ENVIRONMENTAL CONTAMINANTS in WILDLIFE
Interpreting Tissue Concentrations

ENVIRONMENTAL CONTAMINANTS in WILDLIFE
Interpreting Tissue Concentrations

Edited by
W. Nelson Beyer
National Biological Service
Patuxent Environmental Science Center
Laurel, Maryland, USA
Gary H. Heinz
National Biological Service
Patuxent Environmental Science Center
Laurel, Maryland, USA
Amy W. Redmon-Norwood
The Johns Hopkins University
Baltimore, Maryland, USA

SETAC Special Publications Series

Publication sponsored by the Society of Environmental Toxicology and Chemistry (SETAC) and the SETAC Foundation for Environmental Education

Thomas W. La Point, Editor, SETAC Special Publications, Clemson University, Clemson, South Carolina, USA

LEWIS PUBLISHERS

Boca Raton   New York   London   Tokyo

Tennessee Tech Library
Cookeville, TN

The frontispiece of the Eastern screech owl was drawn by Sabra Niebur.

Library of Congress Cataloging-in-Publication Data

Environmental contaminants in wildlife : interpreting tissue concentrations / edited by W. Nelson Beyer, Gary H. Heinz, Amy W. Redmon-Norwood.
    p. cm. -- (SETAC special publications series)
"Publication sponsored by the Society of Environmental Toxicology and Chemistry (SETAC) Foundation for Environmental Education."
    ISBN 1-56670-071-X (alk. paper)
    1. Biological monitoring. 2. Histology, Pathological. 3. Pollution--Environmental aspects. I. Beyer, W. Nelson. II. Heinz, Gary H. III. Redmon-Norwood, Amy W. IV. SETAC (Society) V. SETAC Foundation for Environmental Education. VI. Series.
RA1223.B54E57   1996
615.9′02--dc20   95-45493
                                                                                                                               CIP

    This book contains information obtained from authentic and highly regarded sources. Reprinted material is quoted with permission, and sources are indicated. A wide variety of references are listed. Reasonable efforts have been made to publish reliable data and information, but the author and the publisher cannot assume responsibility for the validity of all materials or for the consequences of their use.
    Neither this book nor any part may be reproduced or transmitted in any form or by any means, electronic or mechanical, including photocopying, microfilming, and recording, or by any information storage or retrieval system, without prior permission in writing from the publisher.
    All rights reserved. Authorization to photocopy items for internal or personal use, or the personal or internal use of specific clients, may be granted by CRC Press, Inc., provided that $.50 per page photocopied is paid directly to Copyright Clearance Center, 27 Congress Street, Salem, MA 01970 USA. The fee code for users of the Transactional Reporting Service is ISBN 1-56670-071-X/96/$0.00+$.50. The fee is subject to change without notice. For organizations that have been granted a photocopy license by the CCC, a separate system of payment has been arranged.
    CRC Press, Inc.'s consent does not extend to copying for general distribution, for promotion, for creating new works, or for resale. Specific permission must be obtained from CRC Press for such copying.
    Direct all inquiries to CRC Press, Inc., 2000 Corporate Blvd., N.W., Boca Raton, Florida 33431.

© 1996 by CRC Press, Inc.
Lewis Publishers is an imprint of CRC Press

No claim to original U.S. Government works
International Standard Book Number 1-56670-071-X
Library of Congress Card Number 95-45493
Printed in the United States of America 1 2 3 4 5 6 7 8 9 0

The SETAC Special Publications Series

The SETAC Special Publications Series was established by the Society of Environmental Toxicology and Chemistry to provide in-depth reviews and critical appraisals on scientific subjects relevant to understanding the impacts of chemicals and technology on the environment. The series consists of single- and multiple-authored or edited books on topics reviewed and recommended by the SETAC Board of Directors for their importance, timeliness, and contribution to multidisciplinary approaches to solving environmental problems. The diversity and breadth of subjects covered in the series reflects the wide range of disciplines encompassed by environmental toxicology, environmental chemistry, and hazard and risk assessment. Despite this diversity, the goals of these volumes are similar; they are to present the reader with authoritative coverage of the literature, as well as paradigms, methodologies, controversies, research needs, and new developments specific to the featured topics. All books in the series are peer reviewed for SETAC by acknowledged experts.

The SETAC Special Publications are useful to environmental scientists in research, research management, chemical manufacturing, regulation, and education, as well as to the students considering careers in these areas. The series provides information for keeping abreast of recent developments in familiar areas and for rapid introduction to principles and approaches in new subject areas.

Environmental Contaminants in Wildlife: Interpreting Tissue Concentrations builds on more than 50 years of collective wisdom concerning contaminants in wildlife. The international authors have provided practical advice and theoretical considerations of contaminants in aquatic and terrestrial animals, from invertebrates to mammals to birds. This book will serve the needs of teachers, scientists, and risk managers as a stand-alone reference work on wildlife contamination near the end of the 20th Century.

Editors, *Environmental Contaminants in Wildlife: Interpreting Tissue Concentrations*
   W. Nelson Beyer and **Gary H. Heinz** are research biologists at the Patuxent Environmental Science Center in Laurel, Maryland, where they study wildlife toxicology. **Amy W. Redmon-Norwood** is a certified Editor in the Life Sciences and senior technical editor of the *American Journal of Epidemiology* at the Johns-Hopkins University in Baltimore, Maryland.

Dedication

This book about interpreting the significance of environmental contaminant residues in fish and wildlife is dedicated to Lucille F. and William H. Stickel. The Stickels pioneered the use of tissue residues analysis as a means of determining whether animals found dead or sick in the wild had been exposed to pesticides or other contaminants. Therefore, a synthesis of methodology and knowledge in contaminant residue interpretation for the use of fish and wildlife biologists provides a fitting opportunity to honor the important contributions of Bill and Lucille Stickel to the field of environmental contaminant research.

The Stickels began their careers at the Patuxent Wildlife Research Center in the early 1940s. During the next four decades, they conducted numerous important studies that provided the basis for the present approaches to the evaluation of the biological and ecological effects of environmental contaminants on wildlife populations and habitats. Until their retirement in 1982, the Stickels were tireless workers for the cause of wildlife conservation, in general, and for the control of environmental pollution, in particular. Their legacy is, in part, the scientific research and scholarly publications that form the foundation for this book.

Lucille was instrumental through her personal research in bringing sharp focus to the effects of pollutants on wildlife and the environment. In addition, she served very capably and effectively as Director of the Patuxent Wildlife Research Center from 1972 to 1981. As a Senior Scientist in the Department of the Interior, she also served on many national and international advisory panels as the U.S. Fish and Wildlife Service expert on environmental contaminants. Lucille was the recipient of many awards, including the Department of the Interior's Distinguished Service Award, The Wildlife Society's Aldo Leopold Award, and the Federal Women's Award.

Bill was also recognized as a pioneer in research on environmental contaminants. He was widely known and respected for his innovative experimental studies, his objectivity in the interpretation of research results, and his development of practical management applications of research findings. Bill also made many behind-the-scenes contributions that fostered contaminant research, such as conducting tours, responding to information requests, managing the reprint collection, and disseminating publications. Bill also devoted countless hours to advising younger members of the staff about projects and pitfalls in contaminant research. He received many professional awards for his communications within the scientific community.

Both Lucille and Bill Stickel dedicated their lives to developing the world-renowned scientific reputation of the Patuxent Wildlife Research Center. Their influence on the selection of scientific staff and on the conduct of high quality research was profound. Many of these environmental contaminants scientists, after over two decades of contaminants research, are still employed at the Center. Others that they hired and trained have continued their contaminant work in other parts of the U.S. Fish and Wildlife Service or in other agencies and academia.

Much of the research conducted by the Stickels relates directly to the purpose of this book, namely, the interpretation of contaminant residues in the tissues of wildlife. They published a long series of articles demonstrating that carefully controlled laboratory studies could be used to establish the brain concentrations of organochlorine pesticides that cause death in birds. Once established, these concentrations could then be compared with the concentrations in the brains of birds found dead in the field. Many references have been made in this book to this pioneering work of the Stickels in establishing what is now known as diagnostic brain residues of organochlorines in birds.

In addition, Bill and Lucille conducted many other studies related to the importance of measuring contaminant residues in wildlife. Some examples are the following:

- Several studies to measure the rate of accumulation and loss of various contaminants from the tissues of birds.
- A study of the effectiveness of different forms of tissue preservation in giving accurate readings of contaminant levels.
- A study to determine how to correct residue readings for loss of moisture occurring in eggs collected in the field.

They also recognized the challenging analytical problems involved in accurate chemical determinations of contaminant residues in animal tissues and fostered the development of a strong analytical chemistry staff and laboratory. Part of this chemistry laboratory has evolved into the present Patuxent Analytical Control Facility, which now serves the needs of the management arm of the U.S. Fish and Wildlife Service to acquire quality data on contaminant residues in fish and wildlife.

Bill and Lucille were practical in their assessments of contaminant threats to wildlife. They recognized that the mere presence of a pesticide or other pollutant in the habitat of a wildlife species did not, by itself, tell anything about the hazard of that contaminant to that species. The chemical had to get into the animal first before there was any chance of harm and, even then, had to accumulate in the animal to the extent that it could cause harm. This concept, which is so fundamental to wildlife toxicology and is the focus of this book, was never lost in their thinking.

Although the Stickels authored numerous scientific papers and technical publications, their individual and collaborative efforts extended significantly beyond to deep personal commitments to natural resource stewardship and the responsible use of pesticides in environmental management. They took great interest in conservation affairs and issues. They financially supported various environmental causes and organizations, especially the Nature Conservancy protection of the Center's diverse wildlife habitats, and they took a keen interest in the plants and animals inhabiting the Center. As Center Director, Lucille added 1760 acres to the Patuxent Wildlife Research Center and designated three large forested tracts as Research Natural Areas.

In recognition of the long-term professional service by the Stickels, the U.S. Fish and Wildlife Service authorized renaming the Chemistry and Physiology Laboratory at the Patuxent Wildlife Research Center as the William H. and Lucille F. Stickel Laboratory. Stickel Laboratory was dedicated on the occasion of the Center's 50th anniversary celebration in 1989. Appropriately, the scientific and administrative offices for the Environmental Contaminants Research Branch were consolidated in Stickel Laboratory. This building also houses the Patuxent Analytical Control Facility.

I had an opportunity to meet and work with Bill and Lucille Stickel, first as a student at Iowa State University and later as a wildlife research biologist for the U.S. Fish and Wildlife Service. When I became Chief of the Service's Division of Wildlife Research, I gained an even deeper appreciation for them personally and professionally. I am grateful for the invitation to write a dedication for this book as a way of recognizing the significant scientific contributions of our friends and colleagues, Bill and Lucille Stickel.

David L. Trauger
Deputy Center Director
Patuxent Environmental
Science Center

Foreword

The handle from a well pump once located outside a modest London pub has come to symbolize for many how thoughtful scientific insight can bring understanding to a collection of observations. A plaque outside the pub reads, "This red granite kerbstone marks the site of the historic Broad Street Pump associated with Dr. John Snow's discovery in 1854 that cholera is conveyed by water." Snow thought about the problem in a new light by considering the geographical distribution of cholera cases in London. By mapping his observations, he showed that the pump was key to the spread of this devasting disease. Persons who drank the well water had a much higher death rate from cholera than those who did not. Moreover, a bottle of water from the pump carried to a widow and her daughter living in an area where no cholera was present appeared to be the cause of infection of both women. Although it was some 40 years before the true bacterial cause of cholera was discovered, Snow's insight and his practical recommendation to remove the pump handle of the contaminated well saved many lives.

We feel the time has come for the field of wildlife toxicology to apply the same thoughtful insight for which Dr. Snow became famous. It is not enough to collect more data, but we need to step back and look at the way in which we think about the problem. It is time to ask objectively what can and cannot be inferred from concentrations of environmental contaminants in tissues of wildlife.

PURPOSE OF THE BOOK

Many books have been written on environmental toxicology and ecotoxicology, but none has focused on the practical question, "How much of a chemical must be in the tissues of a wild animal to cause harm?" This book deals exclusively with that question. The chapters were written by researchers in ecotoxicology for both other researchers and those who regulate or evaluate the hazards of environmental contaminants.

Toxicologists have traditionally based much of their thinking on the concept of dose. As early as the 16th century, a Swiss alchemist, Paracelsus, recognized the importance of dose when he wrote, "What is there that is not poison, all things are poison and nothing without poison. Solely the dose determines that a thing is not a poison" (Deichmann et al., 1986, p. 210). Toxicologists recognize that there is a threshold level of a poison below which harmful effects will not be seen. Generally, this threshold has been expressed as the amount of poison entering an animal. The authors of this book take a different approach and consider the amount of poison in the tissues.

Each year ecotoxicologists analyze many samples of animal tissues for contaminants and then interpret the concentrations in light of seasonal or geographical variation, biomagnification, feeding habits, and various other topics. Authors seldom address the simple question of whether the environmental contaminants measured are harming the animals. We think this question is fundamental for researchers and for those with the task of applying research. Simple questions, such as how to interpret 10 ppm of lead in a duck's liver or 2 ppm of polychlorinated biphenyls (PCBs) in fish, are at the core of applied environmental toxicology.

Claiming that a particular concentration of a chemical in tissue is associated with the onset of harm invites argument among toxicologists for several reasons. Environmental contaminants may cause several biological effects simultaneously. The distinction between harm and effect may be obscure. The background of a researcher is certain to color his or her perception of "harm". Moreover, an animal may experience different degrees of harm from the same concentration of chemical in its body, depending upon age, fat reserves, and breeding condition. Individuals within the same species and, even more so, individuals from different species may react differently to the same concentration of a chemical.

The authors have had to balance their understanding of the usefulness of thresholds with their knowledge of the limitations of generalizing in biology. The authors did not take the easy course of emphasizing the complexity of the topic and the many things not known about interpreting environmental contaminants in tissues. Instead, they did their best to provide useful guidance, while recognizing the shortcomings of their knowledge. This foreword was written to bring out some of the themes and difficulties common to the chapters.

TOPICS AND SCOPE OF THE CHAPTERS

Readers may wonder why some well-known environmental contaminants are not included in this book. Toxicologists have found the concentrations of some chemicals in tissues to be unrelated to harm. For example, they have usually failed when they tried to equate toxicity with the chemical residues of organophosphorous and carbamate pesticides. These pesticides are best investigated by measuring cholinesterase inhibition in the brain or plasma. The gut contents or body may be analyzed for the pesticide to determine the cause of the inhibition.

The first chapter, written by James Keith, describes the history of interpreting environmental contaminants in tissues. Readers will appreciate from his discussion that the reasoning behind associating tissue concentrations of a chemical with harm is still evolving. Much of the early research described was conducted on organochlorine pesticides. Our experiences with these chemicals have influenced our thinking about other kinds of environmental contaminants.

The scope of subsequent chapters is based on the experience of the individual authors and the information available on the topic. This book is a collection of chapters with a common goal, but the chapters were written by authors with different approaches. The authors have studied chemicals having different modes of action, and they based their conclusions on scientific literature emphasizing different disciplines within toxicology. The chapters include a mix of sublethal and lethal results, laboratory and field settings, and observations on individuals and populations. The blend of data from different kinds of studies increases the credibility of the conclusions. Although most authors reasoned along similar lines, others contributed alternative approaches. We cannot single out one approach as best but, as Mark Twain observed wryly, "It were not best that we should all think alike; it is difference of opinion that makes horse-races." Each chapter's conclusions are not indisputable, but they are the best efforts of an authority, using his or her judgment. The authors conclude with a summary, written without the details and references found in the main section of the chapter.

THRESHOLDS AND CRITICAL CONCENTRATIONS

The heart of most chapters is a statement relating a concentration in a tissue to a harmful effect. The authors have used the terms "critical", "diagnostic", or "threshold" or expressions such as "threshold toxic", "toxic effect threshold", and "critical threshold". Other environmental toxicologists have used additional expressions, such as "indicative of acute exposure", "indicative of poisoning", and "toxic". The authors in this book have tried to explain clearly what they mean, knowing that defining the effect carefully is as important as measuring the concentrations accurately. The unexplained use of the term "toxic", for example, may suggest a minor biochemical change to a physiologist, or a dead animal to a population biologist.

In suggesting tissues for analysis, the authors usually selected a target organ of the environmental contaminant. However, the authors were pragmatic. If knowing that the concentration in a liver was useful in determining whether an animal died from a contaminant, then the liver was recommended for analysis, irrespective of whether damage to the liver was thought to have contributed to the death of that animal. Some of the authors discuss threshold concentrations in nonvital tissues, such as hair or feathers, or make inferences about the health of a bird, or even a population, based on concentrations detected in eggs.

Other authors in environmental toxicology have suggested more restrictive definitions and have made cause and effect part of their definitions of "critical levels" and "critical organs". Foulkes (1990, p. 327), for example, in discussing the concept of critical levels of metals in tissues, wrote, "the implicit assumption is usually made that a threshold concentration of specific metals exists in the most sensitive target organ, so an increased frequency of functional lesions will be expected if this threshold is exceeded." Although a definition based on cause and effect is appealing, in practice toxicologists may not agree on the most sensitive target organ and may not be able to apply the criteria to wild animals with unknown exposures. Ecotoxicologists, in contrast to those studying human health, may be less interested in identifying the most sensitive target organ than in understanding the more severe biological effects and how they relate to animal survival and to populations.

Although a "threshold" or "critical" concentration may be estimated from a single controlled laboratory study, most of the authors also evaluated studies on wild animals before reaching a conclusion. The most solid arguments are those based on evidence from both laboratory and field studies. Field data, alone, may be difficult to interpret because wild animals are frequently exposed to many environmental stresses in addition to the contaminant under consideration. Field data are essential when threshold concentrations refer to effects on populations, breeding behavior, or other subjects not suited to laboratory studies. Whenever appropriate, pathological information should accompany chemical analyses of tissues.

DIFFICULTIES IN APPLYING THRESHOLDS

The concept of a threshold concentration is not necessarily useful for all environmental contaminants and all kinds of organisms. For example, because polyaromatic hydrocarbons are metabolized by vertebrates, their tissue concentrations may remain low as exposure increases. Jocelyne Hellou suggests an indirect approach to evaluating these chemicals, using concentrations measured in invertebrates to indicate toxicity to vertebrates in the same waters. In his chapter on metals, Philip Rainbow rejects the use of threshold values for aquatic invertebrates and concludes that whole-body concentrations are not related to toxic effects. He argues, instead, that toxicity occurs when the rate of uptake exceeds the rate of detoxification and excretion. Most of the examples he uses to illustrate his reasoning were taken from studies on zinc, an essential element that may be stored at extremely high concentrations in some species of aquatic invertebrates. He suggests that metabolically active stores of metals such as zinc would have to be differentiated from detoxified stores before the significance of the concentrations could be interpreted.

Variability in tissue concentrations also limits the usefulness of the concept of thresholds. Even under controlled circumstances, variation may be substantial. For example, when birds from six species were given a chronic dose of lead that killed half of the animals in each group

(16 to 30 birds per group), the maximum concentration in the liver was generally about 10 times as great as the minimum for a species (Beyer et al., 1988). The variation in concentrations in wild animals would be expected to be even greater. Age, sex, fat reserves, and many other variables are likely to alter the concentration at which an effect occurs in individual animals. When the variation is too great, calculating thresholds from mean concentrations has little practical value. In a chapter on interpreting fluoride concentrations in birds, Jim Fleming concludes that high variability in the available data precludes determining a threshold concentration.

SELECTING AN APPROPRIATE STATISTICAL ESTIMATOR FOR A THRESHOLD

Thresholds are generally derived from a population and are applied either to an individual or to a population. The intended use of the threshold determines which statistical estimator is appropriate. Suppose, for example, that a biologist is investigating the cause of death of fish found dead by several observers at different locations downstream from a pesticide application. If the pesticide concentrations measured in most of the fish were in the known lethal range, the biologist might reasonably conclude that the pesticide applied nearby killed the fish. The biologist might, for example, compare the concentration in each fish with the fifth percentile of the concentrations detected in experimentally poisoned fish of that or a closely related species. Depending on the situation, the biologist might be justified in calculating a mean for all fish and comparing it with the known mean lethal concentration. Suppose, instead, that the biologist is investigating the cause of death of a single wild bird found dead and has only a list of concentrations of environmental contaminants detected in the bird's liver. Without additional evidence, such as lesions observed by a pathologist, the fifth percentile and probably even the median lethal concentration would be inappropriate statistical estimators.

No matter which statistical estimator is selected as a threshold, some individuals will die at a lower concentration and others will not die at higher concentrations. The choice is a compromise that should depend on the particular circumstances and the questions asked. Generally, toxicologists have been forced to rely on medians or means, rather than on confidence intervals or percentiles of distributions, because they lack the data necessary for estimating their values with adequate precision.

PROVIDING USEFUL ADVICE AMID UNCERTAINTIES

Many variables may affect threshold concentrations. All of the authors agree on this. For example, the chemical form of an environmental contaminant may be relevant. Philip Rainbow explained that the zinc in polychaete mandibles is presumably bound in a nontoxic form and should not be equated with zinc in soft tissues. Threshold effects of mercury may depend on the concentration of selenium, which binds to mercury. Some compounds, such as PCBs and the insecticide toxaphene, are mixtures. Other compounds, such as dioxins, are rarely found alone but are associated with related toxic compounds. Although James Keith refers to a method to sum up the toxic potential of a mixture of organochlorine pesticides present in a tissue, thresholds generally refer to isolated chemicals.

Extrapolating threshold concentrations determined for one species to other species presents another problem. There are no clear guidelines for deciding whether a threshold determined for an environmental contaminant in waterfowl, for example, should be considered appropriate for birds in general, for waterfowl only, or perhaps for only the species tested. Experimental toxicological data are not available for some species, such as marine mammals, and toxicologists must depend on results from distantly related mammals. David Thompson suggests that marine seabirds have been exposed to naturally high concentrations of mercury for so long that they may be able to tolerate concentrations that would be toxic to other species of birds, and that mercury thresholds determined for other avian species may not be applicable to marine seabirds.

Ecotoxicologists would agree that the factors discussed above complicate the interpretation of concentrations of environmental contaminants in wildlife. The authors of the chapters have recognized these difficulties but were not overwhelmed by them. They have done their best to emphasize what is known and to draw the conclusions they deem warranted.

In the tradition of John Snow nearly a century and a half ago, the authors attempt to wrest insight from their observations. The authors have written 22 definitive chapters on interpreting concentrations of environmental contaminants in wildlife. From this well, you, the reader, are invited to drink.

REFERENCES

Beyer, W. N., J. W. Spann, L. Sileo, and J. C. Franson. 1988. Lead poisoning in six captive avian species. *Arch. Environ. Contam. Toxicol.* 17:121–130.

Deichmann, W. B., D. Henschler, B. Holmstedt, and G. Keil. 1986. What is there that is not poison? A study of the *Third Defense* by Paracelsus. *Arch. Toxicol.* 58:207–213.

Foulkes, E. C. 1990. The concept of critical levels of toxic heavy metals in target tissues. *Crit. Rev. Toxicol.* 20:327–339.

W. Nelson Beyer
Gary H. Heinz
Amy W. Redmon-Norwood
March 15, 1995

Reviewers

The authors and editors thank our distinguished reviewers for their assistance:

Daniel W. Anderson
Division of Wildlife and Fisheries
University of California
Davis, California, USA

Gary Atchison
Department of Animal Ecology
Iowa State University
Ames, Iowa, USA

Drew Bodaly
Department of Fisheries and Oceans
Freshwater Institute
Winnipeg, Manitoba, Canada

Don Clark
National Biological Service
Texas A & M University
College Station, Texas, USA

Tracy Collier
National Oceanic and Atmospheric
  Administration
National Marine Fisheries Service
Northwest Fisheries Science Center
Seattle, Washington, USA

Deborah Cory-Slechta
Department of Environmental Medicine
University of Rochester School of Medicine
  and Dentistry
Rochester, New York, USA

Tom Custer
National Biological Service
La Crosse, Wisconsin, USA

John W. Farrington
Woods Hole Oceanographic Institution
Woods Hole, Massachusetts, USA

Norvald Fimreite
Telemark College
Norway

Stephen G. George
NERC Unit of Aquatic Biochemistry
University of Stirling
Stirling, Scotland, United Kingdom

Michael Gilbertson
International Joint Commission
Windsor, Ontario, Canada

Steven Hamilton
National Biological Service
National Fisheries Contaminant Research
  Center
Yankton, South Dakota, USA

John Harwood
Sea Mammal Research Unit
Cambridge, United Kingdom

Barry L. Johnson
National Biological Service
National Fisheries Research Center
La Crosse, Wisconsin, USA

Roy L. Kirkpatrick
Department of Fisheries and Wildlife
  Sciences
Blacksburg, Virginia, USA

Jack Klaverkamp
Freshwater Institute
Winnipeg, Manitoba, Canada

Louis Locke
National Biological Service
National Wildlife Health Research
  Center
Madison, Wisconsin, USA

Carol Meteyer
National Biological Service
National Wildlife Health Research
  Center
Madison, Wisconsin, USA

Greg Mierle
Aquatic Science Section
Dorset Research Centre
Dorset, Ontario, Canada

A. Keith Miles
National Biological Service
University of California
Davis, California, USA

Derek C. G. Muir
Department of Fisheries and Oceans
Freshwater Institute
Winnipeg, Manitoba, Canada

James R. Newman
KBN Engineering and Applied Sciences, Inc.
Gainsville, Florida, USA

Ian Newton
Institute of Terrestrial Ecology
Monks Wood Experiment Station
Cambs, United Kingdom

Ian Nisbet
I.C.T. Nisbet & Company, Inc.
North Falmouth, Massachusetts, USA

John O'Halloran
Department of Zoology
University College
Cork, Ireland

Harry Ohlendorf
CH2M Hill
Sacramento, California, USA

Oliver H. Pattee
National Biological Service
Patuxent Environmental Science Center
Laurel, Maryland, USA

Tony J. Peterle
Ohio Cooperative Fish and Wildlife Research Unit
Columbus, Ohio, USA

Barry Poulton
National Biological Service
National Fisheries Contaminant Research Center
Columbia, Missouri, USA

Robert W. Riseborough
University of California
Berkeley, California USA

Michael K. Saiki
National Biological Service
National Fisheries Contaminant Research Center
Field Research Station - Dixon
Dixon, California, USA

Mark B. Sandheinrich
The Rivers Study Center
University of Wisconsin-La Crosse
La Crosse, Wisconsin, USA

Anton M. Scheuhammer
National Wildlife Research Center
Environment Canada
Hull, Quebec, Canada

Lou Sileo
National Biological Service
National Wildlife Health Lab
Madison, Wisconsin, USA

Mary Walker
University Wisconsin-Madison
School of Pharmacy
Madison, Wisconsin, USA

Ken Walton
School of Biological Sciences
University of Wales
Bangor, Gwynedd LL57 2UW, United Kingdom

Donald H. White
National Biological Service
School of Forest Resources
University of Georgia
Athens, Georgia, USA

Contributors

Lawrence J. Blus
National Biological Survey
Patuxent Environmental Science Center
Corvallis, Oregon

Iain C. Boulton
University of Greenwich
School of Environmental Sciences
London, England

John Cooke
Department of Biology
University of Natal
Dalbridge, South Africa

W. James Fleming
North Carolina Cooperative Fish and
 Wildlife Research Center
National Biological Service
North Carolina State University
Raleigh, North Carolina

J. Christian Franson
National Biological Service
National Wildlife Health Center
Madison, Wisconsin

Robert W. Furness
Institute of Biomedical and Life Sciences
University of Glasgow
Glasgow, Scotland

Gary H. Heinz
U.S. Fish and Wildlife Service
Patuxent Environmental Science Center
Laurel, Maryland

Jocelyne Hellou
Department of Fisheries and Oceans
Toxicology Section
St. John's, Newfoundland
Canada

David J. Hoffman
National Biological Service
Patuxent Environmental Science Center
Laurel, Maryland

Michael S. Johnson
Department of Environmental and
 Evolutionary Biology
University of Liverpool
Liverpool, England

Michael A. Kamrin
Institute for Environmental Toxicology
Michigan State University
East Lansing, Michigan

Jim Keith
Denver Wildlife Research Center
U.S. Department of Agriculture
Denver, Colorado

Timothy J. Kubiak
U.S. Fish and Wildlife Service
Division of Environmental Contaminants
Arlington, Virginia

Robin J. Law
Ministry of Agriculture, Fisheries and
 Food
Fisheries Laboratory
Essex, England

Dennis Lemly
Department of Fisheries and Wildlife
Virginia Tech
Blacksburg, Virginia

Wei-chun Ma
Institute for Forestry and Nature
 Research
Department of Ecotoxicology
Wageningen, The Netherlands

Arthur J. Niimi
Department of Fisheries and Oceans
Canada Center for Inland Waters
Burlington, Ontario, Canada

Antoon Opperhuizen
National Institute for Coastal and Marine
  Management
Ministry of Transport, Public Works, and
  Water Management
The Hague, The Netherlands

Deborah J. Pain
Royal Society for the Protection of Birds
The Lodge
Sandy, Bedfordshire, England

David Peakall
Monitoring and Assessment Research
  Center
King's College, Campden Hill
London, England

Philip S. Rainbow
School of Biological Sciences
Queen Mary and Westfield College
University of London
London, England

Clifford P. Rice
U.S. Department of Agriculture
Agricultural Research Service
Beltsvill, Maryland

Robert K. Ringer
Institute for Environmental Toxicology
Michigan State University
East Lansing, Michigan

Dick T.H.M. Sijm
Research Institute of Toxicology
Environmental Chemistry Group
University of Utrecht
Utrecht, The Netherlands

Douglas J. Spry
Ontario Ministry of Environment and
  Energy
Standards Development Branch
Toronto, Ontario, Canada

David Thompson
Applied Ornithology Unit
Institute of Biomedical and Life
  Sciences
Glasgow University
Glasgow, Scotland

Stanley Wiemeyer
U.S. Fish and Wildlife Service
Reno, Nevada

James Wiener
National Biological Service
Upper Mississippi Science Center
La Crosse, Wisconsin

Contents

1. Residue Analyses: How They were Used to Assess the Hazards of Contaminants to Wildlife 1
   James O. Keith

2. DDT, DDD, and DDE in Birds 49
   Lawrence J. Blus

3. Dieldrin and other Cyclodiene Pesticides in Wildlife 73
   David B. Peakall

4. Other Organochlorine Pesticides in Birds 99
   Stanley N. Wiemeyer

5. PCBs in Aquatic Organisms 117
   Arthur J. Niimi

6. Toxicological Implications of PCB Residues in Mammals 153
   Michael A. Kamrin and Robert K. Ringer

7. PCBs and Dioxins in Birds 165
   David J. Hoffman, Clifford P. Rice, and Timothy J. Kubiak

8. Dioxins: An Environmental Risk for Fish? 209
   Dick T.H.M. Sijm and Antoon Opperhuizen

9. Polycyclic Aromatic Hydrocarbons in Marine Mammals, Finfish, and Molluscs 229
   Jocelyne Hellou

10. Lead in Waterfowl 251
    Deborah J. Pain

11. Interpretation of Tissue Lead Residues in Birds other than Waterfowl 265
    J. Christian Franson

12. Lead in Mammals 281
    Wei-chun Ma

13. Toxicological Significance of Mercury in Freshwater Fish 297
    James G. Wiener and Douglas J. Spry

14. Mercury in Birds and Terrestrial Mammals 341
    David R. Thompson

15.	Metals in Marine Mammals 357
	Robin J. Law
16.	Cadmium in Small Mammals 377
	John A. Cooke and Michael S. Johnson
17.	Cadmium in Birds 389
	Robert W. Furness
18.	Heavy Metals in Aquatic Invertebrates 405
	Philip S. Rainbow
19.	Selenium in Aquatic Organisms 427
	A. Dennis Lemly
20.	Selenium in Birds 447
	Gary H. Heinz
21.	Fluoride in Birds 459
	W. James Fleming
22.	Fluoride in Small Mammals 473
	John A. Cooke, Iain C. Boulton, and Michael S. Johnson
Index 483	

CHAPTER 1

Residue Analyses: How They Were Used to Assess the Hazards of Contaminants to Wildlife

James O. Keith

INTRODUCTION

This chapter describes the history of cooperation between biologists and chemists in measuring and interpreting pesticide residues in fish and wildlife. Considerable effort was spent in gathering the information, but why was it needed and how was it used? More specifically, how did residue data contribute to the clarification of relations among pesticide use, habitat contamination, the death and debility of individual animals, and the decline of animal populations?

Information on the trace amounts of chemicals that occur in materials has been useful in many human activities. An example is that of assaying ore to measure its gold content. Assays help in determining if enough gold is present to make its recovery from ore economical. Assays establish the gold content of ore, but further interpretation is needed. The cost of mining the ore and recovering the gold must be compared with the market value of gold to determine whether the effort would be profitable. Similarly, pesticide residue values required interpretation to establish their significance to the health and well-being of fish and wildlife.

The earliest analyses for chemical residues in animal bodies probably were those requested by coroners to investigate unexpected deaths in humans. When people died under unusual circumstances, coroners had tissues analyzed by chemists to see if poisons were present. The finding of poison residues was intuitively accepted as the cause of death; human tissues should not contain residues unless a person is intentionally poisoned.

The presence of pesticide residues in animals has seldom led to such straightforward interpretations as those obtained from gold assays or even forensic medicine.

It was accepted that wild animals would be in treated fields and that even humans might be exposed to pesticide residues. The significance of residues in humans and wildlife was not always clear, as residues were commonly present in apparently healthy animals. Many years of study by toxicologists, chemists, and biologists were required to document the relative hazards of fungicides, herbicides, insecticides, and other contaminants to wildlife. Chemists learned how to recover and distinguish among the contaminants found in animals and to detect and measure increasingly smaller quantities. At first, wildlife biologists requested analyses of many organisms and substrates in treated areas. With experience, they became more selective and, with the help of the chemists, they focused on more specific questions, problems, and phenomena.

Analyses for residues can verify the presence of insecticides and other contaminants, but even as with the recovery of gold from ore, the recovery of residues from tissues is never 100%. Chemists cannot destroy gold, but they often have inadvertently destroyed pesticide residues. I once collected samples of alfalfa that had been sprayed with endrin for mouse control. I sent split samples to two chemistry laboratories, both respected for their analytical capabilities. One laboratory failed to find any endrin in the sample, while the other reported residues of 80 ppm of endrin. Multilaboratory analyses for residues in split samples seldom gave the same results during the early years of residue chemistry. Chemists had a new challenge in developing methods to recover and quantitate contaminants present in environmental samples. Such analyses are difficult, as isolating residues from animal tissues, fat, eggs, and foods demands complicated procedures. Currently, regulations for quality control and good laboratory practices are in place to ensure that reliable, standardized procedures are used in chemical and biological research on pesticides. Still, residue chemistry and field biology are not exact sciences. Knowledge and experience increase the abilities of both chemists and biologists, but the gifted ones also have had a finely developed intuition that guided them in following leads, forming hypotheses, pursuing creative investigations, and producing reliable results.

The science of residue chemistry developed with the first use of pesticides because of concern about contamination of human foods with poisonous substances. By 1939, about 30 pesticides were in limited use (Moore, 1965a); some, such as pyrethrum, nicotine, and derris root, were of botanical origin and of low toxicity, while others, such as mercury and other heavy metals, offered greater hazards to humans. Lead and calcium arsenate had been used for insect control since the late 19th century. As arsenic was a known poison, some assurance was needed that foods treated for insect control were safe to eat. This concern led to regulations in the early 1900s that established residue tolerances for crops (Rudd and Genelly, 1956). Tolerances are the maximum pesticide residues permitted on foods for human or livestock consumption. Tolerances, expressed as parts per million (ppm) of residues, exist today in the United States for every crop treated or likely to be contaminated with a pesticide. By law, foods containing overtolerance residues can be confiscated and destroyed (Food and Drug Act of 1938 and the Miller Amendment of 1952).

Everything would have been easier for biologists if pesticide tolerances could have been developed and enforced for wildlife foods; however, protection of food quality for wildlife was clearly impossible. A scheme was proposed for pesticide tolerances in wildlife tissues (van Genderen, 1966), and the idea, had it been implemented, could have protected animals on treated areas from lethal exposure. It would also have protected humans who consumed wild game. There have been proposals to establish tolerances for pesticides in waste agricultural water, similar to those for industrial effluents and for human drinking water. Residue data could then be used to ensure that water coming from agricultural lands is safe for use by fish and wildlife. Water quality criteria have been established by the U.S. Environmental Protection Agency, but difficulties in measuring low residues and in enforcing compliance have compromised the program's effectiveness.

Pesticide residues in the environment created many problems for humans. Most were addressed and solutions were found, often at considerable expense to agribusiness. Pesticides also have caused many wildlife problems, but agribusiness usually has offered more resistance than assistance in solving those problems. Pesticide residue data have proved essential in hearings and court proceedings to restrict the use of pesticides that caused damage to fish and wildlife resources.

GENESIS OF PESTICIDE-WILDLIFE STUDIES

Studies of residues in domestic animals preceded those of residues in fish and wildlife. Again, the evaluations resulted from concern about the safety of human foods, but they also helped to solve certain problems created by pesticide use. For example, applications of pesticides sometimes contaminated adjacent crops and pastures and, thereby, animal foodstuffs became contaminated. As a result, poultry, eggs, milk, and beef were confiscated for overtolerance residues. Tissue residue studies helped determine how long poultry, milk cows, and livestock had to be "fed off" on uncontaminated feed before residues in their products decreased below tolerance levels.

By 1945, the first synthetic organic insecticides were in use, and techniques for analysis of their residues in animal tissues were being developed. Dichlorodiphenyltrichloroethane (DDT) was measured in tissues of dogs (Woodard et al., 1945); cows, horses, and sheep (Orr and Mott, 1945); rabbits (Laug, 1946); mice (Woodard and Ofner, 1946); rats (Kunze et al., 1949); and chickens (Bryson et al., 1949; Draper et al., 1950). Residues of DDT and other chlorinated hydrocarbon insecticides were recovered from tissues of cattle and sheep (Diephius and Dunn, 1949) and of dogs and rats (Finnegan et al., 1949), in cows' milk (Carter et al., 1949), and in human milk and fat (Laug et al., 1950), which documented human exposure.

Wildlife biologists were only several years behind in their concern over animal exposure to pesticides, but they often did not have analytical capabilities for pesticide residue analyses within their institutions. Some of the first investigations of pesticide effects were conducted with little or no support from residue data. In England, the deaths of birds and mammals attributable to toxic chemicals were monitored in 1960 and 1961 with few confirming residue analyses (Royal Society for the Protection of Birds, undated, 1962). Linduska and Surber (1948) summarized studies of the U.S.

Fish and Wildlife Service that were conducted in 1947 without residue support. Other investigations were made of vertebrate abundance and reproduction in forests where DDT was sprayed for forest insect control (Kendeigh, 1947; Adams et al., 1949) and for control of Dutch elm disease (Benton, 1951) and in forests that were experimentally treated annually with DDT (Robbins et al., 1951). George and Stickel (1949) looked at the DDT effects on prairie wildlife, Hanson (1952) and Goodrum et al. (1949) evaluated herbicide and insecticide effects on marshes, and Cope et al. (1947) studied insecticide effects on trout and salmon. However, when effects were found, and especially mortality, residues became essential to document exposure and suggest the cause of death. Wildlife agencies were able to take advantage of the residue chemistry experience gained by those working with crops and livestock, and by 1950 chemists had joined several wildlife groups that were investigating pesticide effects. In hindsight, this was one of the more essential and productive marriages in the history of wildlife research; studies demanded the collaborative efforts of these two scientific disciplines.

How, then, did biologists proceed in using residues to assess the impact of contaminants on fish and wildlife? What was the utility of knowing the levels of residues in fish, wildlife, and their environments? At first, the approach was much like that of the coroner striving to determine the cause of death in a person who had been poisoned. Biologists began by using residues to document the insecticide exposure of live and dead animals in the laboratory and in treated areas. Investigations and experience led them into increasingly complex research and a gradual awareness of the invasion of residues into the most basic biological functions of wild animals. Residue analyses provided the description of treatments to which animals were exposed at cellular, organ, individual animal, and population levels. Studies from the 1940s to the present time have shown that birds and fish suffered greater mortality and sublethal effects from environmental contaminants than did other vertebrates. The persistent chlorinated hydrocarbon insecticides have been the worst culprits. Of organochlorines, aldrin, dieldrin, endrin, heptachlor, and toxaphene caused the highest incidences of mortality in fish and wildlife. However, it was the sublethal effects of DDE that were responsible for the critical demise of certain species' populations.

RESIDUES IN ANIMALS ON TREATED AREAS

One of the earliest reports of residues in wildlife was that of Danckwortt and Pfau (1926), who measured traces of arsenic in the livers of deer found dead after forest insect control with calcium arsenate. Kelsall (1950) documented that arsenic also occurred in tissues of birds after apple orchard treatments. Almost 30 years after the use of arsenicals had ceased, Elfving et al. (1979) looked at arsenic residues in old orchard soils and in resident rodents. Soils still contained up to 94 ppm of arsenic, and pine voles (scientific names of animals are given in Appendix Table 1) had whole-body residues of up to 28.0 µg (almost 1.0 ppm) of arsenic.

Among the first reports of organochlorine residues in wildlife were those of DDT found in tissues of birds that were exposed either in the laboratory or to orchard treatments (Barnett, 1950; Mohr et al., 1951). Experimental field applications of DDT caused considerable mortality of nestling songbirds in artificial nest boxes and left residues of 2.6 to 77.0 µg of DDT in whole carcasses of dead young (Mitchell et al., 1953).

Between 1952 and 1955, experimental studies of the relation between the quantities of DDT ingested and tissue residues in wildlife (De Witt et al., 1955) were conducted. Often tissue residues of DDT were more related to the severity of toxic symptoms than to dose or duration of exposure. Genelly and Rudd (1956) looked at the residue levels of DDT, dieldrin, and toxaphene from fat, testes, and livers of pheasants dosed in the laboratory. In those early studies, no measurements were reported of DDE, a principal metabolite of DDT. The existence of DDE was known, and methods were available for its analysis (Schechter and Haller, 1944). However, DDE was of little interest as its bioactivity had not been identified. Only later was it recognized as the most persistent and hazardous metabolite of DDT to occur in animals and the environment.

Post (1952) recovered aldrin residues in brain, liver, and kidney tissues of birds killed during a grasshopper control program. Dieldrin residues were not reported, although ultimately it was learned that aldrin is rapidly converted to dieldrin in the animal body (Bann et al., 1956), in soil (Gannon and Bigger, 1958), and on plants (Gannon and Decker, 1959). Barker (1958) reported residues of DDT as well as those of DDE in tissues from robins and other birds that died after spraying for Dutch elm disease. Dieldrin and heptachlor were recovered from northern bobwhite quail found dead after applications for control of fire ants (Clawson and Baker, 1959). Meanwhile, fishery biologists documented mortality from pesticides in fish (Harrington and Bidlingmayer, 1958; Tarzwell, 1959; U.S. Public Health Service, 1961; Hunt and Linn, 1970) and measured residues in fish exposed to insecticide applications (Bridges, 1961; Cope, 1961).

Scott et al. (1959) conducted an impressive study of wildlife mortality and tissue residues following dieldrin applications for control of Japanese beetles in Illinois. They presented rather clear evidence that the death of animals was associated with dieldrin exposure and residues in tissues. However, the authors were careful scientists and qualified their findings by stating, "There was, of course, no basis for certainty that the amount of toxicant found in the tissues showed conclusively that the animal had consumed enough dieldrin to have caused death." Such was the state of the art in pesticide-wildlife studies in the 1950s. Yet, no aggressive district attorney, with a coroner's report of dieldrin in the liver of a deceased person, would have so qualified his closing remarks to the jury.

Rosene (1965) did a thorough study of heptachlor effects on wildlife following applications to control fire ants. He then carefully made a case relating residues of heptachlor epoxide in birds and their environment to bird mortality and decreases in their abundance; the conversion of heptachlor to its epoxide in the animal body was reported earlier by Davidow and Radomski (1952). Using residue analyses, biologists were making a solid connection between unusual mortalities of wildlife on treated areas and insecticide residues in tissues of dead animals. Thus, residue analyses were critical factors in relating wildlife mortality to insecticide applications. Such findings stimulated a reduction in application rates of organochlorine insecticides and a search for safer replacements. Many persons concerned with agrichemicals challenged the idea that insecticide damage to fish and wildlife was of any great consequence. L. A. McLean of Velsicol Chemical Corporation referred to wildlife studies as "science fiction" and to the biologists as "derogatory writers" reporting "pseudo-scientific" findings. It is difficult to cite his 22-page paper, "The Necessity, Value and Safety of Pesticides," as he failed to put a date on it! Tissue residues were not universally accepted as proof that insecticides caused the death of animals on treated areas. Even as late as 1990, some authors continued to warn about the pitfalls of "...ecological fallacy...the idea that occurrence of an effect in conjunction with a plausible environmental factor proves that factor to be the cause..." (Suter, 1990). Critics became especially cynical as biologists increasingly reported mortalities of fish and wildlife due to insecticides at sites that had never been treated with the chemicals. In the early 1960s, there was, as yet, no perception of how insecticide mortality could occur away from treated areas.

RESIDUES IN ANIMALS AWAY FROM TREATED AREAS

It is unclear why some kinds of insecticide effects were not observed until 15 years or more after the introduction of the synthetic organic insecticides. For instance, it was not until about 1960 that wildlife mortality due to pesticides was reported away from treated areas. Likewise, the first case of wildlife mortality due to chlordane was not documented until the late 1970s (Blus et al., 1983). Fish kills occurred in streams at some distance from forest areas treated with insecticides, but this was reasonable and intuitive as water in the streams had originated in the forests where insecticides were applied (Cope, 1966). Similarly, it was accepted that endrin in an industrial effluent could likely kill fish downstream in the Mississippi River (Mount and Putnicki, 1966). But, how could endrin in industrial effluent or in agricultural runoff water be responsible for the death of brown pelicans and the extirpation of their population in remote coastal areas of Louisiana during the 1960s (King et al., 1977)? How could insecticides cause deaths of birds in isolated areas miles away from points of contamination or treated fields? As Rudd (1964) correctly suggested in his chapter on residues, "Pesticides need not be applied directly to waters to contaminate."

Keith (1966) reported organochlorine residues in dead fish-eating birds found within the vast Tule Lake National Wildlife Refuge in northeastern California. Koeman and van Genderen (1966) recovered organochlorine residues from seals, water birds, and birds of prey found dead or dying along the Netherlands seacoast. Jefferies and Prestt (1966) found organochlorine residues in dead and dying falcons in England, and a recent review by Newton et al. (1992) documented sparrowhawk and kestrel mortalities throughout England due to aldrin and dieldrin between 1963 and

1990. Walker et al. (1967) examined dead birds and found organochlorine residues in 20 of 21 families from terrestrial, aquatic, and marine habitats in England. Moore (1965b) wrote a stimulating discussion of the situation in England in the early 1960s. He concluded that pesticides were widespread in the English fauna at levels high enough to pose hazards. He questioned the source of the residues and stated, "The discovery of residues in bird eggs from unsprayed regions needs explanation."

Investigations were pursued by scientists in several countries to determine the source of residues in dead animals and the mechanism of exposure for second- and third-level consumers in various food chains. Many of these predators simply did not frequent treated areas. Biologists claimed that residues in animals were the cause of death, but others challenged such subjective evidence (Whitten, 1966). The evidence was confusing as investigators often analyzed multiple tissues that had different levels of residues, and levels varied between individual animals. Animals often contained residues of several insecticides and other contaminants. The acute toxicity of insecticides helped interpret the relative contribution of organochlorine residues to mortality. Six of 300 brown pelicans found dead in coastal Louisiana in 1975 were analyzed for pesticide residues. Seven chlorinated hydrocarbon insecticides and polychlorinated biphenyls (PCBs) were found in brain tissue. In interpreting these results, L. F. Stickel stated, "...endrin levels identified in the Louisiana pelicans are sufficient alone to cause death. But we also have reason to believe that there is an additive effect produced by all of the pesticides found in the birds" (Winn, 1975).

Studies uncovered other complicating factors. DDT was shown to stimulate dieldrin metabolism in animals (Street and Chadwick, 1967), while the interacting effects of different organochlorines were found to influence the accumulation of residues in tissues (Street and Blau, 1966). Exposure to a combination of insecticides was determined to have additive effects (Ludke, 1976), and to mimic such exposure, experiments were conducted with combinations of organochlorines, such as the Lake Ontario "soup" used by McArthur et al. (1983).

Analytical confusion also occurred. Some procedures used for analysis of DDT compounds destroyed dieldrin residues (Enderson and Wrege, 1973), while other workers did not try to quantitate DDT and DDE in the presence of PCBs (e.g., Vermeer and Reynolds, 1970). Widespread concern existed briefly when it was determined that DDE and some PCBs eluted from chromatographic columns at the same time (Reynolds, 1969; Risebrough et al., 1969; Switzer et al., 1971). After discovering that PCBs occurred in wildlife tissues, it was logical to assume that some residues reported previously as DDE may have actually been PCBs. However, after determining how to separate the compounds, researchers realized that the highest residues were almost always those of DDE.

DIAGNOSTIC RESIDUE LEVELS

Peakall (1992, p. xvii) observed, "The classical approach to establish the hazard of toxic chemicals to wildlife is to determine the amount of a chemical present and then compare that value with those found to do harm in experimental animals." Such was the approach used to determine residue levels that were diagnostic of death in wildlife and, with time and experience, the tissues examined and the experimental studies became more specific and sophisticated. De Witt et al. (1960) determined the acute and chronic lethal dose for 23 insecticides against northern bobwhite quail and pheasants. They then quantitated tissue residues in experimental animals and in 59 species found dead in the field after insecticide treatments. For DDT, they concluded, "...30 to 40 ppm of DDT in breast muscle of quail, or 20 to 30 ppm in breast muscle of pheasants indicated that the birds had died as a direct result of exposure to this compound...." Barker (1958) analyzed brains from dead birds primarily because they were easy to dissect and fortuitously found that brain residues were the best indicator of DDT debility. He concluded that residues of 60 ppm in brains were indicative of death due to DDT in robins. Wurster et al. (1965) found DDT residues of >30 ppm in carcasses and >50 ppm in brains of robins and other birds that died from DDT treatments to control Dutch elm disease.

Bernard (1963) performed analyses for DDT residues in brains, livers, hearts, kidneys, and breast muscles of house sparrows exposed in the laboratory. Residues varied from 0 to 950 ppm in different tissues. Fat and liver had the highest concentrations, but they were not related to debility and death. Small quantities of DDT were present in the hearts, livers, and kidneys of dying birds. Birds with tremors always had higher residues in brains and breast muscles than did birds that were sacrificed. DDT brain residues in dying house sparrows were usually >65 ppm, while survivors most often had levels of <30 ppm.

Bernard (1966) wrote, "The effects of insecticides on wildlife cannot be assessed until we know the significance of residues recovered from tissues of suspected victims." Subsequently, comprehensive studies defined the diagnostic levels of DDT, DDD, and DDE in the brains and bodies of cowbirds either dying from or surviving DDT exposure (Stickel et al., 1966a), the correlation between tissue residues of both DDE and DDD with mortality in cowbirds (Stickel et al., 1970), and the lethal brain residues of DDE in various species of birds (Stickel et al., 1984a). Similar studies with dieldrin (Stickel et al., 1969) and Aroclor 1254 (Stickel et al., 1984b) described diagnostic levels of those compounds in the brains of dying birds.

These studies gradually provided information on brain residue levels that were associated with mortality in experimental birds. Other workers frequently used those levels to support conclusions that birds found dead in the field had died from insecticide poisoning. Sometimes, however, animals found dead in treated areas had lower brain residues than did birds killed in laboratory studies (Stickel and Stickel, 1969). Heinz and Johnson (1981) determined that both cessation of feeding and the amount of lipid reserves influenced the time to death after exposure, but not the residue level of dieldrin in brains at death.

Rudd and Genelly (1956) noted, "Withdrawal of food from animals on high dosage levels of DDT caused characteristic tremors. Starvation due to sickness produced the same results. Presumably, animals were metabolizing body fat containing stored DDT." Harvey (1967) fed starlings ^{14}C-DDT and showed that less than 25% was absorbed and less than 10% remained in the bird 10 days after feeding stopped. Dying birds had brain residues 3 times higher than did surviving birds.

Dying birds had no body fat, and one half of body residues appeared in brains. Van Velzen et al. (1972) conducted sophisticated experiments on the effects of food deprivation on the lethal mobilization of DDT and metabolites in cowbirds. They also concluded that residues in fat of DDT and its metabolites became hazardous during periods of stress and weight loss.

Stickel et al. (1970) looked at body weight, fat reserves, and muscle weight in relation to the relocation of residues and death of birds. They found that the lightest birds tended to die first, while the heaviest birds died last and retained a greater proportion of their body fat. Weight loss before death represented loss of muscle tissue as well as lipid reserves. Barbehenn and Reichel (1981) used organochlorine residues in bald eagles to explore the hazards to birds of contaminant residues in brain and fat tissues. Other studies were conducted to determine the conditions under which tissue residues of DDT could become lethal following the redistribution of residues (de Freitas et al., 1969; Ecobichon and Saschenbrecker, 1969; Findlay and de Freitas, 1971; Sodergren and Ulfstrand, 1972). All of these investigations helped in assessing how and when residues became hazardous to wild animals and the influence of adverse factors, such as food, cold, and stress, on the lethal activities of residues.

RESIDUES IN FOOD CHAINS

Studies of organochlorine insecticide and PCB residues provided additional facts on ecological relations such as the partitioning of energy in ecosystems, the composition of food webs in contaminated habitats, and the relative importance of organisms in food chains. Much new information on the feeding ecology of animals was generated through study of the bioaccumulation of organochlorines in the environment. Early investigations seldom identified how affected animals were exposed to insecticides, but such knowledge was essential to understand how dead animals found outside treated areas had been killed. Animals on treated areas could have become exposed by inhalation, dermal absorption, or ingestion of insecticides. In contrast, animals dying away from treated areas probably had been exposed only through residues in their foods. But how had the prey of raptors in the arctic and fish and fish-eating birds in wetland refuges become contaminated?

Barker (1958) looked at foliage, litter, and earthworms in environments contaminated with DDT by applications to control Dutch elm disease. From the residue data collected, he was able to conclude that residues in earthworms (up to 100 ppm of DDT plus DDE) caused the death of robins and other birds that died with brain residues as high as 196 ppm of DDE and 65 ppm of DDT. Residues in earthworm food chains in areas treated for Dutch elm disease were also studied by Hickey and Hunt (1960) and Wurster et al. (1965).

De Witt et al. (1960) found that earthworms from Louisiana areas treated with heptachlor for fire ant control contained up to 10 ppm of heptachlor epoxide. Woodcocks wintering in Louisiana ate earthworms and accumulated sublethal residues of heptachlor epoxide. Wright (1960), who was studying woodcock reproduction in

New Brunswick, found heptachlor epoxide in birds breeding there. Thus, insecticide treatment of wintering areas was probably responsible for residues in birds on their breeding grounds. In a similar case, residue analyses helped show that aldrin-treated rice seed was the cause of a snow goose mortality in Texas during March and April of 1972 and 1974 (Flickinger, 1979). In 1974, after initial mortality, geese were hazed from rice fields, and they began their northward migration early. Dead birds with dieldrin residues were found that year along goose migration routes in Missouri (Babcock and Flickinger, 1977). Finally, dieldrin brain residues showed that aldrin exposure in Texas probably resulted in the death of female geese dying on their nesting grounds on the shore of Hudson Bay (Flickinger, 1979).

Anderson et al. (1984) analyzed the carcasses of cackling geese and found large seasonal differences in residue burdens of several organochlorine insecticides. The lowest residues were present in geese collected on Alaskan breeding grounds, but levels increased greatly as the birds migrated into agricultural areas of the Klamath Basin and Central Valley in California.

Korschgen (1970) studied dieldrin residues in soil and wildlife food chains in corn fields treated annually with aldrin. Ground beetles, toads, and snakes accumulated high dieldrin residues, and invertebrate residues were sufficient to be hazardous to quail. Beyer and Krynitsky (1989) followed residues in earthworms annually after a single heavy treatment of plots with either DDT, dieldrin, or heptachlor. Contaminant residues were still present in the soil and found in earthworms 20 years later. Thus, using residue analyses, they were able to document that those insecticides can contaminate earthworms and associated animal food chains for prolonged periods. With residue analyses, other workers demonstrated the transfer of organochlorine contamination in soils to macroinvertebrates and ground-feeding birds (Davis, 1966; Boykins, 1967; Jefferies and Davis, 1968).

In the late 1950s, Hunt and Bischoff (1960) studied a continuing mortality in western grebes following periodic applications of DDD to Clear Lake in California for gnat control. They were the first to describe how an insecticide applied to an aquatic habitat could be accumulated from water into plankton, then into fish, and finally into birds, where residues caused both mortality and reproductive failure. Residue analyses were essential in documenting the cause of grebe debility and the source of their exposure to DDD. Residue analysis was the tool used to show the relations in all of the bioaccumulation studies that followed. Woodwell et al. (1967) measured DDT and DDE residues in an estuary treated for mosquito control. They documented that residue levels increased at each higher trophic level in the estuarine food chain. The residues in birds were about one million times greater than those reported in water. Organisms at lower trophic levels contained a greater proportion of DDT, while those at higher trophic levels contained mostly DDE.

During investigations of a fish-eating bird mortality at the Tule Lake National Wildlife Refuge in northeastern California, Keith (1966) found that lethal organochlorine residues in birds had been transferred and accumulated in a complex food chain. Residues in water were largely carried on organic debris and resulted in contamination of phytoplankton and zooplankton, which were the foods of larger invertebrates. The macroinvertebrates were eaten by fish, which in turn were consumed by the affected birds. The contaminated water that entered the refuge was waste irrigation water, which had been used to irrigate croplands treated with insecticides. This study illustrated for the first time how animals in areas never treated with insecticides could be exposed to lethal levels of residues. Hickey et al. (1966) documented the effects on Lake Michigan herring gulls of insecticides from nearby agricultural areas that had accumulated in one of the lake's food chains.

The accumulation of several organochlorine residues in diverse organisms at the primary, secondary, and tertiary levels of food chains in Louisiana lakes was demonstrated by Niethammer et al. (1984), representing contaminations that had persisted for years; uses of the insecticides in agriculture had been banned in the early 1970s. Norstrom et al. (1988) studied organochlorines in arctic marine food chains and defined their occurrence over space and time in polar bears. The extensive studies of mercury accumulations in terrestrial and aquatic food chains were reviewed by Peterle (1991).

There were a number of experimental field studies where an alternative to chemical residue analyses was used to investigate the fate, persistence, and bioaccumulation of insecticides. These studies used radioactively labeled insecticides and largely were developed and pursued by T. J. Peterle (1966) and his graduate students. After application, the levels of insecticides in diverse environmental substrates were determined by measuring the concentration of the radioactive elements. Meeks (1968) followed ^{36}Cl ring-labeled DDT in water, suspended material, soil, plants, invertebrates, fish, amphibians, reptiles, birds, and mammals for up to 15 months after treatments. At the same time, Dindal (1970) looked at the turnover of tissue residues in pinioned ducks released on the ponds treated for Meeks' study. Giles (1970) followed radiolabeled sulfur from ^{35}S-malathion in insects, birds, and mammals for up to a year after treatment of a forest. Forsyth et al. (1983) followed ^{36}Cl-DDT residues in air, soil, runoff, vegetation, invertebrates, and vertebrates in an old-field terrestrial ecosystem for 6 years.

One serious limitation of pesticide studies has been the cost of chemical residue analyses. In 1 week, a biologist can collect sufficient field samples to keep a chemist busy for several months. The advantages of using isotopes are that samples can be rapidly processed and that sampling capabilities are greatly increased. Numerous samples can be cheaply and rapidly processed with this technique. However, the use of isotopes is feasible only in experimental studies, and the technique measures only the fate of the labeled element and not that of the original insecticide molecule. Isotope studies cannot distinguish between the parent insecticide and its metabolites.

DISPERSAL OF RESIDUE CONTAMINATIONS

After biologists had documented the residue levels in tissues that were diagnostic of wildlife mortality and the bioaccumulation of residues in food chains, certain conclusions became evident. The exposure of most birds and mammals to residues seemed related primarily to their food habits. Fish, however, were exposed directly from water through gill absorption of residues, as well as from their foods (Chadwick and Brocksen, 1969; Jarvinen et al., 1976). Among vertebrates, flesh eaters have been reported to contain higher residue levels of insecticides than do herbivores (Hunt and Bischoff, 1960; Moore and Walker, 1964; Ratcliffe, 1965; Dindal, 1970; Stickel, 1973; Reidinger, 1976). Aquatic habitats seemed to offer greater pesticide exposure to animals than do terrestrial habitats (Moore and Walker, 1964; Keith and Hunt, 1966), except in the cases of animal exposure to seed dressings and poison baits (Rudd and Genelly, 1956; Moore, 1965b; Borg et al., 1969).

Definition of the scope and intensity of pesticide contamination was provided largely by residue data that documented the fate of insecticides in the environment. It became clear that residues were not distributed equally in all habitats or in all animals. These facts led some of us to the conclusion that there were environmental "hot spots," which were the foci of exposure for resident and migrating wildlife. These hot spots were either areas that received repeated treatments, such as Clear Lake, California (Herman et al., 1969), and tidal salt marshes treated for mosquito control (Springer and Webster, 1951), or areas with continuous contaminant input, such as Lake Michigan (Hickey et al., 1966), areas near industrial outfalls (Hom et al., 1974; O'Shea et al., 1980a), and wetland refuges in the western United States (Keith, 1966). It was easy for those of us working in hot spots to envision how fish predators such as white pelicans, bald eagles, and ospreys could be exposed to residue levels high enough to cause mortality and reproductive debility. However, it was more difficult to see how falcons, accipiters, and other raptors with declining populations were being affected by insecticides. The available information suggested that small birds and other prey eaten by these species contained relatively low levels of insecticides, and mortality did not appear to be the cause of decreases in avian predator populations. The source of insecticides to which they were exposed and the route of their exposure were not clear. In addition, the cause of their reproductive debility had not been identified. Again, residue data would slowly contribute to the knowledge essential to resolving these questions.

Studies supported by residue analyses showed that insecticides and PCBs moved from treated areas in surface water (Breidenbach, 1965; Hickey et al., 1966; Keith, 1966; Lichtenstein et al., 1966; Breidenbach et al., 1967; Brown and Huffman, 1976), in air (Antommaria et al., 1965; Abbott et al., 1966; Cohen and Pinkerton, 1966; Risebrough et al., 1968a; Bidleman and Olney, 1974), in rain (Tarrant and Tatton, 1968; Bevenue et al., 1972), in snow (Peterle, 1969), in fog (Glotfelty et al., 1987), in migratory birds (Harvey, 1967; Henny et al., 1982b; Springer et al., 1984; Henny and Blus, 1986), in bats (Reidinger, 1976; Clark, 1981), and in reptiles (Hall, 1980).

This dispersal of residues was responsible for contaminating animals in remote and isolated areas such as the Antarctic (George and Frear, 1966; Sladen et al., 1966; Tatton and Ruzicka, 1967; Brewerton, 1969), the Arctic (Addison and Smith, 1974; Bowes and Jonkel, 1975), Greenland (Braestrup et al., 1974), Iceland (Bengston and Sodergren, 1974), the Atlantic and Pacific oceans (Risebrough, 1969; Heppleston, 1973; Harvey et al., 1974; Thompson et al., 1974), and throughout the global ecosystem (Risebrough et al., 1968b).

Surveys conducted by biologists and chemists of the residues in fish and wildlife showed the kinds of animals exposed and the levels of different contaminants in their tissues. There were reports on the levels of organochlorine insecticides in diverse species of birds from England (Moore, 1965b; Walker et al., 1967), North America (Keith and Gruchy, 1972), Norway (Holt and Sakshaug, 1968), and California (Risebrough et al., 1967). Eggs of many species were collected in England (Moore and Tatton, 1965) and North America (Risebrough et al., 1967; Ohlendorf et al., 1982) and analyzed for organochlorine contaminants. Investigations were conducted of residues in fish (Johnson, 1968; Risebrough, 1969; Skåre et al., 1985); in marine mammals (Jensen et al., 1969; Holden, 1978; O'Shea et al., 1980b); in plankton, pelagic organisms, and fish throughout the Atlantic Ocean (Harvey et al., 1974); in birds and mammals from New Zealand (Lock and Solly, 1976); and in birds, mammals, and fish from California (Keith and Hunt, 1966), the Netherlands (Koeman and van Genderen, 1966), and England (Holmes et al., 1967).

Hall (1980) prepared a comprehensive review of the contaminant residues reported in 36 studies of reptiles, and Clark (1981) reviewed the many studies of residues in bats. Reports appeared on the levels of PCBs (Peakall, 1987) and lead (Stendell et al., 1979; Scanlon et al., 1980; Wickson et al., 1992) in birds; of PCBs in New Zealand birds and mammals (Solly and Shanks, 1976); and of mercury in both fish and fish-eating birds (Fimreite et al., 1971) and in seed-eating birds and their predators (Fimreite et al., 1970).

Critical reviews were published of the residues in fish and wildlife, their environments, and specific ecosystems (De Witt et al., 1960; Moore, 1965a; Dustman and Stickel, 1966, 1969; Stickel, 1968, 1973; Johnson, 1968; Robinson, 1969, 1970; Edwards, 1970). These reviews considered the kinds and amounts of contaminant residues in animal tissues and their possible significance. Sufficient information on the residues, their mode of action, and their effects on animals had accumulated to enable preparation of Edwards' (1973) landmark book, *Environmental Pollution by Pesticides*, with comprehensive reviews by 13 authors on pesticide residues in the environment and in invertebrates, fish, and wildlife. Other residue studies helped biologists to understand the causes of pesticide-related mortalities and the potential threats of pesticides to wildlife populations. The world was certainly contaminated, wild animals were being killed, and their productivity was threatened (see the reviews by Wurster, 1969; Peakall, 1970; and Stickel, 1973).

THREATENED AVIAN POPULATIONS

Residue data were essential to conceptualize and document the phenomenon of bioaccumulation in food chains and the contamination of wildlife and environments throughout the world. Residue information proved equally valuable in resolving the problem of serious reproductive impairment in birds, many of which had become endangered species. Biologists studying declining populations of predaceous birds had found residues of DDE in bird tissues and eggs, but had not identified exactly how DDE might be responsible for reproductive failure and population declines. Affected species included bald eagles (Stickel et al., 1966b), osprey (Ames, 1966), Bermuda petrels (Wurster and Wingate, 1968), brown pelicans (Schreiber and

De Long, 1969; Joanen and Dupuie, 1969), double-crested cormorants (Anderson and Hamerstrom, 1966; Anderson et al., 1969), Scottish golden eagles (Lockie and Ratcliffe, 1964), and peregrine falcons in England (Ratcliffe, 1963), in Alaska (Cade et al., 1968), and worldwide (Hickey, 1969).

The factor causing population declines in birds ultimately was discovered from evidence provided by residue analyses of eggs. In the late 1950s, Ratcliffe (1958) had found broken eggs in eyries of peregrine falcons in England. Then Moore and Ratcliffe (1962) analyzed the contents of a peregrine egg and found organochlorine residues. They stated, "Both of us independently formed the tentative opinion that these substances were the cause of the Peregrine's decline...." Peakall (1974) later extracted remnant lipids from museum eggs and showed that DDE was present in peregrine eggs collected in 1948. In 1965, more information on organochlorine residues in raptor and corvid eggs in England was reported by Ratcliffe (1965). Then Ratcliffe (1967) discovered that shells of raptor eggs from England weighed less (-18.9%) than they did before the DDT era began, which explained why the shells might be weaker and break during incubation.

Hickey and Anderson (1968) were the first to use correlation analyses to show that the shell thickness of eggs was inversely related to residues of DDE in eggs. They demonstrated this with herring gull eggs and presented data showing decreases in shell weights of 18 to 26% for declining and extirpated populations of peregrines, bald eagles, and osprey. Experimental work that followed proved that DDE could cause eggshell thinning in certain species of birds (Heath et al., 1969; Wiemeyer and Porter, 1970). With the insight from these studies, biologists found that reproductive failure was the result of DDE-caused eggshell thinning in British peregrines (Ratcliffe, 1970), Alaskan peregrines and hawks (Cade et al., 1971), brown pelicans (Keith et al., 1970; Risebrough et al., 1970, 1971; Blus et al., 1971), osprey (Wiemeyer et al., 1975), prairie falcons (Fyfe et al., 1969; Enderson and Berger, 1970), great blue heron (Vermeer and Reynolds, 1970), double-crested cormorants (Anderson et al., 1969), and many other species (Anderson and Hickey, 1972). Residue data suggested that peregrine falcons (Nisbet, 1988) and Louisiana brown pelicans (King et al., 1977) also may have suffered direct mortality from other, more toxic, chlorinated hydrocarbon insecticides.

Pitfalls continuously challenged the scientific quality of research on pesticide-wildlife relations. The thickness of eggshells varied over the shell, and it was necessary to standardize measurements among studies. The presence or absence of shell membranes, especially in museum collections, affected shell thickness measurements. Similarly, residue data potentially were subject to extraneous influences. As raptor eggs collected in the field often were addled, the effect of putrefaction on the recovery and degradation of residues had to be determined. Mulhern and Reichel (1970) tested putrefaction effects and found that residues that were recovered from fresh and rotten eggs were not significantly different. The water and lipid contents of eggs vary with the stage of incubation and other factors. Stickel et al. (1973) showed that these differences greatly affected the expression of residues in parts per million of sample and gave methods to compensate for the egg condition.

Some argued that DDE was not responsible for eggshell thinning and population decline in raptors and fish-eating birds (Edwards, 1972; Hazeltine, 1972; Switzer et al., 1972; Beatty, 1973). However, most uses of DDT were banned in the United States in 1972, and the positive response of birds to the ban proved these critics wrong. Residues of DDE declined in songbirds (Johnston, 1974), snakes (Fleet and Plapp, 1978), and some wading birds (Henny et al., 1985). An increase in eggshell thickness and reproductive success accompanied decreases in DDE residues in bald eagles (Grier, 1982; Ohlendorf and Fleming, 1988), ospreys (Spitzer et al., 1978), brown pelicans (Anderson et al., 1975; Anderson and Gress, 1983), British sparrowhawks (Newton and Wyllie, 1992), and certain populations of peregrine falcons (Peakall and Kiff, 1988; Peakall, 1990).

Thus, residue analyses led biologists in their search to understand the causes of population declines that endangered and even extirpated local populations of raptors and fish-eating birds. Residue analyses were essential to proving that DDE was responsible for eggshell thinning in wild birds. Subsequent studies of DDE residues, eggshell thinning, reproductive success, and population status have been reported on a variety of avian species (Table 1).

Contaminant residues in foods, eggs, and tissues have continued to be used to evaluate potential reproductive and population problems in many wild vertebrates. The CD-ROM database, Wildlife Worldwide, lists 1494 citations under the key words, "pesticide residues," for publications between 1935 and 1992. This indicates the extent of research utilizing wildlife residues, most of which were residues of organochlorine insecticides and PCBs. A sampling of recent publications on residues in fish and wildlife are included here to illustrate the scope of this scientific effort; these are in addition to those discussed in the text. Problem assessments supported by residue analyses have been reported for raptors (Table 2), aquatic species of birds (Table 3), mammals (Table 4), and other vertebrates (Table 5). Contaminants that are identified include organochlorine insecticides, polychlorinated biphenyls, heavy metals, dioxins, rodenticides, and other contaminants. The conclusions reached in these studies were based on residue data and have provided a basic perspective on the occurrence and effects of pollutant contaminations in wildlife.

The presence of contaminant residues helped to define problems such as mortality, behavioral aberrations, eggshell thinning, reproductive failure, and population declines in wildlife. Bans in the U.S. on uses of insecticides and regulation of industrial chemicals and effluents have reduced domestic contaminations and exposure of resident wildlife. However, migratory species of both prey and predators often are exposed elsewhere during their travels. Residue analyses again have helped to define the nature and locations of migrant exposure. Henny et al. (1982b) measured the blood levels of DDE in peregrine falcons on the Texas coast as they left and as they returned to the U.S. during their annual migration. They concluded that residues in falcons were largely accumulated in Latin America. Likewise, Springer et al. (1984) compared contaminant profiles (relative amounts of different compounds) among regional populations of peregrine falcons to help evaluate the origin of residues in falcons. Henny and Blus (1986) used radiotelemetry to show that

Table 1 Other Representative Studies of Contaminant Residues, Eggshell Thinning, Reproductive Success, and Population Status in Birds

Species	Contaminants[a]	Area	Ref.
Diverse species	OC	U.K.	Ratcliffe (1970)
Brown pelican	DDE	Mexico	Jehl (1970)
Brown pelican	DDE	U.S.	Blus et al. (1971)
Common tern	DDE/PCB	Alberta	Switzer et al. (1971)
Raptors	DDE	Alaska	Cade et al. (1971)
Great blue heron	DDE	Alberta	Vermeer and Risebrough (1972)
Diverse species	DDE	N. America	Anderson and Hickey (1972)
Aquatic species	OC/PCB/Hg	U.S.	Faber and Hickey (1973)
Prairie falcon	DDE	Colorado	Enderson and Wrege (1973)
Mississippi kite	OC	U.S.	Parker (1976)
Mergansers	OC/PCB/Hg	U.S.	White and Cromartie (1977)
Canvasback	OC/PCB/Hg	N. America	Stendell et al. (1977)
Herring gull	OC/PCB/Hg	Great Lakes	Gilman et al. (1977)
Aquatic species	OC/PCB	E. U.S.	Ohlendorf et al. (1978d)
Loggerhead shrike	DDE	Illinois	Anderson and Duzan (1978)
Anhinga/waders	OC/PCB/Hg	E. U.S.	Ohlendorf et al. (1978a)
Barn owl	OC/PCB	Chesapeake Bay	Klaas et al. (1978)
Aquatic species	OC	Texas	King et al. (1978)
Black-crowned night heron	OC/PCB/HM	E. U.S.	Ohlendorf et al. (1978b)
Osprey	OC/PCB	New Jersey	Wiemeyer et al. (1978)
Common loon	OC	New Hampshire	Sutcliffe (1978)
Woodstork/anhinga	OC	E. U.S.	Ohlendorf et al. (1978c)
Brown pelican	OC/PCB	S.E. U.S.	Blus et al. (1979a)
Peregrine falcon	DDE	Worldwide	Peakall and Kiff (1979)
California condor	DDE	California	Kiff et al. (1979)
Red-tailed hawk	OC	Ohio	Springer (1980a)
Great-horned owl	OC	Ohio	Springer (1980b)
Aquatic species	OC/PCB	E. U.S.	Klaas et al. (1980)
Roseate spoonbill	OC/PCB	Texas	White et al. (1982)
White-tailed eagle	OC/PCB/Hg	Sweden	Helander et al. (1982)
Double-crested cormorant	OC/PCB/Hg	Great Lakes	Weseloh et al. (1983)
Greater snow geese	OC/PCB/Hg	Canada	Longcore et al. (1983)
Black-crowned night heron	OC/PCB	W. U.S.	Henny et al. (1984b)
Bald eagle	OC/PCB/Hg	U.S.	Wiemeyer et al. (1984)
Black-crowned night heron	OC/PCB	Colo./Wyo.	McEwen et al. (1984)
Tree swallow	OC/PCB	Colorado	De Weese et al. (1985)
Brown pelican	OC/PCB/HM	Texas	King et al. (1985)
Caspian/elegant terns	OC/PCB	S. California	Ohlendorf et al. (1985)
Black-crowned night heron	OC/PCB	Idaho	Findholt and Trost (1985)
Aquatic species	OC/PCB	California	Boellstorff et al. (1985)
Long-billed curlew	OC/PCB	Oregon	Blus et al. (1985b)
Aquatic species	OC/PCB/PCS	Texas	King and Krynitsky (1986)
Raptors	OC	S. Africa	Mendelsohn et al. (1988)
Osprey	OC/PCB/Hg	U.S.	Wiemeyer et al. (1988a)
Peregrine falcon	OC/PCB	Worldwide	Cade et al. (1988)
Spanish eagle	OC/PCB/HM	Spain	Gonzalez and Hiraldo (1988)
Black vulture	OC	Mexico	Albert et al. (1989)
White-faced ibis	OC	Texas	Custer and Mitchell (1989)
White-faced ibis	DDE/Se/Hg	Nevada	Henny and Herron (1989)

Table 1 (continued) Other Representative Studies of Contaminant Residues, Eggshell Thinning, Reproductive Success, and Population Status in Birds

Species	Contaminants[a]	Area	Ref.
Wood duck	OC	Mississippi	Ford and Hill (1990)
Bald eagle	OC/PCB	Arizona	Grubb et al. (1990)
Cattle egret	OC/PCB	Baja California	Mora (1991)
Fish eagle	DDT/DDE	Zimbabwe	Douthwaite (1992)
Forster's tern	PCB	Wisconsin	Harris et al. (1993)

[a] OC, organochlorine insecticides; PCB, polychlorinated biphenyls; HM, heavy metals; PCS, polychlorinated styrenes.

Table 2 Representative Studies of Contaminant Residues in Raptors, their Eggs, and their Prey

Species	Contaminants[a]	Area	Ref.
Golden eagle	OC/PCB	Scotland	Lockie et al. (1969)
Bald eagle	OC/PCB/Hg	U.S.	Belisle et al. (1972)
Raptors	OC/PCB	Transvaal	Peakall and Kemp (1976)
Gyrfalcon	OC/PCB	Alaska	Walker (1977)
Eleonora's falcon	OC/PCB	Morocco	Clark and Peakall (1977)
Raptors	OC	Kenya	Frank et al. (1977)
Swainson's hawk	OC/Hg	N.W. U.S.	Henny and Kaiser (1979)
Raptors	OC/PCB	New Zealand	Fox and Lock (1979)
Merlin	OC/PCB/Hg	Canada	Fox and Donald (1980)
Vultures	OC	Africa	Mundy et al. (1982)
Peregrine falcon	DDE	Australia	Olsen and Peakall (1983)
Raptors	Heptachlor	Oregon	Henny et al. (1984a)
Bateleur eagle	OC/PCB	Transvaal	de Kock and Watson (1985)
Snail kite	OC	Florida	Sykes (1985)
Peregrine falcon	OC/PCB	W. U.S.	De Weese et al. (1986)
Sparrowhawk	OC/PCB	U.K.	Newton et al. (1986)
Owls/falcons	OC/PCB/Hg	Norway	Froslie et al. (1986)
Peregrine falcon	OC/PCB	Worldwide	Cade et al. (1988)
California condor	OC/HM/R	California	Wiemeyer et al. (1988b)
Hawks	OC/PCB/Hg	N./S. Dakota	Stendell et al. (1988)
Marsh harrier	OC/PCB	S. Africa	de Kock and Simmons (1988)
Peregrine falcon	OC/PCB	Arizona	Ellis et al. (1989)
Barn owl	R	U.K.	Newton et al. (1990b)
Raptors	OC/PCB/Hg	Canada	Noble and Elliott (1990)
Bald eagle	OC/PCB/HM	Lake Superior	Kozie and Anderson (1991)
Barn owl	Dieldrin[b]	U.K.	Newton et al. (1991)
Osprey	OC/HM	E. U.S.	Steidl et al. (1991)

[a] OC, organochlorine insecticides; PCB, polychlorinated biphenyls; HM, heavy metals; R, rodenticides.
[b] Mortality.

black-crowned night herons carrying high residues and reproducing poorly at Ruby Lake, Nevada, used different wintering grounds than did the less contaminated and more successful herons in other breeding colonies.

Table 3 Representative Studies of Contaminant Residues in Aquatic Bird Species

Species	Contaminants[a]	Area	Ref.
White pelican	OC/Hg	Idaho	Benson et al. (1976a)
Tundra swan	HM	Idaho	Benson et al. (1976b)
Herring gull	OC/PCB/Hg	Great Lakes	Gilman et al. (1977)
Canada geese	Pb	Colorado	Szymczak and Adrian (1978)
White-faced ibis	DDE	Utah	Capen and Leiker (1979)
Snow geese	Pb/Hg	Louisiana	West and Newsom (1979)
Audouin's gull	OC/PCB/HM	Mediterranean	Bijleveld et al. (1979)
Sandhill crane	OC/HM	U.S.	Mullins et al. (1979)
Brown pelican	OC/PCB	Louisiana	Blus et al. (1979b)
Western grebe	OC/PCB	Utah	Lindvall and Low (1979)
Least tern	OC/PCB	S. Carolina	Blus and Prouty (1979)
Herring gull	OC/PCB	Great Lakes	Weseloh et al. (1979)
Red-breasted merganser	OC/PCB/Hg	Lake Michigan	Heinz et al. (1983)
Black-crowned night heron	OC/PCB	E. U.S.	Custer et al. (1983)
Royal tern	OC/PCB/HM	Texas	King et al. (1983)
Black-crowned night heron	HM	E. U.S.	Custer and Mulhern (1983)
Great blue heron	HM[b]	N.W. U.S.	Blus et al. (1985a)
Dipper	OC/PCB	Germany	Monig (1985)
Herring gull	OC/PCB	Finland	Karlin et al. (1985)
Long-billed curlew	OC/PCB[b]	Oregon	Blus et al. (1985b)
Great white egret	HM	Korea	Honda et al. (1985)
Greater scaup	HM	Baltic Sea	Szefer and Falandysz (1986)
Fulvous whistling duck	Dieldrin/endrin[b]	Texas	Flickinger et al. (1986)
Red-necked grebe	OC/PCB	Manitoba	De Smet (1987)
Yellow-legged gull	OC/PCB/HM	Italy	Focardi et al. (1988)
Black-headed gull	OC/PCB/HM	Czechoslovakia	Pellantova et al. (1989)
Common tern	OC/PCB	Great Lakes	Weseloh et al. (1989)
White-faced ibis	OC	Texas	Custer and Mitchell (1989)
Olivaceous cormorant	DDE/PCB	Texas	King (1989a)
Double-crested cormorant	DDE/PCB/HM	Washington	Henny et al. (1989)
Black skimmer	OC/PCB/PCS	Texas	King (1989b)
Gannet	OC/Hg	U.K.	Newton et al. (1990a)
Tundra swan	Pb[b]	Idaho	Blus et al. (1991)
Willet	OC/HM	Texas	Custer and Mitchell (1991)

[a] OC, organochlorine insecticides; HM, heavy metals; PCB, polychlorinated biphenyls; PCS, polychlorinated styrenes.
[b] Mortality.

MONITORING PROGRAMS

The presence and persistence of pesticide residues in soil, water, air, plants, and animals raised concerns about their long-term accumulation and effects. Programs for monitoring pesticide residues were recommended by individuals (Moore, 1966) and official committees (U.S. President's Science Advisory Committee, 1963). In 1967, the Subcommittee on Pesticide Monitoring of the Federal Committee on Pest Control (see *Pesticides Monitoring Journal*, Vol. 1, 1967) developed a program to monitor the "…distribution of pesticides in the various elements of the environment

Table 4 Representative Studies of Contaminant Residue Accumulation in Mammals

Species	Contaminants[a]	Area	Ref.
Bobcat/raccoon	Hg	S.E. U.S.	Cumbie (1975a)
Beaver/otter	OC/PCB	Alabama	Hill and Lovett (1975)
Mink/otter/fish	Hg	Georgia	Cumbie (1975b)
Gray seal	OC/PCB	Nova Scotia	Addison and Brodie (1977)
Fox squirrel	OC	Illinois	Havera and Duzan (1977)
Raccoon	HM	Florida	Hoff et al. (1977)
Raccoon	Pb[b]	Connecticut	Diters and Nielsen (1978)
Raccoon	Kepone	Virginia	Bryant et al. (1978)
Big brown bat	PCB	U.S.	Clark (1978)
Seals	Hg/Se	Canada	Smith and Armstrong (1978)
Deer/antelope	HM	Montana	Munshower and Neuman (1979)
Pine vole	Pb	Virginia	Lochmiller et al. (1979)
Pygmy killer whale	OC/PCB	Florida	Forrester et al. (1980)
Bats	OC	Oregon	Henny et al. (1982a)
Whales	Hg	Tasmania	Munday (1985)
Mink	HM	Virginia	Ogle et al. (1985)
Bats	HM	Florida	Clark et al. (1986)
Striped dolphin	OC/PCB	N. Pacific	Loganathan et al. (1990)
Fur seal	OC/PCB	Australia	Smillie and Waid (1987)
Pinnipeds	OC/PCB	Antarctic	Karolewski et al. (1987)
Mink/otter/fish	OC/PCB/Hg	New York	Foley et al. (1988)
Ringed seal	HM	Canada	Wagemann (1989)
Polecat	OC/PCB/HM	Switzerland	Mason and Webber (1990)
Marten/fisher	OC/PCB	Ontario	Steeves et al. (1991)
Bats	OC/PCB	Europe	Nagel et al. (1991)
Mink	OC/PCB	Great Lakes	Environment Canada (1991)
Seals	OC/PCB	North Sea	Blomkvist et al. (1992)
Otter	OC/PCB	Ireland	Mason and O'Sullivan (1992)

[a] OC, organochlorine insecticides; PCB, polychlorinated biphenyls; HM, heavy metals.
[b] Mortality.

and the changes in these levels with time." The National Pesticide Monitoring Program was established to follow residues in food and feed; humans; fish, wildlife, and estuaries; water; and soil. These activities were assigned to various federal agencies, and most findings were published in the *Pesticides Monitoring Journal* (e.g., birds; Cain, 1981) and later in journals such as *Environmental Monitoring and Assessment* (e.g., birds; Prouty and Bunck, 1986). Johnson et al. (1967) described the monitoring of pesticide residues in fish, wildlife, and estuaries conducted by the U.S. Fish and Wildlife Service. Later, the Service's program was enlarged to include other environmental contaminants. This enlarged effort now is called the National Contaminant Biomonitoring Program and was described by Jacknow et al. (1986). They reported that results for 1980-1981 surveys documented a continuing decline in DDE residues in ducks, but that high PCB residues persisted in ducks from the Atlantic flyway. DDT, PCB, and chlordane residues decreased in fish, but toxaphene residues remained high with regional variations in actual levels. Inorganic contaminants had not yet shown conclusive trends.

Table 5 Representative Studies of Contaminant Residues in Other Species and Groups of Vertebrates

Group	Contaminants[a]	Area	Ref.
Birds			
Everglade kite/whooping crane	OC	U.S.	Lamont and Reichel (1970)
Flicker/bluebird	OC	N.W. U.S.	Henny et al. (1977)
Turkey	HM	Virginia	Scanlon et al. (1979)
Eider/gulls	OC/PCB	E. U.S.	Szaro et al. (1979)
Tree swallow	OC/PCB	Alberta	Shaw (1984)
Egrets/herons	Se/HM	China	Burger and Gochfeld (1993)
Herring gull/common tern	Hg	North Sea	Thompson et al. (1993)
Bird groups			
Aquatic species	OC/PCB	Canada	Vermeer and Reynolds (1970)
Seed eaters/predators	Hg	Canada	Fimreite et al. (1970)
Gulls	DDE/PCB	Norway	Bjerk and Holt (1971)
Fish eaters/fish	Hg	Canada	Fimreite et al. (1971)
Aquatic species	OC/HM	Kenya	Koeman et al. (1972)
Gulls	OC/PCB/Hg	Ontario	Ryder (1974)
Herons	OC/PCB	Transvaal	Peakall and Kemp (1976)
Waterfowl	PCB/TE	New York	Baker et al. (1976)
Waterfowl	Hg	Canada	Pearce et al. (1976)
Seabirds	OC/PCB	Norway	Fimreite et al. (1977)
Diverse species	OC/PCB	France	Mendola and Risebrough (1977)
Fish eaters	HM	Canada	Vermeer and Peakall (1977)
Waterfowl	Cu/Ni	Ontario	Ranta et al. (1978)
Diverse species	OC/PCB	Spain	Baluja and Hernandez (1978)
Seabirds	OC/PCB	Alaska	Ohlendorf et al. (1982)
Shorebirds	OC	Texas	White et al. (1983a)
Diverse species	OC/PCB	Texas	White et al. (1983b)
Coastal species	OC/PCB	S. Africa	de Kock and Randall (1984)
Aquatic species	Various[b]	Lake Michigan	Heinz et al. (1985)
Waterfowl	Se/HM	California	Ohlendorf et al. (1986)
Waterfowl	OC	California/Mexico	Mora et al. (1987)
Aquatic species	OC/PCB/HM	Moravia	Hudec et al. (1988)
Terns/herons	OC/PCB/Hg	California	Ohlendorf et al. (1988)
Aquatic species	OC/PCB	Mississippi	White et al. (1988)
Gulls	OC/PCB	Spain	Gonzalez et al. (1991)
Diverse species	OC/PCB	Mexico	Mora and Anderson (1991)
Shorebirds	Cd/HM	Chile	Vermeer and Castilla (1991)
Aquatic species	OC/PCB	Great Lakes	Environment Canada (1991)
Seabirds	HM	E. Canada	Elliott et al. (1992)
Mammals			
Small mammals	OC/HM	U.K.	Jefferies and French (1976)
Carnivores	Mirex	S.E. U.S.	Hill and Dent (1985)
Fish			
Diverse species	Hg	Canada	Fimreite et al. (1971)
Diverse species	Hg	Georgia	Cumbie (1975b)
Diverse species	OC/PCB	Great Lakes	Frank et al. (1978)
Diverse species	OC	Texas	White et al. (1983a)
Diverse species	OC	Mississippi	Ford and Hill (1991)

Table 5 (continued) Representative Studies of Contaminant Residues in Other Species and Groups of Vertebrates

Group	Contaminants[a]	Area	Ref.
Lake trout	OC/PCB	Canada	Environment Canada (1991)
Reptiles/amphibians			
Turtles	Mirex	Mississippi	Holcomb and Parker (1979)
Diverse species	OC	W. U.S.	Punzo et al. (1979)
Crocodile	OC/PCB	Florida	Hall et al. (1979)
Snapping turtle	OC/PCB	Great Lakes	Environment Canada (1991)
Snakes	OC	Mississippi	Ford and Hill (1991)
General wildlife			
Diverse species	Dioxins	Italy	Fanelli et al. (1980)
Diverse species	HM	Pennsylvania	Beyer et al. (1985)
Diverse species	HM	Missouri	Niethammer et al. (1985)

[a] OC, organochlorine insecticides; HM, heavy metals; PCB, polychlorinated biphenyls; TE, trace elements.
[b] OC/PCB/Hg/PCS (polychlorinated styrenes)/PBB (polybrominated biphenyls).

Most biologists, whether looking at residues in molluscs (Butler, 1966), seabird eggs (Moore, 1966), bird and mammal tissues (De Witt et al., 1960), or tadpoles (Cooke, 1981), had their own ideas on which animals were the best indicator species. Furthermore, small mammals, as a group, were proposed as "monitors of environmental contaminants" (Talmage and Walton, 1991), and animals, in general, as "monitors of environmental quality" (Buck, 1979), and "indicators of pollution" (Bruggemann et al., 1974; National Academy of Sciences, 1979).

Monitoring programs produced information that reinforced conclusions gained in other investigations. Most samples analyzed contained residues. Few samples contained DDT, but most contained DDE and dieldrin, even 10 years after their use was banned. Residues of other organochlorine insecticides occurred at a much lower incidence and usually at lower levels. Monitoring led to the identification of several new problems. Relatively high DDE residues were found in birds near the sites of plants that previously manufactured DDT (O'Shea et al., 1980a). Monitoring pinpointed other areas where wildlife carried high levels of DDE. Subsequent investigations indicated that input of DDT and DDE into the environment of local areas in Arizona and New Mexico was continuing (Clark and Krynitsky, 1983). Likewise, monitoring suggested that there was continued input of DDE and toxaphene into drainages of the lower Rio Grande Valley in Texas (White et al., 1983b; White and Krynitsky, 1986).

BIOMARKERS

Residue data have been essential in identifying and solving wildlife problems created by pesticides. However, they have limitations and, increasingly, use has been made of biomarkers in pollution studies. Biomarkers are changes in cellular and biochemical activities in animals that result from exposure to contaminants. An example would be the inhibition of acetylcholinesterase in animals exposed to organophosphate and carbamate insecticides. Traditional residue analyses are of limited utility in organophosphate and carbamate studies, as residues are rapidly metabolized by birds and mammals following their exposure. However, some effects of exposure, such as cholinesterase inhibition, persist for many days and can be used to measure the intensity of exposure and the degree of hazard posed to wild animals (Hill and Fleming, 1982). Residues of organophosphates and carbamates extracted from water, plants, and other materials can be measured with a bioassay technique that measures cholinesterase inhibition in bovine blood (Ott and Gunther, 1966). Mulla et al. (1966) compared parathion residues in field samples measured by chemical analyses, mosquito larva bioassays, and bovine blood cholinesterase inhibition.

There are numerous other cellular and biochemical indicators that can be used to measure either residues or the intensity of animal exposure to contaminants. Two books have recently appeared that describe the use of various biomarkers in tracking effects on sentinel species and in assessing levels of environmental contamination in wildlife toxicology studies (McCarthy and Shugart, 1990; Peakall, 1992). As pesticide use practices have evolved from organochlorines to organophosphates and carbamates, biomarkers have become increasingly valuable in following the fate and effects of residues in wildlife and the environment. In 1964, 70% of the insecticides used were organochlorines, and 28% were organophosphates and carbamates. By 1982, only 6% were organochlorines, whereas 67% were organophosphates and 18% were carbamates (Osteen, 1993).

WHAT DID RESIDUE STUDIES ACCOMPLISH?

The findings of pesticide residues in animals, especially in dead ones, showed clearly that problems had been created by the introduction of synthetic pesticides into the environment. By the mid-1960s, the somewhat haphazard accumulation of residue data was sufficient for N. Moore (1965b) and L. Stickel (1968) to review the extent and possible hazards of pesticide contaminations in fish and wildlife. A few years later, with additional information, Miller and Berg (1969) and Edwards (1973) edited numerous accounts by experts on the effects of pesticide residues on organisms. Their books remain today the most thorough attempts to interpret pesticide residues found in diverse animals. The most impressive interpretations of residues in a single animal group are the considerations of the effects of DDE and dieldrin on peregrine falcon populations in Cade et al. (1988) and in a special issue of the *Canadian Field Naturalist* (Vol. 104, 1990) on peregrine falcons in the 1980s.

In large part, residue data have been generated by analytical chemists and interpreted by biologists. It was this symbiotic relationship between chemists and biologists that enabled effective studies of the fate of residues in the environment, the effect of residues on animals, and the development of the sciences of wildlife toxicology and ecotoxicology. Some of the most productive scientists were those such as David B. Peakall and Robert W. Risebrough, who were equally skilled as biologists, chemists, and biochemists. Undoubtedly, this combination of abilities contributed to their creativity and excellence in pollution studies.

The findings of pesticide residues in fish and wildlife led scientists to study the toxicity of pesticides to wild animals; to correlate residues in tissues with exposure, death, and debility; to conduct cause-and-effect studies; and to solve population problems in avian predators. The extent of a pesticide's use was important, but it was the persistent residues left behind that caused the greatest problems. The persistence, distribution, and sublethal toxicity of a pesticide often were more important than its acute toxicity.

At first, biologists did not know enough to fully interpret the residues reported by chemists. Often they had not gained sufficient experience to ask the right questions and to collect the most pertinent samples for analyses. Keith and Hunt (1966) reported that, in California, between 1963 and 1965, over 2100 samples of fish, wildlife, and environmental materials were collected and composited into 1200 samples for pesticide residue analysis. This 3-year effort did not identify or solve any specific problems. Results showed only the kinds and amounts of the different insecticides that occurred in tissues and environments of 86 species of fish and wildlife. However, that knowledge led to hypotheses that gave direction to future studies. Wetland wildlife seemed to receive greater exposure than did terrestrial animals. The source of exposure seemed primarily to be residues in foods, and animals at higher trophic levels in food chains carried greater residue levels than did producer organisms or first-level consumers. The authors concluded, "The greatest value of analyses reported...may be in directing the initial course of...future research."

In California, we certainly did not take full advantage of the information we reported. In hindsight, we might ask why we did not follow up immediately on the average DDE residues in fat of 60 ppm in bald eagles, of 38 ppm in Swainson's hawks, and of 25 ppm in California condors. DDE residues in egg yolks were about 56 ppm in red-tailed hawks and 29 ppm in osprey. We missed those clues, but perhaps our data helped orient others who were concerned about raptors. In a somewhat similar case, Risebrough et al. (1967) reported very high residues of DDE in the breast muscle of a brown pelican from California, and Risebrough et al. (1969) showed that DDE residues in anchovies were 50 times higher near pelican breeding colonies at Terminal Island than in San Francisco Bay. As Risebrough was working on residues in birds and fish in the marine environment, he saw the connection between these high DDE residues in anchovies and DDE residues in the pelicans that eat the anchovies. For most of us to make the connection, it took the alert sounded by Schreiber and De Long (1969) that brown pelicans were probably not producing many young; they found few juvenile birds in southern California. Again, it took both the chemical and the biological insights to initially describe the problem of reproductive failure in brown pelicans.

Subsequent studies illustrated that residues were distributed worldwide and were responsible for severe and unexpected effects. Carnivores usually carried greater residues than did herbivores and, ultimately, it was their populations that suffered the greatest effects. Studies of DDE accumulation in food chains and of reproductive problems in endangered populations of avian predators coalesced and led to the discovery of eggshell thinning in predaceous birds and to the inverse relation between increasing DDE residues and decreasing eggshell thickness. The final conclusions were rather straightforward; i.e., the crushing by incubating birds of eggshells thinned by DDE caused reproductive failure and population decreases in many species of birds. Studies in the laboratory and field have used residue data to evaluate the effects of most insecticides, their metabolites, PCBs, mercury, and other heavy metals on fish and wildlife. To date, the effects of DDE on eggshell thinning, decreased reproductive success, and avian population declines remain the most important impacts identified of contaminant residues in the environment.

In large part, it was the data on residues in fish and wildlife that provided the impetus for concern regarding pesticide use practices and environmental pollution. It was the study of the occurrence, levels, and effects of residues that produced the evidence used in court hearings to ban harmful pesticides in the U.S. Certainly the regulations that promulgated worldwide restriction of use of the chlorinated hydrocarbon insecticides were stimulated by wildlife studies that showed the persistence, hazards, and unexpected effects of those insecticides in the environment.

REFERENCES

Abbott, D. C., R. B. Harrison, J. O. Tatton, and J. Thompson. 1966. Organochlorine pesticides in the atmosphere. Nature (Lond.) 211:259-261.

Adams, L., M. G. Hanavan, N. W. Hosley, and D. W. Johnston. 1949. The effects on fish, birds, and mammals of DDT used in the control of forest insects in Idaho and Wyoming. J. Wildl. Manage. 13:245-254.

Addison, R. F., and P. F. Brodie. 1977. Organochlorine residues on maternal blubber, milk, and pup blubber from grey seals *(Halichoerus grypus)* from Sable Island, Nova Scotia. J. Fish. Res. Board Can. 34:937-941.

Addison, R. F., and T. G. Smith. 1974. Organochlorine residue levels in Arctic ringed seals: variation with age and sex. Oikos 25:335-337.

Albert, L. A., C. Barcenas, M. Ramos, and E. Iñigo. 1989. Organochlorine pesticides and reduction of eggshell thickness in a black vulture *Coragyps atratus* population of the Tuxtla Valley, Chiapas, Mexico. p. 473-475. *In* B.-U. Meyburg and R. D. Chancellor (Eds.). Raptors in the modern world. World Working Group on Birds of Prey. London.

Ames, P. L. 1966. DDT residues in the eggs of the osprey in the northeastern United States and their relation to nesting success. J. Appl. Ecol. 3 (Suppl.):87-97.

Anderson, D. W., and F. Gress. 1983. Status of a northern population of California brown pelicans. Condor 85:79-88.

Anderson, D. W., and F. Hamerstrom. 1966. The recent status of Wisconsin cormorants. Passenger Pigeon 29:3-15.

Anderson, D. W., and J. J. Hickey. 1972. Eggshell changes in certain North American birds. Proc. Int. Ornithol. Congr. 15:514-540.

Anderson, D. W., J. J. Hickey, R. W. Risebrough, D. F. Hughes, and R. E. Christensen. 1969. Significance of chlorinated hydrocarbon residues to breeding pelicans and cormorants. Can. Field Nat. 83:91-112.

Anderson, D. W., J. R. Jehl, Jr., R. W. Risebrough, L. A. Woods, Jr., L. R. De Weese, and W. G. Edgecomb. 1975. Brown pelicans: improved reproduction off the southern California coast. Science (Washington, D.C.) 190:806-808.

Anderson, D. W., D. G. Raveling, R. W. Risebrough, and A. M. Springer. 1984. Dynamics of low-level organochlorines in adult cackling geese over the annual cycle. J. Wildl. Manage. 48:1112-1127.

Anderson, W. L., and R. E. Duzan. 1978. DDE residues and eggshell thinning in loggerhead shrikes. Wilson Bull. 90:215-220.

Antommaria, P., M. Corn, and L. De Maio. 1965. Airborne particulates in Pittsburgh: association with p,p'-DDT. Science (Washington, D.C.) 150:1476-1477.

Babcock, K. M., and E. L. Flickinger. 1977. Dieldrin mortality of lesser snow geese in Missouri. J. Wildl. Manage. 41:100-103.

Baker, F. D., C. F. Tumasonis, W. B. Stone, and B. Bush. 1976. Levels of PCB and trace metals in waterfowl in New York State. N.Y. Fish Game J. 23:82-91.

Baluja, G., and L. M. Hernandez. 1978. Organochlorine pesticide and PCB residues in wild bird eggs from the southwest of Spain. Bull. Environ. Contam. Toxicol. 19:655-664.

Bann, J. M., T. J. De Cino, N. W. Earle, and Y. P. Sun. 1956. The fate of aldrin and dieldrin in the animal body. J. Agric. Food Chem. 4:937-941.

Barbehenn, K. R., and W. L. Reichel. 1981. Organochlorine concentrations in bald eagles: brain/body lipid relations and hazard evaluation. J. Toxicol. Environ. Health 8:325-330.

Barker, R. J. 1958. Notes on some ecological effects of DDT sprayed on elms. J. Wildl. Manage. 22:269-274.

Barnett, D. C. 1950. The effect of some insecticide sprays on wildlife. Proc. Annu. Conf. West. Assoc. State Game Fish Comm. 30:125-134.

Beatty, R. G. 1973. The DDT myth. John Day, New York. 201 pp.

Belisle, A. A., W. L. Reichel, L. N. Locke, T. G. Lamont, B. M. Mulhern, R. M. Prouty, R. B. De Wolf, and E. Cromartie. 1972. Residues of organochlorine pesticides, polychlorinated biphenyls, and mercury and autopsy data for bald eagles, 1969 and 1970. Pestic. Monit. J. 6:133-138.

Bengston, S. A., and A. Sodergren. 1974. DDT and PCB residues in airborne fallout and animals in Iceland. Ambio 3:84-86.

Benson, W. W., D. W. Brock, J. Gabica, and M. Loomis. 1976a. Pesticide and mercury levels in pelicans in Idaho. Bull. Environ. Contam. Toxicol. 15:543-546.

Benson, W. W., D. W. Brock, J. Gabica, and M. Loomis. 1976b. Swan mortality due to certain heavy metals in the Mission Lake area, Idaho. Bull. Environ. Contam. Toxicol. 15:171-174.

Benton, A. H. 1951. Effects on wildlife of DDT used for control of Dutch elm disease. J. Wildl. Manage. 15:20-27.

Bernard, R. F. 1963. Studies on the effects of DDT on birds. Mich. State Univ. Mus. Publ. Biol. Ser. 2:160-191.

Bernard, R. F. 1966. DDT residues in avian tissues. J. Appl. Ecol. 3 (Suppl.):193-198.

Bevenue, A., J. N. Ogata, and J. W. Hylin. 1972. Organochlorine pesticides in rainwater, Oahu, Hawaii, 1971-1972. Bull. Environ. Contam. Toxicol. 8:238-241.

Beyer, W. N., and A. J. Krynitsky. 1989. Long-term persistence of dieldrin, DDT, and heptachlor epoxide in earthworms. Ambio 18:271-273.

Beyer, W. N., O. H. Pattee, L. Sileo, D. J. Hoffman, and B. M. Mulhern. 1985. Metal contamination in wildlife living near two zinc smelters. Environ. Pollut. Ser. A Ecol. Biol. 38:63-86.

Bidleman, T. F., and C. E. Olney. 1974. Chlorinated hydrocarbons in the Sargasso Sea atmosphere and sea water. Science (Washington, D.C.) 183:516-518.

Bijleveld, M. F. I. J., P. Goeldlin, and J. Mayol. 1979. Persistent pollutants in Audouin's gull *(Larus audouinii)* in the western Mediterranean: a case study with wide implications? Environ. Conserv. 6:139-142.

Bjerk, J. E., and G. Holt. 1971. Residues of DDE and polychlorinated biphenyls in eggs from herring gull *(Larus argentatus)* and common gull *(Larus canus)* in Norway. Acta Vet. Scand. 12:429-441.

Blomkvist, G., A. Roos, A. Bignert, M. Olsson, and S. Jensen. 1992. Concentration of DDT and PCB in seals from Swedish and Scottish waters. Ambio 21:539-545.

Blus, L., E. Cromartie, L. McNease, and T. Joanen. 1979b. Brown pelican: population status, reproductive success, and organochlorine residues in Louisiana, 1971-1976. Bull. Environ. Contam. Toxicol. 22:128-134.

Blus, L. J., R. G. Heath, C. D. Gish, A. A. Belisle, and R. M. Prouty. 1971. Eggshell thinning in the brown pelican: implication of DDE. Bioscience 21:1213-1215.

Blus, L. J., C. J. Henny, A. Anderson, and R. E. Fitzner. 1985a. Reproduction, mortality, and heavy metal concentrations in great blue herons from three colonies in Washington and Idaho. Colon. Waterbirds 8:110-116.

Blus, L. J., C. J. Henny, D. J. Hoffman, and R. A. Grove. 1991. Lead toxicosis in tundra swans near a mining and smelting complex in northern Idaho. Arch. Environ. Contam. Toxicol. 21:549-555.

Blus, L. J., C. J. Henny, and A. J. Krynitsky. 1985b. Organochlorine-induced mortality and residues in long-billed curlews from Oregon. Condor 87:563-565.

Blus, L. J., T. G. Lamont, and B. S. Neely, Jr. 1979a. Effects of organochlorine residues on eggshell thickness, reproduction, and population status of brown pelicans *(Pelecanus occidentalis)* in South Carolina and Florida, 1969-76. Pestic. Monit. J. 12:172-184.

Blus, L. J., O. H. Pattee, C. J. Henny, and R. M. Prouty. 1983. First records of chlordane-related mortality in wild birds. J. Wildl. Manage. 47:196-198.

Blus, L. J., and R. M. Prouty. 1979. Organochlorine pollutants and population status of least terns in South Carolina. Wilson Bull. 91:62-71.

Boellstorff, D. E., H. M. Ohlendorf, D. W. Anderson, E. J. O'Neill, J. O. Keith, and R. M. Prouty. 1985. Organochlorine chemical residues in white pelicans and western grebes from the Klamath Basin, California. Arch. Environ. Contam. Toxicol. 14:485-493.

Borg, K., H. Wanntorp, K. Erne, and E. Hanko. 1969. Alkyl mercury poisoning in terrestrial Swedish wildlife. Viltrevy (Stockh.) 6:301-379.

Bowes, G. W., and C. J. Jonkel. 1975. Presence and distribution of polychlorinated biphenyls (PCB) in arctic and subarctic marine food chains. J. Fish. Res. Board Can. 32:2111-2123.

Boykins, E. A. 1967. The effects of DDT-contaminated earthworms in the diet of birds. Bioscience 17:37-39.

Braestrup, L., J. Clausen, and O. Berg. 1974. DDE, PCB, and aldrin levels in Arctic birds of Greenland. Bull. Environ. Contam. Toxicol. 11:326-332.

Breidenbach, A. W. 1965. Pesticide residues in air and water. Arch. Environ. Health 10:827-830.

Breidenbach, A. W., C. G. Gunnerson, F. K. Kawahara, J. J. Lichtenberg, and R. S. Green. 1967. Chlorinated hydrocarbon pesticides in major river basins 1957-65. Public Health Rep. 82:139-156.

Brewerton, H. V. 1969. DDT in fats of Antarctic animals. N. Z. J. Sci. 12:194-199.

Bridges, W. R. 1961. Disappearance of endrin from fish and other materials of a pond environment. Trans. Am. Fish. Soc. 90:332-334.

Brown, R. A., and H. L. Huffman, Jr. 1976. Hydrocarbons in open ocean water. Science (Washington, D.C.) 191:847-849.

Bruggemann, J., L. Busch, U. Drescher-Kaden, W. Eisele, and P. Hoppe. 1974. Pesticide residues in organs of wild living animals as an indicator of pollution. Int. Congr. Game Biol. 11:439-449.

Bryant, C. P., R. W. Young, and R. L. Kirkpatrick. 1978. Kepone residues in body tissues of raccoons collected along the James River, east of Hopewell, Virginia. Va. J. Sci. 29:57. (Abstr.)

Bryson, M. J., C. I. Draper, J. R. Harris, C. Biddulph, D. A. Greenwood, L. E. Harris, W. Binns, M. L. Miner, and L. L. Madsen. 1949. DDT in eggs and tissues of chickens fed varying levels of DDT. Proc. Am. Soc. Hortic. Sci. 54:232-236.

Buck, W. B. 1979. Animals as monitors of environmental quality. Vet. Hum. Toxicol. 21:277-284.

Burger, J., and M. Gochfeld. 1993. Heavy metal and selenium levels in feathers of young egrets and herons from Hong Kong and Szechuan, China. Arch. Environ. Contam. Toxicol. 25:322-327.

Butler, P. A. 1966. Pesticides in the marine environment. J. Appl. Ecol. 3 (Suppl.):253-259.

Cade, T. J., J. H. Enderson, C. G. Thelander, and C. M. White (Eds.). 1988. Peregrine falcon populations: their management and recovery. Peregrine Fund, Boise, Id. 949 pp.

Cade, T. J., J. L. Lincer, and C. M. White. 1971. DDE residues and eggshell changes in Alaskan falcons and hawks. Science (Washington, D.C.) 172:955-957.

Cade, T. J., J. L. Lincer, C. M. White, D. G. Roseneau, and L. G. Swartz. 1971. DDE residues and eggshell changes in Alaskan falcons and hawks. Science (Washington, D.C.) 172:955-957.

Cade, T. J., C. M. White, and J. R. Haugh. 1968. Peregrines and pesticides in Alaska. Condor 70:170-178.

Cain, B. W. 1981. Nationwide residues of organochlorine compounds in wings of adult mallards and black ducks, 1979-80. Pestic. Monit. J. 15:128-134.

Capen, D. E., and T. J. Leiker. 1979. DDE residues in blood and other tissues of white-faced ibis. Environ. Pollut. 19:163-171.

Carter, R. H., R. W. Wells, R. D. Radeleff, C. L. Smith, P. E. Hubanks, and H. D. Mann. 1949. The chlorinated hydrocarbon content of milk from cattle sprayed for control of horn flies. J. Econ. Entomol. 42:116-118.

Chadwick, G. G., and R. W. Brocksen. 1969. Accumulation of dieldrin by fish and selected fish-food organisms. J. Wildl. Manage. 33:693-700.

Clark, A. L., and D. B. Peakall. 1977. Organochlorine residues in Eleonora's falcon *Falco eleonorae*, its eggs and its prey. Ibis 119:353-358.

Clark, D. R., Jr. 1978. Uptake of dietary PCB by pregnant big brown bats *(Eptesicus fuscus)* and their fetuses. Bull. Environ. Contam. Toxicol. 19:707-714.

Clark, D. R., Jr. 1981. Bats and environmental contaminants: a review. U. S. Fish Wildl. Serv. Spec. Sci. Rep. Wildl. No. 235, 27 pp.

Clark, D. R., Jr., and A. J. Krynitsky. 1983. DDT: recent contamination in New Mexico and Arizona? Environment (Washington, D.C.) 25:27-31.

Clark, D. R., Jr., A. S. Wenner, and J. F. Moore. 1986. Metal residues in bat colonies, Jackson County, Florida, 1981-1983. Fla. Field Nat. 14:38-45.

Clawson, S. G., and M. F. Baker. 1959. Immediate effects of dieldrin and heptachlor on bobwhites. J. Wildl. Manage. 23:215-219.

Cohen, J. M., and C. Pinkerton. 1966. Widespread translocation of pesticides by air transport and rainout. Adv. Chem. Ser. 60:163-176.

Cooke, A. S. 1981. Tadpoles as indicators of harmful levels of pollution in the field. Environ. Pollut. Ser. A Ecol. Biol. 25:123-133.

Cope, O. B. 1961. Effects of DDT spraying for spruce budworm on fish in the Yellowstone River system. Trans. Am. Fish. Soc. 90:239-251.

Cope, O. B. 1966. Contamination of the freshwater ecosystem by pesticides. J. Appl. Ecol. 3 (Suppl.):33-44.

Cope, O. B., C. M. Gjullin, and A. Storm. 1947. Effects of some insecticides on trout and salmon in Alaska, with reference to blackfly control. Trans. Am. Fish. Soc. 77:160-177.

Cumbie, P. M. 1975a. Mercury in hair of bobcats and raccoons. J. Wildl. Manage. 39:419-425.

Cumbie, P. M. 1975b. Mercury levels in Georgia otter, mink and freshwater fish. Bull. Environ. Contam. Toxicol. 14:193-196.

Custer, T. W., G. L. Hensler, and T. E. Kaiser. 1983. Clutch size, reproductive success, and organochlorine contaminants in Atlantic coast black-crowned night-herons. Auk 100:699-710.

Custer, T. W., and C. A. Mitchell. 1989. Organochlorine contaminants in white-faced ibis eggs in southern Texas. Colon. Waterbirds 12:126-129.

Custer, T. W., and C. A. Mitchell. 1991. Contaminant exposure of willets feeding in agricultural drainages of the lower Rio Grande Valley of south Texas. Environ. Monit. Assess. 16:189-200.

Custer, T. W., and B. L. Mulhern. 1983. Heavy metal residues in prefledgling black-crowned night-herons from three Atlantic coast colonies. Bull. Environ. Contam. Toxicol. 30:178-185.

Danckwortt, P. A., and E. Pfau. 1926. Massenvergiftungen von Tieren durch Arsenbestäubung vom Flugzeug. (In German.) Angew. Chem. 39:1486-1487.

Davidow, B., and J. Radomski. 1952. Metabolite of heptachlor, its analysis, storage, and toxicity. Fed. Proc. 11:336.

Davis, B. N. K. 1966. Soil animals as vectors of organochlorine insecticides for ground-feeding birds. J. Appl. Ecol. 3 (Suppl.):133-139.

de Freitas, A. S. W., J. S. Hart, and H. W. Morley. 1969. Chronic cold exposure and DDT toxicity. p. 361-367. *In* M. W. Miller and G. G. Berg (Eds.) Chemical fallout: current research on persistent pesticides. Charles C Thomas, Springfield, Ill.

de Kock, A. C., and R. M. Randall. 1984. Organochlorine insecticide and polychlorinated biphenyl residues in eggs of coastal birds from the Eastern Cape, South Africa. Environ. Pollut. Ser. A Ecol. Biol. 35:193-201.

de Kock, A. C., and R. Simmons. 1988. Chlorinated hydrocarbon residues in African marsh harrier eggs and concurrent reproductive trends. Ostrich 59:180-181.

de Kock, A. C., and R. T. Watson. 1985. Organochlorine residue levels in bateleur eggs from the Transvaal. Ostrich 56:278-280.

De Smet, K. D. 1987. Organochlorines, predators, and reproductive success of the red-necked grebe in southern Manitoba. Condor 89:460-467.

De Weese, L. R., R. R. Cohen, and C. J. Stafford. 1985. Organochlorine residues and eggshell measurements for tree swallows *(Tachycineta bicolor)* in Colorado. Bull. Environ. Contam. Toxicol. 35:767-775.

De Weese, L. R., L. C. McEwen, G. L. Hensler, and B. E. Petersen. 1986. Organochlorine contaminants in Passeriformes and other avian prey of the peregrine falcon in the western United States. Environ. Toxicol. Chem. 5:675-693.

De Witt, J. B., J. V. Derby, Jr., and G. F. Mangan, Jr. 1955. DDT vs. wildlife. Relationships between quantities ingested, toxic effects, and tissue storage. J. Am. Pharm. Assoc. Sci. Ed. 44:22-24.

De Witt, J. B., C. M. Menzie, V. A. Adomaitis, and W. L. Reichel. 1960. Pesticidal residues in animal tissues. Trans. N. Am. Wildl. Nat. Resour. Conf. 25:277-285.

Diephius, F., and C. L. Dunn. 1949. Toxaphene in tissues of cattle and sheep fed toxaphene-treated alfalfa. p. 22-26. *In* Toxaphene residues. Montana State Coll. Agric. Exp. Station Bull. 461.

Dindal, D. L. 1970. Accumulation and excretion of Cl^{36} DDT in mallard and lesser scaup ducks. J. Wildl. Manage. 34:74-92.

Diters, R. W., and S. W. Nielsen. 1978. Lead poisoning of raccoons in Connecticut. J. Wildl. Dis. 14:187-192.

Douthwaite, R. J. 1992. Effects of DDT on the fish eagle *Haliaeetus vocifer* population of Lake Kariba in Zimbabwe. Ibis 134:250-258.

Draper, C. I., C. Biddulph, D. A. Greenwood, J. R. Harris, W. Binns, and M. L. Miner. 1950. Concentrations of DDT in tissues of chickens fed varying levels of DDT in the diet. Poult. Sci. 29:756. (Abstr.)

Dustman, E. H., and L. F. Stickel. 1966. Pesticide residues in the ecosystem. p. 109-121. *In* Pesticides and their effects on soils and water. Am. Soc. Agronomy Special Publ. No. 8.

Dustman, E. H., and L. F. Stickel. 1969. The occurrence and significance of pesticide residues in wild animals. Ann. N.Y. Acad. Sci. 160:162-172.

Ecobichon, D. J., and P. W. Saschenbrecker. 1969. The redistribution of stored DDT in cockerels under the influence of food deprivation. Toxicol. Appl. Pharmacol. 15:420-432.

Edwards, C. A. 1970. Persistent pesticides in the environment. Crit. Rev. Environ. Control 1:7-67.

Edwards, C. A. 1973. Environmental pollution by pesticides. Plenum Press, New York. 542 pp.

Edwards, J. G. 1972. Cracking the thin shell myth. Agric. Chem. Commer. Fert. 27:20-21, 26.

Elfving, D. C., R. A. Stehn, I. S. Pakkala, and D. J. Lisk. 1979. Arsenic content of small mammals indigenous to old orchard soils. Bull. Environ. Contam. Toxicol. 21:62-64.

Elliott, J. E., A. M. Scheuhammer, F. A. Leighton, and P. A. Pearce. 1992. Heavy metal and metallothionein concentrations in Atlantic Canadian seabirds. Arch. Environ. Contam. Toxicol. 22:63-73.

Ellis, D. H., L. R. De Weese, T. G. Grubb, L. F. Kiff, D. G. Smith, W. M. Jarman, and D. B. Peakall. 1989. Pesticide residues in Arizona peregrine falcon eggs and prey. Bull. Environ. Contam. Toxicol. 42:57-64.

Enderson, J. H., and D. D. Berger. 1970. Pesticides: eggshell thinning and lowered production of young in prairie falcons. Bioscience 20:355-356.

Enderson, J. H., and P. H. Wrege. 1973. DDE residues and eggshell thickness in prairie falcons. J. Wildl. Manage. 37:476-478.

Environment Canada. 1991. Toxic chemicals in the Great Lakes and associated effects. Synopsis. Ottawa. 51 pp.

Faber, R. A., and J. J. Hickey. 1973. Eggshell thinning, chlorinated hydrocarbons, and mercury in inland aquatic bird eggs, 1969 and 1970. Pestic. Monit. J. 7:27-36.

Fanelli, R., M. G. Castelli, G. P. Martelli, A. Noseda, and S. Garattini. 1980. Presence of 2,3,7,8-tetrachlorodibenzo-*p*-dioxin in wildlife living near Seveso, Italy: a preliminary study. Bull. Environ. Contam. Toxicol. 24:460-462.

Fimreite, N., J. E. Bjerk, N. Kveseth, and E. Brun. 1977. DDE and PCBs in eggs of Norwegian seabirds. Astarte 10:15-20.

Fimreite, N., R. W. Fyfe, and J. A. Keith. 1970. Mercury contamination of Canadian prairie seed eaters and their avian predators. Can. Field Nat. 84:269-276.

Fimreite, N., W. N. Holsworth, J. A. Keith, P. A. Pearce, and I. M. Gruchy. 1971. Mercury in fish and fish-eating birds near sites of industrial contamination in Canada. Can. Field Nat. 85:211-220.

Findholt, S. L., and C. H. Trost. 1985. Organochlorine pollutants, eggshell thickness, and reproductive success of black-crowned night-herons in Idaho, 1979. Colon. Waterbirds 8:32-41.

Findlay, O. G., and A. S. W. de Freitas. 1971. DDT movement from adipocyte to muscle cell during lipid utilization. Nature (Lond.) 229:63-65.

Finnegan, J. K., H. B. Haag, and P. S. Larson. 1949. Tissue distribution and elimination of DDD and DDT following oral administration to dogs and rats. Proc. Soc. Exp. Biol. Med. 72:357-360.

Fleet, R. R., and F. W. Plapp, Jr. 1978. DDT residues in snakes decline since DDT ban. Bull. Environ. Contam. Toxicol. 19:383-388.

Flickinger, E. L. 1979. Effects of aldrin exposure on snow geese in Texas rice fields. J. Wildl. Manage. 43:94-101.

Flickinger, E. L., C. A. Mitchell, and A. J. Krynitsky. 1986. Dieldrin and endrin residues in fulvous whistling-ducks in Texas in 1983. J. Field Ornithol. 57:85-90.

Focardi, S., C. Fossi, M. Lambertini, C. Leonzio, and A. Massi. 1988. Long term monitoring of pollutants in eggs of yellow-legged herring gull from Capraria Island (Tuscan Archipelago). Environ. Monit. Assess. 10:43-50.

Foley, R. E., S. J. Jackling, R. J. Sloan, and M. K. Brown. 1988. Organochlorine and mercury residues in wild mink and otter: comparison with fish. Environ. Toxicol. Chem. 7:363-374.

Ford, W. M., and E. P. Hill. 1990. Organochlorine contaminants in eggs and tissue of wood ducks from Mississippi. Bull. Environ. Contam. Toxicol. 45:870-875.

Ford, W. M., and E. P. Hill. 1991. Organochlorine pesticides in soil sediments and aquatic animals in the Upper Steele Bayou watershed of Mississippi. Arch. Environ. Contam. Toxicol. 20:161-167.

Forrester, D. J., D. K. Odell, N. P. Thompson, and J. R. White. 1980. Morphometrics, parasites, and chlorinated hydrocarbon residues of pygmy killer whales from Florida. J. Mammal. 61:356-360.

Forsyth, D. J., T. J. Peterle, and L. W. Bandy. 1983. Persistence and transfer of ^{36}Cl-DDT in the soil and biota of an old-field ecosystem: a six-year balance study. Ecology 64:1620-1636.

Fox, G. A., and T. Donald. 1980. Organochlorine pollutants, nest-defense behavior, and reproductive success in merlins. Condor 82:81-84.

Fox, N. C., and J. W. Lock. 1979. Organochlorine residues in New Zealand birds of prey. N. Z. J. Ecol. 1:118-125.

Frank, L. G., R. M. Jackson, J. E. Cooper, and M. C. French. 1977. A survey of chlorinated hydrocarbon residues in Kenyan birds of prey. East Afr. Wildl. J. 15:295-304.

Frank, R., M. Holdrinet, H. E. Braun, D. P. Dodge, and G. E. Sprangler. 1978. Residues of organochlorine insecticides and polychlorinated biphenyls in fish from Lakes Huron and Superior, Canada — 1968-76. Pestic. Monit. J. 12:60-68.

Froslie, A., G. Holt, and G. Norheim. 1986. Mercury and persistent chlorinated hydrocarbons in owls *Strigiformes* and birds of prey *Falconiformes* collected in Norway during the period 1965-1983. Environ. Pollut. Ser. B Chem. Phys. 91:91-108.

Fyfe, R. W., J. Campbell, B. Hayson, and K. Hodson. 1969. Regional population declines and organochlorine insecticides in Canadian prairie falcons. Can. Field Nat. 83:191-200.

Gannon, N., and J. H. Bigger. 1958. The conversion of aldrin and heptachlor to their epoxides in soil. J. Econ. Entomol. 51:1-2.

Gannon, N., and G. C. Decker. 1959. Insecticide residues as hazards to warm-blooded animals. Trans. N. Am. Wildl. Conf. 24:124-132.

Genelly, R. E., and R. L. Rudd. 1956. Chronic toxicity of DDT, toxaphene, and dieldrin to ring-necked pheasants. Calif. Fish Game 42:5-14.
George, J. L., and D. E. H. Frear. 1966. Pesticides in the Antarctic. J. Appl. Ecol. 3 (Suppl.):155-167.
George, J. L., and W. H. Stickel. 1949. Wildlife effects of DDT dust used for tick control on a Texas prairie. Am. Midl. Nat. 42:228-237.
Giles, R. H., Jr. 1970. The ecology of a small forested watershed treated with the insecticide malathion-S^{35}. Wildl. Monogr. 24:1-81.
Gilman, A. P., G. A. Fox, D. B. Peakall, S. M. Teeple, T. R. Carroll, and G. T. Haymes. 1977. Reproductive parameters and egg contaminant levels of Great Lakes herring gulls. J. Wildl. Manage. 41:458-468.
Glotfelty, D. E., J. N. Seiber, and L. A. Liljedahl. 1987. Pesticides in fog. Nature (Lond.) 325:602-605.
Gonzalez, L. M., and F. Hiraldo. 1988. Organochlorine and heavy metal contamination in the eggs of the Spanish imperial eagle *(Aquila [heliaca] adalberti)* and accompanying changes in eggshell morphology and chemistry. Environ. Pollut. 51:241-258.
Gonzalez, M. J., M. A. Fernandez, and L. M. Hernandez. 1991. Levels of chlorinated insecticides, total PCBs, and PCB congeners in Spanish gull eggs. Arch. Environ. Contam. Toxicol. 20:343-348.
Goodrum, P., W. P. Baldwin, and J. W. Aldrich. 1949. Effect of DDT on animal life of Bull's Island, South Carolina. J. Wildl. Manage. 13:1-10.
Grier, J. W. 1982. Ban of DDT and subsequent recovery of reproduction in bald eagles. Science (Washington, D.C.) 218:1232-1235.
Grubb, T. G., S. N. Wiemeyer, and L. F. Kiff. 1990. Eggshell thinning and contaminant levels in bald eagle eggs from Arizona, 1977 to 1985. Southwest. Nat. 35:298-301.
Hall, R. J. 1980. Effects of environmental contaminants on reptiles: a review. U. S. Fish Wildl. Spec. Sci. Rep. Wildl. No. 228, 12 pp.
Hall, R. J., T. E. Kaiser, W. B. Robertson, Jr., and P. C. Patty. 1979. Organochlorine residues in eggs of the endangered American crocodile *(Crocodylus acutus)*. Bull. Environ. Contam. Toxicol. 23:87-90.
Hanson, W. R. 1952. Effects of some herbicides and insecticides on biota of North Dakota marshes. J. Wildl. Manage. 16:299-308.
Harrington, R. W., Jr., and W. L. Bidlingmayer. 1958. Effects of dieldrin on fishes and invertebrates of a salt marsh. J. Wildl. Manage. 22:76-82.
Harris, H. J., T. C. Erdman, G. T. Ankley, and K. B. Lodge. 1993. Measures of reproductive success and polychlorinated biphenyl residues in eggs and chicks of Forster's terns on Green Bay, Lake Michigan, Wisconsin — 1988. Arch. Environ. Contam. Toxicol. 25:304-314.
Harvey, G. R., H. P. Miklas, V. T. Bowen, and W. G. Steinhauer. 1974. Observations on the distribution of chlorinated hydrocarbons in Atlantic Ocean organisms. J. Mar. Res. 32:103-118.
Harvey, J. M. 1967. Excretion of DDT by migratory birds. Can. J. Zool. 45:629-633.
Havera, S. P., and R. E. Duzan. 1977. Residues of organochlorine insecticides in fox squirrels from south-central Illinois. Trans. Ill. State Acad. Sci. 70:375-379.
Hazeltine, W. 1972. Disagreements on why brown pelican eggs are thin. Nature (Lond.) 239:410-411.
Heath, R. G., J. W. Spann, and J. F. Kreitzer. 1969. Marked DDE impairment of mallard reproduction in controlled studies. Nature (Lond.) 224:47-48.

Heinz, G. H., T. C. Erdman, S. D. Haseltine, and C. Stafford. 1985. Contaminant levels in colonial waterbirds from Green Bay and Lake Michigan, 1975-80. Environ. Monit. Assess. 5:223-236.

Heinz, G. H., S. D. Haseltine, W. L. Reichel, and G. L. Hensler. 1983. Relationships of environmental contaminants to reproductive success in red-breasted mergansers *Mergus serrator* from Lake Michigan. Environ. Pollut. Ser. A Ecol. Biol. 32:211-232.

Heinz, G. H., and R. W. Johnson. 1981. Diagnostic brain residues of dieldrin: some new insights. p. 72-92. *In* D. W. Lamb and E. E. Kenaga (Eds.). Avian and mammalian wildlife toxicology: second conference. ASTM STP 757, Philadelphia, Pa.

Helander, B., M. Olsson, and L. Reutergardh. 1982. Residue levels of organochlorine and mercury compounds in unhatched eggs and the relationships to breeding success in white-tailed sea eagles *(Haliaeetus albicilla)* in Sweden. Holarct. Ecol. 5:349-366.

Henny, C. J., and L. J. Blus. 1986. Radiotelemetry locates wintering grounds of DDE-contaminated black-crowned night-herons. Wildl. Soc. Bull. 14:236-241.

Henny, C. J., L. J. Blus, and C. S. Hulse. 1985. Trends and effects of organochlorine residues on Oregon and Nevada wading birds, 1979-83. Colon. Waterbirds 8:117-128.

Henny, C. J., L. J. Blus, and T. E. Kaiser. 1984a. Heptachlor seed treatment contaminates hawks, owls, and eagles of Columbia Basin, Oregon. Raptor Res. 18:41-48.

Henny, C. J., L. J. Blus, A. J. Krynitsky, and C. M. Bunck. 1984b. Current impact of DDE on black-crowned night-herons in the intermountain west. J. Wildl. Manage. 48:1-13.

Henny, C. J., L. J. Blus, S. P. Thompson, and U. W. Wilson. 1989. Environmental contaminants, human disturbance, and nesting of double-crested cormorants in northwestern Washington. Colon. Waterbirds 12:198-206.

Henny, C. J., and G. B. Herron. 1989. DDE, selenium, mercury, and white-faced ibis reproduction at Carson Lake, Nevada. J. Wildl. Manage. 53:1032-1045.

Henny, C. J., and T. E. Kaiser. 1979. Organochlorine and mercury residues in Swainson's hawk eggs from the Pacific Northwest. Murrelet 60:2-5.

Henny, C. J., C. Maser, J. O. Whitaker, Jr., and T. E. Kaiser. 1982a. Organochlorine residues in bats after a forest spraying with DDT. Northwest Sci. 56:329-337.

Henny, C. J., R. A. Olson, and D. L. Meeker. 1977. Residues in common flicker and mountain bluebird eggs one year after a DDT application. Bull. Environ. Contam. Toxicol. 18:115-122.

Henny, C. J., F. P. Ward, K. E. Riddle, and R. M. Prouty. 1982b. Migratory peregrine falcons, *Falco peregrinus*, accumulate pesticides in Latin America during winter. Can. Field Nat. 96:333-338.

Heppleston, P. B. 1973. Organochlorines in British grey seals. Mar. Pollut. Bull. 4:44-45.

Herman, S. G., R. L. Garrett, and R. L. Rudd. 1969. Pesticides and the western grebe. p. 24-53. *In* M. W. Miller and G. G. Berg (Eds.). Chemical fallout: current research on persistent pesticides. Charles C Thomas, Springfield, Ill.

Hickey, J. J. (Ed.). 1969. Peregrine falcon populations. Their biology and decline. University of Wisconsin Press, Madison, Wisc. 595 pp.

Hickey, J. J., and D. W. Anderson. 1968. Chlorinated hydrocarbons and eggshell changes in raptorial and fish-eating birds. Science (Washington, D.C.) 162:271-273.

Hickey, J. J., and L. B. Hunt. 1960. Initial songbird mortality following a Dutch elm disease control program. J. Wildl. Manage. 24:259-265.

Hickey, J. J., J. A. Keith, and F. B. Coon. 1966. An exploration of pesticides in a Lake Michigan ecosystem. J. Appl. Ecol. 3 (Suppl.):141-154.

Hill, E. F., and W. J. Fleming. 1982. Anticholinesterase poisoning of birds: field monitoring and diagnosis of acute poisoning. Environ. Toxicol. Chem. 1:27-38.

Hill, E. P., and D. M. Dent. 1985. Mirex residues in seven groups of aquatic and terrestrial mammals. Arch. Environ. Contam. Toxicol. 14:7-12.

Hill, E. P., and J. W. Lovett. 1975. Pesticide residues in beaver and river otter from Alabama. Annu. Conf. Southeast. Assoc. Game Fish Comm. (St. Louis, Mo.) 29:365-369.

Hoff, G. L., W. J. Bigler, and J. G. McKinnon. 1977. Heavy metal concentrations in kidneys of estuarine raccoons from Florida. J. Wildl. Dis. 13:101-102.

Holcomb, C. M., and W. S. Parker. 1979. Mirex residues in eggs and livers of two long-lived reptiles *(Chrysemys scripta* and *Terrapene carolina)* in Mississippi, 1970-1977. Bull. Environ. Contam. Toxicol. 23:369-371.

Holden, A. V. 1978. Pollutants and seals — a review. Mammal Rev. 8:53-66.

Holmes, D. C., J. H. Simmons, and J. O. Tatton. 1967. Chlorinated hydrocarbons in British wildlife. Nature (Lond.) 216:227-229.

Holt, G., and J. Sakshaug. 1968. Organochlorine insecticide residues in wild birds in Norway 1965-1967. Nord. Veterinaermed. 20:685-695.

Hom, W., R. W. Risebrough, A. Soutar, and D. R. Young. 1974. Deposition of DDE and polychlorinated biphenyls in dated sediments of the Santa Barbara basin. Science (Washington, D.C.) 184:1197-1199.

Honda, K., B. Yoon Min, and R. Tatsukawa. 1985. Heavy metal distribution in organs and tissues of the eastern great white egret *(Egretta alba modesta)*. Bull. Environ. Contam. Toxicol. 35:781-789.

Hudec, K., F. Kredl, J. Pellantova, J. Svobodnik, and R. Svobodova. 1988. Residues of chlorinated pesticides, PCB, and heavy metals in the eggs of water birds in southern Moravia. Folia Zool. 37:157-166.

Hunt, E. G., and A. I. Bischoff. 1960. Inimical effects on wildlife of periodic DDD applications to Clear Lake. Calif. Fish Game 46:91-106.

Hunt, E. G., and J. D. Linn. 1970. Fish kills by pesticides. p. 97-102. *In* J. W. Gillett (Ed.). The biological impact of pesticides in the environment. Environ. Health Ser. No. 1. Oregon State University, Corvallis, OR.

Jacknow, J., J. L. Ludke, and N. C. Coon. 1986. Monitoring fish and wildlife for environmental contaminants: The National Contaminant Biomonitoring Program. U. S. Fish Wildl. Serv. Fish Wildl. Leafl. No. 4, 15 pp.

Jarvinen, A. W., M. J. Hoffman, and T. W. Thorslund. 1976. Toxicity of DDT food and water exposure to fathead minnows. Duluth Water Qual. Lab., USEPA. EPA-600/3-76-114, 68 pp.

Jefferies, D. J., and B. N. K. Davis. 1968. Dynamics of dieldrin in soil, earthworms, and song thrushes. J. Wildl. Manage. 32:441-456.

Jefferies, D. J., and M. C. French. 1976. Mercury, cadmium, zinc, copper, and organochlorine insecticide levels in small mammals trapped in a wheat field. Environ. Pollut. 10:175-182.

Jefferies, D. J., and I. Prestt. 1966. Post-mortems of peregrines and lanners with particular reference to organochlorine residues. Br. Birds 59:49-64.

Jehl, J. R., Jr. 1970. Is thirty million years long enough? Pac. Discovery 23:16-23.

Jensen, S., A. G. Johnels, M. Olsson, and G. Otterlind. 1969. DDT and PCB in marine animals from Swedish waters. Nature (Lond.) 224:247-250.

Joanen, T., and H. H. Dupuie. 1969. The case of the vanishing brown pelican. Forests and People (Louisiana Forestry Assoc.) 19:23, 24, 38, 40, 41.

Johnson, D. W. 1968. Pesticides and fishes — a review of selected literature. Trans. Am. Fish. Soc. 97:398-424.

Johnson, R. E., T. C. Carver, and E. H. Dustman. 1967. Residues in fish, wildlife, and estuaries. Pestic. Monit. J. 1:7-13.

Johnston, D. W. 1974. Decline of DDT residues in migratory songbirds. Science (Washington, D.C.) 186:841-842.

Karlin, A., P. Rantamki, and R. Lemmetyinen. 1985. Residues of DDT and PCBs in the eggs of the herring gull *Larus argentatus* in the archipelago of southwestern Finland. Ornis Fenn. 62:168-170.

Karolewski, M. A., A. B. Lukowski, and R. Halba. 1987. Residues of chlorinated hydrocarbons in the adipose tissue of the antarctic pinnipeds. Pol. Polar Res. 8:189-197.

Keith, J. A., and I. M. Gruchy. 1972. Residue levels of chemical pollutants in North American birdlife. Proc. Int. Ornithol. Congr. 15:437-454.

Keith, J. O. 1966. Insecticide contaminations in wetland habitats and their effects on fish-eating birds. J. Appl. Ecol. 3 (Suppl.):71-85.

Keith, J. O., and E. G. Hunt. 1966. Levels of insecticide residues in fish and wildlife in California. Trans. N. Am. Wildl. Nat. Resour. Conf. 31:150-177.

Keith, J. O., L. A. Woods, Jr., and E. G. Hunt. 1970. Reproductive failure in brown pelicans on the Pacific coast. Trans. N. Am. Wildl. Nat. Resour. Conf. 35:56-63.

Kelsall, J. P. 1950. A study of bird populations in the apple orchards of the Annapolis Valley, Nova Scotia, with special reference to the effects of orchard sprays upon them. Can. Wildl. Serv. Wildl. Manage. Bull. Ser. 2, No. 1, 69 pp.

Kendeigh, S. C. 1947. Bird population studies in the coniferous forest biome during a spruce budworm outbreak. Ontario, Canada, Dep. Lands and Forest, Biol. Bull. No. 1, 100 pp.

Kiff, L. F., D. B. Peakall, and S. R. Wilbur. 1979. Recent changes in California condor eggshells. Condor 81:166-172.

King, K. A. 1989a. Food habits and organochlorine contaminants in the diet of olivaceous cormorants in Galveston Bay, Texas. Southwest. Nat. 34:338-343.

King, K. A. 1989b. Food habits and organochlorine contaminants in the diet of black skimmers, Galveston Bay, Texas, USA. Colon. Waterbirds 12:109-112.

King, K. A., D. R. Blankinship, E. Payne, A. J. Krynitsky, and G. L. Hensler. 1985. Brown pelican populations and pollutants in Texas 1975-1981. Wilson Bull. 97:201-214.

King, K. A., E. L. Flickinger, and H. H. Hildebrand. 1977. The decline of brown pelicans on the Louisiana and Texas gulf coast. Southwest. Nat. 21:417-431.

King, K. A., E. L. Flickinger, and H. H. Hildebrand. 1978. Shell thinning and pesticide residues in Texas aquatic bird eggs, 1970. Pestic. Monit. J. 12:16-21.

King, K. A., and A. J. Krynitsky. 1986. Population trends, reproductive success, and organochlorine chemical contaminants in waterbirds nesting in Galveston Bay, Texas. Arch. Environ. Contam. Toxicol. 15:367-376.

King, K. A., C. A. Lefever, and B. M. Mulhern. 1983. Organochlorine and metal residues in royal terns nesting on the central Texas coast. J. Field Ornithol. 54:295-303.

Klaas, E. E., H. M. Ohlendorf, and E. Cromartie. 1980. Organochlorine residues and shell thicknesses in eggs of the clapper rail, common gallinule, purple gallinule, and limpkin (Class *Aves*), eastern and southern United States, 1972-74. Pestic. Monit. J. 14:90-94.

Klaas, E. E., S. N. Wiemeyer, H. M. Ohlendorf, and D. M. Swineford. 1978. Organochlorine residues, eggshell thickness, and nest success in barn owls from the Chesapeake Bay. Estuaries 1:46-53.

Koeman, J. H., J. H. Pennings, J. J. M. de Goeij, P. S. Tjioe, P. M. Olindo, and J. Hopcraft. 1972. A preliminary survey of the possible contamination of Lake Nakuru in Kenya with some metals and chlorinated hydrocarbon pesticides. J. Appl. Ecol. 9:411-416.

Koeman, J. H., and H. van Genderen. 1966. Some preliminary notes on residues of chlorinated hydrocarbon insecticides in birds and mammals in the Netherlands. J. Appl. Ecol. 3 (Suppl.):99-106.

Korschgen, L. J. 1970. Soil-food-chain-pesticide wildlife relationships in aldrin-treated fields. J. Wildl. Manage. 34:186-199.

Kozie, K. D., and R. K. Anderson. 1991. Productivity, diet, and environmental contaminants in bald eagles nesting near the Wisconsin shoreline of Lake Superior. Arch. Environ. Contam. Toxicol. 20:41-48.

Kunze, F. M., A. A. Nelson, O. G. Fitzhugh, and E. P. Laug. 1949. Storage of DDT in the fat of the rat. Fed. Proc. 8:311.

Lamont, T., and W. Reichel. 1970. Organochlorine pesticide residues in whooping cranes and everglade kites. Auk 87:158-159.

Laug, E. P. 1946. 2,2,-Bis(p-chlorophenyl)-1,1,1-trichloroethane (DDT) in the tissues, body fluids, and excretia of the rabbit following oral administration. J. Pharmacol. Exp. Ther. 86:332-335.

Laug, E. P., C. S. Prickett, and F. M. Kunze. 1950. Survey analyses of human milk and fat for DDT content. Fed. Proc. 9:294-295.

Lichtenstein, E. P., K. R. Schulz, R. F. Skrentny, and Y. Tsukano. 1966. Toxicity and fate of insecticide residues in water. Arch. Environ. Health 12:199-212.

Linduska, J. P., and E. W. Surber. 1948. Effects of DDT and other insecticides on fish and wildlife. Summary of investigations during 1947. U. S. Dep. Inter., Fish Wildl. Serv. Circ. 15, 19 pp.

Lindvall, M., and J. B. Low. 1979. Organochlorine pesticide and PCB residues in western grebes from Bear River Migratory Bird Refuge, Utah. Bull. Environ. Contam. Toxicol. 22:754-760.

Lochmiller, R. L., II, R. J. Kendall, P. F. Scanlon, and R. J. Kirkpatrick. 1979. Lead concentrations in pine voles trapped from two Virginia orchards. Va. J. Sci. 30:50. (Abstr.)

Lock, J. W., and S. R. B. Solly. 1976. Organochlorine residues in New Zealand birds and mammals. 1. Pesticides. N. Z. J. Sci. 19:43-51.

Lockie, J. D., and D. A. Ratcliffe. 1964. Insecticides and Scottish golden eagles. Br. Birds 57:89-102.

Lockie, J. D., D. A. Ratcliffe, and R. Balharry. 1969. Breeding success and organo-chlorine residues in golden eagles in west Scotland. J. Appl. Ecol. 6:381-389.

Loganathan, B. G., S. Tanabe, H. Tanaka, S. Watanabe, N. Miyazaki, M. Amano, and R. Tatsukawa. 1990. Comparison of organochlorine residue levels in the striped dolphin from western North Pacific, 1978-79 and 1986. Mar. Pollut. Bull. 21:435-439.

Longcore, J. R., J. D. Heyland, A. Reed, and P. LaPorte. 1983. Contaminants in greater snow geese and their eggs. J. Wildl. Manage. 47:1105-1109.

Ludke, J. L. 1976. Organochlorine pesticide residues associated with mortality: additivity of chlordane and endrin. Bull. Environ. Contam. Toxicol. 16:253-260.

Mason, C. F., and W. M. O'Sullivan. 1992. Organochlorine pesticide residues and PCBs in otters *(Lutra lutra)* from Ireland. Bull. Environ. Contam. Toxicol. 48:387-393.

Mason, C. F., and D. Weber. 1990. Organochlorine residues and heavy metals in kidneys of polecats *(Mustela putorius)* from Switzerland. Bull. Environ. Contam. Toxicol. 45:689-696.

McArthur, M. L. B., G. A. Fox, D. B. Peakall, and B. J. R. Philogène. 1983. Ecological significance of behavioral and hormonal abnormalities in breeding ring doves fed an organochlorine chemical mixture. Arch. Environ. Contam. Toxicol. 12:343-353.

McCarthy, J. F., and L. R. Shugart (Eds.). 1990. Biomarkers of environmental contamination. Lewis, Chelsea, MI. 457 pp.

McEwen, L. C., C. J. Stafford, and G. L. Hensler. 1984. Organochlorine residues in eggs of black-crowned night-herons from Colorado and Wyoming. Environ. Toxicol. Chem. 3:367-376.

Meeks, R. L. 1968. The accumulation of ^{36}Cl ring-labeled DDT in a freshwater marsh. J. Wildl. Manage. 32:376-398.

Mendelsohn, J. M., A. C. Butler, and R. R. Sibbald. 1988. Organochlorine residues and eggshell thinning in southern African raptors. p. 439-447. *In* T. C. Cade, J. H. Enderson, C. G. Thelander, and C. M. White (Eds.). Peregrine falcon populations: their management and recovery. Peregrine Fund, Boise, Id.

Mendola, J. T., and R. W. Risebrough. 1977. Contamination de l'avifaune camarguaise par des residus organochlorés. (In French.) Environ. Pollut.13:21-31.

Miller, M. W., and G. G. Berg (Eds.). 1969. Chemical fallout. Charles C Thomas, Springfield, Ill. 531 pp.

Mitchell, R. T., H. P. Blagbrough, and R. C. Van Etten. 1953. The effects of DDT upon the survival and growth of nestling songbirds. J. Wildl. Manage. 17:45-54.

Mohr, R. W., H. S. Telford, E. H. Peterson, and K. C. Walker. 1951. Toxicity of orchard insecticides to birds. Wash. Agric. Exp. Stn. Circ. No. 170, 22 pp.

Monig, R. 1985. Dipper's *(Cinclus c. aquaticus)* quality as bio-indicator analyses of residues of chlorinated hydrocarbons (PCBs) in the eggs of birds living on running waters. Ecol. Birds 7:353-358.

Moore, N. W. 1965a. Environmental contamination by pesticides. p. 221-237. *In* Ecology and the industrial society, fifth symposium of the British Ecological Society, Blackwell, Oxford, England.

Moore, N. W. 1965b. Pesticides and birds — a review of the situation in Great Britain in 1965. Bird Study 12:222-252.

Moore, N. W. 1966. A pesticide monitoring system with special reference to the selection of indicator species. J. Appl. Ecol. 3 (Suppl.):261-269.

Moore, N. W., and D. A. Ratcliffe. 1962. Chlorinated hydrocarbon residues in the egg of a peregrine falcon *(Falco peregrinus)* from Perthshire. Bird Study 9:242-244.

Moore, N. W., and J. O. Tatton. 1965. Organochlorine insecticide residues in the eggs of sea birds. Nature (Lond.) 207:42-43.

Moore, N. W., and C. H. Walker. 1964. Organic chlorine insecticide residues in wild birds. Nature (Lond.) 201:1072-1073.

Mora, M. A. 1991. Organochlorines and breeding success in cattle egrets from the Mexicali Valley, Baja California, Mexico. Colon. Waterbirds 14:127-132.

Mora, M. A., and D. W. Anderson. 1991. Seasonal and geographical variation of organochlorine residues in birds from northwest Mexico. Arch. Environ. Contam. Toxicol. 21:541-548.

Mora, M. A., D. W. Anderson, and M. E. Mount. 1987. Seasonal variation of body condition and organochlorines in wild ducks from California and Mexico. J. Wildl. Manage. 51:132-141.

Mount, D. I., and G. J. Putnicki. 1966. Summary report of the 1963 Mississippi fish kill. Trans. N. Am. Wildl. Nat. Resour. Conf. 31:177-184.

Mulhern, B. M., and W. L. Reichel. 1970. The effect of putrefaction of eggs upon residue analysis of DDT and metabolites. Bull. Environ. Contam. Toxicol. 5:222-225.

Mulla, M. S., J. O. Keith, and F. A. Gunther. 1966. Persistence and biological effects of parathion residues in waterfowl habitats. J. Econ. Entomol. 59:1085-1090.

Mullins, W. H., E. G. Bizeau, and W. W. Benson. 1979. Pesticide and heavy metal residues in greater sandhill cranes. p. 189-195. *In* Proc. 1978 Crane Workshop. International Crane Foundation, Baraboo, WI.

Munday, B. L. 1985. Mercury levels in the musculature of stranded whales in Tasmania. Tasman. Fish. Res. 27:11-13.

Mundy, P. J., K. I. Grant, J. Tannock, and C. L. Wessels. 1982. Pesticide residues and eggshell thickness of Griffon vulture eggs in southern Africa. J. Wildl. Manage. 46:769-773.

Munshower, F. F., and D. R. Neuman. 1979. Metals in soft tissues of mule deer and antelope. Bull. Environ. Contam. Toxicol. 22:827-832.

Nagel, A., S. Winter, and B. Streit. 1991. Residues of chlorinated hydrocarbons in six European bat species. Bat Res. News 32:20-21.

National Academy of Sciences. 1979. Animals as monitors of environmental pollutants. NAS, Natural Resource Council, Washington, D. C.

Newton, I., J. A. Bogan, and P. Rothery. 1986. Trends and effects of organochlorine compounds in sparrowhawk eggs. J. Appl. Ecol. 23:461-478.

Newton, I., M. B. Haas, and P. Freestone. 1990a. Trends in organochlorine and mercury levels in gannet eggs. Environ. Pollut. 63:1-12.

Newton, I., and I. Wyllie. 1992. Recovery of a sparrowhawk population in relation to declining pesticide contamination. J. Appl. Ecol. 29:476-484.

Newton, I., I. Wyllie, and A. Asher. 1991. Mortality causes in British barn owls *Tyto alba*, with a discussion of aldrin-dieldrin poisoning. Ibis 133:162-169.

Newton, I., I. Wyllie, and A. Asher. 1992. Mortality from the pesticides aldrin and dieldrin in British sparrowhawks and kestrels. Ecotoxicology 1:31-44.

Newton, I., I. Wyllie, and P. Freestone. 1990b. Rodenticides in British barn owls. Environ. Pollut. 68:101-117.

Niethammer, K. R., R. D. Atkinson, T. S. Baskett, and F. B. Samson. 1985. Metals in riparian wildlife of the lead mining district of southeastern Missouri. Arch. Environ. Contam. Toxicol. 14:213-223.

Niethammer, K. R., D. H. White, T. S. Baskett, and M. W. Sayre. 1984. Presence and biomagnification of organochlorine chemical residues in oxbow lakes of northeastern Louisiana. Arch. Environ. Contam. Toxicol. 13:63-74.

Nisbet, I. C. T. 1988. The relative importance of DDE and dieldrin in the decline of peregrine falcon populations. p. 351-375. *In* T. J. Cade, J. H. Enderson, C. G. Thelander, and C. M. White (Eds.). Peregrine falcon populations: their management and recovery. Peregrine Fund, Boise, Id.

Noble, D. G., and J. E. Elliott. 1990. Levels of contaminants in Canadian raptors, 1966 to 1988; effects and temporal trends. Can. Field Nat. 104:222-243.

Norstrom, R. J., M. Simon, D. C. G. Muir, and R. E. Schweinsburg. 1988. Organochlorine contaminants in Arctic marine food chains: identification, geographic distribution, and temporal trends in polar bears. Environ. Sci. Technol. 22:1063-1071.

Ogle, M. C., P. F. Scanlon, R. L. Kirkpatrick, and J. V. Gwynn. 1985. Heavy metal concentrations in tissues of mink in Virginia. Bull. Environ. Contam. Toxicol. 35:29-37.

Ohlendorf, H. M., J. C. Bartonek, G. J. Divoky, E. E. Klaas, and A. J. Krynitsky. 1982. Organochlorine residues in eggs of Alaskan seabirds. U. S. Fish Wildl. Serv. Spec. Sci. Rep., Wildl. No. 245, 41 pp.

Ohlendorf, H. M., T. W. Custer, R. W. Lowe, M. Rigney, and E. Cromartie. 1988. Organochlorines and mercury in eggs of coastal terns and herons in California, USA. Colon. Waterbirds 11:85-94.

Ohlendorf, H. M., and W. J. Fleming. 1988. Birds and environmental contaminants in San Francisco and Chesapeake Bays. Mar. Pollut. Bull. 19:487-495.

Ohlendorf, H. M., E. E. Klaas, and T. E. Kaiser. 1978a. Organochlorine residues and eggshell thinning in anhingas and waders. p. 185-195. *In* Proc. 1977 Conf. Colonial Waterbird Group. Department of Biological Science, Northern Illinois University, De Kalb, Ill.

Ohlendorf, H. M., E. E. Klaas, and T. E. Kaiser. 1978b. Environmental pollutants and eggshell thinning in the black-crowned night-heron. p. 63-82. *In* Wading birds. Res. Rep. No. 7. National Audubon Society, New York.

Ohlendorf, H. M., E. E. Klaas, and T. E. Kaiser. 1978c. Organochlorine residues and eggshell thinning in wood storks and anhingas. Wilson Bull. 90:608-618.

Ohlendorf, H. M., R. W. Lowe, P. R. Kelly, and T. E. Harvey. 1986. Selenium and heavy metals in San Francisco Bay diving ducks. J. Wildl. Manage. 50:64-71.

Ohlendorf, H. M., R. W. Risebrough, and K. Vermeer. 1978d. Exposure of marine birds to environmental pollutants. U. S. Fish Wildl. Serv. Wildl. Res. Rep. No. 9, 40 pp.

Ohlendorf, H. M., F. C. Schaffner, T. W. Custer, and C. J. Stafford. 1985. Reproduction and organochlorine contaminants in terns at San Diego Bay. Colon. Waterbirds 8:42-53.

Olsen, P. D., and D. B. Peakall. 1983. DDE in eggs of the peregrine falcon in Australia, 1949-1977. Emu 83:276-277.

Orr, L. W., and L. O. Mott. 1945. The effects of DDT administered orally to cows, horses, and sheep. J. Econ. Entomol. 38:428-432.

O'Shea, T. J., R. L. Brownell, Jr., D. R. Clark, Jr., W. A. Walker, M. L. Gay, and T. G. Lamont. 1980b. Organochlorine pollutants in small cetaceans from the Pacific and South Atlantic Oceans, November 1968-June 1976. Pestic. Monit. J. 14:35-46.

O'Shea, T. J., W. J. Fleming III, and E. Cromartie. 1980a. DDT contamination at Wheeler National Wildlife Refuge. Science (Washington, D.C.) 209:509-510.

Osteen, C. 1993. Pesticide use trends and issues in the United States. p. 307-336. *In* D. Pimentel and H. Lehman (Eds.). The pesticide question. Chapman and Hall, New York.

Ott, D. E., and F. A. Gunther. 1966. Procedure for the analysis of technical grade parathion in waterplants by an anticholinesterase (auto analyzer) method. J. Econ. Entomol. 59:227-229.

Parker, J. W. 1976. Pesticides and eggshell thinning in the Mississippi kite. J. Wildl. Manage. 40:243-248.

Peakall, D. B. 1970. Pesticides and the reproduction of birds. Sci. Am. 222:73-78.

Peakall, D. B. 1974. DDE: its presence in peregrine eggs in 1948. Science (Washington, D.C.) 183:673-674.

Peakall, D. B. 1987. Accumulation and effects in birds. p. 31-47. *In* J. S. Ward (Ed.). PCBs and the environment. CRC Press, Boca Raton, Fla.

Peakall, D. B. 1990. Prospects for the peregrine falcon, *Falco peregrinus*, in the nineties. Can. Field Nat. 104:168-173.

Peakall, D. B. 1992. Animal biomarkers as pollution indicators. Chapman and Hall, New York. 291 pp.

Peakall, D. B., and A. C. Kemp. 1976. Organochlorine residue levels in herons and raptors in the Transvaal. Ostrich 47:139-141.

Peakall, D. B., and L. F. Kiff. 1979. Eggshell thinning and DDE residue levels among peregrine falcons *Falco peregrinus*: a global perspective. Ibis 121:200-204.

Peakall, D. B., and L. F. Kiff. 1988. DDE contamination in peregrines and American kestrels and its effect on reproduction. p. 337-350. *In* T. J. Cade, J. H. Enderson, C. G. Thelander, and C. M. White (Eds.). Peregrine falcon populations: their management and recovery. Peregrine Fund, Boise, Id.

Pearce, P. A., I. M. Price, and L. M. Reynolds. 1976. Mercury in waterfowl from eastern Canada. J. Wildl. Manage. 40:694-703.

Pellantova, J., K. Hudec, F. Kredl, J. Svobodnik, and R. Svobodova. 1989. Organochlorine pesticides, PCB, and heavy metals residues in the eggs of the black-headed gull, *Larus ridibundus*, in Czechoslovakia. Folia Zool. 38:79-86.

Peterle, T. J. 1966. The use of isotopes to study pesticide translocation in natural environments. J. Appl. Ecol. 3 (Suppl.):181-191.
Peterle, T. J. 1969. DDT in Antarctic snow. Nature (Lond.) 224:620.
Peterle, T. J. 1991. Wildlife toxicology. Van Nostrand Reinhold, New York. 322 pp.
Post, G. 1952. The effects of aldrin on birds. J. Wildl. Manage. 16:492-497.
Prouty, R. M., and C. M. Bunck. 1986. Organochlorine residues in adult mallard and black duck wings, 1981-1982. Environ. Monit. Assess. 6:49-57.
Punzo, F., J. Laveglia, D. Lohr, and P. A. Dahm. 1979. Organochlorine insecticide residues in amphibians and reptiles from Iowa and lizards from the southwestern United States. Bull. Environ. Contam. Toxicol. 21:842-848.
Ranta, W. B., F. D. Tomassini, and E. Nieboer. 1978. Elevation of copper and nickel levels in primaries from black and mallard ducks collected in the Sudbury district, Ontario. Can. J. Zool. 56:581-586.
Ratcliffe, D. A. 1958. Broken eggs in peregrine eyries. Br. Birds 51:23-26.
Ratcliffe, D. A. 1963. The status of the peregrine in Great Britain. Bird Study 10:56-90.
Ratcliffe, D. A. 1965. Organo-chlorine residues in some raptor and corvid eggs from northern Britain. Br. Birds 58:65-81.
Ratcliffe, D. A. 1967. Decrease in eggshell weight in certain birds of prey. Nature (Lond.) 215:208-210.
Ratcliffe, D. A. 1970. Changes attributable to pesticides in egg breakage frequency and eggshell thickness in some British birds. J. Appl. Ecol. 7:67-115.
Reidinger, R. F., Jr. 1976. Organochlorine residues in adults of six southwestern bat species. J. Wildl. Manage. 40:677-680.
Reynolds, L. M. 1969. Polychlorobiphenyls (PCB's) and their interference with pesticide residue analysis. Bull. Environ. Contam. Toxicol. 4:128-143.
Risebrough, R. W. 1969. Chlorinated hydrocarbons in marine ecosystems. p. 5-23. *In* M. W. Miller and G. G. Berg (Eds.). Chemical fallout: current research on persistent pesticides. Charles C Thomas, Springfield, Ill.
Risebrough, R. W., J. Davis, and D. W. Anderson. 1970. Effects of various chlorinated hydrocarbons. p. 40-53. *In* J. W. Gillett (Ed.). The biological impact of pesticides in the environment. Environ. Health Sci. Ser. No. 1. Oregon State University, Corvallis, Ore.
Risebrough, R. W., R. J. Huggett, J. J. Griffin, and E. D. Goldberg. 1968a. Pesticides: transatlantic movements in the northeast trades. Science (Washington, D.C.) 159:1233-1236.
Risebrough, R. W., D. B. Menzel, D. J. Martin, Jr., and H. S. Olcott. 1967. DDT residues in Pacific sea birds: a persistent insecticide in marine food chains. Nature (Lond.) 216:589-591.
Risebrough, R. W., P. Reiche, S. G. Herman, D. B. Peakall, and M. N. Kirven. 1968b. Polychlorinated biphenyls in the global ecosystem. Nature (Lond.) 220:1098-1102.
Risebrough, R. W., P. Reiche, and H. S. Olcott. 1969. Current progress in the determination of the polychlorinated biphenyls. Bull. Environ. Contam. Toxicol. 4:192-201.
Risebrough, R. W., F. C. Sibley, and M. N. Kirven. 1971. Reproductive failure of the brown pelican on Anacapa Island in 1969. Am. Birds 25:8-9.
Robbins, C. S., P. F. Springer, and C. G. Webster. 1951. Effects of five-year DDT application on breeding bird population. J. Wildl. Manage. 15:213-216.
Robinson, J. 1969. Organochlorine insecticides and bird populations in Britain. p. 113-173. *In* M. W. Miller and G. G. Berg (Eds.). Chemical fallout: current research on persistent pesticides. Charles C Thomas, Springfield, Ill.
Robinson, J. 1970. Persistent pesticides. Annu. Rev. Pharmacol. 10:353-378.

Rosene, W., Jr. 1965. Effects of field applications of heptachlor on bobwhite quail and other wild animals. J. Wildl. Manage. 29:554-580.

Royal Society for the Protection of Birds. n.d. The deaths of birds and mammals connected with toxic chemicals in the first half of 1960. Rep. No. 1 of the B.T.O.-R.S.P.B. Committee on Toxic Chemicals. London, England. 20 pp.

Royal Society for the Protection of Birds. 1962. Deaths of birds and mammals from toxic chemicals. January-June 1961. Diemer and Reynolds, Bedford, England. 24 pp.

Rudd, R. L. 1964. Pesticides and the living landscape. University of Wisconsin Press, Madison, Wisc. 320 pp.

Rudd, R. L., and R. E. Genelly. 1956. Pesticides: their use and toxicity in relation to wildlife. Calif. Dep. Fish Game Bull. No. 7, 209 pp.

Ryder, J. P. 1974. Organochlorine and mercury residues in gulls' eggs from western Ontario. Can. Field Nat. 88:349-352.

Scanlon, P. F., T. G. O'Brien, N. L. Schauer, J. L. Coggin, and D. E. Steffen. 1979. Heavy metal levels in feathers of wild turkeys from Virginia. Bull. Environ. Contam. Toxicol. 21:591-595.

Scanlon, P. F., V. D. Stotts, R. G. Oderwald, T. J. Dietrick, and R. J. Kendall. 1980. Lead concentrations in livers of Maryland waterfowl with and without ingested lead shot present in gizzards. Bull. Environ. Contam. Toxicol. 25:855-860.

Schechter, M. S., and H. L. Haller. 1944. Colorimetric test for DDT and related compounds. J. Am. Chem. Soc. 66:2129-2130.

Schreiber, R. W., and R. L. De Long. 1969. Brown pelican status in California. Audubon Field Notes 23:57-59.

Scott, T. G., Y. L. Willis, and J. A. Ellis. 1959. Some effects of a field application of dieldrin on wildlife. J. Wildl. Manage. 23:409-427.

Shaw, G. G. 1984. Organochlorine pesticide and PCB residues in eggs and nestlings of tree swallows, *Tachycineta bicolor*, in central Alberta. Can. Field Nat. 98:258-260.

Skåre, J. U., J. Stenersen, N. Kveseth, and A. Polder. 1985. Time trends of organochlorine chemical residues in seven sedentary marine fish species from a Norwegian fjord during the period 1972-1982. Arch. Environ. Contam. Toxicol. 14:33-41.

Sladen, W. J. L., C. M. Menzie, and W. L. Reichel. 1966. DDT residues in Adelie penguins and a crabeater seal from Antarctica: ecological implications. Nature (Lond.) 210:670-673.

Smillie, R. H., and J. S. Waid. 1987. Polychlorinated biphenyls and organochlorine pesticides in the Australian fur seal *(Arctocephalus pusillus doriferus)*. Bull. Environ. Contam. Toxicol. 39:358-364.

Smith, T. G., and F. A. J. Armstrong. 1978. Mercury and selenium in ringed and bearded seal tissues from arctic Canada. Arctic 31:75-84.

Sodergren, A., and S. Ulfstrand. 1972. DDT and PCB relocate when caged robins use fat reserves. Ambio 1:36-40.

Solly, S. R. B., and V. Shanks. 1976. Organochlorine residues in New Zealand birds and mammals. II. Polychlorinated biphenyls. N. Z. J. Sci. 19:53-55.

Spitzer, P. R., R. W. Risebrough, W. Walker II, R. Hernandez, A. Poole, D. Puleston, and I. C. T. Nisbet. 1978. Productivity of ospreys in Connecticut-Long Island increases as DDE residues decline. Science (Washington, D.C.) 202:333-335.

Springer, A. M., W. Walker II, R. W. Risebrough, D. Benfield, D. H. Ellis, W. G. Mattox, D. P. Mindell, and D. G. Roseneau. 1984. Origins of organochlorines accumulated by peregrine falcons, *Falco peregrinus*, breeding in Alaska and Greenland. Can. Field Nat. 98:159-166.

Springer, M. A. 1980a. Pesticide analysis, egg and eggshell characteristics of red-tailed hawk eggs. Ohio J. Sci. 80:206-210.

Springer, M. A. 1980b. Pesticide levels, egg and eggshell parameters of great horned owls. Ohio J. Sci. 80:184-187.

Springer, P. F., and J. R. Webster. 1951. Biological effects of DDT applications on tidal salt marshes. Trans. N. Am. Wildl. Conf. 16:383-397.

Steeves, T., M. Strickland, R. Frank, J. Rasper, and C. W. Douglas. 1991. Organochlorine insecticide and polychlorinated biphenyl residues in martens and fishers from the Algonquin region of south-central Ontario. Bull. Environ. Contam. Toxicol. 46:368-373.

Steidl, R. J., C. R. Griffin, and L. J. Niles. 1991. Contaminant levels of osprey eggs and prey reflect regional differences in reproductive success. J. Wildl. Manage. 55:601-608.

Stendell, R. C., E. Cromartie, S. N. Wiemeyer, and J. R. Longcore. 1977. Organochlorine and mercury residues in canvasback duck eggs, 1972-73. J. Wildl. Manage. 41:453-457.

Stendell, R. C., D. S. Gilmer, N. A. Coon, and D. M. Swineford. 1988. Organochlorine and mercury residues in Swainson's and ferruginous hawk eggs collected in North and South Dakota, 1974-79. Environ. Monit. Assess. 10:37-41.

Stendell, R. C., R. I. Smith, K. P. Burnham, and R. E. Christensen. 1979. Exposure of waterfowl to lead: a nationwide survey of residues in wing bones of seven species, 1972-73. U. S. Fish Wildl. Serv. Spec. Sci. Rep. No. 223.

Stickel, L. F. 1968. Organochlorine pesticides in the environment. U. S. Fish Wildl. Serv. Spec. Sci. Rep., Wildl. No. 119, 32 pp.

Stickel, L. F. 1973. Pesticide residues in birds and mammals. p. 254-312. *In* C. A. Edwards (Ed.). Environmental pollution by pesticides. Plenum Press, New York.

Stickel, L. F., N. J. Chura, P. A. Stewart, C. M. Menzie, R. M. Prouty, and W. L. Reichel. 1966b. Bald eagle pesticide relations. Trans. N. Am. Wildl. Nat. Resour. Conf. 31:190-200.

Stickel, L. F., and W. H. Stickel. 1969. Distribution of DDT residues in tissues of birds in relation to mortality, body condition, and time. Ind. Med. Surg. 38:44-53.

Stickel, L. F., W. H. Stickel, and R. Christensen. 1966a. Residues of DDT in brains and bodies of birds that died on dosage and in survivors. Science (Washington, D.C.) 151:1549-1551.

Stickel, L. F., S. N. Wiemeyer, and L. J. Blus. 1973. Pesticide residues in eggs of wild birds: adjustment for loss of moisture and lipid. Bull. Environ. Contam. Toxicol. 9:193-196.

Stickel, W. H., L. F. Stickel, and F. B. Coon. 1970. DDE and DDD residues correlated with mortality of experimental birds. p. 287-294. *In* W. P. Deichmann (Ed.). Pesticides symposia. Helios and Associates, Miami, FL.

Stickel, W. H., L. F. Stickel, R. A. Dyrland, and D. L. Hughes. 1984a. DDE in birds: lethal residues and loss rates. Arch. Environ. Contam. Toxicol. 13:1-6.

Stickel, W. H., L. F. Stickel, R. A. Dyrland, and D. L. Hughes. 1984b. Aroclor 1254 residues in birds: lethal levels and loss rates. Arch. Environ. Contam. Toxicol. 13:7-13.

Stickel, W. H., L. F. Stickel, and J. W. Spann. 1969. Tissue residues of dieldrin in relation to mortality in birds and mammals. p. 174-204. *In* M. W. Miller and G. G. Berg (Eds.). Chemical fallout: current research on persistent pesticides. Charles C Thomas, Springfield, Ill.

Street, J. C., and A. D. Blau. 1966. Insecticide interactions affecting residue accumulation in animal tissues. Toxicol. Appl. Pharmacol. 8:497-504.

Street, J. C., and R. W. Chadwick. 1967. Stimulation of dieldrin metabolism by DDT. Toxicol. Appl. Pharmacol. 11:68-71.

Sutcliffe, S. A. 1978. Pesticide levels and shell thickness of common loon eggs in New Hampshire. Wilson Bull. 90:637-640.

Suter, G. W. 1990. Use of biomarkers in ecological risk assessment. p. 419-426. *In* J. F. McCarthy and L. R. Shugart (Eds.). Biomarkers of environmental contamination. Lewis, Chelsea, MI.

Switzer, B., V. Lewin, and F. H. Wolfe. 1971. Shell thickness, DDE levels in eggs, and reproductive success in common terns *(Sterna hirundo)* in Alberta. Can. J. Zool. 49:69-73.

Switzer, B. C., F. H. Wolfe, and V. Lewin. 1972. Eggshell thinning and DDE. Nature (Lond.) 240:162-163.

Sykes, P. W., Jr. 1985. Pesticide concentrations in snail kite eggs and nestlings in Florida. Condor 87:438.

Szaro, R. C., N. C. Coon, and E. Kolbe. 1979. Pesticide and PCB of common eider, herring gull, and great black-backed gull eggs. Bull. Environ. Contam. Toxicol. 22:394-399.

Szefer, P., and J. Falandysz. 1986. Trace metals in the bones of scaup ducks *(Aythya marila)* wintering in Gdansk Bay, Baltic Sea, 1982-83 and 1983-84. Sci. Total Environ. 53:193-199.

Szymczak, M. R., and W. J. Adrian. 1978. Lead poisoning in Canada geese in southeast Colorado. J. Wildl. Manage. 42:299-306.

Talmage, S. S., and B. T. Walton. 1991. Small mammals as monitors of environmental contaminants. Rev. Environ. Contam. Toxicol. 119:47-145.

Tarrant, K. R., and J. O. Tatton. 1968. Organochlorine pesticides in rainwater in the British Isles. Nature (Lond.) 219:725-727.

Tarzwell, C. M. 1959. Pollutional effects of organic insecticides. Trans. N. Am. Wildl. Conf. 24:132-142.

Tatton, J. O., and J. H. A. Ruzicka. 1967. Organochlorine pesticides in Antarctica. Nature (Lond.) 215:346-348.

Thompson, D. R., P. H. Becker, and R. W. Furness. 1993. Long-term changes in mercury concentrations in herring gulls *Larus argentatus* and common terns *Sterna hirundo* from the German North Sea coast. J. Appl. Ecol. 30:316-320.

Thompson, N. P., P. W. Rankin, and D. W. Johnston. 1974. Polychlorinated biphenyls and p,p'DDE in green turtle eggs from Ascension Island, South Atlantic Ocean. Bull. Environ. Contam. Toxicol. 11:399-406.

U. S. President's Science Advisory Committee. 1963. Report on the use of pesticides. Commerce Clearing House, Chicago. 22 pp.

U. S. Public Health Service. 1961. Pollution-caused fish kills in 1960. Publ. Health Serv. Publ. 847. USPHS, Washington, D. C. 20 pp.

van Genderen, H. 1966. Tolerances for tissue levels of pesticides in wild animals; a proposal for consideration. J. Appl. Ecol. 3 (Suppl.):271-273.

Van Velzen, A. C., W. B. Stiles, and L. F. Stickel. 1972. Lethal mobilization of DDT by cowbirds. J. Wildl. Manage. 36:733-739.

Vermeer, K., and J. C. Castilla. 1991. High cadmium residues observed during a pilot study in shorebirds and their prey downstream from the El Salvador copper mine, Chile. Bull. Environ. Contam. Toxicol. 46:242-248.

Vermeer, K., and D. B. Peakall. 1977. Toxic chemicals in Canadian fish-eating birds. Mar. Pollut. Bull. 8:205-210.

Vermeer, K., and L. M. Reynolds. 1970. Organochlorine residues in aquatic birds in the Canadian prairie provinces. Can. Field Nat. 84:117-130.

Vermeer, K., and R. W. Risebrough. 1972. Additional information on egg shell thickness in relation to DDE concentrations in great blue heron eggs. Can. Field Nat. 86:384-385.

Wagemann, R. 1989. Comparison of heavy metals in two groups of ringed seals *(Phoca hispida)* from the Canadian Arctic. Can. J. Fish. Aquat. Sci. 46:1558-1563.

Walker, C. H., G. A. Hamilton, and R. B. Harrison. 1967. Organochlorine insecticide residues in wild birds in Britain. J. Sci. Food Agric. 18:123-129.

Walker, W. 1977. Chlorinated hydrocarbon pollutants in Alaskan gyrfalcons and their prey. Auk 94:442-447.

Weseloh, D. V., T. W. Custer, and B. M. Braune. 1989. Organochlorine contaminants in eggs of common terns from the Canadian Great Lakes, 1981. Environ. Pollut. 59:141-160.

Weseloh, D. V., P. Mineau, and D. J. Hallett. 1979. Organochlorine contaminants and trends in reproduction in Great Lakes herring gulls, 1974-1978. Trans. N. Am. Wildl. Nat. Resour. Conf. 44:543-557.

Weseloh, D. V., S. M. Teeple, and M. Gilbertson. 1983. Double-crested cormorants of the Great Lakes: egg-laying parameters, reproductive failure, and contaminant residues in eggs, Lake Huron 1972-1973. Can. J. Zool. 61:427-436.

West, L. D., and J. D. Newsom. 1979. Lead and mercury in lesser snow geese wintering in Louisiana. Proc. Annu. Conf. Southeast. Assoc. Fish Wildl. Agencies 31:180-187.

White, D. H., and E. Cromartie. 1977. Residues of environmental pollutants and shell thinning in merganser eggs. Wilson Bull. 89:532-542.

White, D. H., W. J. Fleming, and K. L. Ensor. 1988. Pesticide contamination and hatching success of waterbirds in Mississippi. J. Wildl. Manage. 52:724-729.

White, D. H., and A. J. Krynitsky. 1986. Wildlife in some areas of New Mexico and Texas accumulate elevated DDE residues, 1983. Arch. Environ. Contam. Toxicol. 15:149-157.

White, D. H., C. A. Mitchell, and E. Cromartie. 1982. Nesting ecology of roseate spoonbills at Nueces Bay, Texas. Auk 99:275-284.

White, D. H., C. A. Mitchell, and T. E. Kaiser. 1983a. Temporal accumulation of organochlorine pesticides in shorebirds wintering on the south Texas coast, 1979-80. Arch. Environ. Contam. Toxicol. 12:241-245.

White, D. H., C. A. Mitchell, H. D. Kennedy, A. J. Krynitsky, and M. A. Ribick. 1983b. Elevated DDE and toxaphene residues in fishes and birds reflect local contamination in the lower Rio Grande Valley, Texas. Southwest. Nat. 28:325-333.

Whitten, J. L. 1966. That we may live. D Van Nostrand Company, New York. 251 pp.

Wickson, R. J., F. I. Norman, G. J. Bacher, and J. S. Garnham. 1992. Concentrations of lead in bone and other tissues of victorian waterfowl. Wildl. Res. 19:221-232.

Wiemeyer, S. N., C. M. Bunck, and A. J. Krynitsky. 1988a. Organochlorine pesticides, polychlorinated biphenyls, and mercury in osprey eggs — 1970-79 — and their relationships to shell thinning and productivity. Arch. Environ. Contam. Toxicol. 17:767-787.

Wiemeyer, S. N., T. G. Lamont, C. M. Bunck, C. R. Sindelar, F. J. Gramlich, J. D. Fraser, and M. A. Byrd. 1984. Organochlorine pesticide, polychlorobiphenyl, and mercury residues in bald eagle eggs — 1969-79 — and their relationships to shell thinning and reproduction. Arch. Environ. Contam. Toxicol. 13:529-549.

Wiemeyer, S. N., and R. D. Porter. 1970. DDE thins eggshells of captive American kestrels. Nature (Lond.) 227:737-738.

Wiemeyer, S. N., J. M. Scott, M. P. Anderson, P. H. Bloom, and C. J. Stafford. 1988b. Environmental contaminants in California condors. J. Wildl. Manage. 52:238-247.

Wiemeyer, S. N., P. R. Spitzer, W. C. Krantz, T. G. Lamont, and E. Cromartie. 1975. Effects of environmental pollutants on Connecticut and Maryland osprey. J. Wildl. Manage. 39:124-139.

Wiemeyer, S. N., D. M. Swineford, P. R. Spitzer, and P. D. McLain. 1978. Organochlorine residues in New Jersey osprey eggs. Bull. Environ. Contam. Toxicol. 19:56-63.

Winn, B. 1975. Pesticides decimate transplanted pelicans. Audubon 77:127-129.

Woodard, G., and R. R. Ofner. 1946. Accumulation of DDT in the fat of rats in relation to dietary level and length of feeding time. Fed. Proc. 5:214.

Woodard, G., R. R. Ofner, and C. M. Montgomery. 1945. Accumulation of DDT in the body fat and its appearance in the milk of dogs. Science (Washington, D.C.) 102:177-178.

Woodwell, G. M., C. F. Wurster, Jr., and P. A. Isaacson. 1967. DDT residues in an East Coast estuary: a case of biological concentration of a persistent insecticide. Science (Washington, D.C.) 156:821-824.

Wright, B. S. 1960. Woodcock reproduction in DDT-sprayed areas of New Brunswick. J. Wildl. Manage. 24:419-420.

Wurster, C. F., Jr. 1969. Chlorinated hydrocarbon insecticides and avian reproduction: How are they related? p. 368-389. In M. W. Miller and G. G. Berg (Eds.). Chemical fallout. Charles C Thomas, Springfield, Ill.

Wurster, C. F., Jr., and D. B. Wingate. 1968. DDT residues and declining reproduction in the Bermuda petrel. Science (Washington, D.C.) 159:979-981.

Wurster, D. H., C. F. Wurster, Jr., and W. N. Strickland. 1965. Bird mortality following DDT spray for Dutch elm disease. Ecology 46:488-499.

Appendix Table 1   List of Species Mentioned in the Text

Birds

American woodcock	*Scolopax minor*
Anhinga	*Anhinga anhinga*
Audouin's gull	*Larus audouinii*
Bald eagle	*Haliaeetus leucocephalus*
Barn owl	*Tyto alba*
Bateleur	*Terathopius ecaudatus*
Bermuda petrel	*Pterodroma cahow*
Black-crowned night heron	*Nycticorax nycticorax*
Black-headed gull	*Larus ridibundus*
Black skimmer	*Rynchops niger*
Black vulture	*Coragyps atratus*
Brown-headed cowbird	*Molothrus ater*
Brown pelican	*Pelecanus occidentalis*
Cackling goose	*Branta canadensis minima*
California condor	*Gymnogyps californianus*
Canada goose	*Branta canadensis*
Canvasback	*Aythya valisineria*
Caspian tern	*Sterna caspia*
Cattle egret	*Bubulcus ibis*
Common eider	*Somateria mollissima*
Common flicker	*Colaptes avratus*
Common loon	*Gavia immer*
Common tern	*Sterna hirundo*
Double-crested cormorant	*Phalacrocorax auritus*
Eastern great white egret	*Egretta alba modesta*
Elegant tern	*Sterna elegans*
Eleonora's falcon	*Falco eleonora*
Everglades kite	*Rostrhamus sociabilis*
Fish eagle	*Haliaeetus vocifer*
Forster's tern	*Sterna forsteri*
Fulvous tree duck	*Dendrocygna bicolor*
Gannet	*Sula bassanus*
Golden eagle	*Aquila chrysaetos*
Great blue heron	*Ardea herodias*
Great-horned owl	*Bubo virginianus*
Great white egret	*Egretta alba*
Greater scaup	*Aythya marila*
Gyrfalcon	*Falco rusticolus*
Herring gull	*Larus argentatus*
House sparrow	*Passer domesticus*
Kestrel	*Falco tinnunculus*
Least tern	*Sterna antillarum*
Loggerhead shrike	*Lanius ludovicianus*
Long-billed curlew	*Numenius americanus*
Marsh harrier	*Circus cyaneus*
Merlin	*Falco columbarius*
Mississippi kite	*Ictinia mississippiensis*
Mountain bluebird	*Sialia currucoides*
Northern bobwhite	*Colinus virginianus*
Olivaceous cormorant	*Phalacrocorax olivaceus*

Appendix Table 1 (Continued)

Osprey	*Pandion haliaetus*
Peregrine falcon	*Falco peregrinus*
Prairie falcon	*Falco mexicanus*
Red-breasted merganser	*Mergus serrator*
Red-necked grebe	*Podiceps grisegena*
Red-tailed hawk	*Buteo jamaicensis*
Ring-necked pheasant	*Phasianus colchicus*
Robin	*Turdus migratorius*
Roseate spoonbill	*Ajaia ajaia*
Royal tern	*Sterna maxima*
Sandhill crane	*Grus canadensis*
Snail kite	*Rostrhamus sociabilis*
Snow goose	*Chen caerulescens*
Spanish eagle	*Aquila adalberti*
Sparrowhawk	*Accipiter nisus*
Swainson's hawk	*Buteo swainsoni*
Tree swallow	*Tachycineta bicolor*
Tundra swan	*Cygnus columbianus*
Turkey	*Meleagris galloparvo*
Western grebe	*Aechmophorus occidentalis*
White-faced ibis	*Plegadis chihi*
White-tailed eagle	*Haliaeetus albicilla*
White-throated dipper	*Cinclus cinclus aquaticus*
White pelican	*Pelecanus erythrorhynchos*
Whooping crane	*Grus americana*
Willet	*Catoptrophorus semipalmatus*
Wood duck	*Aix sponsa*
Woodstork	*Mycteria americana*
Yellow-legged gull	*Larus cachinnans*

Mammals

Antelope	*Antilocapra americana*
Beaver	*Castor canadensis*
Big brown bat	*Eptesicus fuscus*
Bobcat	*Lynx rufus*
Fisher	*Martes pennanti*
Fox squirrel	*Sciurus niger*
Fur seal	*Arctocephalus pusillus doriferus*
Gray seal	*Halichoerus grypus*
Gray squirrel	*Sciurus carolinensis*
Marten	*Martes americana*
Mink	*Mustela vison*
Mule deer	*Odocoileus hemionus*
Pine vole	*Microtus pinetorum*
Polar bear	*Thalarctos maritimus*
Polecat	*Mustela putorius*
Pygmy killer whale	*Feresa attenuata*
Raccoon	*Procyon lotor*
Ringed seal	*Phoca hispida*
River otter	*Lutra canadensis*
Striped dolphin	*Stenella coeruleoalba*

Appendix Table 1 (Continued)

Reptiles

American crocodile — *Orocodylus acutus*
Snapping turtle — *Chelydra serpentina*

Insects

Japanese beetle — *Popillia japonica*
Fire ant — *Solonopis saevissima*

CHAPTER 2

DDT, DDD, and DDE in Birds

Lawrence J. Blus

INTRODUCTION

The organochlorine compound known as dichlorodiphenyltrichloroethane (DDT) was synthesized in 1874. Paul Müller discovered its insecticidal activity in 1939 and subsequently received the Nobel prize for this discovery (Carson, 1962). DDT was used extensively in human health operations during World War II. Agricultural applications started immediately after the war, and the amounts used increased exponentially after that time (Hayes, 1991).

Concern about the effects on wildlife began almost immediately (Cottam and Higgins, 1946). Early field studies with DDT were concerned with the short-term effects after heavy rates of application; for example, 5.6 kg of DDT per hectare (ha) resulted in immediate reductions in the populations of songbirds and invertebrates in an upland hardwood forest (Hotchkiss and Pough, 1946). In contrast, 5.6 kg/ha had no effect on the eggs or nestlings of forest birds, but the DDT spray was limited to an area of only 0.09 m^2 around each nest (Mitchell, 1946). DDT applied to a bottomland hardwood forest at a rate of 2.2 kg/ha had no effect on bird populations when applied for only 1 year (Stewart et al., 1946). We know now that the approximate cause-and-effect relation of DDT for mortality may occur immediately after application, as well as many months or years after application.

The first important step in uncovering long-term relations was dependent on the development of precise and accurate analytical techniques that could detect DDT and its principal metabolites — DDD and DDE — in environmental samples. Heyroth (1950) summarized the early efforts to develop analytical methodology for detecting residues of DDT. Some of the early work was of limited value because DDT and one or more metabolites were lumped together or DDE was not measured. Residue analysis improved with time, vastly progressing with the development of electron-capture gas chromatography, and was essentially perfected with the development of mass spectrometry. With these advances, the residues of DDT and its metabolites in wildlife could be related to lethal as well as sublethal effects, especially eggshell thinning and reduced reproductive success. Technical DDT, the insecticidal formulation applied in the field, consists of several compounds that may be changed or broken down by a number of physical or biological factors in the environment. Of these compounds, only p,p'-DDT (DDT), p,p'-DDD (DDD), and p,p'-DDE (DDE) have been related to adverse environmental effects. The residue data reported here are on a wet-weight basis unless otherwise indicated.

The purposes of this chapter are to summarize the residue levels of these three compounds in birds that are diagnostic for or are associated with mortality and important sublethal effects and to suggest improvements in the design of contemporary field studies that will result in maximum usefulness in interpreting residue data.

INTERPRETING LETHAL RESIDUES

BRAIN

The first experimental attempts to measure lethal levels in animals fed DDT-contaminated diets included analyses of several tissues, including the brain. With the limitations of analytical methodology, some of the studies combined DDT with DDD, or DDE was not detected. Considering DDT and DDD combined, Bernard (1963), Stickel et al. (1966), and Stickel and Stickel (1969) concluded that 30 µg/g in the brain were a useful approximation of the lower level representing serious danger and possible death. Most measurements of DDT + DDD in the brains of birds dying from DDT were above this level, but lethal levels were as low as 25 µg/g in house sparrows *(Passer domesticus)* used in experiments (Bernard, 1963) and 17 µg/g in wild American robins *(Turdus migratorius)* dying with tremors (Hunt, 1968). Stickel et al. (1966) concluded that the relative importance of DDT and DDD was not apparent from these data.

With improvements in analytical methodology, residues of DDT, DDD, and DDE were determined in the brains of experimental and wild animals killed by DDT, and more definitive evaluations of the contribution of the individual compounds to lethality were established. Weighting was necessary, because residues in the brains at death ranged from nearly all DDD to nearly all DDT (Stickel et al., 1970). It was also necessary to evaluate residues in the brains of apparently normal animals exposed to DDT and euthanized at periods when others were dying from accumulated dosage and to evaluate the effects of exposure routes, time to death, age, sex, and various stresses on lethal levels in the brain. It was concluded by Stickel et al. (1970) that there is little or no postmortem breakdown of DDT to DDD. Measurements of the lethal levels of DDE and DDD in animals exposed to these individual compounds also helped in evaluating the relative contributions of each toxicant. By considering the levels of DDT and its major metabolites in the brains of animals on DDT dosage that either died or were euthanized, an excellent relation of residues to lethality was established (Table 1). This separation was possible because residues in the brain increase rapidly shortly before death, and concentrations uniformly meet or exceed the lower lethal limit in relation to exposure routes, time to death, and

Table 1 Residues of DDT, DDD, and DDE in Brains of Birds that Died or were Euthanized while on Experimental DDT Dietary Dosage and in Brains of Wild American Robins that Died in Tremors

Species	Sex/fate[a]	DDT Mean	DDT Range	DDD Mean	DDD Range	DDE Mean	DDE Range	Mean DDT equivalent	Source[b]
Brown-headed cowbird	M/D	39	27-90	59	29-99	7	5-12	51	1
	M/E	7	3-19	9	4-17	1	<1-2	9	1
	F/D	40	27-77	50	27-71	8	6-10	51	1
	F/E	9	6-21	9	6-17	1	<1-3	11	1
House sparrow	B/D	28	18-38	16	8-29	9	5-18	35	2
Northern bobwhite									
Wild	B/D	23	17-29	8	6-14	11	9-13	25	2
Game farm	B/D	25	19-32	3	2-4	9	8-11	26	2
Northern cardinal (Cardinalis cardinalis)	B/D	19	17-24	8	6-10	3	2-3	21	2
Blue jay (Cyanocitta cristata)	B/D	16	12-20	7	6-9	3	2-4	18	2
American robin	B/D	15	NL[c]	39	NL	57	NL	27	3
Clapper rail	M/D	25	19-31	18	13-27	4	3-7	29	4
(Rallus longirostris)	F/D	26	20-31	19	15-23	4	3-5	30	4
	M/E	6	2-10	8	4-13	1	<1-2	8	4

[a] B, both sexes; D, died; E, euthanized, appeared normal.
[b] 1, Stickel and Stickel (1969); 2, Hill et al. (1971); 3, Wurster et al. (1965); 4, Van Velzen and Kreitzer (1975).
[c] NL, not listed.

the other variables mentioned above (Stickel et al., 1970). There was variation in the means of DDT and the major metabolites in the brain at death; for example, DDT ranged from 15 µg/g in American robins (Wurster et al., 1965) to 40 µg/g in brown-headed cowbirds (*Molothrus ater*; Stickel and Stickel, 1969), and DDD ranged from 2 µg/g in northern bobwhite quail (*Colinus virginianus*; Hill et al., 1971) to 99 µg/g in brown-headed cowbirds (Stickel and Stickel, 1969). Because of these variations, Stickel et al. (1970) developed the concept of a DDT equivalent, wherein 1 µg/g of DDT equals 5 µg/g of DDD or 15 µg/g of DDE. Using this weighting system, Stickel et al. (1970) indicated that 10 DDT equivalents in the brain constitute an approximate lower lethal limit. American robins that died in tremors had as little as 10 DDT equivalents in their brains. In Table 1, all the mean DDT equivalents in the brains of animals that died were ≥18, and the mean DDT equivalents in the brains of brown-headed cowbirds that were euthanized on DDT dosage ranged from 8 to 11. Stickel et al. (1970) indicated that the DDT equivalent system was approximate; a 50% margin of error was estimated when all series of data were included. The DDT equivalent weighting system remains a valuable interpretive tool, although the equivalent for DDE probably should be raised to 20 or 25 to reflect the lethal level of DDE alone.

With two notable exceptions, DDE residues in the brains of animals dying from DDT ranged from <1 to 28 µg/g (Table 1). Wurster et al. (1965) reported 57 µg/g of DDE in the brains of wild American robins that died from DDT sprayed for Dutch elm disease; DDE exceeded DDT and DDD in these birds. Another more striking exception was the high mean levels of DDE in the brains of cockerels *(Gallus gallus)* that died after being fed DDT; DDE equaled or exceeded the levels of DDT + DDD in these birds with a high of 227 µg/g in a series of birds fed 250 µg of DDT per day (Ecobichon and Saschenbrecker, 1968). In comparison with the results of other studies, these high levels of DDE seem anomalous, possibly because of problems in analytical methodology or species differences in the metabolism of DDT.

In animals given diets containing DDE during experiments, residues of only DDE were detected in their brains (Table 2). The mean lethal residue of DDE was 499 µg/g in four species of passerine birds, with the lowest individual level of 250 µg/g in a brown-headed cowbird. Stickel et al. (1984) concluded that, for all species tested, residues of DDE in the brain were clearly diagnostic; there was a strong likelihood for death with residues ≥300 µg/g. DDE residues ranged from 52 to 400 µg/g in the brains of birds that were euthanized while receiving dietary levels of DDE that were lethal to other birds. The DDE residue level in the brain of only one euthanized bird overlapped the levels in birds that died (Stickel et al., 1970; Stickel et al., 1984).

Table 2   Residues of DDE and DDD in Brains of Birds that Died or were Euthanized after Experimental Dietary Exposure to DDE or DDD

Species	Sex/ fate[a]	µg/g, Wet weight		Mean DDT equivalent	Source[b]
		Mean	Range		
DDE					
Brown-headed cowbird	M/D	499	250-660	39	1
	M/E	152	67-400	10	1
Passerines[c] (4 species)	B/D	499	305-694	39	2
	B/E	137	52-219	9	2
DDD					
Brown-headed cowbird	M/D	172	86-358	34	1
	M/E	42	19-105	8	1

[a] B, both sexes; D, died; E, euthanized, appeared normal.
[b] 1, Stickel et al. (1970); 2, Stickel et al. (1984).
[c] Combined data for brown-headed cowbird, common grackle *(Quiscalus quiscula)*, red-winged blackbird *(Agelaius phoeniceus)*, and European starling *(Sturnus vulgaris)*.

Regarding DDT equivalents in the brains of birds receiving DDE dosage, the means ranged from 35 to 39 for those that died and from 9 to 10 for those that were euthanized (Table 2). Few possible cases of lethal levels of DDE in the brains of wild birds in the United States exist; these include a bald eagle *(Haliaeetus leucocephalus)* with 385 µg/g (Belisle et al., 1972), a great blue heron *(Ardea herodias)* with 246 µg/g (Call et al., 1976), and a black-crowned night heron *(Nycticorax nycticorax)* with 230 µg/g (Ohlendorf et al., 1981). Also, two experimental American kestrels *(Falco sparverius)* that died after a long period on a low dietary dosage of

DDE (2.8 ppm) had 213 and 301 µg/g of DDE in their brains (Porter and Wiemeyer, 1972). Three other American kestrels died several days after receiving diets containing 160 to 250 ppm of DDE; their brains contained from 230 to 280 µg/g of DDE (Henny and Meeker, 1981). Although the lower lethal limit of 300 µg/g seems to provide a reliable criterion, there is some evidence that lower levels occasionally prove lethal.

Concerning lethal residues of DDD in the brain, birds dying on DDT dosage had mean levels that varied from 3 to 59 µg/g, with individual levels as high as 151 µg/g (Table 1). The brains of birds on experimental dietary dosages of DDD contained an average of 172 µg/g of DDD at death, with an individual lower level of 86 µg/g (Table 2). Birds on DDD dosage that were euthanized had 19 to 105 µg/g in their brains. Stickel et al. (1970) concluded that brain concentrations ≥65 µg/g indicate an increasing likelihood that death was due to poisoning from DDD. DDT equivalents were 34 in those dying on DDD dosage and 8 in those euthanized on that same dosage (Table 2). The only confirmed instance of DDD poisoning in a wild animal was that of a common loon *(Gavia immer)*. Its brain contained 200 µg/g of DDD, 130 µg/g of DDE, and 2 µg/g of DDT; DDT equivalents totaled 41 (Prouty et al., 1975). The most vivid example of the effects of DDD occurred when Clear Lake in California was treated with DDD for several years. Although DDD almost certainly adversely affected western grebe *(Aechmophorus occidentalis)* survival and reproductive success, the only two brains analyzed (both females found moribund and euthanized) had DDD residues of 46 and 48 µg/g (Rudd and Herman, 1972); these levels were less than the lowest individual residues of birds dying on DDD dosage but were slightly greater than the mean level found in birds euthanized (Table 2).

One problem with interpreting residues of the DDT group is that other contaminants including organochlorines are frequently present in the eggs or tissues of wild birds. Regarding organochlorines in the brain, Sileo et al. (1977) assumed straightforward additivity of the toxic effects and developed an "organochlorine index" based on the addition of the proportions of lower lethal levels for all compounds; for example, one half of a lower lethal level contributes 0.5 to the index on the basis of 1 indicating lethality. This index has received little use, one reason being that the lower lethal level of 150 µg/g of DDE ascribed by Sileo et al. (1977) is nearly 100 µg/g less than the accepted lower lethal limit (Table 2). Another reason is that the lethal limits of individual organochlorines in wild birds are usually distinct from one another. Finally, the assumed additivity of organochlorines related to lethality in birds has received little verification from experimental studies, although additivity seems the most common joint action (Smyth et al., 1969).

LIVER

Residues in the livers of animals dying during experiments while receiving dietary dosages of DDT differ from residues in their brains in that DDD constitutes the bulk of the residues (Table 3). Mean levels of DDT ranged from 1 to 35 µg/g, with a range in individual values from <1 to 254 µg/g. Residues of DDE were relatively low in livers except in American robins, where the residues exceeded those of DDD (Wurster et al., 1965); this is the same series that had exceptionally high residues of DDE in their brains. Cockerels dying while receiving a DDT dietary dosage also had levels of DDE in their livers that exceeded the levels of DDD and DDT combined; the same relation held for residues in their brains as previously mentioned (Ecobichon and Saschenbrecker, 1968).

Table 3   Residues of DDT, DDD, and DDE in Livers of Birds That Died or Were Euthanized after Experimental Dietary Exposure to DDT and in Wild American Robins that Died in Tremors

Species	Sex/ fate[a]	μg/g, Wet weight						Source[b]
		DDT		DDD		DDE		
		Mean	Range	Mean	Range	Mean	Range	
Brown-headed cowbird	M/D	34	3-254	768	215-1,640	55	25-104	1
	M/E	5	1-20	58	30-115	3	2-6	1
	F/D	35	9-161	552	292-1,063	53	32-88	1
	F/E	8	4-16	72	61-107	4	2-8	1
American robin	B/D	1	NL[c]	139	NL	165	NL	2
Clapper rail	M/D	3	<1-5	308	75-938	24	7-47	3
	F/D	4	<1-10	229	130-337	19	13-27	3
	M/E	3	<1-7	157	38-352	11	2-36	3

[a] B, both sexes; D, died; E, euthanized, appeared normal.
[b] 1, Stickel and Stickel (1969); 2, Wurster et al. (1965); 3, Van Velzen and Kreitzer (1975).
[c] NL, not listed.

Mean residues of DDE in the livers of brown-headed cowbirds receiving dietary DDE (Table 4) were 3883 μg/g (range, 460 to 11,725 μg/g) in those that died and 523 μg/g (range, 266 to 1560 μg/g) in those that were euthanized (Stickel et al., 1970). In Great Britain, Newton et al. (1992) reported that 23 Eurasian kestrels *(Falco tinnunculus)* and 10 Eurasian sparrowhawks *(Accipiter nisus)* died with lethal levels of DDE in their livers, but the lower lethal limit of 100 μg/g was based on correlative field evidence related to DDT and metabolites (Cooke et al., 1982). According to experimentally derived lethal levels (Stickel et al., 1970), DDE residues in Eurasian sparrowhawk livers (140 to 254 μg/g) were too low to ascribe lethality to DDE, but DDE residues in the livers of at least three Eurasian kestrels (812, 1474, and 1500 μg/g) were within the lethal range. In the United States, the only liver analysis from a wild bird that apparently died from DDE was that of a great blue heron that had DDE residues of 246 μg/g in the brain and 570 μg/g in the liver (Call et al., 1976).

The mean residue levels of DDD in the livers of birds receiving dietary dosages of DDD were 1219 μg/g (range, 79 to 5300 μg/g) in those that died and 521 μg/g (range, 104 to 2854 μg/g) in those that were euthanized (Stickel et al., 1970).

To interpret the lethal levels of DDT and its metabolites in the liver, researchers should devise a weighting system, such as that developed for the brain, to determine the relative contribution of each of the compounds. Also, Stickel et al. (1970) indicated that, for birds killed by DDT, residue levels of DDT and its metabolites in the livers of wild birds were lower than those of laboratory birds. Cooke et al. (1982) indicated that starvation of birds dying from organochlorine pesticides complicated the interpretation of lethal levels in the liver. Bernard (1963) concluded that residues

Table 4 Residues of DDE and DDD in Livers of Male Brown-Headed Cowbirds that Died or were Euthanized after Exposure to DDE or DDD in their Diets (after Stickel et al., 1970)

Fate[a]	μg/g, Wet weight	
	Mean	Range
DDE		
D	3,883	460-11,725
E	523	266-1,560
DDD		
D	1,219	79-5,300
E	521	104-2,854

[a] D, died; E, euthanized, appeared normal.

in the brain were more consistent than those in the liver with regard to the interpretation of lethal residues.

OTHER TISSUES

A number of tissues including blood plasma, carcass remainder, kidney, heart, breast muscle, intestinal tract, skin, and fat have been analyzed in birds that were dying or euthanized while receiving a dosage of DDT (Ecobichon and Saschenbrecker, 1968; Stickel et al., 1970). Although residues in these tissues have not received the same scrutiny as those in the brain during the assessment of diagnostic lethal levels, Stickel et al. (1970) concluded that residue levels of DDD + DDT in the carcasses of birds dying from DDT increased with the time on dietary dosage and that residue levels in those that were euthanized were essentially indistinguishable from those that died on dosage. In contrast, DDE residues, expressed on a lipid basis, in the carcasses of brown-headed cowbirds sacrificed while receiving DDE dietary dosage differed markedly from those that died on that dosage (Stickel et al., 1984). The same authors concluded that residues in carcass lipids accurately predicted lethal brain residues.

INTERPRETING SUBLETHAL RESIDUES

EGGS

Eggshell Thinning

The classic paper by Ratcliffe (1967a) described eggshell thinning in eggs of peregrine falcons *(Falco peregrinus)* and Eurasian sparrowhawks in Great Britain that occurred following the introduction of DDT.

Soon thereafter in the U.S., eggshell thinning was documented in several species of raptorial and fish-eating birds, and the inverse relation between DDE residues in eggs and shell thickness was first established (Hickey and Anderson, 1968). Also, Hickey and Anderson (1968) were the first to document decreases in the mean eggshell thickness over a period of years in relation to population declines. Heath et al. (1969) first documented eggshell thinning and associated lowered reproductive success of experimental birds on DDE diets. Subsequently, there have been a substantial number of experimental and field studies that document eggshell thinning and a smaller number that relate residues in eggs to thinning (Tables 5 and 6).

Table 5 Relation of DDE Residues in Eggs to Eggshell Thinning in Birds on Experimental Dietary Dosages of DDE

Species	Thinning (%)		Residues (μg/g, wet weight)		Source[a]
	Mean	Range	Mean	Range	
Black duck	18	12-29	46	34-63	1
	24	12-32	144	96-219	1
American kestrel	10	1-18	32	17-44	2
Barn owl	20	NL[b]	12	NL	3
	28	NL	41	NL	3

[a] 1, Longcore et al. (1971); 2, Wiemeyer and Porter (1970); 3, Mendenhall et al. (1983).
[b] NL, not listed.

Birds in experiments on dietary dosages of DDE laid eggs that had considerably thinner shells than did birds on "clean" diets without the compound (Table 5). Barn owls *(Tyto alba)* exhibited 20% eggshell thinning when eggs contained 12 μg/g of DDE (Mendenhall et al., 1983), and black ducks *(Anas rubripes)* exhibited 18% thinning when eggs contained 46 μg/g (Longcore et al., 1971). Although there were several studies of eggshell thinning of birds that were given diets containing technical DDT, most did not list residues in the eggs, and DDE — not DDT — comprises most of the dietary exposure of wild birds with significant eggshell thinning (Stickel, 1973). Eggs and tissues of ring-necked pheasants *(Phasianus colchicus)* accumulated high levels of DDT, about 5 times greater than DDE, in areas of intense application of technical DDT (Hunt and Keith, 1963). Domestic chickens given diets containing 300 μg/g of technical DDT showed no effects on eggshell thickness compared with controls, even though the eggs of dosed birds contained mean levels of 10 μg/g of DDE and 87 μg/g of DDT (approximate conversion from egg yolk basis at 14 days of incubation; Waibel et al., 1972). Results of the various studies of DDE relations to eggshell thickness in wild birds indicated extreme interspecific differences in sensitivity. Brown pelicans *(Pelecanus occidentalis)* in California and Baja California (Risebrough, 1972; Jehl, 1973) displayed extreme eggshell thinning and high residues of DDE in 1969 and the early 1970s, with nearly all eggs breaking in the most heavily contaminated colonies. In South Carolina, Florida, and Texas, much lower residues still resulted in mean eggshell thinning of 5 to 17% (Blus et al., 1974, 1979; King et al., 1977). While there was a statistically significant relation between DDE and eggshell thickness or the thickness index, there were some marked intraspecific differences in response. For example, when considering means (Table 6), the

Table 6 DDE Residues in Eggs Associated with Eggshell Thinning in Wild Birds

Species	Area[a]	Mean µg/g, Wet weight	% Thinning	Source[b]
Brown pelican	CA	59[c]	44	1
	BC	66[c]	46	2
	BC	25[c,d]	47	2
	BC	8[c]	26	2
	BC	3[c]	18	2
	SC	5	17	3,4
	SC	3	16	3,4
	SC	1	10	3,4
	FL	1	5	3,4
	FL	2	11	3,4
	TX	3	11	5
American white pelican	CA	2	15	6
(*Pelecanus erythrorhynchos*)	CA	2	10	6
Western grebe	CA	1	1	6
Great blue heron	WA	4	10	7
	WA	5	13	7
Peregrine falcon	AK	2	3[e]	8
	AK	7	8[e]	8
	AK	4	7[e]	8
	AK	44	22[e]	9
	AK	34	17[e]	9
	AK	8	8[e]	9
	AU	18	20[e]	10
Northern gannet (*Sula bassanus*)	QU	19	17	11
Double-crested cormorant	CA[c,f]	32	11	12
	BC[c,f]	24	30	12
	ON	24	15	13
Snowy egret	NV	2	12	14
	NV	1	3	14
White-faced ibis	NV	2	12	14
	NV	1	8	14
	TX	1	3	15
	TX	3	14	15
White-tailed eagle (*Haliaeetus albicilla*)	FI	30	15	16
Black-crowned night heron	CO, WY	4	9	17
	QU	2	<1[e]	18
	MA	4	4	19
	MA	2	0	19
	RI	4	6	19
	RI	1	0	19
Gray heron (*Ardea cinerea*)	GB	6	12[e]	20
	GB	3	9[e]	20
Eurasian sparrowhawk	GB	7	18[e]	21
Black skimmer (*Rynchops niger*)	TX	12	12	22
	TX	3	0	22

Table 6 (continued) DDE Residues in Eggs Associated with Eggshell Thinning in Wild Birds

Species	Area[a]	Mean μg/g, Wet weight	% Thinning	Source[b]
Osprey	CT	9	15	23
	MD	2	12	23
Bald eagle	OR, WA	10	10	24
Golden eagle	GB	0.1	7[e]	25
(Aquila chrysaetos)	GB	0.1	1[e]	25
	GB	0.2	3[e]	25
	GB	0.3	4[e]	25
	GB	0.3	5[e]	25

[a] CA, California; BC, Baja California; SC, South Carolina; FL, Florida; TX, Texas; WA, Washington; AK, Alaska; AU, Australia; QU, Quebec; ON, Ontario; NV, Nevada; FI, Finland; CO, Colorado; WY, Wyoming; MA, Massachusetts; RI, Rhode Island; GB, Great Britain; CT, Connecticut; MD, Maryland; OR, Oregon.

[b] 1, Risebrough (1972); 2, Jehl (1973); 3, Blus et al. (1974); 4, Blus et al. (1979); 5, King et al. (1977); 6, Boellstorff et al. (1985); 7, Fitzner et al. (1988); 8, White et al. (1973); 9, Cade et al. (1971); 10, Pruett-Jones et al. (1980); 11, Elliott et al. (1988); 12, Gress et al. (1973); 13, Weseloh et al. (1983); 14, Henny et al. (1985); 15, King et al. (1980); 16, Koivusaari et al. (1980); 17, McEwen et al. (1984); 18, Tremblay and Ellison (1980); 19, Custer et al. (1983); 20, Cooke et al. (1976); 21, Newton and Bogan (1974); 22, White et al. (1984); 23, Wiemeyer et al. (1975); 24, Anthony et al. (1993); 25, Newton and Galbraith (1991).

[c] Approximate value converted from lipid basis.

[d] Authors suspected residues were too low because of analytical errors.

[e] Percentage of thinning based on thickness index (Ratcliffe, 1967a); all others based on eggshell thickness.

[f] Intact eggs used only for eggshell thickness and residue analysis; the mean eggshell thinning of both crushed and intact eggs was 29% in CA and 38% in BC.

peregrine falcon in Alaska showed 22% thinning at 44 μg/g of DDE (Cade et al., 1971) compared with 20% thinning at 18 μg/g of DDE in Australia (Pruett-Jones et al., 1980). Intact eggs of the double-crested cormorant *(Phalacrocorax auritus)* in California exhibited 11% shell thinning at 32 μg/g compared with 24% thinning at 30 μg/g in Baja California (Gress et al., 1973). Many of the cormorant eggs were crushed in both colonies; overall thinning in collections containing both crushed and intact eggs reached 29% in California and 38% in Baja California.

Eggshell thinning is based on either eggshell thickness or the thickness index. In comparisons of the thickness index with eggshell thickness using museum specimens, the index indicated ≥% thinning 76% of the time (Anderson and Hickey, 1972) with extreme differences of 10% for each measurement. Thus, it is obvious that either of these measurements represents an accurate indication of eggshell thinning, but thickness is probably the measurement of choice in most instances, particularly when shells are cut because of loss of fragments.

The rate of thinning per μg/g of DDE is much greater at lower residues. Using the brown pelican as an example, there is 5 to 10% thinning at 1 μg/g of DDE compared with 44% at 59 μg/g (Table 6). While there is evidence that certain other contaminants and physiological conditions may induce eggshell thinning, the burden of proof overwhelmingly indicates that DDE is the major cause of the eggshell thinning syndrome. There have been attempts to relate DDE residues in the egg to a level of eggshell thinning that is associated with population decline if such thinning persists over a period of years. Initially, Hickey and Anderson (1968) concluded that ≥18% thinning was associated with declining populations; Anderson and Hickey (1972) modified this to "above 15 to 20% for a period of years." Thus, some have taken 15% as an effect level; however, with few exceptions, 18% is probably a more accurate indicator.

One notable exception was that of a declining Eurasian sparrowhawk population in the Netherlands that had poor production, 18% eggshell thinning, and mean DDE residues of 25 µg/g. Even though there was no significant relation between DDE and eggshell thickness, Koeman et al. (1972) suggested that DDE was responsible for the reproductive problems of Eurasian sparrowhawks. Also, Wiemeyer et al. (1972) found a poor correlation between eggshell thinning and DDE concentrations in bald eagle eggs, but there was a significant relation established when additional data were accumulated (Wiemeyer et al., 1988). One of the problems in these relations is that a greater effect per µg/g of DDE occurs at lower levels and, if residues are clumped, particularly on the high side, statistical relations are more difficult to establish. Ideally, a wide spread in residues is optimal for detecting effects on eggshell thickness.

When regression analysis was used to relate DDE levels to 20% eggshell thinning (Table 7), the critical estimates have ranged from 5 µg/g for the California condor (*Gymnogyps californianus*; Kiff et al., 1979) to 60 µg/g (fresh eggs) to 110 µg/g (failed eggs) for the bald eagle (Wiemeyer et al., 1993). Estimates in Table 7 are of value, but they must be interpreted with some caution. For example, the regression equation listed by Cade et al. (1971) indicated that 20% eggshell thinning was associated with 22 µg/g of DDE, whereas their tabular data indicated 17% thinning at 34 µg/g and 22% thinning at 44 µg/g (Tables 6 and 7). Blus (1984) reported that most of the error was related to extending the regression line beyond the range of the data. Thus, the critical level of 19 µg/g for the great blue heron is much too low, and that of 54 µg/g for the black-crowned night heron is much too high. There are wide disparities in the estimates for the common loon, osprey *(Pandion haliaetus)*, and the peregrine falcon. While the California condor is listed as the most sensitive to DDE-induced thinning, the estimated critical level is based on the measurement of eggshell fragments and extraction of DDE from eggshell membranes and then the calculation of residues in the entire egg from these measurements. From work on the extraction of DDE from membranes of intact peregrine falcon eggshells in museums, lower residues are associated with a far greater degree of thinning than are those of intact eggs collected from the field (Peakall and Kiff, 1979). Therefore, the accuracy of this technique requires experimental verification. Although Fox (1979) indicated that the measurement of eggshell thickness was a reliable indicator of the DDE content of the egg in some populations, Blus (1984) concluded that the DDE-thickness relation is not tight enough to do this for individual eggs and that residue analysis is essential for interpretation.

The calculated no-effect level for DDE in eggs related to the effects on eggshell thickness ranged from 0.1 µg/g for the brown pelican (Blus, 1984) to 2 µg/g for the

Table 7 Estimated Residues of DDE in Eggs Associated with Eggshell Thinning of 20% in Wild Birds

Species	DDE (µg/g, Wet weight)	Source[a]
Common loon	14	1
	47	2
California condor	5[b]	3
Peregrine falcon	15-20[c]	4
	22[c]	5
	18[c]	6
	20	7
Brown pelican	8	8
Black-crowned night heron	54	8,9
Prairie falcon	7	7
Great blue heron	19	8
Osprey	9	10
	29[d]	10
	41[d]	11
Eurasian sparrowhawk	10[d]	12
Merlin	16[d]	13
White-faced ibis	7	14
Bald eagle	60-110	15

[a] 1, Price (1977); 2, Fox et al. (1980); 3, Kiff et al. (1979); 4, Peakall et al. (1975); 5, Cade et al. (1971); 6, Pruett-Jones et al. (1980); 7, Enderson and Wrege (1973); 8, Blus (1984); 9, Henny et al. (1984); 10, Wiemeyer et al. (1988); 11, Spitzer et al. (1978); 12, Newton et al. (1986); 13, Newton et al. (1982); 14, Henny and Herron (1989); 15, Wiemeyer et al. (1993).
[b] Based on thickness of eggshell fragments and DDE content of shell membranes.
[c] Percentage of thinning based on thickness index (Ratcliffe, 1967); all others based on eggshell thickness.
[d] Eggs collected after nest failure; all other studies except that of Kiff et al. (1979) included at least some eggs collected while nests were active.

peregrine falcon (Cade et al., 1971). An earlier estimate for the brown pelican was 0.5 µg/g (Blus et al., 1974), but the sample size and range in residues were much smaller than in the subsequent study.

Eggshell Strength

The strength of eggshells, as determined by various mechanical devices, is related to eggshell thickness and, therefore, to DDE residues in the egg. Shell strength decreased more than eggshell thickness per unit of DDE; for example, 8 to 16 µg/g in sample eggs of the white-faced ibis were associated with a decrease of 16% in thickness and 37% in strength (Henny and Bennett, 1990). In addition, productivity of the young was related to DDE, shell thickness, and shell strength. Although shell strength may provide a more sensitive indicator of potential egg failure due to DDE, simple thickness measurements have served very well in that regard, and there are fewer logistical and financial constraints than are required to measure strength.

Productivity

DDT, primarily through its major metabolite DDE, also affects the reproductive success of birds. Eggshell thinning is an important, but not exclusive, factor related to reproductive problems (Blus, 1984). Unfortunately, most experimental studies of the reproductive effects of DDE or DDT did not present residues. This was a loss for interpreting the effects of residues on reproductive success in field studies.

There are several methods of expressing reproductive success relative to residues in birds. One method relates the overall reproductive success of a colony or other breeding group, such as a pen of experimental birds, to the mean residue content in their eggs (Table 8). Black ducks *(Anas rubripes)* receiving dietary dosages of 10 or 30 ppm of DDE had a significantly reduced survival of embryonated eggs or hatchlings to 3 weeks posthatch in relation to controls; DDE averaged 46 and 144 µg/g in eggs of treated birds (Longcore et al., 1971). Barn owls on a diet containing 3 ppm of DDE had hatchling and fledgling rates that were reduced about 75% from control values over 2 years when eggs contained an average of 12 µg/g the first year and 41 µg/g the second year (Mendenhall et al., 1983). The response was approximately the same each year, even though residues in eggs of the barn owls were much higher the second year. Although these studies document the effects from DDE, a narrow part of the relation is presented; for example, there is no indication of the dietary levels or residues in eggs at which problems first appear or where they initially become serious.

Table 8 Residues of DDE in Eggs Related to the Reproductive Success of Birds on Experimental Dietary Dosages of DDE

Species	Source[a]	Treatment	Residues (µg/g, wet weight)		Reproductive success (% of survival to 3 weeks posthatch)	
			Mean	Range	Hatchlings	Embryonated eggs
Black duck	1	Control	0.28	0.14–0.67	91	38
		10 ppm	46	34–63	64[b]	23[b]
		30 ppm	144	96–219	50[b]	9[b]

					Mean per pair	
					Eggs hatched	Young fledged
Barn owl	2	Control (1st year)	0.25	NL[c]	3.2	2.9
		Control (2nd year)	0.40	NL	3.7	3.1
		3 ppm (1st year)	12	NL	1.1[b]	0.7[b]
		3 ppm (2nd year)	41	NL	0.9[b]	0.7[b]

[a] 1, Longcore et al. (1971); 2, Mendenhall et al. (1983).
[b] Significantly different ($P \leq 0.05$) from controls.
[c] NL, not listed.

Table 9 Mean Residues of DDE in Eggs Related to Mean Reproductive Success of Wild Birds

Species	Area[a]	Residues (μg/g, wet weight)	Young produced per active nest	Source[b]
Double-crested cormorant	CA	32[c]	0.0	1
	BC	24[c]	<0.1	1
	ON	14-16	0 to 0.1[d]	2
	ON	5	0.3	2
Black-crowned night heron	QU	2	2.4	3
Common loon	SA	6	0.7	4
Brown pelican	CA	59[e]	<0.1	5
	BC	66[e]	0.0	6
	BC	25[e,f]	<0.1	6
	BC	8[e]	0.1	6
	BC	3[e]	~0.8	6
Bald eagle	OR, WA	10	0.6	7

[a] CA, California; BC, Baja California; ON, Ontario; QU, Quebec; SA, Saskatchewan; OR, Oregon; WA, Washington.
[b] 1, Gress et al. (1973); 2, Weseloh et al. (1983); 3, Tremblay and Ellison (1980); 4, Fox et al. (1980); 5, Risebrough (1972); 6, Jehl (1973); 7, Anthony et al. (1993).
[c] Approximate conversion from dry weight basis.
[d] Five colonies.
[e] Approximate conversion from lipid weight basis.
[f] Authors suspected residues were too low because of analytical errors.

Some of the field data on reproductive success also follow the average residue-average effect design (Table 9). A method that gives more insight into the effects of residues on reproductive success is the sample egg technique (Table 10), whereby one egg is taken from a nest and analyzed, the nest is marked, its fate is monitored through periodic visits, and the residues in eggs are related to nest success (Blus, 1984). This is particularly valuable in the field, where many factors may influence reproductive outcome. Blus (1984) lists advantages and disadvantages of this method. Where nest predation is a problem and where clutch size permits, one egg may be collected for residue analysis and another taken and placed in an incubator.

There are several variations to the sample egg technique as shown in Table 10. One involves work with threatened or endangered species or other special situations where the sample egg is not collected until the fate of the marked nest is determined (Spitzer et al., 1978; Wiemeyer et al., 1984). The major bias with collecting eggs after the fact is that those with thin eggshells and high DDE residues have a greater chance of being crushed or cracked and therefore lost from the population. Other variations relate to statistical analysis of the data, regardless of the time of egg collections. One method ranks young fledged vs. residues (Fyfe et al., 1976; Spitzer et al., 1978), and another method ranks residues or a range of residues vs. young fledged (Blus et al., 1980, 1982; Henny et al., 1984, 1985, 1989; Ambrose et al., 1988; Wiemeyer et al., 1993). Of these two methods, I recommend the second, because it seems to more closely approximate the dependent variable-independent variable relation; however, it should be recognized that these methods have not been subjected to rigid statistical testing.

Table 10 Residues of DDE in Sample Eggs of Wild Birds Related to Reproductive Success

Species	Area[a]	Residues (μg/g, wet weight)	Young produced per active nest[b]	Source[c]
Osprey	CT, NY	23[d-f]	0.0	1
	CT, NY	12[d-f]	1.0	1
	CT, NY	6[d-f]	2.1	1
	ID	14	0.0	2
	ID	6	1.6	2
Peregrine falcon	AK	≤15	1.8	3
	AK	15-30	2.0	3
	AK	>30	1.0	3
Snowy egret	NV	≤1	2.2	4
	NV	1-5	2.4	4
	NV	5-10	1.0	4
	NV	10-20	1.0	4
Prairie falcon	AB	2[d]	0.0	5
	AB	2[d]	1.0	5
	AB	2[d]	2.0	5
	AB	2[d]	3.0	5
	AB	1[d]	4.0	5
Merlin	AB	11[d]	0.0	5
	AB	11[d]	1.0	5
	AB	6[d]	2.0	5
	AB	5[d]	3.0	5
	AB	6[d]	4.0	5
Brown pelican	SC[g]	≤1.5	0.6 (FL),[h] 0.8 (EM)	6
	SC[g]	1.5-3	0.6 (FL), 0.8 (EM)	6
	SC[g]	>3	0.0 (FL), 0.6 (EM)	6
Bald eagle	US[f]	<2.2	1.0	7
	US[f]	2.2-3.5	1.0	7
	US[f]	3.6-6.2	0.5	7
	US[f]	6.3-11.9	0.3	7
	US[f]	≥12	0.2	7
Black-crowned night heron	US	≤1	2.0	8
	US	1-4	1.7	8
	US	4-8	1.5	8
	US	8-12	1.1	8
	US	12-16	1.0	8
	US	16-25	0.8	8
	US	25-50	0.4	8
White-faced ibis	NV	≤1	1.8	9
	NV	1-4	1.8	9
	NV	4-8	1.3	9
	NV	8-16	0.8	9
	NV	>16	0.6	9
Great blue heron	OR, WA	3	1.7-2.0	10

[a] CT, Connecticut; NY, New York; ID, Idaho; AK, Alaska; NV, Nevada; AB, Alberta and nearby areas; SC, South Carolina; US, various locations within the U.S.
[b] Young produced not adjusted for sample egg collected.
[c] 1, Spitzer et al. (1978); 2, Johnson et al. (1975); 3, Ambrose et al. (1988); 4, Henny et al. (1985); 5, Fyfe et al. (1976); 6, Blus (1982); 7, Wiemeyer et al. (1993); 8, Henny et al. (1984); 9, Henny and Herron (1989); 10, Blus et al. (1980).

Table 10 (continued) Residues of DDE in Sample Eggs of Wild Birds Related to Reproductive Success

^d Approximate adjustment from dry weight basis.
^e Elevated levels (17 to 29 μg/g) of polychlorinated biphenyls also present.
^f All or most eggs were collected after the fate of marked nests was determined. Production of young at each nest is based on a 5-year mean.
^g Sample egg either freshly laid or embryonated when collected.
^h FL, freshly laid; EM, embryonated.

Problems with comparing the young fledged per nest with the residue content of sample eggs seemed evident in a study of prairie falcons *(Falco mexicanus)* in Alberta, Canada and surrounding areas, where the mean DDE levels of 1 to 2 μg/g were said to adversely affect fledging success (Table 10; Fyfe et al., 1976). However, differences in the mean residue content were not statistically different, and the egg with the highest level of 11 μg/g was from a nest that fledged five young (Fyfe et al., 1976). In merlins *(Falco columbarius)*, fledging success was significantly related to DDE residues in sample eggs with an effect level of near 10 μg/g but, again, high levels of about 31 and 26 μg/g were found in eggs from nests that fledged one and five young, respectively (Fyfe et al., 1976). On the basis of addled or deserted eggs collected from merlin nests in Great Britain, Newton et al. (1982) indicated a positive correlation between fledging success and DDE residues, with zero young fledged when eggs contained 5 μg/g, increasing to four young fledged at 8 μg/g. Although the lower critical level of DDE that adversely affects the reproductive success of peregrine falcons was considered to be 15 to 20 μg/g (Peakall, 1976), nest success in Great Britain seemed unaffected by DDE, with the highest residues of 25 and 31 μg/g being detected in sample eggs from successful nests (Ratcliffe, 1967b). More recent, albeit limited, evidence from peregrine falcons in Alaska indicated that the effects on nest success occur only when residues exceed 30 μg/g (Ambrose et al., 1988).

Considering DDE residues vs. the young produced per nest, declines in the productivity of brown pelicans (Blus, 1982), bald eagles (Wiemeyer et al., 1993), black-crowned night herons (Henny et al., 1984), snowy egrets *(Egretta thula;* Henny et al., 1985), and the white-faced ibis *(Plegadis chihi;* Henny and Herron, 1989) are obvious (Table 10). One can determine the level at which residues first begin having an adverse effect on the number of young produced, for example, 4 to 8 μg/g in the white-faced ibis (Henny and Herron, 1989), and at which few or no young are produced, for example, ≥15 μg/g in the bald eagle. Black-crowned night herons demonstrate an impressive gradual decline in productivity with an increase in residues; however, a few young are produced even at levels >25 μg/g (Henny et al., 1984). In the brown pelican, Blus (1982) indicated a dramatic effect above 3 μg/g, with no young produced in sample eggs that were freshly laid when collected and a 25% reduction in those that were embryonated when collected (Table 10). The brown pelican is apparently the most sensitive avian species to DDE, with reproductive failure when residues in eggs exceed 3.7 μg/g (Blus, 1982). In South Carolina, a combination of effects from DDE, including eggshell thinning, seemed to adversely affect reproductive success, whereas in California and Baja California nearly every egg collapsed from extreme eggshell thinning in 1969 and several subsequent years (Risebrough, 1972; Jehl, 1973). In order to interpret what a DDE-related reduction in the number of young fledged means in terms of population reduction, one has to know a great deal about the population, e.g., an approximate recruitment standard (number of young that must be fledged per pair of breeding age in order to maintain a stable population) and adult mortality compensating mechanisms, including renesting and other factors. Recruitment standards are 1.2 to 1.5 young for the brown pelican (Henny, 1972), 1.0 young for the bald eagle (Wiemeyer et al., 1993), 1 to 1.3 young (Henny and Wight, 1969) and 0.8 young (Spitzer et al., 1983) for the osprey, 1.9 young for the great blue heron (Henny, 1972), and 2 to 2.1 young for the black-crowned night heron (Henny, 1972). Variations in the recruitment standard for the osprey probably result from whether active nests (Spitzer et al., 1983) or pairs of breeding age (Henny and Wight, 1969) are used in the calculations.

In Table 10, no compensation in the young produced was made for collection of the sample egg. Therefore, most of these productivity data are probably biased low, compared with nests without an egg collected. To measure this bias, Henny and Herron (1989) compared production in white-faced ibis nests with an egg collected to that without an egg collected; sample egg collection was associated with a 30% reduction in the young produced per active nest. The percentage of the reduction, of course, would be influenced by clutch size in the species of interest.

Although DDE was responsible for most reproductive failure in birds, very high levels of DDT in ring-necked pheasant eggs in California may have caused reproductive problems, such as crippling and mortality of young; however, the link between them was never clearly established (Hunt and Keith, 1963).

Domestic chickens given diets containing 300 µg/g of technical DDT showed no significant effects on reproduction compared with controls, even though their eggs contained mean levels of 10 µg/g of DDE and 87 µg/g of DDT (approximate conversion from egg yolk basis at 14 days of incubation; Waibel et al., 1972).

FOOD

The lowest dietary concentration of DDE that resulted in critical eggshell thinning and decreased production in the peregrine falcon was estimated at 1 µg/g (Enderson et al., 1982). A more recent study used 3 µg/g as a critical level, but this was based on dietary levels given to experimental raptors that experienced serious reproductive problems and even adult mortality (DeWeese et al., 1986). For the brown pelican, the lower critical dietary level of DDE was estimated at about 0.1 µg/g on the basis of 31× biomagnification from fish to pelican egg; however, because the chief prey fish also contained DDT at one-half the amount of DDE, the 0.1-µg/g level probably should be raised slightly to account for metabolism from DDT to DDE (Blus et al., 1977). These examples included a highly sensitive species and a moderately sensitive species, so higher estimates for less sensitive species are expected based on experimental studies of the domestic chicken, one of the least sensitive species (Waibel et al., 1972); but lower estimates are unlikely. Because of the lipophilicity and bioaccumulativeness of all three compounds, the highest residues and, depending on species sensitivity, the most extreme effects are found in species at the highest trophic levels, as is evident in most of the studies summarized in this review.

OTHER TISSUES

There are other measurements of the sublethal effects related to residues of DDT and its metabolites in other tissues of animals used in experiments, but most of these data are fragmentary, and few of these measurements have proven useful in field studies.

SUMMARY

Although technical DDT was initially hailed as a tremendous tool in pest control, the environmental problems soon outweighed the positive aspects. As a result, this compound was banned over much of the world; however, use of DDT continues in some countries, especially for control of insects that are disease vectors.

One of the first findings related to the use of technical DDT was the mortality of wildlife after heavy applications, but suitable analytical techniques were required to detect residues because many adverse effects occurred some time after exposure. Residues in tissues, particularly the brain, have proven to be diagnostic of lethality in animals on dietary dosages of DDT, DDD, and DDE in experiments. When used in field investigations, this technique made possible the interpretation of lethality when DDT equivalents (weighting system where an equivalent equals 1 µg/g of DDT, 5 µg/g of DDD, or 15 µg/g of DDE) are as low as 10 in brains; however, most birds or mammals that die from DDT have DDT equivalents >20. Few dead wild birds have been found with lethal levels of DDE or DDD in their brains. Residues in livers also have been used to establish the lethality of DDT in wild animals, but a system for the weighting of the three compounds has not been developed.

Residues in the eggs of birds are a reliable indicator of eggshell thinning and reproductive success. Of the three compounds reviewed in this paper, evidence overwhelmingly indicates that DDE is responsible for most eggshell thinning, reproductive problems, and population reductions. There is a tremendous variation in species sensitivity to these compounds. The brown pelican is the most sensitive, with eggshell thinning and depressed productivity occurring at 3.0 µg/g of DDE in the egg and total reproductive failure when residues exceed 3.7 µg/g. In contrast, adverse effects on the reproductive success of peregrine falcons first occur when DDE residues in the egg are about 10-fold higher, that is, 30 µg/g. Black-crowned night herons demonstrate a different pattern involving a gradual decline in productivity with increasing residues. A few young are still produced at levels >25 µg/g. The domestic chicken is very tolerant of high dietary exposure to technical DDT. By efficient use of the sample egg technique, the effects induced by DDE residues within one colony or breeding area, or compared with a reference colony or area where residues are low, can be quantified and related to the adverse effects on the individual and the population. Techniques for quantifying the relation between residues of DDT and its metabolites and the effects on the biota have been successful, but the process required much time, effort, and financial outlay. Many contemporary field studies are designed inefficiently with regard to quantifying residues, with little or no consideration given to establishing the effects, or less commonly, establishing the effects without evidence from residues. In addition, few experimental studies are directly applicable to the field. Results of experimental and field studies could be made more pertinent to interpretation of field data by changes in the experimental design, so that efficient use can be made of the establishing and measuring of effects induced by residues.

REFERENCES

Ambrose, R. E., C. J. Henny, R. E. Hunter, and J. A. Crawford. 1988. Organochlorines in Alaskan peregrine falcon eggs and their current impact on productivity. p. 385-393. *In* T. J. Cade, J. H. Enderson, C. G. Thelander, and C. M. White (Eds.). Peregrine falcon populations: their management and recovery. Peregrine Fund, Boise, Id.

Anderson, D. W., and J. J. Hickey. 1972. Eggshell changes in certain North American birds. Proc. Int. Ornithol. Congr. 15:514-540.

Anthony, R. G., M. G. Garrett, and C. A. Schuler. 1993. Environmental contaminants in bald eagles in the Columbia River estuary. J. Wildl. Manage. 57:10-19.

Belisle, A. A., W. L. Reichel, L. N. Locke, T. G. Lamont, B. M. Mulhern, R. M. Prouty, R. B. DeWolf, and E. Cromartie. 1972. Residues of organochlorine pesticides, polychlorinated biphenyls, and mercury, and autopsy data for bald eagles, 1969 and 1970. Pestic. Monit. J. 6:133-138.

Bernard, R. F. 1963. Studies of the effects of DDT on birds. Mich. State Univ. Mus. Publ. Biol. Serv. 2:155-192.

Blus, L. J. 1982. Further interpretation of the relation of organochlorine residues in brown pelican eggs to reproductive success. Environ. Pollut. 28:15-33.

Blus, L. J. 1984. DDE in birds' eggs: comparison of two methods for estimating critical levels. Wilson Bull. 96:268-276.

Blus, L. J., A. A. Belisle, and R. M. Prouty. 1974. Relations of the brown pelican to certain environmental pollutants. Pestic. Monit. J. 7:181-194.

Blus, L. J., C. J. Henny, and T. E. Kaiser. 1980. Pollution ecology of breeding great blue herons in the Columbia Basin, Oregon and Washington. Murrelet 61:63-71.

Blus, L. J., T. G. Lamont, and B. S. Neely, Jr. 1979. Effects of organochlorine residues on eggshell thickness, reproduction, and population status of brown pelicans *(Pelecanus occidentalis)* in South Carolina and Florida, 1969-76. Pestic. Monit. J. 12:172-184.

Blus, L. J., B. S. Neely, Jr., T. G. Lamont, and B. M. Mulhern. 1977. Residues of organochlorines and heavy metals in tissues and eggs of brown pelicans, 1969-73. Pestic. Monit. J. 11:40-53.

Boellstorff, D. E., H. M. Ohlendorf, D. W. Anderson, E. J. O'Neill, J. O. Keith, and R. M. Prouty. 1985. Organochlorine chemical residues in white pelicans and western grebes from the Klamath Basin, California. Arch. Environ. Contam. Toxicol. 14:485-493.

Cade, T. J., J. L. Lincer, C. M. White, D. G. Roseneau, and L. G. Swartz. 1971. DDE residues and eggshell changes in Alaskan falcons and hawks. Science (Washington, D.C.) 172:955-957.

Call, D. J., H. J. Shave, H. C. Binger, M. E. Bergeland, B. D. Ammann, and J. J. Worman. 1976. DDE poisoning in wild great blue heron. Bull. Environ. Contam. Toxicol. 16:310-313.

Carson, R. 1962. Silent spring. Houghton Mifflin, Boston.

Cooke, A. S., A. A. Bell, and M. B. Haas. 1982. Predatory birds, pesticides, and pollution. Institute of Terrestrial Ecology, Cambridge, England.

Cooke, A. S., A. A. Bell, and I. Prestt. 1976. Egg shell characteristics and incidence of shell breakage for grey herons *(Ardea cinerea)* exposed to environmental pollutants. Environ. Pollut. 11:59-84.

Cottam, C., and E. Higgins. 1946. DDT: its effects on fish and wildlife. U. S. Fish Wildl. Serv. Circ. 11, 14 pp.

Custer, T. W., C. M. Bunck, and T. E. Kaiser. 1983. Organochlorine residues in Atlantic Coast black-crowned night-heron eggs, 1979. Colon. Waterbirds 6:160-167.

DeWeese, L. R., L. C. McEwen, G. L. Hensler, and B. E. Peterson. 1986. Organochlorine contaminants in Passeriformes and other avian prey of the peregrine falcon in the western United States. Environ. Toxicol. Chem. 5:675-693.

Ecobichon, D. J., and P. W. Saschenbrecker. 1968. Pharmacodynamic study of DDT in cockerels. Can. J. Physiol. Pharmacol. 46:785-794.

Elliott, J. E., R. J. Norstrom, and J. A. Keith. 1988. Organochlorines and eggshell thinning in northern gannets *(Sula bassanus)* from eastern Canada, 1968-1984. Environ. Pollut. 52:81-102.

Enderson, J. H., G. R. Craig, W. A. Burnham, and D. D. Berger. 1982. Eggshell thinning and organochlorine residues in Rocky Mountain peregrines, *Falco peregrinus*, and their prey. Can. Field Nat. 96:255-264.

Fitzner, R. E., L. J. Blus, C. J. Henny, and D. W. Carlile. 1988. Organochlorine residues in great blue herons from the northwestern United States. Colon. Waterbirds 11:293-300.

Fox, G. A. 1979. A simple method of predicting DDE contamination and reproductive success of populations of DDE-sensitive species. J. Appl. Ecol. 16:737-741.

Fox, G. A., K. S. Yonge, and S. G. Sealy. 1980. Breeding performance, pollutant burden, and eggshell thinning in common loons *(Gavia immer)* nesting on a boreal forest lake. Ornis Scand. 11:243-248.

Fyfe, R. W., R. W. Risebrough, and W. Walker, II. 1976. Pollutant effects on the reproduction of the prairie falcons and merlins of the Canadian prairies. Can. Field Nat. 90:346-355.

Gress, F., R. W. Risebrough, D. W. Anderson, L. F. Kiff, and J. R. Jehl, Jr. 1973. Reproductive failures of double-crested cormorants in southern California and Baja California. Wilson Bull. 85:197-208.

Hayes, W. J., Jr. 1991. Introduction. p. 1-37. *In* W. J. Hayes, Jr., and E. R. Laws, Jr. (Eds.). Handbook of pesticide toxicology. Academic Press, San Diego.

Heath, R. G., J. W. Spann, and J. F. Kreitzer. 1969. Marked DDE impairment of mallard reproduction in controlled studies. Nature (Lond.) 224:47-48.

Henny, C. J. 1972. An analysis of the population dynamics of selected avian species — with special reference to changes during the modern pesticide era. U. S. Fish Wildl. Serv. Wildl. Res. Rep. No. 1, 99 pp.

Henny, C. J., and J. K. Bennett. 1990. Comparison of breaking strength and shell thickness as evaluators of white-faced ibis eggshell quality. Environ. Toxicol. Chem. 9:797-805.

Henny, C. J., L. J. Blus, and C. S. Hulse. 1985. Trends and effects of organochlorine residues on Oregon and Nevada wading birds, 1979-1983. Colon. Waterbirds 8:117-128.

Henny, C. J., L. J. Blus, A. J. Krynitsky, and C. M. Bunck. 1984. Current impact of DDE on black-crowned night-herons in the intermountain West. J. Wildl. Manage. 48:1-13.

Henny, C. J., and G. B. Herron. 1989. DDE, selenium, mercury, and white-faced ibis reproduction at Carson Lake, Nevada. J. Wildl. Manage. 53:1032-1045.

Henny, C. J., and D. L. Meeker. 1981. An evaluation of blood plasma for monitoring DDE in birds of prey. Environ. Pollut. Ser. A Ecol. Biol. 25:291-304.

Henny, C. J., and H. M. Wight. 1969. An endangered osprey population: estimates of mortality and production. Auk 86:188-198.

Heyroth, F. F. 1950. The toxicity of DDT. II. A survey of the literature. p. 72-233. Kettering Laboratory, College of Medicine, University of Cincinnati, Cincinnati, OH.

Hickey, J. J., and D. W. Anderson. 1968. Chlorinated hydrocarbons and eggshell changes in raptorial and fish-eating birds. Science (Washington, D.C.) 162:271-273.

Hill, E. F., W. E. Dale, and J. W. Miles. 1971. DDT intoxication in birds: subchronic effects and brain residues. Toxicol. Appl. Pharmacol. 20:502-514.

Hotchkiss, N., and R. H. Pough. 1946. Effect on forest birds of DDT used for gypsy moth control in Pennsylvania. J. Wildl. Manage. 10:202-207.

Hunt, E. G., and J. O. Keith. 1963. Pesticide-wildlife investigations in California, 1962. Proc. 2nd Annu. Conf. on the Use of Agric. Chemicals in California, Davis, Calif., 29 pp.

Hunt, L. B. 1968. Songbirds and Insecticides in a Suburban Elm Environment. Ph.D. thesis. University of Wisconsin, Madison, Wisc.

Jehl, J. R., Jr. 1973. Studies of a declining population of brown pelicans in Northwestern Baja California. Condor 75:69-79.

Johnson, D. R., W. E. Melquist, and G. J. Schroeder. 1975. DDT and PCB levels in Lake Coeur d'Alene, Idaho, osprey eggs. Bull. Environ. Contam. Toxicol. 13:401-405.

Kiff, L. F., D. B. Peakall, and S. R. Wilbur. 1979. Recent changes in California condor eggshells. Condor 81:166-172.

King, K. A., E. L. Flickinger, and H. H. Hildebrand. 1977. The decline of brown pelicans on the Louisiana and Texas Gulf Coast. Southwest. Nat. 21:417-431.

King, K. A., D. L. Meeker, and D. M. Swineford. 1980. White-faced ibis populations and pollutants in Texas, 1969-1976. Southwest. Nat. 25:225-240.

Koeman, J. H., C. F. Van Beusekom, and J. J. M. De Goeij. 1972. Eggshell and population changes in the sparrow-hawk *(Accipiter nisus)*. TNO-Nieuws 27:542-550.

Koivusaari, J., I. Nuuja, R. Palokangas, and M. Finnlund. 1980. Relationships between productivity, eggshell thickness, and pollutant contents of addled eggs in the population of white-tailed eagles *Haliaetus albicilla* L. in Finland during 1969-1978. Environ. Pollut. Ser. A Ecol. Biol. 23:41-52.

Longcore, J. R., F. B. Samson, and T. W. Whittendale, Jr. 1971. DDE thins eggshells and lowers reproductive success of captive black ducks. Bull. Environ. Contam. Toxicol. 6:485-490.

McEwen, L. C., C. J. Stafford, and G. L. Hensler. 1984. Organochlorine residues in eggs of black-crowned night-herons from Colorado and Wyoming. Environ. Toxicol. Chem. 3:367-376.

Mendenhall, V. M., E. E. Klass, and M. A. R. McLane. 1983. Breeding success of barn owls *(Tyto alba)* fed low levels of DDE and dieldrin. Arch. Environ. Contam. Toxicol. 12:235-240.

Mitchell, R. T. 1946. Effects of DDT spray on eggs and nestlings of birds. J. Wildl. Manage. 10:192-194.

Newton, I., and J. Bogan. 1974. Organochlorine residues, eggshell thinning, and hatching success in British sparrowhawks. Nature (Lond.) 249:582-583.

Newton, I., J. Bogan, E. Meek, and B. Little. 1982. Organochlorine compounds and shell-thinning in British merlins *(Falco columbarius)*. Ibis 124:328-335.

Newton, I., J. A. Bogan, and P. Rothery. 1986. Trends and effects of organochlorine compounds in sparrowhawk eggs. J. Appl. Ecol. 23:461-478.

Newton, I., and E. A. Galbraith. 1991. Organochlorines and mercury in the eggs of golden eagles *Aquila chrysaetos* from Scotland. Ibis 133:115-120.

Newton, I., I. Wyllie, and A. Asher. 1992. Mortality from the pesticides aldrin and dieldrin in British sparrowhawks and kestrels. Ecotoxicology 1:31-44.

Ohlendorf, H. M., D. M. Swineford, and L. N. Locke. 1981. Organochlorine residues and mortality of herons. Pestic. Monit. J. 14:125-135.

Peakall, D. B. 1976. The peregrine falcon *(Falco peregrinus)* and pesticides. Can. Field. Nat. 90:301-307.

Peakall, D. B., T. J. Cade, C. M. White, and J. R. Haugh. 1975. Organochlorine residues in Alaskan peregrines. Can. Field. Nat. 104:244-254.

Peakall, D. B., and L. F. Kiff. 1979. Eggshell thinning and DDE residue levels among peregrine falcons *(Falco peregrinus):* a global perspective. Ibis 121:200-204.

Porter, R. D., and S. N. Wiemeyer. 1972. DDE at low dietary levels kills captive American kestrels. Bull. Environ. Contam. Toxicol. 8:193-199.

Price, I. M. 1977. Environmental contaminants in relation to Canadian wildlife. Trans. N. Am. Wildl. Nat. Resour. Conf. 42:382-396.

Prouty, R. M., J. E. Peterson, L. N. Locke, and B. M. Mulhern. 1975. DDD poisoning in a loon and the identification of the hydroxylated form of DDD. Bull. Environ. Contam. Toxicol. 14:385-388.

Pruett-Jones, S. G., C. M. White, and W. B. Emison. 1980. Eggshell thinning and organochlorine residues in eggs and prey of peregrine falcons from Victoria, Australia. Emu 80:281-287.

Ratcliffe, D. A. 1967a. Decrease in eggshell weight in certain birds of prey. Nature (Lond.) 215:208-210.

Ratcliffe, D. A. 1967b. The peregrine situation in Great Britain — 1965-1966. Bird Study 14:238-246.

Risebrough, R. W. 1972. Effects of environmental pollutants upon animals other than man. p. 443-463. *In* Proc. 6th Berkeley Symp. on Mathematical Statistics and Probability. University of California Press, Berkeley, Calif.

Rudd, R. L., and S. G. Herman. 1972. Ecosystemic transferral of pesticides in an aquatic environment. p. 471-485. *In* F. Matsumura, G. Boush, and T. Misato (Eds.). Environmental toxicology of pesticides. Academic Press, New York.

Sileo, L., L. Karstad, R. Frank, M. V. H. Holdrinet, E. Addison, and H. E. Braun. 1977. Organochlorine poisoning of ring-billed gulls in southern Ontario. J. Wildl. Dis. 13:313-322.

Smyth, H. F., Jr., C. S. Weil, J. S. West, and C. P. Carpenter. 1969. An exploration of joint toxic action: twenty-seven industrial chemicals intubated in rats in all possible pairs. Toxicol. Appl. Pharmacol. 14:340-347.

Spitzer, P. R., A. F. Poole, and M. Scheibel. 1983. Initial population recovery of breeding ospreys in the region between New York and Boston. p. 231-241. *In* D. M. Bird (Chief Ed.) Biology and management of bald eagles and ospreys. Harpell Press, Ste. Anne de Bellevue, Quebec, Canada.

Spitzer, P. R., R. W. Risebrough, W. Walker, II, R. Hernandez, A. Poole, D. Puleston, and I. C. T. Nisbet. 1978. Productivity of ospreys in Connecticut-Long Island increases as DDE residues decline. Science (Washington, D.C.) 202:333-335.

Stewart, R. E., J. B. Cope, C. S. Robbins, and J. W. Brainerd. 1946. Effects of DDT on birds at the Patuxent Research Refuge. J. Wildl. Manage. 10:195-201.

Stickel, L. F. 1973. Pesticide residues in birds and mammals. p. 254-312. *In* C. A. Edwards (Ed.). Environmental pollution by pesticides. Plenum Press, London, England.

Stickel, L. F., and W. H. Stickel. 1969. Distribution of DDT residues in tissues of birds in relation to mortality, body condition, and time. Ind. Med. Surg. 38:44-53.

Stickel, L. F., W. H. Stickel, and R. Christensen. 1966. Residues of DDT in brains and bodies of birds that died on dosage and in survivors. Science (Washington, D.C.) 151:1549-1551.

Stickel, W. H., L. F. Stickel, and F. B. Coon. 1970. DDE and DDD residues correlated with mortality of experimental birds. p. 287-294. *In* W. B. Deichmann (Ed.). Pesticides symposia. Seventh Int. Am. Conf. Toxicol. Occup. Med. Helios and Assoc., Miami.

Stickel, W. H., L. F. Stickel, R. A. Dyrland, and D. L. Hughes. 1984. DDE in birds: lethal residues and loss rates. Arch. Environ. Contam. Toxicol. 13:1-6.

Tremblay, J., and L. N. Ellison. 1980. Breeding success of the black-crowned night heron in the St. Lawrence estuary. Can. J. Zool. 58:1259-1263.

Van Velzen, A., and J. F. Kreitzer. 1975. The toxicity of p,p'-DDT to the clapper rail. J. Wildl. Manage. 39:305-309.

Waibel, G. P., G. M. Speers, and P. E. Waibel. 1972. Effects of DDT and charcoal on performance of white leghorn hens. Poult. Sci. 51:1963-1967.

Weseloh, D. V., S. M. Teeple, and M. Gilbertson. 1983. Double-crested cormorants of the Great Lakes: egg-laying parameters, reproductive failure, and contaminant residues in eggs, Lake Huron 1972-1973. Can. J. Zool. 61:427-436.

White, C. M., W. B. Emison, and F. S. L. Williamson. 1973. DDE in a resident Aleutian Island peregrine population. Condor 75:306-311.

White, D. H., C. A. Mitchell, and D. M. Swineford. 1984. Reproductive success of black skimmers in Texas relative to environmental pollutants. J. Field Ornithol. 55:18-30.

Wiemeyer, S. N., C. M. Bunck, and A. J. Krynitsky. 1988. Organochlorine pesticides, polychlorinated biphenyls, and mercury in osprey eggs — 1970-1979 — and their relationships to shell thinning and productivity. Arch. Environ. Contam. Toxicol. 17:767-787.

Wiemeyer, S. N., C. M. Bunck, and C. J. Stafford. 1993. Environmental contaminants in bald eagle eggs — 1980-84 — and further interpretations of relationships to productivity and shell thickness. Arch. Environ. Contam. Toxicol. 24:213-227.

Wiemeyer, S. N., T. G. Lamont, C. M. Bunck, C. R. Sindelar, F. J. Gramlich, J. D. Fraser, and M. A. Byrd. 1984. Organochlorine pesticide, polychlorobiphenyl, and mercury residues in bald eagle eggs — 1969-79 — and their relationships to shell thinning and reproduction. Arch. Environ. Contam. Toxicol. 13:529-549.

Wiemeyer, S. N., B. M. Mulhern, F. J. Ligas, R. J. Hensel, J. E. Mathisen, F. C. Robards, and S. Postupalsky. 1972. Residues of organochlorine pesticides, polychlorinated biphenyls, and mercury in bald eagle eggs and changes in shell thickness, 1969 and 1970. Pestic. Monit. J. 6:50-55.

Wiemeyer, S. N., and R. D. Porter. 1970. DDE thins eggshells of captive American kestrels. Nature (Lond.) 227:737-738.

Wiemeyer, S. N., P. R. Spitzer, W. C. Krantz, T. G. Lamont, and E. Cromartie. 1975. Effects of environmental pollutants on Connecticut and Maryland ospreys. J. Wildl. Manage. 39:124-139.

Wurster, D. H., C. F. Wurster, Jr., and W. N. Strickland. 1965. Bird mortality following DDT spray for Dutch elm disease. Ecology 46:488-499.

CHAPTER **3**

Dieldrin and Other Cyclodiene Pesticides in Wildlife

David B. Peakall

INTRODUCTION

Five cyclodiene pesticides have been used commercially: aldrin, dieldrin, endrin, isodrin, and telodrin. The chemical structures of these compounds are shown in Figure 1. Aldrin is readily converted to dieldrin. The conversion has been documented in mammals and birds (Bann et al., 1956), fish (Stanton and Khan, 1973), and insects (Cohen and Smith, 1961; Ray, 1967), as well as in soil (Bollen et al., 1958). However, Flickinger and Mulhern (1980) found aldrin fairly persistent in the yellow mud turtle *(Kinosternon flavescens)*, finding 4 times as much aldrin than dieldrin in its tissues 14 weeks after exposure to aldrin. Of the cyclodiene pesticides, isodrin has not been widely used and is converted to endrin (Nakatsugawa et al., 1965). Telodrin, likewise, has been used in only small quantities, but it has been implicated in the deaths of terns in Holland (Koeman et al., 1967). A large number of metabolites are known for the cylcodienes (see Menzie, 1969), but the only one of environmental importance is 12-ketoendrin.

In this chapter, the relations between residue levels and biological effects are discussed for the cyclodiene pesticides. By far, the largest data set is available for dieldrin. Aldrin, although widely used, is readily converted to dieldrin; thus, the effects seen in animals exposed to aldrin may be caused by dieldrin. Endrin, although highly toxic, is much less persistent than dieldrin.

DIETARY TOXICITY

Although this chapter concentrates on the relation of residue data to effects, it is useful to summarize the 50% lethal dose (LD_{50}) and 50% lethal concentration

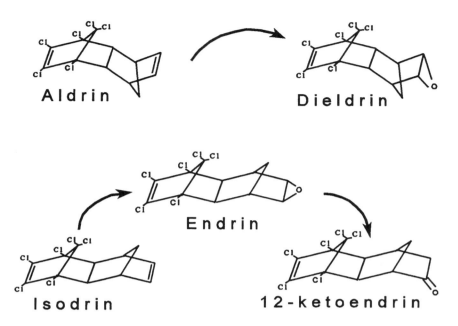

Figure 1 Formula and principal metabolic alterations of the cyclodiene insecticides. (After Menzie, 1969).

(LC_{50}) data for some of the species widely used in toxicity testing so that dietary input can be compared with mortality (Table 1).

In general, the toxicities of aldrin and dieldrin are quite similar, so that the conversion of aldrin to dieldrin in the environment should not result in changes in toxicity. An exception is for fish, where dieldrin is an order of magnitude more toxic than is aldrin. In this case, the conversion of aldrin to dieldrin could result in a major impact. The other points to be made are that endrin is considerably more toxic than either aldrin or dieldrin and that few data appear to be available on isodrin. However, since this pesticide is converted into endrin, estimates of damage can be based on the toxicology of the latter compound. Little information is available on telodrin, although Koeman et al. (1967) note that it is 17 times more toxic than is dieldrin.

In rats, 12-ketoendrin was 5 times as toxic as endrin itself and was considered to be the ultimate cause of death (Bedford et al., 1975). There is considerable interspecies variation in the formation of this metabolite of endrin. Bedford and co-workers found only low levels in cows and rabbits and none in chickens. Studies by Stickel et al. (1979a) showed only low levels in mice and none in specimens from four avian orders heavily exposed to endrin. In studies on the effect of endrin on wildlife (Blus et al., 1983), 12-ketoendrin was found only in voles; no residues were found in birds dying from endrin. In general, for wildlife toxicology purposes, we need not worry about 12-ketoendrin. However, in the case of animals feeding on rodents, the possibility should be borne in mind.

Table 1 Comparative Toxicity of Aldrin, Dieldrin, and Endrin Based on Experimental Studies

Species	LD_{50}/LC_{50}[a] Aldrin	Dieldrin	Endrin	Ref.
Rat (Rattus norvegicus)	54-56 mg/kg	50-55 mg/kg		Spector (1955)
Mule deer (Odocileus hemionus)	19-37 mg/kg	75-150 mg/kg	6-12 mg/kg	Hudson et al. (1984)
Mallard (Anas platyrhynchos)	520 mg/kg	381 mg/kg	5.6 mg/kg	
Bobwhite quail (Colinus virginianus)	6.6 mg/kg			
California quail (Lophortyx california)		8.8 mg/kg	1.2 mg/kg	
Japanese quail (Coturnix coturnix)	62 ppm	60 ppm	17 ppm	Hill and Camardese (1986)
Rainbow trout (Salmo gairdneri)	0.036 ppm	0.0019 ppm	0.0018 ppm	Cope (1965)
Bluegill (Leponis macrochirus)				
24-hr LC_{50}	0.096 ppm	0.0055 ppm	0.00035 ppm	
96-hr LC_{50}	0.013 ppm	0.008 ppm	0.0006 ppm	Henderson et al. (1959)
Fowler's toad (tadpoles) (Bufo woodhousii)	2.0 ppm	1.1 ppm	0.57 ppm	Sanders (1970)

[a] LD_{50}, 50% lethal dose (mg/kg); LC_{50}, 50% lethal concentration (ppm).

EXPERIMENTAL STUDIES OF MORTALITY

A summary of the principal experimental studies to determine the lethal residue levels of dieldrin is given in Table 2, and those for aldrin and endrin are in Table 3. The lethal levels of dieldrin do not vary appreciably among the rather small sample of mammals. In general, mammals are somewhat more sensitive to dieldrin poisoning than are birds. In a recent study of adult clawed frogs, Schuytema et al. (1991) give tissue levels associated with LC_{50} determinations that are lower than those found for either mammals or birds.

There are two major papers on the levels lethal to Japanese quail, that by Robinson and co-workers at Tunstall Research Laboratory of Shell Chemical and that by Stickel and co-workers at the U.S. Fish and Wildlife Service, Patuxent Research Center. While there is a fair measure of agreement between the two papers, there are also some differences. Since these data have been the basis of the accepted lethal level for birds, it seems worthwhile to examine these two studies in some detail.

In both studies, quail were exposed to a range of diets. Robinson et al. (1967) exposed their birds to diets of 10, 20, 30, and 40 ppm as well as to single doses of 50, 75, and 100 mg/kg. The time to death of birds given single doses was much more rapid than those in the dietary studies. Nevertheless, the lethal level in the

Table 2 Experimental Studies Determining Lethal Levels of Dieldrin

Species	Sex	Sample size no.	Organ	Residue level (ppm of wet wt) Mean	Residue level (ppm of wet wt) Range	Ref.
Dog (*Canus familiaris*)	M	6	Brain	5.5	2.4-9.4	Harrison et al., 1963
White-tailed deer (*Odocoileus virginianus*)	M	2	Brain	10.3 and 16.4		Murphy and Korschgen, 1970
Rat	M	6	Brain		2.1-10.8	Hayes, 1974
	n.g.[a]	10	Liver	n.g.	7.8-74.2	Koeman et al., 1967
Short-tailed Shrew (*Blarina brevicauda*)	M	6	Brain	6.8	3.7-12.6	Blus, 1978
	F	8		6.7	4.8-9.3	
Bobwhite quail		?	Brain	12-14		Rudd and Genelly, 1956
	M	11	Brain	12.4	8.5-17.8	Gesell and Robel, 1979
Japanese quail						
Died	M	30	Brain	17.4	15.7-19.3[b]	Robinson et al., 1967
Died	F	35	Brain	17.3	15-20[b]	
Survived	MF	12	Brain	6.9	3.1-15.0	
	M	17[c]	Liver	32.3-41.2		
	F	19[c]	Liver	37.5-57.7		
Died	M	10	Brain	14.8	4.9-44.3[b]	Stickel et al., 1969
Died	F	7	Brain	21.7	10.4-45.4[b]	
Survived	M	8	Brain	2.6	0.1-8.8[b]	
Survived	F	8	Brain	4.1	0.1-18.4[b]	
Pheasant (*Phasianus colchicus*)						
Died	MF	39	Brain	5.8	1.2-27	Linder et al., 1970
Survived	F	10	Brain	0.96	0.2-2.2	
Sharp-tailed grouse (*Pedioecetes phasianellus*)	M	12	Brain	6.9	4.4-10.9	McEwen and Brown, 1966

Species / Condition	Sex	n	Tissue	Concentration	Range/CI	Reference
Pigeon (Columba livia)						Robinson et al., 1967
Died	MF	19	Brain	20	18.1-22.0[b]	
Died	MF	20	Liver	45.6	37.5-55.5[b]	
Survived	M	11	Brain	7.7	2.1-8.5	
Died	MF	2	Brain	27.7	19.2 and 36.2	Jefferies and French, 1972
			Liver	77.2	51.9 and 102.6	
Survived	MF	4	Brain	n.g.	5.4-9.5	
Cowbird (Molothrus ater)						Heinz and Johnson, 1982
Died	M	21	Brain	16.3	9.8-23.5	
Ceased feeding		17		6.8	1.5-11.7	
Red-winged blackbird (Agelaius phoeniceus)						Clark, 1975
Unstressed	M	30	Brain	19.8	(1.7)[d]	
Food stressed	M	30	Brain	22.2	(0.8)[d]	
Leopard frog (Rana pipiens)						Schuytema et al., 1991
Tadpole (chronic, 20 day)		20	Whole body	1.7		
Adult (chronic, 20 day)		8	Skin	5.5		
			Muscle	10.0		
			Liver	13.0		
African clawed frog (Xenopus laevis)						
Tadpole						
Acute, 4 day		20	Whole body	11.0		
Chronic, 20 day		20		1.8		
Juvenile (acute, 4 day)		20		8.0		
Bullfrog (Rana catesbeiana)						
Tadpole						
29 days		20	Whole body	12.0		
50 days		20		5.1		

[a] Not given.
[b] 95% confidence intervals.
[c] Three groups of birds on different dosages. Values are geometric means.
[d] Numbers in parentheses, standard deviation.

Table 3 Experimental Studies Determining Lethal Levels of Endrin

Species	Sex	Sample size (no.)	Organ	Residue level (ppm of wet wt) Mean	Range	Ref.
Mouse (Mus musculus)	MF?	2 pools of 3	Brain	0.7 and 1.0		Stickel et al., 1979[a]
Short-tailed shrew	MF?	2 pools of 3	Brain	0.93 and 0.94		Blus, 1978
Mallard	M	10	Brain	0.88	0.62-1.37	Stickel et al., 1979[b]
Starling	M	7	Brain	1.21	1.03-1.49	
(Sturnus vulgaris)	F	7	Brain	1.03	0.85-1.24	
Cowbird	M	12	Brain	0.99	0.69-1.81	
Red-winged blackbird	F	12	Brain	1.12	0.72-1.66	
Grackle	M	12	Brain	1.28	0.88-1.67	
(Quiscalus quiscula)	F	10	Brain	1.04	0.79-1.35	
Southern leopard frog (Rana sphenocephala)		5 pools of 6 to 8[a]	Whole body	0.43-2.8		Hall and Swineford, 1980
Gizzard shad (Dorosoma cepedianum)		31 20	Blood	0.10[b] 0.15-22[b]		Brungs and Mount, 1967
Channel catfish (Ictalurus punctatus)		67	Blood	0.30[b]		Mount et al., 1966

[a] Each group was exposed to a different concentration of endrin.
[b] Measurement given as μg/g.

brain was independent of treatment. The values obtained for the two sexes were very similar. Stickel et al. (1969) exposed their birds to dietary levels of 2, 10, 50, and 250 ppm of dieldrin. Mortality was rapid at the highest dosage, most males dying within 9 days and females within 14 days. Mortality was less pronounced at 50 ppm, and again there was a pronounced difference between sexes. Some mortality occurred at 10 ppm and none with those birds on the 2-ppm diet. Despite the differences in the time to death, the levels of dieldrin in the brains of those quail that died were independent of dose. There was a statistically significant difference between the sexes, males ranging from 7 to 32 ppm and females from 11 to 33 ppm. The geometric mean values were 14.8 and 21.7 ppm, respectively.

It is in the 95% confidence intervals that there is a marked difference in data. Those of Robinson and co-workers are much smaller than those reported by Stickel and co-workers. Although the mean values are not markedly different, the smaller confidence intervals of Robinson's data enable them to calculate the critical level of dieldrin in the brain to be 10 ppm. Because of the much greater variation of data, Stickel and co-workers conclude that brain residues of 4 to 5 ppm or higher indicate that the animal was in the known danger zone and may have died from dieldrin poisoning. This value appears to be based on the lower limit of the 95% confidence interval (4.9 ppm), and it is below the lowest value (6.23 ppm) of any quail that actually died on dosage.

The value of 4 to 5 ppm of dieldrin in the brain has been widely used, explicitly or implicitly, as the level above which dieldrin poisoning is a reasonable diagnosis. It is hard to say whether this is because people prefer to err on the side of safety, or because they have more confidence in a value determined by a wildlife agency than by industry.

Robinson and co-workers give an overall mean value for the level of dieldrin in the liver associated with mortality as 40 ppm, with a range of values from 20 to 86 ppm. Stickel and co-workers give a mean value of 19.7 ppm, with a range of 5.7 to 52 ppm.

Heinz and Johnson (1982) found that dieldrin at sublethal levels caused cowbirds to stop eating, thereby mobilizing dieldrin to the brain and causing death. They found that the mean level at the time of ceasing to feed was 6.8 ppm, whereas by death the levels had risen to a mean level of 16.3 ppm. While this is an important finding from a mechanistic viewpoint, it does not affect the criteria for lethal levels.

There are few studies relating aldrin levels to toxicity, because of the rapid conversion of aldrin into dieldrin. Hall et al. (1971) examined the effect of a single dose of aldrin that caused the mortality of half the sample of 5-week-old pheasants within 48 hours. The lethal levels of aldrin plus dieldrin (mean, 2.72; range, 2.13 to 4.25 ppm) are considerably lower than those of other avian studies involving dieldrin; either aldrin is, in this case, more toxic than dieldrin, or else young pheasants are particularly susceptible.

The toxicity of endrin to a range of organisms is given in Table 3. Two points can be made: first, endrin is highly toxic; and second, there is little variation of the lethal levels among any of the mammals or birds tested. A review of endrin toxicity in freshwater organisms was made by Grant (1976). Most of the work reviewed refers only to concentrations causing mortality and is, thus, outside the scope of this review.

INVESTIGATIONS OF ENVIRONMENTAL MORTALITY

There is a fundamental difference between laboratory and field studies. In the laboratory, animals are fed dieldrin, and the residue levels are determined in the brain when they die. In animals found dead in the field, residue levels are measured, and a judgment is made whether or not they died from dieldrin poisoning. In some cases, such as mass mortality after the use of the pesticide, field studies are similar to laboratory studies in establishing cause and effect. Usually, however, the relation between mortality and cause is not as clear and must be assigned on the best available evidence on autopsy.

In field studies, there are other variables. The animals may be stressed by food shortage, temperature, and a number of other factors. Another problem is that one chemical can affect the level of another. Street et al. (1966) demonstrated that the administration of dichlorodiphenyltrichloroethane (DDT) markedly reduced the storage of dieldrin in three mammalian species but not in chicken. Field studies are also bedeviled by the fact that most samples contain a mixture of organochlorines. In the case of birds feeding on grain treated with, for example, dieldrin, the amount of other organochlorines (OCs) may be low. In the broader population, this is not the case. This problem is considered in more detail later when the effect of dieldrin on raptorial birds is considered.

Some of the field studies involving mortality caused by cyclodienes are tabulated in Table 4. The levels of dieldrin found in obvious kill situations are similar to, or somewhat higher than, those found experimentally. For example, the levels in brains of rabbits and cotton rats reported by Stickel et al. (1969) are higher than those of most other mammals reported in Table 2. Most of the avian data are in line with experimental work. An exception is the rather low values reported by Koeman et al. (1967) in livers of sandwich terns. In this case, however, considerable amounts of other cyclodienes were also present. In general, the agreement between experimental and field data is good. This gives increased confidence that interspecies variation and variation caused by the various stresses in the wild are not causing serious difficulties in interpretation. Thus, the lethal level concept can be used with some confidence in studies, such as those on raptors, where the relation to pesticide use is not immediate.

WILDLIFE INCIDENTS

Introduction of aldrin and dieldrin as seed dressings in the United Kingdom in 1956 immediately caused considerable mortality, especially among wood pigeons *(Columba livia)* and pheasants *(Phasianus colchicus)*. A Joint Committee of the British Trust for Ornithology and the Royal Society for the Protection of Birds was established and issued a number of reports concerning incidents in the early 1960s. Many of the reports are quite vague, but detailed investigations by J. S. Ash of the Game Research Association (RSPB, 1962) of one incident in Lincolnshire estimated the minimum mortality of four species on 1480 acres as 5668 wood pigeons, 118 stock doves *(Columba oenas)*, 89 pheasants, and 59 rooks *(Corvus frugilegus)*. In another incident in Cambridge, the deaths of over 4000 individuals of eight species were recorded. The numbers of incidents involving mammals were much smaller, the fox *(Vulpes vulpes)* being the species most frequently involved. Analytical chemistry was comparatively unsophisticated at that time, and results in the reports are given as organic chlorine and mercury.

The effects of dieldrin on animal populations are more difficult to determine as many other causes may be involved. Some of these, notably for the peregrine, sparrowhawk, barn owl, and otter *(Lutra lutra)*, are considered below. At the time, declines of the European kestrel, sparrowhawk, and stock dove were attributed to the use of pesticides (RSPB, 1962). A voluntary ban on spring seed dressings in 1962 eased the situation. Nevertheless, the use of dieldrin as a wheat seed dressing continued to be a serious hazard to wildlife during the period from 1963 to 1972 (Van den Heuvel, cited in Stanley and Bunyan [1979]). Aldrin and dieldrin were finally withdrawn as wheat seed treatments in the U.K. at the end of 1975. The effect of this on the percentage of wildlife incidents attributed to dieldrin is shown in Table 5.

Many incidents were also reported in North America. Nisbet (1988) summarizes 29 incidents in the United States between 1949 and 1966. Some of these, notably in North Carolina and Colorado, involved thousands of ducks annually over a period of many years.

Table 4 Field Studies of Dieldrin Mortality

Species	Sample size (no.)	Organ	Residue level (ppm of wet wt) Mean	Residue level (ppm of wet wt) Range	Ref.
Gray bat	8/12[a]	Brain		5-10	Clark et al., 1978
(Myotis grisescens)	2/4	Brain	5.1 and 8.4		
	0/12	Brain			
	10	Brain	7.5	5-10	Clark et al., 1980
	8	Brain	8.6	4.6-13	
Adult	9	Brain	12.1	10.6-13.7[b]	Clark et al., 1983
Juveniles	19		6.5	5.6-7.6[b]	
Cottontail rabbit	5	Brain	13.8	9.3-19.1	Stickel et al., 1969
(Sylvilagus floridanus)		Liver	48.4	28.2-103	
Cotton rat	5	Brain	7.9	5.6-11.1	
(Sigmodon hispidus)		Liver	23.0	15.1-36.8	
Meadowlark	5	Brain	9.3	8.6-12.1	
(Sturnella magna)		Liver	13.1	7.9-15.9	
American Robin	7	Brain	9.6	5.0-17.0	
(Turdus migratorius)					
Pink-footed goose	6	Liver	31	15-48	Stanley and Bunyan, 1979
(Anser brachyrhynchus)					
Snow goose	5/157[a]	Brain	17.3	13.0-24.5	Babcock and Flickinger, 1977
(Chen caerulescens)					
	8/112[a]	Brain	8.2	4.9-14	Flickinger, 1979
Lesser scaup	4	Brain	11.9	7.7-16	Sheldon et al., 1963; cited in Stickel et al., 1969
(Aythya affinis)					
Buzzard (Buteo buteo)	14	Liver	19.7[c]	7.8-31.2	Fuchs, 1967
Lanner falcon	2	Brain	2.0 and 3.3		Jefferies and Prestt, 1966
(Falco biarmicus)					
Peregrine	2	Brain		6.8 and 16.4	Bogan and Mitchell, 1973
(Falco peregrinus)		Liver		17.3 and 53.9	
	1	Brain		5.4[d]	Reichel et al., 1974
	2	Brain		3.5 and 7.8[e]	Jefferies and Prestt, 1966
		Liver		4.0 and 9.3[f]	
European kestrel	84	Liver		6-30[g]	Newton et al., 1992
(Falco tinnunculus)					
Sparrow hawk	25	Liver		5-21[h]	
(Accipiter nisus)					
Owls[i]	10	Brain	15	11-25	Jones et al., 1978
Barn owl	51	Liver	14	6-44	Newton et al., 1991
(Tyto alba)					
Sandwich tern					
(Sterna sandvicensis)					
Chick	6	Liver	5.6	2.4-12[j]	Koeman et al., 1967
Juvenile	8		4.6	1.9-6.6[k]	
Adult	5		5.5	4.7-7.2[l]	
Pigeon	4	Liver		15-24	Turtle et al., 1963; cited in Koeman et al., 1967

[a] Number at each site presumed dead from dieldrin over total number.
[b] 95% confidence limits.
[c] Wet weight assumed.
[d] Also 34 ppm of DDE and 55 ppm of PCBs.

Table 4 (continued)   Field Studies of Dieldrin Mortality

e Also 45 ppm and 44 ppm of DDE.
f Also 70 ppm and 60 ppm of DDE.
g Measurement given as µg/g; a few outliers as high as 99 µg/g.
h Measurement given as µg/g; a few outliers as high as 85 µg/g.
i A number of different species maintained at the London Zoo.
j Also 2.3 ppm of telodrin and 0.47 ppm of endrin.
k Also 0.86 ppm of telodrin and 0.43 ppm of endrin.
l Also 1.0 ppm telodrin and 0.67 ppm of endrin.

Table 5   Wildlife Incidents in the United Kingdom Related to Dieldrin Poisoning[a]

Year	Total no. of incidents	No. attributed to dieldrin	% Involving dieldrin
1973	100	28	28
1974	113	14	12.4
1975	146	18	12.3
1976	179	9	5.0
1977	171	0	0
1978-1979	188	1[b]	0.5

[a] Stanley and Bunyan, 1979; Stanley and Fletcher, 1981.
[b] Plus one incident involving endrin.

An investigation into the effects of the use of endrin on orchards in the state of Washington was carried out by the U.S. Fish and Wildlife Service in the early 1980s (Blus et al., 1983). Incidences where the involvement of pesticides was not immediately obvious in the cause of death, i.e., roadkills, involved three large mammals and 91 birds of 18 species. The brains of 78 of the birds were analyzed for endrin; 46% were found to contain lethal levels of endrin (>0.80 ppm), and a further 4% were in the danger zone (0.60 to 0.79 ppm). The commonest species involved were the California quail *(Callipepla californica)* and the chukar *(Alectoris chukar)*. None of the mammals, except some montane voles *(Microtus montanus)*, had significant levels of endrin in their brains. No systematic searches for carcasses were made, and only those bodies found casually by others were studied. Thus, no idea of the overall mortality can be obtained.

BALD EAGLE *(HALIAEETUS LEUCOCEPHALUS)*

The causes of death of bald eagles found dead in the U.S. by a network of federal, state, and private investigators were determined from 1964 to 1983. Regrettably, there are no published data since 1983. There are obvious biases in such a scheme, but systematic sampling could not be carried out because of the low population and protected status of the species. Nevertheless, the number of bald eagles dying from dieldrin poisoning decreased substantially from 1968 to 1983 (Table 6); data from 1964 to 1965 were omitted because of differences in analytical procedures.

Nisbet (1989) reviewed the data relating the levels of organochlorines to reproductive impairment and population declines in the bald eagle. While he considers that both reproductive impairment caused by 2,2-bis(p-chlorophenyl)-1,1-dichloroethylene (DDE) and excess adult mortality caused by dieldrin appear to have contributed to regional population declines, the evidence is circumstantial at best.

Table 6 Proportion of Deaths of Bald Eagles in the U.S. Caused by Dieldrin

Year	Total examined	No. attributed to dieldrin	% Attributed to dieldrin	Ref.
1966-1968	69	8	11.5	Mulhern et al., 1970
1969-1970	39	6	15.4	Belisle et al., 1972
1971-1972	37	4	10.8	Cromartie et al., 1975
1973-1974	86	4	4.7	Prouty et al., 1977
1975-1977	168	5	3.0	Kaiser et al., 1980
1978-1983	293	5	1.7	Reichel et al., 1984

PEREGRINE

The major impact of dieldrin on raptors has been due to direct mortality, whereas the major impact of DDT has been due to DDE-induced eggshell thinning. Direct mortality, such as the deaths of robins following the heavy use of DDT to control Dutch elm disease (Wallace et al., 1961), has not been observed in raptors. The relative importance of mortality due to dieldrin and impaired reproduction due to DDE-induced eggshell thinning in the decline of the peregrine has been a matter of continuing debate (Nisbet, 1988; Risebrough and Peakall, 1988). Not surprisingly, the number of peregrines found dead and analyzed is small. The main argument has hinged on the timing of the decline, but this is complicated by the fact that DDT acts by causing decreased productivity in raptors and, thus, the effects on the number of breeding pairs are not seen for several years, whereas the cyclodienes, aldrin and dieldrin, introduced several years after DDT, act by direct adult mortality. In the U.K., Ratcliffe (1973) stated, "It is probable that the post-1955 'crash' in numbers resulted primarily from the greatly increased adult mortality, which varied in the different areas; and that the stabilization occurring after 1963 effected a halt in the further spread of this high mortality." However, there is little direct evidence for this claim.

The liver levels of organochlorines in five peregrines found dead in 1963-1964 were 0.6 to 9.3 ppm of dieldrin and 0.1 to 3 ppm of heptachlor epoxide. Brain levels were measured for only two specimens, the values for dieldrin being 3.5 and 7.8 ppm (Jefferies and Prestt, 1966). The authors consider that three of the peregrines may have died from organochlorine poisoning, but only the individual with the highest levels is clearly within the lethal range. Two peregrines with high brain levels of dieldrin (6.8 and 16.8 ppm) were found dead in Stirlingshire in 1973 (Bogan and Mitchell, 1973). In North America, the only case of dieldrin poisoning of a peregrine seems to be one found dead in North Carolina in 1973; its brain contained 5.4 ppm of dieldrin (Reichel et al., 1974).

One of the major difficulties in establishing which pollutant is responsible for which effect is the fact that several different chemicals are usually present, with their concentrations being often strongly intercorrelated. The role of the three main organochlorine pollutants — polychlorinated biphenyls (PCBs), DDE, and dieldrin — on the breeding success of sparrowhawks has been examined by Newton and Bogan (1978). These workers showed, by regression analysis, that eggshell thinning, egg breakage, addling, and hatching failure were correlated with DDE; addling and hatching failure, with PCBs; and none of the aspects of breeding, with dieldrin.

BARN OWL

A survey of the causes of death of barn owls carried out from 1963 to 1977 in the U.K. was compared with a more recent survey carried out in 1987 to 1989 (Newton et al., 1991). In the earlier survey, 25 to 40% of all recorded deaths in some eastern agricultural countries and 8.8% for the entire country were considered to be due to dieldrin poisoning. Birds classed as having died from dieldrin poisoning contained 6 to 44 ppm of dieldrin (geometric mean, 14 ppm) in their livers. In the more recent survey, no deaths from organochlorine poisoning were reported.

BATS

The effects of organochlorines on bats in the large colonies in the U.S. have been studied by Clark and co-workers. These studies have been thoroughly reviewed by Clark (1981). Dieldrin was shown (Clark et al., 1978) to be the cause of death in two of the three incidents of mortality examined in caves in Missouri in 1976. In one cave, the residues in brains of most of the 12 bats analyzed ranged from 5 to 10 ppm (wet weight) and up to 8.4 ppm in another cave. The values from the third cave were all below 1 ppm. The colony in which bats were found with the highest residue levels disappeared in 1979, and the bats were still absent in 1981 (Clark et al., 1983). In this paper, these workers reported that lethal levels of dieldrin were found in the brains of bats in two other colonies in Missouri. Appreciable levels of heptachlor epoxide were also found. Clark (1981) concluded that, although organochlorine chemicals may have played a role in past declines of some populations, it is doubtful that the combined adverse effects of chemicals were as detrimental as the total impact of disturbance, vandalism, and habitat destruction.

OTTERS

Of the causes considered for the decline of the otter in Britain, Chanin and Jefferies (1978) state, "that of pollutants and dieldrin in particular seems to be the only one which corresponds in time, and possibly main area to the decline in the otter populations." However, no recovery was seen after bans on most uses of dieldrin in 1965. Thus, it seems likely that, if dieldrin was the cause of the decline, and this remains unproven, then some other cause has prevented the recovery. Olsson and Sandegren (1991), examining the situation in Sweden, consider PCBs to be the more likely toxic agent and point out that, although PCBs were first commercially synthesized in the 1930s, their use expanded greatly over the critical period. They found that a considerable number of otters had levels in excess of those known to cause reproductive failures in mink (Aulerich and Ringer, 1977). However, these workers did not report levels for dieldrin. Mason et al. (1986) reported organochlorine levels in the liver and muscle of 23 British otters. Values varied widely, with nine having dieldrin levels over 5 ppm, the highest value being 66.4 ppm. The PCB values also varied widely, the highest value being 300 ppm. All values are given on a lipid weight basis. There is, of course, no reason why both compounds should not be involved.

EFFECTS ON REPRODUCTION

Experimental studies of the effects of dieldrin on reproduction are tabulated in Table 7. The major conclusion is that the reproductive processes of mammals and birds are not particularly sensitive to dieldrin. The limited data on amphibians suggest that these animals are even less sensitive. Field studies on the effect of dieldrin on the purple and common gallinules *(Porphyrula martinica* and *Gallinula chloropus)* failed to show any significant effects on either the percentage of eggs hatched or on the survival of young (Fowler et al., 1971). These "no effects" were associated with dieldrin levels of 9.4 ppm (range, 3.2 to 16.4 ppm) in the case of the purple gallinule and even higher levels, namely, 17.5 ppm (range, 4.7 to 28.1 ppm), in the case of the common gallinule.

Eggshell thinning has been investigated by a number of workers, although only one study, that of Winn (1973) on the mallard, relates the degree of thinning to residue levels. While these values are cited in Table 7, it must be stated that they do not fit with the work reported by others and do not appear to have been reported in the open literature. Lehner and Egbert (1969) reported a much smaller reduction in eggshell thickness of 4% associated with a dietary level of 10 ppm of dieldrin. Three other studies, Dahlgren and Linder (1970) on pheasants, Davison and Sell (1972) on chickens, and Hill et al. (1976) on Japanese quail, did not record any significant thinning. However, all of these species are known to be insensitive to DDE-induced eggshell thinning. Wiemeyer et al. (1986), in their experiments feeding a combination of DDT and dieldrin to American kestrels, found correlations with the levels of all four [dieldrin, DDT, DDE, and 2,2-bis(*p*-chlorophenyl)-1,1-dichloroethane (DDD)] contaminants in the egg. However, without experiments using dieldrin alone, it is not possible to be sure whether or not dieldrin causes eggshell thinning in this sensitive species.

Graber et al. (1965) studied the effects of spraying $1/8$ to $1/4$ lb per acre of dieldrin on the reproduction of red-winged blackbirds. They noted that numbers were decreased and that the remaining birds were quiescent after spraying. Eggs containing 5.7 to 6.3 ppm of dieldrin were associated with nest failure, although the possibility exists that disturbance by the observers was a serious factor.

One study indicating an effect at a low level is that on the golden eagle *(Aquila chrysaetos)* (Lockie et al., 1969). These workers found that the proportion of eyries

Table 7 Experimental Studies on the Effect of Dieldrin on Reproduction

Species	Parameter studied	Dosage (ppm)	Finding	Residues (ppm of wet wt.)				Ref.
				Milk	Brain	Liver		
White-tailed deer								Murphy and Korschgen, 1970
	Fawn production	5	No effect	1.50	0.14	3.72		
	Fawn survival	25	No effect	9.18	1.20	16.92		
	Fawn growth		Decreased					
				Egg Yolk				
Mallard	Eggshell thickness	4	Decreased 24%	16.8				Winn, 1972
		10	Decreased 8%	37.4				
		30	Decreased 14%	54.6				
	Egg fertility	10	Decreased 17%	37.4				
		30	Decreased 26%	54.6				
	Chick mortality	4	Increased 29%	16.8				
		10	Increased 37%	37.4				
		30	Increased 62%	54.6				
				Whole Egg[a]				
	Egg production (7 weeks)	10	No effect	23.2				Walker et al., 1969
		20	Reduced 20%	45.2				
		30	Reduced 41%	63.4				
		40	Reduced 100%	92.5				
	Egg fertility (7 weeks)	10	No effect	23.2				
		20	Reduced 34%	45.2				
		30	Reduced 50%	63.2				

DIELDRIN AND OTHER CYCLODIENE PESTICIDES IN WILDLIFE

Species	Endpoint	Dose	Effect	Concentration	Reference
Barn owl *(Tyto alba)*	% of hatchability (7 weeks)	10 20 30	No effect Reduced 41% Reduced 75%	23.2 45.2 63.2	Mendenhall et al., 1983
	Eggs hatched/pair Young fledged/pair	0.5	No effect No effect	**Carcass** M 9.6 (0.17)[b] F 9.2 (0.12)	
	Shell thickness Embryonic mortality		Reduced 5.5% No effect	**Egg** 8.1 + 0.036	
				Egg	**Carcass**	**Liver[a]**
Chicken *(Gallus gallus)*	Eggs per female % of Eggs hatched Survival to 14 days	0.2 2 5	No effect No effect No effect	0.36 1.41 4.20	0.06 0.53 1.91	0.27 2.20 4.56	Graves et al., 1969
				Whole Body			
Frog *(Rana temporaria)*	Abnormalities in tadpoles	0.5	Reversible abnormalities	31.3			Cooke, 1972
Frog *(Buto bufo)*	Abnormalities in tadpoles	0.5	Reversible abnormalities	138			

[a] Wet weight presumed.
[b] Numbers in parentheses, standard deviation.

successfully rearing young in western Scotland increased from 31% in the period 1963 through 1965 to 69% in the period 1966 through 1968. Concurrently, the levels of dieldrin fell from an average of 0.86 ppm to 0.34 ppm. No other environmental factor could be found to explain this result. In a recent analysis of the effects of organochlorines on golden eagles in Scotland, Newton and Galbraith (1991) state that it is "hard to say to what extent, if at all, organochlorines reduced eagle breeding success in the 1950s and 1960s." No experimental studies with dieldrin alone have been carried out on hawks. In experiments using mixtures of DDT and dieldrin on American kestrels *(Falco sparverius)*, it was found that DDE (the major metabolite of DDT) levels were significantly and most closely correlated with fledgling success. The mean levels of dieldrin in the eggs in the various experiments ranged from 4.4 to 7.4 ppm, while the levels of DDE ranged from 12 to 16 ppm. While extrapolation, even within a taxonomic order, can be dangerous, it seems unlikely that the hatchability of eagle eggs is affected at levels as low as 1 ppm of dieldrin.

Blus (1982) examined the role of various organochlorines in affecting the reproduction of the brown pelican *(Pelecanus occidentalis)* in the southeastern U.S. He found that, while several organochlorines appeared to induce adverse effects, the correlation was strongest for DDE. For this compound, a critical egg level was established (3 µg/g associated with substantial impairment, 4 µg/g with total reproductive failure). No definite value was established for dieldrin, but it is stated to exceed 1 µg/g. For endrin, a rough estimate of 0.5 µg/g was given. This study emphasizes the difficulties encountered when multiple residues are present.

The most detailed studies of the effect of endrin on avian reproduction have been carried out by Fleming et al. (1982) on screech owls *(Otus asio)* and by Roylance et al. (1985) and Spann et al. (1986) on mallards. In the screech owl study, it was found that, in comparison to controls, those fed a dietary level of 0.75 ppm laid fewer eggs per day per laying female, hatched fewer eggs per incubated clutch, and produced fewer fledglings per total number of pairs. Overall, the endrin group was only 57% as productive as the control group. This reproductive impairment was associated with endrin levels in eggs of 0.27 ppm (range, 0.12 to 0.46 ppm). Carcass levels were 0.55 ppm (range, 0.36 to 0.80 ppm) in females and 0.36 ppm (range, 0.13 to 0.54 ppm) in males. Regrettably, brain levels were not determined, so that direct comparison with levels causing mortality cannot be calculated.

Roylance and co-workers found no effect of endrin at 0.5 and 3 ppm in the diet on egg production, fertility, hatchability, or hatchling survival. The only parameter affected was embryo survival at the higher dose level. The residue levels associated with the two dosage levels were 0.43 and 2.75 ppm in the eggs, respectively; in fat, the levels were 3 and 16 ppm for males and 3.8 and 22.7 ppm for females. Spann and co-workers studied the effect of 1 and 3 ppm of endrin in the diet. These workers found that birds fed 1 ppm of endrin reproduced as well as, if not better than (the hatching success of the 1-ppm group was greater than either the control group or the 3-ppm group), controls. Mallards fed 3 ppm of endrin appeared to reproduce more poorly than did controls, although the differences were rarely statistically significant. The experiment was bedeviled by a rather poor performance by the control group. The levels of endrin in the fat were 6.4 ppm in males and 7.6 ppm in females at the lower dose and 16 ppm and 11 ppm, respectively, at the higher dose. The egg levels were 1.1 ppm and 2.9 ppm for the two doses.

A prima facie case for endrin's having caused the extirpation of the large breeding population of the brown pelican in Louisiana has been put forward by Blus et al. (1979). The population crashed in the late 1950s and was extirpated by 1963. Endrin was implicated in the die-offs of fish at this time (Mount et al., 1966). Subsequently, some reintroduced pelicans showed poor reproductive success and some mortality. At this time, endrin levels were measured, and those in eggs were within the range that had caused reproductive problems with the screech owl, and brain levels were close to those associated with mortality in experimental studies.

NO-OBSERVED-EFFECT LEVELS

The concept of no-observed-effect levels (NOELs) has increased in popularity in recent years. Newton (1988) examined the large amount of data available on liver levels in sparrowhawks and kestrels and on egg levels of the peregrine falcon and calculated the critical pollutant level of dieldrin for a stable population of these species. The geometric mean values calculated were 1 µg/g in the liver of both the sparrowhawk and kestrel and 0.7 µg/g in the eggs of the peregrine. Levels above these values were associated with population declines.

Nebeker et al. (1992) examined the levels of dieldrin associated with the lowest-observed-adverse-effect level (LOAEL) and no-observed-adverse-effect level (NOAEL) in the mallard duckling. Survival, growth, and behavioral parameters were used. The values obtained for a 24-day LOAEL were 7 µg/g for the liver and 2.5 µg/g for the brain. The values for NOAEL were less than 1 µg/g for both organs.

The same approach has been used on frogs exposed to dieldrin (Schuytema et al., 1991); the criteria used were survival, growth, and teratogenic effects. The values obtained for 24-day tests on tadpoles of *Xenopus leavis* were 1.5 and 0.8 µg/g for the LOAEL and NOAEL, respectively. The values for *Rana pipiens* were even lower, namely, 0.6 and 0.4 µg/g. More acute tests, those carried out over a 4-day period, gave considerably higher values. Nevertheless, these papers indicate that both birds and amphibians show adverse effects at quite low levels.

OTHER PHYSIOLOGICAL PARAMETERS

Although a wide variety of biological changes, "biomarkers," are now available for assessing the impact of chemicals on wildlife (McCarthy and Shugart, 1990; Peakall, 1992), only a small amount of the information is related to residue levels. There are innumerable papers on the effects of certain dietary levels on certain biomarkers, but to put this information in an environmental context is difficult as we are not able to measure exposure in terms of dietary levels. In this section, the limited literature relating residue levels of dieldrin to biomarker changes is considered.

LIVER SIZE AND HEPATIC ENZYME ACTIVITY

An increase in liver weight is known to be a response to exposure to organochlorines. Murphy and Korschgen (1970) found that white-tailed deer exposed to 25 ppm of dieldrin had mean liver weights and liver/body weight ratios significantly higher than either controls or deer exposed to 5 ppm of dieldrin. Residue levels in the liver were 3.72 to 4.15 ppm for the 5-ppm group and 15.80 to 16.92 ppm in the 25-ppm group. Heinz et al. (1980) found a 30% increase in the liver weights of ring doves *(Streptopelia risoria)* exposed to 16 ppm of dietary dieldrin, which corresponded to a brain level of 8.17 ppm. No effect was seen at the 4-ppm dietary level that gave brain levels of 1.06 ppm. Jefferies and French (1972) found no effect on the liver weight of pigeons exposed to diets of 1, 2, or 4 ppm of dieldrin. These corresponded to 6.6, 7.3, and 21.4 ppm in the brain and 27.4, 32.9, and 65.3 ppm in the liver. Some mortality was observed at the highest dose. Sharma et al. (1976) found no significant increase in liver weights in mallards exposed to dietary levels of 4, 10, or 30 ppm of dieldrin. The residue level of dieldrin in the highest group averaged 11.18 ppm. An increase in liver weight seems to be an insensitive indicator of dieldrin exposure. A range of hepatic enzymes, the mixed-function oxidases, is induced as a defense against a wide variety of xenobiotics. Gillett and Arscott (1969) found a doubling of activity of these enzymes in 28-day-old Japanese quail exposed to dieldrin. The dieldrin concentration in the liver was 135 to 144 ng/mg of microsomal protein. If one assumes that microsomal protein is 10% of the liver by weight, then these values can be divided by 10 to give the overall value for the liver. Sharma et al. (1976) found increases of hepatic microsomal enzymes in mallards exposed to dieldrin. In the case of those using *O*-ethyl-*O*-*p*-nitrophenyl benzenethionophosphate, a significant increase was seen at the lowest dose used (4 ppm, liver residue level of 3.08 ppm); another enzyme and DNA and protein content were increased at only the highest dose (30 ppm, 11.18 ppm). The induction of enzymes was 4- to 6-fold at the highest dose.

THYROID STRUCTURE AND FUNCTION

Murphy and Korschgen (1970) found small, but inconsistent increases in thyroid weight in deer exposed to 5 or 25 ppm of dieldrin. Jefferies and French (1972) found that the weight of the thyroid was higher in pigeons fed dieldrin (79.6 mg) compared with that of controls (44.8 mg). However, the dose dependence was not straightforward, as those on dietary levels of 1 and 2 ppm had levels twice as high as did controls, whereas no effect was seen at the highest dose of 4 ppm. Similarly, there was a decrease of mean colloid area, but the largest reduction was at the lowest dose. Residue levels in the animals are given in the preceding paragraph.

BIOGENIC AMINES

Studies of these compounds, which are important in the nervous system, are, by contrast to other biomarkers, well supported by residue data. The effect of dieldrin on the levels of norepinephrine, serotonin, dopamine, and gamma-aminobutyrate in the brains of mallards exposed to diets of 4, 10, and 30 ppm has been studied (Sharma, 1973). Decreases were found for all compounds except gamma-aminobutyrate. The decreases were dose dependent and, at the high dose, levels of the biogenic amines were only 20 to 30% of those of controls. The levels of dieldrin in the brains of mallards in the three groups were 0.109, 0.263, and 0.416 ppm (wet weight). These values seem remarkably low for chronic exposure to dietary levels of 4, 10, and 30 ppm. In a subsequent paper (Sharma et al., 1976), the residue levels in the liver are also given, namely, 2.29, 4.23, and 11.18 ppm. These seem more in line with the dietary levels, but the ratio to the brain levels (which are repeated in this paper) remains an anomaly.

Heinz et al. (1980) exposed ring doves to dietary levels of 1, 4, and 16 ppm of dieldrin. A dose-dependent reduction of norepinephrine and dopamine was found, with significant effects on both compounds at the higher two doses. Reductions at the highest dosage were 62 and 41% of control values, respectively. The brain levels of dieldrin were 0.20, 1.06, and 8.17 ppm, respectively, for the three experimental groups.

BEHAVIORAL CHANGES

The effect on the operant behavior of bobwhites by dieldrin and endrin has been studied by Gesell et al. (1979) and Kreitzer (1980), respectively. The dosages used by Gesell and co-workers were rather high, and consistent behavioral changes were seen only in the three highest dosage groups that had brain levels of 11 to 12 ppm. They conclude, "changes in operant behavior were detected in all bobwhites that had at least 5.73 ppm of dieldrin in their brain tissues." This value is approximately half of the lethal level. Kreitzer exposed quail to 0.1 or 1 ppm of endrin for 20 weeks. He found a significant increase in the number of errors in both endrin groups over that of controls. Surprisingly, the error rate was significantly greater at the lower dose than at the high dose. The residue levels in the brain at the end of the experiment were 0.075 ppm for the low group and 0.35 ppm for the high group. The high error rate of the low group occurred at levels less than a tenth of the lethal, but the ecological significance of changes in operant behavior is virtually impossible to assess.

Changes in several parameters of the breeding behavior of mallards exposed to 10 and 30 ppm of dieldrin were noted by Winn (1973) and by Sharma et al. (1976), whereas 4 ppm did not cause significant alterations. The levels in the brain and liver associated with these three dietary levels were 0.13, 0.26, and 0.42 ppm and 2.29, 4.23, and 11.18 ppm. As already discussed under biogenic amines, the brain levels seem low compared with both the dietary intake and the liver levels. These workers found a decrease of dominant behavior (number of courtship displays, attempted copulations) and, in females, a decrease of nest attentiveness and nest defense.

SUMMARY

The brain levels of dieldrin associated with lethality in mammals are 5 ppm and in birds are 10 ppm. At these levels, one can be reasonably certain that death was due to dieldrin poisoning. The widely accepted value of 4 to 5 ppm for birds allows for a considerable safety margin. For endrin, the value is 0.8 to 1.0 ppm for both mammals and birds. Brain levels are considered to be the best tissue for diagnostic purposes, being less sensitive to such factors as age, sex, species, and dietary concentration. Nevertheless, liver levels have been used by a number of workers.

None of the cyclodienes are known to cause major effects on reproduction at levels well below (i.e., at one half or less) those causing mortality. Few other physiological tests are well supported by residue data. It would greatly add to the environmental usefulness of experimental data if residue levels as well as dosages were available. Within these limitations, it seems that changes to the levels of biogenic amines and induction of hepatic mixed-function oxidases are relatively sensitive to cyclodienes, whereas changes to organ size (liver, thyroid) are relatively insensitive. Some behavioral changes have been observed at low levels, although the ecological significance is difficult to assess. The work on the effect of cyclodienes on the immune system does not seem to be backed by residue data.

ACKNOWLEDGMENTS

The assistance of the library staff of the Central Science Laboratory of the Ministry of Agriculture, Fish, and Food (courtesy of Dr. Peter Stanley) is gratefully acknowledged.

REFERENCES

Aulerich, R. J., and R. K. Ringer. 1977. Current status of PCB toxicity to mink, and effect on their reproduction. Arch. Environ. Contam. Toxicol. 6:279-292.

Babcock, K. M., and E. L. Flickinger. 1977. Dieldrin mortality of lesser snow geese in Missouri. J. Wildl. Manage. 41:100-103.

Bann, J. M., T. J. DeCino, N. W. Earle, and Y.-P. Sun. 1956. The fate of aldrin and dieldrin in the animal body. J. Agric. Food Chem. 4:937-941.

Bedford, C. T., D. H. Hudson, and I. L. Natoff. 1975. The acute toxicity of endrin and its metabolites to rats. Toxicol. Appl. Pharmacol. 33:115-121.

Belisle, A. A., W. L. Reichel, L. N. Locke, T. G. Lamont, B. M. Mulhern, R. M. Prouty, R. B. De Wolf, and E. Cromartie. 1972. Residues of organochlorine pesticides, polychlorinated biphenyls and mercury, and autopsy data for bald eagles, 1969 and 1970. Pestic. Monit. J. 6:133-138.

Blus, L. J. 1978. Short-tailed shrews: toxicity and residue relationship of DDT, dieldrin, and endrin. Arch. Environ. Contam. Toxicol. 7:83-98.

Blus, L. J. 1982. Further interpretation of the relation of organochlorine residues in brown pelican eggs to reproductive success. Environ. Pollut. Ser. A Ecol. Biol. 28:15-33.

Blus, L., E. Cromartie, L. McNease, and T. Joanen. 1979. Brown pelican: population status, reproductive success, and organochlorine residues in Louisiana, 1971-1976. Bull. Environ. Contam. Toxicol. 22:125-135.

Blus, L. J., C. J. Henny, T. E. Kaiser, and R. A. Grove. 1983. Effects on wildlife from use of endrin in Washington State orchards. Trans. N. Am. Wildl. Nat. Resour. Conf. 48:159-174.

Bogan, J. A., and J. Mitchell. 1973. Continuing dangers to peregrines from dieldrin. Br. Birds 66:437-439.

Bollen, W. B., J. E. Roberts, and H. E. Morrison. 1958. Soil properties and factors influencing aldrin-dieldrin recovery and transformation. J. Econ. Entomol. 51:214-219.

Brungs, W. A., and D. I. Mount. 1967. Lethal endrin concentrations in the blood of gizzard shad. J. Fish. Res. Board Can. 24:429-432.

Chanin, P. R. F., and D. J. Jefferies. 1978. The decline of the otter, *Lutra lutra* L., in Britain: an analysis of hunting records and discussion of causes. Biol. J. Linn. Soc. 10:305-328.

Clark, D. R., Jr. 1975. Effect of stress on dieldrin toxicity to male redwinged blackbirds *(Agelaius phoeniceus)*. Bull. Environ. Contamin. Toxicol. 14:250-256.

Clark, D. R., Jr. 1981. Bats and environmental contaminants: a review. U.S. Fish Wildl. Serv. Spec. Sci. Rep. Wildl. No. 235, 27 pp.

Clark, D. R., Jr., C. M. Bunck, E. Cromartie, and R. K. LaVal. 1983. Year and age effects on residues of dieldrin and heptachlor in dead gray bats, Franklin County, Missouri — 1976, 1977, and 1978. Environ. Toxicol. Chem. 2:387-393.

Clark, D. R., Jr., R. K. LaVal, and A. Krynitsky. 1980. Dieldrin and heptachlor residues in dead gray bats, Franklin County, Missouri. 1976 vs. 1977. Pestic. Monit. J. 13:137-140.

Clark, D. R., Jr., R. K. LaVal, and D. M. Swineford. 1978. Dieldrin-induced mortality in an endangered species, the gray bat *(Myotis grisescens)*. Science (Washington, D.C.) 199:1357-1359.

Cohen, A. J., and J. N. Smith. 1961. Fate of aldrin and dieldrin in locusts. Nature (Lond.) 189:600-601.

Cooke, A. S. 1972. The effects of DDT, dieldrin, and 2,4-D on amphibian spawn and tadpoles. Environ. Pollut. 3:51-68.

Cope, O. B. 1965. Sport fishery investigation. p. 51-64. *In* The effect of pesticides on fish and wildlife. U.S. Fish Wildl. Serv. Circ. 226.

Cromartie, E., W. L. Reichel, L. N. Locke, A. A. Belisle, T. E. Kaiser, T. G. Lamont, B. M. Mulhern, R. M. Prouty, and D. M. Swineford. 1975. Residues of organochlorine pesticides and polychlorinated biphenyls and autopsy data for bald eagles, 1971-72. Pestic. Monit. J. 9:11-14.

Dahlgren, R. B., and R. L. Linder. 1970. Eggshell thickness in pheasants given dieldrin. J. Wildl. Manage. 34:226-228.

Davison, K. L., and J. L. Sell. 1972. Dieldrin and p,p'-DDT effects on egg production and eggshell thickness of chickens. Bull. Environ. Contamin. Toxicol. 7:9-18.

Fleming, W. J., M. A. R. McLane, and E. Cromartie. 1982. Endrin decreases screech owl productivity. J. Wildl. Manage. 46:462-468.

Flickinger, E. L. 1979. Effects of aldrin exposure on snow geese in Texas rice fields. J. Wildl. Manage. 43:94-101.

Flickinger, E. L., and B. M. Mulhern. 1980. Aldrin persists in yellow mud turtles. Herpetol. Rev. 11:29-30.

Fowler, J. F., L. D. Newsom, J. B. Graves, F. L. Bonner, and P. E. Schilling. 1971. Effect of dieldrin on egg hatchability, chick survival, and eggshell thickness in purple and common gallinules. Bull. Environ. Contam. Toxicol. 6:495-501.

Fuchs, P. 1967. Death of birds caused by application of seed dressings in the Netherlands. Meded. Rijsfaculteit Landbouwwet. Gent 32:855-859.

Gesell, G. G., R. J. Robel, A. D. Dayton, and J. Frieman. 1979. Effects of dieldrin on operant behavior of bobwhites. J. Environ. Sci. Health Part B Pestic. Food Contam. Agric. Wastes 14:153-170.

Gillett, J. W., and C. H. Arscott. 1969. Microsomal epoxidation in Japanese quail: induction by dietary dieldrin. Comp. Biochem. Physiol. 30:589-600.

Graber, R. R., S. L. Wunderle, and W. N. Bruce. 1965. Effects of a low-level dieldrin application on a red-winged blackbird population. Wilson Bull. 77:168-174.

Grant, B. F. 1976. Endrin toxicity and distribution in freshwater: a review. Bull. Environ. Contamin. Toxicol. 15:283-290.

Graves, J. B., F. L. Bonner, W. F. McKnight, A. B. Watts, and E. A. Epps. 1969. Residues in eggs, preening glands, liver, and muscle from feeding dieldrin-contaminated rice bran to hens and its effect on egg production, egg hatch, and chick survival. Bull. Environ. Contam. Toxicol. 4:375-383.

Hall, J. E., Y. A. Greichus, and K. E. Severson. 1971. Effects of aldrin on young pen-reared pheasants. J. Wildl. Manage. 35:429-434.

Hall, R. J., and D. Swineford. 1980. Toxic effects of endrin and toxaphene on the southern leopard frog, *Rana spenocephala*. Environ. Pollut. Ser. A Ecol. Biol. 23:53-65.

Harrison, D. L., P. E. G. Maskell, and D. F. L. Money. 1963. Dieldrin poisoning of dogs. 2. Experimental studies. N. Z. Vet. J. 11:23-31.

Hayes, W. J., Jr. 1974. Distribution of dieldrin following a single oral dose. Toxicol. Appl. Pharmacol. 28:485-492.

Heinz, G. H., E. F. Hill, and J. F. Contrera. 1980. Dopamine and norepinephrine depletion in ring doves fed DDE, dieldrin, and Aroclor 1254. Toxicol. Appl. Pharmacol. 53:75-82.

Heinz, G. H., and R. W. Johnson. 1982. Diagnostic brain residues of dieldrin: some new insights. p. 72-92. *In* D. W. Lamb and E. E. Kenaga (Eds.). Avian and mammalian wildlife toxicology: second conference, ASTM STP 757. American Society for Testing and Materials, Philadelphia.

Henderson, C., Q. H. Pickering, and C. M. Tarzwell. 1959. Relative toxicity of ten chlorinated hydrocarbon insecticides to four species of fish. Trans. Am. Fish. Soc. 88:23-32.

Hill, E. F., and M. B. Camardese. 1986. Lethal dietary toxicities of environmental contaminants and pesticides to Coturnix. U.S. Fish Wildl. Serv. Fish Wildl. Tech. Rep. 2, 147 pp.

Hill, E. F., R. G. Heath, and J. D. Williams. 1976. Effect of dieldrin and Aroclor 1242 on Japanese quail eggshell thickness. Bull. Environ. Contam. Toxicol. 16:445-453.

Hudson, R. H., R. K. Tucker, and M. A. Haegele. 1984. Handbook of toxicity of pesticides to wildlife. 2nd ed. U.S. Fish Wildl. Serv. Resour. Publ. 153, Washington, D. C., 90 pp.

Jefferies, D. J., and M. C. French. 1972. Changes induced in the pigeon thyroid by p,p'-DDE and dieldrin. J. Wildl. Manage. 36:24-30.

Jefferies, D. J., and I. Prestt. 1966. Post-mortems of peregrines and lanners with particular reference to organochlorine residues. Br. Birds 59:49-64.

Jones, D. M., D. Bennett, and K. E. Elgar. 1978. Deaths of owls traced to insecticide-treated timber. Nature (Lond.) 272:52.

Kaiser, T. E., W. L. Reichel, L. N. Locke, E. Cromartie, A. J. Krynitsky, T. G. Lamont, B. M. Mulhern, R. M. Prouty, C. J. Stafford, and D. M. Swineford. 1980. Organochlorine pesticide, PCB, and PBB residues and necropsy data for bald eagles from 29 states — 1975-77. Pestic. Monit. J. 13:145-149.

Koeman, J. H., A. A. G. Oskamp, J. Veen, E. Brouwer, J. Rooth, P. Zwart, E. van den Brock, and H. van Genderen. 1967. Insecticides as a factor in the mortality of the sandwich tern *(Sterna sandvicensis)*. A preliminary communication. Meded. Rijsfaculteit Landbouwwet. Gent 32:841-854.

Kreitzer, J. F. 1980. Effects of toxaphene and endrin at very low dietary concentrations on discrimination acquisition and reversal in bobwhite quail, *Colinus virginianus*. Environ. Pollut. Ser. A Ecol. Biol. 23:217-230.

Lehner, P. N., and A. Egbert. 1969. Dieldrin and eggshell thickness in ducks. Nature (Lond.) 224:1218-1219.

Linder, R. L., R. B. Dahlgren, and Y. A. Greichus. 1970. Residues in the brain of adult pheasants given dieldrin. J. Wildl. Manage. 34:954-956.

Lockie, J. D., D. A. Ratcliffe, and R. Balharry. 1969. Breeding success and organochlorine residues in golden eagles in West Scotland. J. Appl. Ecol. 6:381-389.

Mason, C. F., T. C. Ford, and N. I. Last. 1986. Organochlorine residues in British otters. Bull. Environ. Contamin. Toxicol. 36:656-661.

McCarthy, J. F., and L. R. Shugart. 1990. Biological markers of environmental contamination. p. 3-14. *In* J. F. McCarthy and L. R. Shugart (Eds.). Biomarkers of environmental contamination. Lewis Publishers, Boca Raton, Fla.

McEwen, L. C., and R. L. Brown. 1966. Acute toxicity of dieldrin and malathion to wild sharp-tailed grouse. J. Wildl. Manage. 30:604-611.

Mendenhall, V. M., E. E. Klaas, and M. A. R. McLane. 1983. Breeding success of barn owls *(Tyto alba)* fed low levels of DDE and dieldrin. Arch. Environ. Contam. Toxicol. 12:235-240.

Menzie, C. M. 1969. Metabolism of pesticides. Bur. Sport Fish. Wildl. Spec. Sci. Rep. 127, 487 pp.

Mount, D. I., L. W. Vigor, and M. L. Schafer. 1966. Endrin: use of concentrations in blood to diagnose acute toxicity to fish. Science (Washington, D.C.) 152:1388-1390.

Mulhern, B. M., W. L. Reichel, L. N. Lockie, T. G. Lamont, A. Belisle, E. Cromartie, G. E. Bagley, and R. M. Prouty. 1970. Organochlorine residues and autopsy data from bald eagles, 1966-68. Pestic. Monit. J. 4:142-244.

Murphy, D. A., and L. J. Korschgen. 1970. Reproduction, growth, and tissue residues of deer fed dieldrin. J. Wildl. Manage. 34:887-903.

Nakatsugawa, T., M. Ishida, and P. A. Dahm. 1965. Microsomal epoxidation of cyclodiene insecticides. Biochem. Pharmacol. 14:1853-1865.

Nebeker, A. V., W. L. Griffis, T. W. Stutzman, G. S. Schuytema, L. A. Carey, and S. M. Scherer. 1992. Effects of aqueous and dietary exposure of dieldrin on survival, growth, and bioconcentration in mallard ducklings. Environ. Toxicol. Chem. 11:687-699.

Newton, I. 1988. Determination of critical pollutant levels in wild populations, with examples from organochlorine insecticides in birds of prey. Environ. Pollut. 55:29-40.

Newton, I., and J. Bogan. 1978. The role of different organochlorine compounds in the breeding of British sparrowhawks. J. Appl. Ecol. 15:105-116.

Newton, I., and E. A. Galbraith. 1991. Organochlorines and mercury in the eggs of golden eagles *Aquila chrysaetos* from Scotland. Ibis 133:115-120.

Newton, I., I. Wyllie, and A. Asher. 1991. Mortality causes in British barn owls *Tyto alba*, with a discussion of aldrin-dieldrin poisoning. Ibis 133:162-169.

Newton, I., I. Wyllie, and A. Asher. 1992. Mortality from the pesticides aldrin and dieldrin in British sparrowhawks and kestrels. Ecotoxicology 1:31-44.

Nisbet, I. C. T. 1988. The relative importance of DDE and dieldrin in the decline of peregrine falcon populations. p. 351-375. *In* T. J. Cade, J. H. Enderson, C. G. Thelander, and C. M. White (Eds.). Peregrine falcon populations: their management and recovery. Peregrine Fund, Boise, Ida.

Nisbet, I. C. T. 1989. Organochlorine, reproductive impairment, and declines in bald eagle *Haliaeetus leucocephalus* populations: mechanisms and dose-response relationships. p. 483-489. *In* B.-U. Meyburg and R. D. Chancellor (Eds.). Raptors in the modern world. World Working Group on Birds of Prey, Berlin.

Olsson, M., and F. Sandegren. 1991. Is PCB partly responsible for the decline of the otter in Europe? p. 223-227. *In* C. Reuther and R. Rochert (Eds.). Proc. 5th Int. Otter Colloq. Hankensbüttel 1989. Gruppe Naturschutz GmbH, Hankensbüttel.

Peakall, D. B. 1992. Animal biomarkers as pollution indicators. Chapman & Hall, London, 290 pp.

Prouty, R. M., W. L. Reichel, L. N. Locke, A. A. Belisle, E. Cromartie, T. E. Kaiser, T. G. Lamont, B. M. Mulhern, and D. M. Swineford. 1977. Residues of organochlorine pesticides and polychlorinated biphenyls and autopsy data for bald eagles, 1973-74. Pestic. Monit. J. 11:134-137.

Ratcliffe, D. A. 1973. Studies of the recent breeding success of the peregrine, *Falco peregrinus*. J. Reprod. Fertil. 19 (Suppl.):377-389.

Ray, J. W. 1967. The epoxidation of aldrin by housefly microsomes and its inhibition by carbon monoxide. Biochem. Pharmacol. 16:99-107.

Reichel, W. L., L. N. Locke, and R. M. Prouty. 1974. Peregrine falcon suspected of pesticide poisoning. Avian Dis. 18:487-489.

Reichel, W. L., S. K. Schmeling, E. Cromartie, T. E. Kaiser, A. J. Krynitsky, T. G. Lamont, B. M. Mulhern, R. M. Prouty, C. J. Stafford, and D. M. Swineford. 1984. Pesticide, PCB, and lead residues and necropsy data for bald eagles from 32 states — 1978-1981. Environ. Monit. Assess. 4:395-403.

Risebrough, R. W., and D. B. Peakall. 1988. The relative importance of several organochlorines in the decline of peregrine falcon populations. p. 449-462. *In* T. J. Cade, J. H. Enderson, C. G. Thelander, and C. M. White (Eds.). Peregrine falcon populations: their management and recovery. Peregrine Fund, Boise, Ida.

Robinson, J., V. K. H. Brown, A. Richardson, and M. Roberts. 1967. Residues of dieldrin (HEOD) in the tissues of experimentally poisoned birds. Life Sci. 6:1207-1220.

Roylance, K. J., C. D. Jorgensen, G. M. Booth, and M. W. Carter. 1985. Effects of dietary endrin on reproduction of mallard ducks *(Anas playrhynchos)*. Arch. Environ. Contamin. Toxicol. 14:705-711.

RSPB (Royal Society for the Protection of Birds). 1962. Deaths of birds and mammals from toxic chemicals January-June 1961, 24 pp.

Rudd, R. L., and R. E. Genelly. 1956. Pesticides: their use and toxicity to wildlife. Calif. Dep. Fish Game Bull. No. 7, 209 pp., cited in McEwen and Brown, 1966.

Sanders, H. O. 1970. Pesticide toxicities to tadpoles of the western chorus frog, *Pseudacris triseriata*, and Fowler's toad, *Bufo woodhousii fowleri*. Copeia 2:246-251.

Schuytema, G. S., A. V. Nebeker, W. L. Griffis, and K. N. Wilson. 1991. Teratogenesis, toxicity, and bioconcentration in frogs exposed to dieldrin. Arch. Environ. Contam. Toxicol. 21:332-350.

Sharma, R. P. 1973. Brain biogenic amines: depletion by chronic dieldrin exposure. Life Sci. 13:1245-1251.

Sharma, R. P., D. S. Winn, and J. B. Low. 1976. Toxic, neurochemical, and behaviorial effects of dieldrin exposure in mallard ducks. Arch. Environ. Contam. Toxicol. 5:43-53.

Spann, J. W., G. H. Heinz, and C. S. Hulse. 1986. Reproduction and health of mallards fed endrin. Environ. Toxicol. Chem. 5:755-759.

Spector, W. S. 1955. Handbook of toxicology. N. A. S.-N. R. C. Wright Air Dev. Cent. Tech. Rep. 55-16. Vol. 1, 408 pp.

Stanley, P. I., and P. J. Bunyan. 1979. Hazards to wintering geese and other wildlife from the use of dieldrin, chlorfenvinphos, and carbophenothion as wheat seed treatments. Proc. R. Soc. Lond. Ser. B Biol. Sci. 205:31-45.

Stanley, P. I., and M. R. Fletcher. 1981. A review of the wildlife incidents investigated from October 1978 to September 1979. Pestic. Sci. 252:55-63.

Stanton, R. H., and M. A. Q. Khan. 1973. Mixed-function oxidase activity towards cylcodiene insecticides in bass and bluegill sunfish. Pestic. Biochem. Physiol. 3:351-357.

Stickel, W. H., T. E. Kaiser, and W. L. Reichel. 1979a. Endrin vs. 12-ketoendrin in birds and rodents. p. 61-68. *In* E. E. Kenaga (Ed.). Avian and mammalian wildlife toxicology, ASTM STP 693. American Society for Testing and Materials, Philadelphia.

Stickel, W. H., W. L. Reichel, and D. L. Hughes. 1979b. Endrin in birds: lethal residues and secondary poisoning. p. 397-406. *In* W. B. Deichman (Org.). Toxicology and occupational medicine. Elsevier, North Holland, New York.

Stickel, W. H., L. F. Stickel, and J. W. Spann. 1969. Tissue residues of dieldrin in relation to mortality in birds and mammals. p. 174-204. *In* M. W. Morton and G. G. Berg (Eds.). Chemical fallout. Charles C Thomas, Springfield, Ill.

Street, J. C., R. W. Chadwick, M. Wang, and R. L. Phillips. 1966. Insecticide interactions affecting residue storage in animal tissues. J. Agric. Food Chem. 14:545-549.

Walker, A. I. T., C. H. Neill, D. E. Stevenson, and J. Robinson. 1969. The toxicity of dieldrin (HEOD) to Japanese quail *(Coturnix coturnix japonica)*. Toxicol. Appl. Pharmacol. 15:69-73.

Wallace, G. J., W. P. Nickell, and R. F. Bernard. 1961. Bird mortality in the Dutch elm disease program in Michigan. Cranbrook Inst. Sci. Bull. 41, 44 pp.

Wiemeyer, S. N., R. D. Porter, G. L. Hensler, and J. R. Maestrelli. 1986. DDE, DDT + dieldrin: residues in American kestrels and relations to reproduction. U.S. Fish Wildl. Serv. Fish Wildl. Tech. Rep. 6, 33 pp.

Winn, D. S. 1973. Effects of Sublethal Levels of Dieldrin on Mallard Breeding Behavior and Reproduction. M. Sci. thesis. Utah State University, Logan, Utah.

CHAPTER 4

Other Organochlorine Pesticides in Birds

Stanley N. Wiemeyer

INTRODUCTION

Organochlorine pesticides were first introduced into the environment in the late 1940s to early 1950s. Their widespread use, lipid solubility, persistence, and biomagnification in many cases resulted in residues of toxicological significance in some bird species. The chemicals discussed in this chapter (heptachlor, chlordane, endosulfan, mirex, chlordecone, hexachlorocyclohexane, hexachlorobenzene, dicofol, methoxychlor, and toxaphene) were often of lesser importance in relation to toxicity, use, or persistence than were others such as dichlorodiphenyltrichloroethane (DDT) and dieldrin. Therefore, the research effort expended on these chemicals was often limited, with most data from experimental studies on captive birds.

Numerous reports have included data on residues of these chemicals in wild birds and their eggs but, most often, these were below effect levels. In many reports other contaminants, especially the DDT metabolite DDE, could have masked the effects of the chemicals discussed. Data from these reports usually were excluded. Data from egg injections or from treatment of the shell surface with the toxicants also were excluded from consideration. The effects in such studies often were not comparable to those of cases where contaminant exposure was via the hen, as shown for mercury exposure (Hoffman and Moore, 1979). Hoffman (1990) reviewed the results of external applications of various toxicants to eggs. Caution is needed in applying cross-species applications of data on residues in eggs associated with reproductive effects, because of known species differences in sensitivity to contaminants (Smith, 1987).

Brief background information is presented for each contaminant, along with essential information regarding metabolites of importance in interpreting residues in tissues or eggs. The information is followed by data useful in interpreting residues in relation to lethality and the effects on reproduction, when available.

HEPTACHLOR

BACKGROUND

Heptachlor (1,4,5,6,7,8,8-heptachloro-3a,4,7,7a-tetrahydro-4,7-methanoindene), a cyclodiene insecticide, was primarily used to control soil pests, including termites (Fairchild, 1976). Most uses of heptachlor were phased out by 1983 in the U.S. Heptachlor is readily metabolized to heptachlor epoxide in vertebrates.

RESIDUES DIAGNOSTIC OF LETHALITY

When red-winged blackbirds *(Agelaius phoeniceus)*, brown-headed cowbirds *(Molothrus ater)*, common grackles *(Quiscalus quiscula)*, and European starlings *(Sturnus vulgaris)* were fed diets containing technical heptachlor (74% heptachlor, 2.5% *trans*-chlordane, and 15% *cis*-chlordane), heptachlor epoxide levels in the brains of those that died were significantly higher than those in survivors that received the same exposure (Stickel et al., 1979). Those that died had 9.2 to 27 ppm (wet weight) of heptachlor epoxide in their brains compared with 2.7 to 7.8 ppm in the brains of survivors. The lethal hazard was estimated to begin near 8 ppm. The birds contained traces of *trans*-nonachlor, but no other organochlorines were detected.

EFFECTS ON REPRODUCTION

Nest success of Canada geese *(Branta canadensis)* in the Columbia Basin, Oregon, was ≥77% when eggs contained ≤1.0, 1.1 to 5.0, and 5.1 to 10.0 ppm (wet weight) of heptachlor epoxide, but declined to 17% when eggs contained >10.0 ppm (Blus et al., 1984). A number of geese died of heptachlor epoxide poisoning. The cause of the poor reproductive success was unknown but could have been due to embryotoxicity or nest desertion.

Field studies were conducted on the effects of heptachlor on raptors. Productivity of American kestrels *(Falco sparverius)* in the Columbia Basin was reduced when eggs contained >1.5 ppm of heptachlor epoxide (Henny et al., 1983). Nest success and the young produced per attempt in relation to the ppm (wet weight) of heptachlor epoxide and sample size follow: ≤0.50 ppm, $n =$ 152, 80%, 2.50 young; 0.51 to 1.50 ppm, $n =$ 40, 78%, 2.48 young; 1.51 to 3.00 ppm, $n =$ 12, 42%, 1.17 young; 3.01 to 6.00 ppm, $n =$ 18, 39%, 1.22 young; >6.00 ppm, $n =$ 2, 0%, 0.00 young. Elevated residues of heptachlor epoxide were detected in eggs of other species of raptors from the same region, but no definite effects on productivity were apparent from the small samples of nests (Henny et al., 1984). They found no significant relation between heptachlor epoxide residues and shell thickness in eggs of Swainson's hawk *(Buteo swainsoni)*. Heptachlor epoxide residues in the eggs of wild prairie falcons *(Falco mexicanus)* and merlins *(Falco columbarius)* between 1 and 2 ppm dry weight (about 0.2 to 0.4 ppm wet weight) seemed not to affect reproduction (Fyfe et al., 1976), in agreement with the findings of Henny et al. (1983) for kestrels.

Chick survival of Japanese quail *(Coturnix japonica)* was reduced by about 50% when eggs contained 14 to 17 ppm (wet weight) of heptachlor epoxide, but hatchability was not affected (Grolleau and Froux, 1973). Chick survival of gray partridges *(Perdix perdix)* was slightly reduced when eggs contained 3 to 7 ppm (wet weight) of heptachlor epoxide, but hatching success was normal (Havet, 1973). Threshold levels of toxicity were not clearly defined in these two studies.

CHLORDANE

BACKGROUND

Chlordane (1,2,4,5,6,7,8,8-octachloro-2,3,3a,4,7,7a-hexahydro-4,7-methano-1H-indene) was the first cyclodiene insecticide used in agriculture (Eisler, 1990). Major restrictions on its use in the U.S. were imposed in 1978, and all uses have recently been banned.

Technical chlordane contains about 45 compounds, of which the major ones are *cis*-chlordane, 19%; *trans*-chlordane, 24%; chlordene isomers, 21.5%; heptachlor, 10%; and *cis*- and *trans*-nonachlor, 7% (Eisler, 1990). Heptachlor can result from the breakdown of *cis*- and *trans*-chlordane, which eventually is oxidized to heptachlor epoxide (Eisler, 1990). Oxychlordane is a metabolite of both *cis*- and *trans*-chlordane (Stickel et al., 1983). The most important toxic products of technical chlordane are oxychlordane and heptachlor epoxide.

RESIDUES DIAGNOSTIC OF LETHALITY

When dietary dosages of oxychlordane were fed to passerines, lethal concentrations in their brains did not differ among species, ages, sexes, dosage levels, or times on dosage (Stickel et al., 1979). The brains of birds that died (range, 5.8 to 16 ppm wet weight; mean, 10 ppm; $n = 36$) had distinctly higher oxychlordane residues than did those of birds that survived (range, 0.73 to 4.1 ppm wet weight; mean, 2.3 ppm; $n = 20$). The lethal hazard zone was estimated to begin near 5 ppm. Residues did not differ in the bodies of those that died and in those that survived.

HCS-3260, an experimental chlordane (not marketed) which contained no heptachlor (70.75% *cis*-chlordane and 23.51% *trans*-chlordane), was fed to red-winged blackbirds, common grackles, brown-headed cowbirds, and European starlings (Stickel et al., 1983). Oxychlordane residues in the brains of all but starlings that died on dosage ranged from 9.4 to 25.0 ppm wet weight, while residues in the same three species that were sacrificed (survivors) ranged from 1.3 to 4.8 ppm; oxychlordane residues in brains provided a clear diagnostic criterion of death. In starlings, those that died had 5.0 to 19.1 ppm (wet weight) of oxychlordane in the brain, which was significantly lower than that in the other three species. Oxychlordane residues in the brains of sacrificed starlings ranged from 1.4 to 10.5 ppm; residues in five of 20 that were sacrificed equalled or exceeded 5 ppm and overlapped the residue levels in those that died. Brain residues in starlings of between 5 and 10 ppm were considered ambiguous, and a diagnosis of poisoning must be supported by necropsy findings. Brains of the four species also contained 1.4 to 13.8 ppm (wet weight) of *cis*-chlordane, whereas sacrificed birds had 1.7 to 5.7 ppm; the overlap was complete. A few birds that died and a few that survived contained low residues of *trans*-chlordane (<0.5 ppm) and heptachlor epoxide (<0.9 ppm), with overlap between groups. Birds that died in this study lost 11 to 32% of their weight, had little or no visible fat, had reduced musculature, nearly always had empty intestines, and had convulsions before death; all signs were generally typical of cyclodiene poisoning. Oxychlordane was lethal to birds fed HCS-3260; lethal residues in the brains of three species, excluding starlings, were significantly higher than when only oxychlordane was fed to the same species.

European starlings and brown-headed cowbirds were fed diets containing a technical standard of nonachlor, which was nearly pure (only 0.3% *trans*-chlordane; Stickel et al., 1983). Nonachlor was reported to metabolize to oxychlordane through *trans*-chlordane. A few birds that died had oxychlordane residues in their brains comparable to those killed by HCS-3260 (see above) and also high levels of nonachlor (166 to 259 ppm wet weight) that clearly exceeded levels in the brains of survivors.

When diets containing chlordane (formulation contained 7% heptachlor) were fed to passerines, heptachlor epoxide residues in the brain were closely related to mortality, although it was clear that oxychlordane also contributed (Stickel et al., 1979). The brain tissue of those that died contained 3.4 to 8.3 ppm (wet weight) of heptachlor epoxide (mean, 5.0 ppm) and 1.1 to 5.0 ppm of oxychlordane (mean, 2.8 ppm), whereas the survivors had 0.87 to 3.2 ppm of heptachlor epoxide (mean, 1.6 ppm) and 0.18 to 1.7 ppm of oxychlordane (mean, 0.72 ppm). Residues of oxychlordane, nonachlor, and unnamed compounds broadly overlapped in brain tissues of both the dead and the survivors. Nonachlor and the unnamed compounds may have contributed to mortality. The levels of heptachlor epoxide and oxychlordane were each about 28% of the concentrations in the brains of birds that died on heptachlor or oxychlordane dosage; an additive effect with other chlordane compounds has been suggested. Others have hypothesized additive effects between different cyclodienes when interpreting lethal concentrations in dosed birds (Robinson, 1969; Ludke, 1976). Deaths of wild birds from chlordane exposure have been reported, with dieldrin a possible contributor in some cases (Blus et al., 1983, 1985; Stone and Okoniewski, 1988). Additive effects among classes of organochlorines (DDE, dieldrin, and polychlorinated biphenyls) were suggested during interpretation of potentially lethal residues in the brains of wild birds (Sileo et al., 1977); however, experimental studies are lacking to support additivity among classes.

EFFECTS ON REPRODUCTION

Technical chlordane at 3 and 15 ppm in the dry feed of northern bobwhite *(Colinus virginianus)* and at 8 ppm in the diet of mallards *(Anas platyrhynchos)* had no effect on reproduction (J. W. Spann, Patuxent Wildlife Research Center, unpublished data); no information was provided on residues in eggs. No data were found on the effects of oxychlordane on the reproduction of birds.

ENDOSULFAN

Endosulfan (6,7,8,9,10,10-hexachloro-1,5,5a,6,9,9a-hexahydro-6,9-methano-2,4,3-benzodioxathiepin 3-oxide) is a member of the cyclodiene group but differs so greatly in properties and effects that it should not be considered one of them (Maier-Bode, 1968). The technical grade consists of two isomers, α and β, at a 70:30 ratio. Endosulfan is rapidly eliminated and does not appear to accumulate in warm-blooded animals. No information was found that interpreted residues in birds or their eggs; therefore, no additional information will be provided for this chemical.

MIREX

BACKGROUND

Mirex (1,1a,2,2,3,3a,4,5,5,5a,5b,6-dodechlorooctahydro-1,3,4-metheno-1H-cyclobuta[cd]pentalene) was primarily used for control of the imported fire ant *(Solenopsis invicta)* in the Southeast but also was used as a fire retardant (Eisler, 1985). Lake Ontario was contaminated with mirex from manufacturing effluents. Mirex is one of the most stable and persistent organochlorines known. Its degradation products include hexachlorobenzene and chlordecone. Mirex was banned from use in the U.S. in 1978 (Eisler, 1985).

RESIDUES DIAGNOSTIC OF LETHALITY

Stickel et al. (1973) administered doses of mirex to four species of passerines. They indicated that, if a bird had ≥177 ppm (wet weight) of mirex in its brain and no more than 2.4% lipid in its carcass, there was a high probability that it died of mirex. Any bird having over 200 ppm (wet weight) of mirex in its brain was either killed or seriously endangered by this chemical. Lethal concentrations of mirex in brain tissue occurred following a reduction in the amount of clean food provided to the birds and mobilization of mirex residues stored in body lipids. Organochlorine-induced mortality may occur in wild birds that have accumulated elevated concentrations of persistent organochlorine pesticides in their bodies and then have subsequently undergone stress that results in mobilization of fat reserves and stored toxicants. The brains of seven northern bobwhites that died after 6 to 12 months on mirex dosage contained 135 to 281 ppm (wet weight) of mirex (mean, 186 ppm), whereas the brains of birds that survived to the end of the study all contained less than 34 ppm (Heath and Spann, 1973). It was not certain that mirex was the cause of death, but mirex residues most likely contributed. The brains of mallards fed a diet containing 100 ppm of mirex in dry feed for 25 weeks had a mean of 28 ppm (wet weight) of mirex with no apparent toxicity when sacrificed (Hyde et al., 1973).

EFFECTS ON REPRODUCTION

Reproduction of northern bobwhites that received a dry diet containing 40 ppm of mirex during the first reproductive season, which was reduced to 10 ppm during the second season because of excessive adult mortality, was unaffected in either year (Heath and Spann, 1973). Eggs collected during the third and fourth weeks of the second season contained 150 ppm (wet weight) of mirex. The reproduction of mallards that received dry diets containing 1 or 10 ppm of mirex for 3 months was unaffected (Heath and Spann, 1973); eggs from birds receiving 10 ppm in the diet contained about 20 ppm (wet weight) of mirex. The reproduction of Japanese quail that received dry diets containing up to 80 ppm of mirex was unaffected (Davison et al., 1975); eggs contained about 450 ppm (dry weight) of mirex (about 90 ppm on a wet-weight basis). The egg residues from the above studies were below levels affecting reproduction.

Reproductive effects were noted when birds were given somewhat higher dietary concentrations of mirex. Mallards were provided a dry diet containing 100 ppm of mirex (Hyde et al., 1973). Although there were no significant effects on hatchability, the survival of ducklings that were maintained on clean feed (73%) was significantly reduced in relation to that of controls (96%). Eggs from the dosed birds that were collected 2 weeks before eggs were taken for the reproductive study contained a mean of 277 ppm (wet weight) of mirex. Ducklings that died contained a mean of 386 ppm of mirex (wet weight on a whole-body basis). Chickens that received dry diets containing 300 or 600 ppm of mirex had normal egg production, and fertility was unaffected, although hens lost weight at the higher concentration (Naber and Ware, 1965). Hatchability of eggs from birds fed 600 ppm of mirex was severely reduced. After 12 weeks on the experimental diet, chicks from hens fed 300 ppm of mirex had reduced survival, and few chicks from hens on the higher concentration survived. Egg yolk residues reached a maximum by the fifth week of the study. Eggs laid 12 weeks after the start of dosage contained 797 ppm (presumed wet weight) of mirex in the yolk for those from the lower dietary concentration group and 1440 ppm in the yolk for those from the higher dietary exposure group. The yolk of a chicken egg makes up 31.9% of the total egg weight (Romanoff and Romanoff, 1949). Therefore, assuming the entire mirex burden in the eggs was in the yolk, the above residues in the yolk would be equivalent to about 255 and 450 ppm wet weight on a whole-egg basis.

OTHER EFFECTS

Structural changes, which included regions of necrosis and other aberrations, were found in the livers of chickens that received 10 ppm or more of mirex in their diet (dry feed) for 12 and 16 weeks in two studies; however, no data were provided on residues in tissues (Davison et al., 1976).

The sperm concentration of American kestrels that received about 8 ppm of mirex wet weight in the diet was reduced, along with a modest increase in semen volume that resulted in a 70% decrease in sperm numbers (Bird et al., 1983). The impact of these changes on egg fertility was unknown. Residues (ppm wet weight) in tissues of the kestrels were 1.6 ppm in liver, 2.0 ppm in muscle, and 4.5 ppm in testes.

CHLORDECONE

BACKGROUND

Chlordecone (decachlorooctahydro-1,3,4-metheno-2H-cyclobuta[cd]pentalen-2-one), also known as Kepone, was registered to control ants and cockroaches, but most production was exported to Central and South America to control the banana root borer (Huggett and Bender, 1980). Effluent from a production plant at Hopewell, Virginia, contaminated the James River. Production in the U.S. ceased in 1975.

PROBABLE SUBLETHAL EFFECTS

Although highly elevated concentrations (extreme of 226 ppm wet weight) of chlordecone were found in the livers of fish-eating birds from the contaminated James River (Stafford et al., 1978; Huggett and Bender, 1980; U.S. Fish and Wildlife Service, 1982), no experimental data were generated to interpret these concentrations. The elevated liver residues might be explained in part by the presence of lipid inclusions in hepatic cells of experimental birds that were given doses of chlordecone (McFarland and Lacy, 1969; Eroschenko and Wilson, 1975).

EFFECTS ON REPRODUCTION

Chlordecone in the dry diet of chickens at 75 and 150 ppm resulted in a significant reduction in the number of eggs produced during the first 12 weeks of the study (Naber and Ware, 1965). Hens fed the higher level also lost body weight. Fertility was unaffected, but hatchability was significantly reduced at 12 weeks for those fed 150 ppm. Most chicks from hens fed 150 ppm exhibited a nervous syndrome characterized by quivering of the legs, wings, and head; none could walk and many could not stand. The survival of chicks from hens receiving 75 ppm of chlordecone was reduced; no chicks survived from hens receiving 150 ppm. Residues of chlordecone in egg yolks reached equilibrium by 5 weeks of dosage. Hens receiving 75 ppm of chlordecone for 6 and 12 weeks laid eggs whose yolks contained 161 and 120 ppm (presumed wet weight) of chlordecone (comparable to about 51 and 38 ppm on a whole-egg wet-weight basis). Hens receiving 150 ppm of chlordecone for 6 and 12 weeks laid eggs whose yolks contained 300 and 232 ppm wet weight (comparable to about 96 and 74 ppm on a whole-egg wet-weight basis). Estrogenic activity was demonstrated in birds fed diets containing as little as 10 ppm of chlordecone (Eroschenko, 1981); however, residues in tissues associated with these effects were not determined. A few eggs of bald eagles *(Haliaeetus leucocephalus)* and ospreys

(Pandion haliaetus) from near the James River, Virginia, contained up to about 5 ppm (wet weight) of chlordecone (Wiemeyer et al., 1984, 1988), but the levels were far below those associated with reproductive effects in chickens.

HEXACHLOROCYCLOHEXANE

BACKGROUND

Hexachlorocyclohexane (1,2,3,4,5,6-hexachlorocyclohexane), also known as benzene hexachloride, occurs as three different isomers, γ, β, and α. The γ isomer, also known as lindane, is the most active insecticide (USDA, 1980). The three major uses are on seed, on hardwood lumber, and on livestock (USDA, 1980). Lindane is readily metabolized and excreted in birds, minimally accumulated in tissues, and without implication as a problem in the field (Blus et al., 1984).

RESIDUES ASSOCIATED WITH EXCESSIVE EXPOSURE

Rock doves *(Columba livia)* were given daily doses of 72 mg/kg of γ-hexachlorocyclohexane for 5 days (Turtle et al., 1963). One bird that died on day 7 had 31.6 ppm (assumed wet weight) in the liver. Birds that were sacrificed during dosing had 18.7 to 67.0 ppm in the liver. Birds sacrificed 15 days after dosing had 4.2 to 37.1 ppm in the liver. While not diagnostic, these residues can be related to dangerous exposure to the toxicant.

EFFECTS ON REPRODUCTION

Hatchability of eggs from ring-necked pheasants *(Phasianus colchicus)* that received a diet containing γ-hexachlorocyclohexane was unaffected; mean γ-hexachlorocyclohexane residues in eggs were about 10 ppm (assumed wet weight; Ash and Taylor, 1964). The nest of an American kestrel, where a sample egg contained 5.5 ppm (wet weight) of β-hexachlorocyclohexane, was successful and fledged three young (Henny et al., 1983). When lindane (γ-hexachlorocyclohexane) was substituted for heptachlor as a seed treatment in the Columbia Basin, Oregon, no hexachlorocyclohexane isomers were found in eggs and tissues of Canada geese nesting nearby (Blus et al., 1984). Reproductive success improved following the substitution.

HEXACHLOROBENZENE

BACKGROUND

Hexachlorobenzene is a fungicide registered for use on seed grains and is an industrial waste product from the manufacture of several chlorinated solvents and pesticides (Courtney, 1979). Hexachlorobenzene is also used in the manufacture of rubber for tires, is a contaminant in the herbicide Dacthal and the fungicide pentachloronitrobenzene, and is persistent in the environment.

RESIDUES ASSOCIATED WITH EXCESSIVE EXPOSURE

Japanese quail that died after receiving a dry diet containing 500 ppm of hexachlorobenzene had a mean of 450 ppm of hexachlorobenzene (wet weight; range, 180 to 850 ppm; $n = 9$) in liver tissue (Vos et al., 1968). Those that received 100 ppm in the diet and died had a mean of 235 ppm in the liver (range, 85 to 720 ppm; $n = 6$). Quail that received 20 ppm in their diet and were sacrificed had a mean of 35.5 ppm in the liver (range, 13.8 to 94 ppm; $n = 6$). Kestrels (presumably *Falco tinnunculus*) were fed a diet containing 200 ppm (wet weight) of hexachlorobenzene (Vos et al., 1972). One male that died after 62 days of exposure had 2600 ppm (wet weight) of hexachlorobenzene in the liver and 1170 ppm in the brain. Two females that were sacrificed after 58 and 80 days of exposure had 348 and 535 ppm in the liver and 465 and 510 ppm in the brain. The females exhibited weight loss, ruffling of feathers, and tremor (started 2 weeks before sacrifice) at the time of sacrifice. Although the tissue concentrations in these two studies were not diagnostic of death, they provided a general indication of levels in severely exposed birds.

EFFECTS ON REPRODUCTION

Japanese quail were given dry diets containing 20 and 100 ppm of hexachlorobenzene (Vos et al., 1968). The hatchability of eggs from the 20-ppm group was significantly depressed from that of controls; few eggs were artificially incubated from the 100-ppm group and none hatched. Four "sterile" and unhatched eggs from the 100-ppm group contained 166 to 217 ppm (presumed wet weight) of hexachlorobenzene. Five chicks from the 20-ppm group contained 30.8 to 58.0 ppm of hexachlorobenzene. There was also evidence of a disturbance of porphyrin metabolism of adults as shown by increased excretion of coproporphyrin in feces in the 20-ppm group. Increased liver weight and fecal coproporphyrin excretion and slight liver lesions were found in Japanese quail that received a dry diet containing 5 ppm of hexachlorobenzene (Vos et al., 1971). Residues in livers ranged from 4.1 to 16.7 ppm of hexachlorobenzene. One ppm of hexachlorobenzene in the diet of Japanese quail for 90 days was considered a no-effect exposure; livers of these birds contained 0.35 to 4.4 ppm of hexachlorobenzene (Vos et al., 1971). In another study in which Japanese quail were given a dry diet containing 20 ppm of hexachlorobenzene for 90 days, survival of chicks was reduced, although other parameters of reproductive success were unaffected (Schwetz et al., 1974). Eggs collected during the last month of the study contained a mean of 6.2 ppm of hexachlorobenzene wet weight.

The hatchability of eggs from Japanese quail given capsule doses of hexachlorobenzene (equivalent to 80 ppm in the diet) for 6 weeks was not significantly affected (Fletcher, 1972). Egg yolks contained about 100 ppm of hexachlorobenzene, which is equivalent to about 35 ppm wet weight, assuming that yolk constitutes 35%

of total egg weight for this species. Adult quail exhibited hyperactivity and increased porphyrin excretion in the last week of the study. The liver, brain, and muscle of males contained about 8, 6, and 7 ppm of hexachlorobenzene (assumed wet-weight basis), whereas those tissues of females contained 20, 7, and 6 ppm, respectively. Hexachlorobenzene residues in chicks that died after hatching were about 6 to 34 ppm in the brain and 52 to 220 ppm in the liver.

Up to 100 ppm of hexachlorobenzene in the dry diet of chickens for 18 to 19 weeks had no adverse effects on the fertility and hatchability of their eggs (Avrahami and Steele, 1972), although sample sizes were small ($n \leq 10$ per treatment). Egg yolks contained 330 ppm of hexachlorobenzene (presumed wet weight) or about 105 ppm on a whole-egg basis. Chicks that hatched had about 100 ppm of hexachlorobenzene in muscle tissue at hatch. The chickens had 50 ppm of hexachlorobenzene in the liver and 12 ppm in muscle tissue after 26 weeks of exposure; the health and general condition of the birds were not affected.

The reproduction of Canada geese (Blus et al., 1984) and American kestrels (Henny et al., 1983) appeared normal when maximum hexachlorobenzene residues in eggs were 2.97 and 2.4 ppm, respectively.

DICOFOL

BACKGROUND

Dicofol [4-chloro-α-(4-chlorophenyl)-α-(trichloromethyl)benzenemethanol] is an organochlorine structurally similar to DDT and is the major constituent of Kelthane (Clark, 1990). Kelthane, primarily a mixture of dicofol, DDT, DDD, DDE, and chloro-DDT, was restricted to contain <0.1% DDT-related contaminants in the U.S. in 1989. Kelthane is a miticide, primarily used on citrus fruit, cotton, apples, and ornamental shrubs. The primary metabolites of dicofol are dechlorodicofol, dichlorobenzophenone, and dichlorobenzhydrol. No information was found regarding dicofol residues in relation to mortality.

EFFECTS ON REPRODUCTION

Eggshells of ringed turtle-doves *(Streptopelia risoria)* fed a dry diet containing 33.4 ppm of dicofol were thinned 8.9%, egg cracking and breakage were increased, and egg production was reduced compared with those of controls (Schwarzbach et al., 1988). Eggs laid during days 39 to 48 of treatment contained about 10 ppm of dicofol, 12 ppm of dichlorobenzophenone, 4 ppm of dechlorodicofol, and 5 ppm of dichlorobenzhydrol (wet weight) in yolk (Schwarzbach, 1991). Assuming that the yolk of a dove egg is 18.1% of total egg weight (Romanoff and Romanoff, 1949) and that all residues are in the yolk, these residues on a whole-egg wet-weight basis would be about 1.8 ppm of dicofol, 2.2 ppm of dichlorobenzophenone, 0.7 ppm of dechlorodicofol, and 0.9 ppm of dichlorobenzhydrol. Eggshells of ringed turtle-doves that received dry diets containing 10 and 32 ppm of dechlorodicofol were not thinned (Schwarzbach, 1991).

American kestrels were fed diets containing 0, 1, 3, 10, and 30 ppm (wet weight) of dicofol containing no detectable DDE or DDT (Clark et al., 1990). The shell thickness and thickness index of the first egg laid in each clutch were significantly reduced from those of controls for birds receiving diets containing 3 ppm (8 and 7%, respectively), 10 ppm (12 and 13%, respectively), and 30 ppm (15 and 17%, respectively). Shell weight was reduced for those receiving 10 ppm (9%) and 30 ppm (16%). Hatchability was reduced for those receiving 10 ppm, but not for those receiving 30 ppm. The mean residues (ppm wet weight) in the first eggs laid (S. N. Wiemeyer, unpublished data) for birds receiving 3 ppm in the diet were 1.2 ppm of dicofol, 0.12 ppm of dechlorodicofol, and 0.19 ppm of dichlorobenzophenone. Egg residues for those receiving 10 ppm were 4.0 ppm of dicofol, 0.70 ppm of dechlorodicofol, and 1.2 ppm of dichlorobenzophenone and, for those receiving 30 ppm, were 15 ppm of dicofol, 2.3 ppm of dechlorodicofol, and 4.0 ppm of dichlorobenzophenone.

Eastern screech owls *(Otus asio)* were also fed a diet containing 10 ppm (wet weight) of dicofol (Wiemeyer et al., 1989). The shell thickness, thickness index, and shell weight were significantly reduced by 11, 15, and 13%, respectively. The residues in eggs were not reported because of problems in the loss of dicofol during chemical analysis; however, one egg, properly analyzed, contained 3.2 ppm of dicofol, 2.2 ppm of dechlorodicofol, and 1.2 ppm of dichlorobenzophenone (Krynitsky et al., 1988).

Mallard hens were provided dry diets containing 0, 3, 10, 30, or 100 ppm of dicofol (Bennett et al., 1990). Hens receiving the highest concentration had an increased incidence of cracked and soft-shelled eggs; and shell strength, thickness, and weight were reduced by 29, 12, and 14%, respectively. These shell parameters were negatively related to dietary concentrations. Eggs that were laid during days 37 to 42 of treatment contained a mean of 151 ppm of dicofol in the yolk. Assuming that the yolk constitutes 30% of total egg weight in mallards (G. H. Heinz, personal communication) and that all residues were in the yolk, about 45 ppm wet weight was present on a whole-egg basis. The residues in eggs were difficult to relate to a given effect because concentrations may not have reached equilibrium by the end of the study and because shell parameter data were collected throughout the study.

METHOXYCHLOR

BACKGROUND

Methoxychlor [1,1'-(2,2,2-trichloroethylidene)-bis(4-methoxybenzene)], the p,p'-methoxy analogue of DDT, is used to control a wide spectrum of insects. A major use in Canada was the control of biting flies (Gardner and Bailey, 1975). Methoxychlor is not persistent in the environment, and bioaccumulation is low. Methoxychlor is readily metabolized in birds and thus not readily accumulated (Gardner and Bailey, 1975). This chemical was a replacement for DDT in many applications. No data were found regarding methoxychlor residues in relation to mortality.

EFFECTS ON REPRODUCTION

The hatchability and other reproductive parameters of chickens fed dry diets containing up to 5000 ppm of methoxychlor for 16 weeks did not differ from those of controls (Lillie et al., 1973). Hens fed diets containing 25, 100, and 250 ppm of methoxychlor accumulated <0.5 ppm in eggs, whereas those receiving 5000 ppm had about 17 ppm in eggs for weeks 9 to 16 of the study. Methoxychlor residues in eggs were rapidly depleted; residues in eggs from all diets were <0.5 ppm after the hens were placed on clean food for 3 weeks.

TOXAPHENE

BACKGROUND

Toxaphene, a complex mixture of chlorinated camphenes, consists of 177 compounds of widely ranging toxicity, most of which have not been identified (Eisler and Jacknow, 1985). These characteristics, in addition to changes in composition from the parent compound in tissues, cause problems in chemical analysis and interpretation of analytical results (Eisler and Jacknow, 1985). At one time, toxaphene was one of the most widely used organochlorine insecticides in the U.S., but most registrations for use in the U.S. were canceled in 1982. The frequency of occurrence of toxaphene in the tissues and eggs of birds is low. Reviews on toxaphene have been prepared by Pollock and Kilgore (1978) and Eisler and Jacknow (1985). Information was not found on residues in relation to mortality.

EFFECTS ON REPRODUCTION

Toxaphene at 100 ppm in the dry diet of chickens did not significantly affect egg production, shell strength, fertility, or hatchability (Bush et al., 1977). Eggs contained a mean of 13.9 ppm (wet weight) of toxaphene, based on three chromatographic peaks.

Haseltine et al. (1980) fed American black ducks *(Anas rubripes)* 0, 10, or 50 ppm of toxaphene through two reproductive seasons. The third egg laid by each female was retained for residue analysis, hens incubated their own eggs, and ducklings were maintained on the same diet as were their parents. No effects on reproduction were found. The mean egg residues (wet-weight basis), based on the area under two chromatographic peaks, for those receiving 50 ppm were 40.5 and 50.2 ppm in the first and second years, respectively. Duckling survival was normal through 12 weeks. Carcasses of those on 50 ppm at 12 weeks of age contained means of 43.1 and 20.1 ppm of toxaphene in the first and second years of the study. Mehrle et al. (1979) examined duckling growth and bone development in birds from the above study.

Duckling growth, as indicated by weight, was significantly depressed 5 and 14 days after hatching for those fed 50 ppm of toxaphene. Backbone development was impaired in those receiving the 50-ppm diet within 14 days after hatch. Collagen was decreased and calcium was increased in ducklings fed 50 ppm. After 12 weeks, collagen was decreased and calcium increased in female ducklings only. Tibia development was normal. Decreased collagen and increased calcium could alter the structural integrity of the vertebral column. Avoidance behavior of 5-day-old ducklings from the above study was not affected by toxaphene exposure (Heinz and Finley, 1978).

SUMMARY

Parent compounds of concern in this chapter are heptachlor, chlordane, endosulfan, mirex, chlordecone, hexachlorocyclohexane, hexachlorobenzene, dicofol, methoxychlor, and toxaphene. Residues of these pesticides or their metabolites have been found in wild birds from various species or in their eggs. However, data were limited on the significance of the concentrations of many of these compounds or their metabolites in birds and their eggs because of limited published research findings.

Threshold concentrations in the brain that may be considered diagnostic of lethality were available for only three toxicants. When heptachlor was fed to passerines, residues of heptachlor epoxide of ≥8 ppm were considered diagnostic of death. A brain concentration of 5 ppm of oxychlordane was the lower concentration associated with death in passerines. Birds that died from exposure to technical chlordane had concentrations of a combination of heptachlor epoxide and oxychlordane that were considered lethal. Birds dying of mirex poisoning had ≥200 ppm of mirex in the brain. Concentrations of other organochlorines in tissues that are associated with excessive exposure were reported, but levels diagnostic of poisoning are unknown.

Threshold residues in eggs that were clearly associated with reductions in productivity have been reported for a few species and chemicals. These data should be used cautiously because species differences in sensitivity may occur. The nest success of wild Canada geese declined sharply when residues of heptachlor epoxide exceeded 10 ppm in eggs. Wild American kestrels, much more sensitive, showed a reduction in productivity when concentrations were >1.5 ppm in eggs. Chick survival of Japanese quail and gray partridges was reduced when eggs contained 14 to 17 and 3 to 7 ppm of heptachlor epoxide, respectively; threshold levels were not defined. Reproduction of northern bobwhites, mallards, and Japanese quail was unaffected when mirex residues in eggs averaged 150 ppm, 20 ppm, and 90 ppm, respectively. Mallard duckling survival was reduced when eggs contained about 280 ppm of mirex; whole bodies of ducklings that died contained about 390 ppm. The hatchability of chicken eggs containing about 450 ppm of mirex was reduced, and chick survival was reduced in eggs containing about 255 ppm of mirex; threshold levels were not defined. Chickens producing eggs containing about 40 to 50 ppm of chlordecone had reduced chick survival; no chicks survived when egg residues were about twice as high. The effects of hexachlorobenzene on the reproduction of Japanese quail, where egg residues were known, were somewhat variable among studies. The hatchability of chicken eggs containing 100 ppm of hexachlorobenzene was normal. The eggshell quality of ringed turtle-doves was reduced when eggs contained about 1.8 ppm of dicofol plus additional residues of various metabolites. The shell quality of American kestrels was clearly reduced when eggs contained 4 ppm of dicofol, 0.7 ppm of dechlorodicofol, and 1.2 ppm of dichlorobenzophenone. The shell quality of mallard eggs was greatly reduced when eggs contained about 45 ppm of dicofol. The hatchability of chicken eggs containing 17 ppm of methoxychlor or 14 ppm of toxaphene was normal. The reproduction of mallards was normal when eggs contained about 50 ppm of toxaphene; however, duckling growth was depressed when carcasses contained 40 ppm of toxaphene.

REFERENCES

Ash, J. S., and A. Taylor. 1964. Further trials on the effects of gamma BHC seed dressing on breeding pheasants. Game Res. Assoc. Annu. Rep. 4:14-20.

Avrahami, M., and R. T. Steele. 1972. Hexachlorobenzene. II. Residues in laying pullets fed HCB in their diet and the effects on egg production, egg hatchability, and on chickens. N. Z. J. Agric. Res. 15:482-488.

Bennett, J. K., S. E. Dominguez, and W. L. Griffis. 1990. Effects of dicofol on mallard eggshell quality. Arch. Environ. Contam. Toxicol. 19:907-912.

Bird, D. M., P. H. Tucker, G. A. Fox, and P. C. Lague. 1983. Synergistic effects of Aroclor® 1254 and mirex on the semen characteristics of American kestrels. Arch. Environ. Contam. Toxicol. 12:633-639.

Blus, L. J., C. J. Henny, and A. J. Krynitsky. 1985. Organochlorine-induced mortality and residues in long-billed curlews from Oregon. Condor 87:563-565.

Blus, L. J., C. J. Henny, D. J. Lenhart, and T. E. Kaiser. 1984. Effects of heptachlor- and lindane-treated seed on Canada geese. J. Wildl. Manage. 48:1097-1111.

Blus, L. J., O. H. Pattee, C. J. Henny, and R. M. Prouty. 1983. First records of chlordane-related mortality in wild birds. J. Wildl. Manage. 47:196-198.

Bush, P. B., J. T. Kiker, R. K. Page, N. H. Booth, and O. J. Fletcher. 1977. Effects of graded levels of toxaphene on poultry residue accumulation, egg production, shell quality, and hatchability in white leghorns. J. Agric. Food Chem. 25:928-932.

Clark, D. R., Jr. 1990. Dicofol (Kelthane®) as an environmental contaminant: a review. U.S. Fish Wildl. Serv. Tech. Rep. 29, 37 pp.

Clark, D. R., Jr., J. W. Spann, and C. M. Bunck. 1990. Dicofol (Kelthane®)-induced eggshell thinning in captive American kestrels. Environ. Toxicol. Chem. 9:1063-1069.

Courtney, K. D. 1979. Hexachlorobenzene (HCB): a review. Environ. Res. 20:225-266.

Davison, K. L., J. H. Cox, and C. K. Graham. 1975. The effect of mirex on reproduction of Japanese quail and on characteristics of eggs from Japanese quail and chickens. Arch. Environ. Contam. Toxicol. 3:84-95.

Davison, K. L., H. H. Mollenhauer, R. L. Younger, and J. H. Cox. 1976. Mirex-induced hepatic changes in chickens, Japanese quail, and rats. Arch. Environ. Contam. Toxicol. 4:469-482.

Eisler, R. 1985. Mirex hazards to fish, wildlife, and invertebrates: a synoptic review. U.S. Fish Wildl. Serv. Biol. Rep. 85(1.1), 42 pp.

Eisler, R. 1990. Chlordane hazards to fish, wildlife, and invertebrates: a synoptic review. U.S. Fish Wildl. Serv. Biol. Rep. 85(1.21), 49 pp.

Eisler, R., and J. Jacknow. 1985. Toxaphene hazards to fish, wildlife, and invertebrates: a synoptic review. U.S. Fish Wildl. Serv. Biol. Rep. 85(1.4), 26 pp.

Eroschenko, V. P. 1981. Estrogenic activity of the insecticide chlordecone in the reproductive tract of birds and mammals. J. Toxicol. Environ. Health 8:731-742.

Eroschenko, V. P., and W. O. Wilson. 1975. Cellular changes in the gonads, livers, and adrenal glands of Japanese quail as affected by the insecticide Kepone. Toxicol. Appl. Pharmacol. 31:491-504.

Fairchild, H. E. 1976. Heptachlor in relation to man and environment. EPA-540/4-76-007. U.S. Environmental Protection Agency, Washington, D.C., 65 pp.

Fletcher, M. R. 1972. Effects of hexachlorobenzene on Japanese quail. Proc. West. Assoc. State Game Fish Comm. 52:374-383.

Fyfe, R. W., R. W. Risebrough, and W. Walker II. 1976. Pollutant effects on the reproduction of the prairie falcons and merlins of the Canadian prairies. Can. Field Nat. 90:346-355.

Gardner, D. R., and J. R. Bailey. 1975. Methoxychlor: its effects on environmental quality. NRCC No. 14102. National Research Council of Canada, Ottawa, 164 pp.

Grolleau, G., and Y. Froux. 1973. Effet de l'heptachlore sur la reproduction de la caille *Coturnix coturnix japonica*. (In French.) Ann. Zool. Ecol. Anim. 5:261-270.

Haseltine, S. D., M. T. Finley, and E. Cromartie. 1980. Reproduction and residue accumulation in black ducks fed toxaphene. Arch. Environ. Contam. Toxicol. 9:461-471.

Havet, P. 1973. Effets de l'heptachlore sur la reproduction de la perdrix rouge *Alectoris rufa* L. et de la perdrix grise *Perdix perdix* L. (In French.) Trans. Int. Congr. Game Biol. 10:175-183.

Heath, R. G., and J. W. Spann. 1973. Reproduction and related residues in birds fed mirex. p. 421-435. *In* W. B. Deichmann (Ed.). Pesticides and the environment: a continuing controversy. Eighth Inter-American Conference on Toxicology and Occupational Medicine. Intercontinental Medical Book Corp., New York.

Heinz, G. H., and M. T. Finley. 1978. Toxaphene does not affect avoidance behavior of young black ducks. J. Wildl. Manage. 42:408-409.

Henny, C. J., L. J. Blus, and T. E. Kaiser. 1984. Heptachlor seed treatment contaminates hawks, owls, and eagles of Columbia Basin, Oregon. Raptor Res. 18:41-48.

Henny, C. J., L. J. Blus, and C. J. Stafford. 1983. Effects of heptachlor on American kestrels in the Columbia Basin, Oregon. J. Wildl. Manage. 47:1080-1087.

Hoffman, D. J. 1990. Embryotoxicity and teratogenicity of environmental contaminants to bird eggs. Rev. Environ. Contam. Toxicol. 115:39-89.

Hoffman, D. J., and J. M. Moore. 1979. Teratogenic effects of external egg applications of methyl mercury in the mallard, *Anas platyrhynchos*. Teratology 20:453-462.

Huggett, R. J., and M. E. Bender. 1980. Kepone in the James River. Environ. Sci. Technol. 14:918-923.

Hyde, K. M., J. B. Graves, A. B. Watts, and F. L. Bonner. 1973. Reproductive success of mallard ducks fed mirex. J. Wildl. Manage. 37:479-484.

Krynitsky, A. J., C. J. Stafford, and S. N. Wiemeyer. 1988. Combined extraction-cleanup column chromatographic procedure for determination of dicofol in avian eggs. J. Assoc. Off. Anal. Chem. 71:539-542.

Lillie, R. J., H. C. Cecil, and J. Bitman. 1973. Methoxychlor in chicken breeder diets. Poult. Sci. 52:1134-1138.

Ludke, J. L. 1976. Organochlorine pesticide residues associated with mortality: additivity of chlordane and endrin. Bull. Environ. Contam. Toxicol. 16:253-260.

Maier-Bode, H. 1968. Properties, effect, residues and analytics of the insecticide endosulfan. Residue Rev. 22:1-44.

McFarland, L. Z., and P. B. Lacy. 1969. Physiologic and endocrinologic effects of the insecticide Kepone in the Japanese quail. Toxicol. Appl. Pharmacol. 15:441-450.

Mehrle, P. M., M. T. Finley, J. L. Ludke, F. L. Mayer, and T. E. Kaiser. 1979. Bone development in black ducks as affected by dietary toxaphene. Pestic. Biochem. Physiol. 10:168-173.

Naber, E. C., and G. W. Ware. 1965. Effect of Kepone and mirex on reproductive performance in the laying hen. Poult. Sci. 44:875-880.

Pollock, G. A., and W. W. Kilgore. 1978. Toxaphene. Residue Rev. 69:87-140.

Robinson, J. 1969. Organochlorine insecticides and bird populations in Britain. p. 113-173. *In* M. W. Miller and G. G. Berg (Eds.). Chemical fallout: current research on persistent pesticides. Charles C Thomas Publisher, Springfield, Ill.

Romanoff, A. L., and A. J. Romanoff. 1949. The avian egg. John Wiley & Sons, New York, 918 pp.

Schwarzbach, S. E. 1991. The role of dicofol metabolites in the eggshell thinning response of ring neck doves. Arch. Environ. Contam. Toxicol. 20:200-205.

Schwarzbach, S. E., L. Shull, and C. R. Grau. 1988. Eggshell thinning in ring doves exposed to p,p'-dicofol. Arch. Environ. Contam. Toxicol. 17:219-227.

Schwetz, B. A., J. M. Norris, R. J. Kociba, P. A. Keeler, R. F. Cornier, and P. J. Gehring. 1974. Reproduction study in Japanese quail fed hexachlorobutadiene for 90 days. Toxicol. Appl. Pharmacol. 30:255-265.

Sileo, L., L. Karstad, R. Frank, M. V. H. Holdrinet, E. Addison, and H. E. Braun. 1977. Organochlorine poisoning of ring-billed gulls in southern Ontario. J. Wildl. Dis. 13:313-322.

Smith, G. J. 1987. Pesticide use and toxicology in relation to wildlife: organophosphorus and carbamate compounds. U.S. Fish Wildl. Serv. Res. Publ. No. 170, 171 pp.

Stafford, C. J., W. L. Reichel, D. M. Swineford, R. M. Prouty, and M. L. Gay. 1978. Gas-liquid chromatographic determination of Kepone in field-collected avian tissues and eggs. J. Assoc. Off. Anal. Chem. 61:8-14.

Stickel, L. F., W. H. Stickel, R. A. Dyrland, and D. L. Hughes. 1983. Oxychlordane, HCS-3260, and nonachlor in birds: lethal residues and loss rates. J. Toxicol. Environ. Health 12:611-622.

Stickel, L. F., W. H. Stickel, R. D. McArthur, and D. L. Hughes. 1979. Chlordane in birds: a study of lethal residues and loss rates. p. 387-396. *In* W. B. Diechmann (Organizer). Toxicology and occupational medicine. Elsevier, North Holland, New York.

Stickel, W. H., J. A. Galyen, R. A. Dyrland, and D. L. Hughes. 1973. Toxicity and persistence of mirex in birds. p. 437-467. *In* W. B. Deichmann (Ed.). Pesticides and the environment: a continuing controversy. Eighth Inter-American Conference on Toxicology and Occupational Medicine. Intercontinental Medical Book Corp., New York.

Stone, W. B., and J. C. Okoniewski. 1988. Organochlorine pesticide-related mortalities of raptors and other birds in New York, 1982-1986. p. 429-438. *In* T. J. Cade, J. H. Enderson, C. G. Thelander, and C. M. White (Eds.). Peregrine falcon populations: their management and recovery. Peregrine Fund, Boise, Ida.

Turtle, E. E., A. Taylor, E. N. Wright, R. J. P. Thearle, H. Egan, W. H. Evans, and N. M. Soutar. 1963. The effects on birds of certain chlorinated insecticides used as seed dressings. J. Sci. Food Agric. 14:567-577.

U.S. Department of Agriculture. 1980. The biologic and economic assessment of lindane. U.S. Dep. Agric. Tech. Bull. 1647, 196 pp.

U.S. Fish and Wildlife Service. 1982. The Chesapeake Bay region bald eagle recovery plan. U.S. Fish and Wildlife Service, Region 5, 81 pp.

Vos, J. G., P. F. Botterweg, J. J. T. W. A. Strik, and J. H. Koeman. 1972. Experimental studies with HCB in birds. TNO Nieuws 27:599-603.

Vos, J. G., H. A. Breeman, and H. Benschop. 1968. The occurrence of the fungicide hexachlorobenzene in wild birds and its toxicological importance. A preliminary communication. Meded. Rijksfac. Landbouwwet. Gent. 33:1263-1269.

Vos, J. G., H. L. van der Maas, A. Musch, and E. Ram. 1971. Toxicity of hexachlorobenzene in Japanese quail with special reference to porphyria, liver damage, reproduction, and tissue residues. Toxicol. Appl. Pharmacol. 18:944-957.

Wiemeyer, S. N., C. M. Bunck, and A. J. Krynitsky. 1988. Organochlorine pesticides, polychlorinated biphenyls, and mercury in osprey eggs — 1970-79 — and their relationships to shell thinning and productivity. Arch. Environ. Contam. Toxicol. 17:767-787.

Wiemeyer, S. N., T. G. Lamont, C. M. Bunck, C. R. Sindelar, F. J. Gramlich, J. D. Fraser, and M. A. Byrd. 1984. Organochlorine pesticide, polychlorobiphenyl, and mercury residues in bald eagle eggs — 1969-79 — and their relationships to shell thinning and reproduction. Arch. Environ. Contam. Toxicol. 13:529-549.

Wiemeyer, S. N., J. W. Spann, C. M. Bunck, and A. J. Krynitsky. 1989. Effects of Kelthane® on reproduction of captive eastern screech-owls. Environ. Toxicol. Chem. 8:903-913.

CHAPTER 5

PCBs in Aquatic Organisms

Arthur J. Niimi

INTRODUCTION

Polychlorinated biphenyls (PCBs) include a group of monochloro- to decachlorinated compounds with a biphenyl nucleus. There are 209 congeners of PCB because of the 10 possible substitution positions on the nucleus with a chlorine or hydrogen atom, although less than 100 congeners would be of environmental or toxicological significance because of their low concentrations (Hutzinger et al., 1974; Hansen, 1987). Commercial PCB mixtures are identified by their trade names, such as Aroclor, Clophen, and Kanechlor, and a number that may indicate their percentage of chlorine content by weight. These mixtures can contain over 140 PCB congeners (Schulz et al., 1989). About 2.2 to 3.3 million tons (equal to 1 to 1.5 million tonnes) of PCBs have been produced for a wide range of industrial applications because of such properties as resistance to breakdown by other chemicals; high thermal stability; and low vapor pressure, flammability, and solubility (WHO, 1976; de Voogt and Brinkman, 1989). At least 660,000 tons (equal to 300,000 tonnes) of PCBs have been released to the environment in North America, although no reliable estimates are available on accidental and incidental releases of PCBs on a worldwide basis (Hansen, 1987). These factors have largely contributed to the ubiquitous distribution of PCBs in the atmospheric, terrestrial, and aquatic environments. The production and use of PCBs were curtailed in the 1970s when they were identified as a persistent chemical in wildlife (Jensen, 1972). These restrictions have resulted in a decline of PCB concentrations in feral animals since the 1970s, when concentrations in excess of 100 mg/kg were reported in fish from highly contaminated systems (e.g., Brown et al., 1985). This decline was most evident during the 1970s and early 1980s and less dramatic thereafter (Brown et al., 1985; Stout, 1986; Borgmann and Whittle, 1991). Recent surveys have reported PCB concentrations in the low-mg/kg range in some aquatic organisms (Gundersen and Pearson, 1992; Williams and Giesy, 1992).

PCBs IN THE AQUATIC ENVIRONMENT

PCB concentrations in water are due to their proximity to a contamination source, but they can be influenced by several factors. PCBs are hydrophobic compounds with octanol water partition coefficients (K_{ow}) that range from log 4.40 for monochloro- to log 8.18 for decachlorobiphenyl (Rapaport and Eisenreich, 1984; Hawker and Connell, 1988). K_{ow} values for commercial mixtures of log 5.58 for Aroclor 1242 and log 6.47 for Aroclor 1254 have been reported (Veith et al., 1979). Aqueous solubilities range from 1 to 5 mg/l for monochlorobiphenyls to the low-µg/l range or less for more highly chlorinated congeners (Opperhuizen et al., 1988; Patil, 1991). Solubilities of 277 µg/l for Aroclor 1242 and 43 µg/l for Aroclor 1254 have been reported (Murphy et al., 1987). It would be unlikely that dissolved concentrations near their solubilities would be found even in highly contaminated systems, because of the hydrophobic behavior of PCBs with their affinity to adsorb to suspended particulate materials like sediment and biota (Hiraizumi et al., 1979; Larsson et al., 1992).

Waterborne PCB concentrations in aquatic ecosystems with no apparent source of local contamination are generally in the low-ng/l range in freshwater and marine coastal waters and in the pg/l range in open oceanic waters (Table 1). Long-range atmospheric transport is the likely source for PCBs that are reported in remote waters and for background concentrations in nearly all environmental matrices (Atlas et al., 1986; Bidleman et al., 1989). PCB deposition in the Great Lakes from the atmosphere may represent about 60% of the total input in Lake Michigan and 85% in Lake Superior (Eisenreich et al., 1981). The highest concentrations of waterborne PCBs are found in rivers that receive point-source discharge with concentrations in the 50- to 500-ng/l range (Tanabe et al., 1989; El-Gendy et al., 1991). The flux of PCBs between the aquatic and atmospheric environments is a dynamic process where contaminated systems can lose more PCBs to the atmosphere through volatilization than are received from aerial deposition (Larsson et al., 1990; Achman et al., 1993). The highest PCB concentrations are often found in riverine and estuarine sediments, where levels can exceed 10 mg/kg, and in hot spots such as the Detroit and Hudson rivers. The New Bedford Harbor can reach 1 g/kg (Brown et al., 1985; Furlong et al., 1988; Lake et al., 1992). Waterborne concentrations of PCBs in marine systems near point-source discharges have been reported in the low-ng/l range or lower because they are generally associated with coastal waters where dilution is rapid.

PCB concentrations reported in aquatic organisms can vary by a factor of 10^5, depending on species and sampling sites (Table 2). PCB concentrations in plankton from relatively uncontaminated waters are in the low-µg/kg range, while those from contaminated water can be higher by a factor of 10 (Kawano et al., 1986; Oliver and Niimi, 1988). Similar differences in PCB concentrations among larger invertebrates are suggested between uncontaminated and more contaminated waters (Ramesh et al., 1990; O'Conner, 1991). The largest differences in PCB concentrations are found among carnivorous fish at the higher trophic levels, where concentrations in the low-µg/kg range are found in fish from uncontaminated waters compared with those in the low-mg/kg range from more contaminated systems (Niimi and Oliver, 1989a; Muir et al., 1992).

Table 1 Polychlorinated Biphenyl (PCB) Concentrations in Water Reported from Relatively Uncontaminated and Contaminated Freshwater and Marine Ecosystems

Location	PCB concentration, mean or range	Ref.
Freshwater Ecosystems with No Known Source of Local Contamination		
Lake Narume, Antarctica	0.048 ng/l	Tanabe et al. (1983)
Hudson River drainage, Quebec	<9 ng/l	Langlois (1987)
Five rivers, northern Ontario	10-14 ng/l	McCrea and Fischer (1986)
Siskiwit Lake, Lake Superior	2 ng/l	Swackhamer et al. (1988)
Marine Ecosystems with No Known Source of Local Contamination		
Antarctic Ocean	42-72 pg/l	Tanabe et al. (1983)
Arctic Ocean	<2-6 pg/l	Hargrave et al. (1992)
North Sea	≤2-≤40 pg/l near- and offshore	Schulz-Bull et al. (1991)
Atlantic Ocean	2-21 pg/l in north Atlantic	Schulz et al. (1988)
Atlantic Ocean	<3 pg/l off U.S. east coast	Sauer et al. (1989)
Gulf of Mexico	<3 pg/l	Sauer et al. (1989)
Pacific Ocean	40-590 pg/l in northwestern Pacific	Tanabe et al. (1984)
Pacific Ocean and Bering Sea	67 and 92 pg/l, respectively	Kawano et al. (1986)
Freshwater Ecosystems with Suspected Source of Local Contamination		
River Em, Sweden	About 5-50 ng/l	Larsson et al. (1990)
River Seine, France	13-190 ng/l at various locations	Chevreuil et al. (1987)
Rio de La Plata, Argentina	<5-75 ng/l	Colombo et al. (1990)
Yodo River, Japan	100-200 ng/l	Tanabe et al. (1989)
Nile River, Egypt	8-650 ng/l at various locations	El-Gendy et al. (1991)
Hudson River, NY	540 ng/l in 1977, 130 ng/l in 1981	Sloan et al. (1983)
Shiawasse River, Mich.	30-1100 ng/l due to dredging	Rice and White (1987)
Lake Butte des Morts, Wis.	1-34 ng/l	Crane and Sonzogni (1992)
Lake Michigan	1 ng/l in offshore waters	Swackhamer and Armstrong (1987)
Lake Superior	1-4 ng/l	Capel and Eisenreich (1985)
Lake Ontario	1 ng/l in offshore waters	Oliver and Niimi (1988)
Marine Ecosystems with Suspected Source of Local Contamination		
Dutch Wadden Sea	0.62 ng/l	Duinker and Hillebrand (1983)
Puget Sound, Wash.	3-22 ng/l at various sites	Pavlou and Dexter (1979)
Mediterranean Sea	<2-11 ng/l in western coastal waters	Marchand et al. (1988)
English Channel and North Sea	<2-39 ng/l in coastal waters	Marchand and Caprais (1985)
New Bedford Harbor, Mass.	~2-70 ng/l in tidal estuary	Connolly (1991)

Table 2 Polychlorinated Biphenyl (PCB) Concentrations in Aquatic Organisms from Relatively Uncontaminated and Contaminate Freshwater and Marine Ecosystems. (Concentrations are Reported on a Wet-Weight, Whole-Body Basis, Unless Noted Otherwise)

System/organism	PCB concentration, mean or range	Ref.
Organisms from Freshwater Ecosystems with No Known Source of Local Contamination		
Algae, Saône River, France	0.29 mg/kg of dry weight in moss	Mouvet et al. (1985)
Fish, Labrador, Canada	<5 µg/kg in muscle among seven species from five lakes	Lockerbie and Clair (1988)
Fish, Northwest Territories, Canada	3-22 µg/kg in whitefish from various waters	Lockhart et al. (1992)
Fish, Alberta, Canada	6-130 µg/kg in fat of three species from three lakes	Chovelon et al. (1984)
Fish, Ontario, Canada	10-40 µg/kg in pike from five northern rivers	McCrea and Fischer (1986)
Fish, Sweden	1-50 µg/kg in pike from 61 lakes	Larsson et al. (1992)
Fish, Finland	42 µg/kg in muscle of pike from 13 northern lakes	Pyysalo et al. (1983)
Organisms from Marine Ecosystems with No Known Source of Local Contamination		
Plankton, Arctic Ocean	~1-23 µg/kg in three size classes	Hargrave et al. (1992)
Plankton, Arctic Ocean	2-27 µg/kg of dry weight in zooplankton	Bidleman et al. (1989)
Plankton, Pacific Ocean and Bering Sea	9 µg/kg in zooplankton	Kawano et al. (1986)
Mollusc, U.S. coastal waters	4-6 µg/kg in mussel from 10 sites	O'Conner (1991)
Mollusc, Pacific Ocean and Bering Sea	17 µg/kg in squid	Kawano et al. (1986)
Crustacea, Arctic Ocean	480-3000 µg/kg of dry weight in amphipod	Bidleman et al. (1989)
Fish, Antarctic Ocean	0.2-0.5 µg/kg in three species	Subramanian et al. (1983)
Fish, Arctic Ocean	1-45 µg/kg in charr	Muir et al. (1992)
Fish, Pacific Ocean and Bering Sea	16 µg/kg in salmon	Kawano et al. (1986)
Fish, Indian Ocean	14-38 µg/kg in muscle of coelacanth	Hale et al. (1991)
Fish, Northwest Atlantic Ocean	~40-90 µg/kg in muscle of tilefish	Steimle et al. (1990)
Organisms from Freshwater Ecosystems with Suspected Source of Local Contamination		
Algae, Lake Geneva	<6 µg/kg in *Cladophora*	Mowrer et al. (1982)
Algae, Lake Huron	92 and 126 µg/kg of dry weight in two species	Anderson et al. (1982)
Plankton, Lake Geneva	17 µg/kg in primarily zooplankton	Mowrer et al. (1982)
Plankton, Lake Huron	1650 µg/kg of dry weight	Anderson et al. (1982)
Plankton, Lake Ontario	50 µg/kg	Oliver and Niimi (1988)

Table 2 (continued) Polychlorinated Biphenyl (PCB) Concentrations in Aquatic Organisms from Relatively Uncontaminated and Contaminate Freshwater and Marine Ecosystems. (Concentrations are Reported on a Wet-Weight, Whole-Body Basis, Unless Noted Otherwise)

System/organism	PCB concentration, mean or range	Ref.
Crustacea, Lake Ontario	790 µg/kg in amphipod	Oliver and Niimi (1988)
Crustacea, Lake Michigan	500 µg/kg of dry weight in *Mysis*	Evans et al. (1991)
Fish, Lake Michigan	23 mg/kg in 1974, 6 mg/kg in 1982 in lake trout	DeVault et al. (1986)
Fish, Hudson River, N.Y.	18 mg/kg in 1978, 5 mg/kg in 1983 in bass fillet	Brown et al. (1985)
Fish, Nagaragawa River, Japan	14 mg/kg in 1968, 0.1 mg/kg in 1986 in gobi	Loganathan et al. (1989)
Fish, Cayuga Lake, N.Y.	1-30 mg/kg in lake trout	Bache et al. (1972)
Fish, Elbe River, Germany	360-590 µg/kg in bream from eight sites	Luckas and Oehme (1990)
Fish, Lake Geneva	1-3 mg/kg in brown trout	Rossel et al. (1987)
Fish, Lake Ontario	2-10 mg/kg in four salmonid species	Niimi and Oliver (1989a)
Fish, Lake Vättem, Sweden	12-19 mg/kg in lipid of salmon	Andersson et al. (1988)
Organisms from Marine Ecosystems with Suspected Source of Local Contamination		
Algae, Adriatic Sea	13-120 µg/kg of dry weight from lagoon of Venice	Pavoni et al. (1990)
Plankton, North Sea	0.2-28 mg/kg of lipid in phyto- and zooplankton	Knickmeyer and Steinhart (1989)
Plankton, Puget Sound, Wash.	2-16 mg/kg of lipid in zooplankton	Pavlou and Dexter (1979)
Mollusc, Northern Atlantic Ocean	1-6 µg/kg in scallop from Wolstenhdme Fjord, Greenland	Kjølholt and Hansen (1986)
Mollusc global monitoring	0.2-530 µg/kg in mussels from 16 international sites	Ramesh et al. (1990)
Crustacea, Dutch Wadden Sea	100 µg/kg in shrimp	Duinker and Hillebrand (1983)
Crustacea, Atlantic coastal waters	91 µg/kg in crab from South Carolina estuaries	Marcus and Renfrow (1990)
Fish, Glomma estuary, Norway	6-8320 µg/kg in liver of two species	Marthinsen et al. (1991)
Fish, San Francisco Bay estuary	2 mg/kg in adult striped bass	Setzler-Hamilton et al. (1988)
Fish, Atlantic coastal waters	2 mg/kg in bluefish fillet from U.S. coast	Sanders and Haynes (1988)
Fish, New Bedford Harbor, Mass.	1-10 mg/kg in flounder	Connolly (1991)
Fish, Long Island Sound, N.Y.	2-15 mg/kg in striped bass	Bush et al. (1989)

The higher PCB concentrations in organisms at higher trophic levels are largely attributable to biomagnification. Laboratory studies have reported bioconcentration factors (BCF) in fish that were greater than log 5.35 for several Aroclors (Nebeker et al.,

1974; DeFoe et al., 1978). Nevertheless, waterborne uptake is not an important pathway for most aquatic organisms because of the low concentrations of dissolved PCBs relative to those in food, where the differences can exceed a factor of 10^5 (Niimi, 1985; Oliver and Niimi, 1988). Contaminant dynamics models and trophodynamic examination of PCB congener distribution patterns in water and organisms at different trophic levels have also indicated that dietary uptake is the more important pathway (Thomann, 1981; Oliver and Niimi, 1988). Some studies also indicate that uptake from contaminated sediments is another pathway for benthic species (Stein et al., 1987).

PCBs AND HEALTH EFFECTS

Mammalian toxicological studies generally conclude that PCBs are not mutagenic or genotoxic (Zeiger, 1989). Most studies indicate that PCBs are not carcinogenic inducers, but are good promoters of hepatocellular carcinomas, neoplastic nodules, and preneoplastic lesions (Silberhorn et al., 1990). PCBs are immunotoxins that affect specific and nonspecific defense mechanisms through thymic and splenic atrophy (Vos and Luster, 1989). There is a reduction of circulating lymphocytes, leukocytes, and natural killer cells; suppression of antibody responses; and increased susceptibility to viral infections (Hansen, 1987). PCBs are teratogens and have other effects on reproduction, such as changes in hormone levels, infertility, increased embryo and fetal mortality, an increased rate of abortion, and lower birth weight (Morrissey and Schwartz, 1989). Other effects include impaired behavioral responses, altered catecholamine levels, numbness in the extremities, gastric hyperplasia and ulceration, decreased respiratory capacity, dermal hyperplasia, and edema (Hansen, 1987).

The increasing use of halogenated chemicals since the 1940s and the increasing frequency of pathological anomalies reported in feral animals have presented strong circumstantial evidence for a chemically based etiology. The effects of some organochlorine pesticides on the reproductive success of avian raptors have been shown through a series of laboratory and field studies (Cooke, 1973). The case for feral aquatic organisms is less clear at this time, because most of the evidence is based on statistical correlations between anomaly frequency and chemical concentrations rather than on studies that examine cause-and-effect relations. This approach can result in a large degree of uncertainty in resolving issues. Nevertheless, many investigators suspect that chemicals are contributing factors to some deleterious effects observed in aquatic animals. Recent studies have indicated that polynuclear aromatic hydrocarbons (PAHs) are important toxicants to benthic organisms (Bender et al., 1988; Myers et al., 1991). This conclusion for PAHs cannot yet be extended to nonbenthic organisms, to other chemical groups found in the sediment, or to other matrices in the aquatic ecosystem.

This study examines the adverse effects of PCBs on aquatic organisms and assesses the toxicological significance of PCB concentrations that are observed in feral organisms.

EVALUATION OF HEALTH EFFECTS IN AQUATIC ORGANISMS

An organism exposed to a chemical above its threshold level may respond through a series of general symptoms, in addition to other biological and physical perturbations. The response could be direct and result in the death of the organism, or it could be less discreet, such as a biochemical dysfunction that can inhibit reproduction and affect the species (Niimi, 1990). To establish which perturbation was responsible is sometimes difficult, because each has the capability to cause these effects. The role of other vectors is sometimes not considered in field studies and cannot be easily rejected in some cases, because that may induce the same or similar anomalies that do not have a chemical etiology (Mix, 1986; Watermann and Kranz, 1992). For example, PAHs have been shown to be potent inducers of carcinomas in fish (Hawkins et al., 1990). Some tumors and papillomas in fish can also be caused by viruses, protozoans, and parasites (Sonstegard, 1977; Grizzle et al., 1981; Möller, 1988; Kirby and Hayes, 1992). Indices associated with poor water quality, such as elevated temperature and low oxygen concentration, may be the cause of an absence of biota from a site rather than chemical contamination. These physical factors can reduce the food source available to fish and increase the incidence of disease in waters where chemicals are also present (Möller, 1990). Hence, the presence of other limiting factors associated with a chemically contaminated environment cannot be ignored.

The effects of PCBs at concentrations that are found in feral aquatic organisms would be difficult to assess directly because of a number of factors. There are no specific clinical symptoms that are associated with PCB-induced toxicity in aquatic organisms. Other natural and anthropogenic organic and inorganic chemicals are invariably present in the organism and its environment. The presence of other chemicals is an important factor that is often recognized but not effectively addressed, because only a few chemicals are monitored in field studies. For example, increased levels of hepatic aryl hydrocarbon hydroxylase activity were reported in trout that corresponded with the level of organic chemical contamination in the Great Lakes (Luxon et al., 1987). A number of chemicals such as PCBs, chlorinated dioxins and furans, and PAHs have similar mechanisms of action that include the activation of aryl hydrocarbon hydroxylase (Niimi, 1994). The presence of these and many other xenobiotic chemicals in Great Lakes Basin biota has been well established (Kuehl et al., 1984). Even when a specific chemical is suspected, its specific role cannot be easily assessed because of chemical interactions. Mammalian studies have shown that PCBs and 2,3,7,8-tetrachlorodibenzo-p-dioxin can act in a synergistic and antagonistic manner (Bannister and Safe, 1987; Bannister et al., 1987). One of the more comprehensive epidemiological studies of the effects of chemicals on fish has been conducted on benthic species in Puget Sound, Washington. Statistically significant correlations have been shown between levels of aromatic hydrocarbons in sediment and their metabolites in fish bile and the prevalence of hepatic lesions (Myers et al., 1990). Increased frequencies of gonadal recrudescence and lower plasma estradiol levels have also been associated with increased concentrations of aromatic hydrocarbons in sediment and tissues (Johnson et al., 1988). These studies illustrate the difficulties in examining cause-and-effect relations for a specific chemical in the natural environment because of the presence of other chemicals. Yet a study that would include monitoring most of the chemicals present may not be feasible, because over 160 compounds have been identified in lake trout from the Great Lakes, and 900 organic chemicals have been found in sediment from Puget Sound (Hesselberg and Seelye, 1977; Malins et al., 1984).

A number of field studies have reported increased mortality, pathological anomalies, and biochemical changes in aquatic organisms from ecosystems where PCBs have been reported (Table 3). These observations include poor survival of eggs and larvae taken from feral organisms and reared in the laboratory (Stauffer, 1979; Monod, 1985). Gross histological anomalies include ovarian atresia and hepatocellular lesions (Cross and Hose, 1988; Gardner et al., 1989; McCain et al., 1992). Other observations include increased oxidase enzyme activity levels and changes in cellular structure (Köhler, 1990; Galgani et al., 1991). Some effects, such as mortality at the early life-stages, are clearly deleterious; others, such as the presence of neoplastic lesions, would be suspected to be deleterious; and the effects of changes at the cellular and biochemical levels would be more difficult to assess.

PCBs are found in some aquatic organisms at concentrations that exceed 2 mg/kg, which is the guideline level for fish and related consumable products set by the Canadian and U.S. regulatory agencies for the protection of human health. Some information is available on the toxicological implications of this concentration on mammals, but little is known on aquatic organisms. Some studies have compiled health-related observations and concentrations of organochlorine chemicals in aquatic life, but the toxicological implications of these observations are speculative because of the presence of other chemicals and other factors (Gilbertson, 1989). Nevertheless, an assessment of the toxicological significance of PCBs in aquatic organisms can be done by comparing PCB concentrations in organisms collected from the natural environment with those from laboratory studies that examine the toxic and chronic effects of PCBs on similar organisms. Measurements of PCBs in water and aquatic organisms have been compiled to indicate the range of concentrations in relatively uncontaminated and contaminated freshwater and marine ecosystems. Some values are dated and not representative of recent levels; they do, however, provide a retrospective view of PCB concentrations in highly contaminated systems. These results are compared with results from short-term studies on acute toxicity and behavioral responses that are often concluded within a few days and from long-term studies that examine chronic effects. Long-term studies would examine whole-organism responses and the effects on growth and reproduction, including probable responses at the cellular and biochemical levels. Laboratory studies may also indicate the threshold waterborne or body residue concentration at which adverse effects occur. Collectively, these results were used to evaluate the toxicological significance of PCB concentrations that are observed in feral organisms.

Table 3 Field and Field-Laboratory Studies Reporting Adverse or Anomalous Effects or Biochemical Changes in Aquatic Organisms that Contain Polychlorinated Biphenyls (PCBs) or Inhabit Waters Known to Contain PCBs (Concentrations of PCBs in the Tissues Reported in the Study, or Estimated from the Data Reported, are Expressed on a Wet-Weight Basis, unless Noted Otherwise)

Response/organism	Observations	Ref.
Reproduction		
Fish	Mortalities of 5-75% reported for Lake Geneva charr eggs with 0.10-0.50 mg/kg of PCBs and 0.04-0.17 mg/kg of DDT; further examination indicated no significant relation between PCB concentration and percentage of mortality.	Monod (1985)
Fish	Viability below 50% in Baltic Sea flounder eggs when PCBs exceeded 120 μg/kg in ovary; further examination indicated hatching was also below 50% among eggs from 18 of 34 ovaries that contained 10-120 μg/kg of PCBs; ovaries also contained 3-92 μg/kg of DDT, and 0.1-4.9 μg/kg of dieldrin.	von Westernhagen et al. (1981)
Fish	Mortality of 75% in rainbow trout fry 30 days posthatch from hatchery stock; eggs contained 2.7 mg/kg of PCBs, and 0.09 mg/kg of DDT; no control fish were reared for comparison.	Hogan and Brauhn (1975)
Fish	Mortality of 80% in Lake Michigan trout fry with 3-10 mg/kg of PCBs and 1-5 mg/kg of DDT, compared with 46% mortality among fish from hatchery; study concluded that these chemicals did not influence fry survival.	Stauffer (1979)
Fish	Mortalities of 3-98% in chinook salmon egg-fry with 3-14 mg/kg of PCBs; no significant correlation shown between PCB concentration and mortality.	Williams and Giesy (1992)
Fish	Mortalities ≥50% among embryos of whiting from North Sea; gonads had 0.1-85 mg/kg of PCBs on a lipid basis and up to 2 mg/kg of DDE.	Cameron et al. (1988)
Fish	Decrease of 17% in weight and 24% in length in flounder larvae with 40 mg/kg of PCBs (dry weight) compared to larvae with 1 mg/kg in northeastern U.S. coastal waters; significant correlation between PCB concentration and fish size content.	Black et al. (1988)
Fish	Survival of striped bass larvae after 30 days about 60% in eggs with 2 mg/kg of PCBs, and 30% among eggs with 9 mg/kg; eggs also had 0.4-1.2 mg/kg of DDT.	Westin et al. (1985)
Histological observations		
Fish	Ovarian atresia in 55-88% of sole from Puget Sound, Wash., with 0.23-0.54 mg/kg of PCBs in ovaries; study concluded that increased PCB concentrations were not associated with decreased reproductive success.	Collier et al. (1992)
Fish	Oocyte atresia increased about 13% in croaker with 2 mg/kg of PCBs in ovary in southern California coastal waters, compared to fish with 0.2 mg/kg from the control site.	Cross and Hose (1988)
Fish	Hepatic neoplastic and nonneoplastic disorder frequencies of 0-79% in flounder from northeastern U.S. coastal waters; causal relation suggested between disorder frequency and degree of sediment contamination with PCBs, polynuclear aromatic hydrocarbons (PAHs), and trace metals.	Gardner et al. (1989)

Table 3 (continued) Field and Field-Laboratory Studies Reporting Adverse or Anomalous Effects or Biochemical Changes in Aquatic Organisms that Contain Polychlorinated Biphenyls (PCBs) or Inhabit Waters Known to Contain PCBs (Concentrations of PCBs in the Tissues Reported in the Study, or Estimated from the Data Reported, are Expressed on a Wet-Weight Basis, unless Noted Otherwise)

Response/organism	Observations	Ref.
Fish	Hepatic lesion frequencies up to 70% in flounder from Elbe estuary, Germany, with ≤3 mg/kg of PCBs in muscle; ultrastructural changes hepatic organelles and cell types related to degree of exposure.	Köhler (1990)
Fish	Hepatic lesion frequencies up to 10% in liver of croaker from San Diego Bay, Calif., with 12 mg/kg of PCBs in liver, compared to fish with no lesions that had 1 mg/kg.	McCain et al. (1992)
Fish	Hepatic lesion frequencies increased in croaker from coastal sites near Los Angeles, Calif., although no statistical correlation was indicated between PCB concentrations of 0.9-23 mg/kg (dry weight) in liver and lesion frequency; fish also had high DDT and PAH concentrations.	Myers et al. (1991)
Fish	Hepatic lesions in tomcod were found in fish with 39 mg/kg of PCBs in liver, while livers of fish without tumors had 33 mg/kg from northeastern U.S. coastal waters.	Smith et al. (1979)

Biochemical observations

Fish	Increased ethoxyresorufin O-deethylase activity in plaice from Seine Bay, France, significantly correlated with concentrations of 0.5-3.4 mg/kg of PCBs.	Galgani et al. (1991)
Fish	Increased aryl hydrocarbons hydroxylase activity in Great Lakes lake trout attributed to increasing concentrations of 0.2-3 mg/kg PCBs.	Luxon et al. (1987)
Fish	Increased aryl hydrocarbon hydroxylase activity in two benthic species with increased PCB content in coastal waters of California.	Spies et al. (1982)

Other observations

Arthropod	Population densities lower, increased frequency of deformities, in *Chironomus* larvae correlated with increases in sediment concentrations of PCBs, PAHs, and heavy metals along the Rhine River.	van Urk et al. (1992)
Fish	Goiter frequencies of 6-80% in coho salmon from Lakes Ontario, Michigan, and Erie not correlated with concentrations of 1-8 mg/kg PCBs. No goiter observed in chinook salmon from these lakes.	Moccia et al. (1977)
Fish	Thyroid hyperplasia in trout reported for Lake Michigan fish compared with those from California.	Robertson and Chaney (1953)
Fish	Thyroid hypoplasia in salmon reported for Lake Michigan fish compared to those from Oregon.	Drongowski et al. (1975)
Fish	Vertebral strength in bass lower among Hudson River fish than hatchery fish with lower PCB content.	Mehrle et al. (1982)
Fish	Erythrocyte micronuclei frequencies increased in two species with elevated PCB and DDT concentrations off southern California.	Hose et al. (1987)
Fish	Fin erosion frequency up to ~20% among three species in San Diego Bay, Calif., in fish with 12 mg/kg of PCBs in liver compared to fish with 1 mg/kg with no erosion. Fish had elevated concentrations of other aromatic compounds.	McCain et al. (1992)

LABORATORY STUDIES ON PCBs

Whole-animal studies on aquatic organisms generally indicate that PCBs can cause adverse effects at low-µg/l waterborne concentrations and low-mg/kg tissue concentrations (Tables 4 and 5). Lethality studies on zooplankton indicate that concentrations of >10 µg/l cause death within a few days and that 1 to 10 µg/l over longer exposure periods can cause death for *Daphnia,* although no lethal tissue concentrations could be estimated (Nebeker and Puglisi, 1974; Branson, 1977). Larger invertebrates, such as shrimp and oysters, appear more tolerant to PCBs where waterborne and tissue concentrations of >10 µg/l and >25 mg/kg may be lethal (Duke et al., 1970; Nimmo et al., 1974). Estimates for fish indicate that waterborne concentrations of >10 µg/l are lethal within a few days and that concentrations of >1 µg/l over longer periods are lethal for minnows (Nebeker and Puglisi, 1974; DeFoe et al., 1978). Long-term waterborne and dietary exposure studies on several species indicate lethal body burdens of >100 mg/kg for young fish and >250 mg/kg for older fish (Hattula and Karlog, 1972; Mayer et al., 1977; Mauck et al., 1978). Acute oral toxicity of PCBs to fish is low where dietary concentrations of 300 mg/kg/day were required to cause death in trout (Mayer et al., 1977).

Growth-related activities in phytoplankton, such as cell division and photosynthesis, can be reduced by <10 µg/l of PCBs (Keil et al., 1971; Moore and Harriss, 1972). Algal species, such as *Chlamydomonas* and *Chlorella,* appear less susceptible than diatoms to PCBs, where somewhat higher concentrations may affect growth (Urey et al., 1976; Christensen and Zielski, 1980). Adverse effects on growth and development among macroinvertebrates occur at 1 to 5 µg/l of PCBs after longer periods of exposure, where PCB concentrations in organisms can exceed 30 mg/kg (Sanders and Chandler, 1972; Nimmo et al., 1975). Long-term exposure studies on trout generally indicate that tissue concentrations of <100 mg of PCBs per kilogram may not affect growth (Lieb et al., 1974; Mayer et al., 1977; Mayer et al., 1985). Some studies have reported a small reduction in growth among trout with >100 mg of PCBs per kilogram, while others reported a 40% reduction among fish fed diets containing >300 mg of PCBs per kilogram for 90 to 365 days (Leatherland and Sonstegard, 1978; Mayer et al., 1985; Cleland et al., 1988a).

Reproduction among zooplankton was not affected by exposure to PCB concentrations up to 1 µg/l (Nebeker and Puglisi, 1974; Dillon et al., 1990). Limited information on congener-specific studies reported that reproduction was not affected in *Daphnia* with tissue concentrations of about 0.4 to 26 mg of PCBs per kilogram (Dillon et al., 1990). Reproduction among larger invertebrates was not markedly inhibited by 1 to 3 µg/l of PCBs among the studies examined, although adverse effects did occur at concentrations >10 µg/l, where tissue concentrations exceeded 30 mg of PCBs per kilogram (Nebeker and Puglisi, 1974; Roesijadi et al., 1976; Borgmann et al., 1990). Some studies on trout reported that PCBs can inhibit spawning and reduce hatching success in fish with tissue concentrations of >30 mg/kg, while others have reported mortality among fry with >125 mg of PCBs per kilogram (Freeman and Idler, 1975; Mauck et al., 1978). Other studies indicate that spawning and hatching success were not affected in minnows that contained >350 mg of PCBs per kilogram (Nebeker et al., 1974; DeFoe et al., 1978).

Table 4 Short-Term Laboratory Studies on the Response of Aquatic Organisms Exposed to Polychlorinated Biphenyls (PCBs)[a]

Response/organism	Observations	Ref.
Lethality		
Zooplankton	48-hour LC_{50}[b] for *Daphnia* reported for hexachlorobiphenyl at 200 µg/l	Bobra et al. (1983)
Zooplankton	48-hour LC_{50}[b] for *Daphnia* reported for capacitor-grade PCB at 21 µg/l	Branson (1977)
Zooplankton	No mortality in *Daphnia* exposed to seven congeners at 0.3-2.8 µg/l after 48 hours	Dillon and Burton (1991)
Crustacea	100% mortality in shrimp for Aroclor 1254 at 100 µg/l after 48 hours	Duke et al. (1970)
Crustacea	96-hour LC_{50} for shrimp for Aroclor 1254 reported between 64-80 µg/l	Roesijadi et al. (1976)
Crustacea	96-hour LC_{50} for shrimp reported for Aroclor 1242 and Aroclor 1254 at 13 and 12 µg/l; threshold concentrations were 6.5 and 0.5 µg/l, respectively	McLeese and Metcalfe (1980)
Crustacea	7-day LC_{50} for shrimp reported for Aroclor 1254 at 9 µg/l; animals had 60 mg/kg of PCBs	Nimmo et al. (1974)
Crustacea	No mortality in crab exposed to Aroclor 1254 at 3-30 µg/l after 96 hours	Stahl (1979)
Fish	No mortality among pinfish exposed to Aroclor 1254 at 1-100 µg/l for 48 hours; fish had 1-17 mg/kg of PCBs	Duke et al. (1970)
Fish	96-hour LC_{50} for minnows reported for Aroclor 1242 and Aroclor 1254 at 15 and 8 µg/l	Nebeker et al. (1974)
Fish	96-hour LC_{50} for scud reported for Aroclor 1242 and Aroclor 1248 at 73 and 29 µg/l	Nebeker and Puglisi (1974)
Fish	96-hour LC_{50} for trout reported for six Aroclors between 1-61 mg/l	Mayer et al. (1977)
Fish	No mortality in minnow exposed to seven congeners at 0.3-2.8 µg/l after 96 hours	Dillon and Burton (1991)
Fish	Acute oral toxicity for trout reported for four Aroclors was >1500 mg/kg of PCBs among fish fed a 300-mg/kg diet per day for 5 days	Mayer et al. (1977)
Growth related		
Protozoa	Reproduction and growth of ameba not affected by Aroclor 1254 at 10 µg/l after 6 days	Prescott et al. (1977)
Phytoplankton	Cell division among three species reduced 50% by Aroclor 1242 at 10-50 µg/l after 5-day exposure; gross morphological changes reported among those exposed to 50 µg/l	Glooschenko and Glooschenko (1975)
Phytoplankton	Growth in two marine species inhibited by Aroclor 1254 at 10-25 µg/l, and at ≥100 µg/l for one marine and two freshwater algae	Mosser et al. (1972)
Phytoplankton	^{14}C uptake in mixed marine species inhibited 50% by Aroclors 1242 and 1254 at 5 and 15 µg/l	Moore and Harriss (1972)
Algae	Growth of *Chlorella* not affected by a tetrachlorobiphenyl at 0.1 and 1 mg/l after 48 hours	Urey et al. (1976)
Algae	26-30% inhibition of ^{14}C uptake by *Scenedesmus* and ≤6% by *Dunaliella* by Aroclor 1254 at 1-10 µg/l after 24 hours	Luard (1973)
Algae	Photosynthesis among seven algal species reduced 0-90% by Aroclor 1254 at 10 µg/l after 48 hours	Harding and Phillips (1978)

Table 4 (continued) Short-Term Laboratory Studies on the Response of Aquatic Organisms Exposed to Polychlorinated Biphenyls (PCBs)[a]

Response/ organism	Observations	Ref.
Coelenterate	Tentacle regeneration of *Hydra* inhibited 50% by Aroclor 1016 and Aroclor 1254 at 1-4 mg/l after 96 hours	Adams and Haileselassie (1984)
Crustacea	Shell growth in oyster inhibited 19-100% by Aroclor 1254 at 1-100 µg/l after 96 hours; oysters had >8 mg/kg of PCBs	Duke et al. (1970)
Arthropod	Metamorphosis into adult mosquito curtailed by Aroclor 1254 at 1.5 µg/l after 7 days; PCB concentration in pupae estimated to be 30 mg/kg	Sanders and Chandler (1972)
Behavior		
Crustacea	Avoidance response by shrimp after exposure to Aroclor 1254 at 10 mg/l, but not at lower concentrations	Hansen et al. (1974)
Crustacea	Melanophore distribution in crab affected by Aroclor 1242 at 8 mg/l after 48 hours	Hanumante et al. (1981)
Fish	Locomotory activity of killifish not affected by exposure to Aroclor 1242 at 4 mg/l after 1 day	Fingerman and Russell (1980)
Fish	Avoidance response by mosquitofish reported for Aroclor 1254 at ≥0.1 mg/l	Hansen et al. (1974)
Fish	Avoidance response by minnow not shown for Aroclor 1254 at 0.001-10 mg/l	Hansen et al. (1974)
Fish	Temperature selection in trout not affected by Aroclor 1254 at 24-100 mg/l exposed for 1 day	Miller and Ogilvie (1975)
Fish	Temperature selection of salmon not affected by Aroclor 1254 at 2 mg/l exposed for 1 day	Peterson (1973)
Biochemical observations		
Fish	Aryl hydrocarbon hydroxylase activity in trout increased by intraperitoneal injection of tetrachlorobiphenyl at 0.6-640 µg/kg	Janz and Metcalfe (1991)
Fish	Glutathione S-transferase activity in flounder increased 1.3-fold by intraperitoneal injection of Aroclor 1254 at 100 mg/kg after 6 days	Scott et al. (1992)
Fish	Ethoxyresorufin- and ethoxycoumarin-O-deethylase and benzo[a]pyrene hydrolase activities in catfish increased 3- to 15-fold by intraperitoneal injection of Aroclor 1254 at 1-100 mg/kg after 8 days; glutathione S-transferase activity not affected by 100 mg/kg of PCB exposure	Ankley et al. (1986)
Fish	Cytochrome P-450 activity in mullet increased 63% by Phenoclor DP6 at 50 mg/kg in diet fed for 8 days	Narbonne and Gallis (1979)
Fish	Interrenal conversion of progesterone to 17α-hydroxyprogesterone, 11-deoxycortisol, and deoxycorticosterone in male and female trout higher after intraperitoneal injection of hexachlorobiphenyl at 1 mg/kg after 5 days, but production of metabolites in fish exposed to 20 mg/kg of PCBs increased in males but decreased in females	Miranda et al. (1992)
Fish	Metabolism of polycyclic aromatic hydrocarbon in salmon increased by intraperitoneal injection of Aroclor 1254 at 5 and 100 mg/kg after 7 days	Collier et al. (1985)

Table 4 (continued) Short-Term Laboratory Studies on the Response of Aquatic Organisms Exposed to Polychlorinated Biphenyls (PCBs)[a]

Response/organism	Observations	Ref.
Fish	In vitro ATPase activity in sunfish kidney reduced 50% by Aroclor 1242 and Aroclor 1254 at 3 mg/kg	Cutkomp et al. (1972)
Fish	Brain neurotransmitters dopamine and norepinephrine in killifish not affected by Aroclor 1242 at 4 mg/l exposure for 1 day	Fingerman and Russell (1980)

[a] The observations recorded indicate the response of exposed organisms compared with that of the controls at the end of the study. PCB concentrations in the organisms reported in the study or estimated from the data presented are expressed on a wet-weight, whole-body basis, unless noted otherwise.
[b] LC_{50}, 50% lethal concentrations.

Disease-challenge studies indicate that resistance to bacterial diseases, such as *Vibrio*, *Aeromonas*, and *Yersinia*, was not compromised in trout exposed to dietary and waterborne PCB levels that were higher than those found in natural systems (Snarski, 1982; Mayer et al., 1985; Cleland et al., 1988b). Other conditions, such as fin erosion, were reported in flagfish and minnows exposed to 3 to 5 µg/l of PCBs, but they were not observed in trout exposed to ≤6 µg/l of PCBs among fish that were exposed for 14 to 90 days (Nebeker et al., 1974; Schimmel et al., 1974; Mayer et al., 1985).

Behavioral responses associated with avoidance, swimming, and temperature selection were not affected by PCB concentrations that approached or exceeded the solubility limits (Hansen et al., 1974; Miller and Ogilvie, 1975). Lipid content was not influenced in fish following prolonged dietary and waterborne exposure where the PCB concentrations in fish were >10 mg/kg (Lieb et al., 1974; Mayer et al., 1985).

The biochemical and cellular observations among PCB-exposed organisms have shown that adverse effects or changes in activity levels can occur at lower concentrations than those reported for whole-organism responses (Tables 4 and 6). The biochemical responses associated with reproduction indicate that possible adverse effects may occur through reduced levels of testosterone, androgen, and vittellogenin (Sivarajah et al., 1978a; Chen et al., 1986; Thomas, 1988). The probable adverse effects on other biochemical observations that were reported are less clear. Thyroid-related products that regulate metabolism were affected in fish exposed to PCBs (Mayer et al., 1977; Folmar et al., 1982). Mixed-function oxidase activities were elevated in fish exposed to waterborne and dietary PCBs (Voss et al., 1982; Nasci et al., 1991). The exposure of fish to PCBs may not alter the lipid content of fish, but can increase lipid peroxidation rates and alter deposition patterns in the liver (Nimmo et al., 1975; Wofford and Thomas, 1988). Changes in ATPase activity, vertebral collagen levels, and vitamin C levels in fish exposed to PCBs were also reported in some studies but not in others (Koch et al., 1972; Mauck et al., 1978).

Changes were observed in cellular structures among organisms whose PCB tissue concentrations were comparable to those reported in feral animals. Most of the

Table 5 Long-Term Laboratory Studies on Whole Organism Responses such as Lethality, Growth, Reproduction and Disease Susceptibility of Aquatic Organisms Exposed to Polychlorinated Biphenyls (PCBs)[a]

Response/organism	Observations	Ref.
Lethality		
Zooplankton	21-day LC_{50}[b] for *Daphnia* reported for Aroclor 1254 at 1 µg/l	Nebeker and Puglisi (1974)
Crustacea	Lethal threshold for *Gammarus* reported for Aroclor 1254 at 1-100 µg/l in colloidal and emulsive solutions after 30-day exposure	Wildish (1970)
Crustacea	72% of shrimp died when exposed to Aroclor 1254 at 5 µg/l during 20-day study; dead shrimp had 16 mg/kg PCBs, while those surviving had 33 mg/kg	Duke et al. (1970)
Crustacea	16-day LC_{50} for shrimp reported for Aroclor 1254 at 12 µg/l; animals had 45 mg/kg of PCBs	Nimmo et al. (1974)
Crustacea	5% of crabs died when exposed to Aroclor 1254 at 5 µg/l during 20-day study; surviving crabs had about 18-27 mg/kg of PCBs	Duke et al. (1970)
Fish	30-day LC_{50} for minnows reported for Aroclor 1248 and Aroclor 1260 at 5 µg/l and 3 µg/l	DeFoe et al. (1978)
Fish	LD_{50}[b] for trout eggs reported for five congeners at 74 to >7000 µg/kg	Walker and Peterson (1991)
Fish	Lethal body burden for goldfish reported for Clophen 50 at 250-324 mg/kg; fish were exposed at 0.5-4 mg/l for 5-21 days	Hattula and Karlog (1972)
Fish	Lethal body burden for trout reported for Aroclor 1254 at 650 mg/kg of PCBs; fish were fed on a 480-mg/kg diet for 260 days	Mayer et al. (1977)
Growth related		
Protozoa	Growth of *Tetrahymena* not affected by exposure to Aroclor 1242 at 0.02-20 mg/l after 22 days	Morgan (1972)
Phytoplankton	Weight, cell count, and chlorophyll index among diatoms significantly lower when exposed to Aroclor 1242 up to 100 µg/l for 14 days; cells had 109 mg/kg of PCBs. RNA synthesis was lower at 10-µg/l treatment, and these cells had 5 mg/kg of PCBs	Keil et al. (1971)
Algae	Growth of *Chlamydomonas* not affected by Aroclor 1242 at 2-20 mg/l after 20 days	Morgan (1972)
Algae	Growth of *Chlamydomonas* reduced by Aroclor 1248 at 11-111 µg/l after about 23 days, but not at 1 µg/l	Christensen and Zielski (1980)
Algae	Biomass of *Chlorella* not affected by Clophen A 50 at 4 µg/l after exposure for 10 days	Södergren (1970)
Mollusc	Growth in oyster reduced by Aroclor 1254 at 5 µg/l exposed for 168 days, but not by those exposed to 1 µg/l for 210 days. PCB concentration in those affected estimated to be 50 mg/kg	Nimmo et al. (1975)
Crustacea	Molting in crabs delayed by exposure to Aroclor 1242 at 8 mg/l for 14 days	Fingerman and Fingerman (1979)
Crustacea	Limb regeneration in crabs adversely affected by exposure to Aroclor 1242 at 8 mg/l for 28 days	Fingerman and Fingerman (1980)
Fish	Growth in trout reduced 40% by Aroclor 1254 at 300 mg/kg in diet fed for 365 days; growth not affected in fish fed a 3-mg/kg diet	Cleland et al. (1988a)

Table 5 (continued) Long-Term Laboratory Studies on Whole Organism Responses such as Lethality, Growth, Reproduction and Disease Susceptibility of Aquatic Organisms Exposed to Polychlorinated Biphenyls (PCBs)[a]

Response/ organism	Observations	Ref.
Fish	Growth in salmon reduced 40% by Aroclor 1242:1254 ratio (1:4) at 500 mg/kg in diet fed for 90 days; growth not affected in fish fed 50-mg/kg diet	Leatherland and Sonstegard (1978)
Fish	Growth in trout reduced 10% by Aroclor 1254:1260 ratio (1:2) at 2.9 µg/l exposed for 90 days; these fish had 120 mg/kg of PCBs, with growth not affected in fish exposed to 0.2-1.5 µg/l of PCBs, these fish having 6-70 mg/kg of PCBs after 90 days	Mayer et al. (1985)
Fish	Growth in minnows not affected by Aroclor 1242 at 0.9-5 µg/l, and Aroclor 1254 at 0.2-5 µg/l, exposed for 260 days; fish had 90-340 mg/kg of Aroclor 1242, and 50-1000 mg/kg of Aroclor 1254	Nebeker et al. (1974)
Fish	Growth in catfish not affected by four aroclors at 2.4-24 mg/kg in diets fed for 193 days; fish had 14-32 mg/kg of PCBs	Mayer et al. (1977)
Fish	Growth in catfish not affected by Aroclor 1242 at 20 mg/kg in a diet fed for 196 days of 242-day study. Fish had 11 mg/kg of PCBs	Hansen et al. (1976)
Fish	Growth in trout not affected by Aroclor 1254 at 15 mg/kg in a diet fed for 224 days; fish had 8 mg/kg of PCBs	Lieb et al. (1974)
Fish	Growth in salmon not affected by Aroclor 1254 at 0.05-480 mg/kg in a diet fed for 260 days; fish had 0.4-645 mg/kg of PCBs	Mayer et al. (1977)
Reproduction		
Zooplankton	Reproduction in *Daphnia* not affected by seven congeners at 0.01 and 1 µg/l exposed for 21 days; congener concentrations estimated to be 0.4-26 mg/kg among those exposed to 1 µg/l	Dillon et al. (1990)
Zooplankton	Reproduction in *Daphnia* not affected by Aroclor 1254 at ≤1.2 µg/l exposed for 21 days	Nebeker and Puglisi (1974)
Crustacea	Reproduction in *Gammarus* inhibited by Aroclor 1242 at 9-234 µg/l exposed for 60 days; animals had >320 mg/kg of PCBs, those exposed to 3 µg/l and reproduced had 76 mg/kg of PCBs	Nebeker and Puglisi (1974)
Crustacea	Survival, growth, and reproduction in *Hyalella* not affected by Aroclor 1242 at 3-30 µg/l exposed for 70 days, but affected animals exposed to ≤100 mg/l; toxic effects observed in animals with 30-100 mg/kg of PCBs	Borgmann et al. (1990)
Crustacea	Larval development period of shrimp not affected by Aroclor 1254 at ≤3 µg/l; those exposed to 16 µg/l died within 11 days	Roesijadi et al. (1976)
Arthropod	Adult emergence of *Tanytarsus* poor after exposed to Aroclor 1254 at 3.5-33 µg/l for 21 days	Nebeker and Puglisi (1974)
Fish	Spawning in minnows inhibited by Aroclor 1242 at 15-51 µg/l and Aroclor 1254 at 4.6-15 µg/l, exposed for 260 days; spawning females exposed to 5 µg/l Aroclor 1242 had 440 mg/kg of PCBs. Spawning females exposed to 1.8 µg/l Aroclor 1254 had 430 mg/kg of PCBs	Nebeker et al. (1974)
Fish	Egg hatch in trout reduced 78% by Aroclor 1254 at 0.2 mg/l exposed for 21 days; fish had 34 mg/kg of PCBs, and 80 mg/kg in eggs	Freeman and Idler (1975)

Table 5 (continued) Long-Term Laboratory Studies on Whole Organism Responses such as Lethality, Growth, Reproduction and Disease Susceptibility of Aquatic Organisms Exposed to Polychlorinated Biphenyls (PCBs)[a]

Response/organism	Observations	Ref.
Fish	Hatching in salmon eggs decreased 8-32% after exposure to Aroclor 1254 at 4-56 µg/l; fry survival decreased 85-15% after 28 days	Halter and Johnson (1974)
Fish	Spawning and hatching in minnows not affected by Aroclor 1248 at 0.1-3 µg/l, and Aroclor 1260 at 0.1-2 µg/l, exposed for 260 days; females had ≤350 mg/kg of Aroclor 1248 and ≤570 mg/kg of Aroclor 1260	DeFoe et al. (1978)
Fish	Hatching in trout eggs not affected when exposed to Aroclor 1254 at 0.4-13 µg/l, but fry mortality was 21-100% at 3-13 µg/l after 114 days; dead fry had >125 mg/kg of PCBs	Mauck et al. (1978)
Fish	Survival of trout progeny not affected by Aroclor 1254 at 200 mg/kg in a diet fed to gravid females for 60 days; eggs had 1.6 mg/kg of PCBs; fry fed a control diet for 365 days	Hendricks et al. (1981)
Disease related		
Fish	Disease resistance in trout not compromised when fed a diet contaminated with PCBs and other organochlorines for 140 days and then exposed to *Vibrio anguillarum*	Cleland et al. (1988b)
Fish	Mortality or time to death in trout not affected when exposed to 5-500 µg/l Aroclor 1254 for 30 days and then to infectious hematopoietic necrosis virus	Spitsbergen et al. (1988)
Fish	Disease resistance in trout increased when exposed to 7 and 15 µg/l of Aroclor 1254 for 30 days and then to *Aeromonas hydropile*	Snarski (1982)
Fish	Disease resistance in trout increased with exposure concentrations to 0.4-6 µg/l Aroclor 1254:1260 ratio (1:2) for 90 days and then to *Yersinia ruckeri;* fish had 6-120 mg/kg of PCBs	Mayer et al. (1985)
Fish	Death and fin erosion in flagfish reported when exposed to Aroclor 1248 at 5-18 µg/l after 14-40 days	Nebeker et al. (1974)
Fish	Fin rot in minnow embryos induced by Aroclor 1254 at 3 µg/l exposed for 21 days; PCB concentrations in fish estimated to be 5-10 mg/kg	Schimmel et al. (1974)
Fish	Fin condition in trout not affected by exposure to an Aroclor 1254:1260 ratio (2:1) at 0.4-6 µg/l for 90 days; fish had 6-120 mg/kg of PCBs	Mayer et al. (1985)
Other observations		
Fish	Lipid content in trout not affected by Aroclor 1242 at 15 mg/kg in a diet fed for 224 days; fish had 8 mg/kg of PCBs	Lieb et al. (1974)
Fish	Lipid content in trout not affected by an Aroclor 1254:1260 ratio (1:2) at 0.2-2.9 µg/l exposed for 90 days; fish had 6-120 mg/kg of PCBs	Mayer et al. (1985)

[a] The observations reported indicate the response of exposed organisms compared with that of the controls at the end of the study. PCB concentrations in exposed organisms reported in the study, or estimated from the data presented, are expressed on a wet-weight whole-body basis, unless noted otherwise.
[b] LC_{50}, 50% lethal concentration; LD_{50}, 50% lethal dose.

Table 6 Long-Term Laboratory Studies on Biochemical and Cellular Responses of Aquatic Organisms Exposed to PCBs[a]

Response/ organism	Observations	Ref.
Reproductive products		
Echinoderm	Testosterone, but not progesterone, levels in ovaries and testes of starfish increased 2- to 3-fold after exposure to >26 mg/kg of PCBs in lipid of food fed for 84 days; PCB concentrations in gonads were 9 mg/kg of lipid	den Besten et al. (1991)
Fish	Plasma androgen, estrogen, and corticoid levels in trout and carp reduced by intraperitoneal injection of Aroclor 1254 at 25 mg/kg once weekly for 4 weeks	Sivarajah et al. (1978a)
Fish	Plasma testosterone levels in croaker reduced 50% by Aroclor 1254 at 3 mg/kg in a diet fed for 30 days	Thomas (1988)
Fish	Sex steroids in cod altered by Aroclor 1242 and Aroclor 1254 at 10-50 µg/kg in a diet fed for 160 days; observations were based on a few fish	Freeman et al. (1982)
Fish	Serum vitellogenin levels in trout reduced 2- to 4-fold by Aroclor 1254 at 3-300 mg/kg of Aroclor 1254 in a diet fed for 180 days	Chen et al. (1986)
Thyroid-related activity		
Fish	Thyroid uptake of ^{125}I in salmon increased 52-119% by Aroclor 1254 at 0.5-480 mg/kg in a diet fed for 260 days (for 193 days, these fish had 0.7-659 mg/kg of PCBs); thyroid activity in catfish was increased 80-120% by Aroclor 1254 at 2.4-24 mg/kg for 193 days, but not in fish fed similar amounts of Aroclors 1232, 1248, and 1260; these fish had 3-32 mg/kg of PCBs	Mayer et al. (1977)
Fish	Thyroid uptake of ^{125}I in salmon increased 119% by Aroclor 1254 at 0.048-480 mg/kg in a diet fed for 260 days (fish had 0.4-650 mg/kg of PCBs)	Mayer et al. (1977)
Fish	Serum triiodothyronine (T_3) levels in salmon reduced by Aroclor 1242:1254 ratio (1:4) at 500 mg/kg in a diet fed for 90 days, but thyroxine (T_4) levels were not affected; fish fed the 50 mg/kg of PCB diet were not affected	Leatherland and Sonstegard (1978)
Fish	Plasma thyroxine and triiodothyronine patterns in salmon altered by intraperitoneal injection of Aroclor 1254 at 150 µg/kg after 42 days	Folmar et al. (1982)
Mixed-function oxidases		
Annelid	Cytochrome P-450 activity in polychaete increased 3-fold by Aroclor 1254 at 500 mg/kg in a diet fed for 56 days	Fries and Lee (1984)
Fish	Cytochrome P-450 activity in goby increased slightly by Aroclor 1254 at 1 µg/l exposed for 30 days	Nasci et al. (1991)
Fish	Ethoxyresorufin- and ethoxycoumarin-O-deethylase and benzo[a]pyrene monooxygenase activities in trout increased 77-fold by Aroclor 1254 at 100 mg/kg in a diet fed for 105 days	Voss et al. (1982)
Cellular observations		
Mollusc	Connective tissue structure of oyster altered by Aroclor 1254 at 5 µg/l exposed for 180 days but recovered when exposed to clean water; PCB concentration in affected animals estimated to be 50 mg/kg	Nimmo et al. (1975)

Table 6 (continued)   Long-Term Laboratory Studies on Biochemical and Cellular Responses of Aquatic Organisms Exposed to PCBs[a]

Response/organism	Observations	Ref.
Crustacea	Hepatopancreas structure of shrimp altered by Aroclor 1254 at 3 μg/l exposed for 30 days. PCB concentrations estimated to be 3 mg/kg	Nimmo et al. (1975)
Fish	Liver histology of trout not affected by Aroclor 1254 at 15 mg/kg in a diet fed for 224 days; fish had 8 mg/kg of PCBs	Lieb et al. (1974)
Fish	Liver histology or size of catfish not affected by Aroclor 1242 at 200 mg/kg in diet fed for 196 of 242-day study; fish had 11 mg/kg of PCBs	Hansen et al. (1976)
Fish	Liver ultrastructure of trout altered by Aroclor 1254 at 10- and 100-mg/kg diets fed for 229-330 days but not changes were seen in fish fed 1 mg/kg of PCBs; affected fish had 2 and 81 mg/kg PCBs	Hacking et al. (1977)
Fish	Ultrastructure of liver, testes, and ovaries of carp and trout altered by Aroclor 1254 in fish receiving 25 mg/kg of PCBs by intraperitoneal injection once weekly for 4 weeks	Sivarajah et al. (1978b)
Fish	Liver of trout enlarged 40% by Aroclor 1254 at 300 mg/kg in a diet fed for 365 days	Cleland et al. (1988a)
Fish	Liver and kidney tissues of minnow degraded by Aroclor 1242 and 1254 in 8 μg/l exposed for 120 days	Koch et al. (1972)
Fish	Fat deposition pattern in liver of spot changed after exposure to Aroclor 1254 at 5 μg/l for >14 days; PCB concentration estimated to be 5 mg/kg	Nimmo et al. (1975)
Other responses		
Fish	Hepatic *in vitro* lipid peroxidation rate of croaker increased 50% by Aroclor 1254 at 5 mg/kg in a diet fed for 17 days of 24-day study	Wofford and Thomas (1988)
Fish	Adrenergic response in gill of trout not affected by exposure to Aroclor 1254 at about 1 g/kg for 3 days and tested after 14-28 days; fish had 6 mg/kg of PCBs in gills and 70 mg/kg in muscle	Kiessling et al. (1983)
Fish	ATPase activity in several minnow tissues affected by Aroclors 1242 and 1254 at 0.3-8.3 μg/l exposed for 120 days although the results appear inconclusive	Koch et al. (1972)
Fish	ATPase activity in salmon gills not affected by intraperitoneal injection of Aroclor 1254 at 150 μg/kg after 42 days	Folmar et al. (1982)
Fish	Vertebral collagen levels in trout reduced by 13-37% by Aroclor 1254 at 0.7-13 μg/l exposed for 38 days; fish had >31 mg/kg of PCBs	Mauck et al. (1978)
Fish	Vertebral collagen levels in trout not affected by an Aroclor 1254:1260 ratio (1:2) at 0.2-2.9 μg/l exposed for 90 days; fish had 6-120 mg/kg of PCBs	Mayer et al. (1985)
Fish	Vitamin C levels in trout reduced by Aroclor 1254 at 1.5-13 μg/l exposed for 38 days; levels not affected at lower exposure concentrations; affected fish had ≥70 mg/kg of PCBs	Mauck et al. (1978)
Fish	Hematocrit levels in trout not affected by an Aroclor 1254:1260 ratio (1:2) at 0.2-2.9 μg/l exposed for 90 days; fish had 6-120 mg/kg of PCBs	Mayer et al. (1985)

Table 6 (continued) Long-Term Laboratory Studies on Biochemical and Cellular Responses of Aquatic Organisms Exposed to PCBs[a]

Response/ organism	Observations	Ref.
Carcinogenicity		
Fish	Hepatic tumors in 120 trout not observed after being fed Aroclor 1254 at 50 mg/kg in a diet for 365 days	Shelton et al. (1984)
Fish	Hepatic tumor in 1 of 80 trout reported after being fed Aroclor 1254 at 100 mg/kg in a diet for 90 days, followed by control diet for 365 days	Shelton et al. (1983)
Fish	Hepatic tumors in trout progeny not observed when Aroclor 1254 at 200 mg/kg in a diet was fed to gravid trout for 60 days; the eggs had 1.6 mg/kg of PCBs. The 60 fish examined were fed a control diet for 365 days	Hendricks et al. (1981)
Protective mechanisms		
Fish	Humoral immodulation in trout not affected by Aroclor 1254 at 3-300 mg/kg in a diet fed for 365 days	Cleland et al. (1988a)
Fish	Natural killer cell activity in trout not affected by Aroclor 1254 at 3-300 mg/kg in a diet fed for 365 days	Cleland and Sonstegard (1987)
Fish	Cortisol levels in trout not affected by an Aroclor 1254:1260 ratio (1:2) at 0.2-2.9 μg/l exposed for 90 days; fish had 5-110 mg/kg of PCBs	Mayer et al. (1985)

[a] The observations reported indicate the response of exposed organisms compared with that of the controls at the end of the study; PCB concentrations in exposed organisms reported in the study or estimated from the data presented are expressed on a wet-weight, whole-body basis, unless noted otherwise.

changes in fish were associated with the nucleus, where components were separated and pseudoinclusions had developed (Hacking et al., 1977). Other changes included alterations of the smooth and rough endoplasmic reticulum (Hacking et al., 1977; Sivarajah et al., 1978b). Changes observed in invertebrates included the irregular structure and infiltration of leukocytes in the vasicular connective tissue of the oyster and formation of crystalloids in the hepatopancreas of shrimp (Nimmo et al., 1975).

PCBs have little or no capability to induce hepatocellular lesions in trout exposed through their diet (Hendricks et al., 1981; Shelton et al., 1984). Exposure of trout to high dietary concentrations of PCBs also did not compromise some protective mechanisms such as humoral immunomodulation and natural killer cell activity (Cleland and Sonstegard, 1987; Cleland et al., 1988a).

EFFECTS OF PCB CONCENTRATIONS IN FERAL ORGANISMS

These laboratory studies generally indicate that waterborne or tissue concentrations of PCBs that would cause adverse effects are higher than those reported from contaminated ecosystems. Waterborne concentrations in the low-μg/l range that were lethal to the more sensitive organisms, such as phytoplankton, were higher than concentrations reported in contaminated ecosystems by a factor of about 10^3 (Table 1). The PCB concentrations reported in dead organisms that approximate lethal tissue concentrations were often at least an order of magnitude higher than concentrations reported in feral organisms. These large differences between lethal and field concentrations would suggest that acute toxicity from PCBs is not an important issue to many organisms in most aquatic ecosystems.

The threshold PCB concentrations at which adverse effects on growth and reproduction occurred in laboratory studies were generally higher than the concentrations observed in natural systems. The PCB concentrations that can impair reproduction among a variety of species in laboratory studies were higher than those reported for comparable feral organisms by a factor of about 10^2 (Table 3). The growth-related effects among fish and larger invertebrates may not be affected by the PCB concentrations reported among comparable feral species, because of similar differences between threshold and field concentrations. A similar conclusion may be suggested for organisms at the lower trophic levels based on waterborne concentrations, even though no threshold tissue concentrations could be determined.

Cytological observations on feral fish have focused on the histopathology of the liver and reproductive organs (Table 3). The increased frequency of hepatocellular lesions among benthic dwelling fish has received considerable attention, because these lesions are one of the few discernible anomalies that may have a chemical etiology (Harshbarger and Clark, 1990). The results of carcinogenic studies on trout are consistent with those of studies on mammals, showing that PCBs have poor initiation properties but are efficacious tumorigenic promotors (Silberhorn et al., 1990). These lesions are often observed in benthic fish from sites with elevated levels of carcinogenic chemicals, such as PAHs (Mix, 1986; Myers et al., 1991). PCBs may not be an important factor in the initiation of hepatocellular lesions, but they could have a promotional effect because of their presence with other chemicals. The ecotoxicological significance of PCBs in this capacity would be difficult to assess because of the interactions with other chemicals. Furthermore, the role of hepatocellular lesions as a lethal or adverse chronic factor has not been clearly demonstrated in fish.

The mixed-function oxygenase (MFO) system is a detoxification mechanism in animals that can enhance the elimination of some hydrophobic chemicals through a series of oxidative reactions. This system is present in most aquatic vertebrates and some invertebrate species, and induction activity is sometimes used to detect chemical exposure (Payne et al., 1987). There are some aspects of this increased activity that are being examined more thoroughly, because of the formation of intermediary or secondary products that may be more toxic than the parent compound (McFarland and Clarke, 1989). Some steroids associated with reproduction may be oxidized or altered by increased MFO activity and affect oogenesis (Freeman et al., 1982; Klotz et al., 1983; Spies and Rice, 1988). Assuming that there is an association between chronic elevated MFO activity and reproductive impairment, PCBs could have an adverse effect on reproduction among feral fish, because field studies have shown that MFO activity levels among fish from PCB-contaminated waters are higher than among fish from control sites (Luxon et al., 1987; Winston et al., 1989).

The toxicological significance of PCBs on MFO activity cannot be assessed at this time, because a cause-and-effect relation has not been established.

PCBs have also been directly or indirectly associated with other responses reported in feral organisms (Table 3). Further evaluation suggests that these responses may be poorly correlated or not directly attributed to PCB exposure. Goiters and thyroid hyperplasia and hypoplasia have been reported in salmonids from the Great Lakes. Laboratory studies reported that thyroid activity had increased and decreased among salmonids fed on a high-PCB diet over a long period (Mayer et al., 1977; Leatherland and Sonstegard, 1978). In another study in which trout were fed a PCB-contaminated diet, the authors did not observe thyroid hyperplasia or significant declines in 3,5,3'-triiodothyronine and thyroxine levels among fish after 168 days (Hilton et al., 1983). The effects of PCB exposure on these thyroid conditions would be speculative.

Collagen levels were lower in feral fish than in those reared in a hatchery (Mehrle et al., 1982). Laboratory studies have reported that collagen levels were reduced in fish exposed to PCBs, while others reported no effect (Mauck et al., 1978; Mayer et al., 1985). Other studies have attributed decreased collagen levels to lower vitamin C levels (Hamilton et al., 1981). Vitamin C levels can be reduced in fish containing ≥70 mg of PCBs per kg, which may not account for the lower collagen levels in feral fish because their PCB concentrations would be considerably lower (Mauck et al., 1978). Increased micronuclei frequencies were also associated with PCB concentrations in feral fish (Hose et al., 1987). Laboratory studies have shown that micronuclei can be induced in fish by mutagenic chemicals such as benzo(a)pyrene; hence, PCBs may not be an important factor in their formation (Metcalfe, 1988). Fin rot was also reported in feral fish and induced in laboratory studies among fish exposed to PCBs (Schimmel et al., 1974). Disease-challenge studies and some immune system activities were not compromised in fish exposed to PCBs in the laboratory (Cleland et al., 1988a; Spitsbergen et al., 1988). Observations such as these on thyroid activity, collagen content, micronuclei formation, and fin rot can result in a large degree of uncertainty when evaluating the role of PCBs in their development in feral fish, even though some have been induced in the laboratory.

Table 7 presents a summary of PCB concentrations in water and tissues that may cause adverse effects or changes in response, among four groups of organisms that are represented by algae, zooplankton, macroinvertebrates, and fish. These estimates represent the threshold concentrations that were derived from a limited information base that may not be representative of the more sensitive species and should be interpreted accordingly. PCB waterborne concentrations of >0.5 to 1 µg/l could be lethal to algae and zooplankton and can impair growth and reproduction. Lethality, growth, and reproductive impairment among macroinvertebrates can occur at PCB tissue concentrations >25 mg/kg. Behavioral changes can occur at >6 µg/l of PCBs, and cellular changes can occur at low-mg/kg tissue concentrations in macroinvertebrates, although it cannot be determined if these changes would have deleterious effects. Fish appear to be more resistant to PCB toxicity, where tissue concentrations >100 mg/kg can be lethal or affect reproduction in females, and where concentrations >50 mg/kg can reduce growth and the survival of progeny. The role of PCBs in

Table 7 Summary of Polychlorinated Biphenyl (PCB) Concentrations in Algae, Zooplankton, Macroinvertebrates, and Fish at Which Adverse and Chronic Effects, Cytological Changes, and Changes in Biochemical Activity Levels May Occur, Based on Short- and Long-Term Laboratory Studies[a]

Response	Algae	Zooplankton	Macroinvertebrates	Fish
Lethality	>0.5–1 µg/l	>0.5 µg/l	>25 mg/kg	>100 mg/kg
Growth	>0.5–1 µg/l	>0.5 µg/l	>25 mg/kg	>50 mg/kg
Reproduction	>0.5–1 µg/l	>0.5 µg/l	>25 mg/kg	
Female				>100 mg/kg
Progeny				>50 mg/kg
Behavior			>100 µg/l	>100 µg/l
Disease				mg/kg range
Cellular changes			Low mg/kg	High µg/kg to low mg/kg
Biochemical changes				High µg/kg to low mg/kg

[a] Estimates for algae include those for phytoplankton, and zooplankton estimates were primarily based on *Daphnia*. A maximum concentration of >100 mg/kg of PCBs in tissues was used because higher concentrations may have limited environmental relevance. Some threshold concentrations are expressed on a waterborne exposure basis, where estimates of tissue concentrations were poorly defined.

enhancing the incidence of bacterial diseases is probably low, based on studies that indicated resistance to disease exposure, and protective mechanisms are not compromised in fish exposed to PCBs. PCBs may contribute indirectly to other disease-related syndromes, such as lesions where mg/kg of tissue concentrations may be required. PCBs can also alter the behavior of fish at >100 µg/l and cause cellular and biochemical changes at tissue concentrations in the high-µg/kg to low-mg/kg range, although the ecotoxicological significance of these changes is largely unknown.

PCBs, as a chemical entity, appear to have only limited adverse effects in feral aquatic organisms, based on the extrapolation of laboratory studies to field observations. This opinion does not consider current views that PCBs and other halogenated organic chemicals can have similar mechanisms of action and induce similar responses (Niimi, 1994). Comparative analyses on the relative toxicity of PCBs and chlorinated dioxins and furans monitored in Great Lakes salmonids using toxic equivalency factors (TEFs) have indicated that the relative toxicity from PCBs is greater; nevertheless, some toxicological effects could be attributed to the cumulative effects of these and other chemicals that have the same mechanism of action (Niimi and Oliver, 1989b; Williams and Giesy, 1992). The toxicological significance of PCBs may assume a greater importance when viewed from this context, because of its ubiquitous distribution and the presence of other chemicals in aquatic ecosystems.

SUMMARY

PCB concentrations that are monitored in feral aquatic organisms and natural waters are not likely to be acutely toxic to most species of the aquatic community, because ambient concentrations are often 10- to 100-fold lower than concentrations shown to be lethal in laboratory studies. These studies have suggested that PCB tissue concentrations of >25 mg/kg in macroinvertebrates and >50 to 100 mg/kg in fish may be required to adversely affect growth and reproduction. Exposure to PCBs can alter biochemical activities and cells at the micro- and ultrastructure levels, notably among macroinvertebrates and fish at tissue concentrations that approximate those of feral organisms from highly contaminated waters. Little is known about the extent of these changes, and even less is known about their effects on the organisms. Current views suggest that chemicals whose mechanism of action is similar to that of PCBs may adversely affect the reproduction of fish and other aquatic organisms, possibly at the biochemical and cellular levels.

REFERENCES

Achman, D. R., K. C. Hornbuckle, and S. J. Eisenreich. 1993. Volatilization of polychlorinated biphenyls from Green Bay, Lake Michigan. Environ. Sci. Technol. 27:75-87.

Adams, J. A., and H. M. Haileselassie. 1984. The effects of polychlorinated biphenyls (Aroclors® 1016 and 1254) on mortality, reproduction, and regeneration in *Hydra oligactis*. Arch. Environ. Contam. Toxicol. 13:493-499.

Anderson, M. L., C. P. Rice, and C. C. Carl. 1982. Residues of PCB in a *Cladophora* community along the Lake Huron shoreline. J. Great Lakes Res. 8:196-200.

Andersson, Ö., C.-E. Linder, M. Olsson, L. Reutergårdh, U.-B. Uvemo, and U. Wideqvist. 1988. Spatial differences and temporal trends of organochlorine compounds in biota from the northwestern hemisphere. Arch. Environ. Contam. Toxicol. 17:755-765.

Ankley, G. T., V. S. Blazer, R. E. Reinert, and M. Agosin. 1986. Effects of Aroclor 1254 on cytochrome P-450-dependent monooxygenase, glutathione *S*-transferase, and UDP-glucuronosyltransferase activities in channel catfish liver. Aquat. Toxicol. (Amst.) 9:91-103.

Atlas, E., T. Bidleman, and C. S. Giam. 1986. Atmospheric transport of PCBs to the oceans. p. 79-100. *In* J. S. Waid (Ed.). PCBs and the environment. CRC Press, Boca Raton, Fla.

Bache, C. A., J. W. Serum, W. D. Youngs, and D. J. Lisk. 1972. Polychlorinated biphenyl residues: accumulation in Cayuga lake trout with age. Science (Washington, D.C.) 177:1191-1192.

Bannister, R., D. Davis, T. Zacharewaki, I. Tizard, and S. Safe. 1987. Aroclor 1254 as a 2,3,7,8-tetrachlorodibenzo-*p*-dioxin antagonist: effects on enzyme induction and immunotoxicity. Toxicology 46:28-42.

Bannister, R., and S. Safe. 1987. Synergistic interactions of 2,3,7,8-TCDD and 2,2',4,4',5,5'-hexachlorobiphenyl in C57BL/6J and DBA/2J mice: role of the Ah receptor. Toxicology 44:159-169.

Bender, M. E., W. J. Hargis, Jr., R. J. Huggett, and M. H. Roberts, Jr. 1988. Effects of polynuclear aromatic hydrocarbons on fishes and shellfish: an overview of research in Virginia. Mar. Environ. Res. 24:237-241.

Bidleman, T. F., G. W. Patton, M. D. Walla, B. T. Hargrave, W. P. Vass, P. Erickson, B. Fowler, V. Scott, and D. J. Gregor. 1989. Toxaphene and other organochlorines in Arctic Ocean fauna: evidence for atmospheric delivery. Arctic 42:307-313.

Black, D. E., D. K. Phelps, and R. L. Lapan. 1988. The effect of inherited contamination on egg and larval winter flounder, *Pseudopleuronectes americanus*. Mar. Environ. Res. 25:45-62.

Bobra, A. M., W. Y. Shiu, and D. Mackay. 1983. A predictive correlation for the acute toxicity of hydrocarbons and chlorinated hydrocarbons to the water flea *(Daphnia magna)*. Chemosphere 12:1121-1129.

Borgmann, U., W. P. Norwood, and K. M. Ralph. 1990. Chronic toxicity and bioaccumulation of $2,5,2',5'$- and $3,4,3',4'$-tetrachlorobiphenyl and Aroclor® 1242 in the amphipod *Hyalella azteca*. Arch. Environ. Contam. Toxicol. 19:558-564.

Borgmann, U., and D. M. Whittle. 1991. Contaminant concentration trends in Lake Ontario lake trout *(Salvelinus namaycush)*: 1977-1988. J. Great Lakes Res. 17:368-381.

Branson, D. R. 1977. A new capacitor fluid — a case study in product stewardship. p. 44-61. In F. L. Mayer and J. L. Hamelink (Eds.). Aquatic toxicology and hazard evaluation. ASTM STP 634. American Society for Testing and Materials, Philadelphia, Pa.

Brown, M. P., M. B. Werner, R. J. Sloan, and K. W. Simpson. 1985. Polychlorinated biphenyls in the Hudson River. Environ. Sci. Technol. 19:656-661.

Bush, B., R. W. Streeter, and R. J. Sloan. 1989. Polychlorinated (PCB) congeners in striped bass *(Morone saxatilis)* from marine and estuarine waters of New York State determined by capillary gas chromatography. Arch. Environ. Contam. Toxicol. 19:49-61.

Cameron, P., V. Dethlefsen, H. von Westerhagen, and D. Janssen. 1988. Chromosomal and morphological investigations on whiting *(Merlangius merlangus)* embryos from the North Sea in relation to organochlorine contamination. Aquat. Toxicol. (Amst.) 11:428-429.

Capel, P. D., and S. J. Eisenreich. 1985. PCBs in Lake Superior, 1978-1980. J. Great Lakes Res. 11:447-461.

Chen, T. T., P. C. Reid, V. Van Beneden, and R. A. Sonstegard. 1986. Effect of Aroclor 1254 and mirex on estradiol-induced vitellogenin production in juvenile rainbow trout *(Salmo gairdneri)*. Can. J. Fish. Aquat. Sci. 43:169-173.

Chevreuil, M., A. Chesterikoff, and E. Letolle. 1987. PCB pollution behaviour in the River Seine. Water Res. 21:427-434.

Chovelon, A., L. George, C. Gulayets, Y. Hoyano, E. McGinness, J. Moore, S. Ramamoorthy, P. Singer, K. Smiley, and A. Wheatley. 1984. Pesticide and PCB levels in fish from Alberta. Chemosphere 13:19-32.

Christensen, E. R., and P. A. Zielski. 1980. Toxicity of arsenic and PCB to a green alga *(Chlaydomonas)*. Bull. Environ. Contam. Toxicol. 25:43-48.

Cleland, G. B., P. J. McElroy, and R. A. Sonstegard. 1988a. The effect of dietary exposure to Aroclor 1254 and/or mirex on humoral immune expression of rainbow trout *(Salmo gairdneri)*. Aquat. Toxicol. (Amst.) 12:141-146.

Cleland, G. B., B. G. Oliver, and R. A. Sonstegard. 1988b. Dietary exposure of rainbow trout *(Salmo gairdneri)* to Great Lakes coho salmon (*Oncorhynchus kisutch* Walbaum). Bioaccumulation of halogenated aromatic hydrocarbons and host resistance studies. Aquat. Toxicol. (Amst.) 13:281-290.

Cleland, G. B., and R. A. Sonstegard. 1987. Natural killer cell activity in rainbow trout *(Salmo gairdneri)*: effect of dietary exposure to Aroclor 1254 and/or mirex. Can. J. Fish. Aquat. Sci. 44:636-638.

Collier, T. K., E. H. Gruger, Jr., and U. Varanasi. 1985. Effect of Aroclor 1254 on the biological fate of 2,6-dimethylnaphthalene in coho salmon *(Oncorhynchus kisutch)*. Bull. Environ. Contam. Toxicol. 34:114-120.

Collier, T. K., J. E. Stein, H. R. Sanborn, T. Hom, M. S. Myers, and U. Varanasi. 1992. Field studies of reproductive success and bioindicators of maternal contaminant exposure in English sole *(Parophtys vetulus)*. Sci. Total Environ. 116:169-185.

Colombo, J. C., M. F. Khalil, M. Arnac, A. C. Horth, and J. A. Catoggio. 1990. Distribution of chlorinated pesticides and individual polychlorinated biphenyls in biotic and abiotic compartments of the Rio de La Plata, Argentina. Environ. Sci. Technol. 24:498-505.

Connolly, J. F. 1991. Application of a food chain model to polychlorinated biphenyl contamination of lobster and winter flounder food chains in New Bedford Harbor. Environ. Sci. Technol. 25:760-770.

Cooke, A. S. 1973. Shell thinning in avian eggs by environmental pollutants. Environ. Pollut. 4:85-152.

Crane, J. L., and W. C. Sonzogni. 1992. Temporal distribution and fractionation of polychlorinated biphenyl congeners in a Wisconsin lake. Chemosphere 24:1921-1941.

Cross, J. N., and J. E. Hose. 1988. Evidence for impaired reproduction in white croaker *(Genyonemus lineatus)* from contaminated areas off southern California. Mar. Environ. Res. 24:185-188.

Cutkomp, L. K., H. H. Yap, D. Desaiah, and R. B. Koch. 1972. The sensitivity of fish ATPases to polychlorinated biphenyls. Environ. Health Perspect. 1:165-168.

DeFoe, D. L., G. D. Veith, and R. W. Carlson. 1978. Effects of Aroclor® 1248 and 1260 on the fathead minnow *(Pimephales promelas)*. J. Fish. Res. Board Can. 35:997-1002.

den Besten, P. J., J. M. L. Elenbaas, J. R. Maas, S. J. Dieleman, H. J. Herwig, and P. A. Voogt. 1991. Effects of cadmium and polychlorinated biphenyls (Clophen A50) on steroid metabolism and cytochrome P-450 monooxygenase system in the sea star *Asterias rubens* L. Aquat. Toxicol. (Amst.) 20:95-110.

DeVault, D. S., W. A. Willford, R. J. Hesselberg, D. A. Nortrupt, E. G. S. Rundberg, A. K. Alwan, and C. Baustita. 1986. Contaminant trends in lake trout *(Salvelinus namaycush)* from the upper Great Lakes. Arch. Environ. Contam. Toxicol. 15:349-356.

de Voogt, P., and U. A. T. Brinkman. 1989. Production, properties, and usage of polychlorinated biphenyls. p. 3-45. In R. D. Kimbrough and A. A. Jensen (Eds.). Halogenated biphenyls, terphenyls, naphthalenes, dibenzodioxins, and related products. Elsevier Scientific Publishers, Amsterdam.

Dillon, T. M., W. H. Benson, R. A. Stackhouse, and A. M. Crider. 1990. Effects of selected PCB congeners on survival, growth, and reproduction in *Daphnia magna*. Environ. Toxicol. Chem. 9:1317-1326.

Dillon, T. M., and W. D. S. Burton. 1991. Acute toxicity of PCB congeners to *Daphnia magna* and *Pimephales promelas*. Bull. Environ. Contam. Toxicol. 46:208-215.

Drongowski, R. A., S. J. Wood, and G. R. Bouck. 1975. Thyroid activity in coho salmon from Oregon and Lake Michigan. Trans. Am. Fish. Soc. 104:349-352.

Duinker, J. C., and M. T. J. Hillebrand. 1983. Composition of PCB mixtures in biotic and abiotic marine compartments (Dutch Wadden Sea). Bull. Environ. Contam. Toxicol. 31:25-32.

Duke, T. W., J. I. Lowe, and A. J. Wilson, Jr. 1970. A polychlorinated biphenyl (Aroclor 1254®) in the water, sediment, and biota of Escambia Bay, Florida. Bull. Environ. Contam. Toxicol. 5:171-180.

Eisenreich, S. J., B. B. Looney, and J. D. Thornton. 1981. Airborne organic contaminants in the Great Lakes ecosystem. Environ. Sci. Technol. 15:30-38.

El-Gendy, K. S., A. A. Abdalla, H. A. Aly, G. Tantawy, and A. H. El-Sebae. 1991. Residue levels of chlorinated hydrocarbon compounds in water and sediment samples from Nile branches in the delta, Egypt. J. Environ. Sci. Health Part B Pestic. Food Contam. Agric. Wastes 26:15-36.

Evans, M. S., G. E. Noguchi, and C. P. Rice. 1991. The biomagnification of polychlorinated biphenyls, toxaphene, and DDT compounds in a Lake Michigan offshore food web. Arch. Environ. Contam. Toxicol. 20:87-93.

Fingerman, S. W., and M. Fingerman. 1979. Comparison of the effects of fourteen-day and chronic exposures to a polychlorinated biphenyl, Aroclor 1242, on molting of the fiddler crab, *Uca pugilator*. Bull. Environ. Contam. Toxicol. 21:352-357.

Fingerman, S. W., and M. Fingerman. 1980. Inhibition by the polychlorinated biphenyl Aroclor 1242 of limb regeneration in the fiddler crab, *Uca pugilator*, in different salinities from which different numbers of limbs have been removed. Bull. Environ. Contam. Toxicol. 25:744-750.

Fingerman, S. W., and L. C. Russell. 1980. Effects of the polychlorinated biphenyl Aroclor 1242 on locomotor activity and on the neurotransmitters dopamine and norepinephrine in the brain of the gulf killifish, *Fundulus grandis*. Bull. Environ. Contam. Toxicol. 25:682-687.

Folmar, L. C., W. W. Dickhoff, W. S. Zaugg, and H. O. Hodgins. 1982. The effects of Aroclor 1254 and No. 2 fuel oil on smoltification and sea-water adaptation of coho salmon *(Oncorhynchus kisutch)*. Aquat. Toxicol. (Amst.) 2:291-299.

Freeman, H. C., and D. R. Idler. 1975. The effect of polychlorinated biphenyl on steroidogenesis and reproduction of brook trout *(Salvelinus fontinalis)*. Can. J. Biochem. 53:666-670.

Freeman, H. C., G. B. Sangalang, and B. Fleming. 1982. The sublethal effects of polychlorinated biphenyl (Aroclor 1254) diet on the Atlantic cod *(Gadus morhua)*. Sci. Total Environ. 24:1-11.

Fries, C. R., and R. F. Lee. 1984. Pollutant effects on the mixed function oxygenase (MFO) and reproductive systems of the marine polychaete *Nereis virens*. Mar. Biol. 79:187-193.

Furlong, E. T., D. S. Carter, and R. A. Hites. 1988. Organic contaminants in sediments from the Trenton Channel of the Detroit River, Michigan. J. Great Lakes Res. 14:489-501.

Galgani, F., G. Bocquene, M. Lucon, D. Grzebyk, F. Letrouit, and D. Claisse. 1991. EROD measurements in fish from the northwest part of France. Mar. Pollut. Bull. 22:494-500.

Gardner, G. R., R. J. Pruell, and L. C. Folmar. 1989. A comparison of both neoplastic and non-neoplastic disorders in winter flounder *(Pseudopleuronectes americanus)* from eight areas in New England. Mar. Environ. Res. 28:393-397.

Gilbertson, M. 1989. Effects on fish and wildlife populations. p. 103-127. *In* R. D. Kimbrough and A. A. Jensen (Eds.). Halogenated biphenyls, terphenyls, naphthalenes, dibenzodioxins, and related products. Elsevier Scientific Publishers, Amsterdam.

Glooschenko, V., and W. Glooschenko. 1975. Effect of polychlorinated biphenyl compounds on growth of Great Lakes phytoplankton. Can. J. Bot. 53:653-659.

Grizzle, J. M., T. E. Schwedler, and A. L. Scott. 1981. Papillomas of black bullheads, *Ictalurus melas* (Rafinesque), living in a chlorinated sewage pond. J. Fish Dis. 4:345-351.

Gundersen, D. T., and W. D. Pearson. 1992. Partitioning of PCBs in the muscle and reproductive tissues of paddlefish, *Polyodon spathula*, at the falls of the Ohio River. Bull. Environ. Contam. Toxicol. 49:455-462.

Hacking, M. A., J. Budd, and K. Hodson. 1977. The ultrastructure of the liver of the rainbow trout: normal structure and modifications after chronic administration of a polychlorinated biphenyl Aroclor 1254. Can. J. Zool. 56:477-491.

Hale, R. C., J. Greaves, J. L. Gunderson, and R. F. Mothershead II. 1991. Occurrence of organochlorine contaminants in tissues of the coelacanth *Latimeria chalumnae*. Environ. Biol. Fishes 32:361-367.

Halter, M. T., and H. E. Johnson. 1974. Acute toxicities of a polychlorinated biphenyl (PCB) and DDT alone and in combination to early life stages of coho salmon *(Oncorhynchus kisutch)*. J. Fish. Res. Board Can. 31:1543-1547.

Hamilton, S. J., P. M. Mehrle, F. L. Mayer, and J. R. Jones. 1981. Mechanical properties of bone in channel catfish as affected by vitamin C and toxaphene. Trans. Am. Fish. Soc. 110:718-724.

Hansen, D. J., S. C. Schimmel, and E. Matthews. 1974. Avoidance of Aroclor® 1254 by shrimp and fishes. Bull. Environ. Contam. Toxicol. 12:253-256.

Hansen, L. G. 1987. Environmental toxicity of polychlorinated biphenyls. p. 15-48. *In* S. Safe and O. Hutzinger (Eds.). Environmental toxin series I. Springer-Verlag, Berlin.

Hansen, L. G., W. B. Wiekhotst, and J. Simon. 1976. Effects of dietary Aroclor® 1242 on channel catfish *(Ictalurus punctatus)* and the selective accumulation of PCB components. J. Fish. Res. Board Can. 33:1343-1352.

Hanumante, M. M., S. W. Fingerman, and M. Fingerman. 1981. Antagonism of the inhibitory effect of the polychlorinated biphenyl preparation, Aroclor 1242, on color changes of the fiddler crab, *Uca pugilator*, by norepinephrine and drugs affecting noradrenergic neurotransmission. Bull. Environ. Contam. Toxicol. 26:479-484.

Harding, L. W., Jr., and J. H. Phillips, Jr. 1978. Polychlorinated biphenyl (PCB) effects on marine phytoplankton photosynthesis and cell division. Mar. Biol. 49:93-101.

Hargrave, B. T., G. C. Harding, W. P. Vass, P. E. Erickson, B. R. Fowler, and V. Scott. 1992. Organochlorine pesticides and polychlorinated biphenyls in the Arctic Ocean food web. Arch. Environ. Contam. Toxicol. 22:41-54.

Harshbarger, J. C., and J. B. Clark. 1990. Epizootiology of neoplasms in bony fish of North America. Sci. Total Environ. 94:1-32.

Hattula, M. L., and O. Karlog. 1972. Toxicity of PCBs to goldfish. Acta Pharmacol. Toxicol. 31:238-240.

Hawker, D. W., and D. W. Connell. 1988. Octanol-water partition coefficients of polychlorinated biphenyl congeners. Environ. Sci. Technol. 22:382-387.

Hawkins, W. E., W. W. Walker, R. M. Overstreet, J. S. Lytle, and T. F. Lytle. 1990. Carcinogenic effects of some polychlorinated aromatic hydrocarbons on the Japanese medaka and guppy in waterborne exposures. Sci. Total Environ. 94:155-167.

Hendricks, J. D., W. T. Scott, T. P. Putnam, and R. O. Sinnhuber. 1981. Enhancement of aflatoxin B_1 hepatocarcinogenesis in rainbow trout *(Salmo gairdneri)* embryos by prior exposure of gravid females to dietary Aroclor 1254. p. 203-214. *In* D. R. Branson and K. L. Dickson (Eds.). Aquatic toxicology and hazard assessment: fourth conference. ASTM STP 737. American Society for Testing and Materials, Philadelphia, Pa.

Hesselberg, R. J., and J. G. Seelye. 1977. Identification of organic compounds in Great Lakes fishes by gas chromatography/mass spectrometry. Great Lakes Fish. Lab. Admin. Rep. No. 82-1. U.S. Fish and Wildlife Service, Ann Arbor, MI.

Hilton, J. W., P. V. Hodson, H. E. Braun, J. L. Leatherland, and S. J. Slinger. 1983. Contaminant accumulation and physiological response in rainbow trout *(Salmo gairdneri)* reared on naturally contaminated diets. Can. J. Fish. Aquat. Sci. 40:1987-1994.

Hiraizumi, Y., M. Takahashi, and H. Nishimura. 1979. Adsorption of polychlorinated biphenyl onto sea bed sediment, marine plankton, and other adsorbing agents. Environ. Sci. Technol. 13:580-583.

Hogan, J. W., and J. L. Brauhn. 1975. Abnormal rainbow trout fry from eggs containing high residues of a PCB (Aroclor 1242). Prog. Fish-Cult. 37:229-230.

Hose, J. E., J. N. Cross, S. G. Smith, and D. Diehl. 1987. Elevated circulating erythrocyte micronuclei in fishes from a contaminated site off southern California. Mar. Environ. Res. 22:167-176.

Hutzinger, O., S. Safe, and V. Zitko. 1974. The chemistry of PCB's. CRC Press, Cleveland, Oh. 269 pp.

Janz, D. M., and C. D. Metcalfe. 1991. Nonadditive interactions of mixtures of 2,3,7,8-TCDD and 3,3',4,4'-tetrachlorobiphenyl on aryl hydrocarbon hydroxylase induction in rainbow trout *(Oncorhynchus mykiss)*. Chemosphere 23:467-472.

Jensen, S. 1972. The PCB story. Ambio 1:123-131.
Johnson, L. L., E. Casillas, T. K. Collier, B. B. McCain, and U. Varanasi. 1988. Contaminant effects on ovarian development in English sole *(Parophrys vetulus)* from Puget Sound, Washington. Can. J. Fish. Aquat. Sci. 45:2133-2146.
Kawano, M., S. Matsushita, T. Inoue, H. Tanaka, and R. Tatsukawa. 1986. Biological accumulation of chlordane compounds in marine organisms from the northern North Pacific and Bering Sea. Mar. Pollut. Bull. 17:512-516.
Keil, J. E., L. E. Priester, and S. H. Sandifer. 1971. Polychlorinated biphenyl (Aroclor 1242®): effects of uptake on growth, nucleic acids, and chlorophyll of a marine diatom. Bull. Environ. Contam. Toxicol. 6:156-159.
Kiessling, A., P. Pärt, O. Ring, and K. Lindahl-Kiessling. 1983. Effects of PCB on the adrenergic response in perfused gills and on levels of muscle glycogen in rainbow trout *(Salmo gairdneri* Rich.). Bull. Environ. Contam. Toxicol. 31:712-718.
Kirby, G. M., and M. A. Hayes. 1992. Significance of liver neoplasia in wild fish: assessment of pathophysiologic responses of a biomonitor species to multiple stress factors. p. 106-116. *In* A. J. Niimi and M. C. Taylor (Eds.). Proc. 18th Annu. Aquat. Toxicity Workshop, Ottawa, Ontario, Canada. September 30-October 3, 1991. Can. Tech. Rep. Fish. Aquat. Sci. No. 1863. Ottawa, Canada.
Kjølholt, J., and M. M. Hansen. 1986. PCBs in scallops and sediments from north Greenland. Mar. Pollut. Bull. 17:432-434.
Klotz, A. V., J. J. Stegmann, and C. Walsh. 1983. An aryl hydrocarbon hydroxylating hepatic cytochrome P-450 from the marine fish *Stenotomus chrysops*. Arch. Biochem. Biophys. 226:578-592.
Knickmeyer, R., and H. Steinhart. 1989. Cyclic organochlorines in plankton from the North Sea in spring. Estuarine Coastal Shelf Sci. 28:117-127.
Koch, R. B., D. Desaiah, H. H. Yap, and L. K. Cutkomp. 1972. Polychlorinated biphenyls: effect of long-term exposure on ATPase activity in fish, *Pimephales promelas*. Bull. Environ. Contam. Toxicol. 7:87-92.
Köhler, A. 1990. Identification of contaminant-induced cellular and subcellular lesions in the liver of flounder *(Platichthys flesus* L.) caught at differently polluted estuaries. Aquat. Toxicol. (Amst.) 16:271-294.
Kuehl, D. W., E. Durham, B. Butterworth, and D. Linn. 1984. Identification of polychlorinated planar chemicals in fishes from major watersheds near the Great Lakes. Environ. Int. 10:45-49.
Lake, J. L., R. J. Pruell, and F. A. Osterman. 1992. An examination of dechlorination processes and pathways in New Bedford Harbor sediments. Mar. Environ. Res. 33:31-47.
Langlois, C. 1987. Étude préliminarie de la qualité des 15 cours d'eau majeurs du Nouveau-Québec. (In French.) Water Pollut. Res. J. Can. 22:530-544.
Larsson, P., L. Collvin, L. Okla, and G. Meyer. 1992. Lake productivity and water chemistry as governors of the uptake of persistent pollutants in fish. Environ. Sci. Technol. 26:346-352.
Larsson, P., L. Okla, S.-O. Ryding, and B. Westöö. 1990. Contaminated sediment as a source of PCBs in a river system. Can. J. Fish. Aquat. Sci. 47:746-754.
Leatherland, J. F., and R. A. Sonstegard. 1978. Lowering of serum thyroxine and triiodothyronine levels in yearling coho salmon, *Oncorhynchus kisutch*, by dietary mirex and PCBs. J. Fish. Res. Board Can. 35:1285-1289.
Lieb, A. J., D. D. Bills, and R. O. Sinnhuber. 1974. Accumulation of dietary polychlorinated biphenyls (Aroclor 1254) by rainbow trout *(Salmo gairdneri)*. J. Agric. Food Chem. 22:638-642.

Lockerbie, D. M., and T. A. Clair. 1988. Organic contaminants in isolated lakes of southern Labrador, Canada. Bull. Environ. Contam. Toxicol. 441:625-632.

Lockhart, W. L., R. Wagemann, B. Tracey, D. Sutherland, and D. J. Thomas. 1992. Presence and implications of chemical contaminants in the freshwaters of the Canadian Arctic. Sci. Total Environ. 122:165-243.

Loganathan, B. G., S. Tanabe, M. Goto, and R. Tastukawa. 1989. Temporal trends of organochlorine residues in lizard goby *Rhinogobius flumineus* from the River Nagaragawa, Japan. Environ. Pollut. 62:237-251.

Luard, E. J. 1973. Sensitivity of *Dunaliella* and *Scenedesmus* (Chlorophyceae) to chlorinated hydrocarbons. Phycologia 12:29-33.

Luckas, B., and M. Oehme. 1990. Characteristic contamination levels for polychlorinated hydrocarbons, dibenzofurans, and dibenzo-*p*-dioxins in bream *(Abramis brama)* from the river Elbe. Chemosphere 21:79-89.

Luxon, P. L., P. V. Hodson, and U. Borgmann. 1987. Hepatic aryl hydroxylase activity of lake trout *(Salvelinus namaycush)* as an indicator of organic pollution. Environ. Toxicol. Chem. 6:649-657.

Malins, D. C., B. B. McCain, D. W. Brown, S.-L. Chan, M. S. Myers, J. T. Landahl, P. G. Prohaska, A. J. Friedman, L. D. Rhodes, D. G. Burrows, W. D. Gronlund, and H. O. Hodgins. 1984. Chemical pollutants in sediments and diseases of bottom-dwelling fish in Puget Sound, Washington. Environ. Sci. Technol. 18:705-713.

Marchand, M., and J. C. Caprais. 1985. Hydrocarbons and halogenated hydrocarbons in coastal waters of the English Channel and the North Sea. Mar. Pollut. Bull. 16:78-81.

Marchand, M., J. C. Caprais, and P. Pignet. 1988. Hydrocarbons and halogenated hydrocarbons in coastal waters of the western Mediterranean (France). Mar. Environ. Res. 25:131-159.

Marcus, J. M., and R. T. Renfrow. 1990. Pesticides and PCBs in South Carolina estuaries. Mar. Pollut. Bull. 21:96-99.

Marthinsen, I., G. Staveland, J. U. Skaare, K. I Ugland, and A. Haugen. 1991. Levels of environmental pollutants in male and female flounder *(Platichthys flesus* L.) and cod *(Gadus morhua* L.) caught during the year 1988 near or in the waterway of Glomma, the largest river in Norway. I. Polychlorinated biphenyls. Arch. Environ. Contam. Toxicol. 20:353-360.

Mauck, W. L., P. M. Mehrle, and F. L. Mayer. 1978. Effects of the polychlorinated biphenyl Aroclor® 1254 on growth, survival, and bone development in brook trout *(Salvelinus fontinalis)*. J. Fish. Res. Board Can. 35:1084-1088.

Mayer, F. J., P. M. Mehrle, and H. O. Sanders. 1977. Residue dynamics and biological effects of polychlorinated biphenyls in aquatic organisms. Arch. Environ. Contam. Toxicol. 5:501-511.

Mayer, K. S., F. L. Mayer, and A. Witt, Jr. 1985. Waste transformer oil and PCB toxicity to rainbow trout. Trans. Am. Fish. Soc. 114:869-886.

McCain, B. B., S.-L. Chan, M. M. Krahn, D. W. Brown, M. S. Myers, J. T. Landahl, S. Pierce, R. C. Clark, Jr., and U. Varanasi. 1992. Chemical contamination and associated fish diseases in San Diego Bay. Environ. Sci. Technol. 26:725-733.

McCrea, R. C., and J. D. Fischer. 1986. Heavy metal and organochlorine contaminants in the five major Ontario rivers of the Hudson Bay lowland. Water Pollut. Res. J. Can. 21:225-234.

McFarland, V. A., and J. U. Clarke. 1989. Environmental occurrence, abundance, and potential toxicity of polychlorinated biphenyl congeners: considerations for a congener-specific analysis. Environ. Health Perspect. 81:225-239.

McLeese, D. W., and C. D. Metcalfe. 1980. Toxicities of eight organochlorine compounds in sediment and seawater to *Crangon septemspinosa*. Bull. Environ. Contam. Toxicol. 25:921-928.

Mehrle, P. M., T. A. Haines, S. Hamilton, J. L. Ludke, F. L. Mayer, and M. A. Ribick. 1982. Relationship between body contaminants and bone development in East-Coast striped bass. Trans. Am. Fish. Soc. 111:231-241.

Metcalfe, C. D. 1988. Induction of micronuclei and nuclear abnormalities in the erythrocytes of mudminnows *(Umbra limi)* and brown bullheads *(Ictalurus nebulosus)*. Bull. Environ. Contam. Toxicol. 40:489-495.

Miller, D. L., and D. M. Ogilvie. 1975. Temperature selection in brook trout *(Salvelinus fontinalis)* following exposure to DDT, PCB, or phenol. Bull. Environ. Contam. Toxicol. 14:545-551.

Miranda, C. L., M. C. Henderson, J.-L. Wang, H-S. Chang, J. D. Hendricks, and D. R. Buhler. 1992. Differential effects of 3,4,5,3',4',5'-hexachlorobiphenyl (HCB) on interrenal steroidogenesis in male and female rainbow trout *Oncorhynchus mykiss*. Comp. Biochem. Physiol. C Comp. Pharmacol. Toxicol. 103:153-157.

Mix, M. C. 1986. Cancerous diseases in aquatic animals and their association with environmental pollutants: a critical literature review. Mar. Environ. Res. 20:1-141.

Moccia, R. D., J. F. Leatherland, and R. A. Sonstegard. 1977. Increasing frequency of thyroid goiters in coho salmon *(Oncorhynchus kisutch)* in the Great Lakes. Science (Washington, D.C.) 198:425-426.

Möller, H. 1988. The problem of quantifying long-term changes in the prevalence of tumours and non-specific growths of fish. J. Cons. Int. Explor. Mer 45:33-38.

Möller, H. 1990. Association between diseases of flounder *(Platichthys flesus)* and environmental conditions in the Elbe estuary, FRG. J. Cons. Int. Explor. Mer 46:187-199.

Monod, G. 1985. Egg mortality of Lake Geneva charr *(Salvelinus alpinus* L.) contaminated by PCB and DDT derivatives. Bull. Environ. Contam. Toxicol. 35:531-536.

Moore, S. A., Jr., and R. C. Harriss. 1972. Effects of polychlorinated biphenyl on marine phytoplankton communities. Nature (Lond.) 240:356-358.

Morgan, J. R. 1972. Effects of Aroclor 1242® (a polychlorinated biphenyl) and DDT on cultures of an alga, protozoan, daphnid, ostracod, and guppy. Bull. Environ. Contam. Toxicol. 8:129-137.

Morrissey, R. E., and B. A. Schwartz. 1989. Reproductive and developmental toxicity in animals. p. 195-225. *In* R. D. Kimbrough and A. A. Jensen (Eds.). Halogenated biphenyls, terphenyls, naphthalenes, dibenzodioxins, and related products. Elsevier Scientific Publishers, Amsterdam.

Mosser, J. L., N. S. Fisher, T.-C. Teng, and C. F. Wurster. 1972. Polychlorinated biphenyls: toxicity to certain phytoplankters. Science (Washington, D.C.) 175:191-192.

Mouvet, C., M. Galoux, and A. Bernes. 1985. Monitoring of polychlorinated biphenyls (PCBs) and hexachlorocyclohexanes (HCH) in freshwater using the aquatic moss *Cinclidotus danubicus*. Sci. Total Environ. 44:253-267.

Mowrer, J., K. Aswald, G. Burgermeister, L. Machado, and J. Tarradellas. 1982. PCB in a Lake Geneva ecosystem. Ambio 11:355-358.

Muir, D. C. G., R. Wagemann, B. T. Hargrave, D. J. Thomas, D. B. Peakall, and R. J. Norstrom. 1992. Arctic marine ecosystem contamination. Sci. Total Environ. 122:75-134.

Murphy, T. J., M. D. Mullin, and J. A. Meyer. 1987. Equilibrium of polychlorinated biphenyls and toxaphene with air and water. Environ. Sci. Technol. 21:155-162.

Myers, M. S., J. T. Landahl, M. M. Krahn, L. L. Johnson, and B. B. McCain. 1990. Overview of studies on liver carcinogenesis in English sole from Puget Sound; evidence for a xenobiotic chemical etiology. I. Pathology and epizootiology. Sci. Total Environ. 94:33-50.

Myers, M. S., J. T. Landahl, M. M. Krahn, and B. B. McCain. 1991. Relationships between hepatic neoplasms and related lesions and exposure to toxic chemicals in marine fish from the U.S. West Coast. Environ. Health Pespect. 90:7-15.

Narbonne, J. F., and J. L. Gallis. 1979. *In vivo* and *in vitro* effect of Phenoclor DP6 on drug metabolizing activity in mullet liver. Bull. Environ. Contam. Toxicol. 23:338-343.

Nasci, C., G. Campesan, V. U. Fossato, L. Tallandini, and M. Turchetto. 1991. Induction of cytochrome P-450 and mixed function oxygenase activity by low concentrations of polychlorinated biphenyls in marine fish *Zoesterissor ophiocephalus* (Pall.). Aquat. Toxicol. (Amst.) 19:281-290.

Nebeker, A. V., and F. A. Puglisi. 1974. Effect of polychlorinated biphenyls (PCB's) on survival and reproduction of *Daphnia, Gammarus,* and *Tanytarus*. Trans. Am. Fish. Soc. 103:722-728.

Nebeker, A. V., F. A. Puglisi, and D. L. DeFoe. 1974. Effect of polychlorinated biphenyl compounds on survival and reproduction of the fathead minnow and flagfish. Trans. Am. Fish. Soc. 103:562-568.

Niimi, A. J. 1985. Use of laboratory studies in assessing the behavior of contaminants in fish inhabiting natural ecosystems. Water Pollut. Res. J. Can. 20:79-88.

Niimi, A. J. 1990. Review of biochemical methods and other indicators to assess fish health in aquatic ecosystems containing toxic chemicals. J. Great Lakes Res. 16:529-541.

Niimi, A. J. 1994. PCBs, PCDDs, and PCDFs. p. 204-243. *In* P. Calow (Ed.). Handbook of ecotoxicology. Vol. 2. Blackwell Scientific Publications, Oxford.

Niimi, A. J., and B. G. Oliver. 1989a. Distribution of polychlorinated biphenyl congeners and other halocarbons in whole fish and muscle among Lake Ontario salmonids. Environ. Sci. Technol. 23:83-88.

Niimi, A. J., and B. G. Oliver. 1989b. Assessment of relative toxicity of chlorinated dibenzo-*p*-dioxins, dibenzofurans, and biphenyls in Lake Ontario salmonids to mammalian systems using toxic equivalent factors (TEFs). Chemosphere 18:1413-1423.

Nimmo, D. R., J. Forester, P. T. Heitmuller, and G. H. Cook. 1974. Accumulation of Aroclor® 1254 in grass shrimp *(Palaemonetes pugio)* in laboratory and field exposures. Bull. Environ. Contam. Toxicol. 11:303-308.

Nimmo, D. R., D. J. Hansen, J. A. Couch, N. R. Cooley, P. R. Parrish, and J. I. Lowe. 1975. Toxicity of Aroclor® 1254 and its physiological activity in several estuarine organisms. Arch. Environ. Contam. Toxicol. 3:22-39.

O'Conner, T. P. 1991. Concentrations of organic contaminants in mollusks and sediments at NOAA national status and trend sites in the coastal and estuarine United States. Environ. Health Perspect. 90:69-73.

Oliver, B. G., and A. J. Niimi. 1988. Trophodynamic analysis of polychlorinated biphenyl congeners and other chlorinated hydrocarbons in the Lake Ontario ecosystem. Environ. Sci. Technol. 22:388-397.

Opperhuizen, A., F. A. P. C. Gobas, and J. M. D. Van der Steen. 1988. Aqueous solubility of polychlorinated biphenyls related to molecular structure. Environ. Sci. Technol. 22:638-646.

Patil, G. S. 1991. Correlation of aqueous solubility and octanol-water partition coefficient based on molecular structure. Chemosphere 22:723-738.

Pavlou, S. P., and R. N. Dexter. 1979. Physical and chemical aspects of the distribution of polychlorinated biphenyls in the aquatic environment. p. 195-211. *In* L. L. Marking and R. A. Kimerle (Eds.). Aquatic toxicology. ASTM STP 667. American Society for Testing and Materials, Philadelphia, PA.

Pavoni, B., C. Calvo, A. Sfriso, and A. A. Orio. 1990. Time trend of PCB concentrations in surface sediments from a hypertrophic, macroalgae populated area of the lagoon of Venice. Sci. Total Environ. 91:13-21.

Payne, J. F., L. L. Fancey, A. D. Rahimtula, and E. L. Porter. 1987. Review and perspective on the use of mixed-function oxygenase enzymes in biological monitoring. Comp. Biochem. Physiol. C 86:233-245.

Peterson, R. H. 1973. Temperature selection of Atlantic salmon *(Salmo salar)* and brook trout *(Salvelinus fontinalis)* as influenced by various chlorinated hydrocarbons. J. Fish. Res. Board Can. 30:1091-1097.

Prescott, L. M., M. K. Kubovec, and D. Tryggestad. 1977. The effects of pesticides, polychlorinated biphenyls, and metals on the growth and reproduction of *Acanthamoeba castellanii.* Bull. Environ. Contam. Toxicol. 18:29-34.

Pyysalo, H., K. Wickström, and R. Litmanen. 1983. A baseline study on the concentrations of chlordane-, PCB-, and DDT-compounds in Finnish fish samples in the year 1982. Chemosphere 12:837-842.

Ramesh, A., S. Tanabe, A. N. Subramanian, D. Mohan, V. K. Venugopalan, and R. Tatsukawa. 1990. Persistent organochlorine residues in green mussels from coastal waters of south India. Mar. Pollut. Bull. 21:587-590.

Rapaport, R. A., and S. J. Eisenreich. 1984. Chromatographic determination of octanol-water partition coefficients (K_{ow}s) for 58 polychlorinated biphenyl congeners. Environ. Sci. Technol. 18:163-170.

Rice, C. P., and D. S. White. 1987. PCB availability assessment of river dredging using caged clams and fish. Environ. Toxicol. Chem. 6:259-274.

Robertson, O. H., and A. L. Chaney. 1953. Thyroid hyperplasia and tissue iodine content in spawning rainbow trout: a comparative study of Lake Michigan and California sea-run trout. Physiol. Zool. 26:328-340.

Roesijadi, G., S. R. Petrocelli, J. W. Anderson, C. S. Giam, and G. E. Neff. 1976. Toxicity of polychlorinated biphenyls (Aroclor 1254) to adult, juvenile, and larval stages of the shrimp *Palaemonetes pugio.* Bull. Environ. Contam. Toxicol. 15:297-304.

Rossel, D., P. Honsberger, and J. Tarradellas. 1987. Bioaccumulative behaviour of some PCB congeners in Lake Geneva brown trout *(Salmo trutta* L.). Int. J. Environ. Anal. Chem. 31:219-233.

Sanders, H. O., and J. H. Chandler. 1972. Biological magnification of a polychlorinated biphenyl (Aroclor® 1254) from water by aquatic invertebrates. Bull. Environ. Contam. Toxicol. 7:257-263.

Sanders, M., and B. L. Haynes. 1988. Distribution pattern and reduction of polychlorinated biphenyls (PCB) in bluefish *Pomatomus saltatrix* (Linnaeus) fillets through adipose tissue removal. Bull. Environ. Contam. Toxicol. 41:670-677.

Sauer, T. C., Jr., G. S. Durell, J. S. Brown, D. Redford, and P. D. Boehm. 1989. Concentrations of chlorinated pesticides and PCBs in microlayer and seawater samples collected in open-ocean waters off the U.S. East Coast and in the Gulf of Mexico. Mar. Chem. 27:235-257.

Schimmel, S. C., D. J. Hansen, and J. Forester. 1974. Effects of Aroclor® 1254 on laboratory-reared embryos and fry of sheepshead minnows *(Cyprinodon variegatus).* Trans. Am. Fish. Soc. 103:582-586.

Schulz, D. E., G. Petrick, and J. C. Duinker. 1988. Chlorinated biphenyls in North Atlantic surface and deep water. Mar. Pollut. Bull. 19:526-531.

Schulz, D. E., G. Pettrick, and J. C. Duinker. 1989. Complete characterization of polychlorinated biphenyl congeners in commercial Aroclor and Clophen mixtures by multidimensional gas chromatography-electron capture detection. Environ. Sci. Technol. 23:852-859.

Schulz-Bull, D. E., G. Petrick, and J. C. Duinker. 1991. Polychlorinated biphenyls in North Sea water. Mar. Chem. 36:365-384.

Scott, K., M. J. Leaver, and S. G. George. 1992. Regulation of hepatic glutathione S-transferase expression in flounder. Mar. Environ. Res. 34:233-236.

Setzler-Hamilton, E. M., J. A. Whipple, and R. B. MacFarlane. 1988. Striped bass populations in Chesapeake and San Francisco Bays: two environmentally impacted estuaries. Mar. Pollut. Bull. 19:466-477.

Shelton, D. W., R. A. Coulombe, C. B. Pereira, J. L. Casteel, and J. D. Hendricks. 1983. Inhibitory effect of Aroclor 1254 on aflatoxin-initiated carcinogenesis in rainbow trout and mutagenesis using a *Salmonella*/trout hepatic activation system. Aquat. Toxicol. (Amst.) 3:229-238.

Shelton, D. W., J. D. Hendricks, R. A. Coulombe, and G. S. Bailey. 1984. Effect of dose on the inhibition of carcinogenesis/mutagenesis by Aroclor 1254 in rainbow trout fed aflatoxin B_1. J. Toxicol. Environ. Health 13:649-657.

Silberhorn, E. M., H. P. Glauert, and L. W. Robertson. 1990. Carcinogenicity of polyhalogenated biphenyls: PCBs and PBBs. Crit. Rev. Toxicol. 20:440-496.

Sivarajah, K., C. S. Franklin, and W. P. Williams. 1978a. The effects of polychlorinated biphenyls on plasma steroid level and hepatic microsomal enzymes in fish. J. Fish Biol. 13:401-409.

Sivarajah, K., C. S. Franklin, and W. P. Williams. 1978b. Some histopathological effects of Aroclor 1254 on the liver and gonads of rainbow trout, *Salmo gairdneri*, and carp, *Cyprinus carpio*. J. Fish Biol. 13:411-414.

Sloan, R. J., K. W. Simpson, R. A. Schroeder, and C. R. Barnes. 1983. Temporal trends toward stability of Hudson River PCB contamination. Bull. Environ. Contam. Toxicol. 31:377-385.

Smith, C. E., T. H. Peck, R. J. Klauda, and J. B. McLaren. 1979. Hepatomas in Atlantic tomcod *Microgadus tomcod* collected in the Hudson River estuary in New York. J. Fish Dis. 2:313-319.

Snarski, V. M. 1982. The response of rainbow trout *Salmo gairdneri* to *Aeromonas hydrophila* after sublethal exposures to PCB and copper. Environ. Pollut. Ser. A Ecol. Biol. 28:219-232.

Södergren, A. 1971. Accumulation and distribution of chlorinated hydrocarbons in cultures of *Chlorella pyrenoidosa* (Chlorophyceae). Oikos 22:215-220.

Sonstegard, R. A. 1977. Environmental carcinogenesis studies in fishes of the Great Lakes of North America. Ann. N.Y. Acad. Sci. 298:261-291.

Spies, R. B., J. S. Felton, and L. Dillard. 1982. Hepatic mixed-function oxidases in California flatfishes are increased in contaminated environments and by oil and PCB ingestion. Mar. Biol. 70:117-127.

Spies, R. B., and D. W. Rice, Jr. 1988. Effects of organic chemicals of the starry flounder *Platichthys stellatus* in San Francisco Bay. II. Reproductive success of fish captured in San Francisco Bay and spawned in the laboratory. Mar. Biol. 98:191-200.

Spitsbergen, J. M., K. A. Schat, J. M. Kleeman, and R. E. Peterson. 1988. Effects of 2,3,7,8-tetrachlorodibenzo-*p*-dioxin (TCDD) or Aroclor 1254 on the resistance of rainbow trout, *Salmo gairdneri* Richardson, to infectious haematopoietic necrosis virus. J. Fish Dis. 11:73-83.

Stahl, R. G., Jr. 1979. Effect of a PCB (Aroclor 1254) on the striped hermit crab, *Clibanarius vittatus* (Anomura: Diogenidae) in static bioassays. Bull. Environ. Contam. Toxicol. 23:91-94.

Stauffer, T. W. 1979. Effects of DDT and PCB's on survival of lake trout eggs and fry in a hatchery and in Lake Michigan, 1973/1976. Trans. Am. Fish. Soc. 108:178-186.

Steimle, R. W., V. S. Zdanowicz, and D. F. Gadbois. 1990. Metals and organic contaminants in northwest Atlantic deep-sea tilefish tissues. Mar. Pollut. Bull. 21:530-535.

Stein, J. E., T. Hom, D. Casillas, A. Friedman, and U. Varanasi. 1987. Simultaneous exposure of English sole *(Parophrys vetulus)* to sediment-associated xenobiotics. Part 2. Chronic exposure to an urban estuarine sediment with added ^{3}H-benzo[*a*]pyrene and ^{14}C-polychlorinated biphenyls. Mar. Environ. Res. 22:123-149.

Stout, V. F. 1986. What is happening to PCBs? p. 163-205. *In* J. S. Waid (Ed.). PCBs and the environment. CRC Press, Boca Raton, Fla.

Subramanian, B. R., S. Tanabe, H. Hidaka, and R. Tatsukawa. 1983. DDTs and PCB isomers and congeners in Antarctic fish. Arch. Environ. Contam. Toxicol. 12:621-626.

Swackhamer, D. L., and D. E. Armstrong. 1987. Distribution and characterization of PCBs in Lake Michigan water. J. Great Lakes Res. 13:24-36.

Swackhamer, D. L., B. D. McDeety, and R. A. Hites. 1988. Deposition and evaporation of polychlorobiphenyl congeners to and from Siskiwit Lake, Isle Royale, Lake Superior. Environ. Sci. Technol. 22:664-672.

Tanabe, S., H. Hidaka, and R. Tatsukawa. 1983. PCBs and chlorinated pesticides in Antarctic atmosphere and hydrosphere. Chemosphere 12:277-288.

Tanabe, S., N. Kannan, M. Fukushima, T. Okamoto, T. Wakimoto, and R. Tatsukawa. 1989. Persistent organochlorines in Japanese coastal waters: an introspective summary from a Far East developed nation. Mar. Pollut. Bull. 20:344-352.

Tanabe, S., H. Tanaka, and R. Tatsukawa. 1984. Polychlorobiphenyls, ΣDDT, and hexachlorocyclohexane isomers in the western North Pacific ecosystem. Arch. Environ. Contam. Toxicol. 13:731-738.

Thomann, R. V. 1981. Equilibrium model of fate of microcontaminants in diverse aquatic food chains. Can. J. Fish. Aquat. Sci. 38:280-296.

Thomas, P. 1988. Reproductive endocrine function in female Atlantic croaker exposed to pollutants. Mar. Environ. Res. 24:179-183.

Urey, J. C., J. C. Kricher, and J. M. Boylan. 1976. Bioconcentration of four pure PCB congeners by *Chlorella pyrenoidosa*. Bull. Environ. Contam. Toxicol. 16:81-85.

van Urk, G., F. C. M. Kerkum, and H. Smit. 1992. Life cycle patterns, density, and frequency of deformities in *Chironomus* larvae (Diptera: Chironomidae) over a contaminated sediment gradient. Can. J. Fish. Aquat. Sci. 49:2291-2299.

Veith, G. D., N. M. Austin, and R. T. Morris. 1979. A rapid method for estimating log P for organic chemicals. Water Res. 13:43-47.

von Westernhagen, H., H. Rosenthal, V. Dethlefsen, W. Ernst, U. Harms, and P.-D. Hansen. 1981. Bioaccumulating substances and reproductive success in Baltic flounder. Aquat. Toxicol. (Amst.) 1:85-99.

Vos, J. G., and M. I. Luster. 1989. Immune alterations. p. 295-322. *In* R. D. Kimbrough and A. A. Jensen (Eds.). Halogenated biphenyls, terphenyls, naphthalenes, dibenzodioxins, and related products. Elsevier Scientific Publishers, Amsterdam.

Voss, S. D., D. W. Shelton, and J. D. Hendricks. 1982. Effects of dietary Aroclor 1254 and cyclopropene fatty acids on hepatic enzymes in rainbow trout. Arch. Environ. Contam. Toxicol. 11:87-91.

Walker, M. K., and R. E. Peterson. 1991. Potencies of polychlorinated dibenzo-*p*-dioxin, dibenzofuran, and biphenyl congeners, relative to 2,3,7,8-tetrachlorodibenzo-*p*-dioxin, for producing early life stage mortality in rainbow trout *(Oncorhynchus mykiss)*. Aquat. Toxicol. (Amst.) 21:219-238.

Watermann, B., and H. Kranz. 1992. Pollution and fish diseases in the North Sea. Some historical aspects. Mar. Pollut. Bull. 24:131-138.

Westin, D. T., C. E. Olney, and B. A. Rogers. 1985. Effects of parental and dietary organochlorines on survival and body burdens of striped bass larvae. Trans. Am. Fish. Soc. 114:125-136.

Wildish, D. J. 1970. The toxicity of polychlorinated biphenyls (PCB) in sea water to *Gammarus oceanicus*. Bull. Environ. Contam. Toxicol. 5:202-204.

Williams, L. L., and J. P. Giesy. 1992. Relationships among concentrations of individual polychlorinated biphenyl (PCB) congeners, 2,3,7,8-tetrachlorodibenzo-*p*-dioxin equivalents (TCDD-EQ), and rearing mortality of chinook salmon *(Oncorhynchus tshawytscha)* eggs from Lake Michigan. J. Great Lakes Res. 18:108-124.

Winston, G. W., S. Narayan, and C. B. Henry. 1989. Induction pattern of liver microsomal alkoxyresorufin *O*-dealkylases of channel catfish *(Ictalurus punctatus)*: correlation with PCB exposure *in situ*. J. Environ. Sci. Health Part B Pestic. Food Contam. Agric. Wastes 24:277-289.

Wofford, H. W., and P. Thomas. 1988. Effect of xenobiotics on peroxidation of hepatic microsomal lipids from striped mullet *(Mugil cephalus)* and Atlantic croaker *(Micropogonias undulatus)*. Mar. Environ. Res. 24:285-289.

World Health Organization. 1976. Polychlorinated biphenyl and terphenyls. WHO, Geneva, 85 pp.

Zeiger, E. 1989. Genetic toxicity. p. 227-237. *In* R. D. Kimbrough and A. A. Jensen (Eds.). Halogenated biphenyls, terphenyls, naphthalenes, dibenzodioxins, and related products. Elsevier Scientific Publishers, Amsterdam.

CHAPTER 6

Toxicological Implications of PCB Residues in Mammals

Michael A. Kamrin and Robert K. Ringer

INTRODUCTION

Polychlorinated biphenyls (PCBs) are a group of aromatic organic chemicals consisting of 209 congeners sharing a common basic two-ring structure. They differ only in the number and placement of chlorine atoms on these rings. The manufacture of PCBs began in 1929 and continued until quite recently in some countries. It is estimated that over one billion pounds of PCBs were produced, mainly for use in electrical equipment. Because these compounds are very persistent and lipid soluble, they have been distributed widely around the globe, and concentrations of PCBs in living organisms have increased steadily over time, at least until the mid-1970s when PCBs were banned in some places.

It is over 25 years since PCBs were first detected in mammals (Jensen, 1972) and over 20 years since a connection between PCB exposure and reproductive toxicity in mink *(Mustela vison)* was established (Ringer et al., 1972; Platonow and Karstad, 1973; Ringer, 1983; Aulerich et al., 1985). As a result of these discoveries, experiments on laboratory or domesticated animals and studies of PCB residues in mammals in the wild were conducted and published during the 1970s and 1980s. Laboratory experiments (Zepp and Kirkpatrick, 1976; Sleight, 1983; Fuller and Hobson, 1987) confirmed the effects found in mink and provided a more complete description of the toxicity of PCBs in a variety of mammals. The experimental literature on the relation between PCB residues and toxicity is limited to studies on standard laboratory animals, except for a number of controlled studies on ranch mink.

Some of the field studies were epidemiological investigations that attempted to assess the relation between PCB residue levels and toxic effects, especially reproductive deficits. In the main, these investigations were performed on marine mammals, especially seals and sea lions. The host of confounding factors in these epidemiological studies makes it difficult to establish the existence of a relation between residue levels and toxicity. These confounders include other chemicals of known toxicity, e.g., dieldrin, 1,1,1-trichloro-2,2-*bis*(*p*-chlorophenyl)ethane (DDT), chlordane, and mercury; pathogenic organisms; and other stressors, such as habitat alterations caused by human activities.

In addition to these confounders, other factors that increase the difficulty in interpreting the results of field studies of mammals include the following:

- The state of the animal at the time of collection. (Was it already dead or was it killed at the time of collection?)
- The treatment of the sample between collection and analysis. (How long was it held and under what conditions?)
- The reproductive status of the animal. (Had the animal just given birth to offspring?)
- The analytical methodology used (the instrumentation selected and the standard chosen for comparison).

As a result of the above considerations, field studies have not provided a scientifically validated quantitative assessment of the PCB residue concentrations associated with adverse effects. Such an assessment is possible at present only for those few mammals studied under controlled experimental conditions, e.g., mink. In light of this, the discussion will focus on mink but will also discuss the state of information about other mammals, information that must be considered tentative because the cause and effect have not been established in these other species. Any use of these values to interpret field data should be done cautiously with the above considerations in mind.

Most residue information is available only as the sum of PCBs; however, there is a limited amount of congener-specific data, which will also be summarized. Results will be reported on a wet-weight basis unless otherwise noted. In addition to this chapter, the reader is directed to recent review articles on PCB residues in mammals (Eisler, 1986; Waid, 1987; Addison, 1989; Talmage and Walton, 1991). These articles provide an indication of the extent of data that are available, including results that are not easily accessible.

TOTAL PCB STUDIES

MINK

The original experimental studies suggesting the link between PCBs and reproductive problems in wildlife were performed on ranch mink. Although field studies on PCB levels in mink in the wild have been performed (Foley et al., 1988), there are no studies on links between reproductive toxicity and PCB levels in these populations. Thus, quantitative assessments of the associations between PCB residues and reproductive or other adverse effects are based solely on data derived from laboratory experiments (Wren, 1991).

It should be mentioned that one epidemiological lethality study was performed that compared a group of ranch mink *(M. vison)* that died in 1969 and 1970 on farms in Massachusetts with healthy animals in that state and in Virginia (Friedman et al., 1977). PCB levels in fat, on a lipid basis, varied from 6 to 60 ppm in the five of eight dead animals that were positive and from 0.3 to 0.5 ppm in the healthy animals, all of which were positive for PCBs. However, pathological analysis of the animals showed effects, e.g., lung congestion, that were distinct from those occurring in animals that died after experimental PCB ingestion. Thus, the authors concluded that the observed correlation between PCB levels and lethality did not represent causation.

Laboratory studies on ranch mink exposed to PCBs have been carried out in two different ways: one involving feeding mink industrial PCB mixtures and the other feeding mink food contaminated with PCBs. There are drawbacks to applying the results of either of these studies directly to animals in the wild. Mink do not ingest industrial PCB formulations in the wild, so the results of this type of study do not mimic the real world. On the other hand, feeding foods, like fish, that are naturally contaminated with PCBs also exposes the mink to other contaminants that are present in the fish.

One of the earliest laboratory studies involved the feeding of meat from cows that had been fed Aroclor 1254 to mink (Platonow and Karstad, 1973). Reproductive failure occurred at the lower administered dose, which resulted in PCB residues in the liver, kidney, muscle, brain, blood, and heart that ranged from 0.87 to 1.33, 1.09 to 1.86, 0.62 to 0.97, 0.33 to 1.36, 0.06 to 0.71, and 1.10 to 1.60 ppm, respectively. At the higher dose, which caused lethality, PCB levels in the liver, kidney, muscle, brain, blood, and heart were 11.99, 7.12, 3.31, 4.72, 1.80, and 8.31 ppm, respectively.

In another early study, mink were fed diets supplemented with industrial PCB mixtures (Ringer et al., 1972). This was lethal to all the adults, and residue levels in the liver, kidney, brain, spleen, lung, muscle, and heart were 4.18, 4.47, 11, 4.79, 4.78, 4.88, and 3.26 ppm, respectively. In a related study in which mink were fed PCB-contaminated fish, lethality occurred at concentrations of 4.2, 4.5, 11, 4.8, 4.8, 4.9, and 3.3 ppm in the liver, kidney, brain, spleen, lung, muscle, and heart, respectively (Aulerich et al., 1973).

Another feeding study using contaminated fish showed that impaired reproduction occurred with a fat concentration of 13.3 ppm and reproductive failure at 24.8 ppm (Hornshaw et al., 1983). In a study in which mink were given PCB-contaminated feed, reproductive success was significantly decreased in animals that had mean PCB concentrations of 86 ppm in fat (lipid basis) (Jensen et al., 1977). The effect was a decrease from 5.1 to 2.9 young born to each female.

In sum, these studies indicate that liver PCB levels above 4 ppm are associated with lethality in mink and that reproductive impairment occurs when PCB fat concentrations on a wet-weight basis exceed about 10 ppm.

BATS

The most comprehensive epidemiological and experimental studies of PCB residues and their reproductive effects in big brown *(Eptisecus fuscus)* and little brown *(Myotis lucifugus)* bats were performed on animals from the United States mid-Atlantic region. The first investigation of PCB residues and reproduction was conducted on big brown bats collected in Maryland in May and June, 1974 (Clark and Lamont, 1976). Only apparently pregnant females were studied, and the animals were kept until the young were born (about a month for the May collection and about a week for the June group). Adult females had mean carcass PCB levels of 1.3 ppm, ranging from 0.8 to 2.9 ppm (May group), and 2.0 ppm, ranging from 1.2 to 3.6 ppm (June group). The young had mean levels of 0.4 ppm, with a range from 0.2 to 3.3 ppm (May group), and 1.2 ppm, with a range from 0.5 to 2.4 ppm (June group).

There was a statistically significant difference in the mean PCB levels in the two of 15 litters from May mothers that contained dead young as opposed to live offspring (2.4 vs. 0.34 ppm). There was also a nonstatistically significant difference in the mean PCB concentrations in two of 11 litters from June mothers with dead as opposed to live young (1.6 vs. 1.1 ppm). There was no clear correlation between dead young and maternal PCB level, although mothers with levels over 2 ppm appeared more likely to produce dead young.

The second study of PCB residue levels and reproduction was performed on a Maryland barn population from which 45 pregnant little brown bats were collected in 1976. The animals were allowed to go to parturition before analyses of mothers and offspring were performed (Clark and Krynitsky, 1978). The PCB levels in the mothers ranged from 3.6 to 24 ppm with a mean of 11.4 ppm and, in the offspring, they ranged from nondetectable to 25 ppm with a mean (of the detects) of 4.2 ppm. Twelve of the 43 mothers produced dead offspring, and the dead young averaged more than twice as much PCB as did the live young (6.7 vs. 3.0 ppm), but the differences were not statistically significant. In addition, females that produced dead young did not contain significantly higher residues than did females giving birth to live young. One additional complication is that the authors suggest that factors associated with first pregnancies accounted for the deaths in a number of young. However, the four dead young with the highest PCB levels (12 to 25 ppm) were born to mothers who were not yearlings.

Following these studies, Clark (1978) fed mealworms *(Tenebrio molitor)* contaminated with PCBs to 18 pregnant big brown bats and clean mealworms to another 18 and followed both groups and their litters through parturition. Measurements were made of whole-body PCB residue concentrations and possible reproductive toxicity as measured by a number of parameters, including the frequency of dead young and litter weight. There was no association between PCB level and toxicity, even though the PCB-dosed females had a mean level of 20.3 ppm (range, 18.3 to 22.6 ppm) and the offspring had a mean concentration of 4.4 ppm (range, 3.6 to 5.3 ppm).

One limited experimental study on lethality, performed by feeding female little brown bats (collected in Maryland) mealworms contaminated with PCBs for 40 days (Clark and Stafford, 1981), showed that, in the two animals that died, brain concentrations were 1300 and 1500 ppm. The corresponding whole-body levels were 5400 and 6100 ppm. These are very high levels and unlikely to occur in field situations.

In sum, the experimental data do not support the link between PCB residue levels and reproductive toxicity in bats suggested in the investigations in which wild bats were collected and studied in captivity. These studies taken in total illustrate the great difficulty in trying to draw firm conclusions from field epidemiological data about relations between toxicity and PCB residue levels.

SEA LIONS

The only research on the connection between PCB residues and possible toxic effects in sea lions was conducted in the field during the early 1970s when there was concern about premature pupping and about the number of sick animals in sea lion populations. All of the work was done on populations living off the western coast of the U.S.

A study of animals collected off the coast of Oregon between 1970 and 1973 compared the PCB levels in various tissues in both sick and healthy sea lions *(Zalophus californianus californianus)* (Buhler et al., 1975). There were no differences between groups, and so the range among all animals was combined. In fat tissue, PCBs varied between 21 and 34 ppm; in muscle, between 0.3 and 0.9 ppm; in the liver, between 2.0 and 4.9 ppm; and in the brain, between 0.5 and 2.8 ppm.

A comparison of PCB levels in premature and live sea lion pups was carried out on a population sampled off the coast of southern California in 1970 (DeLong et al., 1973). In the mothers of premature pups, the PCB concentrations were as follows: blubber, 85 to 145 ppm, mean = 112; liver, 3.4 to 9.7 ppm, mean = 5.7. For the mothers of full-term pups, the corresponding levels were as follows: blubber, 12 to 25 ppm, mean = 17; liver, 0.5 to 2.2 ppm, mean = 1.3. The differences in tissue levels between mothers of premature and full-term pups were statistically significant.

A similar comparison was carried out in the same location in 1972 (Gilmartin et al., 1976). In this research, PCB levels in blubber varied from 33.0 to 92.4 ppm in the mothers of the premature pups and from 4.7 to 39.5 ppm in the mothers of the full-term pups. Maternal liver concentrations ranged from 0.63 to 6.08 ppm for the premature pups and from 0.19 to 1.98 ppm for the full-term sea lion pups. The PCB-level differences between the two groups were statistically significant.

Thus, no relation between PCB levels and adult ill health was found in the single available study, one in which fat levels of PCB were about 20 to 30 ppm. An association was found, however, between the PCB residue concentrations and premature pupping in two sea lion populations. These effects seemed to occur at maternal fat concentrations somewhere above about 30 ppm and liver levels above about 3 ppm. However, because of the presence of confounders, such as DDE and microorganisms, the investigators were not able to determine if there was a causal relation between PCB residues and reproductive problems in sea lions. In addition, no controlled experiments are available to establish such a relation. Thus, associating a specific PCB residue level with toxic effects in sea lions is premature at this time.

SEALS

As mentioned earlier, seals have been studied extensively with respect to PCB residue levels. In the main, studies have focused on the Baltic Sea area, northern

Canada, and the Antarctic. The large decrease in seal population numbers in the Baltic Sea was the impetus for the studies in this area, and the results of a large number of mainly epidemiological studies will be summarized in this section.

RINGED AND GRAY SEALS

Early studies of ringed seals *(Pusa hispida)* in the Gulf of Bothnia (Helle et al., 1976a, 1976b) showed that pregnant females had the lowest PCB fat levels, an average of 73 ppm (lipid basis), while nonpregnant females with occlusions preventing reproduction had an average PCB fat concentration of 110 ppm (lipid basis) and nonpregnant females with normal uteri had levels of 89 ppm (lipid basis). The results were significantly different between the first two groups but not between the first and third groups.

A later study, of gray *(Halichoerus grypus)* and ringed *(Phoca hispida)* seals found dead or shot in the Gulf of Finland between 1976 and 1982, did not show effects corresponding to those of the studies cited above. Of three females with PCB fat concentrations above 250 ppm, two were in the preparturition stage of pregnancy, and one showed no pathological changes in the uterine horns. On the other hand, two females with much lower PCB fat levels, 68 and 79 ppm, had severe pathological changes in the uterine horns (Perttila et al., 1986).

Thus, the studies on ringed and gray seals reveal a complex pattern of relation between PCB residues and reproductive impairment. In addition to the complexity, the data are difficult to interpret because of the presence of other chemical contaminants and other stressors. As a result, the authors of the studies were very reluctant to conclude that there is a causal relation between PCBs alone and reproductive impairment. If such a relation exists, it appears that the effects occur when PCB blubber concentrations are above about 70 ppm (lipid basis).

HARBOR SEALS

There are no epidemiological studies that examine PCB residues and toxicological effects in the same populations. Instead, there are a number of studies conducted in the 1970s on harbor seals *(Phoca vitulina)* from two nearby areas with different seal population histories: the Dutch and German Wadden seas (Drescher et al., 1977; Duinker et al., 1979; Reijnders, 1980; van der Zande and de Ruiter, 1983). A serious decline in numbers has occurred only in the Dutch Wadden Sea harbor seals. Average PCB fat concentrations are also higher in this Dutch population, although there are great variability and overlap in values. The average Dutch Wadden Sea adult seal PCB fat residue levels varied from 89 to 701 ppm (lipid basis), while those from the German Wadden Sea averaged from 76 to 171 ppm (lipid basis) (Reijnders, 1980; van der Zande and de Ruiter, 1983).

There is also one field experiment on common seals *(P. vitulina)* that was conducted by feeding one group a diet of fish low in PCBs and another a diet of fish higher in PCBs and examining the reproductive success of the two groups (Reijnders, 1986). The seals receiving the more highly PCB-contaminated fish showed a statistically significant reduction in reproductive success. A companion study (Boon et al., 1987) of residue levels in these animals revealed that blood PCB concentrations were 25 to 27 ppm (lipid basis) in the affected group and 5.2 to 11 ppm (lipid basis) in the group ingesting lower levels of PCB, differences that were statistically significant.

All of the data concerning harbor seals suffer from the same fundamental defect: lack of control of confounders, including other chemical contaminants. In addition, the measurements were almost all performed on animals found dead, which leaves doubt about the general applicability of such values. Even if there were a relation, the epidemiological population data are very variable, and it is difficult to pick a value at which decreases in population size might be expected. The field experiment is clearer and suggests that, if PCB is the causative factor, a blood PCB concentration of about 25 ppm (lipid basis) is associated with reproductive toxicity.

CONGENER-SPECIFIC ANALYSES

All of the original studies of PCBs, both in the field and in the laboratory, were based on analytical techniques that provided levels of total PCBs in the samples. However, by the early to mid-1980s, it was clear that some PCB congeners are much more toxic than others and that a more accurate assessment of the potential toxicity associated with particular residue levels requires a congener-specific analysis. The available data are of two types: those that report either the congeners present in the greatest concentrations or the congeners considered to be most toxic.

One experimental study (Boon et al., 1987) of the first type compared the concentrations of one of the most common PCB congeners, no. 153, in the blubber from two groups of seals *(P. vitulina)*. One group was fed contaminated fish from the Dutch Wadden Sea, and the other was fed less contaminated fish from the North Atlantic. The results showed that levels of congener no. 153 in the first group were from 6.6 to 6.9 ppm (lipid basis), while those in the second group were 1.5 to 3.5 ppm (lipid basis), a difference that was statistically significant. Thus, ingestion of the contaminated fish resulted in 2 to 5 times higher levels of congener no. 153. A related study showed that reproductive success in the first group was impaired, while that in the second group was not (Reijnders, 1986).

A field study of common seals *(P. vitulina)* in the German North Sea showed that seal blubber levels of congener no. 153 = 10.1 ppm, no. 101 = 0.648 ppm, no. 138 = 7.0 ppm, and no. 180 = 2.2 ppm. Similar levels were found in the Swedish Baltic Sea (congener no. 153 = 10.2 ppm, no. 101 = 0.460 ppm, no. 138 = 6.7 ppm, and no. 180 = 2.7 ppm). The levels were lower in the blubber of seals from less polluted areas, e.g., in Iceland, common seals *(P. vitulina)*, congener no. 153 = 0.639 ppm, no. 101 = 0.125 ppm, no. 138 = 0.405 ppm, and no. 180 = 0.156 ppm. In northern Norway, the levels in ringed seals *(P. hispida)* were as follows: congener no. 153 = 0.302 ppm, no. 101 = 0.120 ppm, no. 138 = 0.205 ppm, and no. 180 = 0.067 ppm. The levels in Weddell seals *(Leptonychotes weddellii)* from the least polluted area, the Antarctic, were as follows: congener no. 153 = 0.009 ppm, no. 101 = 0.003 ppm, no. 138 = 0.006 ppm, and no. 180 = 0.003 ppm (Luckas et al., 1990).

Turning to the most toxic congeners, we found no experimental field or laboratory studies of these toxic congeners and reproduction in marine mammals. There are, however, some data that focus on the levels of three toxic congeners, i.e., no. 77 (3,3',4,4'-PCB), no. 105 (2,3,3',4,4'-PCB), and no. 126 (3,3',4,4',5-PCB), in areas of pollution where reproduction is impaired.

One study of the blubber from Baltic Sea gray *(H. grypus)* and ringed *(P. hispida)* seals showed levels (on a lipid basis) of congener no. 77 = 1.1 to 20 ppb, no. 105 = 42 to 1,100 ppb, and no. 126 = 0.6 to 3.9 ppb (Koistinen, 1990). An investigation of a porpoise and two whales from the North Pacific revealed generally comparable PCB congener levels. The mean blubber concentrations for the female Dall's porpoise *(Phocoenoides dalli)* were as follows: congener no. 77 = 0.930 ppb, no. 126 = 0.092 ppb, and no. 169 = 0.053 ppb. Those for the female Baird's beaked whale *(Berardius bairdii)* were as follows: congener no. 77 = 0.700 ppb, no. 126 = 0.100 ppb, and no. 169 = 0.093 ppb, and for a female killer whale *(Orcinus orca)*, the comparable values were as follows: congener no. 77 = 42 ppb, no. 126 = 4 ppb, and no. 169 = 3.6 ppb (Tanabe et al., 1987).

Although there are differences in congener levels associated with areas with different degrees of pollution, the data are limited to those occurring in highest proportion and not to the most toxic congeners. In addition, even for the most commonly occurring congeners, the data are not sufficient to make any firm conclusions. At best, it can be stated that levels of the most common congeners, nos. 153, 101, 138, and 180, are about 10 ppm, 0.600 ppm, 7 ppm, and 3 ppm, respectively, in highly polluted areas where reproductive problems have been noted in the past.

Some investigators have utilized toxicity equivalency factors in an attempt to assess the combined toxicity of the congeners detected in marine mammal samples (Tanabe et al., 1987; Kannan et al., 1989; Koistinen, 1990). The toxicity equivalency factors are based on *in vitro* studies of enzyme induction in rodent liver cells and provide information about the relative toxicity of each congener in this system (Safe, 1990). The values derived from these techniques cannot be applied to other biological systems to provide an absolute measure of toxicity, since the factor values are both species and effect dependent. Thus, even if more complete congener level information were available, it appears premature to apply this information to the species of concern in this review.

SUMMARY

Although there are many data about PCB residue levels in mammals, there is very little scientifically valid information linking these levels to toxic effects, particularly reproductive impairment, in field populations. The best studies are reproductive toxicity experiments in mink that show a dose-response relation and provide clear evidence that PCBs do impair reproduction. Because there are no studies validating these results in the wild, the applicability of the residue numbers to field situations is uncertain. However, from the available information, it appears that liver concentrations above about 4 ppm are associated with lethality and fat concentrations above 10 ppm are linked to reproductive impairment.

Despite a number of studies, it has not been possible to link reproductive outcome with whole-body PCB residues in bats. In sea lions, the data are quite tenuous and must be treated with great caution. If, however, they are interpreted to show causation, fat concentrations above about 30 ppm and liver levels above about 3 ppm are associated with premature pupping.

Studies on seals comparing animals from areas where pollution is high and reproductive performance is low with seals from less polluted areas provide some data about the relation between PCB residues and reproduction. These epidemiological data are far from those required to establish causation and so should be viewed as tentative. If they do indicate a link, then it appears that PCB blubber concentrations above about 70 ppm (lipid basis) are associated with reproductive impairment in ringed and gray seals and blood levels above about 25 ppm (lipid basis) are associated with reproductive toxicity in harbor seals.

There are even fewer data that provide PCB residue levels on a congener-specific basis, and these are largely limited to seals. As with the total PCB studies, these are mainly epidemiological comparisons of concentrations in animals that live in waters with varying levels of pollutants and show a variety of reproductive problems. Assuming a link between reproductive toxicity and these PCB congener levels, it appears that blubber concentrations of congener no. 153 of about 10 ppm, no. 101 of about 600 ppb, no. 138 of 7 ppm, and no. 180 of 3 ppm are associated with reproductive problems.

To provide a better understanding of the relation between PCB residues and toxicity in mammals in the field, additional studies are needed. Investigations that provide a better link between field studies and laboratory experiments would be very helpful. In view of the large amount of experimental information on ranch mink, studies of mink in the wild would be valuable assets in making this link. Studies of this type should focus on analyses of the PCB congeners of greatest toxicity, although it is possible that improved *in vitro* techniques will enable researchers to assess total toxicity without quantifying individual congeners.

Assessing the role of PCBs will also require better understanding of the impacts of other environmental parameters that may influence toxicity and of possible interactions between these factors and PCBs in producing toxicity. These factors include infectious organisms and physical stress as well as a number of other environmental contaminants, such as heavy metals. It may very well be the case that a single PCB residue value cannot be used as an indicator of toxicity but that, instead, this value will have to be interpreted in the context of other site-specific environmental parameters.

REFERENCES

Addison, R. F. 1989. Organochlorines and marine mammal reproduction. Can. J. Fish. Aquat. Sci. 46:360-368.

Aulerich, R. J., S. J. Bursian, W. J. Breslin, B. A. Olson, and R. K. Ringer. 1985. Toxicological manifestations of 2,4,5-, 2',4',5'-, 2,3,6,2',3',6'-, and 3,4,5,3',4',5'-hexachlorobiphenyl and Aroclor 1254 in mink. J. Toxicol. Environ. Health 15:63-79.

Aulerich, R. J., R. K. Ringer, and S. Iwamoto. 1973. Reproductive failure and mortality in mink fed Great Lakes fish. J. Reprod. Fertil. 19 (Suppl.):365-376.

Boon, J. P., P. J. H. Reijnders, J. Dols, P. Wensvoort, and M. Th. J. Hillebrand. 1987. The kinetics of individual polychlorinated biphenyl congeners in female harbour seals *(Phoca vitulina)*, with evidence for structure-related metabolism. Aquat. Toxicol. 10:307-324.

Buhler, D. R., R. R. Claeys, and B. R. Mate. 1975. Heavy metal and chlorinated hydrocarbon residues in California sea lions *(Zalophus californianus californianus)*. J. Fish. Res. Board Can. 32:2391-2397.

Clark, D. R., Jr. 1978. Uptake of dietary PCB by pregnant big brown bats *(Eptesicus fuscus)* and their fetuses. Bull. Environ. Contam. Toxicol. 19:707-714.

Clark, D. R., Jr., and A. Krynitsky. 1978. Organochlorine residues and reproduction in the little brown bat, Laurel, Maryland — June 1976. Pestic. Monit. J. 12:113-116.

Clark, D. R., Jr., and T. G. Lamont. 1976. Organochlorine residues and reproduction in the big brown bat. J. Wildl. Manage. 40:249-254.

Clark, D. R., Jr., and C. J. Stafford. 1981. Effects of DDE and PCB (Aroclor 1260) on experimentally poisoned female little brown bats *(Myotis lucifugus)*: lethal brain concentrations. J. Toxicol. Environ. Health 7:925-934.

DeLong, R. L., W. G. Gilmartin, and J. G. Simpson. 1973. Premature births in California sea-lions: association with high organochlorine pollutant residue levels. Science (Washington, D.C.) 181:1168-1170.

Drescher, H. E., U. Harms, and E. Huschenbeth. 1977. Organochlorines and heavy metals in the harbour seal *Phoca vitulina* from the German North Sea coast. Mar. Biol. 41:99-106.

Duinker, J. C., M. Th. J. Hillebrand, and R. F. Nolting. 1979. Organochlorines and metals in harbour seals (Dutch Wadden Sea). Mar. Pollut. Bull. 10:360-364.

Eisler, R. 1986. Polychlorinated biphenyl hazards to fish, wildlife, and invertebrates: a synoptic review. Biol. Rep. No. 85 (1.7). Fish and Wildlife Service, U.S. Department of the Interior, Washington, D.C.

Foley, R. E., S. J. Jackling, R. J. Sloan, and M. K. Brown. 1988. Organochlorine and mercury residues in wild mink and otter: comparison with fish. Environ. Toxicol. Chem. 7:363-374.

Friedman, M. A., F. D. Griffith, and S. Woods. 1977. Pathologic analysis of mink mortality in New England mink. Arch. Environ. Contam. Toxicol. 5:457-469.

Fuller, G. B., and W. C. Hobson. 1987. Effect of PCBs on reproduction in mammals. p. 101-125. *In* J. S. Waid (Ed.). PCBs and the environment. Vol. 2. CRC Press, Boca Raton, Fla.

Gilmartin, W. G., R. L. DeLong, A. W. Smith, J. C. Sweeney, B. W. De Lappe, R. W. Risebrough, L. A. Griner, M. D. Dailey, and D. B. Peakall. 1976. Premature parturition in the California sea lion. J. Wildl. Dis. 12:104-115.

Helle, H., M. Olsson, and S. Jensen. 1976a. DDT and PCB levels and reproduction in ringed seal from the Bothnian Bay. Ambio 5:188-189.

Helle, H., M. Olsson, and S. Jensen. 1976b. PCB levels correlated with pathological changes in seal uteri. Ambio 5:261-263.

Hornshaw, T. C., R. J. Aulerich, and H. E. Johnson. 1983. Feeding Great Lakes fish to mink: effects on mink and accumulation and elimination of PCBs by mink. J. Toxicol. Environ. Health 11:933-946.

Jensen, S. 1972. The PCB story. Ambio 1:123-131.

Jensen, S., J. E. Kihlstrom, M. Olsson, C. Lundberg, and J. Orberg. 1977. Effects of PCB and DDT on mink *(Mustela vison)* during the reproductive season. Ambio 6:239.

Kannan, N., S. Tanabe, M. Ono, and R. Tatsukawa. 1989. Critical evaluation of polychlorinated biphenyl toxicity in terrestrial and marine mammals. Increasing impact of non-ortho and mono-ortho coplanar polychlorinated biphenyls from land to ocean. Arch. Environ. Contam. Toxicol. 18:850-857.

Koistinen, J. 1990. Residues of planar polychloroaromatic compounds in Baltic fish and seal. Chemosphere 20:1043-1048.

Luckas, B., W. Vetter, P. Fischer, G. Heidemann, and J. Plotz. 1990. Characteristic chlorinated hydrocarbon patterns in the blubber of seals from different marine regions. Chemosphere 21:13-19.

Perttila, M., O. Stenman, H. Pyysalo, and K. Wickstrom. 1986. Heavy metals and organochlorine compounds in seals in the Gulf of Finland. Mar. Environ. Res. 18:43-59.

Platonow, N. S., and L. H. Karstad. 1973. Dietary effects of polychlorinated biphenyls on mink. Can. J. Comp. Med. 37:391-400.

Reijnders, P. J. H. 1980. Organochlorine and heavy metal residues in harbour seals from the Wadden Sea and their possible effects on reproduction. Neth. J. Sea Res. 14:30-65.

Reijnders, P. J. H. 1986. Reproductive failure in common seals feeding on fish from polluted coastal waters. Nature (Lond.) 324:456-457.

Ringer, R. K. 1983. Toxicology of PCBs in mink and ferrets. p. 227-240. *In* F. D'Itri and M. Kamrin (Eds.). PCBs: human and environmental hazards. Butterworth Publishers, Woburn, Mass.

Ringer, R. K., R. J. Aulerich, and M. Zabik. 1972. Effect of dietary polychlorinated biphenyls on growth and reproduction of mink. Extended abstract. ACS (American Chemical Society) 164th Annu. Meet. 12:149-154.

Safe, S. 1990. Polychlorinated biphenyls (PCBs), dibenzo-*p*-dioxins (PCDDs), dibenzofurans (PCDFs), and related compounds: environmental and mechanistic considerations which support the development of toxic equivalency factors (TEFs). CRC Crit. Rev. Toxicol. 21:51-88.

Sleight, S. D. 1983. Pathologic effects of PCBs in mammals. p. 215-226. *In* F. D'Itri and M. Kamrin (Eds.). PCBs: human and environmental hazards. Butterworth Publishers, Woburn, Mass.

Talmage, S. S., and B. T. Walton. 1991. Small mammals as monitors of environmental contaminants. Rev. Environ. Contam. Toxicol. 119:47-145.

Tanabe, S., N. Kannan, A. Subramanian, S. Watanabe, and R. Tatsukawa. 1987. Highly toxic coplanar PCBs: occurrence, source, persistency, and toxic implications to wildlife and humans. Environ. Pollut. 47:147-163.

van der Zande, T., and E. de Ruiter. 1983. The quantification of technical mixtures of PCBs by microwave plasma detection and the analysis of PCBs in the blubber lipid from harbour seals *(Phoca vitulina)*. p. 133-147. *In* The science of the total environment. Elsevier Science Publishers, Amsterdam.

Waid, J. S. (Ed.). 1987. PCBs and the environment. CRC Press, Boca Raton, Fla.

Wren, C. D. 1991. Cause-effect linkages between chemicals and populations of mink *(Mustala vison)* and otter *(Lutra canadensis)* in the Great Lakes basin. J. Toxicol. Environ. Health 33:549-585.

Zepp, R. L., Jr., and R. L. Kirkpatrick. 1976. Reproduction in cottontails fed diets containing a PCB. J. Wildl. Manage. 40:491-495.

CHAPTER 7

PCBs and Dioxins in Birds

David J. Hoffman, Clifford P. Rice, and Timothy J. Kubiak

INTRODUCTION

Polychlorinated biphenyls (PCBs) are a group of synthetic chlorinated aromatic hydrocarbons that were first synthesized in 1881. Since 1930, PCBs have been in general use, having appeared in commercial products including heat transfer agents, lubricants, dielectric agents, flame retardants, plasticizers, and waterproofing materials (Roberts et al., 1978; Eisler, 1986b). However, their predominant use has been as insulating and cooling agents in closed electrical transformers and capacitors because of their low flammability.

PCB residues have been identified throughout the global ecosystem in rivers and lakes, the atmosphere, and fish, birds, and mammals as well as in human adipose tissue, blood, and breast milk (Roberts et al., 1978; Kimbrough, 1980; Safe, 1984; Eisler, 1986b). Such widespread ecosystem contamination has resulted from industrial discharges, leaks, disposal into municipal sewage and landfills, and incomplete incineration in the atmosphere. Between 1930 and 1975, over 630 million kilograms of PCBs were manufactured in the United States (Safe, 1984). As of 1979, a ban on the manufacture of PCBs in the U.S. was implemented.

Prior commercial production of PCBs consisted of chlorination of biphenyls, resulting in complex mixtures of chlorobiphenyls containing a total of 209 theoretically possible isomers but with some less likely to occur than others (Safe, 1984). Ten possible degrees of chlorination resulted in 10 PCB homolog groups, mono- through decachlorobiphenyl, with various positional isomers (congeners) within each group (Figure 1). Congeners have been assigned identification numbers according to the International Union of Pure and Applied Chemistry (IUPAC).

Commercial PCB formulations were sold under a variety of trade names; e.g., in the U.S., Aroclor was the most common. Aroclors are PCB mixtures that were named according to their chlorine content. For example, Aroclor 1254 contains 54% chlorine by weight, and Aroclor 1260 contains 60%.

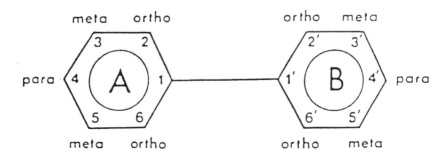

Figure 1 Generalized structure of biphenyl and possible positions for chlorines (ortho, meta, para) on polychlorinated biphenyl (PCB) congeners.

Aroclors vary in their toxicities according to a number of factors including congener composition and chlorine content. PCB congeners with the chlorine atom in positions 2 and 6 (ortho) are generally more readily metabolized, while those with chlorines in positions 4 and 4' (para) or positions 3,4 or 3,4,5 on one or both rings tend to be more toxic and are retained in tissues (Eisler, 1986b). Structures of some of the more active PCB congeners are summarized later in this chapter in Table 4.

Polychlorinated dibenzodioxins (PCDDs), unlike PCBs, have not been purposely manufactured, but rather are present as trace impurities associated with chlorophenols and production of herbicides, such as 2,4,5-trichlorophenoxyacetic acid. PCDDs can also be formed by photochemical and thermal reactions during and after incineration, leading to their presence in fly ash and other products of combustion (Eisler, 1986b). PCDDs, like PCBs, are dispersed throughout the global ecosystem, are chemically stable, and bioaccumulate in animal tissues. The number of chlorine atoms in PCDDs can vary from one to eight per molecule, resulting in 75 positional isomers, with the most toxic and widely studied one being 2,3,7,8-tetra-CDD (2,3,7,8-TCDD) (Kimbrough, 1980; Eisler, 1986b) (see Figure 2).

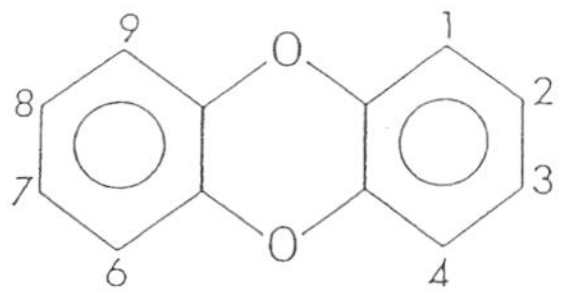

Figure 2 Generalized structure of dibenzo-*p*-dioxin and possible positions for chlorines.

PCBs and PCDDs may be quite biologically toxic, eliciting a number of common responses including but not limited to thymic atrophy (a wasting syndrome), immunotoxic effects, reproductive impairments, porphyria, and related liver damage (Kimbrough, 1980; Safe, 1984, 1990; Safe et al., 1985). The most toxic PCB congeners, including 3,3',4,4',5-penta-CB (PCB 126; IUPAC No. 126), 3,3',4,4'-tetra-CB (PCB 77), and 3,3',4,4',5,5'-hexa-CB (PCB 169), can assume a relatively coplanar conformation generally similar to that of 2,3,7,8-TCDD and are approximate stereo analogs of this compound. Both PCBs and PCDDs are environmentally persistent, resisting bacterial and chemical breakdown, but are readily absorbed from water into the fats of plankton, thereby entering the aquatic food chain. This process continues as fish, and then piscivorous birds including gulls, cormorants, herons, and terns, accumulate progressively higher concentrations (biomagnification) of these compounds as they become deposited in the fat of the body while natural portions of food items are metabolized for energy or excreted. At the top of the aquatic food chain are bald eagles *(Haliaeetus leucocephalus)* and other raptors that consume gulls and other fish-eating birds as well as fish.

In this review, we have mainly incorporated the findings of interpretive toxicological studies that report PCB or dioxin concentrations in avian tissues and eggs as determined in pen and laboratory studies as well as by field observations. All residues in tissues and eggs are reported in terms of the concentration per wet weight unless specified otherwise. Residues in food are reported as either wet weight or dry weight in the text.

PCB PEN AND LABORATORY STUDIES

LETHALITY AND HISTOPATHOLOGY

Adult and juvenile birds of precocial species exhibit varying sensitivities to ingestion of aroclors and other PCB mixtures. In 5-day feeding trials with Aroclor 1254, median lethal concentrations (LC_{50}s) in the diet were 604 ppm (dry weight) for northern bobwhite *(Colinus virginianus)*, 1091 ppm for ring-necked pheasants *(Phasianus colchicus)*, 2697 ppm for mallards *(Anas platyrhynchos)*, and 2895 ppm for Japanese quail *(Coturnix japonica)* (Heath et al., 1972). Dahlgren et al. (1972a) administered capsules daily containing 10, 20, or 210 mg of Aroclor 1254 to 11-week-old hen pheasants until death or sacrifice (Table 1). Birds were killed by PCB ingestion in a dose-dependent manner, with one bird dying in less than 2 days after receiving 210 mg and one bird on 10 mg daily surviving for 8 months. The heaviest birds survived the longest. Birds that died from ingesting 210 mg daily had total PCB brain residues ranging from 320 to 770 ppm (520 ± 110, mean ± SD) wet weight. Liver residues were much more variable and ranged from 390 to 9300 ppm (2500 ± 2000), and muscle levels ranged from 51 to 290 ppm (140 ± 53). Treated birds receiving 210 mg daily that were sacrificed at the same time others were dying on this dose had residue levels in tissues overlapping those of birds that died; however, the least overlap occurred in brain tissue, where the levels in sacrificed birds were 370 ± 65 ppm and were 1900 ± 1300 ppm for liver and 83 ± 17 ppm for muscle. The authors concluded that a brain residue level of 300 to 400 ppm was indicative of death due to PCB toxicosis. Pheasants that died had consistently smaller hearts and very small shrunken spleens due to lymphocyte depletion. Weights of kidney and liver proportional to body weight increased with 10- and 20-mg doses but not with 210-mg doses. Hydropericardium and abdominal and subcutaneous edema were not apparent in this species. Brain residue was the only tissue residue that was independent of a number of other parameters when correlations were made;

Table 1 Effects on Survival, Histopathology and Growth of Aroclors and Other Polychlorinated Biphenyl (PCB) Mixtures

Species, age	Concentration in tissue (ppm wet weight)	Effect	Experimental treatment	Ref.
Pheasants, 11-week-old hens	300 to 400 in brain	Death	Aroclor 1254 daily in capsules of up to 210 mg	Dahlgren et al. (1972a)
Chickens, cockerels (day-old at start)	270 to 420 in brain	Death	Aroclor 1260 in diet, up to 600 ppm	Vos and Koeman (1970)
Chickens, cockerels (day-old at start)	120 in brain, 240 in liver, 410 in kidney, 85 in muscle	Death with severe edema and lesions	Aroclor 1254 in diet, 500 ppm	Platonow et al. (1973)
Chickens, growing chicks	20 in adipose tissue	Death in 30% and decreased growth	Aroclor 1254 in diet, 20 ppm	Bird et al. (1978)
Chickens, growing chicks	325 in fat 100 in fat 21 in fat	Decreased growth Lower hematocrit Lower hemoglobin	Aroclor 1248 in diet, up to 40 ppm	Rehfeld et al. (1972)
Japanese quail	478 in liver	Weight loss and porphyria	Aroclor 1260 oral dose at 100 mg/kg	Vos et al. (1971)
Great cormorants, herons	76–180 in brain, herons with 420–445 in brain	Death	Clophen A60, dosed	Koeman et al. (1973)
Common murres, (Uria aalge)	>25 in brain	Decreased pituitary and thyroid weights	Aroclor 1254, orally dosed up to 400 mg/kg/day	Jefferies and Parslow (1976)
Bengalese finches, adult	290 in brain, 345 in liver	Death with enlarged kidneys, hydropericardium	Aroclor 1254 in diet, up to 440 ppm	Prestt et al. (1970)
Passerine species, (common grackles, red-winged black birds, brown-headed cowbirds, and starlings)	310 in brain	Diagnostic of death; liver hemorrhagic	Aroclor 1254 in diet at 1,500 ppm	Stickel et al. (1984)

these parameters included initial weight, the percentage of weight loss, days to death, and the lipid content in brain, liver, and muscle. Therefore, brain residue is considered a desirable tissue residue for diagnostic purposes. Further studies by these authors (Dahlgren et al., 1972b) revealed that periodic food deprivation increased brain residues more rapidly, leading to more rapid death.

More interpretive PCB studies have been conducted with chickens *(Gallus gallus)* than with any other single avian species. Young cockerels (day-old at start) that died on a dosage of 600 ppm (dry diet) of Aroclor 1260 had 270 to 420 ppm of total PCBs in the brain (Vos and Koeman, 1970). Dosage with other commercial PCB mixtures that also contained 60% chlorine resulted in more variable residues. Residues in the brains of five birds that died on Phenoclor DP6 dosage ranged from

70 to 700 ppm and, in four birds that died on Clophen A60 dosage, from 120 to 380 ppm. Liver residues varied more, ranging from 120 to 2900 ppm among 28 birds that died. These two commercial mixtures were more toxic than Aroclor 1260 and produced greater pathological signs. However, both of these mixtures proved to be contaminated with chlorinated dibenzofurans (Vos et al., 1971). In another study, day-old cockerels fed 500 ppm of Aroclor 1254 died with brain residues of about 80 to 190 ppm (Platonow et al., 1973). Tissue concentrations of PCB increased with the duration of exposure. The PCB ratios between liver, kidney, or muscle and brain were constant throughout the study with the duration of exposure. Half of the birds were dead by 43 days with brain residues averaging about 120 ppm; liver, 240 ppm; kidney, 410 ppm; and muscle, 85 ppm. Ratios of the total PCB concentrations in muscle, liver, and kidney to those in brain were 0.7, 2.0, and 3.5, respectively. Pathological examination of these chickens revealed severe edema and lesions, including hydropericardium and subcutaneous edema as well as edema of the liver and muscles. Hemorrhages, myocarditis, and kidney and liver necrosis were apparent.

Studies conducted with altricial species of birds have included observations with cormorants, herons, finches, and blackbirds. Great cormorants *(Phalacrocorax carbo)* dosed experimentally with Clophen A60 died with total PCB brain residues of 76 to 180 ppm (mean, 130 ppm), whereas herons that died on Clophen dosage contained 420 to 445 ppm, suggesting that cormorants were more sensitive than were herons (Koeman et al., 1973); these authors concluded that the survival time in cormorants was related to the capacity of the birds to store the PCBs in adipose and other tissues apart from the brain, indicating that the total body content is not a good criterion for diagnosis of PCB poisoning. Prestt et al. (1970) fed Aroclor 1254 to adult Bengalese finches, a domesticated form of the sharp-tailed finch *(Lonchura striata)*, at dietary levels ranging from 6 to 440 ppm for 8 weeks. The estimated dose rate for 50% mortality at 56 days was 254 mg/kg/day. The mean liver content was 345 ppm (range, 70 to 697 ppm). All birds dying from PCB ingestion had enlarged kidneys, with several birds also showing hydropericardium. Leg paralysis as well as trembling was apparent. In birds dying from PCB exposure, the ratio of total PCB concentration in the liver to that in the brain was 1.2 ± 0.1 (SD), which was 3 times as high as in those birds on similar diets not dying but sacrificed.

Using Aroclor 1254, Stickel et al. (1984) conducted one of the most comprehensive studies designed to establish lethal brain residues of total PCBs in passerine species. These species included immature male common grackles *(Quiscalus quiscula)*, immature female red-winged blackbirds *(Agelaius phoeniceus)*, adult male brown-headed cowbirds *(Molothrus ater)*, and immature female starlings *(Sturnus vulgaris)*. Aroclor 1254 was selected because chromatographic patterns of PCBs in wild birds were found to resemble those of this mixture closely. Dietary concentrations of 1500 ppm (dry weight) were administered until one half of the birds had died. The 50% mortality point for starlings was reached in 4 days, red-winged blackbirds in 6 days, cowbirds in 7 days, and grackles in 8 days. Signs of PCB poisoning began with birds becoming inactive with tremors. At necropsy, the liver frequently had hemorrhagic areas, and the gastrointestinal tract often had blackish fluid. PCB residues in the brains of birds that died were distinct from those in sacrificed survivors, providing suitable diagnostic criteria. PCB residues varied from 349 to 763 ppm of wet weight in the brains of dead birds and from 54 to 301 ppm in sacrificed birds. The authors considered an approximate level of 310 ppm (3 SDs below the mean) to be diagnostic for a high probability of PCB-induced mortality. PCBs in the brains of dead birds for the three icterine species did not differ significantly from each other (combined mean, 579 ppm), but the average residue levels in starlings (mean, 439 ppm) were significantly lower than those in red-winged blackbirds and grackles. The concentrations in whole bodies and livers were not diagnostic when expressed on a wet-weight basis.

One must be somewhat cautious in utilizing laboratory Aroclor and other commercial PCB mixture feeding studies as absolute predictors of potential avian toxicity in the field, because pattern recognition techniques during analysis of total PCB concentrations and congener patterns suggest that Aroclor mixtures change substantially in the environment and through the food chain (Schwartz and Stalling, 1991).

REPRODUCTIVE EFFECTS

Reproductive impairment has been reported in at least five species of birds that were given experimental doses of PCBs, including chickens, ringed turtledoves *(Streptopelia risoria)*, Japanese quail, mourning doves *(Zenaida macroura)*, and ring-necked pheasants (Table 2). In one study, laying hens were fed 0, 5, or 50 ppm of Aroclor 1254 for up to 39 weeks (Platonow and Reinhart, 1973). With 5 ppm (10% moisture) in the diet, egg production but not hatchability was reduced; however, after 14 weeks, fertility was lower. The hatchability of fertile eggs was not affected when the total PCB concentration in eggs was below 5 ppm. However, when the concentration exceeded 15 ppm in eggs, embryonic mortality was high. In contrast, Cecil et al. (1972) reported that dietary levels of 20 ppm of dry weight for Aroclor 1254 did not affect the hatching success of chickens at the end of 5 weeks with egg residues of 13.2 ppm. Tumasonis et al. (1973) examined the effect of exposing white leghorn hens for 6 weeks to 50 ppm of Aroclor 1254 in drinking water. Egg weight and fertility were not affected but, as total PCB concentrations increased in the yolk, embryonic development was arrested at progressively earlier stages. Within 2 weeks, hatching success dropped to 34%, and the authors concluded that yolk concentrations greater than 10 to 15 ppm (whole egg concentrations above 4 ppm) were required to affect hatching success. Short bowed legs, clenched toes, and neck deformities were present in some of the chicks that were hatched where yolk PCB levels were 10 to 15 ppm. Hemorrhaging and abnormal livers were apparent. In another study, white leghorn hens received diets containing 0 to 80 ppm of Aroclor 1242 for 6 weeks (Britton and Huston, 1973). Egg production, egg weight, shell thickness, and shell weight were not affected, but hatchability was affected within 2 weeks for hens fed as little as 20 ppm of dry weight. Yolks contained 2.4 ppm (expected whole egg concentration, 0.87 ppm). This conversion is based on the chicken yolk being 0.364 of the whole egg content mass (Sotherland and Rahn, 1987). Even 10 ppm of dry weight in the diet caused a small reduction in hatching at the end of 6 weeks (yolk concentration of 3.7 ppm or expected whole egg concentration of 1.3 ppm). Scott

Table 2 Reproductive Effects of Aroclors and Other Polychlorinated Biphenyl (PCB) Mixtures

Species, age	Concentration in tissue (ppm wet weight)	Effect	Experimental treatment	Ref.
Chickens, laying hens	5 in eggs	Hatching reduced	Aroclor 1254 in diet, up to 50 ppm	Platonow and Reinhart (1973)
Chickens, laying hens	13.2 in eggs	Did not affect hatching	Aroclor 1254 in diet, 20 ppm	Cecil et al. (1972)
Chickens, white leghorn hens	Above 4 in eggs	Embryo mortality and teratogenic	Aroclor 1254 in drinking water at 50 ppm	Tumasonis et al. (1973)
Chickens, white leghorn hens	Less than 1 in eggs	Decreased hatching	Aroclor 1242 diet, up to 80 ppm	Britton and Huston (1973)
Chickens, fertile white leghorn eggs	10 in eggs	Embryonic mortality of 64%	Aroclor 1242 injected into air cell of eggs	Blazak and Marcum (1975)
Chickens, laying hens	23 in eggs	Decreased hatching	Aroclor 1248 in diet at 10 ppm	Scott (1977)
Chickens, fertile white leghorn eggs	5 in eggs	Hatching reduced to 17%	Aroclor 1248 injected into yolk sac	Brunstrom and Orberg (1982)
Chickens, fertile eggs	0.05 to 0.1 in eggs	Decreased gluconeogenic enzyme activity	Aroclor 1254 injected into air cell	Srebocan et al. (1977)
Ringed-turtle doves	16 in eggs, 5.5 in adult brain	Embryonic mortality; decreased parental attentiveness	Aroclor 1254 in diet	Peakall and Peakall (1973)
Ringed-turtle doves	2.8 in brain	Depletion of brain dopamine and norepinephrine	Aroclor 1254 in diet	Heinz et al. (1980)
Mallard hens	23 in eggs, 30 in 3-week-old ducklings and 55 in hens	No effects	Aroclor 1254 in diet at 25 ppm	Custer and Heinz (1980)
Mallard hens	105 in eggs	Eggshell thickness decreased; hatching success not affected	Aroclor 1242 in diet	Haseltine and Prouty (1980)
Screech owls	4 to 18 in eggs	No effects	Aroclor 1248 in diet at 3 ppm wet weight	McLane and Hughes (1980)
Atlantic puffins	10 to 81 in eggs, 6 in adults	No effects detected	Aroclor 1254 dosed by implantation of 30-35 mg	Harris and Osborn (1981)

(1977) reported that hatchability was decreased after 4 weeks of Aroclor 1248 at 10 ppm of dry weight in the diet, with egg residues of 22.7 ppm.

Peakall and Peakall (1973) studied the effects of chronic dietary exposure of 10 ppm (dry diet) of Aroclor 1254 on the reproduction of ringed turtledoves. Embryonic mortality was greatly increased when the eggs were incubated by the parents but decreased by artificial incubation. Monitoring of egg temperatures suggested that mortality was increased because of decreased parental attentiveness. The mean PCB residues in dove eggs were 16 ppm (wet weight), with mean levels in adult tissues of 736 ppm in fat, 15 ppm in the liver, 8 ppm in muscle, and 5.5 ppm in the brain. An increase in chromosomal aberrations was apparent in 3- to 6-day-old embryos. Further studies with Aroclor 1254 in this species revealed depletions of brain dopamine and norepinephrine that were negatively correlated with brain residues (Heinz et al., 1980). Brain residues of 2.82 ± 0.29 (SE) ppm of wet weight resulted in significant depletions to levels known to cause behavioral impairments, thus supporting the above findings of Peakall and Peakall (1973).

Other species appear to be less reproductively sensitive to PCB exposure. Neither Aroclor 1254 nor Aroclor 1242 affected the reproductive success of mallards. Aroclor 1254 at 25 ppm of dry weight in the diet was fed to 9-month-old mallard hens for at least a month prior to egg laying with no detrimental effect on reproduction or nest attentiveness (Custer and Heinz, 1980). Hatching of ducklings and survival to 3 weeks were unaffected; the mean total PCB residues in eggs were 23.3 ppm (SE, 1.0 ppm); in 3-week-old ducklings, 29.5 ppm (SE, 1.4 ppm); and in hens, 55.3 ppm (SE, 1.9 ppm). Haseltine and Prouty (1980) fed Aroclor 1242 to mallards for 12 weeks at 150 ppm (dry diet) and did not find any differences in hatching success or nest attentiveness, but they found that eggshell thickness decreased by 8.9%; eggs contained an average of 105 ppm of total PCBs (wet weight). Aroclor 1248 fed to screech owls *(Otus asio)* at 3 ppm (wet weight) failed to affect reproduction with egg residues of 3.9 to 17.8 ppm (McLane and Hughes, 1980). When 108 Atlantic puffins *(Fratercula arctica)* were dosed by implantation under the skin along the ribs with 30 to 35 mg of Aroclor 1254, the PCB concentration in fat rapidly increased 10- to 14-fold, remaining there for 4 to 10 months. Survival and breeding performance were not impaired; the egg concentrations of total PCBs of dosed females ranged from 9.6 to 81.3 ppm compared with a mean concentration of 8.4 ppm (SE, 2.6 ppm) for controls exposed to PCBs from the natural environment (Harris and Osborn, 1981). The mean body burdens of dosed adults were 5.99 ppm (SE, 0.93 ppm) and of controls, 0.58 ppm (SE, 0.08 ppm).

GROWTH AND OTHER SUBLETHAL EFFECTS

Other studies have focused on the sublethal effects, including thyroid and pituitary changes and porphyria. In one study, common murres *(Uria aalge)*, referred to as "guillemots" by the authors, were reared in the laboratory and fed daily doses of Aroclor 1254 for 45 days at dose rates from 12 to 400 mg/kg/day (Jefferies and Parslow, 1976). Brain residues of above 25 ppm were accompanied by dose-related decreases in thyroid weight, follicle size, and colloid area as well as decreased pituitary weight. The authors concluded from these findings that *U. aalge* specimens are at least twice as sensitive to PCBs as are lesser black-backed gulls *(Larus fuscus)*, as judged by their previous findings (Jefferies and Parslow, 1972).

Feeding studies were conducted with Aroclor 1260 in Japanese quail (Vos et al., 1971) and with PCB congeners 2,2',4,4',5,5'-hexa-CB (PCB 153), 2,3,3',4,4'-penta-CB

(PCB 105), and 3,3′,4,4′,5-penta-CB (PCB 126) to assess porphyria in Japanese quail and American kestrels (Elliott et al., 1990, 1991). Aroclor 1260 at a dose rate of 100 mg/kg of body weight per day for 7 days in Japanese quail resulted in a mean liver residue of 478 ppm accompanied by weight loss, porphyria, and an increase of nearly 20-fold in hepatic mitochondrial δ-aminolevulinic acid (ALA) synthetase activity (Vos et al., 1971). However, as low a dose rate as 1 mg/kg/day increased ALA synthetase activity and resulted in a mean liver residue of 1.4 ppm. The mean liver residue levels (± SD) for PCB 105 of 2.6 ± 1.7 ppm and for PCB 126 of only 0.091 ± 0.08 ppm were associated with porphyria after 2 weeks in Japanese quail (Elliott et al., 1990). PCB 153 at 52.6 ppm in the liver had minimal effects. PCB 126 caused a decrease in thymus weight. All three congeners induced mixed-function oxygenase (MFO) activity as detected by the 7-ethoxyresorufin-O-deethylase (EROD) assay. In American kestrels, residue levels in pooled adipose tissue of 182 ppm for PCB 105, 119 ppm for PCB 153, and 3.3 ppm for PCB 126 were not associated with porphyria, but hepatic MFO activities were induced (Elliott et al., 1991).

Aroclor 1254 was more toxic than was Aroclor 1248 and affected growth more readily in chicks. Feeding 10 and 20 ppm of Aroclor 1254 (dry weight) to growing chicks resulted in decreased growth accompanied by 10 and 30% mortality, respectively, by 8 weeks of age, with approximate adipose tissue residues of 10 to 20 ppm (Bird et al., 1978). Feeding 40 ppm of Aroclor 1248 (dry weight) decreased the growth of chicks and resulted in 325 ppm (wet weight) in adipose tissue fat (Rehfeld et al., 1972). Feeding 20 ppm of dry weight decreased the hemoglobin concentration, and 30 ppm decreased hematocrit, with residues of 21 ppm and 100 ppm in the fat, respectively.

Other studies have examined the effects of specific PCB congeners on growth. The effects of feeding 400 ppm (dry weight) of five hexachlorobiphenyl congeners, including 2,2′,4,4′,6,6′-hexa-CB (PCB 155), 2,2′,3,3′,6,6′-hexa-CB (PCB 136), 2,2′,4,4′,5,5′-hexa-CB (PCB 153), 2,2′,3,3′,4,4′-hexa-CB (PCB 128), and 3,3′,4,4′,5,5′-hexa-CB (PCB 169), were examined in growing chicks for 21 days (McKinney et al., 1976). PCB 169 was the most toxic and caused complete mortality, general edema, marked thymic involution, and the highest accumulation (mean, 203 ppm) of all congeners in the liver. Growth was reduced by all congeners except PCB 155. The liver/body weight ratio was increased by all congeners, with the largest increase produced by PCB 155 and the smallest by PCB 136, with mean liver residues of 24 and 9 ppm, respectively. Pathological changes in the liver, including necrosis and fatty infiltration, were marked for PCB 155 and moderate for PCB 128.

Another study examined the effects of PCB congeners 126, 77, and 105 on the growth and development of American kestrel nestlings by daily dosing over the first 10 days posthatching (Hoffman et al., 1993a). Dosing with PCB 126 in the amount of 50 ng/g of body weight per day resulted in a geometric mean liver concentration of 156 ppb (range, 68 to 563 ppb) (wet weight), with pronounced liver enlargement and some mild coagulative necrosis of the liver, some colloid depletion of the thyroid, and lymphoid depletion of the spleen. Other effects at this dose level included a marginal decrease in body and bone lengths. Increasing the dose to 250 ng/g resulted in a mean liver concentration of 380 ppb (218 to 666 ppb); intensification of the above effects was seen, as well as decreased spleen weight and further lymphoid depletion of the spleen and bursa. Dosing at 1000 ng/g resulted in a mean liver concentration of 1098 ppb (652 to 4478 ppb), with decreased bursa weight and body weight in addition to the above effects. Higher doses and resulting liver concentrations of congeners 77 and 105 were required to produce any of the above effects; dosing with 1000 ng/g of PCB 77 resulted in a liver concentration of 892 ppb and the onset of coagulative liver necrosis, whereas dosing with 4000 ng of PCB 105 per gram resulted in a liver concentration of 1677 ppb, with liver necrosis and mild depletion of thyroid colloid.

EGG INJECTION STUDIES

The utility of egg injection studies for predicting potential embryotoxicity of PCBs and tetrachlorodibenzodioxin (TCDD) compares favorably with that of feeding studies. In instances where the same chemicals have been administered by both methods, the egg concentrations and effects are quite similar.

Several egg injection studies have been conducted with Aroclor mixtures, all having used the chicken egg (Table 2). Blazak and Marcum (1975) injected Aroclor 1242 into the air cell of fertile white leghorn eggs and then incubated them for a brief observation period of 4 to 5 days. Both 10 and 20 µg/g of egg (ppm) caused 64 to 67% embryonic mortality but did not result in chromosomal breakage, as had been reported in ringed turtledoves by Peakall et al. (1972). Srebocan et al. (1977) injected Aroclor 1254 (doses of 0, 0.05, 0.1, 0.5, and 5 µg/g of egg; ppm) into the air cell of chicken eggs on day 0 and incubated the eggs until day 14. The effects on survival were not reported, but decreased activity in key gluconeogenic enzymes was found to occur, starting at the lowest dose. Brunstrom and Orberg (1982) injected Aroclor 1248 into the yolk sacs of white leghorn eggs after 4 days of incubation to attain egg concentrations of 0, 1, 5, and 25 ppm. Hatchability was 96, 92, 17, and 0%, respectively, with mortality occurring the earliest in the 25-ppm group (days 6 to 12).

Considerably more egg injection studies have been conducted with specific PCB congeners, studying the effects in chickens and other species (Table 3). Rifkind et al. (1985) injected chicken eggs (unspecified location of injection) at 10 days of incubation with 5 to 1000 nmol per egg for each of three PCB congeners, including 3,3',4,4'-tetra-CB (PCB 77), PCB 169, and PCB 136. PCB 77 caused dose-related decreases in survival from days 10 to 19 of exposure at 100 to 1000 nmol per egg (584 to 5840 ppb, assuming a 50-g constant mass of egg), as did PCB 169 at 500 to 1000 nmol per egg (3610 to 7220 ppb). These decreases in survival were accompanied by decreased thymus weight and increased pericardial and subcutaneous edema in surviving embryos. These authors reported that the dose-response relations for lethality and for hepatic MFO induction, including both aryl hydrocarbon hydroxylase (AHH) and EROD activities, were dissociated and that the maximal induction levels were not correlated with the extent of lethality. Another study by these authors revealed that hepatocyte swelling, the major histopathological change, was apparent

Table 3 Egg Injection and Other Laboratory Studies with Planar Polychlorinated Biphenyls (PCBs) and Dioxin

Species, age	Compound, concentration	Effect	Ref.
Chicken, white leghorn embryo	2,3,7,8-TCDD[a] 10 ppt	2-fold AHH induction	Poland and Glover (1973)
	10-20 ppt	Onset of embryotoxicity	Verrett (1970)
	40-50 ppt	Mortality, edema, hemorrhaging over surface	Verrett (1976)
	63 ppt	ED_{50} for AHH induction	Poland and Glover (1973)
	147 ppt	LD_{50} (air cell injection)	Verrett, 1976
	115 ppt	LD_{50} (yolk sac injection)	Henshel (1993)
	180 ppt	LD_{50} (air cell injection)	Henshel (1993)
	240 ppt	LD_{50} (air cell injection)	Allred and Strange (1977)
	302 ppt	ED_{50} for AHH induction	Sawyer et al. (1986)
	1 ppb	100% mortality	Higginbotham et al. (1968)
	3,3',4,4'5-PeCB (PCB 126) 0.4 ppb	LD_{50} (air cell injection), day 4 through hatching	Hoffman et al. (1995)
	3.1 ppb	LD_{50} (air cell injection), day 7 through day 10	Brunstrom and Andersson (1988)
	3,3'4,4'-TeCB (PCB 77) 2.6 ppb	LD_{50} (air cell injection)	Hoffman et al. (1995)
	8.6 ppb	LD_{50} (air cell injection)	Brunstrom and Andersson (1988)
	40 ppb	LD_{50} (air cell injection)	Vos et al. (1982)
	2,3,3'4,4'-PeCB (PCB 105) 2,200 ppb	LD_{50} (air cell injection)	Brunstrom (1990)
	2,3,3'4,4'5-HxCB (PCB 157) 2,000 ppb	LD_{50} (air cell injection)	Vos et al. (1982)
	1,500 ppb	LD_{50} (air cell injection)	Brunstrom (1990)
Pheasant, embryo	2,3,7,8-TCDD 1.4 ppb	LD_{50} (albumin)	Nosek et al. (1993)
	2.2 ppb	LD_{50} (yolk)	
Bobwhite, embryo	3,3',4,4',5'-PeCB (PCB 126) 24 ppb	LD_{50} (air cell injection), through hatching	Hoffman et al. (1995)
Common tern, embryo	3,3',4,4',5'-PeCB (PCB 126) 45 ppb	35% embryo mortality (air cell injection), through hatching	Hoffman et al. (1995)
American kestrel, embryo	3,3',4,4',5'-PeCB (PCB 126) 65 ppb	LD_{50} (air cell injection), through hatching	Hoffman et al. (1995)
American kestrel, nestling	3,3',4,4',5'-PeCB (PCB 126) 156 ppb (68-563) in liver	Histopathology of liver, thyroid, and spleen	Hoffman et al. (1995)

Table 3 (continued) Egg Injection and Other Laboratory Studies with Planar Polychlorinated Biphenyls (PCBs) and Dioxin

Species, age	Compound, concentration	Effect	Ref.
Mallard, embryo; goldeneye, embryo	3,3'4,4'-TeCB (PCB 77) 5,000 ppb	No effects (air cell injection)	Brunstrom (1988)

[a] 2,3,7,8-TCDD, 2,3,7,8-tetrachlorodibenzodioxin; AHH, aryl hydrocarbon hydroxylase; ED_{50}, 50% effective dose; LD_{50}, 50% lethal dose; 3,3',4,4',5-PeCB, 3,3',4,4',5-pentachlorobiphenyl; 3,3',4,4'-TeCB, 3,3',4,4'-tetrachlorobiphenyl; 2,3,3',4,4'-PeCB, pentachlorobiphenyl; 2,3,3',4,4',5-HxCB, hexachlorobiphenyl.

within 24 hours after doses as low as 5 nmol per egg (29 ppb) for PCB 77 and PCB 169 but only at doses of 5,000 nmol per egg (36,100 ppb) and higher for 2,4,5,2',4',5'-hexa-CB (PCB 153), whereas PCB 136 was inactive (Rifkind et al., 1984). The same relation as for hepatocyte swelling held for induction of MFO (AHH and EROD) assays).

Comparison of the toxicity of PCB congeners injected into the yolk sac of chicken eggs at an earlier stage of development (4 days of incubation) revealed much greater toxicity, where PCB 126 was the most toxic and also the most potent inducer of MFO (EROD) in chick embryo liver (Brunstrom, 1989). The MFO (EROD)-inducing potencies correlated well with embryolethality. With air cell injections on day 7 of incubation, 50% lethal doses (LD_{50}s) (72 hr later) for PCB congeners 126, 77, 169, and 105 were 3.1, 8.6, 170, and 2200 ppb or 9.4, 29, 480, and 6700 nmol/kg, respectively (Brunstrom and Andersson, 1988; Brunstrom, 1990; Brunstrom et al., 1990). When injections were administered at the earlier stage of incubation (day 4) via the yolk sac and eggs were incubated until day 18, the above congeners caused higher embryonic mortality; the congener PCB 126 was approximately fivefold more toxic than was PCB 77, and analogs of PCB 77 (chlorinated at one ortho position) were three to four orders of magnitude less toxic. Hoffman et al. (1995) examined the effects of PCB congeners 126, 77, 105, and 153 in chickens, bobwhite, American kestrels, and common terns through hatching following air cell injections on day 4. The LD_{50}s for these congeners were approximately 0.4 ppb, 2.6 ppb, 3326 ppb, and greater than 14,000 ppb, respectively, in chickens; low-effect levels (10 to 20% embryonic mortality) were 0.2, 2000, and 14,000 ppb, respectively. The difference between these results, especially for PCB 126, and those of Brunstrom and Andersson (1988) is probably due to two important factors: (1) Hoffman et al. (1995) used day 4 of incubation, an earlier and more sensitive stage of embryonic development; and (2) permitted eggs to hatch, also a critical stage for survival; whereas Brunstrom and Andersson (1988) used day 7 embryos for dosing and recorded LD_{50}s 72 hours later. Indeed, when Brunstrom and Danerud (1983) injected PCB 77 into the yolk sac of chicken eggs at 4 days of incubation and permitted the eggs to hatch, they reported that 4 ppb decreased hatching success by 40%, in very close agreement with the findings of Hoffman (1994).

Non-ortho-chlorinated (coplanar) congeners were more potent inhibitors than were mono-ortho-chlorinated congeners of lymphoid development in the embryonic bursa (Brunstrom et al., 1990). Andersson et al. (1991) compared the numbers of lymphoid cells in the thymus and in the bursa after chicken eggs were treated with coplanar PCB congeners by air cell injection on day 13; here, the values for 50% of maximum inhibition (ED_{50}) of bursal development were 4 µg/kg for PCB 126, 50 for PCB 77, and 300 for PCB 169. The most immunotoxic of the mono-ortho-chlorinated analogs of PCB 77 and PCB 126 were about 1000 times less potent than was PCB 126.

Comparative avian egg injection studies by Brunstrom and co-workers have shown that chickens are more sensitive than turkeys *(Meleagris gallopavo)*, pheasants, ducks (mallards and goldeneyes; *Bucephala clangula*), domestic geese *(Anser anser)*, herring gulls, and black-headed gulls *(Larus ridibundus)*, at a dose rate of 20 ppb for PCB 77 mortality in chicken embryos of 70 to 100% that occurred by 18 days of incubation with malformations. Yet, 5000 ppb to ducks and 1000 ppb to geese and herring gulls had no effects (Brunstrom, 1988). However, gallinaceous birds were more sensitive, where 1000 ppb in pheasant eggs resulted in complete mortality (Brunstrom and Reutergardh, 1986). In turkeys, 200 to 1000 ppb caused 17 to 60% mortality (Brunstrom and Lund, 1988). Other species studied have included bobwhite and American kestrels by Hoffman et al. (1995). The LD_{50} for PCB 126 was approximately 0.4 ppb in chickens, whereas the LD_{50} for this congener was 24 ppb for bobwhite and 65 ppb for American kestrels. Forty-five ppb caused a decrease of 35% in hatching success for common terns.

USE OF TOXIC EQUIVALENCY FACTORS FOR "TCDD EQUIVALENTS" IN QUANTIFYING PCB AND DIOXIN TOXICITY

Toxic equivalency factors (TEFs) express the relative potency of dioxin-like compounds including coplanar PCBs relative to 2,3,7,8-TCDD. Related compounds, acting through the same "mode of action," should produce the same effects as actual TCDD but at different concentrations to account for potency differences. The toxicity of commercial PCB mixtures has been associated with the presence of certain PCB congeners having four or more chlorine atoms in both the para and metapositions of the biphenyl rings but no chlorine atoms in the orthoposition (hence, "non-ortho-PCBs"); these congeners are thought to adopt a more coplanar structure than their ortho-substituted analogs, thus attaining more frequent isostereomerism with the highly toxic 2,3,7,8-TCDD. Biological responses are similar between coplanar PCBs (such as 77, 126, and 169) and 2,3,7,8-TCDD; these responses include edema, weight loss, hepatic and thymic changes, embryotoxicity, teratogenicity, and immunotoxicity (Tanabe, 1989). All of these isosteres have high binding affinity to hepatic cytosolic receptor protein (Ah receptor) and can thereby readily induce hepatic microsomal MFO enzymes including AHH and EROD. Collectively, these chemicals are referred to as "planar halogenated hydrocarbons" (PHHs) including the proximate isostereomers for PCBs, PCDDs, and polychlorinated dibenzofurans (PCDFs), as well as certain others.

Because of the generally accepted common mode of action for PHHs, biological potencies have been theoretically calculated for complex mixtures by expressing the potency of individual congeners relative to the most toxic PHH (2,3,7,8-TCDD) (Bradlaw and Casterline, 1979; Safe, 1987) and then summing them. Therefore, one approach has been to perform congener-specific analysis and then to calculate the total potency of the mixture by multiplying the concentration of each congener by its toxic equivalency factor (TEF) and summing the products, assuming an "additive model" (Eadon et al., 1986; Safe, 1987, 1990; Tanabe et al., 1987; Kannan et al., 1988; Kubiak et al., 1989; Kutz et al., 1990; Ahlborg et al., 1992, 1994). The total potency is expressed in units of an equivalent quantity of 2,3,7,8-TCDD (hereinafter referred to as "TCDD-EQs").

A number of TEF schemes have been developed for dioxins and related compounds based on AHH or EROD enzyme induction potency (Safe, 1987; Tanabe et al., 1987; Kannan et al., 1988; Kubiak et al., 1989; Smith et al., 1990) or in vivo and in vitro effects (Kutz et al., 1990, Safe, 1990). Others have attempted to assign TEFs on the basis of acute toxicity (Eadon et al., 1986). TEFs have been derived by several means including the ability of each congener to induce cytochrome P-450-dependent AHH or EROD activity in H4IIE rat hepatoma cell culture (Niwa et al., 1975; Sawyer and Safe, 1982). Mammalian toxicity studies as well as chicken egg injection studies suggest enough correspondence in potency to warrant the use of this mammalian culture system (Kubiak et al., 1989; Ludwig et al., 1993; Tillitt et al., 1993). H4IIE rat hepatoma cells have low basal AHH and EROD enzyme activities, yet they are highly inducible by PHHs. The U.S. Food and Drug Administration (FDA) has used this bioassay to evaluate complex mixtures of PHHs from environmental samples and in food (Bradlaw and Casterline, 1979; Trotter et al., 1982). Kubiak (1991) proposed TEFs for avian embryotoxicity based on either LD_{50} or LD_{85-90} potency values from various chicken egg injection studies. Current understanding of this dioxin-like class of compounds suggests that the suite of acute effects known as the Great Lakes embryo mortality, edema, and deformities syndrome (GLEMEDS) (Gilbertson et al., 1991) are currently the most sensitive endpoints of exposure to the avian embryo. A TEF scheme based on a single endpoint, embryomortality, seems the most realistic. The bases for these TEFs are included in Table 4. Recent work by Nosek et al. (1992a, 1992b, 1993) has shown little toxicological difference related to the source (injection or maternal deposition) of TCDD contamination in pheasant eggs, thus giving further credibility to injection study-derived TEFs. The Netherlands has adopted a TEF scheme for PCB congeners (van Zorge, 1990, reported in Beurskens et al., 1993) that is very similar to that in Table 4. Other in ovo and in vitro studies of the chicken by Vos et al. (1982), Yao et al. (1990), and Bosveld et al. (1992) generally support these TEFs (all potencies are consistent within an order of magnitude). The use of any TEF scheme should be approached with an understanding of the utility and limitations of selecting one scheme over another.

Table 4 Toxic Equivalency Factors (TEFs) for Avian Embryotoxicity Based on Egg Injection Studies[a]

Compound (IUPAC[b] no.)	LD_{50} concentration (pg/g)	LD_{50} TEF (TCDD/PCB ratio)	LD_{85-90} concentration (pg/g)[c]	LD_{85-90} TEF (TCDD/PCB ratio)
2,3,7,8-TCDD	147[d]	1	1,000[c]	1
3,3',4,4',5-PeCB (PCB 126)	3,100[e]	0.05	8,000[e]	0.125
3,3',4,4'-TeCB (PCB 77)	8,600[e]	0.02	22,000[e]	0.045
3,3',4,4',5,5'-HxCB (PCB 169)	170,000[e]	0.001	330,000[e]	0.003
2,3,3',4,4',5-HxCB (PCB 156)	1,400,000[f]	0.0001	2,500,000[f]	0.0001
2,3,3',4,4'-PeCB (PCB 105)	2,200,000[f]	0.00007	2,500,000[f]	0.0001
2,3',4,4',5-PeCB (PCB 118)	>5,000,000[f]	<0.00003	>5,000,000[f]	<0.00001

[a] Adapted from presentation of Kubiak (1991). It should be noted that LD_{50} values for PCB congeners 126 and 77 are lower with treatment on day 4 and survival measured through hatching as shown in Table 3; respective TEFs are 0.25 and 0.04.
[b] IUPAC, International Union of Pure and Applied Chemistry; LD_{50}, 50% lethal dose, TCDD, tetrachlorodibenzodioxin; PCP, polychlorinated biphenyl; LD_{85-90}, 85-90% lethal dose; 3,3',4,4',5-PeCB, pentachlorobiphenyl (other PeCBs defined similarly); 3,3',4,4'-TeCB, tetrachlorobiphenyl; 3,3',4,4',5,5'-HxCB, and 2,3,3',4,4',5-HxCB, hexachlorobiphenyls.
[c] Higginbotham et al. (1968); the LD_{85-90} is defined as the net toxicity over control mortality.
[d] Verrett (1976).
[e] Brunstrom and Andersson (1988).
[f] Brunstrom (1990).

Although analytical techniques for all congeners of these compounds exist, they are extremely time consuming and costly; samples may theoretically contain up to 209 different PCB, 75 PCDD, and 135 PCDF congeners (Safe, 1987). Furthermore, it is presently difficult, if not impossible, to predict the biological effects of these mixtures with any certainty, because of the many possible combinations of congeners with many potential interactions among them that may be synergistic, additive, or antagonistic (Birnbaum et al., 1985; Weber et al., 1985).

Tillitt et al. (1991) examined the overall potencies of PCB mixtures by the ability of each mixture to induce cytochrome P-450-associated EROD activity after it was added to the rat hepatoma cell culture. These overall potencies were compared with potencies derived from summing the components of each mixture as TCDD-EQs. This method was applied to PCB-containing extracts from colonial waterbird eggs collected from the Great Lakes and revealed that the greatest concentrations of TCDD-EQs were found in the most polluted colonies where reproductive impairment was most severe. However, discrepancies between egg extract, rat hepatoma-derived TCDD-EQs, and the "additive model" based upon calculated TCDD-EQs (chemical residue analysis-summed TCDD-EQs) occurred. Hoffman et al. (1995) has been conducting bobwhite and chicken egg injection studies using combinations of two or three congeners, including PCB 126, PCB 105, PCB 77, and PCB 153. These findings suggest less-than-additive and possibly antagonistic interactions. This type of interaction is thought to be due to binding competition for Ah receptors, reducing the receptor-binding probability of congeners that are more active inducers (Safe, 1990). When studying the effects of PHH on the reproductive success of fish-eating birds, Forster's terns *(Sterna forsteri)* in the Great Lakes, Kubiak et al. (1989) converted analytical residue determinations of TCDD and PCB congeners into TCDD-EQs. Here, the summed TCDD-EQ values of individual congener residues resulted in total TCDD-EQs nearly an order of magnitude greater than the egg extract, H4IIE-derived TCDD-EQs (Tillitt et al., 1993). It was thought that these differences may have been due to less-than-additive or antagonistic effects that would only be assessed in the H4IIE assay (Bannister et al., 1987), indicating the merits of the H4IIE method or measured TCDD-EQs over calculated TCDD-EQs. Therefore, the H4IIE extract bioassay, direct injection of dioxin-like extract into fertile eggs of the chicken or other appropriate species, or adult feeding studies with environmentally derived mixtures should be used to measure embryotoxicity and confirm the relative potency of mixture exposure in the environment.

PCB EFFECTS IN THE FIELD

HERRING GULLS

Some of the most thoroughly documented studies linking PCBs to avian mortality, reproductive failure, and population declines have been conducted in the Great Lakes region. PCBs were the probable cause of mortality of many ring-billed gulls *(Larus delawarensis)* that died in southern Ontario in the late summer and early fall of 1969 and 1973 (Sileo et al., 1977) as supported by the laboratory data of Stickel et al. (1984). Among 54 gulls for which no disease-related cause of death could be determined, residues of PCBs in the brain exceeded 300 ppm (310 to 1110 ppm) in 33 specimens and were above 200 ppm in an additional 16. 1,1'-Dichloroethenylidene-bis(4-chlorobenzene) (DDE) residues in all but one of these were well below lethal levels, and dieldrin levels were 5 ppm or higher in only six specimens. Therefore, the concentrations of PCBs alone in most samples were sufficiently high, based upon experimental studies, to have caused mortality (Stickel et al., 1984).

Keith (1966) and Ludwig and Tomoff (1966) reported low reproductive success and eggshell damage in association with high organochlorine residues in Lake Michigan herring gulls *(Larus argentatus)*. Extended studies with Lake Ontario herring gulls documented similar effects, including embryo mortality, reduced hatching success, and high chick mortality (Gilbertson, 1974; Gilbertson and Hale, 1974a, 1974b). These effects were associated with high PCB (550 ppm of dry weight) and DDE (140 ppm) levels that were 10 to 100 times higher than levels in eggs from other North American colonies. Gilbertson and Fox (1977) collected herring gull eggs in 1974 from contaminated colonies on eastern Lake Ontario and from relatively uncontaminated colonies in New Brunswick and Alberta, Canada and incubated them in laboratory incubators. Hatching success for Lake Ontario gulls was 60%

less than that in controls, and subcutaneous edema, hepatomegaly, impaired bone growth, and congenital anomalies were present in hatchlings along with liver porphyria and microsomal AHH induction. Organochlorines may also have reduced reproductive success by altering parental behavior; the total organochlorine (OC) content was correlated with the total time eggs were unattended in the nest by the parent (Fox et al., 1978; Peakall et al., 1980).

Observations of other sublethal effects in herring gulls have included histopathology of the thyroid in adult birds collected from the Great Lakes between 1974 and 1983. Compared with those of a control colony in the Bay of Fundy, the thyroids from Great Lakes gulls had a greater mass and were microfollicular and frequently hyperplastic (Moccia et al., 1986). These effects were found to be consistent with the presence of polyhalogenated hydrocarbons, including PCBs. Porphyria was reported in adult Great Lakes herring gulls collected from 1980 to 1985, with the highest levels in gulls from the lower Green Bay (Lake Michigan), Saginaw Bay (Lake Huron), and Lake Ontario (Fox et al., 1988; Kennedy and Fox, 1990). Concentrations of highly carboxylated porphyrins (HCPs) in herring gulls from Saginaw Bay were significantly correlated with residues of hexachlorobenzene (HCB) ($r = 0.612, P < 0.05$), total PCBs ($r = 0.594, P < 0.05$), and DDE ($r = 0.588, P < 0.05$). DDE is not known to induce porphyria, but concentrations of DDE were highly correlated with total PCBs and PCDD residues in liver.

Gilbertson et al. (1991) reviewed the history of reproductive problems and their classification as GLEMEDS. Some of this material is addressed in more detail below.

TERN STUDIES

In 1983, the reproductive success of a Green Bay colony of state-endangered Forster's terns *(S. forsteri)* was compared with that of a successful inland colony (Hoffman et al., 1987; Kubiak et al., 1989). The hatching success in a laboratory incubator of eggs collected from Green Bay was only 52% of that for eggs from the inland control colony. Green Bay hatchlings weighed less and had an increased ratio of liver weight to body weight, shorter femur length, edema, and malformations. Hepatic MFO activity (AHH) was 3-fold higher in Green Bay hatchlings. Green Bay eggs contained a median concentration of 23 ppm of total PCBs and 37 ppt of 2,3,7,8-TCDD compared with 3.2 ppm and 8 ppt, respectively, for the control colony. The median total PCDD concentrations in eggs from Green Bay and the control area were 102 and 25 ppt, respectively. On the basis of relative AHH induction, Kubiak et al. (1989) estimated the potencies of individual PCB congeners as TCDD-EQs, using the data of Sawyer and Safe (1982); the TEFs used were published in Smith et al. (1990). Kubiak et al. (1989) concluded that two PCB congeners, PCB 105 and PCB 126, accounted for over 90% of the toxicity (2175 ppt vs. 201 ppt of the total median estimated TCDD-EQs). However, studies by Brunstrom et al. (1990) and Hoffman et al. (1995) have revealed the PCB congener 77 to be more embryotoxic than previously thought, thereby accounting for some of the toxicity reported in tern eggs. Additionally, PCB 77 was not recovered efficiently, further diminishing its relative importance (Kubiak et al., 1989). Data from 1982 on concentrations of

PCB 77 in Forster's tern eggs showed this mixture to be of substantially greater importance (Smith et al., 1990). An important extrinsic effect that further reduced the hatching success of field eggs from the Green Bay region was decreased parental nest attentiveness; hatchability was improved when eggs from the Green Bay region were exchanged and incubated by foster parents from the control location. As a sequel to the above study, Harris et al. (1993) reported greater hatching success, number of young fledged, and length of incubation in Forster's terns at Green Bay in 1988. The median total PCB residue (7.3 ppm) was 67% lower in 1988 than in 1983 and corresponded to a 42% reduction in TCDD-EQs from 1983. The authors suggested that contaminant reduction and improved reproductive performance were due to low river flows in 1988 and associated reduced PCB loading to Green Bay. Nevertheless, 42% of the chicks that were monitored died before fledging, and their body weight growth curves deviated from normal, showing signs of a wasting syndrome with loss of soft tissue, primarily pectoral muscle. The young were found to accumulate total PCBs at a rate of 18 μg/day. A no-observable-adverse-effects level (NOAEL) of 40 to 84 μg/kg/day for reproductive success was estimated from the 2-year results.

Schwartz and Stalling (1991) have assessed eggs of Forster's terns from the above study sites through pattern recognition techniques and concluded that total PCB concentrations and congener patterns support the view that Aroclor mixtures change substantially in the environment. It is therefore prudent to be cautious in the use of Aroclor matching as a means for showing a relation between source, biotic exposure, and effects. Technical PCB mixture sources should be linked to biotic exposures and accumulation/biomagnification by use of congener-specific analysis of escape pathways. This will identify which sources contribute to environmental contamination by the specific congeners of interest in organisms.

Ankley et al. (1993) and Jones et al. (1993a, 1993b) have demonstrated the uptake of specific congeners and TCDD equivalents quantified by the H4IIE rat hepatoma cell extract bioassay in Forster's terns by examining the total pollutant mass instead of the concentration. In this way, the effects of growth dilution and metabolism in hatchlings would be minimized in determining uptake/retention of these compounds in chicks and in comparisons with sibling eggs.

The H4IIE extract bioassay (Tillitt et al., 1993) resembled the TEF study approach (Kubiak et al., 1989) in predicting toxic differences between sites, but the bioassay estimated toxicity at only about a tenth of that estimated by the corresponding congener-specific analysis approach using TEFs and an additive model of toxicity used by Kubiak et al. (1989).

Other species of terns studied in the Great Lakes region have included the common tern *(Sterna hirundo)* and the Caspian tern *(Sterna caspia)*. In a study on common terns conducted in 1984 and 1985, the hatching success was determined for eggs from industrialized locations, including Green Bay and Saginaw Bay, and for eggs from several reference locations (Hoffman et al., 1993b). Hatching success was lowest for eggs from the Saginaw Bay (24% for one colony and 60% for another colony), 71% for eggs from Green Bay, and 85% for controls, corresponding to mean total PCB concentrations of 7.6 and 8.5 ppm for Saginaw Bay colonies,

10.0 ppm for Green Bay colonies, and 4.7 ppm for colonies from reference locations. Elevated levels of PCDDs were considered to be a contributing factor to the toxicity at Saginaw Bay. Nevertheless, the log-transformed total PCB content of eggs was related to the femur length/body weight ratio of hatchlings ($r = -0.70$, $P < 0.05$) and to liver microsomal AHH activity ($r = 0.71$, $P < 0.05$).

Concentrations of total PCBs as great as 18.5 to 39.3 ppm in eggs did not seem to reduce the productivity of Caspian terns during 1980 to 1981 (Struger and Weseloh, 1985). This may indicate that Caspian terns are less sensitive to the effects of PCBs than are common and Forster's terns, possibly because of their larger size and slower metabolic rate. However, in the Caspian terns studied at Saginaw Bay colonies, the PCB levels were higher by 2.1-fold as TCDD-EQs (using Brunstrom-derived TEFs) in second-clutch eggs (2800 ppt) than in first-clutch eggs (1300 ppt), and the frequency of deformed embryos was greater in the second clutches (Ludwig et al., 1993; Yamashita et al., 1993). The abnormalities, hatch rate, productivity, and fledging rates all appeared to be related to a large flood in the Saginaw River watershed that mobilized large amounts of contaminated sediment and delivered it to Saginaw Bay (Ludwig et al., 1993). Mora et al. (1993) found that plasma concentrations of total PCBs in adult Caspian terns were greatest in Green Bay and Saginaw Bay, where banding studies showed that fewer birds returned to their natal region, suggesting poorer survival in locations with an increasing plasma PCB concentration. Total PCB concentrations in the plasma of adults were 2.5 to 3.5 ppm in those locations, and only 20% of the birds banded as chicks were observed to return to these sites. At other sites where plasma PCB concentrations were below 1.5 ppm, 70% or more terns returned to the natal sites. Although productivity did not appear to be affected at Green Bay and Saginaw Bay, it is possible that post-fledging survival may have been lower in those locations, affecting the number capable of returning.

CORMORANTS

Double-crested cormorant *(Phalacrocorax auritus)* populations were probably adversely affected by PCBs as well as by DDE; no cormorants were known to have fledged from any of the colonies in the Canadian waters of Lake Ontario from 1954 to 1977 (Price and Wesloh, 1986). Since the late 1970s, there has been a marked expansion of the breeding population. However, congenital anomalies and embryonic death are currently associated with PCBs in certain colonies (Fox et al., 1991b, 1991c).

Fox et al. (1991a) found that the probability of observing a malformed (specifically, bill defects) double-crested cormorant chick on a visit to a colony in Green Bay (Lake Michigan) was 10 to 32 times greater than on a visit to a colony in reference areas located in the Canadian prairies and northwestern Ontario. The prevalence of malformed chicks in the Green Bay region (52.1/10,000) was markedly greater than in all other regions during the 1979 through 1987 period of study. Indeed, data from Tillitt et al. (1992) clearly show that the highest TCDD equivalents determined from the rat hepatoma cell extract bioassay similarly occur in Green Bay and are lowest in the Canadian prairies. In that study, only PCB-fraction extracts from eggs were assessed, not the dioxin or furan fraction.

Yamashita et al. (1993) reported the highest frequencies of deformities in live cormorant embryos from colonies with the highest PCB concentrations in eggs collected in 1988. Total PCB concentrations of 7.3 ppm and TCDD-EQs of 1200 to 1300 ppt were found in eggs from Green Bay and Beaver Islands on Lake Michigan, where the frequency of deformed, live embryos was 6 to 7%. In contrast, on Tahquamenon Island on Lake Superior, the deformity frequency was 2%, the total PCB concentration was 3.6 ppm, and TCDD-EQs were 350 ppt. Embryo mortality was highest in Green Bay colonies (22 to 39%) during 1986 through 1988 (Tillitt et al., 1992). TCDD-EQs derived from the H4IIE rat hepatoma assay using extracts from Green Bay cormorant eggs were 201 ± 13 (SD) to 344 ± 36 ppt, whereas TCDD-EQs using extracts from eggs from Lake Winnipegosis, where embryo mortality was only 8%, were 35 ± 3 ppt. Analysis of data on total PCBs revealed that PCBs at 7 to 9 ppm were associated with approximately 25% embryo mortality. However, the relation for H4IIE-derived TCDD-EQs and embryo mortality was statistically stronger ($r^2 = 0.703$, $P = 0.0003$), where TCDD-EQs at 150 to 250 ppt were associated with 25% embryo mortality.

Work by van den Berg et al. (1992) examined great cormorants *(P. carbo)* from two colonies in the Netherlands. They found yolk sac concentrations of total mono-ortho-PCBs and total PCDD(F)s between 10 and 250 ppm, and 1 to 8 ppb (lipid weight) could produce alterations in EROD liver activity, free plasma thyroxine (T_4) content, head length, size of the yolk sac, and relative liver weight. These effects are consistent with those described as GLEMEDS by Gilbertson et al. (1991). The lipid-basis hatchling yolk sac concentrations reported by van den Berg and co-workers, when converted to fresh wet weight concentrations based on a mean lipid egg content of 5.5% for great cormorants (Sotherland and Rahn, 1987), would yield considerably higher concentrations for PCBs, PCDDs, and PCDFs than reported by Yamashita et al. (1993) for Great Lakes double-crested cormorant eggs exhibiting GLEMEDS effects. Unfortunately, the lipid percentages in hatchling yolk sacs were not reported by van den Berg et al. (1992), so this comparison is illustrative and not directly comparable.

EAGLES

Total PCB concentrations in bald eagle *(H. leucocephalus)* eggs were reported to be as high as 40 to 100 ppm (fresh weight) in 1986 on Lake Erie, Lake Huron, and Lake Michigan (Colborn, 1991). Colborn concluded that, although a strong association between dichlorodiphenyltrichloroethane (DDT)/DDE and impaired bald eagle reproductive success has been provided, the high levels of PCBs and associated toxicity in other species implicate PCBs as possible agents of embryotoxicity in eagles. Kozie and Anderson (1991) reported decreased productivity in bald eagles that were nesting along the Wisconsin shoreline of Lake Superior during 1983 to 1988 compared with inland Wisconsin. PCB concentrations in the brains of nestlings varied from 2.9 to 42 ppm compared with nondetectable to 0.42 ppm for controls from inland locations. DDE concentrations in brains varied from 1.5 to 16 ppm compared with nondetectable to 0.09 ppm for controls from inland locations. Schwartz et al. (1993) reported on congener-specific analysis of an addled bald eagle egg from the Thunder Bay area of Lake Huron, North America. This addled egg contained the highest known concentration of PCB 126 and other PCB congeners in a wild bird egg when normalized to fresh wet weight. The analyzed and reported concentration was 71 ng/g for PCB 126. Normalization to fresh wet weight produced a concentration of approximately 42 ng/g (Bowerman et al., 1994b). This egg was reported to have an embryo with a "beak skewed to the right" that died about day 19-20 of incubation. This concentration was 13.5-fold higher than the LD_{50} in the chicken egg (Brunstrom and Andersson, 1988) and one that causes embryonic mortality in American kestrel embryos (Hoffman et al., 1995). Jones et al. (1993a) cite this egg in their study of biomagnification of dioxin-like compounds in the Thunder Bay ecosystem using the rat hepatoma cell extract bioassay. The egg contained 1,065 pg/g of TCDD-EQs (fresh wet weight), two orders of magnitude above the alewife and smelt concentration of TCDD-EQs from Thunder Bay waters.

Bowerman et al. (1994b) reported on the historic, documented cases of bill defects in bald eagle nestlings of the Great Lakes region. Using banding records to determine sample size ($n = 9444$), these authors found that the prevalence of bill defects in eaglets was comparable to the prevalence of bill deformities in double-crested cormorants in the Great Lakes Basin. During the period of bill defects in the eaglets, concentrations of PCBs in addled eggs ranged from 19 to 98 ppm (fresh wet weight) during 1976 to 1978 (Wiemeyer et al., 1984) and from 3.4 to 119 ppm (fresh wet weight) during 1985 to 1990 (Kubiak and Best, 1991; D. Best, U.S. Fish and Wildlife Service, unpublished data, 1993). Kubiak and Best (1991), Best et al. (1994), and Bowerman et al. (1994b) related nest productivity to concentrations in addled bald eagle eggs from the Great Lakes. Productivity curves developed by Kubiak and Best (1991) look virtually the same as those developed for the white-tailed sea eagle by Helander et al. (1982). Healthy productivity (one young per active nesting pair of adults) associates with 5 to 10 ppm (fresh wet weight) of total PCBs in eagle eggs. Bowerman et al. (1990) analyzed blood plasma from Great Lakes Basin eaglets. They found that plasma PCB concentrations, on an arithmetic mean basis, were greater than 8-fold higher (183 vs. 24 µg/l) in Great Lakes eaglets than in eaglet plasma collected from interior areas away from Great Lakes-contaminated forage.

Anthony et al. (1993) analyzed fresh bald eagle eggs from the Columbia River estuary, North America, and compared the residues to productivity. Productivity averaged 0.56 young/occupied nesting site, and total PCB concentrations averaged 12.7 ppm (fresh wet weight) and ranged from 4.8 to 26.7 ppm. High DDE concentrations in these eggs and both DDE and PCB association with egg shell thinning and productivity did not allow for the putative organochlorine to be clearly determined.

The Baltic Sea is known for significant contamination from PCBs and related compounds (Helander et al., 1982). Tarhanen et al. (1989) analyzed addled white-tailed sea eagle eggs from the Baltic Sea and from interior Lapland. The Baltic Sea eagle egg contained 48.5 ppm of total PCBs, whereas the Lapland eagle egg contained 5.9 ppm. These investigators also determined coplanar PCB congeners in both eggs. Concentrations of coplanar, dioxin-like congeners, PCB 77, PCB 126, and PCB 169, were 21, 20.6, and 6 ppb (fresh wet weight), respectively, for the Baltic Sea eagle egg and 9.8, 0.95, and not detectable for the Lapland eggs. The Baltic Sea egg concentrations of PCBs 77 and 126 were approximately 2- and 7-fold higher, respectively, than the LD_{50} in the domestic chicken egg (Brunstrom and Andersson, 1988), whereas the Lapland egg was equal to the LD_{50} for PCB 77 and 33% of the LD_{50} for PCB 126. Analyses of the muscle and liver of dead Baltic Sea white-tailed sea eagles were also conducted, and the residues were generally of the same order of magnitude as were those in Baltic Sea eggs.

Helander et al. (1982) studied the white-tailed sea eagle of the Baltic Sea coast and Lapland. Total PCB concentrations of approximately 5 to 10 ppm were associated with "healthy" reproductive success of one young produced/active nesting territory. Similar analysis of bald eagle eggs from North America produced similar results. Helander (1983) documented the occurrence of bill deformities in white-tailed sea eagle nestlings in Sweden. These defects were specifically linked to PCB contamination in the Baltic Sea between the study years 1965 and 1978 (Helander et al., 1982). Two nestlings of the 115 Baltic Sea eaglets examined were documented with bill deformities. Although the sample size of Helander and co-workers was small, the prevalence of bill defects per 10,000 observations would have been 173.9, considerably higher than for bald eagles from the Great Lakes region (Bowerman et al., 1994b) or the highest prevalence in double-crested cormorants from the Great Lakes, 52.1 (Fox et al., 1991c). The PCB content of eagle eggs from the Baltic Sea region during the time bill defects were recorded ranged from 18.7 to 159 ppm (fresh wet weight) in comparison with that of eagle eggs from Lapland, containing 8.8 to 11.1 ppm, where no deformities were documented in the 60 nestlings examined.

BLACK-CROWNED NIGHT HERONS

In the San Francisco Bay area (San Francisco Bay National Wildlife Refuge, SFBNWR), Hoffman et al. (1986) revealed a negative correlation ($r = -0.61$, $P < 0.05$) between body weight at hatching (pipping) of black-crowned night herons *(Nycticorax nycticorax)* and log-transformed PCB residues in eggs from the same nest. DDE was not significantly correlated with body weight at hatching. The geometric mean for the total PCB concentration of SFBNWR eggs was 4.1 ppm (range, 0.8 to 52.0 ppm). The yolk-free body weight (internally absorbed yolk sac removed) was lower when SFBNWR embryos were compared with control embryos from a captive colony at the Patuxent Wildlife Research Center. In another study comparing multiple geographical areas, the total PCB concentrations in black-crowned night heron embryos were positively correlated with cytochrome P-450 parameters, including AHH and EROD activities (Rattner et al., 1993); the highest PCB concentrations and MFO induction were found in birds from Green Bay.

Inasmuch as DDE and other "hard organochlorines" do not cause the suite of effects seen in GLEMEDS, the above reproductive problems and presence of bill defects in eagles and other species of birds exhibiting these effects strongly point out the limitations of statistical association of single contaminants in the investigation of chemically induced epizootics involving reproductive impairment or other toxicological endpoints. Future efforts, similar to the GLEMEDS investigations that involved "practical causal inference" (Fox, 1991), will be necessary to more fully interpret field exposures. Highly cocorrelated compounds will have to be assessed relative to their ability to produce specific effects.

LABORATORY STUDIES OF TCDD

Hudson et al. (1984) reported 37-day LD_{50}s of 15,000, >108,000, and >810,000 pg/g for bobwhite quail, mallards, and ringed turtledoves, respectively, following a single oral dose of TCDD. Grieg et al. (1973) reported that chickens given single oral doses of TCDD at 25,000 to 50,000 pg/g died within 12 to 21 days posttreatment. In a more comprehensive study, TCDD was orally administered to 3-day-old white leghorn chickens for 21 days at doses of 0, 10, 100, 1,000, and 10,000 pg/g/day with a NOAEL for mortality of 100 pg/g/day (Schwetz et al., 1973). However, none of the above studies determined residues. Nosek et al. (1992a) treated ring-necked pheasant hens with single doses of TCDD by i.p. injection (6.25, 25, or 100 µg/kg); i.p. injection was favored by the authors over oral dosing for the administration of known quantities of TCDD. The lowest single dose of TCDD to produce a delayed onset of body weight loss and mortality (wasting syndrome) was 25 µg/kg, which resulted in 25% loss in body weight and in approximately 80% mortality at the end of 12 weeks. When hen pheasants were treated weekly with lower doses of TCDD (0.01 to 1.0 µg/kg/week for 10 weeks), signs of the wasting syndrome and mortality were also produced with a cumulative dose of 10 µg/kg at the end of 10 weeks. Egg production at this cumulative dosage level was reduced as was hatchability (egg concentration of approximately 3300 ppt). A cumulative dose of 1 µg/kg resulted in approximately 1% of the cumulative dose being transferred to each of the first 15 eggs laid (Nosek et al., 1992b). The percentage for each egg was not affected by the order in which the eggs were laid. Greater than 99% of all TCDD within the egg was found in the yolk. Injection of 1099 ppt into the egg via the albumin showed a $t_{1/2}$ of 13 days for whole-body elimination of ^{3}H-labeled TCDD in the chicks that hatched. The authors estimated that wild hen pheasants laying two clutches totaling 20 eggs may eliminate 33% of the body burden of TCDD into the eggs.

In another study, Nosek et al. (1993) injected TCDD mixed in 1,4-dioxane into the yolk or the albumin of pheasant eggs prior to incubation. The TCDD doses that were calculated to cause 50% mortality above that in controls (LD_{50}) when injected into the egg albumin or into the yolk were 1354 and 2182 ppt (pg of TCDD/g of egg), respectively. Administration of the vehicle alone resulted in 37.5% mortality when in the albumin and 50% mortality when injected into the yolk. However, the authors stated that this range of mortality was within the historical range for pheasant eggs that had not received injection. Embryo mortality above that in the vehicle control group began to occur at a TCDD dose of 1000 ppt per egg. However, this dose and lower doses had no effect on the growth in hatchlings, edema, or histopathology of the liver, spleen, heart, bursa, or thymus. Cardiac malformations were not apparent in day-old hatchlings, and antibody-mediated immunity was not affected in 28-day-old chicks. The authors concluded that embryo mortality was the most sensitive sign of TCDD toxicity. The LD_{50} of TCDD was similar to the TCDD dose calculated by these authors to be naturally deposited by TCDD-exposed hen pheasants into eggs that failed to hatch (3300 ppt). Induction of hepatic EROD activity in day-old hatchlings was considered to be the most sensitive indicator of exposure. The dose causing half the maximum induction (ED_{50}) was 312 ppt, and the dose causing maximum induction (over 5-fold) was 1000 ppt.

Earlier studies have shown chicken embryos to be considerably more sensitive to TCDD egg injections than are pheasant embryos. Flick et al. (1965) first injected the unsaponifiable fraction of fat-containing PCDDs at 0.9, 1.8, or 4.5 ng per egg, which decreased hatching success by 60 to 100% and resulted in multiple malformations of the brain, legs, and beak, and in stunted growth. Verrett (1970) reported that as little as 10 to 20 ppt of 2,3,7,8-TCDD injected into chicken eggs produced embryonic mortality, edema, and malformations. Verrett (1976) determined LD_{50} concentrations for a variety of dioxins and furans in the chicken egg following injection into the air cell at 96 hours of development through hatching; 2,3,7,8-TCDF was the only compound as toxic as 2,3,7,8-TCDD. The estimated LD_{50} was 0.007 µg of 2,3,7,8-TCDD per egg (personal communication by Verrett in Bradlaw and Casterline, 1979, and in Goldstein, 1980); assuming a 50-g egg content weight, this corresponds to approximately 140 pg/g (ppt). The actual LD_{50} reported by Verrett (1976) was 147 ppt. Allred and Strange (1977) injected TCDD into the air cell of unincubated white leghorn eggs and estimated the LD_{50} by 18 days of incubation to be approximately 240 ppt. An increased liver/egg weight ratio occurred. Higginbotham et al. (1968) determined that 1000 ppt of TCDD injected into the chicken egg produced total mortality. Rifkind et al. (1985) injected TCDD (0.0001 to 12 nmol per egg or 0.65 to 78 ppb) into chicken eggs, with the exact location of injection unspecified, at 10 days of incubation. However, in this study, perhaps because of the late stage of exposure and location of injection, an exceptionally high TCDD concentration was required to affect survival where 6 nmol per egg (39 ppb) caused only 30% embryo mortality. Surviving embryos exhibited decreased thymus weight and increased pericardial and subcutaneous edema. The dose-response relation for lethality and for hepatic MFO induction, including both AHH and EROD activities, was dissociated, and the maximal induction levels were not correlated with the extent of lethality. On day 0, Henshel (1993) injected TCDD into the yolk sac or air cell of chicken eggs and found that the LD_{50}s through hatching were approximately 115 and 180 ppt, respectively. The hatching weight was lower for embryos exposed to higher concentrations. Cheung et al. (1981) injected TCDD doses of 0.009 to 77.5 pmol per egg (0.0585 to 504 ppt) on day 0 into the albumin of white leghorns and examined them on day 14. The subcutaneous edema observed in some embryos was independent of dose. Mortality from treatment was minimal, and control mortality was 21%. However, a dose-dependent increase in cardiovascular malformations was apparent, where 6 ppt caused a 20% increase above the unexpectedly high control incidence of 29% and a doubling at 65 ppt. Other deformities, including malformed legs and crossed beaks associated with microphthalmia, occurred at lower frequencies of 1 to 3%.

Martin et al. (1989) estimated the LD_{50} for eastern bluebirds (*Sialia sialis*) following TCDD injections into the albumin of eggs; the LD_{50} was greater than 1 but less than 10 ng of TCDD/g of egg (ppb), with embryo mortality being the most sensitive manifestation of toxicity. Eye and beak malformations, such as those seen in chicken embryos, were not apparent, nor was there edema or any effects on posthatching growth or histopathology of tissues in 8-day-old bluebird nestlings.

FIELD STUDIES OF TCDD

Retrospective analysis of herring gull eggs from the Great Lakes has revealed that TCDD concentrations averaged 500 ppt during 1974 but were as high as 1200 ppt in earlier years (Gilbertson, 1988), which was probably a factor contributing to reproductive failure in addition to other documented organochlorines, including PCBs and DDT (Gilbertson et al., 1991) (Table 5). These high concentrations of TCDD, as well as of other organochlorines, declined to about 160 ppt by 1976 in gull eggs, and gull reproduction improved dramatically. More recently, Spear et al. (1990), concerned with vitamin A imbalance, examined egg yolk retinoids in this species and found that the molar ratio of retinol to retinyl palmitate was different among colonies from the Great Lakes and correlated with several indices of PCDD and PCDF concentrations in gull eggs from those sites. The molar ratio of retinol to retinyl palmitate was positively correlated with TCDD-EQs of PCDDs and PCDFs ($r = 0.866, P < 0.01$) and with the sum of PCDD and PCDF concentrations ($r = 0.759, P < 0.05$). Excess retinoic acid itself is experimentally teratogenic to avian embryos. The median retinol and retinyl palmitate concentrations in the livers of herring gulls in 1982 were 131 and 231 ppt, respectively, on Lake Ontario; 289 and 377 ppt on Lake Michigan; 382 and 562 ppt on Lake Superior; and 864 and 1737 ppt in New Brunswick (Spear et al., 1986). Corresponding 2,3,7,8-TCDD concentrations in eggs were 90, 10, 13, and 3 ppt, showing an inverse relation between liver retinoid levels and dioxin contamination.

Observations of other sublethal effects in herring gulls have included histopathology of the thyroid in adult birds collected from the Great Lakes between 1974 and 1983. Compared with a control colony in the Bay of Fundy, the thyroids from gulls from the Great Lakes had a greater mass and were microfollicular and frequently hyperplastic (Moccia et al., 1986). These effects were found to be consistent with the presence of polyhalogenated hydrocarbons, including PCBs. Porphyria was reported in adult herring gulls from the Great Lakes collected from 1980 to 1985, with the highest levels in gulls from the lower Green Bay (Lake Michigan), Saginaw Bay (Lake Huron), and Lake Ontario (Fox et al., 1988; Kennedy and Fox, 1990). Concentrations of HCPs in herring gulls from Saginaw Bay were significantly correlated with residues of total 2,3,7,8-substituted tetra- through heptachloro dibenzo-*p*-dioxins ($r = 0.786, P < 0.05$) and DDE ($r = 0.588, P < 0.05$). DDE is not

Table 5 Field Studies Measuring Exposure and Effects Consistent with Planar Polychlorinated Biphenyls (PCBs) and Dioxins

Species, tissue	Concentration	Effects	Ref.
Herring gull,[a] eggs	2,3,7,8-TCDD = 3 > 1200 ppt Total PCBs = 6 > 180 ppm	GLEMEDS consistent,[b] embryo mortality, impaired reproductive success, beak defects, AHH induction, vitamin A depletion, porphyria	Gilbertson et al. (1991); Norstrom et al. (1982); Ellenton et al. (1985); Fox et al. (1988); Spear et al. (1990); Kennedy and Fox (1990)
Forster's tern, eggs	Total PCBs = 6-26 ppm 3,3',4,4',5-PeCB = 540-9,100 ppt 2,3,3',4,4'-PeCB = 330-730 ppb 2,3,7,8-TCDD = 14-105 ppt TCDD equivalents from congener chemistry[d] = 618-7,366 ppt TCDD equivalents from H4IIE extract bioassay = 90-339 ppt	Embryo mortality, impaired reproductive success, subcutaneous edema of head and neck, AHH induction, hard tissue deformities, mostly beaks[b]	Kubiak et al. (1989); Hoffman et al. (1987); Tillitt et al. (1993)
Common tern, eggs	Total PCBs = 5-24 ppm	General reproductive impairment, deformities, edema and AHH induction[b]	Hoffman et al. (1993b)
Caspian tern, eggs	Total PCBs = 4-18 ppm TCDD equivalents from congener chemistry[c] = 1,300-2,800 ppt TCDD equivalents from H4IIE extract bioassay = 50-416 ppt 2,3,7,8-TCDD = 8-22 ppt 3,3',4,4'-TeCB = 15,000-23,000 ppt 3,3',4,4',5-PeCB = 3,300-7,900 ppt 2,3,3',4,4'-PeCB = 140-370 ppb	Multiple deformities, edema, embryo mortality, impaired reproductive success[b]	Ludwig et al. (1993); Yamashita et al. (1993); Tillitt et al. (1991)
Caspian tern, adult blood plasma	Total PCBs = 1-14 ppm	General inverse relationship between population exposure and adult breeder numbers as measured by natal site tenacity suggested lower post-fledging survival to adult breeding age	Mora et al. (1993)
Double-crested cormorant,[a] eggs	TCDD equivalents from H4IIE extract bioassay = 85-413 ppt 2,3,7,8-TCDD = 5.3-22 ppt	Embryonic mortality, beak deformities, club foot[b]	Fox et al. (1991a,b,c); Tillett et al. (1991); Yamashita et al. (1993)

Table 5 (continued) Field Studies Measuring Exposure and Effects Consistent with Planar Polychlorinated Biphenyls (PCBs) and Dioxins

Species, tissue	Concentration	Effects	Ref.
	Total PCBs = 3.6-6.8 ppm 3,3',4,4',5-PeCB = 800-7,900 ppt 2,3,3',4,4'-PeCB = 110-370 ppb TCDD equivalents from congener chemistry[c] = 350-1,300 ppt		
Cormorant, hatchling yolk sac	Total mono-ortho PCBs = 10-250 ppm of lipid weight Total PCDD/PCDF ratio = 1-8 ppb of lipid weight	Concentration-dependent alterations in EROD activity, free T4 plasma content, head length, yolk sac size, and relative liver weight	van den Berg et al. (1992)
Bald eagle, eggs	Total PCBs = 8-77 ppm (All below 1 sample) 3,3',4,4'-TeCB = 25 ppb 3,3',4,4',5-PeCB = 71 ppb Total PCBs = 99.8 ppm TCDD equivalents from H4IIE extract bioassay = 1,065 ppt	General reproductive impairment, deformities of the beak; 50- to 100-fold magnification of TCDD equivalents (H4IIE extract bioassay from small fish to eggs of fish-eating birds; fractional composition of PCB 126 enriched compared with any source Aroclor mixture	Kubiak and Best (1991); Jones et al. (1993a); Schwartz et al. (1993); Bowerman et al. (1994a,b); Bowerman et al. (1995)
White-tailed sea eagle (Sweden), eggs	Total PCBs = 20-159 ppm	General reproductive impairment, deformities of the beak	Helander et al. (1982); Helander (1983)
Black-crowned night heron, eggs	TCDD equivalents from H4IIE extract bioassay = 221 ppt	Hepatomegaly, AHH induction, subcutaneous edema of neck and throat	Hoffman et al. (1993b); Rattner et al. (1993); Tillitt et al. (1991)
Great blue heron, eggs	2,3,7,8-TCDD = 211 ± 34 ppt TCDD equivalents from congener chemistry[c] = 227 ± 36	Altered embryonic growth, shortened beak, scarcity of down follicles, subcutaneous edema, MFO induction, and intercerebral asymmetry	Hart et al. (1991); Bellward et al. (1990); Henshel et al. (1995)
Wood duck, eggs	2,3,7,8-TCDD = 2-482 ppt, 36 ppt (geometric mean) TCDD equivalents from congener chemistry[d] = 3-611 ppt, 52 ppt (geometric mean)	General reproductive impairment, deformities of the beak, subcutaneous edema of head and neck	White and Seginak (1994); White and Hoffman (1995)

Table 5 (continued)   Field Studies Measuring Exposure and Effects Consistent with Planar Polychlorinated Biphenyls (PCBs) and Dioxins

Species, tissue	Concentration	Effects	Ref.
Peregrine falcon, eggs	2,3,7,8-TCDD = 4.7-9 ppt Total PCBs = 1.4-13 ppm 3,3',4,4'-TeCB = 170-3300 ppt 3,3',4,4',5-PeCB = 80-2600 ppt TCDD equivalents from congener chemistry[d] = 120 ppt (geometric mean)	PCB 126 approached the LD_{50} in chicken eggs of 3,100 ppt but considerably lower than the kestrel LD_{50} of 70-100 ppb	Jarman et al. (1993)

[a] Studies cited for certain species involve multiple biological and chemical samples that may not always be of exactly the same year of study or exact site of collection. We have endeavored to screen these citations for temporal and geographic consistency for the identified species. Readers are encouraged to review each citation from an individual species to gain additional appreciation of sample associations.

[b] 2,3,7,8-TCDD, 2,3,7,8-tetrachlorodibenzodioxin; AHH, aryl hydrocarbon hydroxylase; 3,3',4,4',5-PeCB, pentachlorobiphenyl (2,3,3',4,4'-PeCB defined similarly); 3,3',4,4'-TeCB, tetrachlorobiphenyl; PCDD, polychlorinated dibenzodioxin; PCDF, polychlorinated dibenzofuran; EROD, 7-ethoxyresocifin-O-deethylase; MFO, mixed-function oxygenase; LD_{50}, 50% lethal dose.

[c] GLEMEDS, Great Lakes Embryo Mortality, Edema, and Deformities Syndrome as reviewed by Gilbertson et al. (1991).

[d] Readers are cautioned that chemistry-derived TCDD equivalents are a function of the toxic equivalency factors that are used in that study and as such are relative indicators of exposure to actual 2,3,7,8-TCDD, assuming an additive model of toxicity. Various sets of TEFs have been used for this type of data interpretation. Some of these references are identified in the text.

known to induce porphyria, but concentrations of DDE were highly correlated with total PCBs and PCDD residues in liver.

Elliot et al. (1989) reported TCDD concentrations in great blue heron *(Ardea herodias)* eggs from four colonies on the coast of British Columbia. PCDD levels in eggs were significantly elevated at a colony near a Kraft paper mill at Crofton on Vancouver Island in 1986. In 1987, the colony failed to raise any young, with 2,3,7,8-TCDD levels nearly 3 times higher than in 1986. Other contaminants, including PCBs, organochlorine pesticides, and mercury, were generally low. In 1986, the mean 2,3,7,8-TCDD level was 66 ppt (range, 8 to 218 ppt) (wet weight), and that of 2,3,7,8-tetrachlorodibenzofuran (2,3,7,8-TCDF) was 2 ppt (not detectable [ND] to 14 ppt). The total TCDD-EQs calculated according to the method of Mason et al. (1986) were 79 ppt (19 to 272 ppt). In 1987, although the concentrations of 2,3,7,8-TCDD and of total TCDD-EQs had tripled, the authors felt that predation had played a role in the poor productivity. Here, 2,3,7,8-TCDD accounted for 82% of the total TCDD-EQs. In 1988 and further back, eggs were collected and allowed to hatch in a laboratory incubator (Bellward et al., 1990; Hart et al., 1991). Although hatching success was not significantly affected in the incubator with 2,3,7,8-TCDD egg concentrations of 211 ppt, subcutaneous edema was apparent in four of 12 chicks but absent in control chicks. TCDD concentrations in eggs from the same clutch were inversely correlated ($P < 0.01$) with measures of growth, including yolk-free body weight, tibia length, and organ weights. Liver microsomal EROD activity in hatchlings was positively correlated ($r = 0.572$, $P < 0.001$) with 2,3,7,8-TCDD concentrations in eggs from the same clutch. Brains from highly contaminated colonies (Crofton) in 1988 exhibited a high frequency of intercerebral asymmetry that decreased in subsequent years as levels of TCDD decreased (Henshel et al., 1995). The asymmetry was significantly correlated with the level of TCDD in eggs taken from the same nest.

White and Seginak (1994) studied the reproductive success in wood ducks *(Aix sponsa)* in nest boxes downstream from Bayou Meto, a major drainage system in central Arkansas contaminated with PCDDs and PCDFs from a former chemical plant that manufactured the herbicide 2,4,5-trichlorophenoxyacetic acid (2,4,5-T). Based on TEFs, the residues in eggs of wood ducks were 50-fold higher in eggs near the point source than in those from an uncontaminated reference area. The geometric mean and range for 2,3,7,8-TCDD were 36 ppt and 1.6 to 482 ppt, respectively, and were 26 ppt and 2.4 to 244 ppt for 2,3,7,8-TCDF at the most contaminated site. The TCDD-EQs were 52 ppt (range, 3.7 to 611 ppt), with 70% accounted for by 2,3,7,8-TCDD. Overall productivity (nest success, hatching success, and duckling production) was suppressed ($P < 0.05$) at nest sites 9 and 17 km downstream from the source, as was hatching success as far as 58 km downstream compared with that from a reference site. Egg TCDD-EQs were inversely correlated ($P < 0.0001$) with productivity in corresponding nests; i.e., as egg TCDD-EQs decreased, productivity increased. In addition, teratogenesis and oxidative stress were documented at the more contaminated sites (White and Hoffman, 1995). The threshold range of toxicity, based on TCDD-EQs where reduced productivity was evident in wood ducks, was 20 to 50 ppt. Concentrations of PCBs and DDE were virtually absent and therefore not a factor. It was concluded that the wood duck is particularly sensitive to PCDD/PCDF exposure and might serve as a good model for monitoring.

ESTIMATION OF EGG AND BODY RESIDUES FROM FORAGE

Concentrations of PCBs and dioxin-like congeners in target organs, tissues, or whole bodies cannot always be determined at the most opportune time because of numerous factors. Braune and Norstrom (1989) published an eloquent study on apparent biomagnification factors in herring gulls from Lake Ontario. Adult herring gulls of the Great Lakes can be assumed to be at "dynamic steady-state equilibrium" with its main forage food, because herring gull adults do not venture far from their breeding colonies and because they remain on the Lakes virtually year round. Biomagnification factors from forage to egg were generated by Kubiak and Best (1991) from the data of Braune and Norstrom (1989). Because the most sensitive endpoint for PCB- and dioxin-like toxicity appears to be reproductive impairment associated with egg residues, Table 6 depicts the biomagnification factors (BMFs) from alewife, as forage, to herring gull egg, as target organ. Additional BMF data were communicated by R. Norstrom (Environment Canada, 1993) beyond those

Table 6 Biomagnification Factors from Alewife to Herring Gull Egg for Dioxin-like and Other Organochlorine Compounds

Compound	Biomagnification factor
2,3,7,8-TCDD[a]	21
1,2,3,7,8-PeCDD	10
1,2,3,6,7,8-HxCDD	16
1,2,3,4,6,7,8-HpCDD	>6
OCDD	>8
2,3,7,8-TCDF	<0.65
2,3,4,7,8-PeCDF	4
1,2,3,4,7,8-, 1,2,3,4,6,7-HxCDF	>4
1,2,3,6,7,8-HxCDF	>4
Total PCBs	32
2,3,3',4,4'-PeCB	20
2,3',4,4',5-PeCB	31
2,2',3,4,4',5-HxCB	42
3,3',4-TrCB	0.8*
3,3',4,4'-TeCB	1.8*
3,3',4,4',5-PeCB	29*
3,3',4,4',5,5'-HxCB	46*
Hexachlorobenzene	20
DDE	34
Mirex	30
Photomirex	34
β-HCH	10
Octachlorostyrene	8
Oxychlordane	60
trans-Nonachlor	3
cis-Nonachlor	5
DDT	2
Heptachlor epoxide	30
Dieldrin	7

[a] 2,3,7,8-TCDD, 2,3,7,8-tetrachlorodibenzodioxine; 1,2,3,7,8-PeCDD, pentachlorodibenzodioxin; 1,2,3,6,7,8-HxCDD, hexachlorodibenzodioxin; 1,2,3,4,6,7,8-HpCDD, heptachlorodibenzodioxin; OCDD, octachlorodibenzodioxin; 2,3,7,8-TCDF, tetrachlorodibenzofuran; 2,3,4,7,8-PeCDF, pentachlorodibenzofuran; 1,2,3,4,7,8-HxCDF, hexachlorodibenzofuran(1,2,3,4,6,7-HxCDF and 1,2,3,6,7,8-HxCDF defined similarly); total PCBs, total polychlorinated biphenyls; 2,3,3',4,4'-PeCB, pentachlorobiphenyl (2,3',4,4',5-PeCB and 3,3',4,4',5-PeCB defined similarly); 2,2',3,4,4',5-HxCB, hexachlorobiphenyl (3,3',4,4',5,5'-HxCB defined similarly); 3,3',4-TrCB, trichlorobiphenyl; 3,3',4,4'-TeCB, tetrachlorobiphenyl; DDE, 1,1'-(dichloroethenylidane)-bis(4-chlorophenzane); β-HCH, β-benzenehexachloride; DDT, 1,1'-(2,2,2-tetrachloroethylidene)-bis(4-chlorobenzene).

Data adapted from Braune and Norstrom, 1989;* provided by Ross Norstrom, personal communication to T. J. Kubiak.

published. Therefore, it is possible to estimate egg concentrations from adult forage, with the assumption that the forage is the predominant food eaten and that pharmacokinetic differences between carnivorous species are not large. Differences are most likely in species size, which affects metabolism and thus the mass of food ingested per unit of body weight. There may be other differences associated with birds, as specialized feeders are inferior to nonspecialized feeders in their capacities for metabolizing organochlorines (Walker et al., 1987). Nonetheless, this procedure has direct benefit, in that an egg concentration (or lipid concentration in the adult and adult whole-body concentration) can be estimated before the egg is analyzed. This allows for some estimation of the potential risk of exposure, based on the known effect levels discussed in this review. Of course, species-specific and site-specific BMFs would be preferred if available. This approach has been used recently in various ways to estimate the concentrations for risk assessment purposes (Sullivan et al., 1987; Thiel, 1990; Kubiak and Best, 1991; Bowerman et al., 1995). A similar approach using the H4IIE rat hepatoma cell extract bioassay appears to be practical as well, since TCDD-EQs generated from this extract bioassay clearly show biomagnification from forage, such as with the alewife, to predacious fish and fish-eating birds (Jones et al., 1993a, 1993b). The estimated concentrations of TCDD-EQs in eggs could be compared with the threshold effect level of choice or a NOAEL concentration for the actual TCDD in the avian egg.

SUMMARY

PCB residue in the brain was generally found to be the most diagnostic tissue residue for lethality in adult birds. For Aroclor 1254, brain residues of total PCBs of approximately 300 ppm are diagnostic of lethality in pheasants and passerines (Table 7). For Clophen A60, lethal brain residues were 76 to 180 ppm in great cormorants but 420 to 445 ppm in herons. In a field die-off of ring-billed gulls, total PCB brain residues for two thirds of the dead birds exceeded 300 ppm and exceeded 200 ppm for the remainder. However, one must be somewhat cautious in relating Aroclor and other commercial PCB mixture feeding studies to field observations, because pattern recognition techniques during analysis of total PCB concentrations and congener patterns suggest that Aroclor mixtures change substantially in the environment and through the food chain.

The most sensitive functional endpoint for PCB- and dioxin-mediated toxicity appears to be reproductive impairment as associated with egg residues. Chickens are the most sensitive species with respect to PCB effects on reproduction. For Aroclor 1254 and Aroclor 1242, the hatchability of chicken eggs was reduced when residues were above 4 and 1 ppm, respectively. Aroclor 1254 decreased hatching success and parental attentiveness of ringed turtledoves during incubation; the mean total PCB residues in dove eggs were 16 ppm and, in adult brains, 5.5 ppm. Brain residues of 3 ppm resulted in significant depletion of brain dopamine and norepinephrine. Mallards, screech owls, and Atlantic puffins appeared to be more resistant to PCB exposure.

Table 7 Summary of Polychlorinated Biphenyl (PCB) and Tetrachlorodibenzodioxin (TCDD) Effect Levels

Concentration	Effect
20 to 50 ppt of TCDD in eggs	Embryo mortality and teratogenesis in chickens, decreased productivity and teratogenesis for wood ducks
90 to 339 ppt of TCDD equivalents (H4IIE bioassay of egg extract)	Embryotoxicity in Forster's tern
150 to 250 ppt of TCDD in eggs	Decreased embryonic growth, edema in herons
618 to 7,366 ppt of TCDD equivalents (congener chemistry)	Embryotoxicity in Forster's tern
1,000 ppt of TCDD in eggs	Embryo mortality in pheasants
<10,000 ppt of TCDD in eggs	Embryo mortality in bluebirds
1 to 5 ppm of total PCBs in eggs	Decreased hatching success for chickens
8 to 25 ppm of total PCBs in eggs	Decreased hatching success for terns, cormorants, doves, eagles
75 to 300 ppm of total PCBs in brain	Lethality in great cormorants, gulls, passerines, and pheasants

Several developmental and physiological studies have revealed coplanar PCB 126 to be the most toxic of all PCB congeners, causing porphyria in Japanese quail (liver residues under 0.1 ppm) and liver enlargement with coagulative necrosis, colloid depletion of the thyroid, and lymphoid depletion of the spleen in American kestrel nestlings (liver residue of 150 ppb). The utility of studies of injections into eggs for predicting the potential embryotoxicity of PCBs and TCDD compares favorably with feeding studies. In instances in which the same chemicals, aroclors or TCDD, have been administered by both methods, the concentrations in eggs and the effects are quite similar. Comparison of PCB congeners by injection into eggs has revealed chickens to be the most sensitive species and PCB 126 to be the most toxic congener; with air cell injections on day 7 of incubation, LD_{50}s (72 hours later) for PCB congeners 126, 77, 169, and 105 were 3.1, 8.6, 170, and 2200 ppb, respectively, but lower at earlier exposure. For PCB 126, the LD_{50} was between 40 and 70 ppb for bobwhite and between 70 and 100 ppb for American kestrels. For PCB 77 in pheasant eggs, 1000 ppb resulted in complete mortality and, in turkeys, 1000 ppb caused 60% mortality. However, 5000 ppb in ducks and 1000 ppb in geese and herring gulls had no effects. The LD_{50}s for TCDD when injected into the egg albumin or yolk of pheasants were 1354 and 2182 ppt per egg, respectively, and were similar to the dose calculated to be naturally deposited by TCDD-exposed hen pheasants into eggs that failed to hatch (3300 ppt). Chicken embryos are considerably more sensitive to TCDD than are embryos of other species. As little as 10 to 20 ppt of 2,3,7,8-TCDD injected into chicken eggs produced embryonic mortality, edema, and malformations, and LD_{50}s (through hatching) were approximately 150 ppt (yolk sac) and 250 ppt (air cell). For the eastern bluebird, the LD_{50} for injection into egg albumin was between greater than 1000 and 10,000 ppt, with embryo mortality being the most sensitive manifestation of toxicity with little evidence of other effects.

Because of a common mode of action for PHHs including PCBs and PCDDs, biological potencies may be calculated by expressing the potency of individual PCB congeners identified in the sample using a TEF relative to the most toxic PHH (2,3,7,8-TCDD) and then summing them in an "additive" fashion to produce a total potency as TCDD equivalents (TCDD-EQs). For assessing potential avian embryotoxicity, it is best that TEFs be derived from studies of injections into eggs, as presented in Table 4. Another approach assesses the overall potency of PCB-containing extracts from tissues to directly induce cytochrome P-450-dependent EROD activity in H4IIE rat hepatoma cells as compared with the standard 2,3,7,8-TCDD. Discrepancies between the additive model and egg extract, H4IIE-derived TCDD-EQs exist. However, the H4IIE assay has the benefit of directly assessing any nonadditive overall effects that other studies have revealed do occur. Therefore, the H4IIE extract bioassay, direct injection of dioxin-like extract into fertile eggs of the chicken or other appropriate species, or adult feeding studies with environmentally derived mixtures should be used to measure embryotoxicity and confirm the relative potency of mixture exposure in the environment.

When Forster's tern eggs in the Great Lakes region contained a median concentration of 23 ppm of total PCBs and 37 ppt of 2,3,7,8-TCDD (2175-ppt total median-estimated TCDD-EQs), hatching success was only 50%. With a median concentration of 7.3 ppm of total PCBs but approximately 1250 ppt of total TCDD-EQs, 42% of the chicks monitored died before fledging, and their body weight growth curves deviated from normal. Mean total PCB levels of above 7.5 ppm for common tern eggs were associated with decreased hatching success. In Caspian terns, egg concentrations in the 20- to 40-ppm range did not seem to alter productivity, possibly because of the larger size and slower metabolism of this species in contrast to the other terns. In double-crested cormorants, a concentration of 7 to 9 ppm of PCB in eggs was associated with approximately 25% embryo mortality.

In great blue heron eggs collected in the field, TCDD at 211 ppt did not decrease hatching success, but edema was apparent and growth reduced. Mean TCDD-EQs of 52 ppt in wood duck eggs, with 70% accounted for by 2,3,7,8-TCDD, were related to suppressed overall productivity in the field and teratogenesis.

ACKNOWLEDGMENTS

The authors acknowledge Angela Talbert for her helpful assistance in preparing this manuscript.

REFERENCES

Ahlborg, U. G., G. C. Becking, L. S. Birnbaum, A. Brouwer, H. J. G. M. Derks, M. Feeley, G. Golor, A. Hanberg, J. C. Larsen, A. K. D. Liem, S. H. Safe, C. Schlatter, F. Waern, M. Younes, and E. Yrjänheikki. 1994. Toxic equivalency factors for dioxin-like PCBs. Chemosphere 28:1049-1067.

Ahlborg, U. G., A. Brouwer, M. A. Fingerhut, J. L. Jacobson, S. W. Jacobson, S. W. Kennedy, A. A. F. Kettrup, J. H. Koemon, H. Poiger, C. Rappe, S. H. Safe, R. F. Seegal, J. Tuomisto, and M. van den Berg. 1992. Impact of polychlorinated dibenzo-p-dioxins, dibenzofurans, and biphenyls on human health, with special emphasis on application of the toxic equivalency factor concept. European J. Pharmacol. Environ. Toxicol. Pharmacol. Sect. 228:179-199.

Allred, P. M., and J. R. Strange. 1977. The effects of 2,4,5-trichlorophenoxyacetic acid and 2,3,7,8-tetrachlorodibenzo-*p*-dioxin on developing chicken embryos. Arch. Environ. Contam. Toxicol. 5:483-489.

Andersson, L., E. Nikolaidis, B. Brunstrom, A. Bergman, and L. Dencker. 1991. Effects of polychlorinated biphenyls with Ah receptor affinity on lymphoid development in the thymus and bursa of Fabricius of chick embryos *in ovo* and in mouse thymus anlagen *in vitro*. Toxicol. Appl. Pharmacol. 107:183-188.

Ankley, G. T., G. J. Niemi, K. B. Lodge, H. J. Harris, D. L. Beaver, D. E. Tillitt, T. R. Schwartz, J. P. Giesy, P. D. Jones, and C. Hagley. 1993. Uptake of planar polychlorinated biphenyls and 2,3,7,8-substituted polychlorinated dibenzofurans and dibenzo-*p*-dioxins by birds nesting in the Lower Fox River and Green Bay, Wisconsin, USA. Arch. Environ. Contam. Toxicol. 24:332-344.

Anthony, R. G., M. G. Garrett, and C. A. Schuler. 1993. Environmental contaminants in bald eagles in the Columbia River Estuary. J. Wildl. Manage. 57:10-19.

Bannister, R., D. Davis, T. Zacharewski, I. Tizard, and S. Safe. 1987. Aroclor 1254 as a 2,3,7,8-tetrachlorodibenzo-*p*-dioxin antagonist: effects on enzyme induction and immunotoxicity. Toxicology 46:29-42.

Bellward, G. D., R. J. Norstrom, P. E. Whitehead, J. E. Elliott, S. M. Bandiera, C. Dworschak, T. Chang, S. Forbes, B. Cadario, L. E. Hart, and K. M. Cheng. 1990. Comparison of polychlorinated dibenzodioxin and dibenzofuran levels with hepatic mixed-function oxidase induction in great blue herons. J. Toxicol. Environ. Health 30:33-52.

Best, D. A., W. W. Bowerman IV, T. J. Kubiak, S. R. Winterstein, S. Postupalsky, M. Shieldcastle, and J. P. Giesy. 1994. Reproductive impairment of bald eagles *(Haliaeetus leucocephalus)* along the Great Lakes shorelines of Michigan and Ohio. p. 697-702. *In* B. J. Meyburg and R. D. Chancellor (Eds.). Raptor conservation today. World Working Group on Birds of Prey and Pica Press, East Sussex, Great Britain.

Beurskens, J. E. M., G. A. J. Mol, H. L. Barreveld, B. van Munster, and H. J. Winkels. 1993. Geochronology of priority pollutants in a sedimentation area of the Rhine River. Environ. Toxicol. Chem. 12:1549-1566.

Bird, F. H., C. B. Chawan, and R. W. Gerry. 1978. Response of broiler chickens to low level intake of polychlorinated biphenyl isomers. Poult. Sci. 57:538-541.

Birnbaum, L., H. Weber, M. Harris, J. Lamb, and J. McKinney. 1985. Toxic interaction of specific polychlorinated biphenyls and 2,3,7,8-tetrachlorodibenzo-*p*-dioxin: increased incidence of cleft palate in mice. Toxicol. Appl. Pharmacol. 77:292-302.

Blazak, W. F., and J. B. Marcum. 1975. Attempts to induce chromosomal breakage in chicken embryos with Aroclor 1242. Poult. Sci. 54:310-312.

Bosveld, B. A. T. C., M. van den Berg, and R. M. C. Theeien. 1992. Assessment of the EROD inducing potency of eleven 2,3,7,8- substituted PCDD/Fs and three coplanar PCBs in the chick embryo. Chemosphere 25:911-916.

Bowerman, W. W., J. P. Giesy, D. A. Best, and V. J. Kramer. 1995. A review of factors affecting productivity of bald eagles in the Great Lakes region: Implications for recovery. Environ. Health Perspect. 103, suppl. 4:51-59.

Bowerman, W. W., D. A. Best, E. D. Evans, S. Postupalsky, M. S. Martell, K. D. Kozie, R. L. Welch, R. H. Scheel, K. F. Durling, J. C. Rodgers, T. J. Kubiak, D. E. Tillitt, T. R. Schwartz, P. D. Jones, and J. P. Giesy. 1990. PCB concentrations in plasma of nestling bald eagles from the Great Lakes Basin, North America. Vol. 4, p. 212-216. *In* H. Fiedler and O. Hutzinger (Eds.). Proc. 10th Int. Conf. on Organohalogen Compounds. Ecoinforma Press, Bayreuth, Germany.

Bowerman, W. W., IV, J. P. Giesy, Jr., D. A. Best, T. J. Kubiak, and J. G. Sikarskie. 1994a. The influence of environmental contaminants on bald eagle *(Haliaeetus leucocephalus)* populations in the Laurentian Great Lakes, North America. p. 703-707. *In* B. J. Meyburg and R. D. Chancellor (Eds.). Raptor conservation today. World Working Group on Birds of Prey and Pica Press, East Sussex, Great Britain.

Bowerman, W. W., IV, T. J. Kubiak, J. B. Holt, Jr., D. L. Evans, R. G. Eckstein, C. R. Sindelar, D. A. Best, and K. D. Kozie. 1994b. Observed abnormalities in mandibles of nestling bald eagles *Haliaeetus leucocephalus*. Bull. Environ. Contam. Toxicol. 53:450-457.

Bradlaw, J. A., and J. L. Casterline. 1979. Induction of enzymes in cell cultures: a rapid screen for the detection of planar chlorinated organic compounds. J. Assoc. Off. Anal. Chem. 62:904-916.

Braune, B. M., and R. J. Norstrom. 1989. Dynamics of organochlorine compounds in herring gulls. III. Tissue distribution and bioaccumulation in Lake Ontario gulls. Environ. Toxicol. Chem. 8:957-968.

Britton, W. M., and J. M. Huston. 1973. Influence of polychlorinated biphenyls in the laying hen. Poult. Sci. 52:1620-1624.

Brunstrom, B. 1988. Sensitivity of embryos from duck, goose, herring gull, and various chicken breeds to 3,3',4,4'-tetrachlorobiphenyl. Poult. Sci. 67:52-57.

Brunstrom, B. 1989. Toxicity of coplanar polychlorinated biphenyls in avian embryos. Chemosphere 19:765-768.

Brunstrom, B. 1990. Mono-ortho-chlorinated chlorobiphenyls: toxicity and induction of 7-ethoxyresorufin *O*-deethylase (EROD) activity in chick embryos. Arch. Toxicol. 64:188-192.

Brunstrom, B., and L. Andersson. 1988. Toxicity and 7-ethoxyresorufin *O*-deethylase-inducing potency of coplanar polychlorinated biphenyls (PCBs) in chick embryos. Arch. Toxicol. 62:263-266.

Brunstrom, B., L. Andersson, E. Nikolaidis, and L. Dencker. 1990. Non-ortho- and mono-ortho-chlorine-substituted polychlorinated biphenyls — embryotoxicity and inhibition of lymphocyte development. Chemosphere 20:1125-1128.

Brunstrom, B., and P. O. Danerud. 1983. Toxicity and distribution in chick embryos of 3,3',4,4'-tetrachlorbiphenyl injected into the eggs. Toxicology 27:103-110.

Brunstrom, B., and J. Lund. 1988. Differences between chick and turkey embryos in sensitivity to 3,3',4,4'-tetrachlorobiphenyl and in concentration/affinity of the hepatic receptor for 2,3,7,8-tetra-chlorodibenzo-*p*-dioxin. Comp. Biochem. Physiol. C Comp. Pharmacol. Toxicol. 91:507-512.

Brunstrom, B., and J. Orberg. 1982. A method for studying embryotoxicity of lipophilic substances experimentally introduced into hens' eggs. Ambio 11:209-11.

Brunstrom, B., and L. Reutergardh. 1986. Differences in sensitivity of some avian species to the embryotoxicity of a PCB, 3,3',4,4'-tetrachlorobiphenyl, injected into the eggs. Environ. Pollut. Ser. A Ecol. Biol. 42:37-45.

Cecil, H., J. Bitman, G. Fries, L. Smith, and R. Lillie. 1972. PCB's in laying hens. Presented at the American Chemical Society Fall Meeting, p. 86-90.

Cheung, M., E. F. Gilbert, and R. E. Peterson. 1981. Cardiovascular teratogenicity of 2,3,7,8-tetrachlorodibenzo-*p*-dioxin in the chick embryo. Toxicol. Appl. Pharmacol. 61:197-204.

Colborn, T. 1991. Epidemiology of Great Lakes bald eagles. J. Toxicol. Environ. Health 33:395-454.

Custer, T. W., and G. H. Heinz. 1980. Reproductive success and nest attentiveness of mallard ducks fed Aroclor 1254. Environ. Pollut. 21:313-318.

Dahlgren, R. B., R. J. Bury, R. L. Linder, and R. F. Reidinger, Jr. 1972a. Residue levels and histopathology in pheasants given polychlorinated biphenyls. J. Wildl. Manage. 36:524-533.

Dahlgren, R. B., R. L. Linder, and W. L. Tucker. 1972b. Effects of stress on pheasants previously given polychlorinated biphenyls. J. Wildl. Manage. 36:974-978.

Eadon, G., L. Kaminsky, J. Silkworth, K. Aldous, D. Hilker, P. O'Keefe, R. Smith, J. Gierthy, J. Hawley, N. Kim, and A. Decaprio. 1986. Calculation of 2,3,7,8-TCDD equivalent concentrations of complex environmental contaminant mixtures. Environ. Health Perspect. 70:221-227.

Eisler, R. 1986a. Dioxin hazards to fish, wildlife, and invertebrates: a synoptic review. U.S. Fish Wildl. Serv. Biol. Rep. 85 (1.8). 37 pp.

Eisler, R. 1986b. Polychlorinated biphenyl hazards to fish, wildlife, and invertebrates: a synoptic review. U.S. Fish Wildl. Serv. Biol. Rep. 85 (1.7). 72 pp.

Ellenton, J. A., L. F. Brownlee, and B. R. Hollebone. 1985. Aryl hydrocarbon hydroxylase levels in herring gull embryos from different locations on the Great Lakes. Environ. Toxicol. Chem. 4:615-622.

Elliott, J. E., R. W. Butler, R. J. Norstrom, and P. E. Whitehead. 1989. Environmental contaminants and reproductive success of great blue herons *(Ardea herodias)* in British Columbia, 1986-87. Environ. Pollut. 59:91-114.

Elliott, J. E., S. W. Kennedy, D. Jeffrey, and L. Shutt. 1991. Polychlorinated biphenyl (PCB) effects on hepatic mixed function oxidases and porphyria in birds. II. American kestrel. Comp. Biochem. Physiol. C Comp. Pharmacol. Toxicol. 99:141-145.

Elliott, J. E., S. W. Kennedy, D. B. Peakall, and H. Won. 1990. Polychlorinated biphenyl (PCB) effects on hepatic mixed function oxidases and porphyria in birds. I. Japanese quail. Comp. Biochem. Physiol. C Comp. Pharmacol. Toxicol. 96:205-210.

Flick, D. F., R. G. O'Dell, and V. A. Childs. 1965. Studies of the chick edema disease. 3. Similarity of symptoms produced by feeding chlorinated biphenyl. Poult. Sci. 44:1460-1465.

Fox, G. A. 1991. Practical causal inference for ecoepidemiologists. J. Toxicol. Environ. Health 33:359-374.

Fox, G. A., B. Collins, E. Hayaskawa, D. V. Weseloh, J. P. Ludwig, T. J. Kubiak, and T. C. Erdman. 1991a. Reproductive outcomes in colonial fish-eating birds: a biomarker for developmental toxicants in Great Lakes food chains. II. Spatial variation in the occurrence and prevalence of bill defects in young double-crested cormorants in the Great Lakes. J. Great Lakes Res. 17:158-167.

Fox, G. A., M. Gilbertson, A. P. Gilman, and T. J. Kubiak. 1991b. A rationale for the use of colonial fish-eating birds to monitor the presence of developmental toxicants in Great Lakes fish. J. Great Lakes Res. 17:151-152.

Fox, G. A., A. P. Gilman, D. B. Peakall, and F. W. Anderka. 1978. Behavioral abnormalities in nesting Lake Ontario herring gulls. J. Wildl. Manage. 42:477-483.

Fox, G. A., S. W. Kennedy, R. J. Norstrom, and D. C. Wigfield. 1988. Porphyria in herring gulls: a biochemical response to chemical contamination of Great Lakes food chains. Environ. Toxicol. Chem. 7:831-839.

Fox, G. A., D. V. Weseloh, T. J. Kubiak, and T. C. Erdman. 1991c. Reproductive outcomes in colonial fish-eating birds: a biomarker for developmental toxicants in Great Lakes food chains. I. Historical and ecotoxicological perspectives. J. Great Lakes Res. 17:153-157.

Gilbertson, M. 1974. Pollutants in breeding herring gulls in the lower Great Lakes. Can. Field Nat. 88:273-280.

Gilbertson, M. 1988. Epidemics in birds and mammals caused by chemicals in the Great Lakes. p. 133-152. *In* M. S. Evans (Ed.). Toxic contaminants and ecosystem health: a Great Lakes focus. John Wiley & Sons, New York.

Gilbertson, M., and G. A. Fox. 1977. Pollutant-associated embryonic mortality of Great Lakes herring gulls. Environ. Pollut. 12:211-216.

Gilbertson, M., and R. Hale. 1974a. Early embryonic mortality in a herring gull colony in Lake Ontario. Can. Field Nat. 88:354-356.

Gilbertson, M., and R. Hale. 1974b. Characteristics of the breeding failure of a colony of herring gulls on Lake Ontario. Can. Field Nat. 88:356-358.

Gilbertson, M., T. Kubiak, J. Ludwig, and G. Fox. 1991. Great Lakes embryo mortality, edema, and deformities syndrome (GLEMEDS) in colonial fish-eating birds: similarity to chick-edema disease. J. Toxicol. Environ. Health 33:455-520.

Goldstein, J. A. 1980. Structure-activity relationships for the biochemical effects and relationship to toxicity. In R. D. Kimbrough (Ed.). Topics in environmental health. Vol. 4, p. 151-190. Halogenated biphenyls, terphenyls, naphthalenes, dibenzodioxins, and related products. Elsevier/North Holland Biomedical Press, New York.

Grieg, J. B., G. Jones, W. H. Butler, and J. M. Barnes. 1973. Toxic effects of 2,3,7,8-tetrachlorodibenzo-p-dioxin. Food. Cosmet. Toxicol. 11:585-595.

Harris, H. J., T. C. Erdman, G. T. Ankley, and K. B. Lodge. 1993. Measures of reproductive success and PCB residues in eggs and chicks of Forster's tern on Green Bay, Lake Michigan — 1988. Arch. Environ. Contam. Toxicol. 25:304-314.

Harris, M. P., and D. Osborn. 1981. Effect of a polychlorinated biphenyl on the survival and breeding of puffins. J. Appl. Ecol. 18:471-479.

Hart, L. E., K. M. Cheng, P. E. Whitehead, R. M. Shah, R. J. Lewis, S. R. Ruschkowski, R. W. Blair, D. C. Bennett, S. M. Bandiera, R. J. Norstrom, and G. D. Bellward. 1991. Dioxin contamination and growth and development in great blue heron embryos. J. Toxicol. Environ. Health 32:331-344.

Haseltine, S. D., and R. M. Prouty. 1980. Aroclor 1242 and reproductive success of adult mallards *(Anas platyrhynchos)*. Environ. Res. 23:29-34.

Heath, R. G., J. W. Spann, E. F. Hill, and J. F. Kreitzer. 1972. Comparative dietary toxicities of pesticides to birds. U.S. Fish Wildl. Serv. Spec. Sci. Rep. Wildl. 152, 57 pp.

Heinz, G. H., E. F. Hill, and J. F. Contrera. 1980. Dopamine and norepinephrine depletion in ring doves fed DDE, dieldrin, and Aroclor 1254. Toxicol. Appl. Pharmacol. 53:75-82.

Helander, B. 1983. Sea eagle — experimental studies. Artificial incubation of white-tailed sea eagle eggs 1978-1980 and the rearing and introduction to the wild of an eaglet. Natl. Swedish Environ. Protection Board Rep. snv pm 1386.

Helander, B., M. Olsson, and L. Reutergardh. 1982. Residue levels of organochlorine and mercury compounds in unhatched eggs and the relationships to breeding success in white-tailed sea eagles *Haliaeetus albicilla* in Sweden. Holarct. Ecol. 5:349-366.

Henshel, D. S. 1993. LD50 and teratogenicity studies of the effects of TCDD on chicken embryos. Society of Environmental Toxicology and Chemistry Abstracts 14:280.

Henshel, D. S., J. W. Martin, R. Norstrom, P. Whitehead, J. D. Steeves, and K. M. Cheng. 1995. Morphometric abnormalities in brains of great blue heron hatchlings exposed to PCDD's. Environ. Health Perspect., 103, suppl. 4:61-66.

Higginbotham, G. R., A. Huang, D. Firestone, J. Verrett, J. Ress, and A. D. Campbell. 1968. Chemical and toxicological evaluations of isolated and synthetic chloroderivatives of dibenzo-p-dioxin. Nature (Lond.) 220:702-703.

Hoffman, D. J., M. J. Melancon, J. D. Eisemann, and P. N. Klein. 1995. Comparative toxicity of planar PCB congeners by egg injection. Society of Environmental Toxicology and Chemistry Abstracts 16:207.

Hoffman, D. J., B. A. Rattner, C. M. Bunck, A. Krynitsky, H. M. Ohlendorf, and R. W. Lowe. 1986. Association between PCBs and lower embryonic weight in black-crowned night herons in San Francisco Bay. J. Toxicol. Environ. Health 19:383-391.

Hoffman, D. J., B. A. Rattner, L. Sileo, D. Docherty, and T. J. Kubiak. 1987. Embryotoxicity, teratogenicity, and aryl hydrocarbon hydroxylase activity in Forster's terns on Green Bay, Lake Michigan. Environ. Res. 42:176-184.

Hoffman, D. J., C. P. Rice, M. J. Melancon, P. N. Klein, J. D. Eisemann, and R. K. Hines. 1993a. Developmental toxicity of planar PCB congeners in nestling American kestrels *(Falco sparverius)*. Society of Environmental Toxicology and Chemistry Abstracts 14:178.

Hoffman, D. J., G. J. Smith, and B. A. Rattner. 1993b. Biomarkers of contaminant exposure in common terns and black-crowned night herons in the Great Lakes. Environ. Toxicol. Chem. 12:1095-1103.

Hudson, R., R. Tucker, and M. Haegele. 1984. Handbook of toxicity of pesticides to wildlife. 2nd ed. U.S. Fish Wildl. Serv. Resour. Publ. 153, Washington, D.C.

Jarman, W. M., S. A. Burns, R. R. Chang, R. D. Stephens, R. J. Norstrom, M. Simon, and J. Linthicum. 1993. Determination of PCDDS, PCDFS, and PCBS in California peregrine falcons *(Falco peregrinus)* and their eggs. Environ. Toxicol. Chem. 12:105-114.

Jefferies, D. J., and J. L. F. Parslow. 1972. Effect of one polychlorinated biphenyl on size and activity of the gull thyroid. Bull. Environ. Contam. Toxicol. 8:306-310.

Jefferies, D. J., and J. L. F. Parslow. 1976. Thyroid changes in PCB-dosed guillemots and their indication of one of the mechanisms of action of these materials. Environ. Pollut. 10:293-311.

Jones, P. D., G. T. Ankley, D. A. Best, R. Crawford, N. DeGalan, J. P. Giesy, T. J. Kubiak, J. P. Ludwig, J. L. Newsted, D. E. Tillitt, and D. A. Verbrugge. 1993a. Biomagnification of bioassay derived 2,3,7,8-tetrachlorodibenzo-p-dioxin equivalents. Chemosphere 26:1203-1212.

Jones, P. D., J. P. Giesy, J. L. Newsted, D. A. Verbrugge, D. L. Beaver, G. T. Ankley, D. E. Tillitt, K. B. Lodge, and G. J. Niemi. 1993b. 2,3,7,8-Tetrachlorodibenzo-p-dioxin equivalents in tissues of birds at Green Bay, Wisconsin, U.S.A. Arch. Environ. Contam. Toxicol. 24:345-354.

Kannan, N., S. Tanabe, and R. Tatsukawa. 1988. Toxic potential of non-ortho and mono-ortho coplanar PCBs in commercial PCB preparations: "2,3,7,8-TCDD toxicity equivalence factors approach." Bull. Environ. Contam. Toxicol. 41:267-276.

Keith, J. A. 1966. Reproduction in a population of herring gulls *(Larus argentatus)* contaminated by DDT. J. Appl. Ecol. 3:57-70.

Kennedy, S. W., and G. A. Fox. 1990. Highly carboxylated porphyrins as a biomarker of polyhalogenated aromatic hydrocarbon exposure in wildlife: confirmation of their presence in Great Lakes herring gull chicks in the early 1970s and important methodological details. Chemosphere 21:407-415.

Kimbrough, R. D. (Ed.). 1980. Halogenated biphenyls, terphenyls, naphthalenes, dibenzodioxins, and related products. Elsevier/North-Holland, New York.

Koeman, J. H., H. C. W. Van Velzen-Blad, R. de Vries, and J. G. Vos. 1973. Effects of PCB and DDE in cormorants and evaluation of PCB residues from an experimental study. J. Reprod. Fertil. 19 (Suppl.):353-364.

Kozie, K. D., and R. K. Anderson. 1991. Productivity, diet, and environmental contaminants in bald eagles nesting near the Wisconsin shoreline of Lake Superior. Arch. Environ. Contam. Toxicol. 20:41-48.

Kubiak, T. J. 1991. A review of bird egg toxicity studies with planar halogenated hydrocarbons. *In* The Cause Effects Linkages II Symp., Traverse City, Mich. September 27-28, 1991. Michigan Audubon Society, Lansing, Mich.

Kubiak, T. J., and D. A. Best. 1991. Wildlife risks associated with passage of contaminated anadromous fish at federal energy regulatory commission licensed dams in Michigan. Contaminants Program, Ecological Services, East Lansing Field Office, 1405 S. Harrison Rd., East Lansing, Mich.

Kubiak, T. J., H. J. Harris, L. M. Smith, T. R. Schwartz, D. I. Stalling, J. A. Trick, L. Sileo, D. E. Docherty, and T. C. Erdman. 1989. Microcontaminants and reproductive impairment of the Forster's tern on Green Bay, Lake Michigan — 1983. Arch. Environ. Contam. Toxicol. 18:706-727.

Kutz, F. W., D. G. Barnes, E. W. Bretthauer, D. P. Bottimore, and H. Greim. 1990. The international toxicity equivalency factor (I-TEF) method for estimating risks associated with exposures to complex mixtures of dioxins and related compounds. Toxicol. Environ. Chem. 26:99-109.

Ludwig, J. P., H. J. Auman, H. Kurita, M. E. Ludwig, L. M. Campbell, J. P. Giesy, D. E. Tillitt, P. D. Jones, N. Yamashita, S. Tanabe, and R. Tatsukawa. 1993. Caspian tern reproduction in the Saginaw Bay ecosystem following a 100-year flood event. J. Great Lakes Res. 19:96-108.

Ludwig, J., and C. Tomoff. 1966. Reproductive success and insecticide residues in Lake Michigan herring gulls. Jack-Pine Warbler 44:77-84.

Martin, S., J. Duncan, D. Thiel, R. Peterson, and M. Lemke. 1989. Evaluation of the effects of dioxin-contaminated sludges on Eastern bluebirds and tree swallows. Report prepared by Nekoosa Papers, Inc., Port Edwards, Wis.

Mason, G., K. Farrell, B. Keys, J. Piskorska-Pliszczynska, L. Safe, and S. Safe. 1986. Polychlorinated dibenzo-p-dioxins: quantitative in vitro and in vivo structure-activity relationships. Toxicology 41:21-31.

McKinney, J. D., K. Chae, B. N. Gupta, J. A. Moore, and J. A. Goldstein. 1976. Toxicological assessment of hexachlorobiphenyl isomers and 2,3,7,8-tetrachlorodibenzofuran in chicks. Toxicol. Appl. Pharmacol. 36:65-80.

McLane, M. A. R., and D. L. Hughes. 1980. Reproductive success of screech owls fed Aroclor 1248. Arch. Environ. Contam. Toxicol. 9:661-665.

Moccia, R. D., G. A. Fox, and A. Britton. 1986. A quantitative assessment of thyroid histopathology of herring gulls *(Larus argentatus)* from the Great Lakes and a hypothesis on the causal role of environmental contaminants. J. Wildl. Dis. 22:60-70.

Mora, M. A., H. J. Auman, J. P. Ludwig, J. P. Giesy, D. A. Verbrugge, and M. E. Ludwig. 1993. Polychlorinated biphenyls and chlorinated insecticides in plasma of caspian terns: relationships with age, productivity, and colony site tenacity in the Great Lakes. Arch. Environ. Contam. Toxicol. 24:320-331.

Niwa, A., K. Kumaki, and D. W. Nebert. 1975. Induction of aryl hydrocarbon hydroxylase activity in various cell cultures by 2,3,7,8-tetrachlorodibenzo-p-dioxin. Mol. Pharmacol. 11:399-408.

Norstrom, R. J., D. J. Hallett, M. Simon, and M. J. Mulvihill. 1982. Analysis of Great Lakes herring gull eggs for tetrachlorodibenzo-p-dioxins. p. 173-181. *In* O. Hutzinger, R. W. Frei, and E. Merian (Eds.). Chlorinated dioxins and related compounds: impact on the environment. Pergamon Press, New York.

Nosek, J. A., S. R. Craven, J. R. Sullivan, S. S. Hurley, and R. E. Peterson. 1992a. Toxicity and reproductive effects of 2,3,7,8-tetrachlorodibenzo-p-dioxin in ring-necked pheasant hens. J. Toxicol. Environ. Health 35:187-198.

Nosek, J. A., S. R. Craven, J. R. Sullivan, J. R. Olson, and R. E. Peterson. 1992b. Metabolism and disposition of 2,3,7,8-tetrachlorodibenzo-p-dioxin in ring-necked pheasant hens, chicks, and eggs. J. Toxicol. Environ. Health 35:153-164.

Nosek, J. A., J. R. Sullivan, S. R. Craven, A. Gendron-Fitzpatrick, and R. E. Peterson. 1993. Embryotoxicity of 2,3,7,8-tetrachlorodibenzo-*p*-dioxin in the ring-necked pheasant. Environ. Toxicol. Chem. 12:1215-1222.

Peakall, D. B., G. A. Fox, A. D. Gilman, D. J. Hallet, and R. J. Norstrom. 1980. Reproductive success of herring gulls as an indicator of Great Lakes water quality. p. 337-344. *In* B. K. Afghan and D. MacKay (Eds.). Hydrocarbons and halogenated hydrocarbons in the aquatic environment. Plenum Press, New York.

Peakall, D. B., J. L. Lincer, and S. E. Bloom. 1972. Embryonic mortality and chromosomal alterations caused by Aroclor 1254 in ring doves. Environ. Health Perspect. 1:103-104.

Peakall, D. B., and M. L. Peakall. 1973. Effects of a polychlorinated biphenyl on the reproduction of artificially and naturally incubated dove eggs. J. Appl. Ecol. 10:863-868.

Platonow, N. S., L. H. Karstad, and P. W. Saschenbrecker. 1973. Tissue distribution of polychlorinated biphenyls (Aroclor 1254) in cockerels; relation to the duration of exposure and observations on pathology. Can. J. Comp. Med. 37:90-95.

Platonow, N. S., and B. S. Reinhart. 1973. The effects of polychlorinated biphenyls (Aroclor 1254) on chicken egg production, fertility, and hatchability. Can J. Comp. Med. 37:341-346.

Poland, A., and E. Glover. 1973. Chlorinated dibenzo-*p*-dioxins: potent inducers of delta-aminolevulinic acid synthetase and aryl hydrocarbon hydroxylase. Mol. Pharmacol. 9:736-747.

Prestt, I., D. J. Jefferies, and N. W. Moore. 1970. Polychlorinated biphenyls in wild birds in Britain and their avian toxicity. Environ. Pollut. 1:3-26.

Price, I., and D. V. Weseloh. 1986. Increased numbers and productivity of double-crested cormorants, *Phalacrocorax auritus*, on Lake Ontario. Canadian Field Nat. 100:474-482.

Rattner, B. A., M. J. Melancon, T. W. Custer, R. L. Hothem, K. A. King, L. J. LeCaptain, and J. W. Spann. 1993. Biomonitoring environmental contamination with pipping black-crowned night-heron embryos: induction of cytochrome P450. Environ. Toxicol. Chem. 12:1719-1732.

Rehfeld, B. M., R. L. Bradley, Jr., and M. L. Sunde. 1972. Toxicity studies on polychlorinated biphenyls in the chick: biochemical effects and accumulations. Poult. Sci. 51:488-493.

Rifkind, A. B., A. Firpo, Jr., and D. R. Alonso. 1984. Coordinate induction of cytochrome P-448 mediated mixed-function oxidases and histopathologic changes produced acutely in chick embryo liver by polychlorinated biphenyl congeners. Toxicol. Appl. Pharmacol. 72:343-354.

Rifkind, A. B., S. Sassa, J. Reyes, and H. Muschick. 1985. Polychlorinated aromatic hydrocarbon lethality, mixed-function oxidase induction, and uroporphyrinogen decarboxylase inhibition in the chick embryo: dissociation of dose-response relationships. Toxicol. Appl. Pharmacol. 78:268-279.

Roberts, J. R., D. W. Rodgers, J. R. Bailey, and M. A. Rorke. 1978. Polychlorinated biphenyls: biological criteria for an assessment of their effects on environmental quality. Natl. Res. Counc. Can. Assoc. Comm. Sci. Criter. Environ. Qual. Publ. 16077. 172 pp.

Safe, S. 1984. Polychlorinated biphenyls (PCBs) and polybrominated biphenyls (PBBs): biochemistry, toxicology, and mechanism of action. Crit. Rev. Toxicol. 13:319-393.

Safe, S. 1987. Determination of 2,3,7,8-TCDD toxic equivalent factors (TEF): support for the use of the in vitro AHH induction assay. Chemosphere 16:791-802.

Safe, S. 1990. Polychlorinated biphenyls (PCBs), dibenzo-*p*-dioxins (PCDDs), dibenzofurans (PCDFs), and related compounds: environmental and mechanistic considerations which support the development of toxic equivalency factors (TEFs). Crit. Rev. Toxicol. 21:51-58.

Safe, S., S. Bandiera, T. Sawyer, B. Zmudzka, G. Mason, M. Romkes, M. A. Denomme, J. Sparling, A. B. Okey, and T. Fujita. 1985. Effects of structure on binding to the 2,3,7,8-TCDD receptor protein and AHH induction-halogenated biphenyls. Environ. Health Perspect. 61:21-33.

Sawyer, T., D. Jones, K. Rossanoff, G. Mason, J. Piskoska-Pliszczynska, and S. Safe. 1986. The biologic and toxic effects of 2,3,7,8-tetrachlorodibenzo-p-dioxin in chickens. Toxicology 39:197-206.

Sawyer, T., and S. Safe. 1982. PCB isomers and congeners: induction of arylhydrocarbon hydroxylase and ethoxyresorufin O-deethylase enzyme activities in rat hepatoma cells. Toxicol. Lett. (Amst.) 13:57-94.

Schwartz, T. R., and D. L. Stalling. 1991. Chemometric comparison of polychlorinated biphenyl residues and toxicologically active polychlorinated biphenyl congeners in the eggs of Forster's terns *(Sterna fosteri)*. Arch. Environ. Contam. Toxicol. 20:183-199.

Schwartz, T. R., D. E. Tillitt, K. P. Feltz, and P. H. Peterman. 1993. Determination of mono- and non-o,o'-chlorine substituted polychlorinated biphenyls in aroclors and environmental samples. Chemosphere 26:1443-1460.

Schwetz, B. A., J. M. Norris, G. L. Sparschu, V. K. Rowe, P. J. Gehring, J. L. Emerson, and C. G. Gerbig. 1973. Toxicology of chlorinated dibenzo-p-dioxins. Environ. Health Perspect. 5:87-99.

Scott, M. L. 1977. Effects of PCBs, DDT, and mercury compounds in chickens and Japanese quail. Fed. Proc. 36:1888-1893.

Sileo, L., L. Karstad, R. Frank, M. V. H. Holdrinet, E. Addison, and H. E. Braun. 1977. Organochlorine poisoning of ring-billed gulls in southern Ontario. J. Wildl. Dis. 13:313-322.

Smith, L. M., T. R. Schwartz, K. Feltz, and T. J. Kubiak. 1990. Determination and occurrence of AHH-active polychlorinated biphenyls, 2,3,7,8-tetrachloro-p-dioxin and 2,3,7,8-tetrachlorodibenzofuran in Lake Michigan sediment and biota. The question of their relative toxicological significance. Chemosphere 21:1063-1085.

Sotherland, P. R., and H. Rahn. 1987. On the composition of bird eggs. Condor 89:48-65.

Spear, P. A., D. H. Bourbonnais, R. J. Norstrom, and T. W. Moon. 1990. Yolk retinoids (vitamin A) in eggs of the herring gull and correlations with polychlorinated dibenzo-p-dioxins and dibenzofurans. Environ. Toxicol. Chem. 9:1053-1061.

Spear, P. A., T. W. Moon, and D. B. Peakall. 1986. Liver retinoid concentrations in natural populations of herring gulls *(Larus argentatus)* contaminated by 2,3,7,8-tetrachlorodibenzo-p-dioxin and in ring doves *(Streptopelia risoria)* injected with a dioxin analogue. Can. J. Zool. 64:204-208.

Srebocan, V., J. Pompe-Gotal, V. Brmalj, and M. Plazonic. 1977. Effect of polychlorinated biphenyls (Aroclor 1254) on liver gluconeogenic enzyme activities in embryonic and growing chickens. Poult. Sci. 56:732-735.

Stickel, W. H., L. F. Stickel, R. A. Dyrland, and D. L. Hughes. 1984. Aroclor 1254 residues in birds: lethal levels and loss rates. Arch. Environ. Contam. Toxicol. 13:7-13.

Struger, J., and D. V. Weseloh. 1985. Great Lakes Caspian terns: egg contaminants and biological implications. Colon. Waterbirds 8:142-149.

Sullivan, J. R., T. J. Kubiak, T. E. Amundson, R. E. Martini, L. J. Olson, and G. A. Hill. 1987. A wildlife exposure assessment for landspread sludges which contain dioxins and furans. p. 406-415. *In* Proc. 10th Ann. Int. Waste Conf.: Municipal and Industrial Waste, Madison, Wisc.

Tanabe, S. 1989. A need for reevaluation of PCB toxicity. Mar. Pollut. Bull. 20:247-248.

Tanabe, S., N. Kannan, A. Subramanian, S. Watanabe, and R. Tatsukawa. 1987. Highly toxic coplanar PCBs: occurrence, source, persistency, and toxic implications to wildlife and humans. Environ. Pollut. 47:147-163.

Tarhanen, J., J. Koistinen, J. Paasivirta, P. J. Vourinen, J. Koivusaari, I. Nuuga, N. Kannan, and R. Tatsukawa. 1989. Toxic significance of planar aromatic compounds in the Baltic ecosystem — new studies on extremely toxic coplanar PCBs. Chemosphere 18:1067-1077.

Thiel, D. A. 1990. Relating bird egg dioxin concentrations to sludge dioxin exposure. Prepared for the Wisconsin Dioxin Work Group.

Tillitt, D. E., G. T. Ankley, J. P. Giesy, J. P. Ludwig, H. Kurita-Matsuba, D. V. Weseloh, P. S. Ross, C. A. Bishop, L. Sileo, K. L. Stromborg, J. Larson, and T. J. Kubiak. 1992. Polychlorinated biphenyl residues and egg mortality in double crested cormorants from the Great Lakes. Environ. Toxicol. Chem. 11:1281-1288.

Tillitt, D. E., G. T. Ankley, D. A. Verbrugge, J. P. Giesy, J. P. Ludwig, and T. J. Kubiak. 1991. H4IIE rat hepatoma cell bioassay-derived 2,3,7,8-tetrachlorodibenzo-p-dioxin equivalents in colonial fish-eating waterbird eggs from the Great Lakes. Arch. Environ. Contam. Toxicol. 21:91-101.

Tillitt, D. E., T. J. Kubiak, G. T. Ankley, and J. P. Giesy. 1993. Dioxin-like toxic potency in Forster's tern eggs from Green Bay, Lake Michigan, North America. Chemosphere 26:2079-2084.

Trotter, W., S. Young, J. Casterline, Jr., J. Bradlaw, and L. Kamps. 1982. Induction of aryl hydrocarbon hydroxylase activity in cell cultures by aroclors, residues from yusho oil samples, and polychlorinated biphenyl residues from fish samples. J. Assoc. Off. Anal. Chem. 65:838-841.

Tumasonis, C. F., B. Bush, and F. D. Baker. 1973. PCB levels in egg yolks associated with embryonic mortality and deformity of hatched chicks. Arch. Environ. Contam. Toxicol. 1:312-324.

van den Berg, M., B. H. L. J. Craane, T. Sinnige, I. J. Lutke-Schipholt, B. Spenkelink, and A. Brouwer. 1992. The use of biochemical parameters in comparative toxicological studies with the cormorant *(Phalacrocorax carbo)* in the Netherlands. Chemosphere 25:1265-1270.

van Zorge, J. A. 1990. Toxicity equivalency factors for PCBs. (Letter SR/1290255). (In Dutch.) Directorate-General for Environmental Protection, Directorate for Chemicals and Risk Management, Leidschendam, the Netherlands.

Verrett, M. J. 1970. Hearings before the Subcommittee on Energy, Natural Resources, and the Environment of the Committee on Commerce, U.S. Senate. p. 190-203. (Serial 91-60). Government Printing Office, Washington, D.C.

Verrett, M. J. 1976. Investigation of the toxic and teratogenic effects of halogenated dienzo-p-dioxins and dibenzofurans in the developing chicken embryo. *In* Memorandum report. U.S. Food and Drug Administration, Washington, D.C.

Vos, J. G., A. De Liefde, F. L. van Velsen, and the late M. J. van Logten. 1982. Acute toxicity of PCB-isomers in the chick embryo assay. (Rep. 617714001). National Institute of Public Health, Bilthoven, the Netherlands, 15 pp.

Vos, J. G., and J. H. Koeman. 1970. Comparative toxicologic study with polychlorinated biphenyls in chickens with special reference to porphyria, edema formation, liver necrosis, and tissue residues. Toxicol. Appl. Pharmacol. 17:656-668.

Vos, J. G., J. J. T. W. A. Strik, C. W. M. Van Holsteyn, and J. H. Pennings. 1971. Polychlorinated biphenyls as inducers of hepatic porphyria in Japanese quail, with special reference to d-aminolevulinic acid synthetase activity, fluorescence, and residues in the liver. Toxicol. Appl. Pharmacol. 20:232-240.

Walker, C. H., I. Newton, S. D. Hallam, and M. J. J. Ronis. 1987. Activities and toxicological significance of hepatic microsomal enzymes of the kestrel *(Falco tinnunculus)* and sparrowhawk *(Accipiter nisus)*. Comp. Biochem. Physiol. C Comp. Pharmacol. Toxicol. 86:379-382.

Weber, H., M. W. Harris, J. K. Haseman, and L. S. Birnbaum. 1985. Teratogenic potency of TCDD, TCDF, and TCDD-TCDF combinations in C57BL/6N mice. Toxicol. Lett. (Amst.) 26:159-167.

White, D. H., and D. J. Hoffman. 1995. Effects of polychlorinated dibenzo-p-dioxins and dibenzofurans on nesting wood ducks at Bayou Meto, Arkansas. Environ. Health Perspect. 103, suppl. 4:37-39.

White, D. H., and J. T. Seginak. 1994. Dioxins and furans linked to reproductive impairment in wood ducks at Bayou Meto, Arkansas. J. Wildl. Manage. 58:100-106.

Wiemeyer, S. N., T. G. Lamont, C. M. Bunck, C. R. Sindelar, F. J. Gramlich, J. D. Fraser, and M. A. Byrd. 1984. Organochlorine pesticide, polychlorobiphenyl, and mercury residues in bald eagle eggs — 1969-79 — and their relationships to shell thinning and reproduction. Arch. Environ. Contam. Toxicol. 13:529-549.

Yamashita, N., S. Tanabe, J. P. Ludwig, H. Kurita, M. E. Ludwig, and R. Tatsukawa. 1993. Embryonic abnormalities and organochlorine contamination in double-crested cormorants *(Phalacrocorax auritus)* and Caspian terns *(Hydroprogne caspia)* from the upper Great Lakes. Environ. Pollut. 79:163-173.

Yao, C., B. Panigrahy, and S. Safe. 1990. Utilization of cultured chick embryo hepatocytes as in vitro bioassays for polychlorinated biphenyls (PCBs): quantitative structure-induction relationships. Chemosphere 21:1007-1016.

CHAPTER 8

Dioxins: An Environmental Risk for Fish?

Dick T. H. M. Sijm and Antoon Opperhuizen

INTRODUCTION

Polychlorinated dibenzo-p-dioxins (PCDDs) and polychlorinated dibenzofurans (PCDFs) are usually abbreviated as dioxins and furans. These hydrophobic chemicals are found widespread in the environment (Rappe et al., 1987). 2,3,7,8-Tetrachlorodibenzo-p-dioxin (TCDD) is the most toxic congener. Dioxins and furans elicit several responses in fish, such as mortality, decreased survival, growth abnormalities, growth inhibition, immune response effects, blue-sac disease, loss of scale, and enzyme induction (e.g., Kleeman et al., 1988; Mehrle et al., 1988; Spitsbergen et al., 1988a, 1988b, 1991; Cook et al., 1991; Walker and Peterson, 1991; Walker et al., 1991, 1992). Sublethal exposure of rainbow trout fry to TCDD at doses below those causing clinical toxicity did not enhance mortality following challenge with either low or high concentrations of waterborne infectious hematopoietic necrosis virus (Spitsbergen et al., 1988c).

The toxicological properties of dioxins and furans make them likely to be banned. However, a ban for the use and distribution of dioxins and furans is very difficult; various production processes should be controlled to minimize environmental release. Dioxins and furans were formerly produced in the production of herbicides and polychlorinated biphenyls and are formed during municipal waste combustion processes. Furthermore, they are found in chlorine-bleached kraft pulp mill effluents.

Fish are exposed to dioxins and furans by different routes. Dioxins and furans may be taken up by fish from water, food, or sediment (Opperhuizen, 1991). Furthermore, eggs may get their dioxin and furan burdens directly from the mother, probably via vitellogenin transfer (Sijm et al., 1992), and may get additional exposure via the water or sediment-water interface. Studies comparing individual and combined sources revealed that the sediment and food are the most important sources of dioxins and furans for fish (Batterman et al., 1989; Kuehl et al., 1989).

Because of their low solubility in water and their high affinity to sediment, dioxins and furans are believed to be predominantly sorbed (Kuehl et al., 1989). Fish that live near the sediment-water interface may eat some sediment and may thus accumulate dioxins and furans mainly via sediment. In addition, in some cases, PCDD concentrations are found in fish and sediment, whereas the PCDD concentration in water is below detection limits (Cook et al., 1991).

In Lake Ontario sediment, the average TCDD concentration was 68 ng/kg (U.S. EPA, 1990). However, in some sludges near municipal water treatment plants, dioxin and furan levels are as high as 15 µg/kg (Rappe et al., 1989). Octachlorodibenzo-p-dioxin (OCDD), however, usually constitutes 85% to 100% of the congeners. Since OCDD is taken up at very low rates (Opperhuizen and Sijm, 1990), the uptake rate of dioxins and furans from sediment is difficult to estimate and varies widely for different sediments.

Concentrations of dioxins and furans in water have been found at polluted sites at between 25 and 100 pg/l and at cleaner sites at between 1 and 5 pg/l (Rappe et al., 1989; Jobb et al., 1990). OCDD was found to contribute up to 50% or more of the total amount of dioxins and furans in water. TCDD was not detected in any sample (Rappe et al., 1989; Jobb et al., 1990). Since OCDD is not taken up from water by fish (Opperhuizen and Sijm, 1990) and since other congeners are rapidly metabolized (Sijm et al., 1989; Sijm, 1992), only a few dioxins and furans will bioaccumulate in fish. In general, only 2,3,7,8-substituted congeners tend to bioaccumulate in fish as well as in most other animals (Rappe et al., 1987, 1991; Opperhuizen and Sijm, 1990; Zitko, 1992).

To evaluate and estimate the potential risk of dioxins and furans for fish, one has to know the levels that cause toxic responses and the levels of dioxins and furans in fish in the environment. If wild fish have dioxin and furan levels close to the levels that cause toxicological effects, dioxins and furans pose an environmental risk for fish. In addition to fish, there is a risk for human populations, since fish play a significant role as a carrier of dioxins, furans, and related compounds (Tanabe, 1988; Travis and Hattemeyer-Frey, 1991).

ENVIRONMENTAL CONCENTRATIONS

To predict the risk of a complex mixture of PCDDs and PCDFs to fish in the environment, different methods can be used.

1. The simplest method is to sum all the congeners.
2. A less simple method takes into account the different potencies of each congener and expresses the PCDD/PCDF mixture as an equivalent concentration of the most toxic congener, TCDD, i.e., the toxic equivalency concentration (TEC). Calculation of a TEC involves determining the concentration of each TCDD-like dioxin and furan congener and applying congener-specific toxic equivalency factors (TEFs) (Eadon et al., 1986; Walker and Peterson, 1991). A TEF is defined as the ratio of

the 50% effective dose (ED_{50}) of TCDD to the ED_{50} of congener for a specific endpoint, such as in vitro cytochrome P-450IA1 induction (Tillitt et al., 1991) or early life-stage mortality (Walker and Peterson, 1991). The TEFs that are often used in mammalian and wildlife studies to normalize dioxin and furan concentrations on TCDD toxic units (e.g., Tillitt et al., 1991) may not be the same as those used in fish (Cook et al., 1991). The TEFs for different dioxin and furan congeners that have been developed for mammals and rodents are different for fish (Walker and Peterson, 1991; Sijm, 1992). In addition, whereas the largest portion of dioxins and furans is stored in the liver of mammals, the largest portion in fish is stored in other lipid-rich tissues, such as muscle (Sijm et al., 1990; Hektoen et al., 1992). Multiplication of the concentration of a congener in a mixture by its respective TEF yields a TCDD equivalency concentration (TEC) for that congener. The toxic potency of the entire mixture can then be expressed as an equivalent concentration of TCDD by addition of the TECs for all the individual congeners in the mixture. This method assumes that individual congeners interact additively to produce toxicity.

3. The third method calculates a TEC from biological data, which involves extracting the PCDDs and PCDFs, including similar compounds, from a sample and measuring the ability of the extract, relative to TCDD, to produce a TCDD-like response (Tillitt et al., 1991). Cytochrome P-450IA1 enzyme induction in the H4IIE rat hepatoma cell line is the most commonly used TCDD induction. This method assumes that the TCDD-like biochemical response produced in vitro will predict the ability of the extract to produce toxicity in vivo.

The study of Zacharewski et al. (1989) compared methods 1 and 3. They determined the chemical concentrations of different dioxin and furan congeners and the TEC by using the rat hepatoma H4IIE cell line in extracts of fish from the Great Lakes. They found that, in most samples, there was no more than a twofold difference in the bioassay-estimated TECs and the total dioxins and furans determined chemically. The congeners of the total concentration were not normalized for their toxic potencies, whereas TCDF was the most predominant congener in most samples, a congener with a TEF for early life-stage mortality of 0.028. Bioassay-derived values were significantly higher (more than twofold) for the Lake Erie and Lake Ontario fish extracts, which may be due to synergistic interactive effects of the congeners in the bioassay induction response or the presence of "bioassay-active" components, such as bromo/chlorohalogenated aromatics, that are not detected by chemical analysis.

The study of Zacharewski et al. (1989) is an illustrative example of the difficulties scientists meet in interpreting concentrations of a complex mixture of PCDDs, PCDFs, and other compounds. The use of TEFs, according to either Walker and Peterson (1991) or Tillett et al. (1991), would result in at least a 50% decrease of the total concentration of PCDDs and PCDFs, expressed as TEC in the study of Zacharewski et al. (1989). Consequently, these TEF-derived TECs would deviate at least fourfold from the bioassay-derived TECs, whereas the sum of the PCDDs and PCDFs would more properly match the bioassay-derived TECs.

As the study by Eadon et al. (1986) points out, the use of a TEC is not clearly advantageous. While the sum of all the 2,3,7,8-substituted congeners in Binghamton soot was 91.3 mg/kg, the TEC based on TEFs was 22 mg/kg, and the experimentally derived TEC from a 50% lethal dose (LD_{50}) experiment was 58 mg/kg.

It is especially the latter studies (Eadon et al., 1986; Zacharewski et al., 1989) that made us decide to present dioxin and furan levels in wild fish as the sums of the PCDD and PCDF congeners (Table 1). Furthermore, presenting levels of individual dioxin and furan congeners would not facilitate comparability. Consequently, comparison of the environmental concentrations with effect concentrations (see the section, "Effect Concentrations" below) assumes additivity of each dioxin congener. This may overestimate but also underestimate potential toxicity (see, e.g., Zacharewski et al., 1989). However, the fact that environmental concentrations predominantly comprise 2,3,7,8-substituted congeners and the findings that these congeners elicit responses within one order of magnitude except for, e.g., TCDF (Walker and Peterson, 1991), support the decision to present dioxin and furan levels as the sums of the dioxins and furans.

Body burdens of PCDDs in wild fish range between nondetectable levels and 544 ng/kg, and those of PCDFs range between nondetectable levels and 387 ng/kg (Table 1). While most PCDD and PCDF body burdens range between 10 and 100 ng/kg, extremely high concentrations of more than 734 ng/kg of TCDD were found in bass at Newark Bay (Rappe et al., 1991). The facts that high body burdens may result from uptake via the food chain and that high body burdens are found in different geographical areas are shown by the high concentration of 380 ng/kg of PCDFs detected in killer whale, an aquatic mammal (Ono et al., 1987).

In general, concentrations of PCDFs are found to be approximately twice of those of the PCDDs in fish at several sites (Figure 1). This may imply that when the PCDD body burden in fish has been measured, the PCDF body burden can be estimated. However, at sites with mixed sources there is no relationship between PCDD and PCDF body burden.

The distribution of the body burdens of dioxins and furans from Table 1 is shown in Figure 2, which shows that 50% of the reported fish have a body burden of 40 ng/kg or lower and that 90% of the reported fish have a body burden of 100 ng/kg or lower.

EFFECT CONCENTRATIONS

Dioxins and furans elicit several responses in fish. The most typical of these responses in juvenile fish include hepatic cytochrome P-450IA1 induction, delayed mortality, decreased feed consumption, decreased body weight, and epithelial and lymphomyeloid lesions. Early life stages typically exhibit yolk sac edema, hemorrhages, and delayed mortality. Most of these effects have been observed with TCDD, although in some studies other dioxin and furan congeners have been used as well.

The dioxin and furan concentrations that elicit toxicological responses can be divided into external concentrations, such as the lethal dose causing the death of 50% of the population (LC_{50}), and internal concentrations, which are called body burdens. Only the amounts of dioxins and furans that are present in the fish can

Table 1 Concentrations of TCDD[a] ([TCDD]), All PCDD Congeners ([$\sum$PCDDs]), all PCDF Congeners ([$\sum$PCDFs]), and the Sum of All PCDD and PCDF Congeners ($\sum$[PCDDs + PCDFs]) in Wild Fish (ng/kg of Wet Weight)

Species	Site	[$\sum$PCDDs]	[$\sum$PCDFs]	$\sum$[PCDDs + PCDFs]	[TCDD]	Ref.
Carp	Michigan	55			55	Harless et al. (1982)
Channel catfish	Michigan	157			157	
Smallmouth bass	Michigan	8			8	
Sucker	Michigan	11			11	
Lake trout	Michigan	0			0	
Yellow perch	Michigan	14			14	
Brown bullhead	Lake Ontario	4			4	O'Keefe et al. (1983)
Chinook salmon	Lake Ontario	35			35	
Coho salmon	Lake Ontario	22			22	
Lake trout	Lake Ontario	79			79	
Rainbow trout	Lake Ontario	25			25	
Smallmouth bass	Lake Ontario	6			6	
White perch	Lake Ontario	21			21	
White sucker	Lake Ontario	5			5	
Goldfish/carp	Cayuga Creek	87			87	
Northern pike	Cayuga Creek	32			32	
Pumpkinseed	Cayuga Creek	31			31	
Rock bass	Cayuga Creek	12			12	
Goldfish/carp	Lake Erie	<2			<2	
Coho salmon	Lake Erie	1			1	
Smallmouth bass	Lake Erie	1			1	
Walleye pike	Lake Erie	3			3	
Carp	Lake Huron	26			26	
Channel catfish	Lake Huron	20			20	
Lake trout	Lake Huron	21			21	
Sucker	Lake Huron	3			3	
Yellow perch	Lake Huron	<9			<9	
Coho salmon	Lake Michigan	<4			<4	
Rainbow trout	Lake Superior	1			1	
Striped bass	Chesapeake Bay	3	8	11	3	O'Keefe et al. (1984)
Striped bass	Little Neck Bay	16	16	32	16	
Striped bass	Newark Bay	42	28	70	42	
Striped bass	Poughkeepsie	120	76	196	120	
Striped bass	Rhode Island	2	14	16	2	
Striped bass	Rhode Island	4	27	31	4	
Striped bass	Tappan Zee Bridge	32	53	85	32	
Brown trout	Lake Ontario	9			9	Fehringer et al. (1985)
Carp	Saginaw Bay	30			30	
Catfish	Saginaw Bay	60			60	
Lake trout	Lake Huron	6			6	
Lake trout	Lake Ontario	46			46	

Table 1 (continued) Concentrations of TCDD[a] ([TCDD]), All PCDD Congeners ([∑PCDDs]), all PCDF Congeners ([∑PCDFs]), and the Sum of All PCDD and PCDF Congeners (∑[PCDDs + PCDFs]) in Wild Fish (ng/kg of Wet Weight)

Species	Site	[∑PCDDs]	[∑PCDFs]	∑[PCDDs + PCDFs]	[TCDD]	Ref.
Rainbow trout	Lake Ontario	21			21	
Salmon	Lake Ontario	35			35	
White perch	Lake Ontario	25			25	
Eel	The Netherlands	10	5	15	4.0	Van den Berg et al. (1987)
Miscellaneous	Lake Michigan	50	100	150		Baumann and Whittle (1988)
Miscellaneous	Lake Ontario	50	100	150		
Miscellaneous	Lake Huron	50	200	250		
Miscellaneous	Lake Erie	50	25	75		
Miscellaneous	Lake Superior	1	25	26		
Lake trout	Lake Superior	7	21	28	1.0	DeVault et al. (1989)
Lake trout	Lake Huron	27	47	74	8.6	
Lake trout	Saugatuck (Michigan)	22	56	78	3.5	
Lake trout	Sturgeon Bay (Michigan)	24	69	93	3.8	
Lake trout	Kenosha (Michigan)	41	102	143	5.8	
Lake trout	Lake Ontario	65	56	121	48.9	
Walleye	Lake Erie	11	18	29	1.8	
Walleye	Lake St. Clair	18	37	55	6.6	
Miscellaneous	Noncontaminated and contaminated sites in U.S.	0 (70)[b] <5 (19) 5-25 (10) >25 (1)			0 <5 5-25 >25	Kuehl et al. (1989)
Chinook salmon	Fraser River	<36	<12		ND	Rogers et al. (1989)
Chinook salmon	Fraser River	101	387	488	68	
Fish extract	Lake St. Clair	18	36	54	6.6	Zacharewski et al. (1989)
Fish extract	Lake Michigan	55	22	76	3.5	
Fish extract	Lake Ontario	57	61	118	41	
Fish extract	Lake Huron	27	47	72	8.6	
Fish extract	Lake Erie	11	18	29	1.8	
Fish extract	Lake Superior	8	22	30	6.6	
Miscellaneous	U.S.		59			Gardner and White (1990)
Pike	Sweden	2	1	3	0.2	Kjeller et al. (1990)
Bream	Germany (River Elbe)	52	14	66	1.8	Luckas and Oehme (1990)
Brown bullhead	Miscellaneous (U.S.)	1	13	14	1	Smith et al. (1990)
Carp	Miscellaneous (U.S.)	17	24	41	17	
Chinook salmon	Lake Michigan	3	12	15	3	
Fathead minnow	Green Bay, Mich.	<1	7	7	<1	

Table 1 (continued) Concentrations of TCDD[a] ([TCDD]), All PCDD Congeners ($[\sum$PCDDs]), all PCDF Congeners ($[\sum$PCDFs]), and the Sum of All PCDD and PCDF Congeners ($\sum$[PCDDs + PCDFs]) in Wild Fish (ng/kg of Wet Weight)

Species	Site	[$\sum$PCDDs]	[$\sum$PCDFs]	$\sum$[PCDDs + PCDFs]	[TCDD]	Ref.
Lake trout	Lake Michigan	5	35	40	5	
Largemouth bass	Waukegan Harbor, Mich.	2	60	62	2	
Spottail shiner	Fox River, Mich.	0.8	60	61	0.8	
Striped bass	Miscellaneous (U.S.)	10	18	28	10	
Striped bass, fillet	Hudson River, N.J.	29	42	71	29	
Striped bass, ovaries	Hudson River, N.J.	6	31	39	6	
Striped bass, eggs	Hudson River, N.J.	20	100	120	20	
Walleye	Miscellaneous (U.S.)	6	47	53	6	
Miscellaneous	Germany	11	17	28	0.5	Frommberger (1991)
Striped bass	Newark Bay, N.J.	544	260	804	409	Rappe et al. (1991)
Carp	Columbia River	0.3			0.3	Parsons et al. (1991)
Largescale sucker	Columbia River	0.3			0.3	
Salmon	Columbia River	0.1			0.3	
Steelhead trout	Columbia River	0.1			0.1	
White sturgeon	Columbia River	0.6			0.6	
Miscellaneous	Norwich (England)	3	3	6	0.1	Startin et al. (1990)
Brown trout	Lake Ontario	13			13	US EPA (1990)
Lake trout	Lake Ontario	36			36	
Smallmouth bass	Lake Ontario	6			6	
White perch	Lake Ontario	65			65	
Yellow perch	Lake Ontario	5			5	
Bream	Miscellaneous	6	51	57	1.8	Zitko (1992)
Cod	Miscellaneous	10	52	62	4.8	
Eel	Miscellaneous	24	92	116	0.6	
Flounder	Miscellaneous	1	6	7	0.3	
Haddock	Miscellaneous	3	23	26	0.1	
Herring	Miscellaneous	8	32	40	1.0	
Perch	Miscellaneous	13	7	20	12	
Plaice	Miscellaneous	3	10	13	0.3	
Salmon	Miscellaneous	12	43	55	3.6	
Trout	Miscellaneous	28	52	80	12	
Walleye	Miscellaneous	11	26	37	4.2	

[a] TCDD, 2,3,7,8-tetrachlorodibenzo-p-dioxin; PCDD, polychlorinated dibenzo-p-dioxin; PCDF, polychlorinated dibenzofuran; ND, not detected.
[b] Numbers in parentheses, percentage.

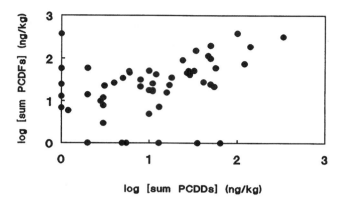

Figure 1 Body burdens of PCDDs ([sum PCDDs]) and of PCDFs ([sum PCDFs]) in wild fish from several geographical sites.

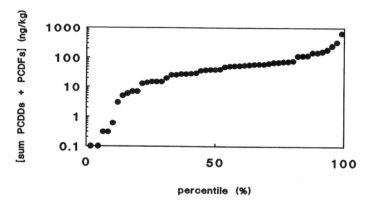

Figure 2 Percentile of body burden distribution of polychlorinated dibenzo-p-dioxins (PCDDs) and of polychlorinated dibenzofurans (PCDFs) in fish from different geographical sites.

elicit toxic effects. Therefore, these environmental body burdens are to be compared to "effect body burdens" in fish, assuming that effect body burdens are the same as critical whole-body doses. The body burdens in wild fish can then be evaluated for the potential risk of dioxins and furans to these fish.

Since environmental body burdens have been reported, while effect body burdens seldom have, the latter have to be estimated. Effect body burdens must be estimated from reported external concentrations, such as the LC_{50}, or from administered doses, such as given intraperitoneally or by food.

To estimate the effect body burden in fish from LC_{50} data, the uptake and elimination kinetics, the bioconcentration factor, and the time of exposition are required. An estimate of the effect body burden is obtained by simply multiplying LC_{50} and the bioconcentration factor (McCarty, 1986). These estimated effect body burdens have been used only for some early life stage effects (Table 2). Intraperitoneal (i.p.) injection and food administration require uptake efficiency and distribution rate constants to estimate the effect body burden. An estimate of the effect body burden is obtained by simply assuming that the entire administrated dose is the effect body burden. In both types of body burdens, resulting from the LC_{50} or from i.p. injection or food administration, the effect body burdens may be overestimated, since the real body burden will be lower because of either non-steady-state conditions for the LC_{50} estimates or less-than-complete absorption of the administered dose from i.p. injection or food.

Since no bioconcentration factors are available for eggs (Opperhuizen and Sijm, 1990), a value of 3 was estimated for TCDD in lake trout and rainbow trout eggs from the two studies by Walker et al. (1991, 1992), who present both aqueous and egg concentrations (egg body burdens). Although it is very likely that this estimated bioconcentration factor is not measured at steady state, it represents the ratio between the concentration in water and that in egg during exposure. For other dioxins also, this bioconcentration factor of 3 was used (Table 2). Differences in the structure of the eggs may influence bioconcentration factors among species. Therefore, estimated body burdens of northern pike may be underestimated (Table 2). Bioconcentration factors for embryos of Japanese medaka were estimated to be 100 from both aqueous and embryo concentrations in the study by Wisk and Cooper (1990b). Bioconcentration factors for older fish are significantly higher and may exceed a value of 10^3 l/kg (Opperhuizen and Sijm, 1990). However, in the present study, no LC_{50} data have been transformed to body burdens for older fish.

Early life stage mortality responses to TCDD are found at body burdens of 0.065 $\mu g/kg_{egg}$ in lake trout and 0.24 or 0.4 $\mu g/kg_{egg}$ in rainbow trout (Table 2). Estimated body burdens of TCDD are 0.04 or 0.9 $\mu g/kg_{egg}$ in Japanese medaka, 0.001 $\mu g/kg_{egg}$ in northern pike, and between 0.006 and 10 $\mu g/kg$ in rainbow trout (Table 2). Other congeners show higher lethal body burdens in eggs, thus making them less potent than TCDD (Table 2).

A number of these body burdens have actually been measured (Spitsbergen et al., 1991; Walker and Peterson, 1991; Walker et al., 1992). LC_{50} data of Japanese medaka (Table 2) when transformed into body burdens fall into the same range, which indicates that the transformation is justified.

The lethal body burdens in the early life stages are similar to the TCDD dose lethal to the most sensitive mammal, the guinea pig. This indicates that fish are very sensitive toward TCDD (Cook et al., in press).

Inhibition of growth at early life stages is found at lower concentrations than those that cause mortality, but such inhibition depends on the study and the species. The lowest observed effect concentration (LOEC) of TCDD for rainbow trout eggs is found at a concentration lower than 0.1 ng/l, which is estimated to give a body burden of <0.0003 $\mu g/kg_{egg}$. Juvenile fish are found to be less affected by TCDD, since growth was inhibited at a concentration of 1.0 ng/l (0.003 $\mu g/kg$), which is higher than that for eggs (Table 2). However, actual body burdens have not been measured in the latter study. Since bioconcentration factors may be different in eggs and juveniles, no comparison can be made between the two life stages, based simply on the external exposure concentrations.

Table 2 Effect Concentrations of PCDDs and PCDFs in Fish

Species and toxicity parameter	Congener	Lowest effect concentration	Body burden[b]	Ref.
Early life-stage/ reproduction				
Japanese medaka				
E.L.S. (LC_{50})	TCDD	13 ng/l	(0.04 $\mu g/kg_{egg}$)	Wisk and Cooper (1990a)
E.L.S. (LC_{50})	TCDF	16 ng/l	(0.05 $\mu g/kg_{egg}$)	
E.L.S. (LC_{50})	1,2,3,7,8-P5CDD	27 ng/l	(0.08 $\mu g/kg_{egg}$)	
E.L.S. (LC_{50})	1,2,3,4,7,8-H6CDD	2,900 μg/l	(9 $\mu g/kg_{egg}$)	
E.L.S. (LC_{50})	1,2,7,8-TCDD	>50 μg/l	(>150 $\mu g/kg_{egg}$)	
E.L.S. (LC_{50})	1,3,6,8-TCDD	>50 μg/l	(>150 $\mu g/kg_{egg}$)	
E.L.S. (LC_{50})	2,3,7-T3CDD	>50 μg/l	(>150 $\mu g/kg_{egg}$)	
E.L.S. (LC_{50})	2,8-DCDD	>50 μg/l	(>150 $\mu g/kg_{egg}$)	
E.L.S. (LC_{50})	2,3-DCDD	>50 μg/l	(>150 $\mu g/kg_{egg}$)	
E.L.S. (LC_{50})	OCDD	>10 μg/l	(>30 $\mu g/kg_{egg}$)	
Japanese medaka				
E.L.S. (LC_{50})	TCDD	9 ng/l	(0.09 $\mu g/kg_{egg}$)	Wisk and Cooper (1990b)
Minor lesions in embryos	TCDD	3.5 ng/l	(0.35 $\mu g/kg$)	
Severe lesions in embryos	TCDD	14 ng/l	(1.4 $\mu g/kg$)	
EC_{50} to prevent hatching	TCDD	14 ng/l	(1.4 $\mu g/kg$)	
Lesions in embryos	TCDD	2.2 ng/l	0.24 $\mu g/kg_{embryo}$	
Lake trout				
LOEC (hatchability)	TCDD	62 ng/l, 48 hr	0.226 $\mu g/kg_{egg}$	Walker et al. (1991)
NOEC (mortality)	TCDD	10 ng/l, 48 hr	0.034 $\mu g/kg_{egg}$	
LOEC (mortality)	TCDD	20 ng/l, 48 hr	0.055 $\mu g/kg_{egg}$	
LD_{50}	TCDD		0.065 $\mu g/kg_{egg}$	
Northern pike				
E.L.S. (LC_{50})	TCDD	1 ng/l	(0.001 $\mu g/kg_{egg}$)	Helder (1980)
E.L.S. (LC_{99})	TCDD	10 ng/l	(0.010 $\mu g/kg_{egg}$)	
Rainbow trout				
Fingerling (LD_{50})	TCDD	Food, 61 days	(1.3 $\mu g/kg$)	Hawkes and Norris (1977)
Fingerling (LD_{88})	TCDD	Food, 71 days	(1.5 $\mu g/kg$)	
E.L.S. (LOEC growth)	TCDD	<0.1 ng/l	(<0.0003 $\mu g/kg_{egg}$)	Helder (1981)
Egg exposure	TCDD	<0.1 ng/l	(<0.0003 $\mu g/kg_{egg}$)	
Yolk sac exposure	TCDD	<0.1 ng/l	(<0.0003 $\mu g/kg_{egg}$)	
Juvenile	TCDD	1.0 ng/l	(0.003 $\mu g/kg$)	
E.L.S. (LC_{50})	TCDD	7.4 ng/l, 10 days	(0.02 $\mu g/kg_{egg}$)	Helder and Seinen (1985)
Fingerling (LD_{20})	TCDD	i.p., 20 days	5 $\mu g/kg$	Spitsbergen et al. (1988b)

Table 2 (continued) Effect Concentrations of PCDDs and PCDFs in Fish

Species and toxicity parameter	Congener	Lowest effect concentration	Body burden[b]	Ref.
Fingerling (LD_{50})	TCDD	i.p., 80 days	10 μg/kg	
E.L.S. (LD_{50})	TCDD	Injection	0.4 μg/kg$_{egg}$	Walker et al. (1992)
E.L.S. (LD_{50})	TCDD	80-500 ng/l	0.4 μg/kg$_{egg}$	
E.L.S. (LD_{50})	TCDD	Injection	0.24 μg/kg$_{egg}$	Walker and Peterson (1991)
E.L.S. (LD_{50})	1,2,3,7,8-P5CDD	Injection	0.57 μg/kg$_{egg}$	
E.L.S. (LD_{50})	1,2,3,4,7,8-H6CDD	Injection	1.43 μg/kg$_{egg}$	
E.L.S. (LD_{50})	TCDF	Injection	8.09 μg/kg$_{egg}$	
E.L.S. (LD_{50})	2,3,4,7,8-P5CDF	Injection	0.70 μg/kg$_{egg}$	
E.L.S. (LD_{50})	1,2,3,7,8-P5CDF	Injection	7.34 μg/kg$_{egg}$	
E.L.S. (LD_{50})	1,2,3,4,7,8-H6CDF	Injection	0.99 μg/kg$_{egg}$	
E.L.S. (25-day LC_{50})	TCDD	1.6-2.7 ng/l	(0.006 μg/kg)	Bol et al. (1989)
Zebrafish (♀)				
LOEC (reproduction)	TCDD	Food	8.3 μg/kg	Wannemacher et al. (1992)
LOEC (oogenesis)	TCDD	Food	8.3 μg/kg	
Older fish				
Bluegill (LD_{50})	TCDD	i.p., 80 days	16 μg/kg	Kleeman et al. (1988)
Bullhead (LD_{50})	TCDD	i.p., 80 days	5 μg/kg	Kleeman et al. (1988)
Carp				
Death	TCDD	62 pg/l, 71 days	2.2 μg/kg	Cook et al. (1991)
LD_{50}	TCDD	i.p., 80 days	3 μg/kg	Kleeman et al. (1988)
Channel catfish (LC_{50})	TCDD	0.01 nmol/l, 20 days		Yockim et al. (1978)
Cod (fin necrosis)	TCDD	Food, 4 weeks	30 μg/kg	Hektoen et al. (1992)
Coho salmon				
LOEC (survival)	TCDD		<0.054 μg/kg	Miller et al. (1973)
NOEC (growth, survival)	TCDD	Food	0.54 μg/kg	Miller et al. (1979)
Fathead minnow				
LD_{100}	TCDD	Food	17-2,042 μg/kg	Adams et al. (1986)
NOEC (survival)	TCDD	<1.7 ng/l		
Guppy				
LOEC (survival)	TCDD	<0.1 μg/l		Norris and Miller (1974)
NOEC (fin necrosis)	TCDD		0.08 μg/kg	Miller et al. (1979)

Table 2 (continued)   Effect Concentrations of PCDDs and PCDFs in Fish

Species and toxicity parameter	Congener	Lowest effect concentration	Body burden[b]	Ref.
NOEC (growth, survival)	TCDD		8 μg/kg	
Largemouth bass (LD_{50})	TCDD	i.p., 80 days	11 μg/kg	Kleeman et al. (1988)
Mosquitofish (LC_{50})	TCDD	3 ng/l, 15 days		Yockim et al. (1978)
Rainbow trout				
Fin necrosis	TCDD	i.p., 2 weeks	10 μg/kg	Spitsbergen et al. (1986)
LD_{50}	TCDD	i.p., 80 days	10 μg/kg	Kleeman et al. (1988)
LOEC (survival, growth)	TCDD	<0.038 ng/l, 28 days	1.0 μg/kg	Mehrle et al. (1988)
NOEC (survival)	TCDF	1.79 ng/l, 28 days	7.6 μg/kg	
NOEC (growth)	TCDF	0.41 ng/l, 28 days	2.5 μg/kg	
LD_{85}	TCDD	i.p., 2-4 weeks	25 μg/kg	Spitsbergen et al. (1988b)
LD_{20}	TCDD	i.p., 11 weeks	5 μg/kg	
LOEC (microscopic lesions)	TCDD	i.p.	10 μg/kg	
LOEC (blood constituents)	TCDD	i.p.	10 μg/kg	
Scale loss	1,2,3,7-TCDD	p.o.	5.2 μg/kg	Sijm et al. (1990)
	1,2,3,4,7-P5CDD	p.o.	11 μg/kg	
EROD induction	1,2,3,7-TCDD	p.o.	5.2 μg/kg	Sijm (1992)
	1,2,3,4,7-P5CDD	p.o.	11 μg/kg	
	2,3,4,7,8-P5CDF	p.o.	2.0 μg/kg	
Liver weight increase	TCDD	i.p., 6 weeks	5 μg/kg	Van der Weiden et al. (1990)
Growth	TCDD	i.p., 11 weeks	5 μg/kg	
EROD induction	TCDD	i.p., 3 weeks	0.5 μg/kg	
AHH induction	TCDD	i.p.	0.6 μg/kg	Janz and Metcalfe (1991)
Fin necrosis	TCDD	Food, 4 weeks	30 μg/kg	Hektoen et al. (1992)
EROD induction	TCDF	Food	1 μg/kg	Muir et al. (1992)
LD_{20}	TCDD	i.p., 12 weeks	3.0 μg/kg	Van der Weiden (1992)
Growth	TCDD	i.p., 6 weeks	3.0 μg/kg	
EROD induction	TCDD	i.p., 3 weeks	0.3 μg/kg	
Scup (EROD induction)	TCDF	i.p.	3.1 μg/kg	Hahn et al. (1989)

Table 2 (continued) Effect Concentrations of PCDDs and PCDFs in Fish

Species and toxicity parameter	Congener	Lowest effect concentration	Body burden[b]	Ref.
Yellow perch				
LD_{50}	TCDD	i.p., 80 days	3 µg/kg	Kleeman et al. (1988)
LD_{80}	TCDD	i.p., 80 days	5 µg/kg	Spitsbergen et al. (1988b)

[a] PCDD, polychlorinated dibenzo-p-dioxin; PCDF, polychlorinated dibenzofuran; E.L.S., early life stage; LC_{50}, lethal concentration causing the death of 50% of a population (LC_{99} defined similarly); TCDD, tetrachlorodibenzo-p-dioxin; TCDF, tetrachlorodibenzofuran; P5CDD, pentachlorodibenzo-p-dioxin; H6CDD, hexachlorodibenzo-p-dioxin; T3CDD, trichlorodibenzo-p-dioxin; DCDD, dichlorodibenzo-p-dioxin; OCDD, octachlorodibenzo-p-dioxin; EC_{50}, concentration causing an effect on 50% of a population; LOEC, lowest observed effect concentration; NOEC, no observed effect concentration; LD_{50}, 50% lethal dose (LD_{88}, LD_{20}, LD_{100}, LD_{85}, and LD_{80} defined similarly); P5CDF, pentachlorodibenzofuran; H6CDF, hexachlorodibenzofuran; EROD, ethoxyresorufin-O-dealkylase; AHH, aryl hydrocarbon hydroxylase.
[b] Values in parentheses are the body burden as estimated by Sijm and Opperhuizen.

Lethal effects at later stages in the development of fish seem to occur at higher concentrations than those at early life stages. Lethal body burdens are found between <0.054 µg/kg and 2042 µg/kg (Table 2). Coho salmon may be very sensitive to TCDD, which may explain their much lower lethal body burden of <0.054 µg/kg compared with those of bluegill (16 µg/kg), bullhead (5 µg/kg), carp (2.2 or 3 µg/kg), fathead minnow (17-2042 µg/kg), largemouth bass (11 µg/kg), rainbow trout (3, 5, 10, or 25 µg/kg), or yellow perch (3 or 5 µg/kg). A high p.o. dose of 30 µg/kg of TCDD, which was administered in a gelatin capsule, however, did not cause death to cod and rainbow trout within 4 weeks from the time of administration (Hektoen et al., 1992). In the latter study, however, the actual body burden of TCDD that was assimilated from the gut and that could cause toxic effects is unknown. A critical body burden for lethality may thus be species dependent; however, at present exact data are lacking. Kleeman et al. (1988) show that the sensitivity of six freshwater juvenile fish species to TCDD-induced lethality varies by fivefold.

Sublethal effects, such as growth inhibition, immunotoxicity, and cytochrome P-450IA1 induction, at more advanced life stages of the fish are found at body burdens between 0.08 µg/kg and 30 µg/kg (Table 2). No effects were observed on the growth of channel catfish and largemouth bass that were exposed to TCDD- and TCDF-containing bleached kraft mill effluent in experimental streams for up to 205 days. Both fish attained maximum body burdens of 20 ng/kg of TCDD and 15 ng/kg of TCDF during the exposure period (NCASI, 1991).

COMPARISON OF ENVIRONMENTAL AND EFFECT CONCENTRATIONS

The body burdens of dioxins and furans that are found in fish in the environment are 0.8 µg/kg and lower, whereas the body burden of TCDD in fish is 0.4 µg/kg or lower (Figure 3). Both concentrations are close to those found to cause lethal or sublethal effects in fish in the laboratory (Figure 3). Lethal effects in older fish are found at body burdens of approximately 0.2 µg/kg and higher, and in fish at early life stages, these effects are found at body burdens between 0.003 and 10 µg/kg. Sublethal effects are observed at body burdens between (estimated) concentrations of <0.0003 and 8.3 µg/kg for early life stages and between <0.054 and 30 µg/kg for older fish (Figure 3). In particular, egg mortality occurs at body burdens as low as an estimated concentration of 0.001 µg/kg$_{egg}$. Since in one study (Smith et al., 1990) it was found that the body burden of dioxin and furan in eggs was 0.12 µg/kg (0.02 µg/kg of TCDD and 0.10 µg/kg of TCDF, which levels were approximately twofold higher than in adult fish [Table 1]), reduced egg survival or sublethal effects at developing stages may be an actual environmental problem.

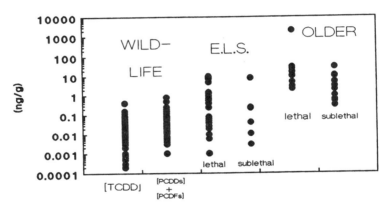

Figure 3 Body burdens of 2,3,7,8-tetrachlorodibenzo-*p*-dioxin ([TCDD]) and of the sum of polychlorinated dibenzo-*p*-dioxins and of polychlorinated dibenzofurans ([PCDDs] + [PCDFs]), which are found in fish in the environment (WILD-LIFE), and which are causing lethal or sublethal effects in laboratory fish at early life stages (E.L.S.) or at older, more developed life stages (OLDER).

In the majority (90%) of the fish sampled at different sites in the United States, body burdens of dioxins and furans are below 0.005 µg/kg. This body burden, however, is also within an order of magnitude of body burdens that elicit toxic responses (Figure 3). So, even at nonpolluted areas, the body burdens of dioxins and furans in fish are close to those that may involve toxicological effects, a conclusion also reported by Cook et al. (1991).

Furthermore, it can be seen that no fish are found in the environment that have body burdens higher than those causing death in laboratory studies (Figure 3). The assumption here is that either fish are not exposed to concentrations causing death, or fish having high levels of dioxins and furans in the environment already died and were not sampled.

Several assumptions that have led to the comparison between environmental and toxic body burdens are given more attention, since the results have great implications.

1. Environmental body burdens usually are measured in the fillet of fish. Since the target for dioxins and furans that cause toxic effects is unknown, it may not be correct to compare fillet concentrations with effect body burdens. In addition, the lipid fraction of fish may greatly determine the tissue distribution of dioxins and furans (Hektoen et al., 1992). This also makes the body burden of the dioxins and furans in fish difficult to compare with body burdens that cause toxic effects. However, environmental body burdens still can be assumed to be within one order of magnitude of effect body burdens.
2. Summing up all the congeners in fish is debatable, since their additivity is not well investigated. Toxic responses have not been measured of either all the individual congeners or the mixtures of congeners. However, there is a rationale for summing the compounds, as mentioned earlier (see the "Environmental Concentrations" section). Other compounds, such as polychlorinated biphenyls or polycyclic aromatic compounds, may contribute to toxicity. The body burdens of solely the dioxins and furans thus possibly underestimate total toxicity. Additivity, synergism, or antagonism of the toxic effects of dioxins, furans, and other chemicals for fish is poorly understood (Janz and Metcalfe, 1991), but may occur (Zacharewski et al., 1989).

SUMMARY

Dioxins and furans elicit several lethal and sublethal effects in fish. These hydrophobic compounds are taken up by fish from water, food, or sediment, of which the latter two are probably the most important sources. Fish eggs may get their dioxin and furan burdens from both ambient water and the mother.

To evaluate and estimate the potential risk of dioxins and furans for fish in the environment, one has to know the levels that cause toxic responses and the levels in fish in the environment. For reasons outlined in the text, the body burdens in wild and laboratory fish are presented as a sum of all the detected dioxin and furan congeners. Measurements are given as wet weight, unless otherwise specified.

Body burdens of polychlorinated dibenzo-p-dioxins (PCDDs) in wild fish range between the nondetectable and 0.804 µg/kg, and those of polychlorinated dibenzofurans (PCDFs) range between the nondetectable and 0.387 µg/kg. Of the reported fish, 50% have a body burden of the sum of dioxins and furans of 40 ng/kg or lower, and 90% have a body burden of 100 ng/kg or lower.

In early life-stage studies, 2,3,7,8-tetrachlorodibenzo-p-dioxin (TCDD) shows the lowest lethal body burdens that are between 0.065 µg/kg$_{egg}$ in lake trout and 0.4 µg/kg$_{egg}$ in rainbow trout. Other dioxin and furan congeners show higher lethal body burdens in eggs, which makes them less potent than TCDD. Lethal body burdens in older fish are found between <0.054 µg/kg in coho salmon and 2042 µg/kg in fathead minnow. These lethal body burdens are higher than those in early life stages and may be species dependent.

Sublethal effects are observed at body burdens estimated at between <0.0003 and 8.3 µg/kg for early life stages and between <0.054 and 30 µg/kg for older fish.

In the environment, dioxin and furan levels in fish are thus within one or two orders of magnitude of the levels that have been found to elicit lethal effects in fish in the laboratory. However, fish that were exposed to TCDD and tetrachlorodibenzofuran (TCDF) in experimental streams attained maximum body burdens of 20 ng/kg of TCDD and 15 ng/kg of TCDF, but they showed no sublethal effects.

The data show that dioxins and furans pose an important environmental risk to fish at early life stages as well as to older fish.

ACKNOWLEDGMENTS

A. O. thanks the Royal Dutch Academy of Sciences.

REFERENCES

Adams, W. J., G. M. DeGraeve, T. D. Sabourin, J. D. Cooney, and G. M. Mosher. 1986. Toxicity and bioconcentration of 2,3,7,8-TCDD to fathead minnow *(Pimephales promelas)*. Chemosphere 15:1503-1511.

Batterman, A. R., P. M. Cook, K. B. Lodge, D. B. Lothenbach, and B. C. Butterworth. 1989. Methodology used for a laboratory determination of relative contributions of water, sediment, and food chain routes of uptake for 2,3,7,8-TCDD bioaccumulation by lake trout in Lake Ontario. Chemosphere 19:451-458.

Baumann, P. C., and D. M. Whittle. 1988. The status of selected organics in the Laurentian Great Lakes: an overview of DDT, PCBs, dioxins, furans, and aromatic hydrocarbons. Aquat. Toxicol. (N.Y.) 11:241-257.

Bol, J., M. van den Berg, and W. Seinen. 1989. Interactive effects of PCDD's, PCDF's, and PCB's as assessed by the E.L.S.-bioassay. Chemosphere 19:899-906.

Cook, P. M., R. D. Johnson, G. T. Ankley, S. P. Bradbury, R. J. Erickson, and R. L. Spehar. In press. Research to characterize ecological risks associated with 2,3,7,8-tetrachlorodibenzo-*p*-dioxin and related chemicals in aquatic ecosystems. *In* Dioxin '92. (Abstr.)

Cook, P. M., D. W. Kuehl, M. K. Walker, and R. E. Peterson. 1991. Bioaccumulation and toxicity of TCDD and related compounds in aquatic ecosystems. p. 143-167. *In* Banbury report. Vol. 35. Biological basis for risk assessment of dioxins and related compounds. Cold Spring Harbor Laboratory Press, Cold Spring Harbor, N.Y.

De Vault, D., W. Dunn, P.-A. Bergqvist, K. Wiberg, and C. Rappe. 1989. Polychlorinated dibenzofurans and polychlorinated dibenzo-*p*-dioxins in Great Lakes fish: a baseline and interlake comparison. Environ. Toxicol. Chem. 8:1013-1022.

Eadon, G., L. Kaminsky, J. Silkworth, K. Aldous, D. Hilker, P. O'Keefe, R. Smith, J. Gierthy, J. Hawley, N. Kim, and A. DeCaprio. 1986. Calculation of 2,3,7,8-TCDD equivalent concentrations of complex environmental contaminant mixtures. Environ. Health Perspect. 70:221-227.

Fehringer, N. V., S. M. Walters, R. J. Ayers, R. J. Kozara, J. D. Ogger, and L. F. Schneider. 1985. A survey of 2,3,7,8-TCDD residues in fish from the Great Lakes and selected Michigan rivers. Chemosphere 14:909-912.

Frommberger, R. 1991. Polychlorinated dibenzo-*p*-dioxins and polychlorinated dibenzofurans in fish from south-west Germany: river Rhine and Neckar. Chemosphere 22:29-38.

Gardner, A. M., and K. D. White. 1990. Polychlorinated dibenzofurans in the edible portion of selected fish. Chemosphere 21:215-222.

Hahn, M. E., B. R. Woodin, and J. J. Stegeman. 1989. Induction of cytochrome P-450E (P-450IA1) by 2,3,7,8-tetrachlorodibenzofuran (2,3,7,8-TCDF) in the marine fish scup *(Stenotomus chrysops)*. Mar. Environ. Res. 28:61-65.

Harless, R. L., E. O. Oswald, R. G. Lewis, A. E. Dupuy, Jr., D. D. McDaniel, and H. Tai. 1982. Determination of 2,3,7,8-tetrachlorodibenzo-*p*-dioxin in fresh water fish. Chemosphere 11:193-198.

Hawkes, C. L., and L. A. Norris. 1977. Chronic oral toxicity of 2,3,7,8-tetrachlorodibenzo-*p*-dioxin (TCDD) to rainbow trout. Trans. Am. Fish. Soc. 106:641-645.

Hektoen, H., K. Ingebrigtsen, E. M. Brevik, and M. Oehme. 1992. Interspecies differences in tissue distribution of 2,3,7,8-tetrachlorodibenzo-*p*-dioxin between cod *(Gadus morhua)* and rainbow trout *(Oncorhynchus mykiss)*. Chemosphere 24:581-587.

Helder, T. 1980. Effects of 2,3,7,8-tetrachlorodibenzo-*p*-dioxin (TCDD) on early life stages of the pike *(Esox lucius* L.). Sci. Total Environ. 14:255-264.

Helder, T. 1981. Effects of 2,3,7,8-tetrachlorodibenzo-*p*-dioxin (TCDD) on early life stages of rainbow trout *(Salmo gairdneri,* Richardson). Toxicology 19:101-112.

Helder, T., and W. Seinen. 1985. Standardization and application of an E.L.S.-bioassay for PCDDs and PCDFs. Chemosphere 14:183-193.

Janz, D. M., and C. D. Metcalfe. 1991. Nonadditive interactions of mixtures of 2,3,7,8-TCDD and 3,3',4,4'-tetrachlorobiphenyl on aryl hydrocarbon hydroxylase induction in rainbow trout *(Oncorhynchus mykiss)*. Chemosphere 23:467-472.

Jobb, B., M. Uza, R. Hunsinger, K. Roberts, H. Tosine, R. Clement, B. Bobbie, G. LeBel, D. Williams, and B. Lau. 1990. A survey of drinking water supplies in the province of Ontario for dioxins and furans. Chemosphere 20:1553-1558.

Kjeller, L-O., S-E. Kulp, S. Bergek, M. Boström, P.-A. Bergqvist, C. Rappe, B. Jonsson, C. de Wit, B. Jansson, and M. Olsson. 1990. Levels and possible sources of PCDD/PCDF in sediment and pike samples from Swedish lakes and rivers. Part one. Chemosphere 20:1489-1496.

Kleeman, J. M., J. R. Olson, and R. E. Peterson. 1988. Species differences in 2,3,7,8-tetrachlorodibenzo-*p*-dioxin toxicity and biotransformation in fish. Fundam. Appl. Toxicol. 10:206-213.

Kuehl, D. W., B. C. Butterworth, A. McBride, S. Kroner, and D. Bahnick. 1989. Contamination of fish by 2,3,7,8-tetrachlorodibenzo-*p*-dioxin: a survey of fish from major watersheds in the United States. Chemosphere 18:1997-2014.

Luckas, B., and M. Oehme. 1990. Characteristic contamination levels for polychlorinated hydrocarbons, dibenzofurans, and dibenzo-*p*-dioxins in bream *(Abramis brama)* from the river Elbe. Chemosphere 21:79-89.

McCarty, L. S. 1986. Relationship between toxicity and bioconcentration for some organic chemicals. Environ. Toxicol. Chem. 5:1071-1080.

Mehrle, P. M., D. R. Buckler, E. E. Little, L. M. Smith, J. D. Petty, P. H. Peterman, D. L. Stalling, G. M. De Graeve, J. J. Coyle, and W. J. Adams. 1988. Toxicity and bioconcentration of 2,3,7,8-tetrachlorodibenzo-*p*-dioxin and 2,3,7,8-tetrachlorodibenzofuran in rainbow trout. Environ. Toxicol. Chem. 7:47-62.

Miller, R. A., L. A. Norris, and C. L. Hawkes. 1973. Toxicity of 2,3,7,8-tetrachlorodibenzo-*p*-dioxin (TCDD) in aquatic organisms. Environ. Health Perspect. 5:177-186.

Miller, R. A., L. A. Norris, and B. R. Loper. 1979. The response of Coho salmon and guppies to 2,3,7,8-tetrachlorodibenzo-*p*-dioxin (TCDD) in water. Trans. Am. Fish. Soc. 108:401-407.

Muir, D. C. G., A. L. Yarechewski, D. A. Metner, and W. L. Lockhart. 1992. Dietary 2,3,7,8-tetrachlorodibenzofuran in rainbow trout: accumulation, disposition, and hepatic mixed function oxidase enzyme induction. Toxicol. Appl. Pharmacol. 117:65-74.

NCASI. 1991. Observations on the bioaccumulation of 2,3,7,8-TCDD and 2,3,7,8-TCDF in channel catfish and largemouth bass and their survival or growth during exposure to biologically treated bleached kraft mill effluent in experimental streams. National Council of the Paper Industry for Air and Stream Improvement Tech. Bull. No. 611.

Norris, L. A., and R. A. Miller. 1974. The toxicity of 2,3,7,8-tetrachlorodibenzo-*p*-dioxin (TCDD) in guppies (*Poecilia reticulatus* Peters). Bull. Environ. Contam. Toxicol. 12:76-80.

O'Keefe, P., D. Hilker, C. Meyer, K. Aldous, L. Shane, R. Donnelly, R. Smith, R. Sloan, L. Skinner, and E. Horn. 1984. Tetrachlorodibenzo-*p*-dioxins and tetrachlorodibenzofurans in Atlantic coast striped bass and in selected Hudson River fish, waterfowl, and sediments. Chemosphere 13:849-860.

O'Keefe, P., C. Meyer, K. Aldous, B. Jelus-Tyror, K. Dillon, R. Donnelly, E. Horn, and R. Sloan. 1983. Analysis of 2,3,7,8-tetrachlorodibenzo-*p*-dioxin in Great Lakes fish. Chemosphere 12:325-332.

Ono, M., N. Kannan, T. Wakimoto, and R. Tatsukawa. 1987. Dibenzofurans a greater global pollutant than dioxins? Evidence from analyses of open ocean killer whale. Mar. Pollut. Bull. 18:640-643.

Opperhuizen, A. 1991. Bioaccumulation kinetics: experimental data and modelling. p. 61-70. *In* G. Angeletti and A. Bjørseth (Ed.). Organic micropollutants in the aquatic environment. Proceedings of the sixth European symposium held in Lisbon, Portugal. Kluwer Academic Publishers, Dordrecht.

Opperhuizen, A., and D. T. H. M. Sijm. 1990. Bioaccumulation and biotransformation of polychlorinated dibenzo-*p*-dioxins and dibenzofurans in fish. Environ. Toxicol. Chem. 9:175-186.

Parsons, A. H., S. L. Huntley, E. S. Ebert, E. R. Algeo, and R. E. Keenan. 1991. Risk assessment for dioxin in Columbia River fish. Chemosphere 23:1709-1717.

Rappe, C., R. Andersson, P.-A. Bergqvist, C. Brohede, M. Hansson, L.-O. Kjeller, G. Lindström, S. Marklund, M. Nygren, S. E. Swanson, M. Tysklind, and K. Wiberg. 1987. Overview on environmental fate of chlorinated dioxins and dibenzofurans. Sources, levels, and isomeric pattern in various matrices. Chemosphere 16:1603-1618.

Rappe, C., P.-A. Bergqvist, L.-O. Kjeller, S. Swanson, T. Belton, B. Ruppel, K. Lockwood, and P. C. Kahn. 1991. Levels and patterns of PCDD and PCDF contamination in fish, crabs, and lobsters from Newark Bay and the New York Bight. Chemosphere 22:239-266.

Rappe, C., L.-O. Kjeller, and R. Andersson. 1989. Analyses of PCDDs and PCDFs in sludge and water samples. Chemosphere 19:13-20.

Rogers, I. H., C. D. Levings, W. L. Lockhart, and R. J. Norstrom. 1989. Observations on overwintering juvenile Chinook salmon (*Oncorhynchus tshawytscha*) exposed to bleached kraft mill effluent in the upper Fraser River, British Columbia. Chemosphere 19:1853-1868.

Sijm, D. T. H. M. 1992. Influence of Biotransformation on Bioaccumulation and Toxicity of Chlorinated Aromatic Compounds in Fish. Ph.D. thesis. University of Utrecht, Utrecht, The Netherlands.

Sijm, D. T. H. M., W. Seinen, and A. Opperhuizen. 1992. A life-cycle study on biomagnification of PCBs in guppy. Environ. Sci. Technol. 26:2162-2174.

Sijm, D. T. H. M., H. Wever, and A. Opperhuizen. 1989. Influence of biotransformation on the accumulation of PCDDs from fly-ash in fish. Chemosphere 19:475-480.

Sijm, D. T. H. M., A. L. Yarechewski, D. C. G. Muir, G. R. B. Webster, W. Seinen, and A. Opperhuizen. 1990. Biotransformation and tissue distribution of 1,2,3,7-tetrachlorodibenzo-*p*-dioxin, 1,2,3,4,7-pentachlorodibenzo-*p*-dioxin, and 2,3,4,7,8-pentachlorodibenzofuran in rainbow trout. Chemosphere 21:845-866.

Smith, L. M., T. R. Schwartz, K. Feltz, and T. J. Kubiak. 1990. Determination and occurrence of AHH-active polychlorinated biphenyls, 2,3,7,8-tetrachloro-*p*-dioxin, and 2,3,7,8-tetrachlorodibenzofuran in Lake Michigan sediment and biota. The question of their relative toxicological significance. Chemosphere 21:1063-1085.

Spitsbergen, J. M., J. M. Kleeman, and R. E. Peterson. 1988a. Morphologic lesions and acute toxicity in rainbow trout *(Salmo gairdneri)* treated with 2,3,7,8-tetrachlorodibenzo-*p*-dioxin. J. Toxicol. Environ. Health 23:333-358.

Spitsbergen, J. M., J. M. Kleeman, and R. E. Peterson. 1988b. 2,3,7,8-Tetrachlorodibenzo-*p*-dioxin toxicity in yellow perch *(Perca flavescens)*. J. Toxicol. Environ. Health 23:359-383.

Spitsbergen, J. M., K. A. Schat, J. M. Kleeman, and R. E. Peterson. 1986. Interactions of 2,3,7,8-tetrachlorodibenzo-*p*-dioxin (TCDD) with immune response of rainbow trout. Vet. Immunol. Immunopathol. 12:263-280.

Spitsbergen, J. M., K. A. Schat, J. M. Kleeman, and R. E. Peterson. 1988c. Effects of 2,3,7,8-tetrachlorodibenzo-*p*-dioxin (TCDD) or Aroclor 1254 on the resistance of rainbow trout, *Salmo gairdneri* Richardson, to infectious haematopoietic necrosis virus. J. Fish Dis. 11:73-83.

Spitsbergen, J. M., M. K. Walker, J. R. Olson, and R. E. Peterson. 1991. Pathologic alterations in early life stages of lake trout, *Salvelinus namaycush*, exposed to 2,3,7,8-tetrachlorodibenzo-*p*-dioxin as fertilized eggs. Aquat. Toxicol. (Amst.) 19:41-72.

Startin, J. R., M. Rose, C. Wright, I. Parker, and J. Gilbert. 1990. Surveillance of British foods for PCDDs and PCDFs. Chemosphere 20:793-798.

Tanabe, S. 1988. Dioxin problems in the aquatic environment. Mar. Pollut. Bull. 19:347-348.

Tillitt, D. E., J. P. Giesy, and G. T. Ankley. 1991. Characterization of the H4IIE rat hepatoma cell bioassay as a tool for assessing toxic potency of planar halogenated hydrocarbons in environmental samples. Environ. Sci. Technol. 25:87-92.

Travis, C. C., and H. A. Hattemeyer-Frey. 1991. Human exposure to dioxin. Sci. Total Environ. 104:97-127.

U.S. EPA. 1990. Lake Ontario TCDD bioaccumulation study. Final report. U.S. Environmental Protection Agency, New York State Department of Environmental Conservation, New York State Department of Health, Occidental Chemical Corporation.

Van den Berg, M., F. Blank, C. Heeremans, H. Wagenaar, and K. Olie. 1987. Presence of polychlorinated dibenzo-*p*-dioxins and polychlorinated dibenzofurans in fish-eating birds and fish from The Netherlands. Arch. Environ. Contam. Toxicol. 16:149-158.

Van der Weiden, M. E. J., J. van der Kolk, R. Bleumink, W. Seinen, and M. van den Berg. 1992. Concurrence of P-450 1A1 induction and toxic effects after administration of a low dose of 2,3,7,8-tetrachlorodibenzo-*p*-dioxin (TCDD) to the rainbow trout *(Oncorhynchus mykiss)*. Aquat. Toxicol. (Amst.) 24:23-142.

Van der Weiden, M. E. J., J. van der Kolk, A. H. Penninks, W. Seinen, and M. van den Berg. 1990. A dose/response study with 2,3,7,8-TCDD in the rainbow trout *(Oncorhynchus mykiss)*. Chemosphere 20:1053-1058.

Walker, M. K., L. C. Hufnagle, Jr., M. K. Clayton, and R. E. Peterson. 1992. An egg injection method for assessing early life stage mortality of polychlorinated dibenzo-*p*-dioxins, dibenzofurans, and biphenyls in rainbow trout *(Oncorhynchus mykiss)*. Aquat. Toxicol. (Amst.) 22:15-38.

Walker, M. K., and R. E. Peterson. 1991. Potencies of polychlorinated dibenzo-*p*-dioxin, dibenzofuran, and biphenyl congeners, relative to 2,3,7,8-tetrachlorodibenzo-*p*-dioxin, for producing early life stage mortality in rainbow trout *(Oncorhynchus mykiss)*. Aquat. Toxicol. (Amst.) 21:219-238.

Walker, M. K., J. M. Spitsbergen, J. R. Olson, and R. E. Peterson. 1991. 2,3,7,8-Tetrachlorodibenzo-*p*-dioxin (TCDD) toxicity during early life stage development of lake trout *(Salvelinus namaycush)*. Can. J. Fish. Aquat. Sci. 48:875-883.

Wannemacher, R., A. Rebstock, E. Kulzer, D. Schrenk, and K. W. Bock. 1992. Effects of 2,3,7,8-tetrachlorodibenzo-*p*-dioxin on reproduction and oogenesis in zebrafish *(Brachydanio rerio)*. Chemosphere 24:1361-1368.

Wisk, J. D., and K. R. Cooper. 1990a. Comparison of the toxicity of several polychlorinated dibenzo-*p*-dioxins and 2,3,7,8-tetrachlorodibenzofuran in embryos of the Japanese medaka *(Oryzias latipes)*. Chemosphere 20:361-377.

Wisk, J. D., and K. R. Cooper. 1990b. The stage specificity of 2,3,7,8-tetrachlorodibenzo-*p*-dioxin in embryos of the Japanese medaka *(Oryzias latipes)*. Environ. Toxicol. Chem. 9:1159-1169.

Yockim, R. S., A. R. Isensee, and G. E. Jones. 1978. Distribution and toxicity of TCDD and 2,4,5-T in an aquatic model ecosystem. Chemosphere 7:215-220.

Zacharewski, T., L. Safe, S. Safe, B. Chittim, D. DeVault, K. Wiberg, P.-A. Bergqvist, and C. Rappe. 1989. Comparative analysis of polychlorinated dibenzo-*p*-dioxin and dibenzofuran congeners in Great Lakes fish extracts by gas chromatography-mass spectrometry and in vitro enzyme induction activities. Environ. Sci. Technol. 23:730-735.

Zitko, V. 1992. Patterns of 2,3,7,8-substituted chlorinated dibenzodioxins and dibenzofurans in aquatic fauna. Sci. Total Environ. 111:95-108.

CHAPTER 9

Polycyclic Aromatic Hydrocarbons in Marine Mammals, Finfish, and Molluscs

Jocelyne Hellou

INTRODUCTION

This chapter examines the fate of polycyclic aromatic hydrocarbons (PAHs) in marine organisms to determine if the PAH concentrations observed in tissues of various animals can be related to biological effects. In the first section, the source, fate, and some of the characteristic PAHs present in the aquatic environment are discussed. Then, the different chromatographic and spectroscopic methods available for the analysis of total or individual PAHs are briefly outlined to give the reader an appreciation of the various methods and reference materials used in the quantification of PAHs. Next, data regarding the bioaccumulation of PAHs in marine vertebrates are presented in relation to the possible fates awaiting PAHs taken up by mammals and finfish and in relation to our limited knowledge of these outcomes. The case of molluscs is then investigated as well as the recent research concerning sediment quality criteria. Subsequently, we attempt to relate PAH levels in sediments to PAH levels in marine molluscs.

PAHs IN THE AQUATIC ENVIRONMENT

PAHs are a group of chemicals containing carbon and hydrogen atoms with more than one benzene ring in their structure. PAHs are ubiquitous in nature and, in the marine environment, they originate mainly from anthropogenic sources. They can enter the aquatic environment directly, via petroleum spills, discharges from ships, oil seepage, runoff from roads, effluents from industrial processes, forest fires, and atmospheric transport, after the incomplete combustion of fossil fuels (Neff, 1990; Baek et al., 1991).

Once in the aquatic environment, hydrocarbons undergo a series of weathering (physical, chemical, and biological) processes, such as evaporation, photochemical oxidation, microbial degradation, dispersion and dissolution in the water, and deposition on sediments, which can have variable importance depending on the location (Neff, 1979). In general, higher hydrocarbon concentrations are found in sediments (>1000 times) than in the water column, while surface waters have the highest PAH concentrations within the water column. Because hydrocarbons are hydrophobic and lipophilic, they are also present in higher concentrations in suspended material than in the water itself. Typically, in the same geographical region, colder waters and near-shore and finer sediments will display higher hydrocarbon concentrations than warmer waters and offshore and coarser sediments display.

Many individual PAHs have carcinogenic (White, 1986) or mutagenic potential (Pahlmann and Pelkonen, 1987). Baumann (1989) and Moore et al. (1989) have reviewed the biological effects on finfish and shellfish caused by the presence of PAHs in the environment. The effect of oil spills on the health of marine mammals has also been discussed (Geraci and St. Aubin, 1990). Various aspects relating to the presence of PAHs in the marine environment have previously been reviewed (Corner et al., 1976; Malins, 1977; Neff, 1979; Cooke and Dennis, 1980; Bjorseth, 1983; Bjorseth and Ramdahl, 1985; Varanasi, 1989; Dunn, 1991; Farrington, 1991). In this chapter, the emphasis is on the interpretation of monitoring results.

A more recent example illustrating the importance of monitoring PAHs in the environment is found in the study of chemical pollution in Puget Sound, Washington, where a significant correlation was observed between the concentration of PAHs in sediments and the prevalence of hepatic neoplasms in English sole (Malins et al., 1988; Stein et al., 1990).

Sixteen PAHs (Figure 1) have been recommended as priority pollutants by the World Health Organization, the European Economic Community, and the U.S. Environmental Protection Agency. These priority pollutants are all parental compounds (only rings, no alkyl substituents) and major constituents of a pyrolytic source of PAHs. Characteristic differences in the PAHs originating from a combustion or petroleum source are listed in Table 1.

Upon weathering, the PAH composition described in Table 1 will be altered. Weathering processes and the presence of more than one source of PAH for a certain location will add to the difficulty of assigning a source to an environmental PAH extract containing hundreds of components. Other differences due to the hydrocarbon's source can be observed in the saturated fraction or the aromatic heterocyclic components of an environmental extract.

Once animals are exposed to PAHs in their environment, several fates can await these xenobiotics, depending on the species considered. In general, the process is more complex in vertebrates than in invertebrates.

CONSIDERATIONS REGARDING ANALYSES

The difficulty in interpreting and comparing the results of various PAH monitoring studies resides not only in the number of components present in an extract

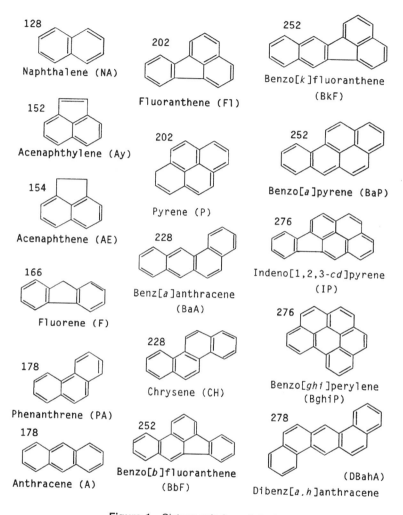

Figure 1 Sixteen priority pollutants.

but also in the different extraction and purification steps used in the isolation of PAHs, especially in the spectroscopic and/or chromatographic instrumentation and standards used to report concentrations. Although intercalibration exercises using sediments and biota (Law and Portman, 1982; Farrington et al., 1986, 1988; MacLeod et al., 1988; Law, 1989; Law and Nicholson, 1990) have shown a decreasing variability in the last 20 years between the results of different laboratories, it remains that the concentrations obtained for total PAHs, i.e., ultraviolet/fluorescence (uv/f) or gas chromatography (GC), or individual PAHs, i.e., GC-mass spectrometry (GC-MS) or high-performance liquid chromatography (HPLC), are not always unequivocal.

Table 1 Content of Aromatic Fraction Depending on Source of Polycyclic Aromatic Hydrocarbon (PAH)

	Aromatics phenanthrene/ anthracene	No. of rings	Alkylated/ parental	Methylpyrene/ pyrene	PA[a] and C-1 and C-2 PA	Most abundant
Petroleum	Present/not present	1 to >9	Higher/lower concentration	>2	C-1 and C-2 PA > PA	Alkylated PA
Combustion	Present/present	2 to 6	Lower/higher concentration	<1	PA > C-1 PA > C-2 PA	Fluoranthene-pyrene

[a] PA, phenanthrene; C-1 PA, methyl PA; C-2 PA, dimethyl PA.

As is the case with all chemical analyses, results have to be accompanied by enough information regarding the quality assurance and quality control protocols (details regarding blanks, recoveries, and results of duplicate analyses) to allow the reader to evaluate the precision and accuracy of results. These types of data have unfortunately not always been provided in the past. Also, results are reported in terms of dry or wet weight, even though dry weight values allow a better comparison of PAH results in tissues of different animal species. Since hydrocarbons are hydrophobic, the percentage of water in tissues is variable and the determination of dry weights simple.

When total PAHs are analyzed, the reference material used has to be chosen carefully. Even though chrysene has been nominated as the standard uv/f material by the World Health Organization, a second analytical approach is very useful as a check on the significance of the results. Overestimation and underestimation of the actual PAH concentrations and the choice of a standard, when using uv/f, have been discussed by Mason (1987). The technique of synchronous fluorescence spectroscopy has been reviewed (Vo-Dinh, 1981).

The importance of assessing the purity of an extract quantified by uv/f, using an alternate analytical method (GC-MS), has to be stressed. It has been shown that the presence of an unsaturated compound, such as squalene, in a hydrocarbons extract obtained from muscle tissue will not affect the uv/f absorbance (at the wavelengths used in PAH analyses) as much as the presence of leftover lipids (Hellou and Upshall, 1992). As could be expected, this effect becomes negligible as the ratio of PAH to interfering material increases. However, since there is an effect, especially in the case of "cleaner" environmental samples, the importance of the chromatographic purification prior to quantification needs to be emphasized, as well as the investigation of the composition of the extracts (Boehm and Quinn, 1978; Farrington et al., 1986). Fluorometric analysis can be regarded as a preliminary screening method that should be followed up by further investigations if elevated concentrations of fluorescing compounds are determined.

There are no specific guidelines regarding the analysis of total PAHs by GC other than the presence, in the standard, of an unresolved complex mixture in the same range of retention times as in the extracts. The application of this chromatographic technique to the analysis of PAHs has been reviewed (Bartle, 1985) and compared with uv/f (e.g., Mason, 1987). It should be pointed out that, when analysis of an unresolved complex mixture is done by uv/f or GC, other fluorescing nonpolar components such as heterocyclic aromatic compounds (non-PAHs) are included in the quantification.

When individual PAHs are analyzed, they must be representative of the total content of the extract. GC-MS single-ion monitoring or multiple-ion monitoring allows separation, partial identification, and quantification of the components of interest in a mixture. It is the best spectroscopic technique presently available for the routine analysis of the PAH components of an extract. However, it has not always been demonstrated that the target compounds are appropriate.

Wise (1985) reviewed the use of HPLC (with uv or uv/f detectors) in the analyses of PAHs. Depending on the selectivity of the method (number of wavelengths used), the identity of the components in representative extracts might still need to be ascertained, i.e., liquid chromatography-MS or GC-MS. It is understood that even when the last two techniques are applied, foolproof identity of PAHs can only be accomplished when concentrations allow MS analysis in the total-ion chromatogram mode.

The above analytical methods, as well as infrared spectroscopy, more useful in the analysis of aliphatic hydrocarbons, have been previously discussed from a comparative point of view (Farrington and Medeiros, 1975; Farrington and Tripp, 1975; Lake et al., 1980; Ehrhardt et al., 1991). Ideally, results for total and individual PAHs should be compared (e.g., Boehm and Hirtzer, 1982; Cretney et al., 1987; Hellou et al., 1990, 1991; Law and Whinnett, 1992).

An example of the detection limits expected when analyzing a single PAH or the aromatic fraction of a petroleum oil is shown in Table 2. The detection limits obtained with an HPLC as well as with a uv/f instrument will vary with the wavelengths chosen. Different instruments display different detection limits; the values reported below were obtained in our laboratory. It is also important to consider detection limits when choosing an analytical technique.

Table 2 Expected Detection Limits of a Single Polycyclic Aromatic Hydrocarbon or of the Aromatic Fraction of a Petroleum Oil

Chemical and technique	Detection limits (ng/μl)
Chrysene	
uv/f[a] (310/360 nm)	0.05
GC or GC-MS-TIC	0.5
GC-MS-SIM	0.1
Venezuelan crude oil (aromatic fraction	
uv/f (335/360 nm)	0.1
GC or GC-MS-TIC	20
GC-MS-MIM	1

[a] uv/f, ultraviolet/fluorescence; GC, gas chromatography; MS, mass spectrometry; TIC, total-ion chromatogram; SIM, single-ion monitoring; MIM, multiple-ion monitoring.

FINFISH AND MAMMALS

There are several possible fates for PAHs taken up by vertebrates. The major difference from those of invertebrates is due to the activity of mixed-function oxygenase (MFO) enzymes. These enzymes (phase I and phase II) are involved in the biotransformation of PAHs (oxidation and conjugation) to more polar, water-soluble metabolites that can, in turn, be retained or excreted. This difference leads to the following possible fates for organic xenobiotics:

- Excretion as a free xenobiotic;
- Retention as a free xenobiotic;
- Oxidation and excretion (e.g., alcohols, quinones, epoxides);
- Oxidation and retention (e.g., alcohols, quinones, epoxides);
- Oxidation, conjugation, and excretion; and
- Oxidation, conjugation, and retention.

Oxidation can yield more than one product per PAH, while conjugation can take place with several types of molecules (e.g., amino acids, sulfate groups, glucuronic acid, and/or DNA). Conjugation usually takes place after oxidation of the PAH. In the case of DNA-adducts, one electron-mediated oxidation can also lead to the binding between a PAH and a purine or pyrimidine moiety (Cavalieri et al., 1990). These different outcomes can take place at different rates depending on the PAH considered and species examined. Therefore, analyzing PAHs in vertebrates could mean examining free hydrocarbons and/or various metabolites. Consequently, to the difficulty of comparing analytical techniques is added the choice of which partial (or all) fate and tissue to consider. This chapter emphasizes free unmetabolized PAHs.

Examples of PAH concentrations obtained from feral fish species along with locations are presented in Table 3. Data on feral marine mammals are more scarce, and only muscle tissue has been analyzed. Hellou et al. (1990) examined four species of seals and six species of whales collected in the Northwest Atlantic Ocean. PAH concentrations ranged between 0.10 and 1.21 ppm (dry weight, chrysene equivalents). A study of PAH levels in fetal, juvenile, and older male and female harp seals indicated that the lowest levels were present in fetal muscle (0.29 ppm for females, 0.44 ppm for males) and the highest in juvenile muscle (2.08 ppm for females, 3.04 ppm for males), with no significant difference between sexes (Hellou et al., 1991). Law and Whinnett (1992) investigated PAHs in the muscle of harbor porpoises collected in waters from the United Kingdom. The concentrations ranged between 0.55 and 2.80 ppm (dry weight, chrysene equivalents).

The concentration of individual PAHs (vs. the sum or value of not detected), results from single fish (vs. the mean), the number of fish analyzed, general morphometric data, and the quality assurance and quality control are not always provided. These variables are needed to gain a better insight into the results. Taking this into consideration, one can deduce a general distribution of PAHs in tissues from the results presented in Table 3. In most cases, muscle tissue displayed lower free PAH concentrations than did the liver, gonads, stomach, or gall bladder, while the liver displayed higher concentrations than did muscle tissue. The mean liver/muscle tissue concentration ratios (for the total or sum of PAHs) obtained from relatively less to more contaminated areas follow the order (Table 3):

North Atlantic Ocean, >25 → Spain, 2 to 66 →
Turkey, 2 to 75 → Arabian Gulf, 1 to 2

This suggests a trend toward lower ratios at more contaminated sites. When the ratio is taken for a particular PAH (mean or single values that can be compared between

Table 3 Comparison of Polycyclic Aromatic Hydrocarbon (PAH) Concentrations Obtained in Tissues of Feral Fish

Species	Sample	Concentration of PAH (dry weight)	Type of PAH[a]	Location	Ref.
Hake	Muscle	5 ppm	Total PAH (GC)[b]	Georges Bank	Boehm and Hirtzer (1982)
	Liver	127 ppm			
Four-spot flounder	Muscle	38 ppm			
	Liver	885 ppm			
Hake	Muscle	14 ppb	$\sum$>13 PAH[c]		
	Liver	204 ppb			
Four-spot flounder	Muscle	18 ppb			
	Liver	902 ppb			
Parophrys vetulus	Muscle	ND	$\sum$23 PAH	Puget Sound (concentration at 2 sites)	Malins et al. (1984, 1985)
	Liver	989, 72 ppb			
	Stomach	67, 140, 692 ppb			
Stizostedion lucioperca	Muscle	85-150 ppb	$\sum$14 PAH	Finnish Archipelago	Rainio et al. (1986)
	Liver	590 ppb			
	Gall bladder	237 ppb			
Lota lota	Muscle	130 ppb			
	Liver	2,225 ppb			
	Gall bladder	313 ppb			
Mullus barbatus	Muscle	7, 8, 1[d] ppm	Total PAH (CH)	Turkey	Salihoglu et al. (1987)
	Liver	18, 75, 75 ppm			
Solea solea	Muscle	1, 6, 1 ppm			
	Liver	13, 55, 5 ppm			
Epinephelus aenus	Muscle	7, 3, 3 ppm			
	Liver	40, 10, 10 ppm			
M. barbatus	Muscle	4, 9, 11[c] ppm	Total PAH (Kuwait crude)	Spain (concentration at 3 sites)	Albaiges et al. (1987a, 1987b)
	Liver	132, 602, 125 ppm			
	Gonads	16, 57 ppm			
Trachurus trachurus	Muscle	4, 11, 4 ppm			
	Liver	34, 43, 8 ppm			
	Gonads	19 ppm			
Mugil capito	Muscle	ND			
Cyprinus carpito	Muscle	ND-3.0 ppm			
Argyrops spinifer	Muscle	66-200 ppm	Total PAH (CH)	Arabian Gulf (concentration range)	El Deeb and El Ebiary (1988)
	Liver	82-176 ppm			
Mylio bifasciatus	Muscle	114-689 ppm			
	Liver	76-677 ppm			
Mullus barbatus	June muscle	0.02, 0.17[e] ppm	Total PAH (CH)	Adriatic Sea	Dujmov and Sucevic (1989)
	July muscle	0.06, 0.09 ppm			
Tilefish	Muscle	1.96, 3.95[e] ppb	$\sum$24 PAH	Middle Atlantic Bight, 2 sites	Steimle et al. (1990)
	Liver	21.96, 12.80 ppb			
	Gonad	7.01, 23.53 ppb			
Gailus morhua	Muscle	ND[f]	$\sum$27 PAH	NW Atlantic (concentration range)	Hellou et al. (1993)
	Liver	ND-585 ppb			
	Gonad	ND-336 ppb			

Table 3 (continued) Comparison of Polycyclic Aromatic Hydrocarbon (PAH) Concentrations Obtained in Tissues of Feral Fish

[a] Reference material is indicated in parentheses, CH represents chrysene.
[b] GC, gas chromatography; ND, not determined; CH, chrysene.
[c] Includes dibenzothiophenes.
[d] Three values represent concentrations in March, August, and November.
[e] Two values represent concentrations at sites 1 and 2, respectively.
[f] Dibenzothiophenes were analyzed.

studies), this trend is not so clear, most likely because different PAHs undergo different fates to different extents and because different PAHs can have different origins. For example, the following is observed for phenanthrene:

Northwest Atlantic Ocean, >20 to >50 → Mid-Atlantic Bight, 4 to 6 → North Atlantic Ocean, 10 to 12 → Finnish Archipelago, 4 to 6

"Free PAHs" refer to the PAHs that have "escaped" metabolism and, therefore, reflect the MFO enzyme activity and PAH exposure. Liver concentrations appear to provide information regarding short-term exposure, while muscle concentrations inform about long-term bioaccumulation. It is also generally held that a threshold level of exposure to contaminants is needed before induction of MFO enzymes, although there can also be a saturation of the enzymatic activity at high exposure. The liver/muscle ratio of concentrations would reflect the ability of the vertebrates to eliminate and/or biotransform PAHs, a lower ratio possibly indicating a lower efficiency.

Exceptions to this observation are apparent (Table 3). It could be argued that this difference is due to the actual composition of the extracts or that the sum or the total PAHs are not really representative of the actual components within the mixture. Many factors could also affect the observed variation in PAH concentrations between muscle tissue and other organs. Possible factors might include the fish species, age, sex, reproductive state, feeding behavior, type of habitat, type and amount of PAHs and/or other contaminants present in the environment, temperature, and mixed-function oxygenase activity. Elucidating the role played by these interacting variables still presents a long-term challenge.

The presence of PAHs in gonads is of great concern, since it is the early developmental stages of fish that are more affected by contaminants (Solbakken et al., 1984; Falk-Petersen and Kjorsvik, 1989; Castello and Gamble, 1992).

Additional observations regarding the fate of PAHs in tissues can be made from the results obtained in laboratory experiments involving fish and mammals (Engelhardt et al., 1977; Melancon and Lech, 1978; Nava and Engelhardt, 1980; Solbakken et al., 1983; Stein et al., 1984). Higher concentrations of polar metabolites have been observed in bile than in liver and muscle. Higher or longer exposure to PAHs has resulted in higher concentrations of PAHs in tissues. The relative concentration of free PAHs in tissues varied with the PAHs and fish species examined. After a short-term exposure, concentrations persisted in muscle for a longer time than in the other organs. The relative lipid content of liver and muscle has also been proposed to play a role in the distribution of PAHs in tissues (Solbakken and Palmork, 1980).

Varanasi et al. (1989) and Varanasi and Stein (1991) have pointed out the tendency for the accumulation of the relatively lower molecular weight PAHs (two and three rings) under the free form. The higher molecular weight PAHs (four to six rings) with higher chemical reactivity and carcinogenic potential appear to undergo metabolism. This observation has led to the investigation of other biochemical fates of PAHs, such as the presence of conjugated metabolites and of DNA-adducts in tissues. To date, a limited number of publications deal with bile metabolites (e.g., Collier and Varanasi, 1991; Gronlund et al., 1991; Krahn et al., 1992), metabolites in tissues (Krone et al., 1992), and DNA-adducts (e.g., Kurelec et al., 1989; MacCubbin et al., 1990; Liu et al., 1991) obtained from feral fish. Even fewer publications deal with these from marine mammal populations (Martineau et al., 1988; Ray et al., 1992). These two approaches are also potentially capable of clarifying the problem of whether finfish and marine mammals have been exposed to PAHs and to how much.

MOLLUSCS

In the past, molluscs were thought to lack mixed-function oxygenase enzymes (Lee et al., 1972). Subsequently, cytochrome P-450 and associated enzyme activities have been detected in at least 23 species of molluscs (Livingstone et al., 1989). However, even though mixed-function oxygenase activities have been detected, the metabolic rates are much lower in molluscs than in vertebrates (Stegeman and Lech, 1991). This phenomenon leads to a major difference in the fate of PAHs in molluscs (simpler) compared with fish and mammals and has made molluscs the sentinel organisms of choice in monitoring studies (Mix, 1984; James, 1989).

The bioaccumulation of PAHs (the concentration of chemical in animal tissue over the concentration in the environment) is also influenced by many biochemical and environmental factors (McElroy et al., 1989; Farrington, 1991), which have to be taken into consideration when examining monitoring results. Table 4 gives some examples of the correlations observed between a change in a species-related variable and in PAH concentrations. These results were obtained in either laboratory or field studies and do not represent an exhaustive listing.

The chronological sequence presented in Table 4 attempts to reflect the insight gained by the scientific community in the past 20 years regarding the variables affecting the accumulation of PAHs in molluscs. For example, in 1973, Stegeman and Teal associated the amount of lipid present in tissue with the PAH content of oysters. A relationship was therefore expected with the reproductive cycle of molluscs (Joseph, 1982, reviewed lipids in molluscs). Later, Nasci and Fossato (1982) and Friocourt et al. (1985), respectively, examined the accumulation of PAHs in mussels and in scallops and found no relation with gametogenesis or spawning. Later on, Jovanovich and Marion (1987) analyzed many variables in clams (lipids, proteins, carbohydrates, and gonad index) in relation to the rate of uptake and

POLYCYCLIC AROMATIC HYDROCARBONS

Table 4 Variables Influencing the Concentration of Polycyclic Aromatic Hydrocarbons (PAHs) in Molluscs

Species	Variable	Effect on PAH	Ref.
Oysters	Higher lipid	Higher concentration	Stegeman and Teal (1973)
Clams	Lower exposure	Higher bioconcentration	Clement et al. (1980)
Mussels	Gametogenesis	Not related	Nasci and Fossato (1982)
Mussels	Filter feeder	Surface PAH	Boehm et al. (1982)
Clams	Deposit feeder	Sediment PAH	
Clams	Fall-winter	Lower concentration	Mix and Schaffer (1983)
	Spring-summer	Higher concentration	
Scallops	Not only maximum gonad index	Maximum concentration in muscle, gonads	Friocourt et al. (1985)
Mussels vs. clams	Respiration or pumping rate	Faster vs. slower uptake	Broman and Ganning (1986)
Clams	Maximum gonad index	Highest concentration	Jovanovich and Marion (1987)
Three mussel species	Fall-winter	Lower concentration	Mason (1988)
	Spring-summer	Higher concentration	
Three mussel species	Lower exposure	Higher bioconcentration	Mason (1988)
Clams	Larger animals	Higher concentration	Tanacredi and Cardenas (1991)

elimination of PAHs. They found a correlation between the concentration of PAHs in tissues and the peak in the gonad index, but not with the lipids concentration. Therefore, from a combination of studies, it would appear that, in some species, such as clams, a maximum gonad index can be correlated with the maximum bioaccumulation of PAHs, while in all cases there seems to be a seasonal variation, perhaps due to temperature, which can be related to metabolic rates and/or PAH concentrations in the environment.

Another point brought out by Table 4 is that different species of molluscs can take up different types or levels of PAHs from their environment (Boehm et al., 1982). Different concentrations of PAHs in mollusc species have also been observed in other studies (Grahl-Nielsen et al., 1978; Cretney et al., 1987; Bender and Huggett, 1989), although generally the concentration in various bivalves differs only by a factor of 2 or 3. Our present knowledge concerning PAHs in molluscs points to season, species, and animal size as important variables to consider in planning, comparing, or interpreting a monitoring study.

Taking the above considerations into account, it remains that the concentration of PAHs in molluscs reflects the concentration of PAHs in the environment. A few examples obtained from laboratory and field studies are presented in Table 5, which illustrate the change in the PAH concentration of molluscs with PAH exposure.

The results presented in Table 5 are expressed in terms of different standards (reference materials) and show that the concentration of total and individual PAHs generally increases when molluscs from polluted sites are compared with those from unpolluted sites. Some important points should be made in relation to this table. First, at best, the total PAHs (uv/f) will represent the sum of biogenic olefins, heterocyclic aromatic compounds, and anthropogenic compounds. Second, as illustrated by Mix

Table 5 Polycyclic Aromatic Hydrocarbon (PAH) Concentrations Reported for Whole Molluscs Compared with Water

Species	Concentration (dry weight)[a]	Conc in water[b]	Type of PAH[c]	Ref.
Mussels	0.013	Unpolluted	BaP[b]	Fossato et al.
Mytilus sp.	0.116	Polluted		(1979)
Mussels	0.055/0.035/0.010	Unpolluted	Fl/P/BaP	Mackie et al.
M. edulis	1.345/1.415/1.645	Polluted		(1980)
Clams	42	Control	Total PAH	Clement et al.
Macoma balthica	405	0.03	(Prudhoe Bay	(1980)
	650	0.3	crude)[e]	
	1,750	3 ppb		
Clams	0.4-2.2	pre-spill	Total PAH	Boehm et al.
M. balthica	84-435	post-spill 1	(Lagomedia	(1982)
	115-860	post-spill 2	crude)[f]	
Mussels	0.71-2.01	Unpolluted	Σ15 PAH[g]	Mix and
M. edulis	3.37-6.62	Polluted		Schaffer (1983)
Mussels	70	0.02	Total PAH	Mason (1988)
(3 spp.)	990	0.3	(Quatar	
	2440	1.5 ppb	light crude)	
Oysters	4.3-74.7	0.2-2.4	Total PAH	Badawy and
Saccostrea	6.8-74.5	0.4-6.5	(chrysene)	Al-Harthy (1991)
cuculata	0.03-13.0	0.1-23.6		
	1.2-13.9	0.3-66 ppb		

[a] Assumed 80% water in tissues to transform wet weights into dry weights (ppm).
[b] Actual concentration not always reported.
[c] Concentrations reported in terms of the reference material indicated in parentheses, using uv/f spectroscopy. Letters refer to specific PAH listed in Figure 1.
[d] BaP, benzo(a)pyrene; Fl, fluoranthene; P, pyrene.
[e] Values obtained after an exposure of 180 days.
[f] Values taken for Mya truncata and Macoma calcarea, from Bays 9 and 10. Post-spill-1 and 2 refer to 3 days and 2 to 3 weeks after the spill.
[g] Individual PAHs were from 2 to 20 times higher, when comparing sites at each of the 12 sampling dates.

and Schaffer (1983), the concentration of individual PAHs can increase by 2 to 20 times, depending on the PAH examined. This variation is related to the PAH source in the particular environment examined, to the mollusc species analyzed and, especially, to the physical-chemical properties of PAH. Finally, molluscs are usually analyzed whole, because of the difficulty in separating the organs of small animals and because they are often eaten as such (scallops being one exception). However, it has been demonstrated that contaminants are not evenly distributed in invertebrate organs (e.g., Friocourt et al., 1985), levels being more elevated in the visceral mass.

The range of concentrations that can be expected for total low-molecular-weight (two and three rings, Σ12) and high-molecular-weight (four to six rings, Σ12) PAHs in mussels compared with sediments is well presented by Robertson et al. (1991). Under the National Status and Trends Program (NST) (NOAA, 1985, 1988, 1989; O'Connor, 1991), PAH concentrations in mussels from 95 sites and sediments from 217 sites in the United States were compared. This program showed a range of concentrations between 2 and 2000 ppb for low-molecular-weight PAHs and between

0.2 and 200 ppb for high-molecular-weight PAHs in mussels, compared with values between 1 and 6000 ppb for low-molecular-weight PAHs and between 7 and 10,000 ppb for high-molecular-weight PAHs in sediments.

SEDIMENT QUALITY CRITERIA

It is important to recognize that 1 ppm of a PAH (dry weight) is quite different from 1 ppm of another PAH or of total PAHs expressed in terms of various standards or of the sum of a different number of PAH components in sediments, present in combination with other contaminants or not. The physical-chemical form of hydrophobic compounds plays a major role in their bioavailability (Farrington, 1991). Factors such as the molecular weight, water solubility, octanol-water partition coefficient, sorption among three phases (water or dissolved-phase colloids, organics in water particulates, or solid phase), concentration of dissolved organic carbon or dissolved organic matter, salinity, and water temperature will influence the bioavailability of PAHs (McElroy et al., 1989). Synergistic or antagonistic effects can take place, and differences according to variables such as animal species, time of year, age, feeding behavior, reproductive stage, and activity of mixed-function oxygenase enzymes will occur.

Long (1992) and Long and Morgan (1990) have analyzed the studies aimed at discovering the threshold level of a variety of contaminants in sediments. The data considered by these authors covered laboratory experiments and field or modeling studies, using a variety of organisms and the following four different approaches:

- Sediment-water, equilibrium partitioning,
- Spiked-sediments bioassays,
- Screening-level contaminants, and
- Apparent-effect threshold.

They concluded that, with the present state of our knowledge, which leaves ample space for uncertainty, 1 ppm (dry weight) of a PAH (typically fluoranthene) in sediments would represent a threshold value between biological effects and no effects.

The specific case of PAHs examined using the results of the above four approaches was presented by Chapman (1986). This author concluded that 2 to 12 ppm (dry weight) of a PAH in sediments would represent a threshold range.

A few more studies have appeared since the publication by Chapman (1986). These include the exposure of flounder to a crude oil in sediments (Payne et al., 1988), demonstrating physiological and biochemical sublethal effects at a concentration of 1 ppm (dry weight, $\sum 18$ PAHs). A study by Casillas et al. (1992) showed impaired growth in sand dollars exposed to environmental sediments containing a variety of contaminants, with PAH concentrations between 0.825 and 2.220 ppm ($\sum 16$ PAHs). An investigation by Stein et al. (1992) of bioindicators in three species of benthic fish exposed to sediments of similar composition to those of Casillas et al.

(1992) reported sublethal effects at very low PAH concentrations (0.061 ppm, $\sum$18 PAHs, wet weight). The above studies illustrate the rate at which our knowledge is increasing and the uncertainty behind sediment quality criteria (discussed by Gray, 1989; Malins, 1989; Long, 1992).

HOW MUCH OF WHAT, WHY?

Can we propose a threshold value for PAHs in vertebrate tissues where biological effects can be predicted? Can we come up with a PAH level in marine molluscs corresponding to the threshold level obtained through studies of sediments' quality criteria? We can answer "no" and a qualified "yes."

Our present knowledge allows us to translate a PAH concentration in sediments to a PAH concentration in water (to a certain degree, we can approximate, by ppm almost becoming ppb, an oversimplification). From recent publications (Payne et al., 1988; Casillas et al., 1992; Stein et al., 1992) using a multidisciplinary approach in the investigation of the effect of contaminants in sediments on marine animals, we can deduce a PAH concentration in sediments, 0.1 to 1 ppm (dry weight, sum of individual PAHs obtained by GC-MS), where biochemical and/or physiological changes are observed in vertebrates and/or invertebrates.

The sediment concentrations in the above section are expressed in terms of the sum of 16 or 18 PAHs. The data presented by Mason (1988) and Clement et al. (1980) concerning the bioaccumulation of PAHs in molluscs are expressed in terms of total PAHs (Table 5, uv/f, sum of biogenic and anthropogenic hydrocarbons). The data from the NST Program are expressed in terms of 12 low-molecular-weight and 12 high-molecular-weight PAHs, and final conclusions are not available yet. We, therefore, have to make a generalization concerning these different quantities. It is recognized that the sum of PAHs in sediments (GC-MS) represents only a fraction of the total PAHs (uv/f). In the case of animal tissues, different sums can represent from 0.1 to 4% of the total, depending on the various reference materials used and the degree of contamination (Boehm and Hirtzer, 1982; Farrington et al., 1986, 1988; Cretney et al., 1987; Hellou et al., 1991; Law and Whinnett, 1992). From these studies and others, we also know that uv/f results can be from 1.5 to more than 10 times more elevated or lower, depending on the choice of standard. In the case of certain oils (Venezuelan crude, crankcase, and no. 2 fuel oil), the sum of 22 or 26 PAHs can represent up to 10% of the total PAHs expressed in terms of chrysene units. The case of sediments, which is of interest presently, will again vary according to the type of contamination and appropriateness of the reference materials used. In the most optimistic estimate, the sum of PAHs could represent 20% (to less than 1%) of the total measured (Prudhoe Bay crude, Clement et al., 1980; Quatar crude, Mason, 1988).

We can, therefore, attempt to transform the above sediment concentration into a concentration of 5 to 0.5 ppb in water (total PAHs) and from the results listed in Table 5 (using power regression with original data from Mason, 1988; using linear regression with data from Clement et al., 1980) into a concentration in mussels (including *Mytilus edulis*) and clams *(Macoma balthica)*. From these latter publications, we extrapolate a bioaccumulation factor of 1720 to 2240 and of 580 to 1070 in mussels and clams, respectively, for the above range of water concentrations (5 to 0.5 ppb). Therefore, we can deduce in our most cautious estimate that a concentration of 8620 to 1120 and of 2900 to 540 ppm of PAH (dry weight, total PAHs, analyzed by uv/f) in mussels and clams, respectively, would represent a threshold value where physiological or biochemical effects would be produced.

Higher bioconcentration factors have been reported by Murray et al. (1991), Pruell et al. (1986), and Burns and Smith (1981) for total, aromatic, or biogenic hydrocarbons, as well as for linear alkyl benzenes or specific PAHs in mussels. Using these values ($\pm 2 \times 10^4$ to 4×10^5, dry weight, relative to water concentrations) would lead to the prediction of higher concentrations in tissues of molluscs.

A few major assumptions have been made in relation to the different standards used in the analysis of PAHs. These are at the basis of the complexity of the PAH question: how to interrelate these various quantities? Another assumption concerns the use of one animal species in a particular test and of a different one in the application of the results. These suppositions constitute major shortcomings in our attempt to answer important questions in a straightforward manner. The results of the NST Program will hopefully clarify the case of the concentration of individual PAHs by relating a PAH level in mussels to a PAH level in sediments. The challenging case of vertebrates is slowly being untangled.

SUMMARY

In this chapter, we have attempted to answer a number of questions related to polycyclic aromatic hydrocarbons, in order to interpret the meaning of a PAH concentration found in tissues of marine animals.

We have presented information concerning the presence of polycyclic aromatic hydrocarbons in the environment and how a PAH fingerprint can be related to its source. We have briefly discussed the various spectroscopic and chromatographic methods used in the quantification of PAHs, summarizing the type of information obtained with each method. The PAH levels found in marine molluscs, finfish, and mammals, in relation to PAH concentrations in the environment, have been reviewed to understand how much PAH can accumulate in marine animals. The environmental and physiological variables thought to affect PAH concentrations in animal tissues have been outlined to help in the interpretation of results from monitoring studies. The recent literature (up to 1992) covering the subject of sediment quality criteria has been examined to assess the threshold level of PAHs in sediments to distinguish between biochemical and/or physiological effects and no effects in marine organisms.

What is the meaning of a PAH concentration obtained from a tissue of molluscs, finfish, or marine mammals? This main question found only a partial, complicated answer.

It is still premature to attempt to relate PAH levels in sediments to PAH levels in vertebrates, because of the restricted amount of information available and the complexity of the subject. Laboratory experiments suggest that, after a single exposure, the concentration of free PAHs persists longer in the muscle tissue of vertebrates. Some of the data regarding PAH concentrations in fish tissues point to a decreased liver/muscle ratio with continuous exposure to elevated levels of PAHs. The lower ratio could reflect the inability of the enzymatic system to efficiently metabolize xenobiotics. There are too many unknowns behind the present observation and, therefore, for the time being, a PAH concentration in a tissue of a vertebrate species cannot be properly deciphered without a combination of other biochemical and physiological analyses. Other fates for PAHs in vertebrates (oxidation and conjugation) have been briefly documented.

The case of invertebrates has found a very partial answer. From recent (1988 to early 1992) multidisciplinary studies, 0.1 to 1 ppm of PAHs in sediments (dry weight, sum of individual PAHs obtained by GC-MS) would correspond to a concentration range where biochemical and/or physiological changes are observed in vertebrates and/or invertebrates. This level tentatively translates into 1120 to 8620 ppm of PAHs (dry weight, total PAHs, uv/f) in mussels *(Mytilus edulis)* and 540 to 2900 ppm in clams *(Macoma balthica)*.

In spite of the uncertain state of our knowledge, progress in the PAH field is being made at a rapid rate. The author believes that laboratory experiments with controlled variables, with a multispectroscopic and multidisciplinary approach, will help to determine when and how the fate and effect of PAHs on biota in the environment would be of concern.

ACKNOWLEDGMENTS

The author thanks Dr. Jerry Payne for introducing her to this field of research and for providing continuous support since the author's arrival at the Department of Fisheries and Oceans, St. John's, Newfoundland.

REFERENCES

Albaiges, J., J. Algaba, P. Arambarri, F. Cabrera, G. Baluja, L. M. Hernandez, and J. Castroviejo. 1987a. Budget of organic and inorganic pollutants in the Donano Natural Park (Spain). Sci. Total Environ. 63:13-28.

Albaiges, J., A. Farran, M. Soler, and A. Gallifa. 1987b. Accumulation and distribution of biogenic and pollutant hydrocarbons, PCBs, and DDT in tissues of western Mediterranean fishes. Mar. Environ. Res. 22:1-18.

Badawy, M. I., and F. Al-Harthy. 1991. Hydrocarbons in sea water, sediments, and oysters from the Omani coastal waters. Bull. Environ. Contam. Toxicol. 47:386-391.

Baek, S. O., R. A. Field, M. E. Goldstone, P. W. Kirk, J. N. Lesterand, and R. Perry. 1991. A review of atmospheric polycyclic aromatic hydrocarbons: sources, fate, and behavior. Water Air Soil Pollut. 60:279-300.

Bartle, K. D. 1985. Recent advances in the analyses of polycyclic aromatic compounds by gas chromatography. p. 193-236. *In* A. Bjorseth and T. Ramdahl (Eds.). Handbook of polycyclic aromatic compounds. Vol. 2. Marcel Dekker, New York.

Baumann, P. C. 1989. PAH metabolites and neoplasia in feral fish populations. p. 269-290. *In* U. Varanasi (Ed.). Metabolism of polycyclic aromatic hydrocarbons in the aquatic environment. CRC Press, Boca Raton, Fla.

Becher, G., and A. Bjorseth. 1985. Determination of occupational exposure to PAH by analysis of body fluids. p. 237-252. *In* A. Bjorseth and T. Ramdahl (Eds.). Handbook of polycyclic aromatic hydrocarbons. Vol. 2. Emission sources and recent progress in analytical chemistry. Marcel Dekker, New York.

Bender, M. E., and R. J. Huggett. 1989. Polynuclear aromatic hydrocarbon residues in shellfish: species variations and apparent intraspecific differences. p. 226-234. *In* H. E. Kaiser (Ed.). Comparative aspects of tumor development. Kluver Academic Publishers, Dordrecht, The Netherlands.

Bjorseth, A. 1983. Handbook of polycyclic aromatic hydrocarbons. Vol. 1. Marcel Dekker, New York, 727 pp.

Bjorseth, A., and T. Ramdahl. 1985. Handbook of polycyclic aromatic hydrocarbons. Vol. 2. Emission sources and recent progress in analytical chemistry. Marcel Dekker, New York, 416 pp.

Boehm, P. D., and P. Hirtzer. 1982. Gulf and Atlantic survey for selected organic pollutants in fish. NOAA (Natl. Oceanic Atmos. Adm.) Tech. Rep. NMFS (Natl. Mar. Fish. Serv.) Circ. NEC-13, 101 pp.

Boehm, P. D., and J. G. Quinn. 1978. Benthic hydrocarbons of Rhode Island Sound. Estuarine Coastal Mar. Sci. 6:471-494.

Boehm, P. H., D. L. Fiest, P. Hirtzer, L. Scott, R. Norstrom, and R. Engelhardt. 1982. A biogeochemical assessment of the BIOS experimental spills: transport, pathways, and fates of petroleum in benthic animals. p. 581-618. *In* Proc. Arct. Mar. Oil Spill Program Tech. Semin., Edmonton, Alberta, Canada, June 15-17, 1982.

Broman, D., and B. Ganning. 1986. Uptake and release of petroleum hydrocarbons by two brackish water bivalves, *Mytilus edulis* and *Macoma balthica* L. Ophelia 25:49-57.

Burns, K. A., and J. L. Smith. 1981. Biological monitoring of ambient water quality: the case for using bivalves as sentinel organisms for monitoring petroleum pollution in coastal waters. Estuarine Coastal Shelf Sci. 13:433-443.

Casillas, E., D. Weber, C. Haley, and S. Sol. 1992. Comparison of growth and mortality in juvenile sand dollars *(Dendraster excenticus)* as indicators of contaminated marine sediments. Environ. Toxicol. Chem. 11:559-569.

Castello, J. J., and J. C. Gamble. 1992. Effects of sewage sludge on marine fish embryo and larvae. Mar. Environ. Res. 33:49-74.

Cavalieri, E. L., E. G. Rogan, P. D. Devanesan, P. Cremonesi, R. L. Cerny, M. L. Gross, and W. J. Bodell. 1990. Binding of benzo[a]pyrene to DNA cytochrome P-450 catalyzed one-electron oxidation in rat liver microsomes and nuclei. Biochemistry 29:4820-4827.

Chapman, P. M. 1986. Sediment quality criteria from the sediment quality triad: an example. Environ. Toxicol. Chem. 5:957-964.

Clement, L. E., M. S. Stekoll, and D. G. Shaw. 1980. Accumulation, fractionation, and release of oil by the intertidal clam *Macoma balthica*. Mar. Biol. (N.Y.) 57:41-50.

Collier, T. K., and U. Varanasi. 1991. Hepatic activities of xenobiotic metabolizing enzymes and biliary levels of xenobiotics in English sole *(Parophrys vetulus)* exposed to environmental contaminants. Arch. Environ. Contam. Toxicol. 20:462-473.

Cooke, M., and A. J. Dennis. 1980. Polynuclear aromatic hydrocarbons. Batelle Press, Columbus, Ohio, 343 pp.

Corner, E. D. S., R. P. Harris, K. J. Whittle, and P. R. Mackie. 1976. Hydrocarbons in marine zooplankton and fish. p. 70-105. *In* A. P. M. Lockwood (Ed.). Effects of pollutants on aquatic organisms. Vol. 2. Society for Environmental Biology Seminar Series, Cambridge University Press, Cambridge, United Kingdom.

Cretney, W. J., D. R. Green, B. R. Fowler, B. Humphrey, F. R. Engelhardt, R. J. Norstrom, M. Simon, D. L. Fiest, and P. D. Boehm. 1987. Hydrocarbon biogeochemical setting of the Baffin Island oil spill experimental sites. III. Biota. Arctic 40 (Suppl. 1):71-79.

Dujmov, J., and P. Sucevic. 1989. Contents of polycyclic aromatic hydrocarbons in the Adriatic Sea determined by uv-fluorescence spectroscopy. Mar. Pollut. Bull. 20:405-409.

Dunn, B. P. 1991. Carcinogen adducts as an indicator for the public health risks of consuming carcinogen-exposed fish and shellfish. Environ. Health Perspect. 90:111-116.

Ehrhardt, M., J. Klungskoyr, and R. J. Law. 1991. Hydrocarbons: review of methods for analysis in sea water, biota, and sediments. Int. Counc. Explor. Sea (ICES) Coop. Res. Rep. No. 12, 47 pp.

El Deeb, K. Z., and E. H. El Ebiary. 1988. Total aromatic hydrocarbon content in the muscle and liver lipid extracts of two seabream fishes from the Arabian Gulf. Arab Gulf J. Sci. Res. Agric. Biol. Sci. 86:139-151.

Engelhardt, F. R., J. R. Geraci, and T. G. Smith. 1977. Uptake and clearance of petroleum hydrocarbons in the ringed seal *Phoca hispida*. J. Fish. Res. Board Can. 34:1143-1147.

Falk-Petersen, I.-B., and E. Kjorsvik. 1989. Histological and ultrastructural effects due to hydrocarbon exposure in larval cod (*Gadus morhua* L.). *In* J. H. C. Blaxter, J. C. Gamble, and H. von Westernhagen (Eds.). The early life history of fish. 3rd ICES Symp., Bergen, October 3-5, 1988. Vol. 191, 493 pp.

Farrington, J. W. 1991. Biogeochemical processes governing exposure and uptake of organic pollutant compounds in aquatic organisms. Environ. Health Perspect. 90:75-84.

Farrington, J. W., A. C. Davis, N. M. Frew, and A. Knap. 1988. ICES/IOC intercomparison exercise on the determination of petroleum hydrocarbons in biological tissues (mussel homogenate). Mar. Pollut. Bull. 19:372-380.

Farrington, J. W., A. C. Davis, J. B. Livramento, C. Hovey Clifford, N. M. Frew, A. Knap, J. F. Uthe, C. J. Musial, and G. R. Sirota. 1986. Reports on the ICES/IOC intercomparison exercise on the determination of petroleum hydrocarbons in biological tissues (2/HC/BT) and the ICES intercomparative study (3/HC/BT) on PAH in biological tissues. Int. Counc. Explor. Sea Coop. Res. Rep. No. 141, 85 pp.

Farrington, J. W., and G. C. Medeiros. 1975. Evaluation of some methods of analysis of petroleum hydrocarbons in marine organisms. p. 115-121. *In* Proc. 1975 Conf. Prev. Control Oil Pollut. American Petroleum Institute, Washington, D.C.

Farrington, J. W., and B. W. Tripp. 1975. A comparison of analysis methods for hydrocarbons in surface sediments. *In* T. M. Church (Ed.). Marine chemistry in the coastal environment. American Chemical Society, Washington, D.C., 710 pp.

Farrington, J. W., S. G. Wakeham, J. B. Livramento, B. W. Tripp, and J. M. Teal. 1986. Aromatic hydrocarbons in New York Bight polychaetes: ultraviolet fluorescence analyses and gas chromatography/gas chromatography-mass spectrometry analyses. Environ. Sci. Technol. 20:69-72.

Fossato, V. U., C. Nasci, and F. Dolci. 1979. 3,4-Benzopyrene and perylene in mussels, *Mytilus* sp., from the Laguna Veneta, North-East Italy. Mar. Environ. Res. 2:47-53.

Friocourt, M., P. G. Bodennec, and F. Berthou. 1985. Determination of polyaromatic hydrocarbons in scallops (*Pecten maximus*) by UV (fluorescence and HPLC combined with UV) and fluorescence detectors. Bull. Environ. Contam. Toxicol. 34:228-238.

Geraci, J. R., and D. J. St. Aubin. 1990. Sea mammals and oil: confronting the risks. Academic Press, San Diego, Calif., 282 pp.

Grahl-Nielsen, O., J. T. Staveland, and S. Wilhelmsen. 1978. Aromatic hydrocarbons in benthic organisms from coastal areas polluted by Iranian crude oil. J. Fish. Res. Board Can. 35:615-623.

Gray, J. S. 1989. Do bioassays adequately predict ecological effects of pollutants? Hydrobiologia 188/189:397-402.

Gronlund, W. D., S. L. Chan, B. B. McCain, R. C. Clark, Jr., M. S. Myers, J. E. Stein, D. W. Brown, J. T. Landahl, M. M. Krahn, and U. Varanasi. 1991. Multidisciplinary assessment of pollution at three sites in Long Island Sound. Estuaries 14:299-305.

Hellou, J., J. F. Payne, and C. Hamilton. 1993. GC-MS analysis of polycyclic aromatic compounds in cod *(Gadus morhua)* from the Northwest Atlantic. Environ. Pollut. 85:197-202.

Hellou, J., G. Stenson, I. H. Ni, and J. F. Payne. 1990. Polycyclic aromatic hydrocarbons in muscle tissue of marine mammals from the Northwest Atlantic. Mar. Pollut. Bull. 21:469-473.

Hellou, J., and C. Upshall. 1992. Of uv/f, biogenic and anthropogenic unsaturated compounds. Int. J. Environ. Anal. Chem. 53:249-259.

Hellou, J., C. Upshall, I. H. Ni, J. F. Payne, and Y. S. Huang. 1991. Polycyclic aromatic hydrocarbons in harp seals *(Phoca groenlandica)* from the Northwest Atlantic. Arch. Environ. Contam. Toxicol. 21:135-140.

James, M. O. 1989. Biotransformation and disposition of PAH in aquatic invertebrates. p. 69-92. *In* U. Varanasi (Ed.). Metabolism of polycyclic aromatic hydrocarbons in the aquatic environment. CRC Press, Boca Raton, Fla.

Joseph, J. D. 1982. Lipid composition of marine and estuarine invertebrates. Part II. Mollusca. Prog. Lipid Res. 21:109-153.

Jovanovich, M. C., and K. R. Marion. 1987. Seasonal variation, uptake, and depuration of anthracene by the brackish water clam *Rangia cuneata*. Mar. Biol. (N.Y.) 95:395-403.

Krahn, M. M., D. G. Burrows, G. M. Yitalo, D. W. Brown, C. A. Wigren, T. K. Collier, S.-L. Chan, and U. Varanasi. 1992. Mass spectrometric analysis for aromatic compounds in bile of fish sampled after the Exxon Valdez oil spill. Environ. Sci. Technol. 26:116-126.

Krone, C. A., J. E. Stein, and U. Varanasi. 1992. Estimation of levels of metabolites of aromatic hydrocarbons in fish tissues by HPLC/fluorescence analysis. Chemosphere 24:497-510.

Kurelec, B., A. Garg, S. Krca, and R. C. Gupta. 1989. DNA adducts as biomarkers in genotoxic risk assessment in the aquatic environment. Mar. Environ. Res. 28:317-321.

Lake, J. L., C. W. Dimock, and C. B. Norwood. 1980. A comparison of methods for the analysis of hydrocarbons in marine sediments. Adv. Chem. Ser. 185:343-360.

Law, R. J. 1989. The fourth round hydrocarbon intercomparison programme: report of progress to April 1989. Int. Counc. Explor. Sea Coop. Res. Rep. E-15, 8 pp.

Law, R. J., and M. D. Nicholson. 1990. The fourth round hydrocarbon intercomparison programme: preliminary assessment of Stage 1. Int. Counc. Explor. Sea (ICES) Coop. Res. Rep. E-14, 14 pp.

Law, R. J., and J. E. Portman. 1982. Report on the first ICES intercomparison exercise on petroleum hydrocarbon analyses in marine samples. Int. Counc. Explor. Sea (ICES) Coop. Res. Rep. 117, 55 pp.

Law, R. J., and J. A. Whinnett. 1992. Polycyclic aromatic hydrocarbons in muscle tissue of harbour porpoises *(Phocoena phocoena)* from UK waters. Mar. Pollut. Bull. 24:550-553.

Lee, R., R. Sauerhebel, and A. A. Benson. 1972. Petroleum hydrocarbons: uptake and discharge by the marine mussel *Mytilus edulis*. Science (Washington, D.C.) 177:344-346.

Liu, T.-Y., S-L. Cheng, T.-H. Ueng, Y.-F. Ueng, and C.-W. Chi. 1991. Comparative analysis of aromatic DNA adducts in fish from polluted and unpolluted areas by the ^{32}P-postlabeling analysis. Bull. Environ. Contam. Toxicol. 47:783-789.

Livingstone, D. R., M. A. Kirchin, and A. Wiseman. 1989. Cytochrome P-450 and oxidative metabolism in molluscs. Xenobiotica 19:1041-1062.

Long, E. R. 1992. Ranges in chemical concentrations in sediments associated with adverse biological effect. Mar. Pollut. Bull. 24:38-45.

Long, E. R., and L. G. Morgan. 1990. The potential for biological effects of sediment-sorbed contaminants tested in the National Status and Trends Program. NOAA (Natl. Oceanic Atmos. Adm.) Tech. Rep. NOS OMA52. National Oceanic and Atmospheric Administration, Seattle, Wash.

MacCubbin, A. E., J. J. Black, and B. P. Dunn. 1990. ^{32}P-Postlabeling detection of DNA adducts in fish from chemically contaminated waterways. Sci. Total Environ. 94:89-104.

Mackie, P. R., R. Hardy, K. J. Whittle, C. Bruce, and A. S. McGill. 1980. p. 379-394. In A. Bjorseth and A. J. Dennis (Eds.). Polynuclear hydrocarbons: chemistry and biological effects. Battelle Press, Columbus, Ohio.

MacLeod, W. D., A. J. Friedman, and D. W. Brown. 1988. Improved interlaboratory comparisons of polycyclic aromatic hydrocarbons in marine sediment. Mar. Environ. Res. 26:209-221.

Malins, D. C. 1977. Effects of petroleum on Arctic and Subarctic marine environments and organisms. Vol. 1. Nature and fate of petroleum. Academic Press, New York, 321 pp.

Malins, D. C. 1989. The use of environmental assays for impact assessment. Hydrobiologia 188/189:87-91.

Malins, D. C., M. M. Krahn, M. S. Myers, L. D. Rhodes, D. W. Brown, C. A. Krone, B. B. McCain, and S-L. Chan. 1985. Toxic chemicals in sediment and biota from a creosote-polluted harbor: relationships with hepatic neoplasms and other hepatic lesions in English sole *(Parophys vetulus)*. Carcinogenesis (Lond.) 6:1463-1469.

Malins, D. C., B. B. McCain, D. W. Brown, S.-L. Chan, M. S. Myers, J. T. Landahl, P. G. Prohaska, A. J. Friedman, L. D. Rhodes, D. G. Burrows, W. D. Gronlund, and H. O. Hodgins. 1984. Chemical pollutants in sediments and diseases of bottom-dwelling fish in Puget Sound, Washington. Environ. Sci. Technol. 18:705-713.

Malins, D. C., B. B. McCain, M. S. Landahl, M. M. Krahn, D. W. Brown, S.-L. Chen, and W. T. Roubal. 1988. Neoplastic and other diseases in fish in relation to toxic chemicals: an overview. Aquat. Toxicol. (N.Y.) 11:43-67.

Martineau, D., A. Lagacé, P. Béland, R. Higgins, D. Armstrong, and L. R. Shugart. 1988. Pathology of stranded beluga whales *(Delphinapterus leucas)* from the St. Lawrence Estuary, Québec, Canada. J. Comp. Pathol. 98:287-311.

Mason, R. P. 1987. A comparison of fluorescence and GC for the determination of petroleum hydrocarbons in mussels. Mar. Pollut. Bull. 18:528-533.

Mason, R. P. 1988. Hydrocarbons in mussels around the Cape Peninsula, South Africa. S. Afr. J. Mar. Sci. 7:139-151.

McElroy, A. E., J. W. Farrington, and J. M. Teal. 1989. Bioavailability of polyaromatic hydrocarbons in the aquatic environment. p. 1-40. *In* U. Varanasi (Ed.). Metabolism of polycyclic aromatic hydrocarbons in the aquatic environment. CRC Press, Boca Raton, Fla.

Melancon, M. J., and J. J. Lech. 1978. Distribution and elimination of naphthalene and 2-methylnaphthalene in rainbow trout during short- and long-term exposures. Arch. Environ. Contam. Toxicol. 7:207-220.

Mix, M. C. 1984. Polycyclic aromatic hydrocarbons in the aquatic environment: occurrence and biological monitoring. Rev. Environ. Toxicol. 50:51-102.

Mix, M. C., and R. L. Schaffer. 1983. Concentrations of unsubstituted polynuclear aromatic hydrocarbons in Bay mussels *(Mytilus edulis)* from Oregon, USA. Mar. Environ. Res. 9:193-209.

Moore, M. N., D. R. Livingstone, and J. Widdons. 1989. Hydrocarbons in marine molluscs: biological effects and ecological consequences. p. 291-328. *In* U. Varanasi (Ed.). Metabolism of polycyclic aromatic hydrocarbons in the aquatic environment. CRC Press, Boca Raton, Fla.

Murray, A. P., B. J. Richardson, and C. F. Gibbs. 1991. Bioconcentration factors for petroleum hydrocarbons, PAHs, LABs, and biogenic hydrocarbons in blue mussel. Mar. Pollut. Bull. 22:595-603.

Nasci, C., and V. U. Fossato. 1982. Studies on physiology of mussels and their ability in accumulating hydrocarbons and chlorinated hydrocarbons. Environ. Technol. Lett. 3:273-280.

National Oceanic and Atmospheric Administration (NOAA). 1985. National estuarine inventory data atlas. *In* Vol. 1. Physical and hydrologic characteristics. National Oceanic and Atmospheric Administration, Rockville, Md., 103 pp.

National Oceanic and Atmospheric Administration (NOAA). 1988. National Status and Trends Program for Marine Environmental Quality: progress report. A summary of selected data on chemical contaminants in sediments collected during 1984, 1985, 1986, and 1987. NOAA (Natl. Oceanic Atmos. Adm.) Tech. Rep. NOS OMA 44. National Oceanic and Atmospheric Administration, Rockville, Md., 15 pp. (plus Appendices A-D).

National Oceanic and Atmospheric Administration (NOAA). 1989. National Status and Trends Program for Marine Environmental Quality; progress report. A summary of data on tissue contamination from the first three years (1986-1988) of the Mussel Watch Project. NOAA (Natl. Oceanic Atmos. Adm.) Tech. Rep. NOS OMA 49. National Oceanic and Atmospheric Administration, Rockville, Md., 22 pp. (plus Appendices A-C).

Nava, M. E., and F. R. Engelhardt. 1980. Compartmentalization of ingested labelled petroleum in tissues and bile of the American eel *(Anguilla rostrata)*. Bull. Environ. Contam. Toxicol. 24:879-885.

Neff, J. J. 1990. Composition and fate of petroleum and spill treating agents in the marine environment. p. 1-33. *In* J. R. Geraci and D. J. St. Aubin (Eds.). Sea mammals and oil: confronting the risks. Academic Press, San Diego, Calif.

Neff, J. M. 1979. Polycyclic aromatic hydrocarbons in the aquatic environment: sources, fates, and biological effects. Applied Science Publishers, London, 262 pp.

O'Connor, T. P. 1991. Concentrations of organic contaminants in molluscs and sediments at NOAA National Status and Trends sites in the coastal and estuarine United States. Environ. Health Perspect. 90:69-73.

Pahlmann, R., and O. Pelkonen. 1987. Mutagenicity studies of different polycyclic aromatic hydrocarbons. The significance of enzymatic factors and molecular structures. Carcinogenesis (Lond.) 8:773-778.

Payne, J. F., J. Kiceniuk, L. L. Fancey, U. Williams, G. L. Fletcher, A. Rahimtula, and B. Fowler. 1988. What is a safe level of polycyclic aromatic hydrocarbons for fish: subchronic toxicity study on winter flounder *(Pseudopleuronectes americanus)*. Can. J. Fish. Aquat. Sci. 45:1983-1993.

Pruell, R. J., J. L. Lake, W. R. Davis, and J. G. Quinn. 1986. Uptake and depuration of organic contaminants by blue mussels *(Mytilus edulis)* exposed to environmentally contaminated sediment. Mar. Biol. (N.Y.) 91:497-507.

Rainio, K., R. R. Linko, and L. Ruotsila. 1986. Polycyclic aromatic hydrocarbons in mussel and fish from the Finnish Archipelago Sea. Bull. Environ. Contam. Toxicol. 37:337-343.

Ray, S., B. P. Dunn, J. F. Payne, L. Fancey, R. Helbig, and P. Béland. 1992. Aromatic DNA-carcinogen adducts in beluga whales from the Canadian Arctic and Gulf of St. Lawrence. Mar. Pollut. Bull. 22:329-396.

Robertson, A., B. W. Gottholm, D. D. Turgeon, and D. A. Wolfe. 1991. A comparative study of contaminant levels in Long Island Sound. Estuaries 14:290-298.

Salihoglu, I., C. Saydam, and A. Yilmaz. 1987. Long term impact of dissolved dispersed petroleum hydrocarbons (DDPH) in Gulf of Iskenderun. Chemosphere 16:381-394.

Solbakken, J. E., and K. H. Palmork. 1980. Distribution of radioactivity in the Chondrichthyes *Squalus acanthias* and the Osteichthyes *Salmo gairdneri* following intragastric administration of (9-^{14}C)-phenanthrene. Bull. Environ. Contam. Toxicol. 25:902-908.

Solbakken, J. E., M. Solberg, and K. H. Palmork. 1983. A comparative study on the disposition of three aromatic hydrocarbons in flounder *(Platichthys flesus)*. Fiskedir. Skr. Ser. Havunders. 17:473-481.

Solbakken, J. E., S. Tilseth, and K. H. Palmork. 1984. Uptake and elimination of aromatic hydrocarbons and chlorinated biphenyl in eggs and larvae of cod *Gadus morhua*. Mar. Ecol. 16:297-301.

Stegeman, J. J., and J. J. Lech. 1991. Cytochrome P-450 monooxygenase systems in aquatic species: carcinogen metabolism and biomarkers for carcinogen and pollutant exposure. Environ. Health Perspect. 90:101-109.

Stegeman, J. J., and J. M. Teal. 1973. Accumulation, release, and retention of petroleum hydrocarbons by the oyster *Crassostrea virginica*. Mar. Biol. (N.Y.) 22:37-44.

Steimle, F. W., V. S. Zdauowicz, and D. F. Gadbois. 1990. Metals and organic contaminants in Northwest Atlantic deep-sea tilefish tissues. Mar. Pollut. Bull. 21:530-535.

Stein, J. E., T. K. Collier, W. L. Reichert, E. Casillas, T. Hom, and U. Varanasi. 1992. Bioindicators of contaminant exposure and sublethal effects: studies with benthic fish in Puget Sound. Environ. Toxicol. Chem. 11:701-714.

Stein, J. E., T. Hom, and U. Varanasi. 1984. Simultaneous exposure of English sole *(Parophrys vetulus)* to sediment-associated xenobiotics. Part 1. Uptake and disposition of ^{14}C-polychlorinated biphenyls and ^{3}H-benzo[a]pyrene. Mar. Environ. Res. 13:97-119.

Stein, J. E., W. L. Reichert, M. Nishimoto, and U. Varanasi. 1990. Overview of studies on liver carcinogenesis in English sole from Puget Sound: evidence for xenobiotic chemical etiology. II. Biochemical studies. Sci. Total Environ. 94:51-69.

Tanacredi, J. T., and R. R. Cardenas. 1991. Biodepuration of polynuclear aromatic hydrocarbons from a bivalve mollusc, *Mercenaria merceneria* L. Environ. Sci. Technol. 25:1453-1461.

Varanasi, U. 1989. Metabolism of polycyclic aromatic hydrocarbons in the aquatic environment. CRC Press, Boca Raton, Fla., 341 pp.

Varanasi, U., and J. E. Stein. 1991. Disposition of xenobiotic chemicals and metabolites in marine organisms. Environ. Health Perspect. 90:93-100.

Varanasi, U., J. E. Stein, and M. Nishimoto. 1989. Biotransformation and disposition of polycyclic aromatic hydrocarbons in fish. p. 93-150. *In* U. Varanasi (Ed.). Metabolism of polycyclic aromatic hydrocarbons in the aquatic environment. CRC Press, Boca Raton, Fla.

Vo-Dinh, T. 1981. Synchronous excitation spectroscopy. p. 167-192. *In* E. L. Wehry (Ed.). Modern fluorescence spectroscopy. Vol. 4. Plenum Press, New York.

White, K. L. 1986. An overview of immunotoxicology and carcinogenic polycyclic aromatic hydrocarbons. Environ. Carcinogen. Rev. 4:163-202.

Wise, S. A. 1985. Recent progress in the determination of PAH by high performance liquid chromatography. p. 113-192. *In* A. Bjorseth and T. Ramdahl (Eds.). Handbook of polycyclic aromatic compounds. Vol. 2. Marcel Dekker, New York.

CHAPTER 10

Lead in Waterfowl

Deborah J. Pain

INTRODUCTION

Lead is a highly toxic heavy metal that acts as a nonspecific poison affecting all body systems. There is no known biological requirement for lead. Absorption of low concentrations may result in a wide range of sublethal effects in animals, and higher concentrations may result in mortality (Demayo et al., 1982).

Lead has been mined and smelted by humans for centuries, but the use of lead-based products increased greatly following the Industrial Revolution. Consequently, lead today is ubiquitous in air, water, and soil, in both urban and rural environments, in far above background concentrations. Terrestrial vertebrates are exposed to lead mainly via inhalation and ingestion. A proportion of lead entering the body will be absorbed into the bloodstream and will subsequently become distributed among body tissues, primarily the blood, liver, kidney, and bone. As a result of anthropogenic activities, most animals have higher tissue lead concentrations than in preindustrialized times. Although even very low tissue lead concentrations have some measurable physiological effects, the kinds of concentrations usually encountered in the wider environment (i.e., distant from lead emission sources) have not generally been considered to directly affect survival in most forms of wildlife.

However, waterfowl and certain other avian species may be exposed to large amounts of lead in a very particular way through the ingestion of spent lead gunshot. Gunshot, ingested by waterfowl, presumably as grit or food particles, has resulted in widespread waterfowl mortality in the United States, Europe, and elsewhere for over a century (Bellrose, 1959; Sanderson and Bellrose, 1986; Pain, 1992). A similar problem has arisen in some areas (primarily Britain) when swans (*Cygnus* spp.) and other waterfowl have ingested anglers' lead weights (Sears, 1988).

This focus of this chapter is the interpretation of tissue lead concentrations resulting from exposure to lead shot, although conclusions relate equally to the ingestion of anglers' lead weights.

Research into the lead poisoning of waterfowl, under both experimental and field conditions, has been conducted since the 1950s, when Bellrose (1959) completed the first comprehensive study of this condition. This chapter draws upon the extensive research carried out and covers the distribution of lead within the body, factors that influence tissue lead concentrations, and the effects of lead poisoning. The concentrations of lead found in unexposed and lead-poisoned birds are described, along with the threshold tissue concentrations that have previously been considered to indicate excessive lead absorption or poisoning.

The chapter concludes with suggested thresholds for tissue lead concentrations indicative of different levels of exposure and poisoning and discusses their interpretation and the limitations associated with their use at both individual and population levels.

A NOTE ON TERMINOLOGY

To avoid confusion, the terminology used in this chapter is defined as follows.

1. Natural environmental concentrations no longer exist, because lead resulting from anthropogenic emissions is ubiquitous. Concentrations in the wider environment far from lead emission sources have consequently been described as "background" and exposure to such concentrations as "background exposure." "Elevated exposure" is that resulting from exposure to above-background concentrations.
2. Exposure to lead has been defined as "acute" (exposure to high lead concentrations over a short period of time) or "chronic" (sustained exposure to lower lead concentrations).
3. The lowest measurable lead concentrations affect biological systems. Consequently, "no effect" tissue concentrations cannot be defined. The terminology used in this chapter to help interpret the significance of lead concentrations in waterfowl tissues is as follows. "Subclinical poisoning" suggests that physiological effects are present but not sufficient to severely impair normal biological functioning, and there are no external signs of poisoning. "Clinical poisoning" suggests that normal biological functions may be impaired and that external signs of poisoning may be present. "Severe clinical poisoning" suggests that the effects may be directly life threatening.

LEAD DISTRIBUTION AMONG WATERFOWL TISSUES

Ingested shot may be retained along with grit in the muscular gizzard of waterfowl and mechanically ground down. The acidic conditions (pH 2.5) in waterfowl stomachs facilitate the dissolution of lead, and the resultant toxic lead salts are absorbed into the bloodstream. Shot may be evacuated immediately following ingestion, thus resulting in little or no absorption of lead, or retained until it is completely eroded and absorbed, approximately 6 weeks post ingestion. The average retention times are 18 to 21 days (Jordan and Bellrose, 1951). Some of the lead in the bloodstream is deposited rapidly into soft tissues, primarily the liver and kidney, and into bone. Lead in blood and soft tissues retains a fairly mobile equilibrium and usually remains elevated from several weeks to several months following exposure (when exposure is not prolonged), in relation to the initial amount absorbed. Lead in bone is relatively immobile, loss is very slow, and lead accumulates in bone throughout the lifetime. Excepting acute lead exposure, bone lead concentrations are positively correlated with age in waterfowl populations (Stendell et al., 1979; Clausen et al., 1982).

Table 1 illustrates the distribution of lead in different tissues in an experiment in which drake mallard ducks *(Anas platyrhynchos)* were given doses of one number four lead shot and subsequently died (Longcore et al., 1974a). Tissue lead concentrations increased over time (shot was retained) and were highest in the tibia and kidney, followed by the liver and blood.

The distribution of lead among tissues may be different between females and males during the prebreeding and breeding seasons, with females depositing comparatively less lead in the liver and more in high medullary content bones than males (Finley and Dieter, 1978; Rocke and Samuel, 1991). The storage of higher proportions of lead in bone prior to and during breeding by females could result in sex differences in the toxic effects of exposure to a similar amount of lead during this period; for example, Rocke and Samuel (1991) noted an increased immunosuppressive effect of lead in males compared with females during the prebreeding season.

The chronicity of exposure to lead has an important influence upon tissue lead concentrations and effects. Most waterfowl that die of lead poisoning die in small numbers following chronic exposure, several weeks after the ingestion of a small number of shot. Under these conditions, the highest lead concentrations are often found in bone, followed by those in the kidney, liver, and blood (Mautino and Bell, 1986, 1987). However, less usually, waterfowl die following acute exposure shortly after the ingestion of many shot and the absorption of very high lead concentrations. Under these circumstances, liver and kidney lead concentrations may exceed those in bone. In a die-off of lesser scaup *(Aythya affinis)*, acute lead exposure and rapid poisoning were suspected as body weights were positively correlated with the number of shot in the gizzard, indicating that birds died before the effects of lead poisoning resulted in weight loss (Anderson, 1975). Tissue lead concentrations were consistent with this hypothesis, with mean lead concentrations being 46 ppm of wet weight (approximately 160 ppm of dry weight) in the liver, 66 ppm of wet weight (approximately 230 ppm of dry weight) in the kidney, and 40 ppm of dry weight in the bone.

However, as lead is lost far more slowly from bone than from soft tissues, few birds sampled from wild populations have higher lead concentrations in soft tissues than bone. Clausen et al. (1982) found only 7% of mute swans *(Cygnus olor)* that died of lead poisoning with high liver but low bone lead concentrations, whereas high bone and low liver lead concentrations occur relatively frequently (31% of birds sampled; Clausen and Wolstrup, 1979).

The bone lead concentration is the least useful indicator of recent lead exposure and absorption. The tissues usually chosen are blood, liver, and occasionally kidney.

Table 1 Lead Concentrations in Selected Tissues of Male Game Farm Mallard that Died of Lead Poisoning Following Experimental Shot Ingestion[a]

Tissue	Lead Concentrations (ppm wet weight)	
	Range	Arithmetic mean
Tibia		
Group I	87-167	112
Group II	102-191	140
Group III	130-209	157
Kidney		
Group I	53-299	158
Group II	69-205	124
Group III	98-408	259
Liver		
Group I	32-63	51
Group II	43-83	58
Group III	37-72	64
Blood		
Group I	4-20	12
Group II	5-18	9
Group III	3-18	9
Brain		
Group I	2-5	3
Group II	2-6	4
Group III	4-11	6
Breast muscle		
Group I	0.7-1.1	0.9
Group II	0.6-3.2	1.4
Group III	1.3-2.5	1.8

[a] Group I, 5-8 days after shot ingestion ($n = 6$); Group II, 10-14 days after shot ingestion ($n = 6$); Group III, 16-21 days after shot ingestion ($n = 6$).

From Longcore, J. R., R. Andrews, L. N. Locke, G. E. Bagley, and L. T. Young. 1974. Toxicity of lead and proposed substitute shot to mallards. U.S. Fish Wildl. Serv. Spec. Sci. Rep. No. 183.

FACTORS INFLUENCING CONCENTRATIONS OF LEAD IN TISSUES

It is often difficult to directly relate exposure to lead (i.e., shot ingestion) to tissue lead concentrations, as many factors influence shot retention (e.g., the amount and characteristics of food and grit ingested) and lead absorption and deposition into body tissues. The influence of level and duration of exposure (chronic or acute) to shot upon tissue lead concentrations has already been discussed. Absorption of lead and deposition within body tissues are influenced by many factors including age, sex, and diet, as discussed below.

Finley et al. (1976a) found that laying mallards dosed with one shot (196 mg) accumulated significantly higher liver, kidney, and bone lead concentrations than did males. While lead concentrations remained low (means of <2 ppm of wet weight) in soft tissues, mean bone lead concentrations were 112 ppm of dry weight in females compared with 10 ppm of dry weight in males. Subsequent studies recorded a similar result and additionally found femur lead concentrations to be 4 times higher in laying than nonlaying females (Finley and Dieter, 1978). This effect is thought to be a competitive lead/calcium effect during periods of active calcium metabolism (such as egg laying). This is supported by evidence from studies of wild birds. Anderson (1975) reported significantly higher lead concentrations in the wing bones of immature than adult lesser scaup that died of lead poisoning. As bone lead concentrations generally increase with age, the above result suggests either an increased resistance to lead poisoning or a faster deposition rate of lead into the bones of immature ducks, possibly while calcium deposition into bones is very rapid. Similar results have been found under experimental conditions (Elder, 1954). The age, sex, and physiological condition appear to be important factors influencing the absorption and/or deposition of lead within the body, particularly into bone. However, there is little evidence of differential tissue lead accumulation rates between the sexes outside the breeding season.

It appears that lead is deposited to different extents in different types of bone. Finley and Dieter (1978) recorded 2 to 3 times more lead accumulation in bones of high medullary content (femur and sternum) than in those with low medullary content (ulna-radius or wing bones) from lead shot-dosed mallards. Medullary bone is produced almost exclusively by females (Simkiss, 1961).

The single most important factor influencing lead absorption and deposition into tissues is probably diet, and both the quantity and quality (chemical and physical) of diet are important (Jordan and Bellrose, 1951; Longcore et al., 1974a; Sanderson and Irwin, 1976; Koranda et al., 1979; Sanderson and Bellrose, 1986). Koranda et al. (1979) dosed mallards with one, three, or six shots (approximately 220 mg each, $n = 8$ for each treatment). In each dosage group, half of the birds were kept on a diet rich in calcium, phosphorus, and protein, and half were kept on a low-nutrient diet. Mallards on the high-nutrient diet were sacrificed after 36 days. Liver lead concentrations were 3 to 18 ppm of dry weight for birds fed one shot, 1 to 15 ppm of dry weight for those fed three shots, and 7 to 24 ppm of dry weight in the six-shot group. Nineteen days after dosage (or at death if sooner), birds on the low-nutrient diet had liver lead concentrations of 4 to 186, 132 to 219, and 79 to 290 ppm of dry weight, respectively, at the one-, three-, and six-shot dosages. Although the effects of diet are widely recognized, the exact mechanisms by which they occur have not been fully explained. Much of the protective effect of nutrient-rich diets appears to occur in the digestive system (Sanderson, 1992) and, when lead is ingested along with food, certain chemical groups in food components have a ligand effect, binding lead in a nonsoluble form in the intestine (Morton et al., 1985).

These examples serve to illustrate that tissue lead concentrations are not always directly related to exposure.

TISSUE LEAD CONCENTRATIONS AND SUBLETHAL EFFECTS

Tissue lead "threshold" concentrations for lead poisoning can be defined according to the tissue lead concentrations at which measurable effects occur. Lead absorption may result in a range of sublethal effects or in mortality. Lead acts at the molecular level affecting the hematological, muscular, behavioral, nervous, and reproductive systems. The signs and effects of lead poisoning in waterfowl have been described by many authors (Forbes and Sanderson, 1978; Sanderson and Bellrose, 1986; Friend, 1987; Eisler, 1988) and will not be covered in great detail here.

Because of the dynamics of lead uptake and retention, lead concentrations in blood and soft tissues are more easily related to effects than are bone lead concentrations. Waterfowl frequently accumulate significant lead concentrations in bone with no apparent signs of lead toxicity (Mautino and Bell, 1987). The effects of blood lead upon hematological parameters have been extensively studied in waterfowl. Blood lead studies provide useful information, as blood lead concentrations can be related to effect over time, with the extent and duration of lead exposure experimentally controlled.

Lead inhibits the activities of several enzymes necessary for the synthesis of heme, e.g., delta-aminolevulinic acid dehydratase (ALAD) and heme synthetase. Heme is incorporated into hemoglobin and mitochondrial cytochromes and is part of cytochrome P-450 that is required in the liver for certain detoxification processes (Sassa et al., 1975; Dieter and Finley, 1979). Heme synthetase is responsible for the incorporation of ferrous iron into protoporphyrin IX (PPIX) at the last stage of heme formation. Inhibition of heme synthetase activity results in an accumulation of PPIX in the blood. PPIX fluoresces when exposed to specific wavelengths, and this fluorescence has been used for the quantitative estimation of the PPIX concentration in waterfowl blood and, consequently, as an indicator of lead contamination (Roscoe et al., 1979; O'Halloran et al., 1988b). However, the first measurable biochemical change resulting from lead absorption appears to be the inhibition of erythrocyte ALAD activity (Hernberg et al., 1970; Tola et al., 1973). The inhibition of erythrocyte ALAD activity by blood lead is described in some detail below and illustrates the difficulty of interpreting the biological significance of sublethal effects, even when they can be directly related to tissue lead concentrations.

Inhibition of erythrocyte ALAD activity persists for several weeks to several months following lead absorption, in relation to blood lead concentrations (Finley et al., 1976b; Dieter and Finley, 1978; Pain, 1987). Inhibition of ALAD activity in mallard blood has been reported at blood lead concentrations of <5 µg/dl (Pain, 1989). An inhibition in ALAD activity of 50% has been associated with a blood lead concentration of 15 to 20 µg/dl (Finley et al., 1976b, 1976c), and a blood lead concentration of approximately 100 µg/dl has resulted in 75 to 95% inhibition in ALAD activity (Dieter and Finley, 1978; Pain, 1989).

Waterfowl appear to be able to tolerate a certain reduction in erythrocyte ALAD activity without showing signs of reduced hematocrit or hemoglobin concentration (anemia). Anemia may occur following severe (e.g., >75%) inhibition in ALAD activity in cases of acute exposure and poisoning (Pain and Rattner, 1988) or lower level inhibition sustained over a long period. However, although the significance of reduced erythrocyte ALAD activity is not always easily determined, ALAD activity is also inhibited in other body tissues. One month after administering doses of one lead shot to mallards, Dieter and Finley (1979) recorded a 75% reduction in erythrocyte ALAD activity (blood lead, 98 µg/dl), a 42% inhibition in liver ALAD activity (liver lead, 2.24 ppm of wet weight), a 50% reduction in ALAD activity in the cerebellum, and a 35% reduction in the cerebral hemisphere (brain lead, 0.43 ppm of wet weight). ALAD activity was correlated with the lead concentration in all tissues but was more sensitive to lead in the brain than in the liver, where some lead may possibly be bound in a biologically inactive form. The authors also recorded a significant increase in butylcholinesterase (a marker enzyme for glial or supportive cells) activity in the brain, suggesting possible brain damage. The results of this study suggest that blood lead concentrations of 100 µg/dl may be associated with pathological changes in waterfowl brains. However, the examination of the effects of lead upon the central nervous system (CNS) has been largely neglected in birds, and histological lesions found in the CNS have been minor (Wobester, 1981).

It is important to note that for lead, as for many other contaminants, toxic effects may depend upon factors other than just the concentrations in tissues. These factors include the form in which the contaminant is stored and the level and duration of exposure. The biological significance of a tissue lead concentration may, therefore, be very difficult to determine if the history of the bird is not known. The chronicity of exposure is particularly important. Birds exposed to a relatively low lead level on a sustained basis may suffer similar effects, but with lower tissue lead concentrations, to birds acutely exposed to lead for a short period of time.

ESTABLISHMENT OF THRESHOLD LEAD CONCENTRATIONS IN WATERFOWL TISSUES

As discussed in the last two sections, there are difficulties associated with relating exposure to tissue lead concentrations and with relating tissue lead concentrations to effect. However, for wildlife management and the establishment of emission standards for contaminants, it is important to be able to give simple information concerning the maximally acceptable levels of contaminants in wildlife. These are based upon the tissue lead concentrations in unexposed wild birds and the concentrations at which sublethal effects and mortality may occur.

Blood lead concentrations in unexposed waterfowl are generally very low, usually <30 µg/dl and frequently <10 µg/dl (Dieter et al., 1976; Birkhead, 1983; Pain, 1989). Waterfowl with no history of lead poisoning usually have liver lead concentrations of <2 ppm of wet weight and frequently of <1 ppm of wet weight (Bagley and Locke, 1967; Irwin, 1975; Clausen and Wolstrup, 1979; Spray and Milne, 1988; Kingsford et al., 1989). Bone lead concentrations are more difficult to interpret, and concentrations tend to be higher because of accumulation. Different values for background concentrations have been proposed, but concentrations of up to 10 to 20 ppm of dry weight are generally considered as background (Longcore et al., 1974b; Moore, 1978; Szymczak and Adrian, 1978). The liver is the tissue most frequently analyzed from dead birds in suspected cases of lead exposure. Table 2 illustrates liver lead concentrations recorded from lead-poisoned waterfowl under both experimental conditions and in the wild.

Table 2 Lead Concentrations in the Livers of Lead-poisoned Waterfowl from Representative Studies

Species and Conditions	Range	Mean Pb concentrations (ppm wet weight)[a]	Ref.
Experimental			
Summary of seven studies (1994-1971)[b]	5-80	12-51	Longcore (1974a)
Mallard (Anas platyrhynchos) male birds, n = 18	32-83	51-64	Longcore (1974a)
Free-ranging birds			
Summary of seven die-offs (1944-1967)[b]	9-53	12-28	Longcore (1974a)
Lesser scaup (Aythya affinis)		46 (±4)	Anderson (1975)
Mallard, n = 52		41	Clausen and Wolstrup (1979)
Mute swans (Cygnus olor, n = 69)		33	Clausen and Wolstrup (1979)
Four species, two geese, two duck, n = 995	43-137[c]	15-31	Zwank et al. (1985)
Swans (Cygnus spp., n = 5)	42-145	87	O'Halloran et al. (1988b)

[a] Wet weight values have been presented, as is common. However, dry weight values for soft tissue analysis are more reliable and consistent (Adrian and Stevens, 1979). For comparative purposes, 1 ppm wet weight = approximately 3 to 4 ppm dry weight.
[b] Because of inadequate sensitivity and contamination problems, tissue lead determinations were less precise 20 years ago. Values from early studies may therefore be elevated.
[c] Maximum values.

Tissue concentrations of lead used by different authors as thresholds to indicate abnormal exposure and absorption and to determine toxic effect levels are presented in Table 3. Although within the same range, different threshold concentrations have been proposed by different authors and have often been chosen to reflect slightly different things, with no standard definitions.

RECOMMENDATIONS FOR THE INTERPRETATION OF TISSUE LEAD CONCENTRATIONS IN WATERFOWL

The interpretative difficulties described and the different lead threshold levels previously used suggest that ranges of tissue lead concentrations might be more appropriate to evaluate the relative levels of contamination. Tissue lead concentrations

Table 3 Tissue Lead Concentrations Used as Thresholds to Indicate Lead Exposure and Poisoning in Waterfowl

Tissue lead	Threshold concentration	Ref.
Blood (µg/dl)	>20 (elevated)	Friend (1985); U.S. Fish and Wildlife Service (1986)
	>40 (indicative of poisoning)	Cook and Trainer (1966); Birkhead (1983); Spray and Milne (1988)
	>50 (acute toxicity)	Mauser et al. (1990)
Liver (ppm of wet weight)	>2 (elevated)	Friend (1985); U.S. Fish and Wildlife Service (1986)
	>6 (indicative of poisoning)	Clausen et al. (1982)
	>6-20 (acute exposure and absorption)	Longcore et al. (1974b)
	>7 (poisoning)	Clausen and Wolstrup (1979)
	>8 (consistent with signs of poisoning)	Friend (1985)
	>12.5 (poisoning)	Clarke and Clarke (1975); O'Halloran et al. (1988b)
Bone lead (ppm of dry weight)	>20 (excessive exposure and absorption)	White and Stendell (1977); Stendell et al. (1979)

Table 4 Suggested Interpretations of Tissue Lead Concentrations in Waterfowl

Tissue	Background	Subclinical poisoning	Clinical poisoning	Severe clinical poisoning
Blood (µg/dl)	<20	20 < 50	50-100	>100
Liver (ppm of wet weight)	<2	2 < 6	6-15	>15
Bone (ppm of dry weight)	<10	10-20		>20

corresponding to background contamination, subclinical poisoning, clinical poisoning, and severe clinical poisoning are presented in Table 4. These values provide broad guidelines. Appropriate uses and limitations are discussed below.

INDIVIDUAL BIRDS

It is often difficult to interpret lead concentrations in tissues from individual birds. A bird that has been chronically exposed to lead may die of lead poisoning with far lower blood and liver lead concentrations than a bird that has died following acute exposure. Sequential blood lead analyses from an individual bird give a much clearer picture of the significance of contamination as chronicity can be established. In addition, hematological measurements such as ALAD activity, protoporphyrin IX concentration, hematocrit, and hemoglobin concentrations will indicate biochemical damage.

Single bone lead values are the least useful indices of poisoning due to accumulation. However, high liver and low bone lead concentrations may suggest severe clinical poisoning following acute exposure.

While tissue lead concentrations are the single most reliable diagnostic feature of lead poisoning in waterfowl, a variety of clinical signs and pathological changes are associated with poisoning and provide supporting evidence in the diagnosis of lead poisoning. These have been described in detail by Friend (1987) and will only be briefly summarized here.

External signs of lead poisoning following chronic exposure include emaciation, lethargy, wing and tail droop, green-stained vent, and green diarrhea. Gross lesions observed at necropsy include the following: wasting of breast muscles, reduced amounts of visceral fat, impactions of the esophagus or proventriculus, distended gallbladder, dark discolored gizzard lining, wasting of internal organs, pale flabby heart, and pale internal organs and muscle tissue. These, and the presence of worn disk-like lead fragments within the gizzard, provide presumptive evidence of lead poisoning. However, these lesions alone are not diagnostic, and definitive diagnosis requires additional toxicological evidence (usually tissue lead analysis). In cases where birds die rapidly following acute exposure to lead, many of these lesions may be absent.

POPULATIONS

Although less useful than other tissues in individual cases, bone lead concentrations have been used to determine geographical patterns of lead poisoning in populations (Stendell et al., 1979). In addition, at a population level, bone and liver lead concentrations may be correlated (Anderson, 1975), and tissue lead concentrations may be correlated with exposure, as measured by the presence of shot in gizzards (White and Stendell, 1977). Pain et al. (1992) found a positive correlation between exposure to shot (measured as the percentage of shot ingestion in eight species of waterfowl and shore birds) and liver and bone lead concentrations. Exposure and tissue lead concentrations are not always associated in individual birds because of the varying retention time of shot in the gizzard and the uptake/retention dynamics of lead in tissues.

Although most cases of lead poisoning in waterfowl result from shot ingestion, tissue levels will sometimes become elevated because of exposure via other pathways. This may occur if birds are occupying or feeding in an area near a point emission source, such as a lead-mining area (Chupp and Dalke, 1964; Blus et al., 1991), or near an industrial effluent outfall (Bull et al., 1983). Although tissue lead concentrations give no direct information concerning the source of contamination, an examination of the distribution of blood or liver lead concentrations can indicate the type of exposure. Distributions of blood and soft tissue lead concentrations from populations in which some birds have ingested shot tend to be very skewed or have distinct outliers (Dieter, 1979). Distributions are likely to be closer to normality from populations exposed to a more general source of lead, e.g., in water or the atmosphere. Bone lead distributions are more difficult to interpret as population age (and sex) structures may influence results.

SUMMARY

Lead is a nonessential, highly toxic heavy metal that affects all body systems. Waterfowl are directly exposed to high lead concentrations through the ingestion of spent lead gunshot. This has resulted in widespread waterfowl mortality in the U.S., Europe, and elsewhere for more than a century.

Following absorption into the bloodstream, lead becomes deposited into soft tissues (primarily liver and kidney) and bone. Lead has different retention times in these tissues, with soft tissue lead concentrations reflecting recent absorption and bone lead concentrations reflecting long-term absorption. Many factors influence lead absorption and distribution within the body, including age, sex, physiological condition, diet, and exposure level and duration.

The first measurable effects of lead are in the bloodstream, where the activities of heme-biosynthetic enzymes are inhibited. There does not appear to be a no-effect level for lead, and the activities of certain enzymes are inhibited at blood lead concentrations of <5 µg/dl. The biological significance of tissue lead concentrations is related to the exposure regimen and duration, as well as to the tissue and form in which the lead is deposited.

A range of tissue lead concentrations have been used as threshold indicators for elevated lead absorption and lead poisoning. The suggested threshold limits for blood and liver lead concentrations, respectively, are as follows: <20 µg/dl, <2 ppm of wet weight (background); 20 to <50 µg/dl, 2 to <6 ppm of wet weight (subclinical poisoning); 50 to 100 µg/dl, 6 to 15 ppm of wet weight (clinical poisoning); and >100 µg/dl, >15 to 20 ppm of wet weight (severe clinical poisoning). Bone lead concentrations of >20 ppm of dry weight are considered to suggest excessive exposure.

Because of the rapid uptake and slow release of lead from bone, bone lead concentrations are the least useful index of recent lead poisoning, although bone lead concentrations can be used to determine geographical patterns of poisoning in populations. An examination of the distributions of bone or liver lead concentrations from populations may help to determine the source of lead to which they have been exposed.

REFERENCES

Adrian, W. J., and M. L. Stevens. 1979. Wet vs. dry weights for heavy metal toxicity determinations in duck liver. J. Wildl. Dis. 15:125-126.

Anderson, D. R. 1975. Lead poisoning in waterfowl at Rice Lake, Illinois. J. Wildl. Manage. 39:264-70.

Bagley, G. E., and L. N. Locke. 1967. The occurrence of lead in tissues of wild birds. Bull. Environ. Contam. Toxicol. 2:297-305.

Bellrose, F. C. 1959. Lead poisoning as a mortality factor in waterfowl populations. Ill. Nat. Hist. Surv. Bull. 27:235-288.

Birkhead, M. 1983. Lead levels in the blood of mute swans *Cygnus olor* on the River Thames. J. Zool. (Lond.) 199:59-73.

Blus, L. J., C. J. Henny, D. J. Hoffmann, and R. A. Grove. 1991. Lead toxicosis in tundra swans near a mining and smelting complex in northern Idaho. Arch. Environ. Contam. Toxicol. 21:549-555.

Bull, K. R., W. J. Every, P. Freestone, J. R. Hall, D. Osborn, A. S. Cooke, and T. Stowe. 1983 Alkyl lead pollution and bird mortalities on the Mersey estuary, UK, 1979-1981. Environ. Pollut. Ser. A Ecol. Biol. 31:239-259.

Chupp, N. R., and P. D. Dalke. 1964. Waterfowl mortality in the Coeur d'Alene River Valley, Idaho. J. Wildl. Manage. 28:692-702.

Clarke, E. G. C., and M. L. Clarke. 1975. Veterinary toxicology. Bailliere Tindall, London.

Clausen, B., K. Elvestad, and O. Karlog. 1982. Lead burden in mute swans from Denmark. Nord. Veterinaermed. 34:83-91.

Clausen, B., and C. Wolstrup. 1979. Lead poisoning in game from Denmark. Dan. Rev. Game Biol. 11:1-22.

Cook, R. S., and D. O. Trainer. 1966. Experimental lead poisoning in Canada geese. J. Wildl. Manage. 30:1-8.

Demayo, A., M. C. Taylor, K. W. Taylor, and P. V. Hodson. 1982. Toxic effects of lead and lead compounds on human health, aquatic life, wildlife, plants, and livestock. Crit. Rev. Environ. Control 12:257-305.

Dieter, M. P. 1979. Blood delta-aminolevulinic acid dehydratase (ALAD) to monitor lead contamination in canvasback ducks *(Aythya valisineria)*. p. 177-191. In S. W. G. Nielsen, G. Migaki, and D. G. Scarpelli (Eds.). Animals as monitors of environmental pollutants. (Symp., Storrs, Conn., 1977.) National Academy of Sciences, Washington, D.C., 421 pp.

Dieter, M. P., and M. T. Finley. 1978. Erythrocyte delta-aminolevulinic acid dehydratase activity in mallard ducks: duration of inhibition after lead shot dosage. J. Wildl. Manage. 42:621-625.

Dieter, M. P., and M. T. Finley. 1979. Delta-aminolevulinic acid dehydratase enzyme activity in blood, brain, and liver of lead-dosed ducks. Environ. Res. 19:127-135.

Dieter, M. P., M. C. Perry, and B. M. Mulhern. 1976. Lead and PCB's in canvasback ducks: relationship between enzyme levels and residues in blood. Arch. Environ. Contam. Toxicol. 5:1-13.

Eisler, R. 1988. Lead hazards to fish, wildlife, and invertebrates: a synoptic review. Biol. Rep. 85(1.14). U.S. Fish and Wildlife Service, Washington, D.C., 134 pp.

Elder, W. H. 1954. The effect of lead poisoning on the fertility and fecundity of domestic mallard ducks. J. Wildl. Manage. 18:315-323.

Finley, M. T., and M. P. Dieter. 1978. Influence of laying on lead accumulation in bone of mallard ducks. J. Toxicol. Environ. Health 4:123-129.

Finley, M. T., M. P. Dieter, and L. N. Locke. 1976a. Lead in tissues of mallard ducks dosed with two types of lead shot. Bull. Environ. Contam. Toxicol. 16:261-269.

Finley, M. T., M. P. Dieter, and L. N. Locke. 1976b. Delta-aminolevulinic acid dehydratase: inhibition in duck dosed with lead shot. Environ. Res. 12:243-249.

Finley, M. T., M. P. Dieter, and L. N. Locke. 1976c. Sublethal effects of chronic lead ingestion in mallard ducks. J. Toxicol. Environ. Health 1:929-937.

Forbes, R. M., and G. C. Sanderson. 1978. Lead toxicity in domestic animals and wildlife. p. 225-277. In J. O. Nriagu (Ed.). The biogeochemistry of lead in the environment. Part B. Biological effects. Elsevier/North Holland Biomedical Press, Amsterdam, 397 pp.

Friend, M. 1985. Interpretation of criteria commonly used to determine lead poisoning problem areas. Fish Wildl. Leaflet 24. U.S. Fish and Wildlife Service, Washington, D.C.

Friend, M. (Ed.). 1987. Field guide to wildlife diseases. Fish Wildl. Resour. Publ. No. 167. U.S. Fish and Wildlife Service, Washington, D.C., 225 pp.

Hernberg, S., J. Nikkanen, G. Mellin, and H. Lilius. 1970. Delta-aminolevulinic acid dehydratase as a measure of lead exposure. Arch. Environ. Health 21:140-145.
Irwin, J. C. 1975. Mortality factors in whistling swans at Lake St. Clair, Ontario. J. Wildl. Dis. 11:8-12.
Jordan, J. S., and F. C. Bellrose. 1951. Lead poisoning in wild waterfowl. Ill. Nat. Hist. Surv. Biol. Notes 26, 27 pp.
Kingsford, R. T., J. Flanjak, and S. Black. 1989. Lead shot and ducks on Lake Cowal. Aust. Wildl. Res. 16:167-172.
Koranda, J., K. Moore, M. Stuart, and C. Conrado. 1979. Dietary effects on lead uptake and trace element distribution in mallard ducks dosed with lead shot. (Report). UCID-18044. Lawrence Livermore Laboratory, Environmental Sciences Division, Livermore, Calif., 41 pp.
Longcore, J. R., R. Andrews, L. N. Locke, G. E. Bagley, and L. T. Young. 1974a. Toxicity of lead and proposed substitute shot to mallards. U.S. Fish Wildl. Serv. Spec. Sci. Rep. No. 183.
Longcore, J. R., L. N. Locke, G. E. Bagley, and R. Andrews. 1974b. Significance of lead residues in mallard tissues. U.S. Fish Wildl. Serv. Spec. Sci. Rep. No. 182.
Mauser, D. M., T. E. Rocke, J. G. Mensik, and C. J. Brand. 1990. Blood lead concentrations in mallards from Delevan and Colusa national wildlife refuges. Calif. Fish Game 76:132-136.
Mautino, M., and J. U. Bell. 1986. Experimental lead toxicity in ring-necked duck. Environ. Res. 41:538-545.
Mautino, M., and J. U. Bell. 1987. Hematological evaluation of lead intoxication in mallards. Bull. Environ. Contam. Toxicol. 38:78-85.
Moore, K. C. 1978. Investigations of waterfowl lead poisoning in California. Trans. West. Sec. Wildl. Soc. Annu. Mtg., pp. 209-220.
Morton, A. P., S. Partridge, and J. A. Blair. 1985. The intestinal uptake of lead. Chem. Br. Oct:923-927.
O'Halloran, J., P. F. Duggan, and A. A. Myers. 1988a. Biochemical and haematological values for mute swans *(Cygnus olor)*: effects of acute lead poisoning. Avian Pathol. 17:667-678.
O'Halloran, J., A. A. Myers, and P. F. Duggan. 1988b. Blood lead levels and free red blood cell protoporphyrin as a measure of lead exposure in mute swans. Environ. Pollut. 52:19-38.
Pain, D. J. 1987. Lead Poisoning in Waterfowl: Sources and Screening Techniques. Ph.D. thesis. Oxford University, Oxford, U.K.
Pain, D. J. 1989. Haematological parameters as predictors of blood lead poisoning in the black duck *(Anas rubripes)*. Environ. Pollut. 60:67-81.
Pain, D. J. (Ed.). 1992. Lead poisoning in waterfowl. *In* Proc. International Waterfowl and Wetlands Research Bureau (IWRB) workshop, Brussels, Belgium, 1991. IWRB Spec. Publ. 16. IWRB, Slimbridge, U.K., 105 pp.
Pain, D. J., C. Amiard-Triquet, and C. Sylvestre. 1992. Tissue lead concentrations and shot ingestion in nine species of waterbird from the Camargue (France). Ecotoxicol. Environ. Saf. 24:217-233.
Pain, D. J., and B. A. Rattner. 1988. Mortality and hematology associated with the ingestion of one number four lead shot in black ducks, *Anas rubripes*. Bull. Environ. Contam. Toxicol. 40:159-164.
Rocke, T. E., and M. D. Samuel. 1991. Effects of lead shot ingestion on selected cells of the mallard immune system. J. Wildl. Dis. 27:1-9.

Roscoe, D. E., S. W. Nielsen, A. A. Lamola, and D. Zuckermann. 1979. A simple quantitative test for erythrocyte protoporphyrin in lead poisoned ducks. J. Wildl. Dis. 15:127-136.

Sanderson, G. 1992. Lead poisoning mortality. p. 14-18. In D. J. Pain (Ed.). Lead poisoning in waterfowl. Proc. IWRB Workshop, Brussels, Belgium, 1991. IWRB Spec. Publ. 16. IWRB, Slimbridge, U.K., 105 pp.

Sanderson, G. C., and F. C. Bellrose. 1986. A review of the problem of lead poisoning in waterfowl. Ill. Nat. Hist. Surv. Spec. Publ. No. 4, 33 pp.

Sanderson, G. C., and J. C. Irwin. 1976. Effects of various combinations and numbers of lead:iron pellets dosed in wild-type captive mallards. Final Rep. Contract No. 14-16-0008-914. U.S. Fish and Wildlife Service and Illinois Natural History Survey, 67 pp.

Sassa, S., S. Granick, and A. Kappas. 1975. Effect of lead and genetic factors on heme biosynthesis in the human red cell. Ann. N.Y. Acad. Sci. 244:419-440.

Sears, J. 1988. Regional and seasonal variations in lead poisoning in the mute swan *Cygnus olor* in relation to the distribution of lead and lead weights in the Thames area, England. Biol. Conserv. 46:115-134.

Simkiss, K. 1961. Calcium metabolism in avian reproduction. Biol. Rev. Camb. Philos. Soc. 36:321-367.

Spray, C. J., and H. Milne. 1988. The incidence of lead poisoning among whooper and mute swans *Cygnus cygnus* and *C. olor* in Scotland. Biol. Conserv. 44:265-281.

Stendell, R. C., R. I. Smith, K. P. Burnham, and R. E. Christensen. 1979. Exposure of waterfowl to lead: a nationwide survey of residues in wing bones of seven species, 1972-73. U.S. Fish Wildl. Serv. Spec. Sci. Rep. Wildl. 223:1-12.

Szymczak, M. R., and W. J. Adrian. 1978. Lead poisoning in Canada geese in southeast Colorado. J. Wildl. Manage. 42:299-306.

Tola, S., S. Humberg, S. Asp, and J. Nikkanen. 1973. Parameters indicative of absorption and biological effect in new lead exposure: a prospective study. Br. J. Ind. Med. 30:134.

U.S. Fish and Wildlife Service. 1986. Use of lead shot for hunting migratory birds in the United States. In Final supplement environmental impact statement. U.S. Fish and Wildlife Service, Washington, D.C.

White, D. H., and R. C. Stendell. 1977. Waterfowl exposure to lead and steel shot on selected hunting areas. J. Wildl. Manage. 41:469-475.

Wobester, G. A. 1981. Diseases of wild waterfowl. p. 153-161. Plenum Press, New York.

Zwank, P. J., V. L. Wright, P. M. Shealy, and J. D. Newsom. 1985. Lead toxicosis in waterfowl on two major wintering areas in Louisiana. Wildl. Soc. Bull. 13:17-26.

CHAPTER 11

Interpretation of Tissue Lead Residues in Birds Other Than Waterfowl

J. Christian Franson

INTRODUCTION

Lead poisoning has been reported in at least 30 species of birds other than waterfowl but, because of behavioral and environmental factors that contribute to conceal this disease, the true number of species affected may be higher (U.S. Fish and Wildlife Service, 1986; Eisler, 1988; Pain, 1991; Locke and Friend, 1992). Mortality has resulted from direct consumption of spent lead shot, consumption of lead shot or bullet fragments embedded in food items, ingestion of lead fishing weights, and exposure to lead-based paint. Lead exposure of non-waterfowl avian populations also occurs in urban environments (Ohi et al., 1974; Ohi et al., 1981; Grue et al., 1986), in locations surrounding metal smelters (Beyer et al., 1985), and in areas contaminated with industrial effluents (Bull et al., 1983).

The diagnosis of a disease or alteration in normal health is made when sufficient data (including field observations, objective physical or functional evidence of disease, and laboratory results) are available and consistent with that condition. A diagnosis of lead poisoning as the cause of death is established through a synthesis of necropsy observations, pathological findings, and tissue lead residues. Lead poisoning as a cause of death should be distinguished from the diagnosis of lead poisoning as a disease or poisoning syndrome. Death from lead poisoning cannot be diagnosed on the basis of tissue residues only, but the disease of lead poisoning can, if sufficient data exist from related species showing a relation between illness and residue levels.

Some of the gross and microscopic lesions in waterfowl with lead poisoning (Coburn et al., 1951; Cook and Trainer, 1966; Karstad, 1971; Clemens et al., 1975; Hunter and Wobeser, 1980) have been reported for other avian species (Locke and Bagley, 1967; Hunter and Haigh, 1978; Pattee et al., 1981; Beyer et al., 1988; Langelier et al., 1991). The presence of these lesions, in conjunction with elevated tissue residues, can be used to support a diagnosis of lead poisoning as the cause of death without the definition of a "toxic" tissue lead level in each species. The toxic level suggested for waterfowl (8 ppm wet weight in liver; Friend, 1985) can be used as a guide because, if gross or microscopic lesions are consistent with lead poisoning, it is the presence, not the magnitude, of elevated tissue lead that is critical to a diagnosis.

The focus of this chapter is the interpretation of tissue lead residues from avian species other than Anseriformes, in the absence of clinical and pathological observations or when no lesions characteristic of lead poisoning are seen. In these situations, a prediction can be made regarding the degree to which birds were affected by the lead-poisoning syndrome but, technically, a diagnosis of lead poisoning as the cause of death is not possible. Enough data exist for Falconiformes, Columbiformes, and Galliformes to categorize lead residues in the blood, liver, and kidney according to increasing severity of effects (Table 1): (1) subclinical, a range of residues reported to cause physiological effects only, such as the inhibition of delta-aminolevulinic acid dehydratase (ALAD); (2) toxic, an approximate threshold level marking the initiation of clinical signs (pathological manifestations of physiological effects) such as anemia, microscopic lesions in tissues, weight loss, muscular incoordination, green diarrhea, and anorexia; and (3) compatible with death, an approximate threshold value associated with death in field, captive, and/or experimental cases of lead poisoning.

I suggest that Table 1 be used as follows. Consider residues below the subclinical range as "background," i.e., evidence of normal environmental exposure. Residues in the subclinical category are indicative of potential physiological injury from which the bird would probably recover if lead exposure were terminated. Toxic residues could result in severe physiological effects, leading to clinical signs and probable death if lead exposure were to continue. Residues above the compatible-with-death threshold are consistent with lead-poisoning mortality. Table 1 is meant to provide general guidance in the assessment of residues, with an awareness that numerous factors may contribute to the substantial overlap of residue values in the three categories. Among these factors are species variability in response to lead exposure (Beyer et al., 1988), the immunosuppressive effects of lead (Franson, 1986; Rocke and Samuel, 1991), and the potential interactions between lead and other disease agents (Wobeser, 1986). In addition, a prediction of the impact of residue levels depends on the overall health of the individual or population, the extent of physiological damage already done, the likelihood that lead exposure will continue, and the status of tissue lead kinetics at the time the bird was sampled. In other words, a given tissue residue may be subclinical in one situation and toxic in another, just as birds may die of lead poisoning with tissue residues below the compatible-with-death threshold.

Residues reviewed in the text are reported as they appear in the literature, as the wet or dry weight basis for soft tissue and bone and as ppm or μg/dl in whole blood. To facilitate interpretation, residue data in Table 1 and the Summary are expressed as ppm of wet weight. Where necessary, the estimates of wet weight residues were calculated from dry weight data using the moisture content of mallard tissues reported by Scanlon (1982). Blood lead expressed on a volumetric basis was

Table 1 Interpretation[a] of Tissue Lead Residues (ppm wet weight) in Falconiformes, Columbiformes, and Galliformes

Order	Blood	Liver	Kidney	Ref.
Falconiformes				
Subclinical	0.2-1.5	2-4	2-5	Franson et al. (1983), Custer et al. (1984), Henny et al. (1991), Redig et al. (1991)
Toxic	>1	>3	>3	Redig et al. (1980), Hoffman et al. (1981), Lumeij et al. (1985)
Compatible with death	>5	>5	>5	Kaiser et al. (1980), Pattee et al. (1981), MacDonald et al. (1983), Wiemeyer et al. (1988), Langelier et al. (1991)
Columbiformes				
Subclinical	0.2-2.5	2-6	2-20	Ohi et al. (1974), Cory-Slechta et al. (1980), Hutton (1980), Hutton and Goodman (1980), Kendall and Scanlon (1982), Kendall et al. (1982), DeMent et al. (1987), Scheuhammer and Wilson (1990)
Toxic	>2	>6	>15	Cory-Slechta et al. (1980), Anders et al. (1982), Kendall and Scanlon (1982), Boyer et al. (1985)
Compatible with death	>10	>20	>40	Locke and Bagley (1967), Barthalmus et al. (1977), Dietz et al. (1979), Cory-Slechta et al. (1980), Anders et al. (1982), Boyer et al. (1985)
Galliformes				
Subclinical	0.2-3	2-6	2-20	Salisbury et al. (1958), Stone and Soares (1976), Stone et al. (1977, 1979), Gjerstad and Hanssen (1984), Yamamoto et al. (1993)
Toxic	>5	>6	>15	Franson and Custer (1982)
Compatible with death	>10	>15	>50	McConnell (1967), Stone and Butkas (1978), Keymer and Stebbings (1987), Beyer et al. (1988)

[a] Subclinical, physiological effects (e.g., ALAD depression) only, with no overt clinical signs; toxic, clinical signs, such as muscle wasting, green diarrhea, weakness, anemia, muscular incoordination; compatible with death, residues associated with mortality in field reports of lead poisoning or in experimental dosing studies.

converted to ppm (e.g., µg/dl divided by 100 = estimated ppm), and blood lead data reported in the literature as ppm or µg/g were assumed to be on a wet weight basis. Principles of the tissue kinetics of lead and of the use of ALAD as an indicator of lead exposure are similar for all avian species; these are discussed in the waterfowl chapter in this volume.

FALCONIFORMES

More is known about lead poisoning and exposure in Falconiformes than in any other group of wild non-waterfowl species. A wide range of liver and kidney residues have been reported from raptors dying of lead poisoning, but levels are often in the range of 5 to 40 ppm wet weight. The published records of 36 wild bald eagles *(Haliaeetus leucocephalus)* that died of lead poisoning report liver lead residues of 5 to 61 ppm wet weight and of 5 and 12 ppm wet weight in the two kidneys tested (Mulhern et al., 1970; Kaiser et al., 1980; Reichel et al., 1984; Frenzel and Anthony, 1989; Craig et al., 1990; Langelier et al., 1991). The unpublished records of lead poisoning mortality in wild bald eagles include 174 cases in which necropsy evaluations and toxicology were conducted at the National Wildlife Health Research Center. The mean liver lead concentration in these eagles was 31 ppm wet weight (standard deviation = 15). Of three captive bald eagles that died of lead poisoning, two had liver lead residues of 23 and 15 ppm and kidney residues of 11 ppm each (wet weight presumed but not stated) (Jacobson et al., 1977; Redig et al., 1980). The third had 26 ppm (wet weight) of lead in the liver and 9 ppm wet weight in the kidney (Janssen et al., 1979). In an experimental study of lead shot poisoning, four eagles died with mean lead residues of 17 ppm wet weight in the liver, 6 ppm wet weight in the kidney, 1.4 ppm wet weight in the brain, and 10 ppm dry weight in the bone (Pattee et al., 1981). A fifth eagle that became blind and was euthanized had lead concentrations of 3 ppm wet weight in the liver and kidney, 2 ppm wet weight in the brain, and 13 ppm dry weight in the bone.

Liver lead concentrations in three California condors *(Gymnogyps californianus)* that died of lead poisoning were 6, 23, and 35 ppm wet weight (Wiemeyer et al., 1988). The relatively low liver lead concentration in the first bird was probably the result of chelation therapy administered before it died (Mautino, 1990). Reports of lead poisoning mortality in captive condors and vultures include an Andean condor *(Vultur gryphus)* with a liver lead level of 38 ppm wet weight (Locke et al., 1969) and two captive king vultures *(Sarcoramphus papa)* with liver lead concentrations of 63 and 7 ppm and kidney residues of 71 and 25 ppm wet weight (Decker et al., 1979).

A wild red-tailed hawk *(Buteo jamaicensis)* that died of lead poisoning had a liver lead concentration of 71 ppm (wet weight presumed but not stated) and eight ingested lead shot (Sikarskie, 1977). MacDonald et al. (1983) reported lead poisoning mortality, probably caused by the presence of metallic lead in food items, in captive Falconiformes including a common buzzard *(Buteo buteo)*, Eurasian sparrowhawk *(Accipiter nisus)*, two peregrine falcons *(Falco peregrinus)*, and a lagger falcon *(Falco jugger)*. Liver lead concentrations were 36 to 175 ppm dry weight, and kidney residues were 31 to 221 ppm dry weight. Two prairie falcons *(Falco mexicanus)* that died of lead poisoning after being fed parts of animals shot by hunters had lead residues of 17 and 57 ppm in the liver and 6 and 78 ppm (wet weight presumed but not stated) in the kidney (Benson et al., 1974; Redig et al., 1980). Three raptors that died after receiving daily doses of lead acetate for several weeks had mean lead residues of 11 ppm dry weight in the liver, 10 ppm in the kidney, and 37 ppm in bone (Reiser and Temple, 1981). Nestling American kestrels *(Falco sparverius)* dosed with metallic lead powder exhibited reduced ALAD activity and anemia with liver and kidney lead concentrations of 4 and 7 ppm wet weight, while survivors of a dosage that killed 40% of the treatment group had lead residues of 6 and 16 ppm wet weight in the liver and kidney (Hoffman et al., 1985a; Hoffman et al., 1985b). Franson et al. (1983) reported a mean liver lead concentration of 2.4 ppm dry weight associated with inhibited ALAD, but no anemia, in adult American kestrels fed 50 ppm of lead powder. Kestrels fed a diet containing about 450 ppm dry weight of biologically incorporated lead had mean tissue residues of 10 ppm dry weight in the liver, 15 in the kidney, 2 in the brain, and 18 in bone with no effects on body weight, hematocrit, hemoglobin concentration, or red blood cell count (Custer et al., 1984).

As blood lead concentrations in Falconiformes near the 5-ppm level, clinical signs of toxicity can be expected. Raptors exhibited anemia, anorexia, and bile-stained feces when blood lead residues reached 5 to 8 ppm (Reiser and Temple, 1981). After 14 days of exposure to lead shot, five bald eagles had a mean blood lead concentration of 5.4 ppm and exhibited 80% ALAD depression and a 20 to 25% reduction in hematocrit and hemoglobin concentration (Hoffman et al., 1981). A California condor that was captured in a weakened condition and later died had a blood lead concentration of 420 µg/dl (Janssen et al., 1986). Two turkey vultures *(Cathartes aura)* with clinical signs consistent with lead toxicosis had blood lead residues of 320 µg/dl and 11 ppm, but both recovered (Janssen et al., 1979, Reiser and Temple, 1981). A prairie falcon with weakness, weight loss, and anemia had 11 ppm of lead in its blood but recovered after chelation therapy (Redig et al., 1980). A honey buzzard *(Pernis apivorus)* with a relatively low blood lead concentration of 80 µg/dl exhibited clinical signs of green diarrhea, muscle wasting, and weakness and had one lead shot in its stomach (Lumeij et al., 1985). The bird was released to the wild 2 weeks after removal of the lead shot, when the blood lead concentration was 16 µg/dl. Falconiformes have been reported to tolerate blood lead concentrations of about 1.5 ppm without the development of overt clinical signs. American kestrels fed lead-contaminated diets had blood lead levels of 1.69 ppm with no resultant anemia (Custer et al., 1984). Redig et al. (1991) reported no mortality or clinical signs of toxicity, based on gross observations and body weight, in red-tailed hawks with blood lead concentrations of up to 1.58 µg/ml. In a survey of blood lead concentrations in 162 wild golden eagles *(Aquila chrysaetos)*, Pattee et al. (1990) considered 36% of the birds to have been exposed to lead, but the overall mean blood lead concentration was 0.25 ppm. Ospreys *(Pandion haliaetus)* from a mining and smelting area had a mean blood lead concentration of 0.20 µg/g and inhibited ALAD activity, but there was no effect on hemoglobin or hematocrit (Henny et al., 1991).

COLUMBIFORMES

Soft tissue lead residues associated with mortality in doves and pigeons tend to be in the 20- to 60-ppm wet weight range or even higher, as in the case of a wild mourning dove *(Zenaidura macroura)* that died of lead poisoning with 72 ppm wet weight of lead in the liver (Locke and Bagley, 1967). Clausen and Wolstrup (1979) reported liver and kidney lead residues of 48 and 200 ppm (wet weight presumed but not stated) in a wood pigeon *(Columba palumbus)* that died of lead poisoning. Mourning doves that died after dosage with lead shot had mean lead residues of

80 to 93 µg/g dry weight in the liver, 230 to 300 µg/g in the kidney, and 116 to 192 µg/g in bone (Buerger et al., 1986). In another lead shot dosing study, a mourning dove that died had lead residues of 267 ppm dry weight in the liver, 1901 ppm in the kidney, 11 ppm in the brain, and 403 ppm in bone (Kendall et al., 1983). Mourning doves in that study that were euthanized after 9 days of lead shot exposure had microscopic lesions of lead poisoning and lead concentrations of 150 to 179 ppm dry weight in the liver, 1182 to 1298 ppm in the kidney, 11 to 12 ppm in the brain, and 473 to 528 ppm in bone. Ringed turtle-doves *(Streptopelia risoria)* given doses of lead shot and euthanized 9 days later had liver lead residues of 24 to 128 µg/g and kidney residues of 633 to 2384 µg/g dry weight (Kendall et al., 1981). Several of these birds had seizures, and all had microscopic lesions in the kidney and liver, but none died during the course of the experiment. Ringed turtle-doves that received lead acetate in drinking water for 90 days had mean kidney and liver residues of 900 and 8 µg/g dry weight, respectively (Kendall and Scanlon, 1981). Although no birds died, cellular necrosis and lead inclusions were noted in the kidneys. Reduced ALAD activity occurred in ringed turtle-doves that had liver and kidney lead residues of 1 and 6 ppm wet weight (Scheuhammer and Wilson, 1990). In other studies with ringed turtle-doves, ALAD activity was inhibited with tissue lead levels of 4 to 9 ppm dry weight in the brain, 9 to 19 ppm in the liver, and 84 to 839 ppm in the kidney (Kendall and Scanlon, 1982; Kendall et al., 1982). The hemoglobin concentration was reduced when tissue residues reached 28 ppm dry weight in the liver, 12 ppm in the brain, and 457 ppm in the kidney (Kendall and Scanlon, 1982).

ALAD activity was reduced in feral pigeons *(Columba livia)* with lead residues of 6 to 10 ppm dry weight in the liver, 10 to 49 ppm in the kidney, 3 to 6 ppm in the brain, and 108 to 282 ppm in bone (Ohi et al., 1974; Hutton, 1980; Hutton and Goodman, 1980). Clinical signs of lead poisoning as well as ALAD inhibition were reported in feral pigeons, with 14 to 22 ppm (dry weight) of lead in the liver, 189 to 321 ppm in the kidney, 12 to 20 ppm in the brain, and 245 to 669 ppm in bone (Hutton, 1980; Hutton and Goodman, 1980; Johnson et al., 1982). Two clinically normal feral pigeons with lead shot in their gizzards had liver lead concentrations of 5 and 13 ppm and kidney residues of 14 and 81 ppm wet weight (DeMent et al., 1987). Domestic pigeons that became anemic after treatment with lead acetate had lead residues of 8 to 20 ppm wet weight in the liver, 33 to 603 ppm in the kidney, 0.9 to 2.3 ppm in the brain, and 57 to 501 ppm in bone (Anders et al., 1982).

Blood lead concentrations associated with death in Columbiformes can be extremely high. Two domestic pigeons treated with lead acetate had blood lead residues of 4000 and 2320 µg/dl shortly before they died (Dietz et al., 1979; Anders et al., 1982). Other pigeons that were treated with lead acetate developed blood lead concentrations of 250 to 440 µg/dl, resulting in anemia and lead inclusions in the kidneys, but survived the dosage regimen for up to 64 weeks (Anders et al., 1982). Domestic pigeons that died after experimental lead acetate exposure had blood lead levels of 569 to 1235 and 1245 µg/dl, crop stasis occurred at 220 to 820 and 450 to 1100 µg/dl, but there were no observable clinical signs when blood lead was less than 200 µg/dl (Barthalmus et al., 1977; Cory-Slechta et al., 1980; Boyer et al., 1985). ALAD activity was inhibited in ringed turtle-doves with blood lead concentrations of 21, 81 to 122, and 142 to 245 μg/dl, and hemoglobin concentration was reduced, with mean blood lead concentrations of 395 μg/dl (Kendall and Scanlon, 1982; Kendall et al., 1982; Scheuhammer and Wilson, 1990). Depression of ALAD in feral pigeons has been associated with lead residues of 15 to 45 μg/dl in the blood (Ohi et al., 1974; Hutton, 1980; Hutton and Goodman, 1980), and clinical signs of lead poisoning were reported with 101 μg/dl (Hutton, 1980; Hutton and Goodman, 1980). Two urban pigeons with lead shot in their gizzards had blood lead concentrations of 95 and 1870 μg/dl, but they were not anemic or emaciated (DeMent et al., 1987).

GALLIFORMES

Galliforms poisoned by lead also tend to have high tissue lead residues. A wild turkey *(Meleagris gallopavo)* that died of lead poisoning had a liver lead concentration of 17 ppm wet weight (Stone and Butkas, 1978). A wild ring-necked pheasant *(Phasianus colchicus)* found dead with 29 lead shot in its gizzard had 168 ppm (wet weight presumed but not stated) of lead in the liver (Hunter and Rosen, 1965). Keymer and Stebbings (1987) reported lead poisoning as the cause of death in a gray partridge *(Perdix perdix)* with 40 ppm (wet weight) of lead in the liver and 100 ppm wet weight in the kidney. An emaciated gray partridge had lead residues of 130 ppm in the liver and 440 in the kidney (wet weight presumed but not stated) with 34 lead pellets in the gizzard (Clausen and Wolstrup, 1979). Gjerstad and Hanssen (1984) administered doses to willow ptarmigan *(Lagopus lagopus)* of one, three, or six lead shot. Three ptarmigan that died had liver lead residues of 64, 134, and 274 ppm wet weight. Birds given doses of one lead shot survived with no clinical signs and had mean liver lead residues of about 3 ppm wet weight 15 days after dosing. When northern bobwhite *(Colinus virginianus)* were fed increasing amounts of lead acetate until half of the birds died, lead residues in the liver and kidney were 21 to 277 and 85 to 500 ppm wet weight, respectively (Beyer et al., 1988). Chickens *(Gallus domesticus)* that died after being fed lead-containing grit had liver lead residues of up to 54 ppm wet weight, while liver lead in chickens that exhibited no clinical signs and were euthanized was 1 to 6 ppm wet weight (Salisbury et al., 1958). ALAD was inhibited but there was no effect on hemoglobin or packed cell volume in Japanese quail *(Coturnix coturnix japonica)* that had liver lead concentrations of about 0.5, 4, and 6 ppm and kidney concentrations of 6, 20, and 30 ppm wet weight (Stone and Soares, 1976; Stone et al., 1977; Stone et al., 1979). Northern bobwhite given doses of lead shot had a pooled blood lead concentration of 4.3 mg/100 g 9 days postexposure (McConnell, 1967). Weakness, lethargy, and loss of weight were seen in 19% of the quail and 10% died. Chickens fed lead acetate exhibited severely inhibited ALAD activity, weight loss, and mild anemia with blood lead concentrations of 322 to 832 μg/dl (Franson and Custer, 1982). Mean liver and kidney lead residues were 17 and 56 ppm dry weight after 28 days. Japanese quail given doses of lead shot had lead residues of 7 ppm wet weight in the blood, 3 ppm in the liver, and 5 ppm in the kidney (Yamamoto et al., 1993). ALAD was inhibited, but no other clinical signs were reported.

PASSERIFORMES

Two cuckoos *(Coccyzus americanus)* collected near a zinc smelter had liver lead concentrations of 18 and 25 ppm wet weight and kidney residues of 21 and 14 ppm wet weight (Beyer et al., 1985). ALAD was inhibited, but there was no anemia or gross or microscopic lesions of lead poisoning. Getz et al. (1977) measured tissue lead residues in songbirds collected in urban and rural areas. House sparrows *(Passer domesticus)*, starlings *(Sturnus vulgaris)*, common grackles *(Quiscalus quiscala)*, and robins *(Turdus migratorius)* had maximum liver lead residues of 10 to 16 ppm dry weight and kidney residues of 14 to 98 ppm dry weight.

Starlings given doses of trialkyl lead died within 6 days of treatment, with mean liver and kidney residues of 40 and 20 ppm wet weight, respectively (Osborn et al., 1983). Beyer et al. (1988) fed lead acetate to red-winged blackbirds *(Agelaius phoeniceus)*, brown-headed cowbirds *(Molothrus ater)*, and common grackles until half of the birds in each group died. Median liver lead concentrations were 20 to 50 ppm wet weight (range, 4 to 97 ppm), and median kidney residues were 22 to 160 ppm wet weight (range, 2 to 740 ppm).

CHARADRIIFORMES

Bull et al. (1983) reported liver lead levels of 8 to 31 ppm wet weight in three dunlin *(Calidris alpina)* and two redshank *(Tringa totanus)* found dead, apparently after consuming prey contaminated with lead. Dunlin found sick had mean liver and kidney lead concentrations of 9 and 7 ppm wet weight, respectively ($n = 6$). Locke et al. (1991) reported a liver lead concentration of 52 ppm wet weight in a marbled godwit *(Limosa fedoa)* that died of lead poisoning after ingesting lead shot. Seven herring gull *(Larus argentatus)* chicks that received a one-time injection of lead nitrate survived 45 days but exhibited slower growth than did controls, with mean lead concentrations of 197 µg/dl in the blood, 21 ppm dry weight in the liver, and 41 ppm dry weight in the kidney (Burger and Gochfeld, 1990). Apparently healthy adult laughing gulls *(Larus atricilla)* collected in a lead-contaminated area had mean liver lead residues of 4 to 5 ppm and kidney residues of 2 ppm wet weight (Munoz et al., 1976; Hulse et al., 1980). Adult royal terns *(Thalasseus maximus)* collected in the same region had liver and kidney residues of 0.4 and 1.1 ppm wet weight, while sandwich terns *(T. sandvicensis)* had liver and kidney residues of 0.5 and 0.8 ppm wet weight (Maedgen et al., 1982). Herring gulls feeding at a dump site had liver and kidney lead residues of 3 and 13 ppm dry weight, respectively (Leonzio et al., 1986). Black-headed gulls *(Larus ridibundus)* collected at the same location had liver and kidney residues of 8 and 31 ppm dry weight, respectively.

GRUIFORMES

Windingstad et al. (1984) reported on two wild sandhill cranes *(Grus canadensis)* that died of lead poisoning with parts of fishing sinkers in their gizzards. One bird had liver and kidney lead concentrations of 23 and 30 ppm wet weight, while the other had liver and kidney lead residues of 259 and 113 ppm dry weight. A captive sandhill crane that died after ingesting two 0.22-caliber rifle cartridges had 30 ppm (wet weight) of lead in its liver, and another dead captive sandhill crane with an ingested copper-coated penny had a liver lead concentration of 24 ppm wet weight (Windingstad et al., 1984). A sandhill crane that died after exposure to lead-based paint in a zoo facility had liver and kidney lead concentrations of 29 and 19 ppm (wet weight presumed but not stated), respectively (Kennedy et al., 1977). Four additional cranes in this facility developed clinical signs of lead poisoning with blood lead concentrations of 146 to 378 µg/100 ml, but they recovered following chelation treatment. A whooping crane *(Grus americana)* had a blood lead concentration of 5.6 ppm shortly before it died and was found to have liver and kidney concentrations of 24 and 10 ppm wet weight (Snyder et al., 1991). Apparently healthy sora rails *(Porzana carolina)* collected with lead shot in their gizzards had mean liver lead residues of up to 3 ppm wet weight and up to 11 ppm dry weight in bone (Stendell et al., 1980).

CICONIFORMES

Liver lead concentrations of 17 Caribbean flamingoes *(Phoenicopterus ruber ruber)* that died of lead poisoning after consumption of lead shot ranged from 128 to 771 ppm dry weight (Schmitz et al., 1990). Apparently healthy cattle egrets *(Bubulcus ibis)* collected in an industrialized area had liver lead concentrations of 0.07 to 1.3 ppm and kidney concentrations of 0.08 to 3.5 ppm wet weight (Hulse et al., 1980).

GAVIFORMES

Locke et al. (1982) reported liver lead residues of 21, 46, and 38 ppm wet weight in three common loons *(Gavia immer)* that died of lead poisoning. Two of the three had lead fishing sinkers in gizzard contents. Liver lead concentrations were 5 to 38 ppm wet weight in 15 additional cases of lead-poisoned loons, 13 of which had ingested lead fishing weights (Pokras and Chafel, 1992; Franson and Cliplef, 1992).

STRIGIFORMES

Two captive snowy owls *(Nyctea scandiaca)* that died of lead poisoning had 45 and 204 ppm (dry weight) of lead in the liver and 68 and 146 ppm dry weight in the kidney (MacDonald et al., 1983). Exposure was thought to have resulted from consumption of bullet fragments in food items. Eastern screech owls *(Otus asio)* fed lead acetate until half the birds died had median liver and kidney lead concentrations of 22 and 33 ppm wet weight (Beyer et al., 1988).

PROCELLARIIFORMES

Lead poisoning from consumption of paint chips contributed to epizootic mortality in Laysan albatross *(Diomedea immutabilis)* on Midway Atoll (Sileo and Fefer, 1987). Blood lead concentrations in 10 sick albatross ranged from 0.03 to 4.8 ppm (mean, 1.0 ppm), and liver lead residues in eight sick birds were 6 to 110 ppm dry weight (mean, 53 ppm).

SUMMARY

Lead exposure in birds results from direct consumption of spent lead shot, consumption of shot or bullet fragments in food items, ingestion of lead fishing weights and lead-based paints, and environmental contamination of urban and industrial areas. Lead-poisoning mortality has occurred in over 30 species of birds other than waterfowl. Blood, liver, and kidney residues are useful for determining the prevailing status of lead exposure and/or toxicity in birds because they are relatively labile. The bone lead concentration is an indicator of chronic exposure and, thus, less useful for the evaluation of toxicity at the time of sampling. Tissue lead residues associated with physiological injury, clinical signs, and death due to lead poisoning vary greatly among species, and assessments of toxicosis should be done by comparison with data from phylogenetically related groups.

Considerable toxicity data exist for Falconiformes, Columbiformes, and Galliformes, allowing categorization of lead residues according to the increasing severity of effects: (1) subclinical, the range of residues associated with physiological effects only; (2) toxic, the approximate threshold concentration consistent with the development of clinical signs of lead toxicosis; and (3) compatible with death, the approximate threshold concentration associated with mortality in field or laboratory cases of lead poisoning. Lead residues, in parts per million wet weight, suggested to be subclinical, toxic, and compatible with death for Falconiformes are the following: blood, 0.2 to 1.5, >1, >5; liver, 2 to 4, >3, >5; kidney, 2 to 5, >3, >5; for Columbiformes: blood, 0.2 to 2.5, >2, >10; liver, 2 to 6, >6, >20; kidney, 2 to 20, >15, >40; and for Galliformes: blood, 0.2 to 3, >5, >10; liver, 2 to 6, >6, >15; kidney, 2 to 20, >15, >50. Birds may die of lead poisoning with tissue residues well below the threshold level compatible with death, but death is not likely to result from residues below the toxic threshold.

Few data are available for other orders of birds, but the following liver lead residues, in parts per million wet weight, have been reported in cases of morbidity or mortality in the laboratory and field: for Passeriformes: 20 to 40; for Charadriiformes: 8 to 31; for Gruiformes: 30; for Ciconiformes: 40 to 250; for Gaviformes: 20 to 50; for Strigiformes: 15 to 67; and for Procellariiformes: 2 to 36.

Care should be exercised when comparing tissue lead residues from experimental studies with field data. Wild birds with inadequate diets and exposed to ambient environmental conditions may be more susceptible to lead toxicosis than are birds in controlled situations. A clinical evaluation of health and the determination of lead residues in sequential blood samples will enhance the assessment of lead exposure and toxicity in live birds. Although assumptions can be made regarding the role of lead as a contributing factor in the death of birds, a definitive diagnosis of lead poisoning as the cause of death cannot be made on the basis of lead residues alone. Interpretation of tissue lead residues in dead birds should be accompanied, when possible, by examination of carcasses for gross and microscopic lesions of lead poisoning.

REFERENCES

Anders, E., D. D. Dietz, C. R. Bagnell, Jr., J. Gaynor, M. R. Krigman, D. W. Ross, J. D. Leander, and P. Mushak. 1982. Morphological, pharmacokinetic, and hematological studies of lead-exposed pigeons. Environ. Res. 28:344-363.

Barthalmus, G. T., J. D. Leander, D. E. McMillan, P. Mushak, and M. R. Krigman. 1977. Chronic effects of lead on schedule-controlled pigeon behavior. Toxicol. Appl. Pharmacol. 42:271-284.

Benson, W. W., B. Pharaoh, and P. Miller. 1974. Lead poisoning in a bird of prey. Bull. Environ. Contam. Toxicol. 11:105-108.

Beyer, W. N., O. H. Pattee, L. Sileo, D. J. Hoffman, and B. M. Mulhern. 1985. Metal contamination in wildlife living near two zinc smelters. Environ. Pollut. Ser. A Ecol. Biol. 38:63-86.

Beyer, W. N., J. W. Spann, L. Sileo, and J. C. Franson. 1988. Lead poisoning in six captive avian species. Arch. Environ. Contam. Toxicol. 17:121-130.

Boyer, I. J., D. A. Cory-Slechta, and V. DiStefano. 1985. Lead induction of crop dysfunction in pigeons through a direct action on neural or smooth muscle components of crop tissue. J. Pharmacol. Exp. Ther. 234:607-615.

Buerger, T. T., R. E. Mirarchi, and M. E. Lisano. 1986. Effects of lead shot ingestion on captive mourning dove survivability and reproduction. J. Wildl. Manage. 50:1-8.

Bull, K. R., W. J. Every, P. Freestone, J. R. Hall, D. Osborn, A. S. Cooke, and T. Stowe. 1983. Alkyl lead pollution and bird mortalities on the Mersey Estuary, UK, 1979-1981. Environ. Pollut. Ser. A Ecol. Biol. 31:239-259.

Burger, J., and M. Gochfeld. 1990. Tissue levels of lead in experimentally exposed herring gull *(Larus argentatus)* chicks. J. Toxicol. Environ. Health 29:219-233.

Clausen, B., and C. Wolstrup. 1979. Lead poisoning in game from Denmark. Dan. Rev. Game Biol. 11:1-22.

Clemens, E. T., L. Krook, A. L. Aronson, and C. E. Stevens. 1975. Pathogenesis of lead shot poisoning in the mallard duck. Cornell Vet. 65:248-285.

Coburn, D. R., D. W. Metzler, and R. Treichler. 1951. A study of absorption and retention of lead in wild waterfowl in relation to clinical evidence of lead poisoning. J. Wildl. Manage. 15:186-192.

Cook, R. S., and D. O. Trainer. 1966. Environmental lead poisoning of Canada geese. J. Wildl. Manage. 30:1-8.

Cory-Slechta, D. A., R. H. Garman, and D. Seidman. 1980. Lead-induced crop dysfunction in the pigeon. Toxicol. Appl. Pharmacol. 52:462-467.

Craig, T. H., J. W. Connelly, E. H. Craig, and T. L. Parker. 1990. Lead concentrations in golden and bald eagles. Wilson Bull. 102:130-133.

Custer, T. W., J. C. Franson, and O. H. Pattee. 1984. Tissue lead distribution and hematologic effects in American kestrels *(Falco sparverius* L.) fed biologically incorporated lead. J. Wildl. Dis. 20:39-43.

Decker, R. A., A. M. McDermid, and J. W. Prideaux. 1979. Lead poisoning in two captive king vultures. J. Am. Vet. Med. Assoc. 175:1009.

DeMent, S. H., J. J. Chisolm, Jr., M. A. Eckhaus, and J. D. Strandberg. 1987. Toxic lead exposure in the urban rock dove. J. Wildl. Dis. 23:273-278.

Dietz, D. D., D. E. McMillan, and P. Mushak. 1979. Effects of chronic lead administration on acquisition and performance of serial position sequences by pigeons. Toxicol. Appl. Pharmacol. 47:377-384.

Eisler, R. 1988. Lead hazards to fish, wildlife, and invertebrates: a synoptic review. U.S. Fish Wildl. Serv. Biol. Rep. No. 85 (1.14), 134 pp. U.S. Fish and Wildlife Service, Washington, D.C.

Franson, J. C. 1986. Immunosuppressive effects of lead. p. 106-109. *In* J. S. Feierabend and A. B. Russell (Eds.). Lead poisoning in wild waterfowl — a workshop. National Wildlife Federation, Washington, D.C.

Franson, J. C., and D. J. Cliplef. 1992. Causes of mortality in common loons. p. 2-12. *In* L. Morse, S. Stockwell, and M. Pokras (Eds.). Proc. 1992 Conf. on the Loon and Its Ecosystem; Status, Management, and Environmental Concerns. U.S. Fish and Wildlife Service, Concord, N.H.

Franson, J. C., and T. W. Custer. 1982. Toxicity of dietary lead in young cockerels. Vet. Hum. Toxicol. 24:421-423.

Franson, J. C., L. Sileo, O. H. Pattee, and J. F. Moore. 1983. Effects of chronic dietary lead in American kestrels *(Falco sparverius)*. J. Wildl. Dis. 19:110-113.

Frenzel, R. W., and R. G. Anthony. 1989. Relationship of diets and environmental contaminants in wintering bald eagles. J. Wildl. Manage. 53:792-802.

Friend, M. 1985. Interpretation of criteria commonly used to determine lead poisoning problem areas. U.S. Fish Wildl. Serv. Leaflet 2, 4 pp. U.S. Fish and Wildlife Service, Washington, D.C.

Getz, L. L., L. B. Best, and M. Prather. 1977. Lead in urban and rural songbirds. Environ. Pollut. 12:235-238.

Gjerstad, K. O., and I. Hanssen. 1984. Experimental lead poisoning in willow ptarmigan. J. Wildl. Manage. 48:1018-1022.

Grue, C. E., D. J. Hoffman, W. N. Beyer, and L. P. Franson. 1986. Lead concentrations and reproductive success in European starlings *(Sternus vulgaris)* nesting within highway roadside verges. Environ. Pollut. Ser. A Ecol. Biol. 42:157-182.

Henny, C. J., L. J. Blus, D. J. Hoffman, R. A. Grove, and J. S. Hatfield. 1991. Lead accumulation and osprey production near a mining site on the Coeur d'Alene River, Idaho. Arch. Environ. Contam. Toxicol. 21:415-424.

Hoffman, D. J., J. C. Franson, O. H. Pattee, C. M. Bunck, and A. Anderson. 1985a. Survival, growth, and accumulation of ingested lead in nestling American kestrels *(Falco sparverius)*. Arch. Environ. Contam. Toxicol. 14:89-94.

Hoffman, D. J., J. C. Franson, O. H. Pattee, C. M. Bunck, and H. C. Murray. 1985b. Biochemical and hematological effects of lead ingestion in nestling American kestrels *(Falco sparverius)*. Comp. Biochem. Physiol. 80C:431-439.

Hoffman, D. J., O. H. Pattee, S. N. Wiemeyer, and B. Mulhern. 1981. Effects of lead shot ingestion on delta-aminolevulinic acid dehydratase activity, hemoglobin concentration, and serum chemistry in bald eagles. J. Wildl. Dis. 17:423-431.

Hulse, M., J. S. Mahoney, G. D. Schroder, C. S. Hacker, and S. M. Pier. 1980. Environmentally acquired lead, cadmium, and manganese in the cattle egret, *Bubulcus ibis*, and the laughing gull, *Larus atricilla*. Arch. Environ. Contam. Toxicol. 9:65-78.

Hunter, B., and J. C. Haigh. 1978. Demyelinating peripheral neuropathy in a guinea hen associated with subacute lead intoxication. Avian Dis. 22:344-349.

Hunter, B., and G. Wobeser. 1980. Encephalopathy and peripheral neuropathy in lead-poisoned mallard ducks. Avian Dis. 24:169-178.

Hunter, B. F., and M. N. Rosen. 1965. Occurrence of lead poisoning in a wild pheasant *(Phasianus colchicus)*. Calif. Fish Game 51:207.
Hutton, M. 1980. Metal contamination of feral pigeons *Columbia livia* from the London area: Part 2. Biological effects of lead exposure. Environ. Pollut. Ser. A Ecol. Biol. 22:281-293.
Hutton, M., and G. T. Goodman. 1980. Metal contamination of feral pigeons *Columbia livia* from the London area: Part 1. Tissue accumulation of lead, cadmium, and zinc. Environ. Pollut. Ser. A Ecol. Biol. 22:207-217.
Jacobson, E., J. W. Carpenter, and M. Novilla. 1977. Suspected lead toxicosis in a bald eagle. J. Am. Vet. Med. Assoc. 171:952-954.
Janssen, D. L., J. E. Oosterhuis, J. L. Allen, M. P. Anderson, D. G. Kelts, and S. N. Wiemeyer. 1986. Lead poisoning in free-ranging California condors. J. Am. Vet. Med. Assoc. 189:1115-1117.
Janssen, D. L., P. T. Robinson, and P. K. Ensley. 1979. Lead toxicosis in three captive avian species. p. 40-42. *In* Proc. 1979 Annu. Meet. Am. Assoc. Zoo Vet., Denver, Colo., Oct. 1979.
Johnson, M. S., H. Pluck, M. Hutton, and G. Moore. 1982. Accumulation and renal effects of lead in urban populations of feral pigeons, *Columbia livia*. Arch. Environ. Contam. Toxicol. 11:761-767.
Kaiser, T. E., W. L. Reichel, L. N. Locke, E. Cromartie, A. J. Krynitsky, T. G. Lamont, B. M. Mulhern, R. M. Prouty, C. J. Stafford, and D. M. Swineford. 1980. Organochlorine pesticide, PCB, and PBB residues and necropsy data for bald eagles from 29 states — 1975-77. Pestic. Monit. J. 13:145-149.
Karstad, L. 1971. Angiopathy and cardiopathy in wild waterfowl from ingestion of lead shot. Conn. Med. 35:355-360.
Kendall, R. J., and P. F. Scanlon. 1981. Chronic lead ingestion and nephropathy in ringed turtle doves. Poult. Sci. 60:2028-2032.
Kendall, R. J., and P. F. Scanlon. 1982. The toxicology of ingested lead acetate in ringed turtle doves *Streptopelia risoria*. Environ. Pollut. Ser. A Ecol. Biol. 27:255-262.
Kendall, R. J., P. F. Scanlon, and R. T. Di Giulio. 1982. Toxicology of ingested lead shot in ringed turtle doves. Arch. Environ. Contam. Toxicol. 11:259-263.
Kendall, R. J., P. F. Scanlon, and H. P. Veit. 1983. Histologic and ultrastructural lesions of mourning doves *(Zenaida macroura)* poisoned by lead shot. Poult. Sci. 62:952-956.
Kendall, R. J., H. P. Veit, and P. F. Scanlon. 1981. Histological effects and lead concentrations in tissues of adult male ringed turtle doves that ingested lead shot. J. Toxicol. Environ. Health 8:649-658.
Kennedy, S., J. P. Crisler, E. Smith, and M. Bush. 1977. Lead poisoning in sandhill cranes. J. Am. Vet. Med. Assoc. 171:955-958.
Keymer, I. F., and R. St. J. Stebbings. 1987. Lead poisoning in a partridge *(Perdix perdix)* after ingestion of gunshot. Vet. Rec. 120:276-277.
Langelier, K. M., C. E. Andress, T. K. Grey, C. Wooldridge, R. J. Lewis, and R. Marchetti. 1991. Lead poisoning in bald eagles in British Columbia. Can. Vet. J. 32:108-109.
Leonzio, C., C. Fossi, and S. Focardi. 1986. Lead, mercury, cadmium, and selenium in two species of gull feeding on inland dumps, and in marine areas. Sci. Total Environ. 57:121-127.
Locke, L. N., and G. E. Bagley. 1967. Lead poisoning in a sample of Maryland mourning doves. J. Wildl. Manage. 31:515-518.
Locke, L. N., G. E. Bagley, D. N. Frickie, and L. T. Young. 1969. Lead poisoning and aspergillosis in an Andean condor. J. Am. Vet. Med. Assoc. 155:1052-1056.

Locke, L. N., and M. Friend. 1992. Lead poisoning of avian species other than waterfowl. p. 19-22. *In* D. J. Pain (Ed.). Lead poisoning in waterfowl. Int. Waterfowl Wetlands Res. Bur. Spec. Publ. No. 16. Slimbridge, Gloucester, U.K.

Locke, L. N., S. M. Kerr, and D. Zoromski. 1982. Lead poisoning in common loons *(Gavia immer)*. Avian Dis. 26:392-396.

Locke, L. N., M. R. Smith, R. M. Windingstad, and S. J. Martin. 1991. Lead poisoning of a marbled godwit. Prairie Nat. 23:21-24.

Lumeij, J. T., W. T. C. Wolvekamp, G. M. Bron-Dietz, and A. J. H. Schotman. 1985. An unusual case of lead poisoning in a honey buzzard *(Pernis apivorus)*. Vet. Q. 7:165-168.

MacDonald, J. W., C. J. Randall, H. M. Ross, G. M. Moon, and A. D. Ruthven. 1983. Lead poisoning in captive birds of prey. Vet. Rec. 113:65-66.

Maedgen, J. L., C. S. Hacker, G. D. Schroder, and F. W. Weir. 1982. Bioaccumulation of lead and cadmium in the royal tern and sandwich tern. Arch. Environ. Contam. Toxicol. 11:99-102.

Mautino, M. 1990. Avian lead intoxication. p. 245-247. Proc. 1990 Annu. Conf. Assoc. Avian Vet., Phoenix, Ariz., Sept. 1990.

McConnell, C. A. 1967. Experimental lead poisoning of bobwhite quail and mourning doves. p. 208-219. Proc. 21st Annu. Conf. Southeast. Assoc. Game Fish Commiss., New Orleans, La., Sept. 1967.

Mulhern, B. M., W. L. Reichel, L. N. Locke, T. G. Lamont, A. Belisle, E. Cromartie, G. E. Bagley, and R. M. Prouty. 1970. Organochlorine residues and autopsy data from bald eagles 1966-68. Pestic. Monit. J. 4:141-144.

Munoz, R. V., Jr., C. S. Hacker, and T. F. Gesell. 1976. Environmentally acquired lead in the laughing gull, *Larus atricilla*. J. Wildl. Dis. 12:139-142.

Ohi, G., H. Seki, K. Akiyama, and H. Yagyu. 1974. The pigeon, a sensor of lead pollution. Bull. Environ. Contam. Toxicol. 12:92-98.

Ohi, G., H. Seki, K. Minowa, M. Ohsawa, I. Mizoguchi, and F. Sugimori. 1981. Lead pollution in Tokyo — the pigeon reflects its amelioration. Environ. Res. 26:125-129.

Osborn, D., W. J. Every, and K. R. Bull. 1983. The toxicity of trialkyl lead compounds to birds. Environ. Pollut. Ser. A Ecol. Biol. 31:261-275.

Pain, D. J. 1991. Lead poisoning in birds: an international perspective. p. 2343-2351. *In* Acta XX Congr. Int. Ornithol. New Zealand Ornithological Congress Trust Board, Wellington, N.Z.

Pattee, O. H., P. H. Bloom, J. M. Scott, and M. R. Smith. 1990. Lead hazards within the range of the California condor. Condor 92:931-937.

Pattee, O. H., S. N. Wiemeyer, B. M. Mulhern, L. Sileo, and J. W. Carpenter. 1981. Experimental lead-shot poisoning in bald eagles. J. Wildl. Manage. 45:806-810.

Pokras, M. A., and R. Chafel. 1992. Lead toxicosis from ingested fishing sinkers in adult common loons *(Gavia immer)* in New England. J. Zoo Wildl. Med. 23:92-97.

Redig, P. T., E. M. Lawler, S. Schwartz, J. L. Dunnette, B. Stephenson, and G. E. Duke. 1991. Effects of chronic exposure to sublethal concentrations of lead acetate on heme synthesis and immune function in red-tailed hawks. Arch. Environ. Contam. Toxicol. 21:72-77.

Redig, P. T., C. M. Stowe, D. M. Barnes, and T. D. Arent. 1980. Lead toxicosis in raptors. J. Am. Vet. Med. Assoc. 177:941-943.

Reichel, W. L., S. K. Schmeling, E. Cromartie, T. E. Kaiser, A. J. Krynitsky, T. G. Lamont, B. M. Mulhern, R. M. Prouty, C. J. Stafford, and D. M. Swineford. 1984. Pesticide, PCB, and lead residues and necropsy data for bald eagles from 32 states — 1978-81. Environ. Monit. Assess. 4:395-403.

Reiser, M. H., and S. A. Temple. 1981. Effects of chronic lead ingestion on birds of prey. p. 21-25. *In* J. E. Cooper and A. G. Greenwood (Eds.). Recent advances in the study of raptor diseases. Chiron Publications, West Yorkshire, England.

Rocke, T. E., and M. D. Samuel. 1991. Effects of lead shot ingestion on selected cells of the mallard immune system. J. Wildl. Dis. 27:1-9.

Salisbury, R. M., E. L. J. Staples, and M. Sutton. 1958. Lead poisoning of chickens. N. Z. Vet. J. 6:2-7.

Scanlon, P. F. 1982. Wet and dry weight relationships of mallard *(Anas platyrhynchos)* tissues. Bull. Environ. Contam. Toxicol. 29:615-617.

Scheuhammer, A. M., and L. K. Wilson. 1990. Effects of lead and pesticides on delta-aminolevulinic acid dehydratase of ring doves *(Streptopelia risoria)*. Environ. Toxicol. Chem. 9:1379-1386.

Schmitz, R. A., A. A. Aguirre, R. S. Cook, and G. A. Baldassarre. 1990. Lead poisoning of Caribbean flamingos in Yucatan, Mexico. Wildl. Soc. Bull. 18:399-404.

Sikarskie, J. 1977. The case of the red-tailed hawk. Intervet 8:4.

Sileo, L., and S. I. Fefer. 1987. Paint chip poisoning of Laysan albatross at Midway Atoll. J. Wildl. Dis. 23:432-437.

Snyder, S. B., M. J. Richard, R. C. Drewien, N. Thomas, and J. P. Thilsted. 1991. Diseases of whooping cranes seen during annual migration of the Rocky Mountain flock. p. 74-80. *In* R. E. Junge (Ed.). 1991 Proc. Am. Assoc. Zoo Vet., Calgary, Alberta, Canada, Sept. 1991.

Stendell, R. C., J. W. Artmann, and E. Martin. 1980. Lead residues in sora rails from Maryland. J. Wildl. Manage. 44:525-527.

Stone, C. L., M. R. S. Fox, A. L. Jones, and K. R. Mahaffey. 1977. Delta-aminolevulinic acid dehydratase — a sensitive indicator of lead exposure in Japanese quail. Poult. Sci. 56:174-181.

Stone, C. L., K. R. Mahaffey, and M. R. S. Fox. 1979. A rapid bioassay system for lead using young Japanese quail. J. Environ. Pathol. Toxicol. 2:767-779.

Stone, C. L., and J. H. Soares, Jr. 1976. The effect of dietary selenium level on lead toxicity in the Japanese quail. Poult. Sci. 55:341-349.

Stone, W. B., and S. A. Butkas. 1978. Lead poisoning in a wild turkey. N.Y. Fish Game J. 25:169.

U.S. Fish and Wildlife Service. 1986. Final supplemental environmental impact statement on the use of lead shot for hunting migratory birds in the United States, 535 pp. U.S. Fish and Wildlife Service, Washington, D.C.

Wiemeyer, S. N., J. M. Scott, M. P. Anderson, P. H. Bloom, and C. J. Stafford. 1988. Environmental contaminants in California condors. J. Wildl. Manage. 52:238-247.

Windingstad, R. M., S. M. Kerr, L. N. Locke, and J. J. Hurt. 1984. Lead poisoning of sandhill cranes *(Grus canadensis)*. Prairie Nat. 16:21-24.

Wobeser, G. 1986. Interaction between lead and other disease agents. p. 109-112. *In* J. S. Feierabend and A. B. Russell (Eds.). Lead poisoning in wild waterfowl — a workshop. Natl. Wildl. Fed., Washington, D.C.

Yamamoto, K., M. Hayashi, M. Yoshimura, H. Hayashi, A. Hiratsuka, and Y. Isii. 1993. The prevalence and retention of lead pellets in Japanese quail. Arch. Environ. Contam. Toxicol. 24:478-482.

CHAPTER 12

Lead in Mammals

Wei-chun Ma

INTRODUCTION

Lead poisoning in mammals has been often described in domestic livestock upon accidental ingestion of lead-based paint or of feed contaminated with lead shot. Domestic and wild mammals are also exposed to lead when grazing in polluted areas near metal smelters, motorways, and mine tailings. An additional important pathway of lead exposure is through inhalation of lead-containing dust. Several reviews on the exposure of mammals, including humans, to lead are available, e.g., National Research Council (1972), Underwood (1977), Mahaffey (1978), and Forbes and Sanderson (1978).

Lead affects a wide range of physiological systems, including the central nervous system, the kidneys, the hematopoietic system, the cardiovascular system, and the gastrointestinal tract. Lead toxicity in mammals is difficult to diagnose because of the multifaceted nature of the symptoms involved, which may vary from clear-cut clinical symptoms to subtle biochemical changes without overt signs. Clinically manifest signs of lead poisoning in mammals include encephalopathy, peripheral neuropathy, blindness, anemia, hypertension, nephropathy (renal damage and dysfunction), excessive salivation, vomiting, intestinal colic (constipation, diarrhea, abdominal pains), weight loss, stillbirths, and abortions (Rom, 1976; Edwards and Clay, 1977; Posner et al., 1978; Tachon et al., 1983; Kisseberth et al., 1984; de Kort et al., 1987; Sharp et al., 1987; Verschoor et al., 1987). Acute lead poisoning may further cause such abnormal behavior patterns as insomnia, ataxia, eye blinking, rapid ear movement, and teeth grinding as well as loss of appetite and lassitude (Baars et al., 1990).

Apart from overt clinical symptoms due to high-level lead exposure, lead may cause subclinical (asymptomatic) effects in mammals that are chronically exposed to relatively low levels of lead. Such exposure conditions, which are often encountered in environments with lead-polluted soils, may have significant effects on the behavior of animals, particularly if the exposure includes the early developmental stages. Chronic low-level exposure during the prenatal and postnatal stages of mammals may thus cause subnormal physical growth and irreversible disturbed brain development, resulting in neurobehavioral deficits.

The establishment of meaningful dose-effect relations for lead in mammals is very complex because of the multitude of factors involved, e.g., those related to the specific exposure pathway and dietary factors affecting lead absorption. Absorption rates of lead in mammals, for instance, may vary from 10 to 50% for inhalation and from 2 to 20% for ingestion. The confounding effect of a multifactor-mediated exposure on dose-effect relations may be circumvented by applying a unified approach by normalizing the exposure to internal dose levels rather than to environmental or oral intake. Determining internal dose levels in target tissues and in risk assessment with regard to critical levels is useful as a diagnostic approach to metal pollution biomonitoring, even though such an approach obviously is a simplified one, as it does not account for kinetic and dynamic aspects (Ma, 1993).

The objectives of this chapter are the following: (1) to determine the association of internal dose levels of lead in target tissues with (sub)clinical adverse health effects and specific biochemical or morphological responses (biomarkers) indicative of lead exposure in mammals, and (2) to define the critical thresholds for internal lead dose with respect to lead effects. The effects of inorganic lead rather than of organic speciation forms will be addressed. Although organolead compounds such as alkyllead are more toxic than inorganic forms, internal dose levels do not seem to have a clinical significance for the exposure to organic lead. The blood lead level, for instance, is not a reliable index of organolead exposure (Jensen, 1984).

INTERPRETING LEAD CONCENTRATIONS IN MAMMALIAN TISSUES

BLOOD

The usual and generally the best measure of current lead absorption in mammals is blood lead concentration. This is especially true under conditions of long-term exposure when the mean blood lead concentration (PbB) reaches a steady state. Under steady-state conditions, lead in the blood is about 99% confined to the erythrocytes, where it is bound to hemoglobin. In human adults, it is estimated that every 100 µg of long-term daily intake of dietary lead contribute about 10 µg of lead per 100 ml of blood (Lee, 1983). The half-life of lead in the blood is usually about 2 to 4 weeks. An average half-life of 10.5 days has been calculated in acutely lead-poisoned cattle (Baars et al., 1990).

Normal baseline values of PbB in mature mammals are about 2 to 6 µg/dl in cattle (Prpic-Majic et al., 1990), 4 µg/dl in laboratory rats (Barrett and Livesey, 1985), 8 µg/dl in wood mice and 4 µg/dl in bank voles living in uncontaminated areas (Ma, unpublished data), 5 µg/dl or less in monkeys (Laughlin et al., 1983), and 3 to 5 µg/dl in humans from remote nonindustrialized areas (Piomelli et al.,

1980). PbB levels may be higher in immature animals than in mature adults. Neuman and Dollhopf (1992) found that immature cattle from 2 to 12 weeks of age have significantly elevated PbB concentrations compared with mature animals of more than 1 year of age from the same herd. Average background levels of PbB measured in a herd from an uncontaminated area were 7 µg/dl in immature cattle and 5 µg/dl in mature animals. These values rose to 78 and 21 µg/dl, respectively, in a herd grazing in a lead-polluted area close to a metal smelter.

Several clinical symptoms of lead intoxication in mammals are associated with elevated levels of blood lead. Anemia, weight loss, and renal necrosis have been observed in dogs with a PbB of 170 µg/dl due to feeding on a lead-enriched diet (Forbes and Sanderson, 1978). The diet had low calcium and phosphorus contents, which are known to promote lead absorption and retention in mammals (Mahaffey, 1978). Clinical lead toxicity has been observed in cattle with a PbB of 80 µg/dl (Dollahite et al., 1978; Osweiler and Van Gelder, 1983). Impaired kidney function in metal workers has been associated with a PbB of about 60 to 70 µg/dl (Lilis et al., 1979; Verschoor et al., 1987). A disturbed spermatogenesis has been reported in lead workers with a PbB of 50 to 75 µg/dl (Lancranjan et al., 1975).

Clinical neuropathological effects with demyelination and axonal degeneration of motor nerves are also associated with relatively high blood lead levels. The effects of lead on peripheral nerve conduction velocities appear at a PbB of about 40 to 50 µg/dl or more (WHO, 1980). Clinical encephalopathy, with convulsions, stupor, and coma in the advanced form, appears especially in children who have a PbB above 60 to 70 µg/dl. It may develop in adults only at a higher PbB of above 80 µg/dl (WHO, 1977). Fatigue, a frequent symptom of occupational lead intoxication in humans, has been encountered in 50% of a metal worker population with a PbB of about 65 µg/dl (Lilis et al., 1979).

There are indications that lead exposure may also have adverse effects on reproduction in mammals, which include effects on implantation, embryonic development, and male reproductive organs (Rom, 1976; Lee, 1983). Male rats exposed to sufficient lead to produce a PbB of greater than 39 µg/dl showed prostatic hyperplasia, impaired sperm motility, reduced testicular weight, seminiferous tubular damage, and spermatogenic cell arrest (Hildebrand et al., 1973). Male mice exposed to sufficient lead in the drinking water to produce a PbB of 32 µg/dl were shown to have a reduced fertility compared with control mice with a PbB of <0.5 µg/dl, possibly because of disturbed sperm maturation including an impairment of sperm motility (Johansson and Wide, 1986). Prenatal lead exposure at a PbB as low as 10 to 15 µg/dl, and possibly lower, is linked to reduced gestational age, lowered birth weight, and other adverse effects on early development and growth (Davis and Svendsgaard, 1987). It is unknown whether these observed effects are sufficiently severe to reduce the reproductive capacity of wild mammals exposed to environmental lead.

The lowest reported blood lead levels that have been associated with clinical lead toxicosis in mammals are 35 µg/dl, observed in both cattle and horses (Hammond and Aronson, 1964; Buck, 1975; Osweiler and Van Gelder, 1983; Baars et al., 1990). In addition, the highest level of PbB in cattle at which clinical signs of lead toxicosis have not been noted is 29 µg/dl (Sharma et al., 1982). In summary, clinical symptoms of lead poisoning in mammals are associated with a PbB of 35 µg/dl, or higher, and this range in PbB can therefore be taken as a chemical biomarker of acute toxic lead exposure in mammals.

Compared with acute toxicity effects, about which much has been written, relatively few studies have been concerned with subclinical effects of lead in mammals. The most important subclinical effect of lead is associated with the interference with the central nervous system after the passage of lead through the blood-brain barrier and, in the prenatal stage, after crossing of the placenta (Barltop, 1969; Davis, 1990). Subclinical effects are of special concern in the prenatal and early postnatal stages because of their greater gastrointestinal absorption compared with that in adults. The absorption of oral lead in newborn rats is as high as 90%, but decreases to 15% within 20 to 30 days of age (Forbes and Reina, 1972). Young children have about 50% absorption of dietary lead compared with values of 6 to 13% in adults (Rabinowitz et al., 1976; Rosen and Sorrell, 1978). As the placenta is a poor barrier to lead, the PbB in cord blood is almost the same as the maternal blood lead level (Hubermont et al., 1978). Weanlings are exposed to lead upon excretion in milk (Vreman et al., 1986). Lead in milk is incorporated into casein micelles and is absorbed in such protein-bound form by the small intestine of infant rats (Beach and Henning, 1988). Blood lead levels in suckling infants have indeed been associated with lead in milk (Rabinowitz et al., 1985). The life stages most sensitive to low lead exposure are therefore those of the fetus and the developing young.

Minimal impaired brain function has been suggested to occur at a PbB of 50 to 60 µg/dl in children and 60 to 70 µg/dl in adults. Rhesus monkeys chronically exposed by adding lead acetate to the daily milk formula to maintain a PbB of 40 to 60 µg/dl from birth to 1 year of age exhibited learning deficits while showing normal food consumption and weight gain. The effects persisted for many years beyond the termination of lead exposure and were possibly permanent (Laughlin et al., 1983). Similarly, newly born macaque monkeys, given 100 µg/kg of body weight of lead as acetate in milk-feed for 200 days and which at the end of the administration period had a PbB of 25 µg/dl, showed an impaired ability to perform motor discrimination reversal tasks when tested at 3 years of age.

Altered electrophysiological auditory perceptual processing has been observed in animals with a PbB of 35 to 45 µg/dl. Subclinical effects on the peripheral nervous system including a disturbed motor nerve conduction velocity have also been reported at the lower PbB of 30 µg/dl (Seppalainen et al., 1983). These studies also showed that impaired color discrimination persisted down to the lowest dose of 50 µg/kg of body weight, which corresponded to a PbB of 15.4 µg/l. Lambs exposed in utero to lead have been found to exhibit impaired visual discrimination and learning behavior while showing a PbB of 35 µg/dl. Rat weanlings exposed for 50 days to 25 mg/l of lead as acetate in drinking water showed subclinical behavioral deficits at a PbB of 15 to 20 µg/dl, measured at day 86 (Cory-Slechta et al., 1985). Mice with elevated blood lead levels have been found to exhibit inhibition of isolation-induced aggression (Ogilvie and Martin, 1982).

Subclinical toxicity with regard to brain development and functioning may occur in children at a PbB of 10 to 15 µg/dl (Bellinger et al., 1987) compared with 30 to

40 µg/dl in adults (Hogstedt et al., 1983). Neurobehavioral symptoms of lead toxicity in children as well as increased blood pressure in adults may extend all the way down to a PbB of 7 to 8 µg/dl (Pocock et al., 1988). Rabinowitz et al. (1992) investigated the effect of lead exposure, as measured from incisor teeth, on the intelligence quotient (IQ) deficit of young children. The IQ deficit is defined as the difference between a child's observed IQ and the IQ predicted from available information about the child aside from lead. An upper boundary on the threshold was found at a dentine lead level of 3.5 µg/g, which corresponded to a PbB of approximately 8 µg/dl by association.

It is unknown whether the action of lead on the developing brain may affect the survival fitness of wild mammals. However, if toxicological endpoints are considered at the individual level, the available evidence indicates that the safe threshold for lead in immature mammals should be set at a PbB below 15 µg/dl.

LIVER, KIDNEY, BONE, AND OTHER TISSUES

In addition to blood lead concentration, other useful chemical biomarkers of lead exposure in mammals include the concentration of lead in the liver, kidneys, brain, and bone. Lead in soft tissues, including in trabecular bone, has a half-life in the order of weeks and therefore indicates relatively recent exposure. Baars et al. (1990) calculated a half-life of 7.4 days for lead in the liver and kidneys of acutely lead-poisoned cattle.

Among soft tissues, the liver and kidneys contain the highest lead concentration in mammals. Muscle is only a poor indicator of lead burden. The tissue distribution of lead in mammals has the following order: bone > kidneys > liver > brain > muscle. The skeleton, which in adults consists of about 80% cortical bone and 20% trabecular bone, contains more than 90% of the total lead body burden (Barry, 1975). There is no indication to suggest that the concentration of lead in the soft tissues of mammals bears any relation to the concentration in bone (Barry, 1975). The range of conversion factors that can be used to calculate lead concentrations from fresh weight to dry weight or vice versa in mammals is three to five for liver, kidney, and bone tissue (Barry, 1975; Ulrich, 1978; Ma, 1989).

Normal baseline or background values of tissue lead in cattle include a liver Pb concentration of about 0.16 µg/g of fresh weight and a kidney Pb level of 0.46 µg/g of fresh weight (Vreman et al., 1986). The corresponding transfer factors, defined as the concentration of lead in tissue divided by the concentration of lead in the animal's feed, are 0.064 and 0.184 for the liver and kidney, respectively. Considerable variations have been noted in the level of tissue lead in small mammals. Liver and kidney Pb levels have been reported in rodents (mice and voles) from areas used as noncontaminated reference sites in pollution studies. The lower range of background levels is 1 µg/g of dry weight or below (Mierau and Favara, 1975; Welch and Dick, 1975; Getz et al., 1977; Chmiel and Harrison, 1981; Anderson et al., 1982; Anthony and Kozlowski, 1982; Hegstrom and West, 1989; Ma, 1989; Clark et al., 1992). Small mammalian predators (Insectivora) such as shrews, however, have higher baseline values which can be largely attributed to a higher dietary exposure (Andrews et al., 1989; Ma et al., 1991). Values of liver Pb levels in shrews measured at reference sites are 2 μg/g of dry weight, or lower, with kidney Pb concentrations of about 5 μg/g of dry weight (Getz et al., 1977; Chmiel and Harrison, 1981; Hegstrom and West, 1989; Ma, 1989). Insectivora thus have a greater risk of lead intoxication than do rodents, which is due to the fact that the invertebrate prey, notably earthworms, contain relatively high levels of lead through internal accumulation and surface contamination (Ma et al., 1991).

In a systematic study of the normal background values of lead in soft tissues of small mammals, it has been found that the values vary with the dietary habit of the species concerned (Ma et al., 1992). Thus, in areas situated on peat or clay soils, the average background kidney Pb concentrations are 0.2 to 0.6 μg/g of dry weight in mice and voles and 3 to 11 μg/g of dry weight in shrews. In areas on sandy soil, however, the background values are higher, i.e., 0.4 to 1.5 μg in mice and voles and 13 to 19 μg in shrews. Moles in sandy soil have average reference levels similar to those of shrews, i.e., a liver Pb concentration of 10 μg/g of dry weight and a kidney Pb concentration of about 20 μg/g of dry weight (Ma, 1987). Large mammalian predators (Carnivora) are not expected to have a high dietary lead exposure because muscle is a poor accumulator of lead. Polecats have indeed been reported to contain reference kidney Pb values of about 0.4 μg/g of fresh weight (Mason and Weber, 1990). Polar bears have liver Pb levels of 0.5 μg/g of wet weight and kidney Pb levels of 0.6 μg/g (Norheim et al., 1992). Considering the much longer life span of the animals, these background values of lead are relatively low compared with those of small mammals. It is clear that the assessment of lead exposure in wildlife requires a differentiation according to the type of dietary exposure (Ma, 1993).

Lead in hard tissues has a half-life in the order of 10 to 20 years. The lead content of teeth and bones therefore represents a long-term cumulative lead exposure. Normal background concentrations of bone Pb are 2 to 3 μg/g of dry weight or lower in mice and voles from reference sites (Jefferies and French, 1972; Kisseberth et al., 1984; Ma, 1989). In shrews, however, baseline values of bone Pb are higher, i.e., 12 to 55 μg/g of dry weight (Getz et al., 1977; Chmiel and Harrison, 1981; Andrews et al., 1989; Ma, 1989). Lead concentrations in the bone of occupationally nonexposed workers are about 3 μg/g of fresh weight (Ulrich, 1978; Lindh et al., 1980). The lowest bone lead levels in humans are 1.5 μg/g of dry weight in the age group of 0 to 11 months and 3 μg/g in the age group of 12 to 19 years (Samuels et al., 1989). Age may have a clear impact on bone Pb concentrations. Concentrations of Pb in the bone of humans decline with age from about 50 years on, because of bone resorption (Drasch et al., 1987). Bone resorption is especially important in postmenopausal stages and may cause a mobilization of lead sequestered in bone. Similar age effects have also been observed in laboratory rats; i.e., the lowering of lead levels in bone upon advanced age is accompanied by increased concentrations of lead in the brain, liver, and kidneys (Cory-Slechta, 1990).

The toxicity of lead relative to the internal dose level in target organs has been less well documented than relations with the level of blood lead. Studies on rat weanlings have shown that exposure till age 50 days to drinking water containing 25 mg of Pb per liter as lead acetate causes behavioral deficits at day 86 that are accompanied by an elevated brain Pb level of 0.07 µg/g of fresh weight vs. 0.03 µg/g in the controls (Cory-Slechta et al., 1985). Dogs with symptoms of acute lead poisoning by feeding for 12 weeks on a low-calcium/phosphorus diet with 100 µg of Pb per gram showed tissue lead levels of 23 µg/g of fresh weight in the liver and 32 µg/g in the kidneys (Forbes and Sanderson, 1978). The animals also had 1.2 µg of Pb per gram of fresh weight in the brain, 16 µg of Pb per gram of fresh weight in the spleen, and 735 µg of Pb per gram in the bone. Plumbism in cattle has been reported to be associated with tissue lead levels of 26 µg/g of fresh weight in the liver and 50 µg/g in the kidneys (Osweiler and Van Gelder, 1983). Equivalent values in horses were 18 µg/g of fresh weight in the liver and 16 µg/g in the kidney.

Donald et al. (1987) found that pregnant mice showed a normal reproductive success when chronically exposed to lead acetate in drinking water (12 µmol/l) from conception onward. However, birthweight was decreased and growth was retarded when the pups continued to receive lead in drinking water. As food ingestion remained normal, the retarded postnatal development was not related to malnutrition. Social behavioral changes were observed in males and females when tested at ages of up to 16 weeks. When sacrificed at age 20 weeks, females showed an average lead content in femur bone of 1326 µg/g of ash weight, and males showed a much lower content of 227 µg/g. Brain lead levels were 270 and 21 µg/g in females and males, respectively.

Tissue lead levels must be very high before being associated with body weight loss and death. Body weight loss in rats has been found to be associated with a renal lead concentration of 120 µg/g of dry weight (Goyer et al., 1970). Acute poisoning with some deaths has been reported in cattle due to the ingestion of lead shot-contaminated corn silage (Dollahite et al., 1978). The animals showed a liver Pb level of 16 µg/g of fresh weight and a kidney Pb level of >32 µg/g of fresh weight. Kwatra et al. (1986) reported deaths and neurological disorders in cattle consuming vegetation near a lead battery recycling plant and showing very high lead levels of 338 µg/g of dry weight in the liver and 552 µg/g in the kidney. Baars et al. (1990) reported lower mean lead concentrations of 32 µg/g of dry weight in the liver, 87 µg/g in the kidneys, and 125 µg/l in milk of cattle that had died from acute lead poisoning due to the accidental ingestion of lead-contaminated feed. In surviving animals, lead decreased according to a first-order elimination process with half-time values of 7.4 days in the liver and kidneys and of 4.6 days in milk.

On the other hand, talpid moles in areas near metal smelters appear to survive while showing liver Pb levels as high as 40 µg/kg of dry weight and kidney Pb levels of 400 µg/g of dry weight, or above (Ma, 1987). Similar high values of lead in body tissues have been measured in shrews living in a lead shot-polluted area (Ma, 1989). It is unknown whether moles or shrews have a greater resistance to lead than do other mammals or whether selective adaption to lead toxicity has occurred in the population living in the area. Furthermore, these animals did show some effects of lead poisoning (see below).

Lead toxicity in mammals is not noted below a tissue level of 5 µg/g of dry weight in the liver or 10 µg/g of dry weight in the kidneys. Mierau and Favara (1975) thus did not find significant effects on reproductive success, i.e., birth rate and litter survival, or other signs of plumbism for that matter in a roadside population of deer mice showing mean lead levels of 3.3 µg/g of dry weight in the liver and 8.5 µg/g in the kidneys. Lead levels in the bone were 52 µg/g of dry weight and in the brain, 0.84 µg/g of dry weight.

An increased somatic organ index, defined as an increased organ weight relative to the total body weight, is indicative of edema and may be used as an indicator of toxic lead exposure (Goyer et al., 1970; Bankowska and Hine, 1985). Evidence of an increased somatic kidney index has been reported in lead-exposed shrews and bank voles living in a lead shot-contaminated area (Ma, 1989). The shrews had a highly elevated liver Pb level of 16 µg/g of dry weight, a kidney Pb level of 270 µg/g of dry weight, and a Pb level in femur bone of 550 µg/g of dry weight. The somatic kidney index was also found to be increased in bank voles, even though these animals had considerably lower tissue lead levels than did shrews, i.e., 5.1 µg of Pb per gram of dry weight in the liver, 16 µg of Pb per gram of dry weight in the kidney, and 26 µg of Pb per gram of dry weight in the bone. The tissue lead levels were lowest in wood mice, i.e., 2.7 µg/g in the liver, 5.9 µg/g in the kidneys, and 14 µg/g in the bone. These animals did not show a significantly increased somatic kidney index (Ma, 1989). An increased somatic kidney index has also been reported in wood mice living at a lead/zinc mine site (Roberts et al., 1978). The animals showed mean tissue lead levels of 12 µg/g of dry weight in the liver, 47 µg/g of dry weight in the kidneys, 352 µg/g of dry weight in the bone, and 13 µg/g of dry weight in the brain.

In summary, clinical signs of lead toxicosis in mammals appear to be associated with a range of concentrations of lead in the liver and kidneys that extend down to a relatively low level. The evidence thus suggests that liver Pb levels above 5 µg/g of dry weight and kidney Pb levels above 15 µg/g of dry weight can be taken as a chemical biomarker of toxic exposure to lead in mammals.

BIOCHEMICAL BIOMARKERS

There are several biochemical and structural changes that correlate with low-level lead exposure in mammals and therefore can be used as biomarkers. The most sensitive biochemical correlate of blood lead concentration is the inhibition of the enzymes porphobilinogen synthetase (δ-aminolevulinic acid dehydratase or ALAD) and pyrimidine-5'-nucleotidase (Py-5-N) in erythrocytes. The inhibition of both enzymes has a PbB threshold of 5 to 10 µg/dl (Schutz and Skerfving, 1976; WHO, 1977; Angle et al., 1982). ALAD is a cytosolic enzyme that converts δ-aminolevulinic acid (ALA) to porphobilinogen, causing an accumulation of ALA and coproporphyrin in blood plasma and subsequently in urine (ALAU). Normal baseline values for ALAD in cattle have been reported as 4 to 9 units per liter of erythrocytes and about 1.5 to 2.2 µmol/l of erythrocytes for erythrocyte protoporphyrin (EP) values (Prpic-Majic et al., 1990).

Another useful biochemical biomarker of lead exposure is the inhibition of ferrochelatase (heme synthetase), an enzyme that catalyzes the incorporation of the ferrous ion into protoporphyrin IX to form heme. The result is a rise in the formation of EPs, i.e., free erythrocyte protoporphyrin (FEP) and zinc protoporphyrin (ZPP).

Like ALAD activity, EP levels are a reliable indicator of lead dose only under steady-state conditions of exposure. Normal levels of ZPP in wood mice living in uncontaminated areas are about 2.0 to 2.5 mg/l of erythrocytes (Ma, unpublished data). The variability of ALAD and FEP in cattle can be rather large regardless of breed or age (Ruhr, 1984).

Erythrocyte ALAD activity in the blood begins to decrease rapidly when PbB levels are above 10 µg/dl and is almost completely inhibited when PBb levels are in excess of 70 to 80 µg/dl. ALAU increases at PbB levels above about 40 µg/dl (WHO, 1977). The inhibition is a subcritical effect, as ALAD activity has no functional significance in the mature erythrocyte where heme synthesis no longer occurs. Beyer et al. (1985) reported reduced ALAD activity in wild populations of mice near a lead-polluted smelter, but packed cell volumes and hemoglobin values of the blood remained normal. Increased EP and urinary excretion of ALA and coproporphyrin are not detrimental to the health of mammals as the amount of hemoglobin is normally not seriously affected. Decreased hemoglobin levels and the consequent occurrence of anemia are observed only in severe cases of lead poisoning when the reduction of heme synthesis occurs concomitantly with a shortened life span of the erythrocytes. A reduced hemoglobin content occurs at a PbB above 50 µg/dl in human adults and 40 µg/dl in children, and a reduced erythrocyte survival time occurs at more than 60 µg/dl (Hernberg, 1980). The occurrence of anemia has indeed been associated with a PbB of about 60 to 80 µg/l and above (Goyer et al., 1970; WHO, 1977). Anemia is thus a rather insensitive indicator of lead exposure.

The EP level is a useful screening parameter to indicate lead exposure in mammals when used in conjunction with measurements of blood lead. Erythrocyte FEP has a PbB threshold of about 25 to 30 µg/dl in women and children and of about 45 to 50 µg/dl in men (WHO, 1977). According to other studies, a detectable rise in FEP or ZPP in erythrocytes is estimated at a PbB of 15 to 25 µg/dl in young children and 20 to 35 µg/dl in adults (Hammond et al., 1985). The sensitivity of EP as an indicator of lead absorption is lower than that of ALAD (Parsons et al., 1991). In contrast to the immediate response of ALAD inhibition, the effect on EP may take a longer time to manifest itself. Experiments with laboratory mice have shown that FEP and ZPP remain unaltered when treated with lead acetate in drinking water for 6 months, during which period the animals did show an increase in PbB from 4.5 to 60 µg/dl and a decrease in ALAD activity from 21.5 to 6.8 units per liter (Torra et al., 1989). Other experiments done with laboratory rats suggest that it may take a prolonged exposure period for FEP or ZPP to rise to significantly elevated levels in erythrocytes and, moreover, the elevation may take longer in younger than in older rats (Cory-Slechta, 1990).

In any case, the blood lead level correlating with elevated FEP or ZPP values is higher than the very low levels of blood lead associated with decreased ALAD activity. On the basis of the present evidence, ferrochelatase inhibition may serve as a screening parameter for lead exposure in the range associated with clinical lead poisoning, whereas ALAD inhibition is clearly a more sensitive biochemical biomarker of total lead exposure.

MORPHOLOGICAL BIOMARKERS

Lead exposure in mammals may be specifically indicated from detailed examination for structural alterations in the kidneys. An example is the accumulation of nucleotides and basophilic stippling in the erythrocytes due to the inhibition of Py-5-N, an enzyme that mediates the phosphorolysis of pyrimidine nucleotides. Basophilic stippling reflects aggregates of incompletely degraded ribosomal material. The formation of acid-fast-staining intranuclear inclusion bodies in the renal proximal convoluted tubule cells is a characteristic index of lead exposure that is more sensitive than urinary ALA excretion (Goyer et al., 1970; Kisseberth et al., 1984). The pathological significance of the inclusion bodies or their correlation with detrimental health effects is unclear. Renal edema, interstitial nephritis, and tubular and glomerular damage are less sensitive and may be observed at a more advanced stage of lead exposure (Goyer et al., 1970).

In wild small mammals, acid-fast intranuclear inclusion bodies in the kidney have been found in a shrew showing a high kidney lead level of 280 $\mu g/g$ of dry weight (Beyer et al., 1985). No inclusion bodies were found in mice containing a lead level of about 9 to 19 $\mu g/g$ in the liver. Inclusion bodies and mitochondrial alterations have been detected in rats with kidney Pb levels as low as 30 $\mu g/g$ of dry weight (Goyer et al., 1970). Hegstrom and West (1989) could not find any structural symptoms in the kidney of a shrew with a kidney Pb level of 4 $\mu g/g$ of dry weight.

SUMMARY

Although lead can be taken up from the intake of air and water, the more common pathway of exposure of mammals is the transfer of lead via the food chain. In this respect, the major source of chronic exposure to environmental lead is essentially similar for both humans and mammals. Chronic low-level lead exposure of mammals and humans is of greater concern than is acute poisoning, because of the irreversible and diffuse nature of environmental contamination.

Chronic lead poisoning has an insidious and slowly progressive character, with a continuum of biochemical and clinical effects. Most clinical phenomena of lead intoxication in mammals are nonspecific, e.g., anemia, weakness, high blood pressure, fatigue, or changes in nervous system and gastrointestinal tract functioning. A causal relation with lead should therefore always be investigated by demonstrating an increased blood lead level, preferably in conjunction with a biochemical or morphological biomarker. ALAD and Py-5-N inhibition already occurs at very small increases in the PbB below those known to have any physiological effect, whereas ferrochelatase inhibition is related to higher blood lead levels when signs of clinical health effects of lead poisoning become prevalent.

The PbB in conjunction with EP gives the best overall prediction of the risk of clinical lead effects in mammals, as they correlate with most toxicity endpoints of either neurological, renal, hematopoietic, or musculoskeletal nature. Additional determination of bone lead, if feasible, is useful to obtain historical information on the contribution of recent exposure to the blood lead level. In cases of clinical lead poisoning, the PbB in mammals is never less than 35 μg/dl and typically is greater than 60 μg/dl. From the available evidence, it is suggested that a PbB of 35 μg/dl is the lower limit for overt symptoms of clinical lead intoxication in mammals. A PbB of 20 μg/dl can be regarded as the upper limit at which clinical effects of lead exposure are not encountered.

Concerning the significance of lead levels in body tissues other than blood, concentrations of liver Pb of 30 μg/g of dry weight or of kidney Pb of 90 μg/g of dry weight and above appear to be associated with clinical signs of lead poisoning in mammals. Histopathological changes can be detected at three times lower lead levels in the liver than the kidney. Allowing for the differences due to species variation in susceptibility, the present evidence suggests that a liver Pb level above 10 μg/g of dry weight or a kidney Pb level above 25 μg/g of dry weight can be applied for diagnosing acute lead poisoning in mammalian wildlife. These values agree with earlier recommendations made by veterinary pathologists with regard to the diagnosis of lead poisoning in cattle (Fenstermacher et al., 1946; Garner and Papworth, 1967).

In comparison with the abundant studies on clinical lead poisoning, there is much less information available on the adverse effects of low-level lead exposure in mammals. Relatively few studies on mammals are concerned with the effect of exposures that give rise to blood lead levels of less than 35 μg/dl. There is some indication that low-level exposure may produce reproductive impairment, such as decreased fertility and retarded development of the fetus. More studies should be pursued to elucidate the significance of lead on reproduction in mammals.

With regard to impaired brain function, developing stages including the fetus, infants, and young children are more at risk than are adults because of a greater lead absorption efficiency and larger sensitivity in dose-effect relations. The adverse effects on cognitive and behavioral performance occur in early life at asymptomatic levels of lead intoxication. A lead exposure capable of producing a PbB of 10 to 15 μg/dl and above is already very likely to affect the mental and physical development in infants and children. The definition of lead poisoning in children has therefore been lowered from an action level of concern of 25 μg/dl in 1985 to 10 μg/dl in 1991 (Centers for Disease Control, 1991).

Assuming that the susceptibility of wild or domestic mammals to lead does not differ essentially from that of humans, it is suggested that environmental management steps should be considered to reduce lead exposure if the PbB in neonates or pregnant females is 20 μg/dl or above. The paucity of data on the subclinical health effects of lead in mammals renders it impossible to establish an upper range for safe or no-adverse-effect internal concentrations. Obviously more research is needed to establish the range of lead in blood or tissues that gives a no-adverse-effects diagnosis.

The consequences of adverse neurobehavioral effects in early life for the performance of mammals in later life are unknown. However, it can be speculated that diminished learning capability during certain critical developmental phases may not be entirely without consequences to long-term survival potential. It would thus be interesting to know whether changes in motor functions and deficits in reaction time, attentional performance, learning of tasks, and visual-motor discrimination abilities are likely to have any long-term adverse effects on predator-prey relationships and the competitiveness of mammalian wildlife.

REFERENCES

Anderson, T. J., G. W. Barrett, C. S. Clark, V. J. Elia, and V. A. Majeti. 1982. Metal concentrations in tissues of meadow voles from sewage sludge-treated fields. J. Environ. Qual. 11:272-277.

Andrews, S. M., M. S. Johnson, and J. A. Cooke. 1989. Distribution of trace element pollutants in a contaminated grassland ecosystem established on metalliferous fluorspar taillings. I. Lead. Environ. Pollut. 58:73-85.

Angle, C. R., M. S. McIntire, M. S. Swanson, and S. J. Stohs. 1982. Erythrocyte nucleotides in children — increased blood lead and cytidine triphosphate. Pediatr. Res. 16:331-334.

Anthony, R. G., and R. Kozlowski. 1982. Heavy metals in tissues of small mammals inhabiting waste-water-irrigated habitats. J. Environ. Qual. 11:20-22.

Baars, A. J., H. van Beek, I. J. R. Visser, G. Vos, W. van Delft, G. Fennema, G. W. Lieben, K. Lautenbag, J. H. M. Nieuwenhuijs, P. A. de Lezenne Coulander, F. H. Pluimers, G. van der Haar, T. Jorna, L. G. M. T. Tuinstra, P. Zandstra, and B. Bruins. 1990. Lead intoxication in cattle in the northern part of the Netherlands. (In Dutch.) Tijdschr. Diergeneeskd. 115:882-890.

Bankowska, J., and C. Hine. 1985. Retention of lead in the rat. Arch. Environ. Contam. Toxicol. 14:621-629.

Barltop, D. 1969. Transfer of lead to the human fetus. p. 135-151. *In* D. Barltop and W. L. Barland (Eds.). Mineral metabolism in pediatrics. Davis, Philadelphia.

Barrett, J., and P. J. Livesey. 1985. Low level lead effects on activity under varying stress conditions in the developing rat. Pharmacol. Biochem. Behav. 22:107-118.

Barry, P. S. I. 1975. A comparison of concentrations of lead in human tissues. Br. J. Ind. Med. 32:119-139.

Beach, J. R., and S. J. Henning. 1988. The distribution of lead in milk and the fate of milk lead in the gastrointestinal tract of suckling rats. Pediatr. Res. 23:58-62.

Bellinger, D., A. Leviton, C. Waternaux, H. Needleman, and M. Rabinowitz. 1987. Longitudinal analyses of prenatal and postnatal lead exposure and early cognitive development. N. Engl. J. Med. 316:1037-1043.

Beyer, W. N., O. H. Pattee, L. Sileo, D. J. Hoffman, and B. M. Mulhern. 1985. Metal contamination in wildlife living near two zinc smelters. Environ. Pollut. Ser. A Ecol. Biol. 38:63-86.

Buck, W. B. 1975. Toxic materials and neurologic disease in cattle. J. Am. Vet. Med. Assoc. 166:222-226.

Centers for Disease Control. 1991. Preventing lead poisoning in young children. U.S. Department of Health and Human Services, Atlanta, Ga.

Chmiel, K. M., and R. M. Harrison. 1981. Lead content of small mammals at a roadside site in relation to the pathways of exposure. Sci. Total Environ. 17:145-154.

Clark, D. R., Jr., K. S. Foerster, C. M. Marn, and R. L. Hothem. 1992. Uptake of environmental contaminants by small mammals in pickleweed habitats at San Francisco Bay, California. Arch. Environ. Contam. Toxicol. 22:389-396.

Cory-Slechta, D. A. 1990. Lead exposure during advanced age: alterations in kinetics and biochemical effects. Toxicol. Appl. Pharmacol. 104:67-78.

Cory-Slechta, D. A., B. Weiss, and C. Cox. 1985. Performance and exposure indices of rats exposed to low concentrations of lead. Toxicol. Appl. Pharmacol. 71:342-352.

Davis, J. M. 1990. Risk assessment of the developmental neurotoxicity of lead. Neurotoxicology (Little Rock) 11:285-291.

Davis, J. M., and D. J. Svendsgaard. 1987. Lead and child development. Nature (Lond.) 329:297-300.

de Kort, W. L., M. A. Verschoor, A. A. Wibowo, and J. J. van Hemmen. 1987. Occupational exposure to lead and blood pressure: a study in 105 workers. Am. J. Ind. Med. 11:145-156.

Dollahite, J. W., R. L. Younger, H. R. Crookshank, L. P. Jones, and H. D. Petersen. 1978. Chronic lead poisoning in horses. Am. J. Vet. Res. 39:961-964.

Donald, J. M., M. G. Cutler, and M. R. Moore. 1987. Effects of lead in the laboratory mouse. Development and social behaviour after lifelong exposure to 12 µM lead in drinking fluid. Neuropharmacology 26:391-399.

Drasch, G. A., J. Bohm, and C. Baur. 1987. Lead in human bones: investigations on an occupationally non-exposed population in southern Bavaria (FRG). I. Adults. Sci. Total Environ. 64:303-315.

Edwards, W. C., and B. R. Clay. 1977. Reclamation of rangeland following a lead poisoning incident in livestock from industrial airborne contamination of forage. Vet. Hum. Toxicol. 19:247-249.

Fenstermacher, R., B. S. Pomeroy, M. H. Roepke, and W. L. Boyd. 1946. Lead poisoning of cattle. J. Am. Vet. Med. Assoc. 108:1-12.

Forbes, R. M., and J. C. Reina. 1972. Effect of age on gastrointestinal absorption (Fe, Sr, Pb) in the rat. J. Nutr. 102:647-659.

Forbes, R. M., and G. C. Sanderson. 1978. Lead toxicity in domestic animals and wildlife. p. 225-277. In J. O. Nriagu (Ed.). The biogeochemistry of lead in the environment. Part B. Biological effects. Elsevier/North Holland Biomedical Press, Amsterdam.

Garner, R. J., and D. S. Papworth. 1967. Garner's veterinary toxicology. Williams & Wilkins, Baltimore, Md.

Getz, L. L., L. Verner, and M. Prather. 1977. Lead concentrations in small mammals living near highways. Environ. Pollut. 13:151-157.

Goyer, R. A., D. L. Leonard, J. F. Moore, B. Rhyne, and M. R. Krigman. 1970. Lead dosage and the role of the intranuclear inclusion body. Arch. Environ. Health 20:705-711.

Hammond, P. B., and A. C. Aronson. 1964. Lead poisoning in cattle and horses in the vicinity of a smelter. Ann. N.Y. Acad. Sci. 111:595-611.

Hammond, P. B., R. L. Bornschein, and P. Succop. 1985. Dose-effect and dose-response relationships of blood lead to erythrocytic protoporphyrin in young children. Environ. Res. 38:187-196.

Hegstrom, L. J., and S. D. West. 1989. Heavy metal accumulation in small mammals following sewage sludge application to forests. J. Environ. Qual. 18:345-349.

Hernberg, S. 1980. Biochemical and clinical effects and responses as indicated by blood lead concentration. p. 367-399. In R. L. Singhal and J. Thomas (Eds.). Lead toxicity. Urban and Schwarzenberg, Baltimore, Md.

Hildebrand, D. C., H. Der, W. Griffin, and F. S. Fahim. 1973. Effect of lead acetate on reproduction. Am. J. Obstet. Gynecol. 115:1058-1065.

Hogstedt, C., M. Hane, A. Agrell, and L. Bodin. 1983. Neuropsychological test results and symptoms among workers with well-defined long-term exposure to lead. Br. J. Ind. Med. 40:99-105.

Hubermont, G., J. P. Buchet, H. Roels, and R. Lauwerys. 1978. Placental transfer of lead, mercury, and cadmium in women living in a rural area. Int. Arch. Occup. Environ. Health 41:117-124.

Jefferies, D. J., and M. C. French. 1972. Lead concentrations in small mammals trapped on roadside verges and field sites. Environ. Pollut. 3:147-156.

Jensen, A. A. 1984. Metabolism and toxicokinetics. p. 259-266. *In* P. Grandjean (Ed.). Biological effects of organolead compounds. CRC Press, Boca Raton, Fla.

Johansson, L., and M. Wide. 1986. Long-term exposure of the male mouse to lead: effect on fertility. Environ. Res. 41:481-487.

Kisseberth, W. C., J. P. Sundberg, R. W. Nyboer, J. D. Reynolds, S. C. Kasten, and V. R Beasley. 1984. Industrial lead contamination of an Illinois wildlife refuge and indigenous small mammals. J. Am. Vet. Med. Assoc. 185:1309-1313.

Kwatra, M. S., B. S. Gill, R. Singh, and M. Singh. 1986. Lead toxicosis in buffaloes and cattle in Punjab. Indian J. Anim. Sci. 56:412-413.

Lancranjan, T., H. Popescu, O. Gavenescu, J. Klepsch, and M. Serbanescu. 1975. Reproductive ability of workmen occupationally exposed to lead. Arch. Environ. Health 300:396-401

Laughlin, N. K., E. E. Bowman, E. D. Levin, and P. J. Bushnell. 1983. Neurobehavioral consequences of early exposure to lead in rhesus monkeys: effects on cognitive behaviors. p. 497-515. *In* T. W. Clarkson, G. F. Nordberg, and P. R. Sager (Eds.). Reproductive and developmental toxicity of metals. Plenum Press, New York.

Lee, I. P. 1983. Effects of environmental metals on male reproduction. p. 253-279. *In* T. W. Clarkson, G. F. Nordberg, and P. R. Sager (Eds.). Reproductive and developmental toxicity of metals. Plenum Press, New York.

Lilis, R., J. Valciukas, and A. Fischbein. 1979. Renal function impairment in secondary lead smelter workers: correlations with zinc protoporphyrin and blood lead levels. J. Environ. Pathol. Toxicol. 2:1447-1474.

Lindh, U., D. Brune, G. Nordberg, and P. O. Wester. 1980. Levels of antimony, arsenic cadmium, copper, lead, mercury, selenium, tin, and zinc in bone tissues of industrially exposed workers. Sci. Total Environ. 16:109-116.

Ma, W. C. 1987. Heavy metal accumulation in the mole, *Talpa europea*, and earthworms as an indicator of metal bioavailability in terrestrial environments. Bull. Environ. Contam. Toxicol. 39:933-938.

Ma, W. C. 1989. Effect of soil pollution with metallic lead pellets on lead bioaccumulation and organ/body weight alterations in small mammals. Arch. Environ. Contam. Toxicol. 18:617-622.

Ma, W. C. 1993. Methodological principles of using small mammals for ecological hazard assessment of chemical soil pollution, with examples on cadmium and lead. p. 357-371. *In* M. H. Donker, H. Eysackers, and F. Heimbach (Eds.). Ecotoxicology of soil pollution. Lewis Publishers, Boca Raton, Fla.

Ma, W. C., W. Denneman, and J. Faber. 1991. Hazardous exposure of ground-living small mammals to cadmium and lead in contaminated terrestrial ecosystems. Arch. Environ. Contam. Toxicol. 20:266-270.

Ma, W. C., H. van Wezel, and D. van den Ham. 1992. Background concentrations of fifteen metal elements in soil, vegetation, and soil fauna of twelve nature areas in the Netherlands. (In Dutch.) Rep. No. 92/11. Institute for Forestry and Nature Research, the Netherlands, 67 pp.

Mahaffey, K. R. 1978. Environmental exposure to lead. p. 1-36. *In* J. O. Nriagu (Ed.). The biogeochemistry of lead in the environment. Elsevier/North-Holland Biomedical Press, Amsterdam.

Mason, C. F., and D. Weber. 1990. Organochlorine residues and heavy metals in kidneys of polecats *(Mustela putorius)* from Switzerland. Bull. Environ. Contam. Toxicol. 45:689-696.

Mierau, G. W., and B. E. Favara. 1975. Lead poisoning in roadside populations of deer mice. Environ. Pollut. 8:55-64.

National Research Council. 1972. Lead: airborne lead in perspective. National Academy of Sciences, Washington, D.C.

Neuman, D. R., and D. J. Dollhopf. 1992. Lead levels in blood from cattle near a lead smelter. J. Environ. Qual. 21:181-184.

Norheim, G., J. U. Skaare, and O. Wiig. 1992. Some heavy metals, essential elements, and chlorinated hydrocarbons in polar bears *(Ursus maritimus)* at Svalbard. Environ. Pollut. 77:51-57.

Ogilvie, D. M., and A. H. Martin. 1982. Aggression and open-field activity of lead-exposed mice. Arch. Environ. Contam. Toxicol. 11:249-252.

Osweiler, G. D., and G. A. Van Gelder. 1983. Epidemiology of lead poisoning in animals. p. 143-177. *In* F. W. Oehme (Ed.). Toxicity of heavy metals in the environment. Part I. Marcel Dekker, New York.

Parsons, P. J., A. A. Reilly, and A. Hussain. 1991. Observational study of erythrocyte protoporphyrin as a screening test for detecting lead exposure in children: impact of lowering the blood lead action threshold. Clin. Chem. 37:216-225.

Piomelli, S., L. Corash, M. B. Corash, C. Seaman, P. Mushak, B. Glover, and R. Padgett. 1980. Blood lead concentrations in a remote Himalayan population. Science (Washington, D.C.) 210:1135-1137.

Pocock, S. J., A. G. Shaper, D. Ashby, T. Delves, and B. E. Clayton. 1988. The relationship between blood lead, blood pressure, stroke, and heart attacks in middle-aged British men. Environ. Health Perspect. 78:28-30.

Posner, H. S., T. Damstra, and J. O. Nriagu. 1978. Human health effects of lead. p. 173-223. *In* J. O. Nriagu (Ed.). The biogeochemistry of lead in the environment. Elsevier/North-Holland Biomedical Press, Amsterdam.

Prpic-Majic, D., V. Karacic, and L. Skender. 1990. A follow-up study of lead absorption in cows as an indicator of environmental lead pollution. Bull. Environ. Contam. Toxicol. 45:19-24.

Rabinowitz, M., A. Leviton, and H. Needleman. 1985. Lead in milk and infant blood: a dose-response model. Arch. Environ. Health 40:283-286.

Rabinowitz, M. B., J. D. Wang, and W. T. Soong. 1992. Apparent threshold of lead's effect on child intelligence. Bull. Environ. Contam. Toxicol. 48:688-695.

Rabinowitz, M. B., G. W. Wetherill, and J. D. Kopple. 1976. Kinetic analysis of lead metabolism in healthy humans. J. Clin. Invest. 58:260-270.

Roberts, R. D., M. S. Johnson, and M. Hutton. 1978. Lead contamination of small mammals from abandoned metalliferous mines. Environ. Pollut. 15:61-69.

Rom, W. N. 1976. Effects of lead on the female and reproduction: a review. Mt. Sinai J. Med. 43:544-552.

Rosen, J. F., and M. Sorell. 1978. The metabolism and subclinical effects of lead in children. p. 151-172. *In* J. O. Nriagu (Ed.). The biogeochemistry of lead in the environment. Elsevier/North-Holland Biomedical Press, Amsterdam.

Ruhr, L. P. 1984. Blood lead, delta-aminolevulinic dehydratase, and free erythrocyte porphyrins in normal cattle. Vet. Hum. Toxicol. 26:105-107.

Samuels, E. R., J. C. Meranger, B. L. Tracy, and K. S. Subramanian. 1989. Lead concentrations in human bones from the Canadian population. Sci. Total Environ. 89:261-269.

Schutz, A., and S. Skerfving. 1976. Effect of a short, heavy exposure to lead dust upon blood lead level, erythrocyte delta-aminolevulinic acid, and coproporphyrin. Scand. J. Work Environ. Health 3:176-184.

Seppalainen, A. M., S. Hernberg, R. Vesanto, and B. Kock. 1983. Early neurotoxic effects of occupational lead exposure: a prospective study. Neurotoxicology (Little Rock) 4:181-192.

Sharma, R. P., J. C. Street, J. L. Shupe, and D. R. Boureier. 1982. Accumulation and depletion of cadmium and lead in tissues and milk of lactating cows fed small amounts of the metals. J. Dairy Sci. 65:972-979.

Sharp, D. S., C. E. Becker, and A. H. Smith. 1987. Chronic low-level lead exposure. Its role in the pathogenesis of hypertension. Med. Toxicol. 2:210-232.

Tachon, P., A. Laschi, J. P. Briffaux, and G. Brain. 1983. Lead poisoning in monkeys during pregnancy and lactation. Sci. Total Environ. 30:221-229.

Torra, M., M. Rodamilans, J. To-Figueras, I. Hornos, and J. Corbella. 1989. Delta aminolevulinic dehydratase and ferrochelatase activities during chronic lead exposure. Bull Environ. Contam. Toxicol. 42:476-481.

Ulrich, L. 1978. Untersuchungen über den Bleigehalt in Wirbeln und Rippen. (In German.) Arch. Toxicol. 41:133-148.

Underwood, E. J. 1977. Trace elements in humans and animal nutrition, 4th ed. Academic Press, New York.

Verschoor, M., A. Wibowo, R. Herber, J. van Hemmen, and R. Zielhuis. 1987. Influence of occupational low-level exposure on renal parameters. Am. J. Ind. Med. 12:341-351.

Vreman, K., N. G. van der Veen, E. J. van der Molen, and W. G. de Ruig. 1986. Transfer of cadmium, lead, mercury, and arsenic from feed into milk and various tissues of dairy cows — chemical and pathological data. Neth. J. Agric. Sci. 34:129-144.

Welch, W. R., and D. L. Dick. 1975. Lead concentrations in tissues of roadside mice. Environ Pollut. 8:15-21.

WHO. 1977. Environmental health criteria. 3. Lead. World Health Organization, Geneva, Switzerland.

WHO. 1980. Recommended health-based limits in occupational exposure to heavy metals 3. Inorganic lead. p. 36-80. World Health Organization, Geneva, Switzerland.

CHAPTER 13

Toxicological Significance of Mercury in Freshwater Fish

James G. Wiener and Douglas J. Spry

"Fish became contaminated with methylmercury to levels dangerous to consumers. There is little evidence to show what levels of mercury contamination in the fish may be lethal to the creatures themselves."

F. A. J. Armstrong (1979)

INTRODUCTION

Mercury (Hg) is a toxic metal with no known essential function in vertebrate organisms. The potential consequences of mercury contamination of aquatic food chains were first recognized in the 1950s and 1960s in Minamata and Niigata, Japan, where human consumers of methylmercury-contaminated fish were severely poisoned (Nomura, 1968; Tsubaki and Irukayama, 1977; Davies, 1991; Fujiki and Tajima, 1992). Humans and birds were also poisoned by eating seed grain treated with alkylmercury fungicides in the 1960s (Clarkson, 1992). These tragedies prompted widespread reductions in the industrial discharges of mercury into surface waters and in the agricultural use of alkylmercury fungicides. The concentrations of methylmercury declined in fishes inhabiting affected surface waters after industrial discharges of mercury were reduced, although concentrations in fish have remained unacceptably high in some waters (Honma, 1977; Rada et al., 1986; Parks and Hamilton, 1987; D'Itri, 1991; Lodenius, 1991; Fujiki and Tajima, 1992). Recent reports of high mercury concentrations in fish, particularly in newly flooded reservoirs (Hecky et al., 1991; Jackson, 1991; Verdon et al., 1991) and in low-alkalinity lakes (Wiener and Stokes, 1990; Spry and Wiener, 1991), have renewed concerns about mercury in the environment.

The regulatory and scientific focus on mercury in aquatic ecosystems has been motivated largely by the health risks of consuming contaminated fish, because human exposure to methylmercury is almost wholly due to consumption of fish (Clarkson, 1990, 1992; Wiener and Stokes, 1990; Fitzgerald and Clarkson, 1991). Consequently, much effort has been expended to quantify mercury concentrations in key sport fishes and in commercial fishery products (Schreiber, 1983; Håkanson et al., 1988; McMurtry et al., 1989; Lathrop et al., 1991). The environmental fate, bioaccumulation, and mammalian toxicity of mercury have also received considerable study. However, there has been comparatively little study of the potential toxicological effects of mercury contamination on fish populations (Armstrong, 1979; Spry and Wiener, 1991).

In this chapter, we summarize the state of our knowledge of the uptake, tissue distribution, retention, and toxicity of methylmercury in freshwater fish. We identify environmental conditions or situations often associated with high mercury levels in fish and quantify the concentrations typically present in piscivorous fishes under such conditions. Our primary objective is to define the toxicological significance of mercury residues in fish tissues at the organismal and population levels. For consistency, mercury concentrations in fish are expressed as micrograms of total mercury per gram of wet tissue ($\mu g/g$ wet weight, equivalent to parts per million wet weight) throughout the chapter.

MERCURY IN FISH

Nearly all (95 to 99%) of the mercury in fish is methylmercury (Huckabee et al., 1979; Grieb et al., 1990; Bloom, 1992), even though very little of the total mercury in the waters and sediments of freshwater ecosystems exists as methylmercury (Bloom et al., 1991; Bubb et al., 1991; Lee and Iverfeldt, 1991; Wilken and Hintelmann, 1991; St. Louis et al., 1994; Watras et al., 1994). Fish do not methylate inorganic mercury within their tissues (Pennacchioni et al., 1976; Huckabee et al., 1978; Birge et al., 1979), although methylation does occur within the gut (Rudd et al., 1980). Fish obtain methylmercury from their diet and, to a lesser extent, from water passed over the gills (Olson et al., 1973; Phillips and Buhler, 1978; Huckabee et al., 1979; Rodgers, 1994). Inorganic mercury is absorbed much less efficiently across the gut and gills and is eliminated more rapidly than is methylmercury (Huckabee et al., 1979; Ribeyre and Boudou, 1984; Boudou and Ribeyre, 1985; Niimi and Kissoon, 1994).

Mercury in natural waters can exist in many forms, including elemental mercury (Hg^0), dissolved and particulate ionic forms, and dissolved and particulate methylmercury (Gill and Bruland, 1990; Vandal et al., 1991; Mason and Fitzgerald, 1993). The production of methylmercury by methylation of inorganic Hg(II) in the environment, which is primarily a microbial process, is a key mechanism affecting the quantity of mercury accumulated in fish (Rudd et al., 1983; Winfrey and Rudd, 1990; Gilmour and Henry, 1991; Gilmour et al., 1992; Bodaly et al., 1993). Methylmercury can be produced within a water body by methylation of inorganic mercury in sediments and the water column (Winfrey and Rudd, 1990; Gilmour and Henry, 1991). A water body can also import methylmercury in wet deposition or in surface inflows from the catchment (Bloom and Watras, 1989; Lee and Iverfeldt, 1991).

UPTAKE FROM FOOD AND WATER

Abundances of methylmercury and total mercury in fresh waters are much lower than previously believed (Gill and Bruland, 1990; Porcella et al., 1992), even though all water bodies in the Northern Hemisphere are probably contaminated because of long-range transport and deposition of mercury from anthropogenic sources (Evans, 1986; Rada et al., 1989; Verta et al., 1989; Norton et al., 1990; Swain et al., 1992; Rognerud and Fjeld, 1993). Recent applications of reliable methods for quantifying methylmercury in water (Bloom, 1989) and of clean techniques for collecting and handling samples (Gill and Fitzgerald, 1987; Fitzgerald and Watras, 1989; Gill and Bruland, 1990; Porcella et al., 1992) have shown that methylmercury concentrations in oxic surface waters typically range from about 0.01 to 0.8 ng of Hg per liter (Table 1). Concentrations of methylmercury seem to be higher in surface waters that drain wetland areas (e.g., 0.18 to 0.63 ng of Hg per liter; St. Louis et al., 1994) than in many other waters. In nearly pristine or lightly contaminated lakes and streams, concentrations of total mercury in water (unfiltered samples) are in the range of 0.6 to 4 ng/l (Table 1) but may seasonally be much higher in very humic streams (Johansson et al., 1991; Mierle and Ingram, 1991). In lakes and streams with direct sources of mercury, the concentrations of total mercury in water vary from about 5 to 100 ng/l and are often in the range of 10 to 40 ng/l (Table 1).

Empirical evidence indicates that the diet is the primary route of methylmercury uptake by fish in natural waters (Phillips et al., 1980; MacCrimmon et al., 1983; Mathers and Johansen, 1985; Cope et al., 1990; Harrison et al., 1990; Andersson and Håkanson, 1991; Rask and Metsälä, 1991; Francesconi and Lenanton, 1992; Futter, 1994), probably contributing more than 90% of the methylmercury accumulated (Spry and Wiener, 1991; Rodgers, 1994). The assimilation efficiency for the uptake of dietary methylmercury in fish is probably 65 to 80% or greater (Rodgers, 1994). Phillips and Gregory (1979) estimated that piscivorous northern pike *(Esox lucius)* assimilated about 20% of the methylmercury present in naturally contaminated forage fish; however, bioenergetic-based models show that an assimilation efficiency of 20% is too small to account for the observed methylmercury accumulation in fish (Rodgers, 1994). About 7 to 12% of the waterborne methylmercury passing over the gills of fish is assimilated (Phillips and Buhler, 1978; Rodgers and Beamish, 1981). In temperate waters, the accumulation of mercury by fish seems to be most rapid in summer (Uthe et al., 1973; Weis et al., 1986), when the feeding and metabolic rates of fish and the microbial production of methylmercury are greatest (Korthals and Winfrey, 1987; Bodaly et al., 1993; Ramlal et al., 1993).

Although laboratory studies have shown that fish can accumulate high concentrations of methylmercury directly from water (Reinert et al., 1974; Olson et al., 1975; Wobeser, 1975a; McKim et al., 1976; Phillips and Buhler, 1978; Niimi and Kissoon, 1994), the *minimum* exposure concentrations in such studies have typically been in the range of 20 to 1000 ng of Hg per liter or greater. Even the *lowest* exposure concentrations of methylmercury in these laboratory studies were at least 10 to 10^3 times greater than the *highest* concentration of methylmercury reported for fresh waters (2.0 ng of Hg per liter in grossly polluted Onondaga Lake, New York;

Table 1 Range of Concentrations of Total Mercury, Dissolved Mercury, and Methylmercury in Oxic Fresh Waters, from Studies in which Trace-Metal-Free Protocols were used during Handling and Analyses of Samples

Location	Number and type of surface waters sampled	Concentration (Hg/l)			Ref.
		Total Hg	Dissolved Hg[a]	Methyl Hg	
Lakes and Reservoirs					
Northern Manitoba, Canada	Flooded Southern Indian Lake	0.7-3.2		0.014-0.045	Ramsey (1990b)
	Granville Lake	0.7-2.4		0.013-0.054	Ramsey (1990b)
	Stephens Lake	0.8-2.7		0.027-0.060	Ramsey (1990b)
	Notigi Reservoir	1.1		0.046	Ramsey (1990b)
Northern Wisconsin, U.S.	4 seepage lakes	0.9-1.9			Fitzgerald and Watras (1989)
	6 seepage lakes	0.7-2.1		0.05-0.33	Watras et al. (1994)
Sweden	3 lakes	1.4-15		0.06-0.41	Lee and Hultberg (1990)
	8 drainage lakes			0.04-0.8	Lee and Iverfeldt (1991)
	25 forest lakes	1.9-7.8			Meili et al. (1991)
New York State, U.S.	Onondaga Lake (Hg polluted)	7-19	2-10 (0.2 μm)	0.4-2.0	Bloom and Effler (1990)
California, U.S.	Silver Lake (pristine)	0.6	0.4		Gill and Bruland (1990)
	Clear Lake (Hg-polluted tailings)	3.6-104	1.1-1.5		Gill and Bruland (1990)
	Davis Creek Reservoir (Hg mine tailings)	2.7-13	1.9-4.0		Gill and Bruland (1992)
Nevada, U.S.	Mono Lake (desert lake)	20-22	11-13		Gill and Bruland (1990)
	Pyramid Lake (desert lake)	1.9	0.9		Gill and Bruland (1990)
N. American Great Lakes	Lake Ontario	0.9	0.7		Gill and Bruland (1990)
	Lake Erie	3.6	1.8		Gill and Bruland (1990)

Location	Description				Reference
Guangxi Province, People's Republic of China	Nan Hu (urban lake)	3.4-12		0.07-0.12	Bloom et al. (1994)
Streams and rivers					
Wisconsin (statewide), U.S.	Rivers and streams (39 sites)	0.7-43	0.3-24	<0.02-0.87	Hurley et al. (1995)
Northern Manitoba, Canada	Burntwood River	2.1		0.027	Ramsey (1990b)
Southeastern Ontario, Canada	6 streams	1.6-3.2[b]			Mierle (1990)
	5 humic streams	1-28[c]			Mierle and Ingram (1991)
Northwestern Ontario, Canada	5 streams	1.7-13[d]		0.03-0.63[d]	St. Louis et al. (1994)
Sweden	13 streams	1.0-6.5			Iverfeldt and Johansson (1988)
West Coast, U.S.	Sacramento River (at Sacramento, Calif.)	4.6	0.9		Gill and Bruland (1990)
	Columbia River (at Dodson, Ore.)	2.4	0.4		Gill and Bruland (1990)
	Davis Creek, Calif. (abandoned Hg mine)	12-34	12		Gill and Bruland (1990)
Central Italy	3 rivers (cinnabar deposits)	7-270	2-10 (0.45 μm)		Ferrara et al. (1991)
Brazil	Madeira River and tributaries (Hg polluted)	20-33	10-17		Nriagu et al. (1992)

[a] Filter-pore size of 0.4 μm, unless otherwise indicated.
[b] Range of volume-weighted mean concentrations for the six streams.
[c] Range of concentrations in individual samples of stream water.
[d] Range of mean concentrations for the five streams.

Table 1). Laboratory and field tests examining bioaccumulation in fish exposed to environmentally realistic concentrations of waterborne and dietary methylmercury are needed.

TISSUE DISTRIBUTION AND RETENTION

In fish, methylmercury rapidly penetrates and is cleared from the gut and the gills, binds to red blood cells, and is rapidly transported to all organs, including the brain (Giblin and Massaro, 1973; McKim et al., 1976; Olson et al., 1978; Harrison et al., 1990). The route of uptake (gut vs. gill) has little influence on the distribution of methylmercury among most internal organs and tissues, except that concentration in the gills are much higher after waterborne (than dietary) exposure and concen trations in the intestines are higher after dietary exposure (McKim et al., 1976 Huckabee et al., 1979; Boudou and Ribeyre, 1983; Harrison et al., 1990). Concen trations of methylmercury in exposed fish are typically greatest in the blood, spleen kidney, and liver in both laboratory tests (McKim et al., 1976; Olson et al., 1978 Boudou and Ribeyre, 1983, 1985; Ribeyre and Boudou, 1984) and field studies (Harrison et al., 1990; Niimi and Kissoon, 1994).

There is a dynamic internal redistribution of assimilated methylmercury among the tissues and organs of fish exposed to methylmercury in laboratory and field studies. The concentrations and burdens (masses) in the blood, spleen, kidney, liver and brain decrease after exposure to either waterborne or dietary methylmercury ceases, and skeletal muscle is the primary "receiver" of the redistributed methylm ercury (Boudou and Ribeyre, 1983; Ribeyre and Boudou, 1984; Harrison et al. 1990). Most of the methylmercury in the body eventually accumulates in muscle bound to sulfhydryl groups in protein, even though concentrations are usually les in muscle than in other tissues (Giblin and Massaro, 1973; McKim et al., 1976 Olson et al., 1978; Huckabee et al., 1979; Boudou and Ribeyre, 1983, 1985; Ribeyre and Boudou, 1984; Harrison et al., 1990).

Fish eliminate methylmercury very slowly relative to its rate of uptake (Lockhart et al., 1972; Ruohtula and Miettinen, 1975; Laarman et al., 1976; McKim et al. 1976). Possible excretory pathways include the feces, kidney, and gill (Giblin and Massaro, 1973).

ELEVATED MERCURY CONCENTRATIONS IN FISH POPULATIONS

The accumulation of methylmercury in aquatic organisms is influenced by many biotic, physicochemical, ecological, and human processes (Verta, 1990; Boudou et al., 1991; Meili, 1991; Spry and Wiener, 1991; Miskimmin et al., 1992). We briefly review selected papers to illustrate some of the factors that are associated with the accumulation of high concentrations of methylmercury in fish. These factors include piscivorous feeding habits, biomagnification of methylmercury in food chains, fish age and longevity, anthropogenic discharges of mercury to the environment, wate temperature, low acid-neutralizing capacity of surface waters, atmospheric deposition of mercury to lakes, and flooding of new impoundments.

The cycling of mercury in the global environment has been altered significantly by human activities. Humans have used mercury for thousands of years for many applications (Clarkson, 1994) and, since 1900, anthropogenic emissions of mercury to the environment have increased greatly (Nriagu and Pacyna, 1988; Fitzgerald and Clarkson, 1991). On a global scale, recent human-related emissions of mercury to the atmosphere may equal or exceed emissions from natural sources (Nriagu, 1989). Slemr and Langer (1992) estimated that atmospheric concentrations of total gaseous mercury in the Northern Hemisphere increased about 1.5% per year during 1977 through 1990, which they attributed to anthropogenic releases. Moreover, spatial gradients in the mercury content of air and precipitation (Iverfeldt, 1991), surface mineral soil and organic litter (Nater and Grigal, 1992), lake sediments (Rognerud and Fjeld, 1993), and fish (Johansson et al., 1991; Lathrop et al., 1991; Fjeld and Rognerud, 1993) parallel spatial gradients in atmospheric mercury deposition, sulfate deposition, and human activity, indicating a link between mercury deposition and anthropogenic activities.

Mercury is readily dispersed and transported through the atmosphere, and the aquatic and global cycling of mercury is strongly affected by exchange processes across the air-water interface (Fitzgerald et al., 1991; Mason and Fitzgerald, 1993). Mercury in air exists primarily as elemental mercury (Hg^0) and, to a much lesser extent, in the particle phase (Schroeder et al., 1991). Elemental mercury can be oxidized to mercuric ion, Hg(II), which readily absorbs to inorganic and organic particles and is removed from the atmosphere in wet and dry deposition (Fitzgerald et al., 1991). Further discussion of the mercury cycle is beyond the scope of this chapter. However, several features of the biogeochemical cycling of mercury have been examined recently, including global fluxes (Fitzgerald and Clarkson, 1991), atmospheric transformations (Schroeder et al., 1991), speciation and fluxes in oceans (Mason and Fitzgerald, 1993) and inland lakes (Meili et al., 1991; Watras et al., 1994), methylation and demethylation (Winfrey and Rudd, 1990; Gilmour and Henry, 1991), and complexation with sulfur (Dyrssen and Wedborg, 1991).

DIET, TROPHIC STRUCTURE, AND LONGEVITY

Feeding habits and the food-chain structure influence the methylmercury uptake in fish, and piscivorous fishes usually contain higher concentrations than do coexisting fishes of lower trophic levels (Phillips et al., 1980; MacCrimmon et al., 1983; Wren et al., 1983; Francesconi and Lenanton, 1992). Methylmercury biomagnifies in aquatic food chains (Wren et al., 1983; Francesconi and Lenanton, 1992; Watras and Bloom, 1992). Moreover, the fraction of total mercury that exists as methylmercury in aquatic organisms increases progressively from primary producers to fish (Hildebrand et al., 1980; May et al., 1987; Francesconi and Lenanton, 1992; Watras and Bloom, 1992).

The trophic structure of a waterbody can influence mercury concentrations in fish, particularly in species that are typically piscivorous. For example, lake trout

(Salvelinus namaycush) have higher mercury concentrations when forage fish, such as rainbow smelt *(Osmerus mordax)*, are present (Akielaszek and Haines, 1981; Futter, 1994). Similarly, mercury concentrations in northern pike in a Finnish lake lacking forage fish were about one fourth of those in northern pike in nearby, similar lakes with forage fish (Rask and Metsälä, 1991).

Lake morphometry and temperature also affect the bioaccumulation of mercury in fish. In northwestern Ontario, for example, Bodaly et al. (1993) found that the *mean* concentrations in axial muscle of walleyes *(Stizostedion vitreum)* and northern pike ranged about 0.7 to 1.1 µg/g wet weight in small (89- to 706-hectare [ha]) lakes, but were less than 0.4 µg/g in nearby, larger (2219- to 34,690-ha) and colder lakes. Specific rates of mercury methylation in the lakes were positively correlated with water temperature, whereas specific rates of methylmercury demethylation (by microbes) were negatively correlated with temperature. Bodaly et al. (1993) attributed the differences in mercury concentrations in fish among the lakes to the temperature-related variation in microbial production of methylmercury in the epilimnia.

Whole-lake experiments indicate that mercury enters food chains more rapidly in small, shallow lakes with high littoral/pelagic area ratios than in large, deep lakes. Harrison et al. (1990), who added inorganic ^{203}Hg to the epilimnion of oligotrophic Lake 224 in northwestern Ontario, observed rapid accumulation of ^{203}Hg (presumably after methylation) by fathead minnows *(Pimephales promelas)* in the epilimnion. Lake trout inhabiting the hypolimnion of the same lake accumulated the added ^{203}Hg much more slowly and had lower concentration factors than did the fathead minnows.

Mercury concentrations in tissues of a fish species within a given water body generally increase with increasing age or body size (Phillips and Buhler, 1978; Huckabee et al., 1979; Lange et al., 1993), because of the very slow rate of elimination of methylmercury relative to the rate of uptake. In addition, the methylmercury content of the diet of some fishes, particularly those that are partly or totally piscivorous as adults, increases as the fish grows larger (Mathers and Johansen, 1985). The rate of methylmercury accumulation in lake trout, for example, increases greatly when the fish becomes large enough to change from a diet of invertebrates to forage fish (MacCrimmon et al., 1983). Consequently, mercury concentrations in axial muscle sometimes exceed 0.5 or 1.0 µg/g wet weight (values widely used as criteria in fish-consumption advisories) in long-lived, piscivorous fish inhabiting waters that lack both direct sources of anthropogenic mercury and conditions, such as low pH, that enhance the methylation or bioavailability of the metal (Sumner et al., 1972; Bodaly et al., 1993).

POINT-SOURCE POLLUTION

High concentrations of methylmercury have been widely reported for fish from surface waters contaminated by mercury from point sources (Table 2), including chloralkali plants (Hildebrand et al., 1980; Rudd et al., 1983; Effler, 1987; Bloom and Effler, 1990; Mukherjee, 1991), pulp and paper mills (Rada et al., 1986; Lodenius, 1991; Mukherjee, 1991), and other industrial facilities (Turner et al., 1984; Mukherjee, 1991; Francesconi and Lenanton, 1992). Concentrations in axial muscle in *individual* piscivorous fishes taken from these contaminated waters were often in the range of 1 to 10 µg/g wet weight (Sumner et al., 1972; Effler, 1987; Parks and Hamilton, 1987; Francesconi and Lenanton, 1992), with mean concentrations in piscivorous species often in the range of 1 to 7 µg/g wet weight (Sumner et al., 1972; Armstrong and Scott, 1979; Effler, 1987; Parks and Hamilton, 1987; Lodenius, 1991; Francesconi and Lenanton, 1992). High concentrations of methylmercury have also been observed in omnivorous fishes inhabiting such contaminated waters; Hildebrand et al. (1980), for example, found about 1.2 to 2.5 µg/g wet weight in axial muscle of rock bass *(Ambloplites rupestris)* and northern hogsucker *(Hypentelium nigricans)* at sites just downstream from an inactive chloralkali plant in the North Fork Holston River, Virginia.

Point-source discharges of mercury to surface waters declined in many industrialized countries since the 1960s and 1970s. Such reductions were generally followed by decreased mercury concentrations in aquatic biota. During 1977 through 1988, for example, whole-body concentrations of mercury in lake trout, rainbow smelt, and slimy sculpin *(Cottus cognatus)* from Lake Ontario declined about 50% (Borgmann and Whittle, 1991, 1992). In some waters, however, concentrations in fish continued to exceed 0.5 or 1.0 µg/g wet weight for several years after mercury inputs were reduced (Parks and Hamilton, 1987; Lodenius, 1991) or after industrial-source plants were inactivated (Hildebrand et al., 1980).

Gold-mining operations that used the mercury-amalgamation process have caused long-term contamination of sediment and fish in certain rivers (Bycroft et al., 1982; Cooper and Vigg, 1984). Recent gold-mining activities caused substantial mercury pollution in the Madeira River in the Amazon River basin of South America (Martinelli et al., 1988; Pfeiffer et al., 1991; Nriagu et al., 1992). Mercury concentrations in axial muscle of fish from contaminated sites in the Madiera River frequently exceeded 1.0 µg/g wet weight (Martinelli et al., 1988; Pfeiffer et al., 1991).

ATMOSPHERIC DEPOSITION TO LOW-pH AND HUMIC LAKES

Piscivorous fish in waters with low acid-neutralizing capacity (≤60 µeq/l), low pH (≤6.7), or high humic content often contain mercury concentrations in axial muscle in the range of 0.5 to 2.0 µg/g wet weight (Table 2), even in waters far from anthropogenic sources of the metal (Mannio et al., 1986; Verta, 1990; Spry and Wiener, 1991; Wren et al., 1991). This is a geographically widespread pattern, observed in surveys of largemouth bass *(Micropterus salmoides)* (Lange et al., 1993), smallmouth bass *(M. dolomieu)* (Suns et al., 1987; McMurtry et al., 1989), walleye (Scheider et al., 1979; Wiener et al., 1990b; Lathrop et al., 1991), and northern pike (Håkanson, 1980; Björklund et al., 1984; Heiskary and Helwig, 1986; Verta et al., 1990; Wren et al., 1991). This pattern is also evident in forage fishes, such as yellow perch *(Perca flavescens)* (Cope et al., 1990; Suns and Hitchin, 1990). The greater accumulation of methylmercury in fish in low-pH waters has been attributed in part to greater in-lake microbial production of methylmercury (Xun et al., 1987; Wiener et al., 1990a, 1990b; Winfrey and Rudd, 1990; Gilmour and Henry, 1991; Gilmour

Table 2 Mercury Concentrations in Axial Muscle Tissue of Largely Piscivorous Fish in Surface Waters in Relation to Sources of Anthropogenic Mercury or Environmental Conditions that Enhance the Bioavailability of the Metal

Probable source of Hg or environmental condition and location of surface waters	Fish species	n	Wet weight (kg) of fish analyzed Mean	Wet weight (kg) of fish analyzed Range	Total length (cm) of fish analyzed Mean	Total length (cm) of fish analyzed Range	Hg concentration (μg/g wet weight) Mean	Hg concentration (μg/g wet weight) Range	Ref.
Chloralkali plant									
Saskatchewan River, Saskatchewan, Canada	Northern pike	3					4.8	2.6-6.1	Sumner et al. (1972)
Clay Lake, northwestern Ontario, Canada	Walleye				50		4-15[a]		Parks and Hamilton (1987)
	Northern pike				60		3-6[b]		Parks and Hamilton (1987)
Ball Lake, northwestern Ontario (1971)	Walleye	60	1.05		44[c]		2.0		Armstrong and Scott (1979)
	Northern pike	20	1.52		59[c]		5.1		Armstrong and Scott (1979)
Onondaga Lake, New York, U.S.	Walleye	122					2.4	0.37-7.9	Effler (1987)
	Smallmouth bass	325				18-44	1.0	0.14-2.2	Effler (1987)
Superphosphate plant									
Princess Royal Harbour, southwestern Australia (marine embayment)	Long-headed flathead[d]	16			51	40-60	6.7	2.6-9.0	Francesconi and Lenanton (1992)
	Blue-spotted flathead[d]	29			46	28-63	3.8	0.11-9.1	Francesconi and Lenanton (1992)
Pulp and paper mill									
Lake Kirkkojärvi, Southwestern Finland	Northern pike	294					1.5[a]		Lodenius (1991)
Wisconsin River, Wisconsin, U.S.	Walleye	76	0.40	0.03-2.9	31	15-65	0.35	0.05-2.0	Rada et al. (1986)
Atmospheric deposition to low-pH lakes									
Northern Wisconsin, U.S., 8 lakes, pH 5.0-6.7	Walleye	34	1.01	0.45-1.67	46	37-55	0.93	0.41-1.7	Wiener et al. (1990b)
Ontario, Canada, lakes with pH ≤7.0									G. Mierle, personal communication[g]
62 lakes with mean pH 6.8[f]	Walleye	1,021	0.96	0.34-2.22	43	30-59	0.75	0.35-1.55	
63 lakes with mean pH 6.8[f]	Northern pike	822	1.13	0.44-2.56	54	39-71	0.58	0.30-1.09	
14 lakes with mean pH 6.5[f]	Largemouth bass	162	0.50	0.21-1.09	31	23-40	0.46	0.27-0.87	

Location	Species	n							Reference
72 lakes with mean pH 6.4[f]	Smallmouth bass	1,099	0.49	0.18-1.14	30	22-41	0.53	0.26-1.08	Allard and Stokes (1989)
Southern Ontario, Canada, 7 lakes (pH 6.0-6.6)	Smallmouth bass	169					0.63	0.10-2.9	
Finland, 17 lakes (pH 4.9-6.6)	Northern pike	223		0.002-6.1		6-97	0.71[h]	0.02-2.1	Rask and Metsälä (1991)
Florida, U.S., 25 lakes (pH 3.6-6.7)	Largemouth bass	344	0.81	0.04-4.6	36	15-66	0.76	0.15-2.0	Lange et al. (1993)
Newly flooded lakes and reservoirs									
Southern Indian Lake, northern Manitoba, Canada (21% increase in area after flooding)	Northern pike[j]	531			55[f]		0.7	0.05-2.6	Bodaly et al. (1984)
	Walleye[j]	289			38[f]		0.6	0.06-2.2	Bodaly et al. (1984)
Notigi Reservoir, northern Manitoba, Canada (282% increase in area)	Northern pike[f]	120			66[i]		1.5	0.24-2.8	Bodaly et al. (1984)
	Walleye[i]	155			43[f]		1.8	0.19-3.5	Bodaly et al. (1984)
La Grande 2 Reservoir, Québec, Canada	Northern pike[j]		2.1		70		3.0		Verdon et al. (1991)
	Walleye[k]		0.63		40		2.8		Verdon et al. (1991)
Six reservoirs (≤10 years old) in Finland	Northern pike	192					1.4	0.5-4.1	Verta et al. (1986)
Lake Jocassee, South Carolina, U.S.	Largemouth bass	41	0.9	0.2-2.3	29	18-44	2.9	1.3-5.7	Abernathy et al. (1975)
Golf courses (treated with Hg fungicides)									
Seven golf-course lakes, Missouri, U.S.	Largemouth bass	17	1.2	0.05-2.55			2.2	0.43-7.1	Koirtyohann et al. (1974)
Gold mining (Hg-amalgamation process)									
Lahontan Reservoir, Carson River, northern Nevada, U.S.	White bass	8	0.16	0.04-0.26			2.5	0.97-4.0	Cooper (1983)

[a] Range of concentrations in 50-cm walleye, estimated for each year from 1970 to 1983.
[b] Range of concentrations in 60-cm northern pike, estimated for each year from 1975 to 1983.
[c] Fork length, rather than total length, is given.
[d] The long-headed flathead is *Platycephalus inops*; the blue-spotted flathead is *Platycephalus speculator*.
[e] Estimated value for 1-kg fish during 1971-1974.
[f] The minimum and maximum values given for fish weight, total length, and mercury concentration are the mean minimum values and the mean maximum values, respectively, for the group of lakes sampled.
[g] Data from the Ontario Ministry of Environment and Energy, Sport Fish Contaminant Survey.
[h] Estimated grand mean for 1-kg northern pike in the 17 lakes.
[i] Survey data only (samples from commercial fisheries not included); fork length, rather than total length, is given.
[j] Data for all variables are estimates for fish having a standardized total length of 70 cm, sampled 9 years after reservoir creation.
[k] Data for all variables are estimates for fish having a standardized total length of 40 cm, sampled 9 years after reservoir creation.

et al., 1992; Miskimmin et al., 1992). In addition, the efflux of the metal (as elemental mercury, Hg^0) to the atmosphere seems to be much greater from high-pH lakes than from low-pH lakes (Fitzgerald et al., 1991; Vandal et al., 1991; Watras et al., 1994), leading to greater retention of mercury in low-pH aquatic systems (Rada et al., 1993).

In regions of Sweden, Finland, Canada, and the United States that have many low-alkalinity and humic waters, much of the mercury in fish in remote or semiremote lakes seems to be derived from atmospheric deposition (Björklund et al., 1984; Johnson, 1987; Mierle, 1990; Verta, 1990; Wiener et al., 1990a; Fitzgerald et al., 1991; Johansson et al., 1991). The surficial sediments in lakes (Evans, 1986; Rada et al., 1989; Verta et al., 1989; Norton et al., 1990; Swain et al., 1992; Rognerud and Fjeld, 1993) and the surficial strata of peat cores in ombrotrophic bogs (raised, precipitation-dominated bogs; Jensen and Jensen, 1991; Steinnes and Andersson, 1991) in the Northern Hemisphere are enriched with mercury, a pattern indicating recent increases in atmospheric deposition of the metal. In the north-central U.S., the rate of atmospheric deposition of mercury has probably increased by a factor of 2 to 4 since the mid-1800s (Rada et al., 1989; Norton et al., 1990; Swain et al., 1992). In northern Wisconsin, the annual atmospheric input (wet plus dry deposition) of mercury averaged about 0.1 g/ha during 1988 through 1990, an input sufficient to account for the mass of mercury accumulating in fish in semiremote seepage lakes, which lack surface inflows (Wiener et al., 1990a; Fitzgerald et al., 1991). Drainage lakes receive significant influxes of atmospheric mercury indirectly in streamflow from their catchments as well as direct inputs in wet and dry deposition (Iverfeldt and Johansson, 1988; Lee and Hultberg, 1990; Mierle, 1990; Johansson et al., 1991; Mierle and Ingram, 1991).

NEWLY FLOODED RESERVOIRS

Piscivorous fish in newly flooded reservoirs and impounded lakes also contain high concentrations of methylmercury (Verta et al., 1986; Hecky et al., 1991; Jackson, 1991; Verdon et al., 1991; Yingcharoen and Bodaly, 1993). In newly flooded temperate and subarctic reservoirs, concentrations in axial muscle of piscivorous fishes often average 0.6 to 3.0 µg/g wet weight; maximum concentrations of 2 to 6 µg/g (Table 2) can, in some cases, equal or exceed concentrations in fishes from waters heavily contaminated by direct industrial discharges (Surma-Aho et al., 1986; Hecky et al., 1991). For comparison, the mean concentrations in northern pike and walleye were typically in the range of 0.20 to 0.35 µg/g in existing surface waters before flooding by the Churchill River diversion in northern Manitoba, Canada (Bodaly et al., 1984). Nine years after creation of the La Grande 2 reservoir (part of the La Grande hydroelectric complex) in the Canadian province of Québec, standardized concentrations of mercury were 3.0 µg/g wet weight in 70-cm northern pike and 2.8 µg/g in 40-cm walleye (Verdon et al., 1991); concentrations were even higher (3.5 µg/g in 70-cm northern pike) farther downstream, in an unimpounded section of the La Grande River. In comparison, concentrations in fish from 29 reference lakes near the La Grande complex were 0.6 µg/g wet weight in 70-cm northern pike and 0.7 µg/g in 40-cm walleye (Verdon et al., 1991).

The rapid increase in bioaccumulation of methylmercury after flooding is due to the enhanced microbial methylation of inorganic mercury present in the inundated terrestrial habitats (Bodaly et al., 1984; Ramsey, 1990a, 1990b; Hecky et al., 1991). In subarctic reservoirs, the magnitude of the increase in fish mercury concentration after flooding is positively related to the ratio of the newly flooded area to the preimpoundment lake area (Bodaly et al., 1984). Mercury burdens in fish in subarctic reservoirs can be reasonably predicted with multiple-regression models containing two independent variables: (1) the ratio of flooded (terrestrial) area to water volume in the reservoir and (2) the ratio of flooded area upstream to water volume upstream of the reservoir (Johnston et al., 1991).

In northern Canada, the mercury concentrations in fish may remain elevated for decades after flooding (Ramsey, 1990a, 1990b; Hecky et al., 1991; Verdon et al., 1991; Bodaly and Johnston, 1992). In Finnish reservoirs, mercury concentrations in fish seem to be most elevated during the first 10 years after flooding (Verta et al., 1986).

RESIDUES IN MERCURY-INTOXICATED ADULT FISH

Our assessment of organismal toxicity and associated methylmercury residues in wild adult fish is based on three underlying assumptions:

1. That the expected mode of toxic action in adult fishes inhabiting natural waters will be on internal organs (e.g., the brain), not on organ systems exposed directly to waterborne mercury (e.g., the gill);
2. That any toxic effects of mercury on *adult* fish in natural fresh waters will result from methylmercury obtained from dietary exposure; and
3. That the route of uptake does not affect the toxicological significance of methylmercury concentrations in tissues of adult fish.

The first assumption, that the expected mode of toxic action in adult wild fishes is on internal organs, is valid because the concentrations of total mercury and methylmercury in fresh waters (Table 1) are much lower (10^4- to 10^5-fold less) than are the concentrations that harm the gills (Drummond et al., 1974; Wobeser, 1975a; Lock et al., 1981). In rainbow trout *(Oncorhynchus mykiss)* exposed to waterborne methylmercury, a 24- to 96-h exposure to 40 to 135 µg of Hg per liter damaged the gills, causing asphyxiation and death (Wobeser, 1975a), and a 7-day exposure to 5 µg of Hg per liter disturbed ionoregulation (Lock et al., 1981). In brook trout *(Salvelinus fontinalis)*, exposure to 3 µg of Hg per liter of methylmercury causes coughing, which is presumably done to clear the gills after a noxious stimulus (Drummond et al., 1974). Wild fishes are not likely to encounter methylmercury concentrations of this magnitude.

The second assumption, that any toxic effects of mercury on adult wild fishes will result from dietary exposure to methylmercury, is supported by three general observations. First, 95 to 99% of the mercury in the tissues of adult freshwater fish is methylmercury. Second, the methylmercury accumulated in wild adult fishes is almost entirely from dietary exposure. Third, concentrations of mercury (total or methyl) in fresh waters (Table 1), even in mercury-contaminated systems, are not high enough to be directly toxic to fish (Figure 1; also see reviews by Armstrong [1979], Birge et al. [1979], Eisler [1987], and Mance [1987]).

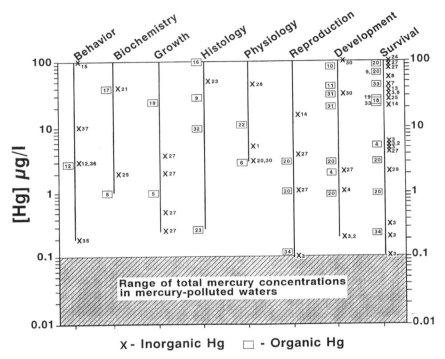

Figure 1 Concentrations of waterborne mercury affecting fish behavior, biochemistry, growth, histology, physiology, reproduction, development (teratogenic effects), and survival, shown in relation to concentrations of total mercury in surface waters. Effect concentrations are indicated for inorganic mercury (denoted by "X"; primarily mercuric chloride) and organic mercury (denoted by open squares; primarily methylmercuric chloride). Effects concentrations exceeding 100 µg of mercury per liter were not included. Numbers on the graph indicate the source of data, as follows: 1, Armstrong (1979); 2, Birge (1978); 3, Birge et al. (1979); 4, Birge et al. (1983); 5, Christensen (1975); 6, Christensen et al. (1977); 7, Curtis et al. (1979); 8, Deshmukh and Marathe (1980); 9, Devlin and Mottet (1992); 10, Dial (1978a); 11, Dial (1978b); 12, Drummond et al. (1974); 13, Hale (1977); 14, Heisinger and Green (1975); 15, Hilmy et al. (1987); 16, Kirubagaran and Joy (1988); 17, Kirubagaran and Joy (1990); 18, Lock and van Overbeeke (1981); 19, Matida et al. (1971); 20, McKim et al. (1976); 21, Menezes and Qasim (1984); 22, O'Connor and Fromm (1975); 23, Olson et al. (1973); 24, Rehwoldt et al. (1972); 25, Sastry and Rao (1984); 26, Sastry et al. (1982); 27, Snarski and Olson (1982); 28, Verma and Tonk (1983); 29, Walczak et al. (1986); 30, Weis and Weis (1977a); 31, Weis and Weis (1977b); 32, Wester (1991); 33, Wobeser (1975a); 34, Mount (1974), cited in Birge et al. (1979); 35, Black and Birge (1980); 36, Weir and Hine (1970); 37, Morgan (1979).

The third assumption, that the route of uptake does not influence the toxicological significance of methylmercury concentrations in tissues of adult fish, is supported by two general observations. First, methylmercury is rapidly transported via the circulatory system from the site of uptake to all other internal tissues and organs (Giblin and Massaro, 1973; McKim et al., 1976; Olson et al., 1978). Second, the mode of uptake has little influence on the distribution of methylmercury among the internal organs and tissues (McKim et al., 1976; Boudou and Ribeyre, 1983; Niimi and Kissoon, 1994). Given this assumption, we can use data from laboratory studies of fish accumulating waterborne methylmercury to make inferences about the toxicological significance of methylmercury concentrations in wild fishes, which obtain methylmercury mostly through the diet. This is pertinent because many toxicological studies with fish have involved aqueous, rather than dietary, exposure to methylmercury.

MODE OF ACTION

Neurotoxicity seems to be the most probable chronic response of wild adult fishes to dietary methylmercury, even though other effects have been observed in laboratory studies. In birds and mammals (including humans), methylmercury is primarily neurotoxic, damaging the central nervous system (Chang, 1979; Clarkson et al., 1984; Scheuhammer, 1991; Clarkson, 1992, 1994). In fish, long-term dietary exposure to methylmercury can cause incoordination, inability to feed, and diminished responsiveness (Matida et al., 1971; Scherer et al., 1975). In Minamata Bay, Japan, methylmercury-poisoned fish were sluggish (could easily be captured by hand), exhibited abnormal movements, were emaciated, and had lesions in the brain (Takeuchi, 1968). Thus, the central nervous system is the probable site of the most harmful toxic action in fish exposed to methylmercury.

DETOXIFICATION MECHANISMS

Detoxification processes that decrease the retention and accumulation of methylmercury have been well studied in mammals (reviewed by the World Health Organization, 1989; Clarkson, 1992, 1994). We summarize mammalian detoxification processes and compare them with those in fish.

Mammals assimilate almost 100% of the methylmercury present in the diet. Once in the blood, methylmercury binds to red blood cells and is rapidly transported to internal tissues. After administration of a single dose, for example, peak concentrations occur in 3 days in the brain and in 2 days in most tissues. Inorganic mercury is found in most tissues of methylmercury-exposed mammals, indicating that demethylation occurs in the tissues. Mammals eliminate mercury mostly in the feces, although the incorporation of methylmercury in growing hair can remove considerable amounts from some mammals. Both methylmercury and inorganic mercury are secreted from the liver into the bile. Glutathione, a small tripeptide rich in sulfur, is the most important ligand binding methylmercury and inorganic mercury. Because of microbial demethylation of methylmercury in the gut, much of the mercury in the feces is inorganic mercury, which is poorly absorbed and readily excreted, whereas methylmercury in the feces is reabsorbed very efficiently. Mammalian elimination of methylmercury, therefore, relies partly on microbial demethylation in the gut to interrupt the cycle of absorption and excretion of methylmercury. In fact, it is unclear whether mammals excrete any organic mercury in the feces. This excretory process in mammals results in half-lives of 7 days in mice, 20 to 70 days or more in humans, and longer in some other primates.

The processing of methylmercury in fish has not been studied as rigorously, but seems to differ from that in mammals with respect to assimilation, demethylation, and excretion. Uptake across the gut may be less efficient in fish than in mammals, with assimilation efficiencies in fish probably ranging from 65 to 80% or greater (Rodgers, 1994). The presence of inorganic mercury in methylmercury-exposed fish in the laboratory has been interpreted as evidence of demethylation in the tissues (Olson et al., 1978). Bloom (1992), however, confirmed that nearly all (99%) of the mercury in the axial muscle of fish is methylmercury. Bloom (1992) concluded that earlier reports of significant amounts of inorganic mercury in axial muscle probably resulted from analytical errors and handling contamination, which adds *inorganic* mercury to the sample. Methylation of mercury can occur in the gut of fish (Rudd et al., 1980), but demethylation in the gut has not been examined in detail. If quantitatively significant demethylation were occurring in fish, excretion should be faster at higher metabolic rates (Huckabee et al., 1979); however, Sharpe et al. (1977) found that the metabolic rate (as determined by ambient temperature) did not affect the rate of excretion of methylmercury in goldfish *(Carassius auratus)*. Although demethylation clearly occurs in mammals and is a crucial stage in the mammalian excretion of mercury, there is no convincing evidence that methylmercury is demethylated in fish. A simultaneous analysis of methylmercury and total mercury in other tissues and organs of fish (handled with ultraclean techniques to prevent sample contamination), similar to that done for axial muscle by Bloom (1992), would be useful in assessing whether methylmercury is demethylated in fish.

Fish do not readily excrete methylmercury. Some authors have reported that there is no measurable excretion of methylmercury from fish (Laarman et al., 1976; McKim et al., 1976), whereas others have reported measurable — but very slow — rates of excretion (Lockhart et al., 1972; Giblin and Massaro, 1973; Huckabee et al., 1975; Ruohtula and Miettinen, 1975; deFreitas, 1979). The estimated half-retention times of methylmercury ranged from about 200 to 516 days in rainbow trout (Giblin and Massaro, 1973; Ruohtula and Miettinen, 1975), exceeded 400 days in mosquitofish *(Gambusia affinis)* (Huckabee et al., 1975), and were about 2 years in northern pike (Lockhart et al., 1972).

Metal-binding proteins, such as metallothioneins, may detoxify metals in fish, mammals, and many other animals by binding free metal ions, thereby reducing the availability of metal ions to sites of potentially toxic action (Klaverkamp et al., 1984; Roesijadi, 1992). Although metallothioneins have a high affinity for inorganic mercury (Dutton et al., 1993), methylmercury neither induces metallothionein nor binds to existing metallothionein with much affinity (Chen et al., 1973; Piotrowski et al., 1973, cited in Weis, 1984; Carty and Malone, 1979). Any inorganic mercury formed through demethylation in the tissues of fish would presumably be strongly bound to metallothionein (Stegeman et al., 1992; Dutton et al., 1993). In mummichog *(Fundulus heteroclitus),* Weis (1984) found a metallothionein-like protein that bound copper but not mercury, presumably because the mercury was in the methyl form. He also found that metallothionein had little, if any, effect on mercury tolerance in mummichog. Olson et al. (1978) found radiolabeled methylmercury associated with presumptive metallothionein, suggesting binding of methylmercury; however, the weight of evidence indicates that metallothionein does little to detoxify methylmercury. Recent reviews of metallothioneins as bioindicators of metal pollution in aquatic organisms (Roesijadi, 1992; Stegeman et al., 1992) discuss the binding of inorganic mercury to metallothioneins, but they do not mention methylmercury.

Fish have evolved in environments that have always contained mercury. Fish possess mechanisms to protect against inorganic mercury but seem to have fewer defenses against methylmercury. Methylmercury crosses the organism's biological barriers (gills, intestines, and internal cellular membranes) much more readily than does inorganic Hg(II) (Boudou et al., 1991). Unlike inorganic mercury, methylmercury in fish is neither effectively excreted nor bound to metallothioneins. Storage in the muscle, which seems to be less sensitive to methylmercury than other tissues and organs (Niimi and Kissoon, 1994), may function as the primary detoxification mechanism for methylmercury in fish. The binding of assimilated methylmercury to proteins in the skeletal muscle, even if incidental, clearly reduces the exposure of the brain to methylmercury.

The toxicity of methylmercury is reduced by selenium (reviewed by Cuvin-Aralar and Furness, 1991). In the tissues of marine mammals, for example, very high concentrations of mercury may be tolerated if selenium concentrations are also high, although the mechanism by which selenium protects against mercury toxicity is not understood (Cuvin-Aralar and Furness, 1991). Field studies have shown that adding small amounts of selenium to water reduces the bioaccumulation of methylmercury in fish (Turner and Swick, 1983; Paulsson and Lundbergh, 1991). Because of its potential toxicity, however, the addition of selenium to surface waters is not a practical approach for reducing mercury concentrations in sportfish (Hodson, 1988; Lemly, 1993).

TOXICITY AND ASSOCIATED RESIDUES

Laboratory Studies

Scherer et al. (1975) studied the effects of dietary methylmercury on 1-year-old walleyes, which were fed shredded northern pike in two treatments. Walleyes in the high-mercury treatment group were provided a diet with a mean concentration of 7.9 µg of Hg per gram wet weight, composed of northern pike from industrially contaminated Clay Lake, Ontario. Test fish in the low-mercury treatment group, termed "controls" by Scherer et al. (1975), were fed northern pike with a mean concentration of 0.41 µg of Hg per gram. The daily food consumption by individual test fish was described as "somewhat less" than 2 g (Scherer et al., 1975). Thus, test fish were presumably ingesting about 10 to 15 μg of Hg per day in the high-mercury treatment group and about 0.5 to 0.8 μg of Hg per day in the low-mercury treatment group (as methylmercury in both treatment groups). Walleyes in the high-mercury treatment group became emaciated and exhibited greatly diminished locomotor activity, coordination, escape behavior, and response to light, relative to fish in the low-mercury treatment group.

The mortality of test fish was much greater in the high-mercury treatment group than in the low-mercury treatment group. Scherer et al. (1975) did not report the tissue concentrations of mercury that coincided with the specific toxic symptoms observed; however, they did measure and graphically summarize the mercury concentrations in tissues of fish that died during the bioassay. In the high-mercury treatment group, 20 fish died between days 42 and 63 of the test, presumably because of methylmercury intoxication, because only two fish died in the low-mercury treatment group in the same period. The concentrations of total mercury in the dead walleyes from the high-mercury treatment group during this 21-day interval were about 3 to 6 μg/g wet weight in brain tissue, 6 to 14 μg/g in liver, and 5 to 8 μg/g in axial muscle (Table 3). The mortality of test fish in the high-mercury treatment group continued throughout the 314-day test, whereas no mortality occurred after day 120 in the low-mercury treatment group. At the end of the test, the cumulative mortality was 88% in the high-mercury treatment group and 27% in the low-mercury treatment group. The concentrations of mercury in walleyes from the high-mercury treatment group that died after day 240 were about 15 to 40 μg/g wet weight in the brain, 18 to 50 μg/g in the liver, and 15 to 45 μg/g in axial muscle. In comparison, the concentrations of mercury in fish from the low-mercury treatment group were usually less than 2 μg/g wet weight in brain tissue, 3 μg/g in liver, and 2.5 μg/g in axial muscle.

McKim et al. (1976) examined the toxicity and bioaccumulation of waterborne methylmercury, administered as methylmercuric chloride, in three generations of chronically exposed brook trout. Most (88%) of the fish exposed to 2.93 μg of Hg per liter died, after exhibiting loss of appetite, muscle spasms, and deformities. The concentrations (μg/g wet weight) in these brook trout at the time of death averaged 155 in the spleen, 58 in liver, 147 in kidney, 48 in gill, 42 in brain, 32 in gonad, and 24 in axial muscle and the whole body (McKim et al., 1976); the concentration in the brain in these dead fish varied less among individuals than did those concentrations in the other tissues, regardless of the duration of exposure. Many brook trout exposed to 0.93 μg of Hg per liter for 39 weeks exhibited sluggish behavior, deformities, reduced growth, and mortality; the concentrations (μg/g wet weight) in fish exposed to 0.93 μg of Hg per liter averaged 39 in the spleen, 24 in liver, 27 in kidney, 22 in gill, 17 in brain, 12 in gonad, 10 in axial muscle, and 9 in the whole body (McKim et al., 1976). In this study, whole-body concentrations of 5 to 7 μg/g wet weight or greater eventually caused toxic symptoms and death, whereas no deaths or abnormalities were observed in fish with whole-body concentrations of 2.7 μg/g or less (McKim et al., 1976). The no-observed-effect concentration for methylmercury in water was 0.29 μg of Hg per liter; the concentrations (μg/g wet weight) in fish from this treatment averaged 12 in the spleen, 8 in liver, 9 in kidney, 6 in gill, 5 in brain, 3 in gonad, and 5 in axial muscle.

Table 3 Total Mercury Concentrations in Tissues of Fish Exhibiting Symptoms of Methylmercury Toxicity

Type of study and species of fish	Location or mode and duration of exposure	Hg concentration (μg/g wet weight)				Toxic effect(s)[a]	Ref.
		Brain	Liver	Muscle	Whole body		
Laboratory studies							
Walleye	Diet (42–63 days)	3–6	6–14	5–8		Onset of mortality Fish emaciated (–) locomotor activity (–) coordination (–) appetite	Scherer et al. (1975)
Rainbow trout	Diet (240–314 days)	15–40	18–50	15–45		(+) mortality (–) growth	Scherer et al. (1975) Wobeser (1975b)
Rainbow trout	Diet (105 days)			12–23		Darkened skin	
Rainbow trout	Diet (84 days)				30–35	Lethargic behavior	Rodgers and Beamish (1982)
	Diet (84 days)				10–30	(–) appetite (–) growth	Rodgers and Beamish (1982)
Rainbow trout	Diet (270 days)	16–30	26–68	20–28	19	(–) appetite (–) activity (–) visual acuity (–) growth Darkened skin Loss of equilibrium	Matida et al. (1971)
Rainbow trout	Water (4 μg/l)[b] (30–98 days)	7–32	32–114	9–52		Death, preceded by (–) appetite and (–) activity	Niimi and Kissoon (1994)
Rainbow trout	Water (9 μg/l)[b] (12–33 days)				4–27	Death, preceded by (–) appetite and (–) activity	Niimi and Kissoon (1994)

Table 3 (continued) Total Mercury Concentrations in Tissues of Fish Exhibiting Symptoms of Methylmercury Toxicity

Type of study and species of fish	Location or mode and duration of exposure	Hg concentration (µg/g wet weight)				Toxic effect(s)[a]	Ref.
		Brain	Liver	Muscle	Whole body		
Brook trout	Water (2.9 µg/l)[b] (273 days)	42	58	24	24	Mortality, preceded by loss of appetite, muscle spasms, and deformities	McKim et al. (1976)
	Water (0.93 µg/l)[b] (273 days)	17	24	10	5–7	(+) mortality (–) growth Sluggish behavior Deformities	McKim et al. (1976)
	Water (0.29 µg/l)[b] (273 days)	5	8	5	3	None observed	McKim et al. (1976)
Field studies							
Nibea schlegeli	Minamata Bay			8–15		Fish enfeebled	Kitamura (1968)
Latcolabrax japonicus	Minamata Bay			17		Fish enfeebled	Kitamura (1968)
Sparus macrocephalus	Minamata Bay			24		Fish enfeebled	Kitamura (1968)
Scomberomorus niphonicus	Minamata Bay		15	9		Fish enfeebled	Kitamura (1968)
Striped mullet (Mugil cephalus)	Minamata Bay			11		Fish enfeebled	Kitamura (1968)
Cardinalfish (Apogon sp.)	Minamata Bay			19		Fish enfeebled	Kitamura (1968)
Northern pike	Clay Lake, Ontario			6–16		Fish emaciated, with biochemical symptoms of starvation and (–) immunity	Lockhart et al. (1972)

[a] An increase is indicated by (+); a decrease is indicated by (–).
[b] Concentration of mercury in water, administered as methylmercuric chloride.

The effects of dietary methylmercury were much less severe in fingerling rainbow trout (weight range, 17 to 52 g) fed a commercial trout food spiked with methylmercuric chloride to yield nominal treatment concentrations of 4, 8, 16, and 24 μg of Hg per gram (Wobeser, 1975b). In control fish (no added methylmercuric chloride in the diet), the concentrations of mercury in muscle were less than 0.2 μg/g wet weight. In each of the four treatment groups, the concentrations in muscle increased throughout the test and were slightly higher than the nominal dietary concentration after the 105-day test. No differences in mortality, appetite, avoidance of capture (netting), or apparent visual acuity were noted between the treatment groups and controls. The skin of fish fed the 24-μg/g ration darkened slightly (relative to controls) during days 50 to 70 of the test but subsequently returned to normal. Test fish were offered excess food (daily rate of 3 to 4% of body weight), yet mean weight gains were less in the 16-μg/g and 24-μg/g treatment groups during days 70 to 105 than in controls; these reductions in growth were associated with mean concentrations in muscle of about 12 and 23 μg/g wet weight in the two treatment groups (Wobeser, 1975b). There were no neurological lesions in exposed fish; observed histopathological lesions were minor and limited to the gills and posterior kidney. Subadult rainbow trout, exposed to waterborne methylmercury (150 ng of Hg per liter, administered as methylmercuric chloride) for 75 days, accumulated whole-body concentrations of 12 μg/g without apparent effects on survival or hematological variables (Niimi and Lowe-Jinde, 1984).

The effects of dietary methylmercury on rainbow trout were also examined by Rodgers and Beamish (1982). For 84 days, test fish were fed a commercial trout food spiked with methylmercuric chloride to yield nominal treatment concentrations of 0, 25, 45, 75, and 95 μg of Hg per gram. About 20 to 30% of the rainbow trout fed the two most contaminated diets had darkened skin and were lethargic; the concentrations of total mercury in whole fish in these treatments were about 30 to 35 μg/g wet weight. The appetites of fish provided food with 25, 45, and 95 μg of Hg per gram declined during the 84-day test, whereas the appetites of controls remained stable; whole-body concentrations in fish with reduced appetites were about 10 to 30 μg/g wet weight during days 56 to 84 of the test. The growth rates of rainbow trout fed ad libitum were inversely related to the methylmercury concentration in the diet, an effect attributed to the reduced appetite and to the lower apparent digestibility of the diet in methylmercury-exposed fish (Rodgers and Beamish, 1982). The other response variables measured (serum osmolality, sodium, and potassium; histology of the gill, liver, kidney, and intestine; and mortality) did not differ between test fish and controls.

Matida et al. (1971) fed ground soft tissues of mercury-contaminated shellfish *(Horumomura mutabilis)* from Minamata Bay, Japan, to rainbow trout for 270 days. The shellfish diet contained 48 μg of Hg per gram dry weight, one third of which was methylmercury. The exposed fish exhibited several symptoms of methylmercury intoxication, including loss of appetite, reduced activity, darkened skin, and loss of equilibrium; these symptoms were not observed in control fish that were fed shellfish from an uncontaminated bay. Methylmercury-exposed fish also seemed to suffer reduced visual acuity, as indicated by their collisions with tank walls and reduced ability to catch food. The exposed fish also grew more slowly than did controls. Corresponding concentrations of mercury in exposed rainbow trout were 19 µg/g wet weight in the whole body, 20 to 28 µg/g in muscle, 26 to 68 µg/g in liver, and 16 to 30 µg/g in the brain (Matida et al., 1971).

Niimi and Kissoon (1994), who quantified mercury in subadult rainbow trout killed by exposure to waterborne methylmercuric chloride, questioned the validity of the critical tissue-concentration concept. In their laboratory tests, mercury concentrations in the kidney, liver, spleen, brain, and muscle of rainbow trout increased with increasing exposure time, suggesting that death did not occur in response to a critical threshold concentration in the internal tissues. Fish died faster at higher exposure concentrations, with time to death averaging 1, 8, 22, 24, and 58 days, respectively, in test fish exposed to 34, 13, 10, 9, and 4 µg of Hg per liter. At death, fish had accumulated less mercury at higher exposure concentrations than at lower exposure concentrations; for example, fish exposed to 4, 10, 13, and 34 µg of Hg per liter had concentrations averaging 19, 13, 7.7, and 1.1 µg/g wet weight in the brain and 89, 72, 51, and 6.4 in the spleen, respectively. Niimi and Kissoon (1994) suggested that critical concentrations of methylmercury in fish might be more appropriate if based on whole-body residues than on tissue residues. Whole-body concentrations, measured in fish exposed to 9 µg of Hg per liter until death, averaged 11 µg/g wet weight and ranged from 4 to 27 µg/g in their study.

Multiple modes of toxic action on multiple target sites were presumably involved in the study by Niimi and Kissoon (1994). The concentrations of mercury in the brains of rainbow trout killed by exposure to waterborne methylmercury varied 50-fold among individuals (range, 0.6 to 32 µg/g wet weight). The three higher concentrations of waterborne methylmercury applied by Niimi and Kissoon (10, 13, and 34 µg of Hg per liter) were probably sufficient to cause severe — perhaps lethal — adverse effects on the gills (Wobeser, 1975a; Lock et al., 1981). Mercury concentrations in the gills, in contrast to the internal tissues, varied little among and within treatments, averaging 66, 51, 64, and 56 µg/g wet weight in fish exposed to 4, 10, 13, and 34 µg of Hg per liter, respectively; it is not known if these represent critical tissue concentrations for lethal effects on the gill. At lower exposure concentrations (specific treatments not defined), test fish became lethargic and gradually reduced their food intake, symptoms suggesting a generalized stress response.

Thus, the critical-tissue-concentration concept may not apply to situations involving acute, lethal exposures and multiple modes of toxic action. The results of Niimi and Kissoon (1994), however, do not invalidate the critical-tissue-concentration concept for fish exposed to environmental concentrations of methylmercury. Indeed, the results of long-term chronic exposures support the critical-tissue-concentration concept. McKim et al. (1976), for example, found that mercury concentrations in the brains of brook trout killed by exposure to 2.9 µg of Hg per liter of waterborne methylmercury varied little, regardless of the duration of exposure.

Olson et al. (1975) analyzed whole fathead minnows exposed to waterborne methylmercuric chloride for 336 days. Their test included controls (<10 ng of Hg per liter) and five treatment concentrations ranging from 18 to 247 ng of Hg per liter. The mean concentrations of mercury in whole fathead minnows after the test increased with exposure concentration, ranging from 0.21 µg/g wet weight in controls to 11 µg/g in the 247-ng/l treatment. Olson et al. (1975) observed no readily apparent effects of mercury on the survival, behavior, growth, or general appearance of the test fish.

Field Studies

Coincident with the methylmercury poisoning of humans and other organisms near Minamata Bay, Japan, fish from the bay exhibited several symptoms of methylmercury intoxication (Kitamura, 1968; Takeuchi, 1968) seen in subsequent laboratory studies (Scherer et al., 1975; McKim et al., 1976; Rodgers and Beamish, 1982). Takeuchi (1968, p. 212) reported that "a number of fish died" and that fish in the bay "frequently could easily be captured by hand," indicating severely diminished locomotor activity and escape behavior. Moreover, fish from the bay were commonly emaciated and had lesions in the brain (Takeuchi, 1968). Mercury concentrations in axial muscle of "enfeebled fishes" found floating in seawater in the area (Kitamura, 1968) averaged 15 µg/g wet weight and ranged from 8.4 to 24 µg/g in six species (Table 3).

Lockhart et al. (1972) examined the physiological and biochemical condition of mercury-contaminated northern pike from grossly polluted Clay Lake in the English-Wabigoon River system, northwestern Ontario. The concentrations of mercury ranged from 6 to 16 µg/g wet weight in the axial muscle of the fish examined, which were from 3 to 8 years old. Compared with northern pike from a relatively uncontaminated reference lake, fish from Clay Lake were emaciated, had low hepatic fat stores, exhibited symptoms of starvation (low levels of total protein, glucose, and alkaline phosphatase in blood serum), and had low serum cortisol levels. Lockhart et al. (1972) also transplanted contaminated northern pike from Clay Lake into the reference lake. After 1 year, transplanted fish had serum concentrations of total protein, alkaline phosphatase, and cortisol intermediate to those in Clay Lake fish and reference fish, suggesting partial recovery. Transplanted northern pike eliminated methylmercury slowly, losing only 30% of their body burden in 1 year. Although Lockhart et al. (1972) did not attribute the poor condition of Clay Lake fish directly to the effects of methylmercury, Scherer et al. (1975) subsequently reported similar symptoms in methylmercury-intoxicated fish in a laboratory study.

Bidwell and Heath (1993) studied the physiological condition of rock bass in the South River, Virginia, in relation to concentrations of total mercury in axial muscle and liver tissues. Concentrations in rock bass from a contaminated reach averaged 1.4 µg/g wet weight in muscle and 2.9 µg/g in liver. The concentrations in fish from an upstream reference site were much lower, averaging 0.17 µg/g in muscle and 0.10 µg/g in liver. Physiological differences in fish between the contaminated and reference sites were few, based on comparison of several physiological and biochemical variables. Bidwell and Heath (1993) concluded that the mercury concentrations in rock bass from the contaminated area did not adversely affect their well-being.

We suspect that wild fishes are more severely affected by methylmercury than are fish in laboratory tests. Incoordination, diminished responsiveness, reduced appetite, and reduced growth have been documented in intoxicated fish in field studies and in laboratory tests, where fish were provided with excess food. The neurotoxic effects of methylmercury may severely impede the ability of wild fish to locate, capture, handle, and ingest prey and to avoid predation.

Critical Tissue Concentrations

Defining critical tissue concentrations for adult freshwater fish is not straightforward because of the variation in tissue levels associated with toxic effects of methylmercury (e.g., Scherer et al., 1975). Some of this variation may stem from the fact that the brain, rather than muscle or whole-body tissue, is the primary site of toxicity. In brook trout killed by exposure to 2.9 µg of Hg per liter of waterborne methylmercury, for example, mercury concentrations in most tissues (including axial muscle) varied among individuals, whereas concentrations in the brain varied little, regardless of the duration of exposure before death (McKim et al., 1976).

Interspecific differences in sensitivity to methylmercury may also contribute to the variation in fish-tissue concentrations associated with toxicity. In laboratory tests, for example, mercury concentrations in tissues of walleye and brook trout at the onset of mortality were less than those in tissues of rainbow trout suffering sublethal effects (Table 3).

The rate of accumulation and exposure time seem to significantly affect the toxicity of methylmercury to fish. These factors probably account for some of the observed variation in tissue concentrations associated with toxicity (McKim et al., 1976). In the laboratory, a large single dose of methylmercury is often very toxic to fish, yet fish can accumulate and tolerate equivalent or larger burdens of methylmercury if accumulated over a much longer time (Giblin and Massaro, 1973; Wobeser, 1975b; Niimi and Kissoon, 1994). This is probably due to the dynamic internal redistribution and binding of assimilated methylmercury to proteins in the skeletal muscle, a process that reduces the exposure of the central nervous system to methylmercury. In adult wild fish, the symptoms of methylmercury toxicity have involved grossly contaminated food chains and surface waters, such as Minamata Bay, Japan (Kitamura, 1968; Takeuchi, 1968), and Clay Lake, Ontario (Lockhart et al., 1972) — situations presumably involving rapid bioaccumulation and less effective detoxification.

Given the above caveats, we offer the following guidance for interpreting methylmercury residues in *adult* freshwater fish.

Brain Tissue

Fish with mercury concentrations of 7 to 15 µg/g wet weight or greater in the brain are probably suffering significant, potentially lethal effects. For brook trout, a concentration of 5 µg/g in the brain can be regarded as a no-observed-effect concentration, based on the work of McKim et al. (1976). In walleyes, which seem to be more sensitive to methylmercury, concentrations of 3 µg/g wet weight or greater in the brain are probably indicative of significant toxic effects, including death. A value of 1.5 µg/g in the brain could be used as a no-observed-effect concentration for walleyes.

Muscle Tissue

Muscle is the only tissue for which data have been obtained on methylmercury-intoxicated fish in both field and laboratory studies. Field studies indicate that fish with residues in muscle in the range of 6 to 20 µg/g wet weight are adversely affected. The range for laboratory studies is similar, with sublethal effects or mortality associated with muscle tissue concentrations of 5 to 8 µg/g in walleyes and 10 to 20 µg/g in salmonids. For salmonids, a concentration of 5 µg/g in muscle can be regarded as a no-observed-effect concentration, based on the work of McKim et al. (1976).

Whole Body

In brook trout, whole-body concentrations of about 5 µg/g wet weight or greater are indicative of probable toxic effects, whereas a value of 3 µg/g could be regarded as a no-observed-effect concentration. In rainbow trout, whole-body concentrations of about 10 µg/g wet weight or greater seem to be associated with sublethal or lethal toxic effects.

Symptoms of methylmercury toxicity, coincident with high concentrations in fish tissues, would strengthen a diagnosis of mercury toxicity. Such symptoms include emaciation and other evidence of starvation, lack of appetite or feeding (indicated by lack of food in stomachs), reduced growth, abnormal or reduced locomotor activity, poor coordination, diminished escape behavior, deformities, and perhaps darkened skin.

Given the extreme neurotoxicity of methylmercury, behavioral studies might show that the behavior of adult fish is affected at tissue concentrations much lower than those indicated previously and in Table 3. Many fish behaviors are sensitive and ecologically relevant indicators of contaminant toxicity that are affected at exposure concentrations much lower than those causing direct mortality (Little and Finger, 1990; Sandheinrich and Atchison, 1990). Kania and O'Hara (1974), for example, found that the ability of mosquitofish to avoid predation by largemouth bass was greatly diminished by direct exposure to 10, 50, and 100 µg of Hg per liter (administered as mercuric chloride), concentrations that otherwise did not influence mortality. Mean whole-body concentrations of 0.7 to 5.4 µg/g wet weight in mosquitofish were associated with diminished predator-avoidance behavior, yet no direct mercury-induced mortality was evident in these same fish. Predator-avoidance behavior was not diminished in mosquitofish with a mean whole-body concentration of 0.4 µg/g. Behavioral studies of methylmercury-exposed fish are few; most prior work has involved inorganic mercury (Figure 1).

EFFECTS ON REPRODUCTION AND EARLY LIFE STAGES

The embryos of vertebrate organisms are very sensitive to mercury. In mammals, the prenatal life stage is most sensitive to methylmercury, and all prenatal effects seem to be irreversible because they involve developing neural pathways (Clarkson et al., 1984; Clarkson, 1992). Similarly, the effects of methylmercury on birds are much more severe in embryos and chicks than in adults (Fimreite, 1979; Scheuhammer, 1991). Low-level dietary exposures to methylmercury that cause no measurable adverse effects in adult birds can significantly impair egg fertility, hatchling survival, and overall reproductive success (Scheuhammer, 1991). Moreover, the dietary concentrations of methylmercury needed to significantly impair avian reproduction are about one fifth those needed to produce overt toxicity in the adult bird (Scheuhammer, 1991). In fish, the early life stages are also typically more sensitive than adults to both methylmercury and inorganic mercury, with regard to developmental toxicity and mortality (Wobeser, 1975a; McKim et al., 1976; Christensen et al., 1977; McKim, 1977; Birge et al., 1979; Weis and Weis, 1991).

Teratogenic effects of mercury in fish embryos and larvae have been studied intensively in the mummichog, an estuarine fish (Weis and Weis, 1977a, 1977b, 1991), and have been examined in freshwater fishes, including medaka *(Oryzias latipes)* (Dial, 1978a, 1978b), rainbow trout (Birge et al., 1979), brook trout (McKim et al., 1976), and fathead minnow (Snarski and Olson, 1982). Reported effective concentrations for teratogenic effects in fish exposed to waterborne mercury range from 0.2 to 100 µg of Hg per liter (Figure 1). Mercurials cause three general effects: craniofacial (including cyclopia or fusion of the eyes), cardiovascular, and skeletal flexures (Birge et al., 1979; Weis and Weis, 1991). Moreover, there are developmental stages when fish are particularly susceptible to mercury (Weis and Weis, 1991; Devlin and Mottet, 1992).

Fish embryos in natural waters are presumably exposed to methylmercury primarily via the yolk (Weis and Weis, 1991), which is synthesized during oogenesis by the adult female. Nicoletto and Hendricks (1988) found higher concentrations of mercury in reproducing female sunfishes (Centrarchidae) than in males of the same species and age, apparently because females ate more to meet the energy and nutritional needs of reproduction. Exposure of the developing fish embryo to waterborne inorganic mercury and methylmercury seems to be inhibited somewhat by the chorion, the outermost membrane on the fertilized egg (Christensen, 1975; Weis and Weis, 1991).

The gonads of mercury-exposed female fishes contain smaller concentrations of mercury than most other tissues and organs (McKim et al., 1976; Olson et al., 1978). In wild fishes, a small quantity of mercury is transferred from the female to the eggs during oogenesis (Niimi, 1983; Weis and Weis, 1991), although data are few. The mean concentrations of total mercury in eggs from gravid females of five species from lakes Ontario and Erie, for example, varied from 0.004 µg/g wet weight in white bass *(Morone chrysops)* to 0.011 µg/g in rainbow trout (Niimi, 1983), representing 0.3 to 2.3% of the whole-body burden of mercury in females of the five species. Concentrations in female carcasses without eggs and gastrointestinal tract contents, in comparison, ranged from 0.067 to 0.28 µg/g wet weight for the five species (Niimi, 1983).

The mercury content of fish eggs reflects the exposure history of the parental female. Weis and Weis (1984), for example, found that mummichog inhabiting a mercury-contaminated environment produced eggs with higher mercury content (mean, 0.022 µg/g wet weight) than did gravid females from a less contaminated environment (mean, 0.005 µg/g). In the laboratory, McKim et al. (1976) found that mercury concentrations in eggs of brook trout increased concomitantly with the waterborne methylmercury concentration (administered as methylmercuric chloride) to which the parental females had been exposed.

Laboratory bioassays have shown that survival of fish embryos can be substantially reduced by a seemingly minute quantity of either inorganic mercury or methylmercury within the fertilized egg, whether from waterborne exposure (Birge et al., 1979) or maternal transfer (McKim et al., 1976; Birge et al., 1979). Fertilized eggs of rainbow trout suffered 100% mortality after 8-day exposures to 100 ng/l of inorganic mercury in flow-through bioassays; the survival of controls was about 85% (Birge et al., 1979). In these bioassays, the mean concentration (± standard error) of mercury in rainbow trout eggs was 0.019 (±0.004) µg/g wet weight before exposure to mercury, 0.068 (±0.023) µg/g after 4 days of exposure, and 0.097 (±0.006) µg/g after 7.5 days. Survival ranged from 79 to 83% in the three bioassays after 4 days of exposure to inorganic mercury and from 23 to 26% after 7 days (Birge et al., 1979). Thus, embryonic mortality probably coincided with inorganic mercury concentrations of about 0.07 to 0.10 µg/g wet weight in the fertilized eggs, concentrations that are less than 1% of the tissue residues (10 to 30 µg/g wet weight) associated with overt mercury toxicity in adult rainbow trout (Table 3). Furthermore, the residues in the exposed eggs were only 6 to 9 times those in eggs obtained from rainbow trout in Lake Ontario (Niimi, 1983).

Reproductive impairment occurred in fathead minnows exposed to inorganic mercury in water (Snarski and Olson, 1982). Fathead minnows chronically exposed to 1020, 2010, and 3690 ng of Hg per liter, administered as mercuric chloride, did not mature sexually. Spawning and total egg production were substantially reduced, relative to controls, in fish exposed to 260 and 500 ng of Hg per liter.

Either inorganic mercury or methylmercury transferred from the female to the eggs during oogenesis can adversely affect developing fish embryos. Birge et al. (1979) exposed adult rainbow trout to two concentrations of waterborne inorganic mercury (210 to 240 ng of Hg per liter and 700 to 790 ng of Hg per liter, administered as mercuric chloride) and reared the fertilized eggs from these fish in clean water. At 4 days after hatching, the survival of larvae from adults exposed to 700 to 790 ng/l was only 14 to 44%, and 25% of the alevins produced were deformed. Survival of larvae from adults exposed to 210 to 240 ng/l was 26 to 68%, and 27% of the larvae were deformed. The mean concentrations in eggs, taken at spawning from females exposed to waterborne mercury for 400 days, were 0.26 µg/g wet weight in the 210- to 240-ng/l treatment group and 3.67 µg/g in the 700- to 790-ng/l treatment group. Nearly all (92%) of the mercury in the tissues of exposed, adult rainbow trout was inorganic (Birge et al., 1979).

McKim et al. (1976), who exposed three generations of brook trout to waterborne methylmercuric chloride, found that all second-generation embryos exposed maternally to 2930 ng of Hg per liter were badly deformed and died. The mean concentration in these eggs, stripped from females, was 12.5 µg/g wet weight. Third-generation embryos maternally exposed to 930 ng of Hg per liter were also deformed; the mean concentration in these eggs was 2.2 µg/g.

The early life stages of fish do not seem to be measurably affected by *waterborne* exposure to the concentrations of inorganic mercury and methylmercury typically present in natural waters (Table 1) except in the most contaminated lakes and rivers. The waterborne concentrations of inorganic mercury and methylmercury that are known to cause lethal or sublethal (e.g., teratogenic) effects in young fish (McKim, 1977; Birge et al., 1979; Weis and Weis, 1991), embryos (Christensen, 1975; McKim et al., 1976; Birge et al., 1979; Klaverkamp et al., 1983; Perry et al., 1988; Weis and Weis, 1991), or sperm (McIntyre, 1973) are typically orders of magnitude greater than recently measured concentrations in fresh waters. The fertilization success of sperm of steelhead rainbow trout, for example, was not reduced by a 30-min exposure to methylmercury concentrations of 100 µg of Hg per liter (McIntyre, 1973). Similarly, the survival of embryos of brook trout was not measurably affected by exposure to 1 µg of Hg per liter, administered as methylmercuric chloride, for 16 to 17 days (Christensen, 1975).

Mercury levels common in wild adult walleyes may affect reproductive success. Whitney (1991) examined the effects of maternally transmitted mercury on the hatching success, survival, growth, and behavior of embryo and larval walleyes from two lakes in northern Wisconsin. Eggs were taken from six walleyes from Escanaba Lake (range in total length, 38 to 54 cm; range in age, 5 to 7 years) and from six fish from nearby Trout Lake (48 to 59 cm; 5 to 7 years). These fish were probably exposed to low concentrations of waterborne mercury because nearby lakes contain 0.7 to 2.1 ng of Hg per liter of total mercury and 0.05 to 0.33 ng of Hg per liter of methylmercury (Fitzgerald and Watras, 1989; Watras et al., 1994). The mercury content of stripped eggs ranged from 0.002 to 0.013 µg/g wet weight for the six fish from Escanaba Lake and from 0.015 to 0.058 µg/g for the six fish from Trout Lake. Mercury concentrations in eggs from the Trout Lake fish were positively correlated with those in maternal skin-on fillets, which ranged from 0.20 to 0.57 µg/g wet weight. In contrast, concentrations in eggs from the Escanaba Lake fish were negatively correlated with those in maternal skin-on fillets, which ranged from 0.14 to 0.36 µg/g. Hatching success varied from 81 to 100% for the 12 maternal groups of eggs and was negatively correlated with the mercury concentration in eggs ($r = -0.76$, $P < 0.01$). Survival of embryos varied from 70 to 99% for the 12 groups and was negatively correlated with the mercury concentration in eggs ($r = -0.70$, $P < 0.02$). The growth, survival, and behavior of larval walleyes were not correlated with the mercury concentration in eggs. Walleyes in many surface waters contain much higher concentrations of mercury (Table 2) than those studied by Whitney (1991), indicating that mercury may impair reproductive success in some walleye populations.

Research on the toxicological significance of maternal methylmercury, obtained from the diet of the parental female, in embryo-larval stages of fish is clearly needed.

The effect of maternally derived methylmercury on the behavior of larval fish also merits scrutiny, given the extreme neurotoxicity of this compound and the sensitivity of fish behaviors to toxic substances. Such work would better define the relation between mercury concentrations in fish eggs and the toxic effects seen in embryonic and larval stages.

SUMMARY

In the last decade, our understanding of the biogeochemistry of mercury has advanced greatly. In addition, the concentrations of mercury in edible fish tissues have been monitored widely because of the health concerns about eating contaminated fish. Yet relatively scant progress has been made in defining fish-tissue residues of mercury associated with toxic effects and in assessing the toxicological significance of methylmercury at environmentally relevant exposure levels. Armstrong's statement about the lack of information on the toxicological significance of mercury in fish, quoted at the beginning of this chapter, remains applicable 15 years later.

The abundances of methylmercury and total mercury in fresh waters are much lower than was believed even a decade ago, even though all surface waters probably contain anthropogenic mercury. Methylmercury concentrations in most oxic surface waters range about 0.01 to 0.8 ng of Hg per liter. Ranges for total mercury (unfiltered samples) are 0.6 to 4 ng/l for lightly contaminated lakes and streams and about 5 to 100 ng/l for directly contaminated waters.

The concentrations of mercury in most surface waters are probably much too low to cause direct toxic effects at the gill surface, either in adult fish or in the more sensitive early life stages. Rather, fish in natural waters obtain methylmercury almost entirely from the diet. Methylmercury accumulates in internal tissues and organs and exerts its most harmful effects on the central nervous system. In the laboratory, long-term dietary exposure to methylmercury has caused incoordination, inability to feed, diminished responsiveness, and starvation. These symptoms were also observed at grossly polluted Minamata Bay, Japan, where severely poisoned adult fish had residues of 8 to 24 µg/g wet weight in axial muscle.

We offer the following guidance for interpreting mercury residues in *adult* fish. In the brain, concentrations of 7 µg/g wet weight or greater probably cause severe, potentially lethal effects. In mercury-sensitive species, such as the walleye, brain tissue concentrations of 3 µg/g wet weight or greater probably indicate significant toxic effects. For axial muscle tissue, field studies indicate that residues of 6 to 20 µg/g wet weight are associated with toxicity. The range for laboratory studies is similar, with sublethal effects or death associated with muscle tissue residues of 5 to 8 µg/g in walleyes and 10 to 20 µg/g in salmonids. Whole-body concentrations associated with sublethal or lethal toxic effects are about 5 µg/g wet weight for brook trout and 10 µg/g for rainbow trout. Estimated no-observed-effect concentrations in salmonids are 3 µg/g for the whole body and 5 µg/g for brain or axial muscle tissue.

Methylmercury residues that would impair fish behavior, thereby reducing growth and survival, have been little studied. Given the neurotoxicity of methylmercury, the residue levels causing behavioral effects may be much lower than the concentration ranges given in the preceding paragraph, which were associated with overt toxicity. The neurotoxic effects of methylmercury may impede the ability of wild fish to locate, capture, handle, and ingest prey, causing starvation, a condition seen in mercury-contaminated northern pike from polluted Clay Lake, Ontario. Furthermore, the consequences of chronic methylmercury exposure may be underestimated in laboratory tests, in which fish are typically offered excess food and are not exposed to predators.

The internal toxicity of methylmercury in fish is influenced by factors that contribute uncertainty to our estimates of critical tissue residues. First, interspecific and intraspecific variation in sensitivity to methylmercury is apparent. Second, the rate of accumulation in fish seems to affect the toxicity of methylmercury. If accumulated slowly, fish can tolerate higher tissue concentrations of mercury, presumably because of the internal transfer and binding of methylmercury to proteins in skeletal muscle (the primary storage site), which decreases exposure of the central nervous system. In addition, the toxicity of methylmercury may be influenced by environmental factors, such as the abundance of selenium.

Many conditions can lead to the bioaccumulation of high concentrations of methylmercury in fish, including anthropogenic discharges of mercury to surface waters, flooding of new impoundments, and atmospheric deposition of mercury to low-pH and humic waters. Methylmercury biomagnifies in food chains; hence, piscivorous fishes accumulate higher concentrations and are at greater risk than coexisting fishes. The most contaminated piscivorous fish, with maximum concentrations in axial muscle of about 5 to 15 $\mu g/g$ wet weight, have been associated with point-source dischargers, such as chloralkali plants; these residues are within the range of concentrations associated with toxic effects in field and laboratory studies. Piscivorous fish from newly flooded impoundments have maximum concentrations in muscle of 3 to 4 $\mu g/g$ wet weight or greater; the toxicological significance of these residues is unclear. In piscivores from low-pH or humic lakes, axial muscle generally contains less than 2 $\mu g/g$ wet weight, seemingly below residue values associated with overt toxicity to adult fish.

The developing fish embryo can be severely affected by a small quantity of methylmercury or inorganic mercury. Methylmercury derived from the adult female, however, probably poses greater risk than waterborne mercury for embryos in natural waters, even though the amount of mercury transferred to the eggs during oogenesis is small. In laboratory bioassays, maternally derived mercury (both inorganic and methyl) can adversely affect the survival, hatching, and development of embryos. The mercury content of eggs reflects the maternal exposure history, with the concentration in the egg increasing concomitantly with parental exposure and tissue concentrations.

We conclude that the primary effect of mercury on fish populations — if any, at existing exposure levels — would be reduced reproductive success resulting from toxicity of maternally derived mercury to embryonic and larval stages. Sublethal and lethal effects on fish embryos are associated with mercury residues in eggs that are much lower than (perhaps 1 to 10% of) the residues associated with toxicity in adult fish. In rainbow trout, for example, the mortality of embryos coincided with total mercury concentrations of 0.07 to 0.10 $\mu g/g$ wet weight in the eggs, values less than 1% of the tissue residues (10 to 30 µg/g) associated with toxicity in adults. Furthermore, the limited data imply that, for some fish populations, the margin of safety between harmful and existing mercury residue levels may be much less for embryo-larval stages than for adults. The reproductive success of some walleye populations may be impaired by existing levels of mercury exposure.

Mercury surveillance programs, which usually analyze axial muscle or skin-on fillets of fish, could be modified and closely coordinated with research to assess the ecotoxicological effects of methylmercury. Analyses of brain tissue, the target organ for methylmercury toxicity, would facilitate assessment of potential effects on adult fish. Sampling and analysis protocols for fish-surveillance programs could be expanded to ensure that potential symptoms of methylmercury intoxication are recorded; these include emaciation and other evidence of starvation, lack of appetite or feeding, slow growth, diminished locomotor activity, poor coordination, diminished escape behavior, and deformities. The presence of such symptoms, coincident with high concentrations of mercury in sampled fish, would strengthen any diagnoses of methylmercury toxicity. Analyses of eggs from gravid females could facilitate assessment of potential reproductive effects of mercury on fish populations. The toxic effects of maternally derived methylmercury on embryo-larval stages of piscivorous fishes clearly merit further study.

ACKNOWLEDGMENTS

Preparation of this chapter was facilitated by the support of the National Biological Service, Upper Mississippi Science at La Crosse (Wisconsin), and the Ontario Ministry of Environment and Energy, Standards Development Branch (Toronto).

We thank Gregory Mierle (Ontario Ministry of Environment and Energy, Aquatic Sciences Section, Dorset Research Centre) for providing statistical summaries of data from the Ministry's Sport Fish Contaminant Survey and Ted Lange (Florida Game and Freshwater Fish Commission, Eustis) for providing data on mercury in largemouth bass from low-pH Florida lakes. We are grateful to Gary Atchison, Drew Bodaly, Barry Johnson, Jack Klaverkamp, Gregory Mierle, and Mark Sandheinrich for providing constructive reviews of the manuscript. Michelle Bartsch prepared the figure, and William Swink and Patricia Watts edited the draft manuscript.

REFERENCES

Abernathy, A. R., D. B. Cox, and C. R. Carter. 1975. Mercury concentrations in fish in lakes Keowee and Jocassee, South Carolina. Completion Rep. No. B-07-09-191-00-573. Clemson University, Clemson, S.C. 51 pp.

Akielaszek, J. J., and T. A. Haines. 1981. Mercury in the muscle tissue of fish from three northern Maine lakes. Bull. Environ. Contam. Toxicol. 27:201-208.

Allard, M., and P. M. Stokes. 1989. Mercury in crayfish species from thirteen Ontario lakes in relation to water chemistry and smallmouth bass *(Micropterus dolomieui)* mercury. Can. J. Fish. Aquat. Sci. 46:1040-1046.

Andersson, T., and L. Håkanson. 1991. Time resolution of mercury dose and lake sensitivity related to mercury content of fish. Water Air Soil Pollut. 56:169-186.

Armstrong, F. A. J. 1979. Effects of mercury compounds on fish. p. 657-670. *In* J. O. Nriagu (Ed.). Biogeochemistry of mercury in the environment. Elsevier/North-Holland Biomedical Press, New York.

Armstrong, F. A. J., and D. P. Scott. 1979. Decrease in mercury content of fishes in Ball Lake, Ontario, since imposition of controls on mercury discharges. J. Fish. Res. Board Can. 36:670-672.

Bidwell, J. R., and A. G. Heath. 1993. An *in situ* study of rock bass *(Ambloplites rupestris)* physiology: effect of season and mercury contamination. Hydrobiologia 264:137-152.

Birge, W. J. 1978. Aquatic toxicology of trace elements of coal and fly ash. p. 219-240. *In* J. H. Thorp and J. W. Gibbons (Eds.). Energy and environmental stress in aquatic systems. CONF-771114. National Technical Information Service, Springfield, Va.

Birge, W. J., J. A. Black, A. G. Westerman, and J. E. Hudson. 1979. The effects of mercury on reproduction of fish and amphibians. p. 629-655. *In* J. O. Nriagu (Ed.). Biogeochemistry of mercury in the environment. Elsevier/North-Holland Biomedical Press, New York.

Birge, W. J., J. A. Black, A. G. Westerman, and B. A. Ramey. 1983. Fish and amphibian embryos — a model system for evaluating teratogenicity. Fundam. Appl. Toxicol. 3:237-242.

Björklund, I., H. Borg, and K. Johansson. 1984. Mercury in Swedish lakes — its regional distribution and causes. Ambio 13:118-121.

Black, J. A., and W. J. Birge. 1980. An avoidance response bioassay for aquatic pollutants. Univ. Ky. Water Resour. Res. Inst. Rep. 123:1-34.

Bloom, N. S. 1989. Determination of picogram levels of methylmercury by aqueous phase ethylation, followed by cryogenic gas chromatography with cold vapour atomic fluorescence detection. Can. J. Fish. Aquat. Sci. 46:1131-1140.

Bloom, N. S. 1992. On the chemical form of mercury in edible fish and marine invertebrate tissue. Can. J. Fish. Aquat. Sci. 49:1010-1017.

Bloom, N. S., and S. W. Effler. 1990. Seasonal variability in the mercury speciation of Onondaga Lake (New York). Water Air Soil Pollut. 53:251-265.

Bloom, N. S., L. Liang, Z. Q. Xie, and S. S. Wang. 1994. Distribution and speciation of mercury in the water and fish of Nan Hu (South Lake), Guangxi Province, People's Republic of China. p. 51-56. *In* C. J. Watras and J. W. Huckabee (Eds.). Mercury pollution: integration and synthesis. Lewis Publishers, Boca Raton, Fla.

Bloom, N. S., and C. J. Watras. 1989. Observations of methylmercury in precipitation. Sci. Total Environ. 87/88:199-207.

Bloom, N. S., C. J. Watras, and J. P. Hurley. 1991. Impact of acidification on the methylmercury cycle of remote seepage lakes. Water Air Soil Pollut. 56:477-491.

Bodaly, R. A., R. E. Hecky, and R. J. P. Fudge. 1984. Increases in fish mercury levels in lakes flooded by the Churchill River diversion, northern Manitoba. Can. J. Fish. Aquat. Sci. 41:682-691.

Bodaly, R. A., and T. A. Johnston. 1992. The mercury problem in hydro-electric reservoirs with predictions of mercury burdens in fish in the proposed Grande Baleine Complex, Québec. *In* Hydro-electric development environmental impacts. Paper No. 3. James Bay Publ. Ser., Montreal, Québec. 15 pp.

Bodaly, R. A., J. W. M. Rudd, R. J. P. Fudge, and C. A. Kelly. 1993. Mercury concentrations in fish related to size of remote Canadian Shield lakes. Can. J. Fish. Aquat. Sci. 50:980-987.

Borgmann, U., and D. M. Whittle. 1991. Contaminant concentration trends in Lake Ontario lake trout *(Salvelinus namaycush)*: 1977 to 1988. J. Great Lakes Res. 17:368-381.

Borgmann, U., and D. M. Whittle. 1992. DDE, PCB, and mercury concentration trends in Lake Ontario rainbow smelt *(Osmerus mordax)* and slimy sculpin *(Cottus cognatus)*: 1977 to 1988. J. Great Lakes Res. 18:298-308.

Boudou, A., M. Delnomdedieu, D. Georgescauld, F. Ribeyre, and E. Saouter. 1991. Fundamental roles of biological barriers in mercury accumulation and transfer in freshwater ecosystems (analysis at organism, organ, cell, and molecular levels). Water Air Soil Pollut. 56:807-821.

Boudou, A., and F. Ribeyre. 1983. Contamination of aquatic biocenoses by mercury compounds: an experimental ecotoxicological approach. p. 73-116. *In* J. O. Nriagu (Ed.). Aquatic toxicology. John Wiley & Sons, New York.

Boudou, A., and F. Ribeyre. 1985. Experimental study of trophic contamination of *Salmo gairdneri* by two mercury compounds — $HgCl_2$ and CH_3HgCl — analysis at the organism and organ levels. Water Air Soil Pollut. 26:137-148.

Bubb, J. M., T. Rudd, and J. N. Lester. 1991. Distribution of heavy metals in the River Yare and its associated broads. I. Mercury and methylmercury. Sci. Total Environ. 102:147-168.

Bycroft, B. M., B. A. W. Coller, G. B. Deacon, D. J. Coleman, and P. S. Lake. 1982. Mercury contamination of the Lerderderg River, Victoria, Australia, from an abandoned gold field. Environ. Pollut. Ser. A Ecol. Biol. 28:135-147.

Carty, A. J., and S. F. Malone. 1979. The chemistry of mercury in biological systems. p. 433-480. *In* J. O. Nriagu (Ed.). Biogeochemistry of mercury in the environment. Elsevier/North-Holland Biomedical Press, New York.

Chang, L. W. 1979. Pathological effects of mercury poisoning. p. 519-580. *In* J. O. Nriagu (Ed.). Biogeochemistry of mercury in the environment. Elsevier/North-Holland Biomedical Press, New York.

Chen, R. W., H. E. Ganther, and W. G. Hoekstra. 1973. Studies on the binding of methylmercury by thionein. Biochem. Biophys. Res. Commun. 51:383-390.

Christensen, G. M. 1975. Biochemical effects of methylmercuric chloride, cadmium chloride, and lead nitrate on embryos and alevins of the brook trout, *Salvelinus fontinalis*. Toxicol. Appl. Pharmacol. 32:191-197.

Christensen, G., E. Hunt, and J. Fiandt. 1977. The effect of methylmercuric chloride, cadmium chloride, and lead nitrate on six biochemical factors of the brook trout *(Salvelinus fontinalis)*. Toxicol. Appl. Pharmacol. 42:523-530.

Clarkson, T. W. 1990. Human health risks from methylmercury in fish. Environ. Toxicol. Chem. 9:957-961.

Clarkson, T. W. 1992. Mercury: major issues in environmental health. Environ. Health Perspect. 100:31-38.

Clarkson, T. W. 1994. Toxicology of mercury and its compounds. p. 631-641. *In* C. J. Watras and J. W. Huckabee (Eds.). Mercury pollution: integration and synthesis. Lewis Publishers, Boca Raton, Fla.

Clarkson, T. W., J. Cranmer, D. J. Sivulka, and R. Smith. 1984. Mercury health effects update: health issue assessment. Rep. EPA-600/8-84-019F. U.S. Environmental Protection Agency, Washington, D.C. 149 pp.

Cooper, J. J. 1983. Total mercury in fishes and selected biota in Lahontan Reservoir, Nevada: 1981. Bull. Environ. Contam. Toxicol. 31:9-17.

Cooper, J. J., and S. Vigg. 1984. Extreme mercury concentrations of a striped bass, *Morone saxatilis*, with a known residence time in Lahontan Reservoir, Nevada. Calif. Fish Game 70:190-192.

Cope, W. G., J. G. Wiener, and R. G. Rada. 1990. Mercury accumulation in yellow perch in Wisconsin seepage lakes: relation to lake characteristics. Environ. Toxicol. Chem. 9:931-940.

Curtis, M. W., T. L. Copeland, and C. H. Ward. 1979. Acute toxicity of 12 industrial chemicals to freshwater and saltwater organisms. Water Res. 13:137-141.

Cuvin-Aralar, M. L. A., and R. W. Furness. 1991. Mercury and selenium interaction: a review. Ecotoxicol. Environ. Saf. 21:348-364.

Davies, F. C. W. 1991. Minamata disease: a 1989 update on the mercury poisoning epidemic in Japan. Environ. Geochem. Health 13:35-38.

deFreitas, A. S. W. 1979. Uptake and retention of mercury by aquatic organisms. p. 201-211. *In* Effects of mercury in the Canadian environment. Rep. NRCC 16739. National Research Council of Canada, Ottawa, Canada.

Deshmukh, S. S., and V. B. Marathe. 1980. Size related toxicity of copper and mercury to *Lebistes reticulatus* (Peter), *Labeo rohita* (Ham.), and *Cyprinus carpio* (L.). Indian J. Exp. Biol. 18:421-427.

Devlin, E. W., and N. K. Mottet. 1992. Embryotoxic action of methyl mercury on coho salmon embryos. Bull. Environ. Contam. Toxicol. 49:449-454.

Dial, N. A. 1978a. Methylmercury: some effects on embryogenesis in the Japanese medaka, *Oryzias latipes*. Teratology 17:83-92.

Dial, N. A. 1978b. Some effects of methylmercury on development of the eye in medaka fish. Growth 42:309-318.

D'Itri, F. M. 1991. Mercury contamination — what we have learned since Minamata. Environ. Monit. Assess. 19:165-182.

Drummond, R. A., G. F. Olson, and A. R. Batterman. 1974. Cough response and uptake of mercury by brook trout, *Salvelinus fontinalis*, exposed to mercuric compounds at different hydrogen-ion concentrations. Trans. Am. Fish. Soc. 103:244-249.

Dutton, M. D., M. Stephenson, and J. F. Klaverkamp. 1993. A mercury saturation assay for measuring metallothionein in fish. Environ. Toxicol. Chem. 12:1193-1202.

Dyrssen, D., and M. Wedborg. 1991. The sulphur-mercury(II) system in natural waters. Water Air Soil Pollut. 56:507-519.

Effler, S. W. 1987. The impact of a chlor-alkali plant on Onondaga Lake and adjoining systems. Water Air Soil Pollut. 33:85-115.

Eisler, R. 1987. Mercury hazards to fish, wildlife, and invertebrates: a synoptic review. Biol. Rep. 85(1.10). U.S. Fish and Wildlife Service, Laurel, Md. 90 pp.

Evans, R. D. 1986. Sources of mercury contamination in the sediments of small headwater lakes in south-central Ontario, Canada. Arch. Environ. Contam. Toxicol. 15:505-512.

Ferrara, R., B. E. Maserti, and R. Breder. 1991. Mercury in abiotic and biotic compartments of an area affected by a geochemical anomaly (Mt. Amiata, Italy). Water Air Soil Pollut. 56:219-233.

Fimreite, N. 1979. Accumulation and effects of mercury on birds. p. 601-627. *In* J. O. Nriagu (Ed.). Biogeochemistry of mercury in the environment. Elsevier/North-Holland Biomedical Press, New York.

Fitzgerald, W. F., and T. W. Clarkson. 1991. Mercury and monomethylmercury: present and future concerns. Environ. Health Perspect. 96:159-166.

Fitzgerald, W. F., R. P. Mason, and G. M. Vandal. 1991. Atmospheric cycling and air-water exchange of mercury over mid-continental lacustrine regions. Water Air Soil Pollut. 56:745-767.

Fitzgerald, W. F., and C. J. Watras. 1989. Mercury in surficial waters of rural Wisconsin lakes. Sci. Total Environ. 87/88:223-232.

Fjeld, E., and S. Rognerud. 1993. Use of path analysis to investigate mercury accumulation in brown trout *(Salmo trutta)* in Norway and the influence of environmental factors. Can. J. Fish. Aquat. Sci. 50:1158-1167.

Francesconi, K. A., and R. C. J. Lenanton. 1992. Mercury contamination in a semi-enclosed marine embayment: organic and inorganic mercury content of biota, and factors influencing mercury levels in fish. Mar. Environ. Res. 33:189-212.

Fujiki, M., and S. Tajima. 1992. The pollution of Minamata Bay by mercury. Water Sci. Technol. 25:133-140.

Futter, M. N. 1994. Pelagic food web structure influences probability of mercury contamination in lake trout *(Salvelinus namaycush)*. Sci. Total Environ. 145:7-12.

Giblin, F. J., and E. J. Massaro. 1973. Pharmacodynamics of methyl mercury in the rainbow trout *(Salmo gairdneri):* tissue uptake, distribution, and excretion. Toxicol. Appl. Pharmacol. 24:81-91.

Gill, G. A., and K. W. Bruland. 1990. Mercury speciation in surface freshwater systems in California and other areas. Environ. Sci. Technol. 24:1392-1400.

Gill, G. A., and K. W. Bruland. 1992. Mercury speciation and cycling in a seasonally anoxic freshwater system: Davis Creek Reservoir. Final report to Electric Power Research Institute, Palo Alto, Calif.

Gill, G. A., and W. F. Fitzgerald. 1987. Picomolar mercury measurements in seawater and other materials using stannous chloride reduction and two-stage gold amalgamation with gas phase detection. Mar. Chem. 20:227-243.

Gilmour, C. C., and E. A. Henry. 1991. Mercury methylation in aquatic systems affected by acid deposition. Environ. Pollut. 71:131-169.

Gilmour, C. C., E. A. Henry, and R. Mitchell. 1992. Sulfate stimulation of mercury methylation in freshwater sediments. Environ. Sci. Technol. 26:2281-2287.

Grieb, T. M., C. T. Driscoll, S. P. Gloss, C. L. Schofield, G. L. Bowie, and D. B. Porcella. 1990. Factors affecting mercury accumulation in fish in the upper Michigan peninsula. Environ. Toxicol. Chem. 9:919-930.

Håkanson, L. 1980. The quantitative impact of pH, bioproduction, and Hg-contamination on the Hg-content of fish (pike). Environ. Pollut. Ser. B Chem. Phys. 1:285-304.

Håkanson, L., Å. Nilsson, and T. Andersson. 1988. Mercury in fish in Swedish lakes. Environ. Pollut. 49:145-162.

Hale, J. G. 1977. Toxicity of metal mining wastes. Bull. Environ. Contam. Toxicol. 17:66-73.

Harrison, S. E., J. F. Klaverkamp, and R. H. Hesslein. 1990. Fates of metal radiotracers added to a whole lake: accumulation in fathead minnow *(Pimephales promelas)* and lake trout *(Salvelinus namaycush)*. Water Air Soil Pollut. 52:277-293.

Hecky, R. E., D. J. Ramsey, R. A. Bodaly, and N. E. Strange. 1991. Increased methylmercury contamination in fish in newly formed freshwater reservoirs. p. 33-52. *In* T. Suzuki et al. (Eds.). Advances in mercury toxicology. Plenum Press, New York.

Heisinger, J. F., and W. Green. 1975. Mercuric chloride uptake by eggs of the ricefish and resulting teratogenic effects. Bull. Environ. Contam. Toxicol. 14:665-673.

Heiskary, S. A., and D. D. Helwig. 1986. Mercury levels in northern pike, *Esox lucius*, relative to water chemistry in northern Minnesota lakes. p. 33-37. *In* Lake and reservoir management. Vol. 2. North American Lake Management Society, Washington, D.C.

Hildebrand, S. G., R. H. Strand, and J. W. Huckabee. 1980. Mercury accumulation in fish and invertebrates of the North Fork Holston River, Virginia and Tennessee. J. Environ. Qual. 9:393-400.

Hilmy, A. M., N. A. El Domiaty, A. Y. Daabees, and F. I. Moussa. 1987. Short-term effects of mercury on survival, behaviour, bioaccumulation, and ionic pattern in the catfish *(Clarias lazera)*. Comp. Biochem. Physiol. C Comp. Pharmacol. Toxicol. 87:303-308.

Hodson, P. V. 1988. The effect of metal metabolism on uptake, disposition, and toxicity in fish. Aquat. Toxicol. (Amst.) 11:3-18.
Honma, Y. 1977. Concentration of methylmercury by organisms, especially fish, in the Agano River. p. 288-310. *In* T. Tsubaki and K. Irukayama (Eds.). Minamata disease: methylmercury poisoning in Minamata and Niigata, Japan. Kodansha, Tokyo, Japan.
Huckabee, J. W., J. W. Elwood, and S. G. Hildebrand. 1979. Accumulation of mercury in freshwater biota. p. 277-302. *In* J. O. Nriagu (Ed.). Biogeochemistry of mercury in the environment. Elsevier/North-Holland Biomedical Press, New York.
Huckabee, J. W., R. A. Goldstein, S. A. Janzen, and S. E. Woock. 1975. Methylmercury in a freshwater food chain. p. 199-216. *In* Proc. First Int. Conf. on Heavy Metals in the Environment, Toronto, Ontario.
Huckabee, J. W., S. A. Janzen, B. G. Blaylock, Y. Talmi, and J. J. Beauchamp. 1978. Methylated mercury in brook trout *(Salvelinus fontinalis):* absence of an in vivo methylating process. Trans. Am. Fish. Soc. 107:848-852.
Hurley, J. P., J. M. Benoit, C. L. Babiarz, M. M. Shafer, A. W. Andren, J. R. Sullivan, R. Hammond, and D. A. Webb. 1995. Influences of watershed characteristics on mercury levels in Wisconsin Rivers. Environ. Sci. Technol. 29:1867-1875.
Iverfeldt, Å. 1991. Occurrence and turnover of atmospheric mercury over the Nordic countries. Water Air Soil Pollut. 56:251-265.
Iverfeldt, Å., and K. Johansson. 1988. Mercury in run-off water from small watersheds. Verh. Int. Verein. Limnol. 23:1626-1632.
Jackson, T. A. 1991. Biological and environmental control of mercury accumulation by fish in lakes and reservoirs of northern Manitoba, Canada. Can. J. Fish. Aquat. Sci. 48:2449-2470.
Jensen, A., and A. Jensen. 1991. Historical deposition rates of mercury in Scandinavia estimated by dating and measurement of mercury in cores of peat bogs. Water Air Soil Pollut. 56:769-777.
Johansson, K., M. Aastrup, A. Andersson, L. Bringmark, and Å. Iverfeldt. 1991. Mercury in Swedish forest soils and waters — assessment of critical load. Water Air Soil Pollut. 56:267-281.
Johnson, M. G. 1987. Trace element loadings to sediments of fourteen Ontario lakes and correlations with concentrations in fish. Can. J. Fish. Aquat. Sci. 44:3-13.
Johnston, T. A., R. A. Bodaly, and J. A. Mathias. 1991. Predicting fish mercury levels from physical characteristics of boreal reservoirs. Can. J. Fish. Aquat. Sci. 48:1468-1475.
Kania, H. J., and J. O'Hara. 1974. Behavioral alterations in a simple predator-prey system due to sublethal exposure to mercury. Trans. Am. Fish. Soc. 103:134-136.
Kirubagaran, R., and K. P. Joy. 1988. Toxic effects of three mercurial compounds on survival and histology of the kidney of the catfish *Clarias batrachus* (L.). Ecotoxicol. Environ. Saf. 15:171-179.
Kirubagaran, R., and K. P. Joy. 1990. Changes in brain monoamine levels and monoamine oxidase activity in the catfish, *Clarias batrachus*, during chronic treatments with mercurials. Bull. Environ. Contam. Toxicol. 45:88-93.
Kitamura, S. 1968. Determination on mercury content in bodies of inhabitants, cats, fishes, and shells in Minamata District and in the mud of Minamata Bay. p. 257-266. *In* Minamata disease. Study Group of Minamata Disease, Kumamoto University, Japan.
Klaverkamp, J. F., W. A. Macdonald, D. A. Duncan, and R. Wagemann. 1984. Metallothionein and acclimation to heavy metals in fish: a review. p. 99-113. *In* V. W. Cairns, P. V. Hodson, and J. O. Nriagu (Eds.). Contaminant effects on fisheries. John Wiley & Sons, New York.

Klaverkamp, J. F., W. A. Macdonald, W. R. Lillie, and A. Lutz. 1983. Joint toxicity of mercury and selenium in salmonid eggs. Arch. Environ. Contam. Toxicol. 12:415-419.

Koirtyohann, S. R., R. Meers, and L. K. Graham. 1974. Mercury levels in fishes from some Missouri lakes with and without known mercury pollution. Environ. Res. 8:1-11.

Korthals, E. T., and M. R. Winfrey. 1987. Seasonal and spatial variations in mercury methylation and demethylation in an oligotrophic lake. Appl. Environ. Microbiol. 53:2397-2404.

Laarman, P. W., W. A. Willford, and J. R. Olson. 1976. Retention of mercury in the muscle of yellow perch *(Perca flavescens)* and rock bass *(Ambloplites rupestris)*. Trans. Am. Fish. Soc. 105:296-300.

Lange, T. R., H. E. Royals, and L. L. Connor. 1993. Influence of water chemistry on mercury concentration in largemouth bass from Florida lakes. Trans. Am. Fish. Soc. 122:74-84.

Lathrop, R. C., P. W. Rasmussen, and D. R. Knauer. 1991. Mercury concentrations in walleyes from Wisconsin (USA) lakes. Water Air Soil Pollut. 56:295-307.

Lee, Y.-H., and H. Hultberg. 1990. Methylmercury in some Swedish surface waters. Environ. Toxicol. Chem. 9:833-841.

Lee, Y.-H., and Å. Iverfeldt. 1991. Measurement of methylmercury and mercury in run-off, lake, and rain waters. Water Air Soil Pollut. 56:309-321.

Lemly, A. D. 1993. Metabolic stress during winter increases the toxicity of selenium to fish. Aquat. Toxicol. 27:133-158.

Little, E. E., and S. E. Finger. 1990. Swimming behavior as an indicator of sublethal toxicity in fish. Environ. Toxicol. Chem. 9:13-19.

Lock, R. A. C., P. M. J. M. Cruijsen, and A. P. van Overbeeke. 1981. Effects of mercuric chloride and methylmercuric chloride on the osmoregulatory function of the gills in rainbow trout, *Salmo gairdneri* Richardson. Comp. Biochem. Physiol. C Comp. Pharmacol. Toxicol. 68:151-159.

Lock, R. A. C., and A. P. van Overbeeke. 1981. Effects of mercuric chloride and methylmercuric chloride on mucus secretion in rainbow trout, *Salmo gairdneri* Richardson. Comp. Biochem. Physiol. C Comp. Pharmacol. Toxicol. 69:67-73.

Lockhart, W. L., J. F. Uthe, A. R. Kenney, and P. M. Mehrle. 1972. Methylmercury in northern pike *(Esox lucius)*: distribution, elimination, and some biochemical characteristics of contaminated fish. J. Fish. Res. Board Can. 29:1519-1523.

Lodenius, M. 1991. Mercury concentrations in an aquatic ecosystem during twenty years following abatement of the pollution source. Water Air Soil Pollut. 56:323-332.

MacCrimmon, H. R., C. D. Wren, and B. L. Gots. 1983. Mercury uptake by lake trout, *Salvelinus namaycush*, relative to age, growth, and diet in Tadenac Lake with comparative data from other Precambrian Shield lakes. Can. J. Fish. Aquat. Sci. 40:114-120.

Mance, G. 1987. Pollution threat of heavy metals in aquatic environments. Elsevier, London. 362 pp.

Mannio, J., M. Verta, P. Kortelainen, and S. Rekolainen. 1986. The effect of water quality on the mercury concentration of northern pike *(Esox lucius* L.) in Finnish forest lakes and reservoirs. p. 32-43. Publ. No. 65. Water Research Institute, National Board of Waters, Helsinki, Finland.

Martinelli, L. A., J. R. Ferreira, B. R. Forsberg, and R. L. Victoria. 1988. Mercury contamination in the Amazon: a gold rush consequence. Ambio 17:252-254.

Mason, R. P., and W. F. Fitzgerald. 1993. The distribution and biogeochemical cycling of mercury in the equatorial Pacific Ocean. Deep-Sea Res. 40:1897-1924.

Mathers, R. A., and P. H. Johansen. 1985. The effects of feeding ecology on mercury accumulation in walleye *(Stizostedion vitreum)* and pike *(Esox lucius)* in Lake Simcoe. Can. J. Zool. 63:2006-2012.

Matida, Y., H. Kumada, S. Kumura, Y. Saiga, T. Nose, M. Yokote, and H. Kawatsu. 1971. Toxicity of mercury compounds to aquatic organisms and accumulation of the compounds by the organisms. Bull. Freshwater Fish. Res. Lab. (Tokyo) 21:197-227.

May, K., M. Stoeppler, and K. Reisinger. 1987. Studies in the ratio total mercury/methylmercury in the aquatic food chain. Toxicol. Environ. Chem. 13:153-159.

McIntyre, J. D. 1973. Toxicity of methyl mercury for steelhead trout sperm. Bull. Environ. Contam. Toxicol. 9:98-99.

McKim, J. M. 1977. Evaluation of tests with early life stages of fish for predicting long-term toxicity. J. Fish. Res. Board Can. 34:1148-1154.

McKim, J. M., G. F. Olson, G. W. Holcombe, and E. P. Hunt. 1976. Long-term effects of methylmercuric chloride on three generations of brook trout *(Salvelinus fontinalis)*: toxicity, accumulation, distribution, and elimination. J. Fish. Res. Board Can. 33:2726-2739.

McMurtry, M. J., D. L. Wales, W. A. Scheider, G. L. Beggs, and P. E. Dimond. 1989. Relationship of mercury concentrations in lake trout *(Salvelinus namaycush)* and smallmouth bass *(Micropterus dolomieui)* to the physical and chemical characteristics of Ontario lakes. Can. J. Fish. Aquat. Sci. 46:426-434.

Meili, M. 1991. Fluxes, pools, and turnover of mercury in Swedish forest lakes. Water Air Soil Pollut. 56:719-727.

Meili, M., Å. Iverfeldt, and L. Håkanson. 1991. Mercury in the surface water of Swedish forest lakes — concentrations, speciation, and controlling factors. Water Air Soil Pollut. 56:439-453.

Menezes, M. R., and S. Z. Qasim. 1984. Effects of mercury accumulation on the electrophoretic patterns of the serum, haemoglobin, and eye lens proteins of *Tilapia mossambica* (Peters). Water Res. 18:153-161.

Mierle, G. 1990. Aqueous inputs of mercury to Precambrian Shield lakes in Ontario. Environ. Toxicol. Chem. 9:843-851.

Mierle, G., and R. Ingram. 1991. The role of humic substances in the mobilization of mercury from watersheds. Water Air Soil Pollut. 56:349-357.

Miskimmin, B. M., J. W. M. Rudd, and C. A. Kelly. 1992. Influence of dissolved organic carbon, pH, and microbial respiration rates on mercury methylation and demethylation in lake water. Can. J. Fish. Aquat. Sci. 49:17-22.

Morgan, W. S. G. 1979. Fish locomotor behavior patterns as a monitoring tool. J. Water Pollut. Control Fed. 51:580-589.

Mount, D. I. 1974. Chronic toxicity of methylmercuric chloride to fathead minnow. Testimony in the matter of proposed toxic pollutant effluent standards for aldrin-dieldrin et al. Federal Water Pollution Control Act Amendments (307) Docket No. 1, Exhibit No. 4.

Mukherjee, A. B. 1991. Industrial emissions of mercury in Finland between 1967 and 1987. Water Air Soil Pollut. 56:35-49.

Nater, E. A., and D. F. Grigal. 1992. Regional trends in mercury distribution across the Great Lakes states, north central USA. Nature (Lond.) 358:139-141.

Nicoletto, P. F., and A. C. Hendricks. 1988. Sexual differences in accumulation of mercury in four species of centrarchid fishes. Can. J. Zool. 66:944-949.

Niimi, A. J. 1983. Biological and toxicological effects of environmental contaminants in fish and their eggs. Can. J. Fish. Aquat. Sci. 40:306-312.

Niimi, A. J., and G. P. Kissoon. 1994. Evaluation of the critical body burden concept based on inorganic and organic mercury toxicity to rainbow trout *(Oncorhynchus mykiss)*. Arch. Environ. Contam. Toxicol. 26:169-178.

Niimi, A. J., and L. Lowe-Jinde. 1984. Differential blood cell ratios of rainbow trout *(Salmo gairdneri)* exposed to methylmercury and chlorobenzenes. Arch. Environ. Contam. Toxicol. 13:303-311.

Nomura, S. 1968. Epidemiology of Minamata disease. p. 5-35. *In* Minamata disease. Study Group of Minamata Disease, Kumamoto University, Japan.

Norton, S. A., P. J. Dillon, R. D. Evans, G. Mierle, and J. S. Kahl. 1990. The history of atmospheric deposition of Cd, Hg, and Pb in North America: evidence from lake and peat bog sediments. p. 73-102. *In* S. E. Lindberg, A. L. Page, and S. A. Norton (Eds.). Sources, deposition, and canopy interactions. Vol. 3. Acidic precipitation. Springer-Verlag, New York.

Nriagu, J. O. 1989. A global assessment of natural sources of atmospheric trace metals. Nature (Lond.) 338:47-49.

Nriagu, J. O., and J. M. Pacyna. 1988. Quantitative assessment of worldwide contamination of air, water, and soils by trace metals. Nature (Lond.) 333:134-139.

Nriagu, J. O., W. C. Pfeiffer, O. Malm, C. M. M. Souza, and G. Mierle. 1992. Mercury pollution in Brazil. Nature (Lond.) 356:389. (Letter).

O'Connor, D. V., and P. O. Fromm. 1975. The effect of methyl mercury on gill metabolism and blood parameters of rainbow trout. Bull. Environ. Contam. Toxicol. 13:406-411.

Olson, G. F., D. I. Mount, V. M. Snarski, and T. W. Thorslund. 1975. Mercury residues in fathead minnows, *Pimephales promelas* Rafinesque, chronically exposed to methylmercury in water. Bull. Environ. Contam. Toxicol. 14:129-134.

Olson, K. R., H. L. Bergman, and P. O. Fromm. 1973. Uptake of methyl mercuric chloride and mercuric chloride by trout: a study of uptake pathways into the whole animal and uptake by erythrocytes in vitro. J. Fish. Res. Board Can. 30:1293-1299.

Olson, K. R., K. S. Squibb, and R. J. Cousins. 1978. Tissue uptake, subcellular distribution, and metabolism of $^{14}CH_3HgCl$ and $CH_3^{203}HgCl$ by rainbow trout, *Salmo gairdneri*. J. Fish. Res. Board Can. 35:381-390.

Parks, J. W., and A. L. Hamilton. 1987. Accelerating recovery of the mercury-contaminated Wabigoon/English River system. Hydrobiologia 149:159-188.

Paulsson, K., and K. Lundbergh. 1991. Treatment of mercury contaminated fish by selenium addition. Water Air Soil Pollut. 56:833-841.

Pennacchioni, A., R. Marchetti, and G. F. Gaggino. 1976. Inability of fish to methylate mercuric chloride in vivo. J. Environ. Qual. 5:451-454.

Perry, D. M., J. S. Weis, and P. Weis. 1988. Cytogenetic effects of methylmercury in embryos of the killifish, *Fundulus heteroclitus*. Arch. Environ. Contam. Toxicol. 17:569-574.

Pfeiffer, W. C., O. Malm, C. M. M. Souza, L. Drude de Lacerda, E. G. Silveira, and W. R. Bastos. 1991. Mercury in the Madeira River ecosystem, Rondônia, Brazil. For. Ecol. Manage. 38:239-245.

Phillips, G. R., and D. R. Buhler. 1978. The relative contributions of methylmercury from food or water to rainbow trout *(Salmo gairdneri)* in a controlled laboratory environment. Trans. Am. Fish. Soc. 107:853-861.

Phillips, G. R., and R. W. Gregory. 1979. Assimilation efficiency of dietary methylmercury by northern pike *(Esox lucius)*. J. Fish. Res. Board Can. 36:1516-1519.

Phillips, G. R., T. E. Lenhart, and R. W. Gregory. 1980. Relation between trophic position and mercury accumulation among fishes from the Tongue River Reservoir, Montana. Environ. Res. 22:73-80.

Piotrowski, J. K., B. Trojanowska, J. M. Wisniewska-Knypl, and W. Bolanowska. 1973. Further investigations on binding and release of mercury in the rat. p. 247-263. *In* M. W. Miller and T. W. Clarkson (Eds.). Mercury, mercurials, and mercaptans. Charles C Thomas, Springfield, Ill.

Porcella, D. B., C. J. Watras, and N. S. Bloom. 1992. Mercury species in lake water. p. 127-138. *In* The deposition and fate of trace metals in our environment. Gen. Tech. Rep. NC-150. U.S. Forest Service, North Central Forest Experiment Station, St. Paul, Minn.

Rada, R. G., J. E. Findley, and J. G. Wiener. 1986. Environmental fate of mercury discharged into the upper Wisconsin River. Water Air Soil Pollut. 29:57-76.

Rada, R. G., D. E. Powell, and J. G. Wiener. 1993. Whole-lake burdens and spatial distribution of mercury in surficial sediments in Wisconsin seepage lakes. Can. J. Fish. Aquat. Sci. 50:865-873.

Rada, R. G., J. G. Wiener, M. R. Winfrey, and D. E. Powell. 1989. Recent increases in atmospheric deposition of mercury to north-central Wisconsin lakes inferred from sediment analyses. Arch. Environ. Contam. Toxicol. 18:175-181.

Ramlal, P. S., C. A. Kelly, J. W. M. Rudd, and A. Furutani. 1993. Sites of methyl mercury production in remote Canadian Shield lakes. Can. J. Fish. Aquat. Sci. 50:972-979.

Ramsey, D. J. 1990a. Experimental studies of mercury dynamics in the Churchill River diversion, Manitoba. Collect. Environ. Géol. 9:147-173.

Ramsey, D. J. 1990b. Measurements of methylation balance in Southern Indian Lake, Granville Lake, and Stephens Lake, Manitoba, 1989. Ecol. Rep. 90-3. Environment Canada, Fisheries and Oceans, Northern Flood Agreement, Winnipeg, Manitoba. 89 pp.

Rask, M., and T.-R. Metsälä. 1991. Mercury concentrations in northern pike, *Esox lucius* L., in small lakes of Evo area, southern Finland. Water Air Soil Pollut. 56:369-378.

Rehwoldt, R., L. W. Menapace, B. Nerrie, and D. Allessandrello. 1972. The effect of increased temperature upon the acute toxicity of some heavy metal ions. Bull. Environ. Contam. Toxicol. 8:91-96.

Reinert, R. E., L. J. Stone, and W. A. Willford. 1974. Effect of temperature on accumulation of methylmercuric chloride and p,p'DDT by rainbow trout *(Salmo gairdneri)*. J. Fish. Res. Board Can. 31:1649-1652.

Ribeyre, F., and A. Boudou. 1984. Bioaccumulation et repartition tissulaire du mercure — $HgCl_2$ et CH_3HgCl — chez *Salmo gairdneri* apres contamination par voie directe. (In French.) Water Air Soil Pollut. 23:169-186.

Rodgers, D. W. 1994. You are what you eat and a little bit more: bioenergetics-based models of methylmercury accumulation in fish revisited. p. 427-439. *In* C. J. Watras and J. W. Huckabee (Eds.). Mercury pollution: integration and synthesis. Lewis Publishers, Boca Raton, Fla.

Rodgers, D. W., and F. W. H. Beamish. 1981. Uptake of waterborne methylmercury by rainbow trout *(Salmo gairdneri)* in relation to oxygen consumption and methylmercury concentration. Can. J. Fish. Aquat. Sci. 38:1309-1315.

Rodgers, D. W., and F. W. H. Beamish. 1982. Dynamics of dietary methylmercury in rainbow trout, *Salmo gairdneri*. Aquat. Toxicol. 2:271-290.

Roesijadi, G. 1992. Metallothioneins in metal regulation and toxicity in aquatic animals. Aquat. Toxicol. 22:81-114.

Rognerud, S., and E. Fjeld. 1993. Regional survey of heavy metals in lake sediments in Norway. Ambio 22:206-212.

Rudd, J. W. M., A. Furutani, and M. A. Turner. 1980. Mercury methylation by fish intestinal contents. Appl. Environ. Microbiol. 40:777-782.

Rudd, J. W. M., M. A. Turner, A. Furutani, A. L. Swick, and B. E. Townsend. 1983. The English-Wabigoon River system. I. A synthesis of recent research with a view towards mercury amelioration. Can. J. Fish. Aquat. Sci. 40:2206-2217.

Ruohtula, M., and J. K. Miettinen. 1975. Retention and excretion of ^{203}Hg-labelled methylmercury in rainbow trout. Oikos 26:385-390.

Sandheinrich, M. B., and G. J. Atchison. 1990. Sublethal toxicant effects on fish foraging behavior: empirical vs. mechanistic approaches. Environ. Toxicol. Chem. 9:107-119.

Sastry, K. V., and D. R. Rao. 1984. Effect of mercuric chloride on some biochemical and physiological parameters of the freshwater murrel, *Channa punctatus*. Environ. Res. 34:343-350.

Sastry, K. V., D. R. Rao, and S. K. Singh. 1982. Mercury induced alterations in the intestinal absorption of nutrients in the fresh water murrel, *Channa punctatus*. Chemosphere 11:613-619.

Scheider, W. A., D. S. Jeffries, and P. J. Dillon. 1979. Effects of acidic precipitation on Precambrian freshwaters in southern Ontario. J. Great Lakes Res. 5:45-51.

Scherer, E., F. A. J. Armstrong, and S. H. Nowak. 1975. Effects of mercury-contaminated diet upon walleyes, *Stizostedion vitreum vitreum* (Mitchill). Can. Fish. Mar. Serv. Resour. Dev. Branch Winnipeg Tech. Rep. No. 597. 21 pp.

Scheuhammer, A. M. 1991. Effects of acidification on the availability of toxic metals and calcium to wild birds and mammals. Environ. Pollut. 71:329-375.

Schreiber, W. 1983. Mercury content of fishery products: data from the last decade. Sci. Total Environ. 31:283-300.

Schroeder, W. H., G. Yarwood, and H. Niki. 1991. Transformation processes involving mercury species in the atmosphere — results from a literature survey. Water Air Soil Pollut. 56:653-666.

Sharpe, M. A., A. S. W. deFreitas, and A. E. McKinnon. 1977. The effect of body size on methylmercury clearance by goldfish. Environ. Biol. Fishes 2:177-183.

Slemr, F., and E. Langer. 1992. Increase in global atmospheric concentrations of mercury inferred from measurements over the Atlantic Ocean. Nature (Lond.) 355:434-437.

Snarski, V. M., and G. F. Olson. 1982. Chronic toxicity and bioaccumulation of mercuric chloride in the fathead minnow *(Pimephales promelas)*. Aquat. Toxicol. 2:143-156.

Spry, D. J., and J. G. Wiener. 1991. Metal bioavailability and toxicity to fish in low-alkalinity lakes: a critical review. Environ. Pollut. 71:243-304.

Stegeman, J. J., M. Brouwer, R. T. Di Giulio, L. Forlin, B. A. Fowler, B. M. Sanders, and P. A. Van Veld. 1992. Molecular responses to environmental contamination: enzyme and protein systems as indicators of chemical exposure and effect. p. 235-335. *In* R. J. Huggett, R. A. Kimerle, P. M. Mehrle, Jr., and H. L. Bergman (Eds.). Biomarkers: biochemical, physiological, and histological markers of anthropogenic stress. Lewis Publishers, Chelsea, Mich.

Steinnes, E., and E. M. Andersson. 1991. Atmospheric deposition of mercury in Norway: temporal and spatial trends. Water Air Soil Pollut. 56:391-404.

St. Louis, V. L., J. W. M. Rudd, C. A. Kelly, K. G. Beaty, N. S. Bloom, and R. J. Flett. 1994. Importance of wetlands as sources of methyl mercury to boreal forest ecosystems. Can. J. Fish. Aquat. Sci. 51:1065-1076.

Sumner, A. K., J. G. Saha, and Y. W. Lee. 1972. Mercury residues in fish from Saskatchewan waters with and without known sources of pollution — 1970. Pestic. Monit. J. 6:122-125.

Suns, K., and G. Hitchin. 1990. Interrelationships between mercury levels in yearling yellow perch, fish condition, and water quality. Water Air Soil Pollut. 50:255-265.

Suns, K., G. Hitchin, B. Loescher, E. Pastorek, and R. Pearce. 1987. Metal accumulations in fishes from Muskoka-Haliburton lakes in Ontario (1978-1984). Tech. Rep. Ontario Ministry of the Environment, Rexdale, Ontario. 38 pp.

Surma-Aho, K., J. Paasivirta, S. Rekolainen, and M. Verta. 1986. Organic and inorganic mercury in the food chain of some lakes and reservoirs in Finland. Chemosphere 15:353-372.

Swain, E. B., D. R. Engstrom, M. E. Brigham, T. A. Henning, and P. L. Brezonik. 1992. Increasing rates of atmospheric mercury deposition in midcontinental North America. Science (Washington, D.C.) 257:784-787.

Takeuchi, T. 1968. Pathology of Minamata disease. p. 141-228. *In* Minamata disease. Study Group of Minamata Disease, Kumamoto University, Japan.

Tsubaki, T., and K. Irukayama (Eds.). 1977. Minamata disease: methylmercury poisoning in Minamata and Niigata, Japan. Kodansha, Tokyo, Japan.

Turner, M. A., and A. L. Swick. 1983. The English-Wabigoon River system. IV. Interactions between mercury and selenium accumulated from waterborne and dietary sources by northern pike *(Esox lucius)*. Can. J. Fish. Aquat. Sci. 40:2241-2250.

Turner, R. R., C. R. Olsen, and W. J. Wilcox, Jr. 1984. Environmental fate of Hg and ^{137}Cs discharged from Oak Ridge facilities. p. 329-338. *In* D. D. Hemphill (Ed.). Trace substances in environmental health. Vol. 18. University of Missouri Press, Columbia, Mo.

Uthe, J. F., F. M. Atton, and L. M. Royer. 1973. Uptake of mercury by caged rainbow trout *(Salmo gairdneri)* in the South Saskatchewan River. J. Fish. Res. Board Can. 30:643-650.

Vandal, G. M., R. P. Mason, and W. F. Fitzgerald. 1991. Cycling of volatile mercury in temperate lakes. Water Air Soil Pollut. 56:791-803.

Verdon, R., D. Brouard, C. Demers, R. Lalumiere, M. Laperle, and R. Schetagne. 1991. Mercury evolution (1978-1988) in fishes of the La Grande hydroelectric complex, Quebec, Canada. Water Air Soil Pollut. 56:405-417.

Verma, S. R., and I. P. Tonk. 1983. Effect of sublethal concentrations of mercury on the composition of liver, muscles, and ovary of *Notopterus notopterus*. Water Air Soil Pollut. 20:287-292.

Verta, M. 1990. Mercury in Finnish forest lakes and reservoirs: anthropogenic contribution to the load and accumulation in fish. p. 5-34. Publ. No. 6. Water and Environment Research Institute, National Board of Waters and the Environment, Helsinki.

Verta, M., J. Mannio, P. Iivonen, J.-P. Hirvi, O. Järvinen, and S. Piepponen. 1990. Trace metals in Finnish headwater lakes — effects of acidification and airborne load. p. 883-908. *In* P. Kauppi et al. (Eds.). Acidification in Finland. Springer-Verlag, Berlin.

Verta, M., S. Rekolainen, and K. Kinnunen. 1986. Causes of increased fish mercury levels in Finnish reservoirs. p. 44-58. Publ. No. 65. Water Research Institute, National Board of Waters, Helsinki, Finland.

Verta, M., K. Tolonen, and H. Simola. 1989. History of heavy metal pollution in Finland as recorded by lake sediments. Sci. Total Environ. 87/88:1-18.

Walczak, B. Z., U. T. Hammer, and P. M. Huang. 1986. Ecophysiology and mercury accumulation of rainbow trout *(Salmo gairdneri)* when exposed to mercury in various concentrations of chloride. Can. J. Fish. Aquat. Sci. 43:710-714.

Watras, C. J., and N. S. Bloom. 1992. Mercury and methylmercury in individual zooplankton: implications for bioaccumulation. Limnol. Oceanogr. 37:1313-1318.

Watras, C. J., N. S. Bloom, R. J. M. Hudson, S. A. Gherini, R. Munson, S. A. Claas, K. A. Morrison, J. Hurley, J. G. Wiener, W. F. Fitzgerald, R. Mason, G. Vandal, D. Powell, R. Rada, L. Rislove, M. Winfrey, J. Elder, D. Krabbenhoft, A. W. Andren, C. Babiarz, D. B. Porcella, and J. W. Huckabee. 1994. Sources and fates of mercury and methylmercury in Wisconsin lakes. p. 153-177. *In* C. J. Watras and J. W. Huckabee (Eds.). Mercury pollution: integration and synthesis. Lewis Publishers, Boca Raton, Fla.

Weir, P. A., and C. H. Hine. 1970. Effects of various metals on behavior of conditioned goldfish. Arch. Environ. Health 20:45-51.

Weis, J. S., and P. Weis. 1984. A rapid change in methylmercury tolerance in a population of killifish, *Fundulus heteroclitus*, from a golf course pond. Mar. Environ. Res. 13:231-245.

Weis, P. 1984. Metallothionein and mercury tolerance in the killifish, *Fundulus heteroclitus*. Mar. Environ. Res. 14:153-166.

Weis, P., and J. S. Weis. 1977a. The effects of heavy metals on embryonic development of the killifish, *Fundulus heteroclitus*. J. Fish Biol. 11:49-54.

Weis, P., and J. S. Weis. 1977b. Methylmercury teratogenesis in the killifish, *Fundulus heteroclitus*. Teratology 16:317-326.

Weis, P., and J. S. Weis. 1991. The developmental toxicity of metals and metalloids in fish. p. 145-169. *In* M. C. Newman and A. W. McIntosh (Eds.). Metal ecotoxicology — concepts and applications. Lewis Publishers, Boca Raton, Fla.

Weis, P., J. S. Weis, and J. Bogden. 1986. Effects of environmental factors on release of mercury from Berry's Creek (New Jersey) sediments and its uptake by killifish *Fundulus heteroclitus*. Environ. Pollut. Ser. A Ecol. Biol. 40:303-315.

Wester, P. W. 1991. Histopathological effects of environmental pollutants beta-HCH and mercury on reproductive organs in freshwater fish. Comp. Biochem. Physiol. C Comp. Pharmacol. Toxicol. 100:237-239.

Whitney, S. D. 1991. Effects of Maternally Transmitted Mercury on the Hatching Success, Survival, Growth, and Behavior of Embryo and Larval Walleye *(Stizostedion vitreum vitreum)*. M.S. thesis. University of Wisconsin-La Crosse, La Crosse, Wisc. 80 pp.

Wiener, J. G., W. F. Fitzgerald, C. J. Watras, and R. G. Rada. 1990a. Partitioning and bioavailability of mercury in an experimentally acidified Wisconsin lake. Environ. Toxicol. Chem. 9:909-918.

Wiener, J. G., R. E. Martini, T. B. Sheffy, and G. E. Glass. 1990b. Factors influencing mercury concentrations in walleyes in northern Wisconsin lakes. Trans. Am. Fish. Soc. 119:862-870.

Wiener, J. G., and P. M. Stokes. 1990. Enhanced bioavailability of mercury, cadmium, and lead in low-alkalinity waters: an emerging regional environmental problem. Environ. Toxicol. Chem. 9:821-823.

Wilken, R.-D., and H. Hintelmann. 1991. Mercury and methylmercury in sediments and suspended particles from the River Elbe, North Germany. Water Air Soil Pollut. 56:427-437.

Winfrey, M. R., and J. W. M. Rudd. 1990. Environmental factors affecting the formation of methylmercury in low pH lakes. Environ. Toxicol. Chem. 9:853-869.

Wobeser, G. 1975a. Acute toxicity of methyl mercury chloride and mercuric chloride for rainbow trout *(Salmo gairdneri)* fry and fingerlings. J. Fish. Res. Board Can. 32:2005-2013.

Wobeser, G. 1975b. Prolonged oral administration of methyl mercury chloride to rainbow trout *(Salmo gairdneri)* fingerlings. J. Fish. Res. Board Can. 32:2015-2023.

World Health Organization. 1989. Mercury — environmental aspects. WHO Environmental Criteria 86, Geneva, Switzerland. 115 pp.

Wren, C. D., H. R. MacCrimmon, and B. R. Loescher. 1983. Examination of bioaccumulation and biomagnification of metals in a Precambrian Shield lake. Water Air Soil Pollut. 19:277-291.

Wren, C. D., W. A. Scheider, D. L. Wales, B. W. Muncaster, and I. M. Gray. 1991. Relation between mercury concentrations in walleye *(Stizostedion vitreum vitreum)* and northern pike *(Esox lucius)* in Ontario lakes and influence of environmental factors. Can. J. Fish. Aquat. Sci. 48:132-139.

Xun, L., N. E. R. Campbell, and J. W. M. Rudd. 1987. Measurements of specific rates of net methyl mercury production in the water column and surface sediments of acidified and circumneutral lakes. Can. J. Fish. Aquat. Sci. 44:750-757.

Yingcharoen, D., and R. A. Bodaly. 1993. Elevated mercury levels in fish resulting from reservoir flooding in Thailand. Asian Fish. Sci. 6:73-80.

CHAPTER **14**

Mercury in Birds and Terrestrial Mammals

David R. Thompson

INTRODUCTION

Mercury is a biologically nonessential heavy metal that can exist in a range of inorganic and organic forms with varying degrees of stability and toxicity. It is generally accepted that methylmercury is the most stable and toxic to wildlife, being absorbed efficiently from the diet and ultimately attacking the nervous system. Typical poisoning symptoms include loss of coordination, reduction in the field of vision, numbness of the extremities, and decline in mental activity and awareness.

Mercury is emitted to the environment from a number of natural sources, including volcanic activity, continental particulate and volatile matter, and fluxes from the marine environment (Nriagu, 1989). In addition, a considerable amount of mercury is released through human activities such as industrial and domestic coal combustion, nonferrous metal production, waste incineration, chemical production processes, and the dumping of sewage sludge (Nriagu and Pacyna, 1988); in total, these have been estimated to produce over twice the amount released from natural sources (Nriagu, 1989).

The toxicological history of mercury is a long one, with documented human fatalities dating back several centuries. More recently, there have been several pronounced poisoning incidents. Around Minamata Bay in Japan (Kurland et al., 1960), several tens of human deaths were recorded as the result of industrial discharge of mercury to the environment and its subsequent accumulation through the food chain via fish. Similar outbreaks of mercury poisoning in Iraq — where several hundreds of people died (Bakir et al., 1973), Guatemala, and Pakistan have, in part, meant that mercury pollution has a relatively high profile and has been the subject of much investigation. Similarly, the widespread poisoning of wildlife in Sweden following the application of organomercury seed dressing (Borg et al., 1969; Johnels and Westermark, 1969) illustrated that accumulation of mercury through the food chain is pronounced, a characteristic for which little, if any, evidence exists for other metals (Bryan, 1979).

Although much work has focused upon mercury as an environmental contaminant, it is clear that the results of laboratory-based studies can be difficult to extrapolate to natural populations, partly since laboratory studies have usually involved mercury at concentrations resulting in acute challenges unlikely to be encountered in most investigations of wild populations. Furthermore, relatively few studies, both laboratory based and field based, have considered the interactive effects of combinations of pollutants or elements. The apparent amelioratory effect of selenium upon the toxicity of mercury, for example, has been well documented (Stoewsand et al., 1978; Magos and Webb, 1980; Cuvin-Aralar and Furness, 1991) and may lead to problems of interpretation of relatively high mercury concentrations if the latter were to be considered in isolation. Although an important area of metal toxicity, such interactions are beyond the scope of this review that will focus upon the effects of mercury only. I will deal with mercury toxicity in free-living individuals or populations and in laboratory-based studies separately and draw relevant conclusions from both approaches where appropriate. Mercury concentrations refer to "total" mercury unless otherwise indicated.

MERCURY TOXICITY IN BIRDS

STUDIES OF FREE-LIVING POPULATIONS AND INDIVIDUALS

Within this category, some studies have failed to detect any detrimental effect of mercury. Such studies are included in this review, however, since they may highlight the problems of this type of work or set a "baseline" of mercury contamination that may be tolerable by the particular species studied. Investigations into the effects of mercury on wild birds have generally fallen into three broad categories as outlined below.

First, mercury concentrations have been reported in individual birds that were found dead and that were considered to have succumbed directly to the toxic effects of mercury, dying and exhibiting symptoms consistent with chronic/acute mercury poisoning, or in birds of a single population or subpopulation exhibiting some gross effect(s) due to mercury poisoning.

Henriksson et al. (1966) noted mercury concentrations of 4.6 to 27.1 and 48.6 to 123.1 mg/kg of wet weight in the liver and kidney, respectively, of white-tailed eagles *(Haliaaetus albicilla)* found dead in Finland; the authors concluded that mercury poisoning was the cause of death. Similarly, Koeman et al. (1972) recorded a mercury concentration of 48.2 mg/kg of wet weight in the liver of a white-tailed eagle found dead in northern Germany. Oehme (1981) concluded that mercury poisoning was the cause of death in seven white-tailed eagles found dead in northern Germany with median mercury concentrations of 91 and 120 mg/kg of wet weight in the liver and kidney, respectively, while Falandysz (1984, 1986) and Falandysz et al. (1988) recorded mercury concentrations of 30, 11, and 33 mg/kg of wet weight in the liver of three white-tailed eagles thought to have died from mercury poisoning. Koeman et al. (1969) measured mercury concentrations of 73 to 125, 85 and 93, and 68 mg/kg of wet weight in the kidneys of kestrels *(Falco tinnunculus)*, buzzards *(Buteo buteo)*, and a long-eared owl *(Asio otus)*, respectively, which were either dead or exhibiting symptoms of mercury poisoning. Fimreite (1974) suggested that mercury contamination of Ball Lake, Ontario, resulting in mercury concentrations as high as 90.5 mg/kg of wet weight in the liver (mean, 51.9 mg/kg) of common loons *(Gavia immer)*, was "responsible" for the absence of chicks of this species at this site. Furthermore, mercury concentrations of 40 to 121 mg/kg of wet weight in the liver of turkey vultures *(Cathartes aura)* were thought to be associated with impaired wing movement (Fimreite, 1974).

Nicholson and Osborn (1983) noted kidney lesions and other nephrotoxic effects in puffins *(Fratercula arctica)* (mean mercury concentration in the kidney, 5.02 mg/kg of dry weight), fulmars *(Fulmarus glacialis)* (13.4 mg/kg), and Manx shearwaters *(Puffinus puffinus)* (4.67 mg/kg) similar to, though less severe than, those in dosed starlings *(Sturnus vulgaris)* with mercury concentrations of 20 to 40 mg/kg of dry weight in the kidney. However, cadmium was thought to play a more important role in this respect (Nicholson and Osborn, 1983; see Chapter 17 by R. W. Furness, this volume).

Second, comparisons have been made between mercury concentrations in two or more populations of birds that show differences in some aspect of reproduction. Fimreite (1974) found low hatching success (approximately 27%) in common tern *(Sterna hirundo)* eggs from Ball Lake (mean total mercury concentration, 3.65 mg/kg of wet weight), while at Wabigoon Lake (mean, 1.0 mg/kg) hatching success was high. However, the influence of mercury upon hatching success was far from clear-cut when mercury concentrations were compared among clutches of one, two, or three eggs (Fimreite, 1974). In another study of common terns, it was concluded that the low reproductive success of birds at Hamilton Harbour, compared with that in birds from Great Gull Island, was not related to mercury concentration (mean, 2.28; maximum, 4.74 mg/kg of wet weight in the liver of Hamilton Harbour birds; Connors et al., 1975), while Gilman et al. (1977) could find "no toxicological significance" in mercury concentrations in eggs of herring gulls *(Larus argentatus)* from the Great Lakes (means, 0.22 to 0.51 mg/kg of wet weight, depending on location).

Furness et al. (1989) recorded a trend of decreasing reproductive success with increasing mercury concentration in feather samples of golden eagles *(Aquila chrysaetos)* from different regions of Scotland (maximum mean, approximately 8 mg/kg of fresh weight in feathers from eagles from Rum, western Scotland), a pattern confirmed in golden eagle eggs by Newton and Galbraith (1991). Both studies cautioned against linking mercury concentrations directly with reproductive success, since there was a high likelihood that both variables could be correlated with a range of other parameters, most notably diet. In an extensive study of bald eagles *(Haliaeetus leucocephalus)*, Wiemeyer et al. (1984) found a negative correlation between the mercury concentration in eggs and the mean 5-year productivity, although there was much intercorrelation between contaminants and although the significance of the above relation was removed in a multiple regression analysis when the 2,2-bis(*p*-chlorophenyl)-1,1-dichloroethylene (DDE) concentration emerged as the most important variable. Mercury concentrations in eggs that failed to hatch were rarely greater than 1.0 mg/kg of wet weight, with most geometric mean concentrations being 0.4 mg/kg of wet weight or less (Wiemeyer et al., 1984). King et al. (1991) found no correlation between the mercury concentration in eggs of Forster's tern *(Sterna forsteri)* (mean, 0.40 mg/kg of wet weight) or of black skimmers *(Rynchops niger)* (mean, 0.46 mg/kg of wet weight) and hatching success at what was considered a contaminated site in Texas.

Third, several studies have investigated the relation between mercury concentration and some aspect of reproduction and/or survival on an individual basis within a particular population. Barr (1986) found a negative correlation between the breeding success in common loons and the degree of environmental contamination with mercury. It was suggested that mercury concentrations of 0.3 to 0.4 mg/kg of wet weight in prey were sufficient to impair territorial fidelity and egg laying (Barr, 1986). For the most part, these studies have used eggs as a monitoring tissue. Vermeer et al. (1973) noted that the hatching success of herring gull eggs was unaffected by mercury concentrations in the first-laid egg of a given clutch, values ranging from 2.3 to 15.8 mg/kg of wet weight. Fyfe et al. (1976) found no significant relation between the mercury concentration in eggs of the prairie falcon *(Falco mexicanus)* (mean, 0.35 mg/kg of dry weight) or Richardson's merlin *(Falco columbarius richardsonii)* (mean, 0.66 mg/kg of dry weight) and a range of reproductive parameters. Similarly, Odsjo and Sondell (1977) found that mercury concentrations in eggs of marsh harriers *(Circus aeruginosus)* were not related to the number of young in a particular clutch (means, 1 to 2 mg/kg of dry weight for nests of 0 to 4 young; mean, less than 1.0 mg/kg of dry weight for nests of five young). Koivusaari et al. (1980) found that mercury concentrations of 1.31 to 4.16 mg/kg of dry weight (0.33 to 2.25 mg/kg of wet weight) in eggs of white-tailed eagles were not associated with changes (actually increases over the period of study) in reproductive success, and Helander et al. (1982) found no correlation between the mercury concentration in eggs of the same species from the Baltic Sea coastal area of Sweden (mean, 4.6 mg/kg of dry weight) and reproductive success.

Newton and Haas (1988) suggested that mercury concentrations in excess of 3 mg/kg of dry weight in merlin *(F. columbarius)* eggs impaired productivity (brood size), although the relation between brood size and mercury concentration actually improved if eggs with the highest mercury concentrations were removed from the regression. In a study of mercury concentrations in eggs of the peregrine falcon *(Falco peregrinus)*, brood size was not significantly associated with mercury concentration alone (geometric mean from inland sites, 0.21 mg/kg of dry weight; that from coastal sites, 1.27 mg/kg of dry weight), although mercury may have reduced brood sizes by augmenting the effects of DDE (Newton et al., 1989). Henny and Herron (1989) found that mercury concentrations in eggs (generally less than 1.0 mg/kg of dry weight) had no effect upon the reproduction of the white-faced ibis *(Plegadis chihi)*.

In a study using feathers as monitoring tissue, Thompson et al. (1991) found no relation between the mercury concentration and a range of reproductive parameters and survival in great skuas *(Catharacta skua)* thought to be exposed to considerable stress during the breeding season through acute food shortage (Hamer et al., 1991). Mercury concentrations ranged from 1.2 to 32.4 mg/kg of fresh weight in body feathers and from 4.5 to 33.4 mg/kg of dry weight in the liver, while organic (methyl)mercury concentrations in the liver ranged from 1.8 to 11.1 mg/kg of dry weight (Thompson et al., 1991).

In summary, it seems reasonable to conclude that mercury has been responsible for the death and/or poisoning of a range of bird species, but it is more difficult to be absolutely unambiguous with respect to the particular concentration at which these effects become manifest. Birds "found dead" are prone to alterations in tissue structure and form postmortem, and mercury concentrations are likely to alter accordingly, masking the actual mercury concentration that caused death or poisoning. This potential problem is compounded further by the fact that mercury intoxication can suppress the appetite, leading to weight loss and an increase in mercury concentration. From the published data, however, mercury concentrations in the liver and kidney in excess of 30 mg/kg of wet weight would appear to be harmful (lethal) to a range of species of birds of prey. It is worth highlighting that, in many of the studies quoted, mercury concentrations in the kidney were higher than those in the liver, a pattern that has been suggested as being indicative of mercury poisoning (Lewis and Furness, 1990).

It should be noted that mercury concentrations far in excess of the 30-mg/kg level have been recorded, particularly in seabirds, which were apparently healthy and reproducing normally (Muirhead and Furness, 1988; Honda et al., 1990; Lock et al., 1992); however, the form of mercury measured in the seabirds in the above studies was predominantly inorganic, suggesting that biotransformation of ingested methylmercury is an important mechanism by which long-lived and slow-moulting seabirds avoid the toxic effects of accumulating large quantities of methylmercury, for which the opportunities of elimination from the body are limited (Thompson and Furness, 1989; Honda et al., 1990). It is true to say that, in general, seabirds exhibit higher mercury concentrations than terrestrial birds because of the higher mercury burdens encountered in marine ecosystems. It follows, therefore, that seabirds are more likely to be able to tolerate higher concentrations of mercury before toxic effects become evident.

Interpopulation studies often appear inconclusive, since the effects of some other pollutants seem at least as significant as those of mercury, and it has been difficult to tease apart the most important contaminants with respect to some difference in reproduction. It seems reasonable to conclude, however, that mercury concentrations in eggs of up to approximately 0.5 mg/kg of wet weight (roughly 2.5 mg/kg of dry weight, assuming the water content of eggs to be approximately 80%) appear to have little detrimental effect on reproduction. The use of eggs as a monitor tissue for studies of mercury on individuals within a population obviously reduces the scope for following breeding success if viable eggs are used or may introduce bias if infertile or addled eggs are analyzed. However, this approach is more rigorous than that of comparing different populations and should allow any chronic effects of mercury to be detected; feathers are particularly apt for studies of this kind, since their use is relatively noninvasive and permits a given breeding attempt to be monitored fully. Although the majority of investigations on an individual basis yielded negative results, this should not preclude their potential to elucidate any possible effects of mercury upon the breeding success and survival of a free-living population, especially since sublethal effects of mercury are likely to occur at concentrations far lower than those required to produce more pronounced pathological effects (Scheuhammer, 1987).

CONTROLLED STUDIES OF CAPTIVE BIRDS

A wide range of deleterious effects, resulting from the administration of mercury, have been reported in dosing experiments of captive birds. These have included, in extreme cases, death through to impairment of some aspect of reproduction.

Finley et al. (1979) recorded the time for five individuals of a species to succumb to a diet containing 40 mg/kg of mercury as methylmercury dicyandiamide in Morsodren ad libitum. They found that this dosage was lethal to all study species, with 6, 8, 10, and 11 days required for five grackles *(Quiscalus quiscula)*, starlings, cowbirds *(Molothrus ater)*, and red-winged blackbirds *(Agelaius phoeniceus)* to succumb, respectively. The mean mercury concentrations in the liver ranged from 54.5 mg/kg of wet weight in grackles to 126.5 mg/kg of wet weight in blackbirds. Gardiner (1972) found that a diet containing 33 mg/kg of methylmercury for 35 days produced 90% mortality in pheasants *(Phasianus colchicus)* with progressively reduced mortality in ducks and chickens. Kestrels were found to succumb to a diet of mice dosed with methylmercury dicyandiamide to an overall body concentration of 13.3 mg/kg of wet weight. Mercury concentrations in the liver of poisoned kestrels ranged from 49 to 122 mg/kg of wet weight (Koeman et al., 1971).

Borg et al. (1970) fed chicks contaminated with methylmercury dicyandiamide to levels of 10 and 40 mg/kg of wet weight in the muscle and liver, respectively, to juvenile goshawks *(Accipiter gentilis)*. The overall mercury concentrations in chick samples provided to the experimental hawks were approximately 10 and 13 mg/kg of wet weight, depending upon treatment. The goshawks showed a reduction in body weight and food consumption, with death ensuing in 30 to 47 days. Internally, dosed birds were found to be in poor condition with major reductions in fat and muscle. Total mercury (predominantly methylmercury) concentrations in dosed hawks ranged from 103 to 144 and from 121 to 138 mg/kg of wet weight in the liver and kidney, respectively. In a similar study, Fimreite and Karstad (1971) dosed 1-year-old, red-tailed hawks *(Buteo jamaicensis)* with chicks contaminated with methylmercury as methylmercury dicyandiamide for 4 weeks. Mercury concentrations in chick livers were either 3.9, 7.2, or 10.0 mg/kg of wet weight, depending upon the dosing regimen. A slight weight loss was observed in the hawks of the higher dose group, together with loss of coordination and ultimately death. Mercury concentrations in the livers of poisoned hawks ranged from 16.7 to 20.0 mg/kg of wet weight, and lesions in nerve tissue were recorded.

Adult ring-necked pheasants were killed by a diet containing mercury at 12.5 mg/kg of wet weight as ethyl mercury p-toluene sulfonanilide, but their survival was unaffected by a reduced dose of 4.2 mg/kg of mercury. However, the lower dose caused a reduction in egg production and survival of third-week embryos; the mean mercury concentration in eggs from the second experiment was 1.5 mg/kg of wet weight (Spann et al., 1972). Reduced hatchablity was recorded in the same species by Borg et al. (1969) associated with mercury concentrations in eggs of 1.3 to 2.0 mg/kg of wet weight. Similarly, Fimreite (1971) found that adult pheasants fed a diet containing 2 to 3 mg/kg of methylmercury as methylmercury dicyandiamide for 12 weeks were generally unaffected; there was no adult mortality, and food consumption was normal as were mating behavior and egg production. However, this dosage was sufficient to cause an increase in eggs without shells, a decrease in egg weight, decreased hatchability, and an increase in unfertilized eggs. Mercury concentrations in unhatched eggs ranged from 0.5 to 1.5 mg/kg of wet weight. The liver of adult pheasants contained approximately 2 mg/kg of mercury, wet weight.

Similar sublethal effects of mercury have been reported by Heinz (1974) in mallard ducks *(Anas platyrhynchos)*. Breeding adults were fed a diet containing up to 3 mg/kg of methylmercury, dry weight, as methylmercury dicyandiamide over a 12-month period and appeared healthy throughout the study. However, the period of egg output was curtailed, hatching success was reduced, and there was an increase in the early mortality of ducklings from parents fed mercury; ducklings were found to possess brain lesions (Heinz and Locke, 1976). The mercury concentrations in eggs from this study were, in some cases, in excess of 5 mg/kg of wet weight (Heinz, 1974). Comparable, though less severe, findings were reported from an extension of the above study (Heinz, 1976). During a second breeding year, egg production and hatching success were similar to those of controls, and mercury concentrations in eggs were reduced by 0.5 to 1.0 mg/kg of wet weight (Heinz, 1976). Black ducks *(Anas rubripes)* were subjected to feeding experiments in which adults were dosed with 3 mg/kg of methylmercury, dry weight, as methylmercury dicyandiamide in Morsodren. These birds showed no adverse effects of this feeding regimen, but the clutch size and egg hatchability were reduced compared with those of controls. Furthermore, duckling survival was also impaired, and lesions in nerve tissue were found in ducklings from mercury-fed parents. The mean mercury concentrations in eggs of treated birds were 5.53 and 4.70 mg/kg of wet weight in the 2 years of the study (Finley and Stendell, 1978).

Scott et al. (1975) found little, if any, effect of inorganic mercury (mercuric sulfate or mercuric chloride) upon egg weight, fertility, or hatchability in chickens, even at concentrations of up to 200 mg/kg added to a basal diet. However, organic mercury, as methylmercury chloride, added to a basal diet at concentrations of 10 and 20 mg/kg produced a marked reduction in the above parameters, together with severe ataxia in laying hens and misshapen and truncated eggs; a reduction in shell strength was recorded in such eggs. Young chickens exhibited depressed growth and increased mortality when supplied with drinking water containing mercuric chloride at concentrations up to 250 mg/kg (Parkhurst and Thaxton, 1973). Similarly, Japanese quail *(Coturnix coturnix japonica)* showed a reduction in mating attempts, egg fertility, egg hatchability, ova development, and testes weight when supplied with drinking water containing 125 mg/kg of mercuric chloride (Thaxton and Parkhurst, 1973).

Damage to liver littoral cells has been recorded in quail fed up to 8 mg/kg of mercury, wet weight, as methylmercury chloride for 6 weeks (Gullvag et al., 1978), while Nicholson and Osborn (1984) found nephrotoxic lesions in starlings that had inadvertently been exposed to 1.1 mg/kg of mercury, dry weight, in their diet, resulting in mercury concentrations of 36.3 and 6.55 mg/kg, dry weight, in the kidney and liver, respectively.

Stoewsand et al. (1971) found thinning of eggshells in quail fed a diet containing 8 mg/kg of mercury as mercuric chloride. However, eggshell thinning was not recorded by Peakall and Lincer (1972) in either American kestrel *(Falco sparverius)* or ring doves *(Streptopelia risoria)* dosed with 10 mg/kg of dimethylmercury or methylmercury chloride either orally or via injection, and mercury has been shown to be ineffectual with respect to eggshell thinning in other studies (Spann et al., 1972; Heinz, 1974; Hill and Shaffner, 1976).

In summary, laboratory-based studies of birds of prey tend to confirm the findings of field studies; dosed birds that succumbed to mercury intoxication were found to contain 30 mg/kg and above of mercury, wet weight, in the liver and kidney. Dietary mercury concentrations of approximately 10 mg/kg of wet weight were lethal to birds of prey, passerines, and pheasants. Such high concentrations in the diet are unlikely to be encountered in any but the most locally contaminated environment. However, lower dietary mercury concentrations (up to approximately 3 mg/kg of dry weight) caused impaired reproduction with little effect to breeding adults. The marine environment, where mercury concentrations tend to be highest, would seem to offer the only realistic possibility of such dietary mercury concentrations being encountered, and then only by top predators. There is a degree of agreement between dosing experiments in that mercury concentrations in eggs of 0.5 to 2.0 mg/kg of wet weight in pheasants, with higher concentrations in duck eggs, caused impairment of hatching or survival of embryos/hatchlings. Mercury concentrations in eggs similar to these have been measured in wild populations (see this Chapter), but they had apparently little or no effect upon hatching in some studies (Vermeer et al., 1973, for example), while some authors suggested some impairment of hatching (Fimreite, 1974, for example), perhaps indicating species-specific responses. Overall, however, mercury concentrations in excess of 2.0 mg/kg of wet weight in eggs would seem to have some detrimental effect.

MERCURY TOXICITY IN MAMMALS

The reader is directed to Wren (1986) for a more comprehensive appraisal of mercury in mammals; the aim of this review will be to focus upon the deleterious effects of mercury to essentially nonmarine mammals.

STUDIES OF FREE-LIVING POPULATIONS AND INDIVIDUALS

There have been rather few documented studies of mercury poisoning in wild mammals. Borg et al. (1969) reported mercury concentrations in mammals found exhibiting symptoms consistent with mercury poisoning in Sweden. A fox *(Vulpes vulpes)* behaving erratically had 30 mg/kg of mercury, wet weight, in a liver-kidney homogenate, while the mercury concentration in a similarly distressed marten *(Martes martes)* was 40 mg/kg. Wren (1985) noted mercury concentrations of 96 and 58 mg/kg of wet weight in the liver and kidney, respectively, in an otter *(Lutra canadensis)* found dead at Clay Lake, Ontario, and thought to have died from mercury poisoning. Wobeser and Swift (1976) reported the case of a mink *(Mustela vison)* found behaving erratically near South Saskatchewan River. Within 2 hours of capture, the animal became moribund and died. Lesions were found most notably in the cerebral cortex and cerebellum, and mercury concentrations of 58.2 and 31.9 mg/kg of wet weight were measured in the liver and kidney, respectively.

Takeuchi et al. (1977) reported behavioral abnormalities in a cat fed entrails and scraps of fish from the English River, Ontario; the river system was contaminated with mercury from a chloralkali plant. The cat was found to jump into the air, growl, and froth at the mouth. With time, it became ataxic and anorexic. Pathological investigation revealed severe deterioration of brain cells, particularly of the cerebellum cortex. The cat was found to contain 67.1, 16.4, and 13.4 mg/kg of mercury, wet weight, in the liver, brain, and kidney, respectively (Takeuchi et al., 1977).

In a study of otters *(Lutra lutra)* in Shetland, Scotland, Kruuk and Conroy (1991) suggested that relatively high concentrations of mercury (up to 65 mg/kg of dry weight in the liver) may have had some detrimental effect on otter survival over and above that of periods of natural food shortage, during which peak numbers of nonviolent otter deaths were recorded.

Roelke et al. (1991) recorded a mercury concentration of 110 mg/kg of wet weight in the liver of a Florida panther *(Felis concolor coryi)* found dead and presumed to have died from mercury poisoning. Furthermore, mercury accumulation from nonungulate prey (mainly raccoons [*Procyon lotor*] and alligators [*Alligator mississippiensis*]) in female panthers was thought to result in reduced kitten survival when mercury concentrations in the blood exceeded 0.5 mg/kg (mean, 0.17 kittens per female year) compared with blood levels ≤0.25 mg/kg (mean, 1.46 kittens per female year; Roelke et al., 1991).

LABORATORY-BASED STUDIES

Albanus et al. (1972) investigated the effects of feeding a fish homogenate with a mean mercury concentration of approximately 6.0 mg/kg of wet weight, the mercury present as methylmercury, to cats; the daily dose of methylmercury was equivalent to 0.45 mg of mercury/kg of body weight. After approximately 50 days, the cats developed behavioral and locomotory abnormalities, followed later by convulsions.

The mean terminal mercury concentrations were 39, 31, and 18 mg/kg of wet weight in the liver, kidney, and brain, respectively. Eaton et al. (1980) fed cats ringed seal *(Phoca hispida)* liver tissue that contained a mean mercury concentration of 26.2 mg/kg of wet weight; however, only 3% was methylmercury (mean concentration, 0.79 mg/kg of wet weight). Three experimental groups were fed either 25, 50, or 100 g of seal liver per day; in addition, a fourth group was fed beef liver to which was added methylmercury chloride, equivalent to 0.25 mg of mercury/kg of body weight per day. After 90 days, there were no signs of mercury poisoning in the cats fed seal liver, but after only 68 days, cats fed beef liver plus methylmercury chloride exhibited convulsions; the mean survival time for the latter group was 78 days. The mean mercury concentrations in the liver of the groups fed seal liver were 1.08, 2.27, and 3.66 mg/kg of wet weight for the 25-, 50-, and 100-g treatments, respectively. A mean mercury concentration of 40.3 mg/kg of wet weight was recorded in the group fed beef liver augmented with methylmercury chloride.

Hanko et al. (1970) fed chickens, contaminated through methylmercury-dressed wheat feed, to ferrets (polecat [*Mustela furo*] × ferret [*M. putorius*] cross) at dietary concentrations of either 5 or 7 mg/kg of wet weight. Decreased appetite and other symptoms of mercury poisoning developed in 2 weeks in the higher dose group and in 3 weeks in the lower dose group. Survival was similarly linked to dose; ferrets succumbed after 35 or 36 days in the higher dose group and after 58 days in the lower dose group. Mean mercury concentrations of 53.7 and 69.0 mg/kg of wet weight were recorded in the liver and kidney, respectively (Hanko et al., 1970).

In a study of adult mink, 5 mg of methylmercury, wet weight, per kg in the diet were found to be lethal in approximately 30 days, whereas 10 mg of mercuric chloride, wet weight, per kg in the diet had no effect upon survival and reproduction after 5 months. Mink fed the diet containing methylmercury developed signs of mercury poisoning after 24 days; a mean mercury concentration of 55.6 mg/kg of wet weight was recorded in the livers of these animals, compared with 3.16 mg/kg of wet weight in those fed the diet containing mercuric chloride (Aulerich et al., 1974). Wobeser et al. (1976a) found that a mean mercury concentration of 0.44 mg/kg of wet weight in a fish diet fed to mink for 145 days had no detectable detrimental effects. However, in a second study (Wobeser et al., 1976b), mink were fed mercury at 1.1, 1.8, 4.8, 8.3, or 15.0 mg/kg of wet weight in the diet as methyl mercuric chloride. Animals in the dietary groups with 1.1 to 15.0 mg/kg of mercury developed symptoms of clinical mercury intoxication, the onset of which was negatively correlated with dose, such that 1 or 2 days were sufficient for symptoms to develop in mink fed 15.0 mg/kg of mercury, while 70 days elapsed before signs of mercury poisoning developed in mink fed 1.8 mg/kg of mercury. Mink that died contained mean mercury concentrations of 24.3, 23.1, and 11.9 mg/kg of wet weight in the liver, kidney, and brain, respectively. In contrast to the above studies, Jernelov et al. (1976) could demonstrate no adverse effects of feeding mink with fish that contained, on average, 5.7 mg/kg of methylmercury, wet weight (5.8 mg/kg of total mercury). After 100 days, mercury concentrations of 65 and 46.5 mg/kg of wet weight were recorded in the liver and kidney, respectively. Selenium may have ameliorated the effects of mercury in this study, as concentrations of selenium increased along with those of mercury in the liver and kidney (Jernelov et al., 1976).

O'Connor and Nielsen (1980) investigated the feeding of otters diets containing mercury at 2, 4, or 8 mg/kg of wet weight as methylmercury hydroxide (equivalent to 9.3, 17.0, and 37.0 mg of mercury per kilogram of body weight per day, respectively). The mean survival time of the three groups was 184, 117, and 54 days, with the onset of mercury-poisoning symptoms occurring first in the group with the highest dose. Lesions in the central nervous system were found; terminal mercury concentrations were similar in all dose groups. Overall mean mercury concentrations of 33.0, 39.0, and 19.0 mg/kg of wet weight were recorded in the liver, kidney, and brain, respectively.

In summary, from the limited data available from studies of mercury in wild mammals, a mercury concentration of approximately 30 mg/kg of wet weight in the liver or kidney would seem lethal or at least harmful. This suggestion has to be tentative, given the relative paucity of information, but is strengthened by the findings of dosing experiments. Terminal mercury concentrations in mammals that succumbed to mercury poisoning in laboratory studies were generally of a similar magnitude, approximately 25 mg/kg of wet weight, and above. Furthermore, there was general agreement among laboratory-based studies that dietary mercury concentrations of 2 to 6 mg/kg of wet weight (methylmercury) were sufficient to cause mercury intoxication.

SUMMARY

Mercury is a biologically nonessential heavy metal occurring naturally in the environment but one that is augmented by significant anthropogenic emissions. The most stable and toxic form of mercury is methylmercury, and it is to this form that most top avian and mammalian predators are exposed through their diet.

Nonmarine birds with mercury concentrations in the liver and kidney in excess of approximately 20 to 30 mg/kg of wet weight were likely to suffer toxic effects and ultimately even death. Poisoning symptoms, including altered and erratic behavior, weight loss, appetite suppression, and ataxia, leading to death, were induced in laboratory studies by dietary mercury concentrations of approximately 10 mg/kg of wet weight. Lower dietary concentrations of up to approximately 3 mg/kg of dry weight were insufficient to affect adult survival, but reproductive success was impaired, as measured by reduced egg production, egg viability and hatchability, embryo survival, and chick survival. Mercury concentrations of 0.5 to 2.0 mg/kg of wet weight in eggs were sufficient to induce the above effects. The concentrations of mercury in seabirds are more difficult to interpret from a toxicological viewpoint. It is clear that much higher concentrations (an order of magnitude greater than the 20- to 30-mg/kg range quoted above in some species) have apparently little, if any, effect. It seems reasonable to conclude that pelagic seabirds that accumulate the highest mercury concentrations, compared with other birds, have yet to be exposed to sufficiently high burdens of mercury to induce measurable effects on reproduction or survival.

Nonmarine mammals with mercury concentrations in the liver and kidney in excess of approximately 30 mg/kg of wet weight were likely to suffer mercury intoxication. The results of laboratory studies supported this value and indicated that a dietary methylmercury concentration of approximately 2 to 6 mg/kg of wet weight produced mercury poisoning in feeding experiments using a range of species of mammals.

ACKNOWLEDGMENTS

I should like to thank Dr. Sharon Lewis for help with literature searches.

REFERENCES

Albanus, L., L. Frankenberg, C. Grant, U. von Haartman, A. Jernelov, G. Nordberg, M. Rydalv, A. Schutz, and S. Skerfving. 1972. Toxicity for cats of methylmercury in contaminated fish from Swedish lakes and of methylmercury hydroxide added to fish. Environ. Res. 5:425-442.

Aulerich, R. J., R. K. Ringer, and J. Iwamoto. 1974. Effects of dietary mercury on mink. Arch. Environ. Contam. Toxicol. 2:43-51.

Bakir, F., S. F. Damluji, L. Amin-Zaki, M. Murtadha, A. Khalidi, N.Y. Al-Rawi, S. Tikriti, H. I. Dhahir, T. W. Clarkson, J. C. Smith, and R. A. Doherty. 1973. Methylmercury poisoning in Iraq. Science (Washington, D.C.) 181:230-241.

Barr, J. F. 1986. Population dynamics of the common loon *(Gavia immer)* associated with mercury-contaminated waters in northwestern Ontario. Can. Wildl. Serv. Occas. Pap. No. 56, 23 pp.

Borg, K., K. Erne, E. Hanko, and H. Wanntorp. 1970. Experimental secondary methylmercury poisoning in the goshawk *(Accipiter g. gentilis* L.). Environ. Pollut. 1:91-104.

Borg, K., H. Wanntorp, K. Erne, and E. Hanko. 1969. Alkyl mercury poisoning in Swedish wildlife. Viltrevy (Stockh.) 6:301-379.

Bryan, G. W. 1979. Bioaccumulation of marine pollutants. Philos. Trans. R. Soc. Lond. B Biol. Sci. 286:483-505.

Connors, P. G., V. C. Anderlini, R. W. Risebrough, M. Gilbertson, and H. Hays. 1975. Investigations of heavy metals in common tern populations. Can. Field Nat. 89:157-162.

Cuvin-Aralar, M. L. A., and R. W. Furness. 1991. Mercury and selenium interaction: a review. Ecotoxicol. Environ. Saf. 21:348-364.

Eaton, R. D. P., D. C. Secord, and P. Hewitt. 1980. An experimental assessment of the toxic potential of mercury in ringed seal liver for adult laboratory cats. Toxicol. Appl. Pharmacol. 55:514-521.

Falandysz, J. 1984. Metals and organochlorines in a female white-tailed eagle from Uznam Island, southwestern Baltic Sea. Environ. Conserv. 11:262-263.

Falandysz, J. 1986. Metals and organochlorines in adult and immature males of white-tailed eagle. Environ. Conserv. 13:69-70.

Falandysz, J., B. Jakuczun, and T. Mizera. 1988. Metals and organochlorines in four female white-tailed eagles. Mar. Pollut. Bull. 19:521-526.

Fimreite, N. 1971. Effects of dietary methylmercury on ring-necked pheasants with special reference to reproduction. Can. Wildl. Serv. Occas. Pap. No. 9, 39 pp.

Fimreite, N. 1974. Mercury contamination of aquatic birds in northwestern Ontario. J. Wildl. Manage. 38:120-131.

Fimreite, N., and L. Karstad. 1971. Effects of dietary methylmercury on red-tailed hawks. J. Wildl. Manage. 35:293-300.

Finley, M. T., and R. C. Stendell. 1978. Survival and reproductive success of black ducks fed methylmercury. Environ. Pollut. 16:51-64.

Finley, M. T., W. H. Stickel, and R. E. Christensen. 1979. Mercury residues in tissues of dead and surviving birds fed methylmercury. Bull. Environ. Contam. Toxicol. 21:105-110.

Furness, R. W. 1996. Cadmium in birds. In W. N. Beyer, G. H. Heinz, and A. W. Redmon (Eds.). Interpreting environmental contaminants in animal tissues. Lewis Publishers, Boca Raton, Fla.

Furness, R. W., J. L. Johnston, J. A. Love, and D. R. Thompson. 1989. Pollutant burdens and reproductive success of golden eagles *Aquila chrysaetos* exploiting marine and terrestrial food webs in Scotland. p. 495-500. In B.-U. Meyburg and R. D. Chancellor (Eds.). Raptors in the modern world. WWGBP, Berlin.

Fyfe, R. W., R. W. Risebrough, and W. Walker. 1976. Pollutant effects on the reproduction of the prairie falcons and merlins of the Canadian prairies. Can. Field Nat. 90:346-355.

Gardiner, E. E. 1972. Differences between ducks, pheasants, and chickens in tissue mercury retention, depletion, and tolerance to increasing levels of dietary mercury. Can. J. Anim. Sci. 52:419-423.

Gilman, A. P., G. A. Fox, D. B. Peakall, S. M. Teeple, T. R. Carroll, and G. T. Haymes. 1977. Reproductive parameters and egg contaminant levels of Great Lakes herring gulls. J. Wildl. Manage. 41:458-468.

Gullvag, B. M., J. P. Aagdal, and B. Eskeland. 1978. A fine structural study of liver (littoral) cells of methyl-mercury fed Japanese quail *(Coturnix coturnix japonica)*. Acta Pharmacol. Toxicol. 43:93-98.

Hamer, K. C., R. W. Furness, and R. W. G. Caldow. 1991. The effects of changes in food availability on the breeding ecology of great skuas *Catharacta skua* in Shetland. J. Zool. (Lond.) 223:175-188.

Hanko, E., E. Erne, H. Wanntorp, and K. Borg. 1970. Poisoning in ferrets by tissues of alkyl mercury-fed chickens. Acta Vet. Scand. 11:268-282.

Heinz, G. H. 1974. Effects of low dietary levels of methylmercury on mallard reproduction. Bull. Environ. Contam. Toxicol. 11:386-392.

Heinz, G. H. 1976. Methylmercury: second-year feeding effects on mallard reproduction and duckling behavior. J. Wildl. Manage. 40:82-90.

Heinz, G. H., and L. N. Locke. 1976. Brain lesions in mallard ducklings from parents fed methylmercury. Avian Dis. 20:9-17.

Helander, B., M. Olsson, and L. Reutergardh. 1982. Residue levels of organochlorine and mercury compounds in unhatched eggs and the relationships to breeding success in white-tailed sea eagles *Haliaeetus albicilla* in Sweden. Holarct. Ecol. 5:349-366.

Henny, C. J., and G. B. Herron. 1989. DDE, selenium, mercury, and white-faced ibis reproduction at Carson Lake, Nevada. J. Wildl. Manage. 53:1032-1045.

Henriksson, K., E. Karppanen, and M. Helminen. 1966. High residue of mercury in Finnish white-tailed eagles. Ornis Fenn. 43:38-45.

Hill, E. F., and C. S. Shaffner. 1976. Sexual maturation and productivity of Japanese quail fed graded concentrations of mercuric chloride. Poult. Sci. 55:1449-1459.

Honda, K., J. E. Marcovecchio, S. Kan, R. Tatsukawa, and H. Ogi. 1990. Metal concentrations in pelagic seabirds from the north Pacific Ocean. Arch. Environ. Contam. Toxicol. 19:704-711.

Jernelov, A., A. Johansson, L. Sorenson, and A. Svenson. 1976. Methylmercury degradation in mink. Toxicology 6:315-321.

Johnels, A. G., and T. Westermark. 1969. Mercury contamination of the environment in Sweden. p. 221-239. *In* M. W. Millar and G. G. Berg (Eds.). Chemical fallout: current research on persistent pesticides. Charles C Thomas, Springfield, Ill.

King, K. A., T. W. Custer, and J. S. Quinn. 1991. Effects of mercury, selenium, and organochlorine contaminants on reproduction of Forster's terns and black skimmers nesting in a contaminated Texas bay. Arch. Environ. Contam. Toxicol. 20:32-40.

Koeman, J. H., J. Garssen-Hoekstra, E. Pels, and J. J. M. de Goeij. 1971. Poisoning of birds of prey by methylmercury compounds. Meded. Rijksfac. Landbouwwet. Gent. 36:43-49.

Koeman, J. H., R. H. Hadderingh, and M. F. I. J. Bijlveld. 1972. Persistent pollutants in the white-tailed eagle *(Haliaeetus albicilla)* in the Federal Republic of Germany. Biol. Conserv. 4:373-377.

Koeman, J. H., J. A. J. Vink, and J. J. M. de Goeij. 1969. Causes of mortality in birds of prey and owls in the Netherlands in the winter of 1968-1969. Ardea 57:67-76.

Koivusaari, J., I. Nuuja, R. Palokangas, and M. Finnlund. 1980. Relationships between productivity, eggshell thickness, and pollutant contents of addled eggs in the population of white-tailed eagles *Haliaeetus albicilla* L. in Finland during 1969-1978. Environ. Pollut. Ser. A Ecol. Biol. 23:41-52.

Kruuk, H., and J. W. H. Conroy. 1991. Mortality of otters *(Lutra lutra)* in Shetland. J. Appl. Ecol. 28:83-94.

Kurland, L. T., S. N. Faro, and H. Siedler. 1960. Minamata disease: the outbreak of a neurological disorder in Minamata, Japan, and its relationship to the ingestion of seafood contaminated by mercuric compounds. World Neurol. 1:370-395.

Lewis, S. A., and R. W. Furness. 1991. Mercury accumulation and excretion in laboratory reared black-headed gull *Larus ridibundus* chicks. Arch. Environ. Contam. Toxicol. 21:316-320.

Lock, J. W., D. R. Thompson, R. W. Furness, and J. A. Bartle. 1992. Metal concentrations in seabirds of the New Zealand region. Environ. Pollut. 75:289-300.

Magos, L., and M. Webb. 1980. The interactions of selenium with cadmium and mercury. CRC Crit. Rev. Toxicol. 8:1-42.

Muirhead, S. J., and R. W. Furness. 1988. Heavy metal concentrations in the tissues of seabirds from Gough Island, south Atlantic Ocean. Mar. Pollut. Bull. 19:278-283.

Newton, I., J. A. Bogan, and M. B. Haas. 1989. Organochlorines and mercury in the eggs of British peregrines *Falco peregrinus*. Ibis 131:355-376.

Newton, I., and E. A. Galbraith. 1991. Organochlorines and mercury in the eggs of golden eagles *Aquila chrysaetos* from Scotland. Ibis 133:115-120.

Newton, I., and M. B. Haas. 1988. Pollutants in merlin eggs and their effects on breeding. Br. Birds 81:258-269.

Nicholson, J. K., and D. Osborn. 1983. Kidney lesions in pelagic seabirds with high tissue levels of cadmium and mercury. J. Zool. (Lond.) 200:99-118.

Nicholson, J. K., and D. Osborn. 1984. Kidney lesions in juvenile starlings *Sturnus vulgaris* fed on a mercury-contaminated synthetic diet. Environ. Pollut. Ser. A Ecol. Biol. 33:195-206.

Nriagu, J. O. 1989. A global assessment of natural sources of atmospheric trace metals. Nature (Lond.) 338:47-49.

Nriagu, J. O., and J. M. Pacyna. 1988. Quantitative assessment of worldwide contamination of air, water, and soils by trace metals. Nature (Lond.) 333:134-139.
O'Connor, D. J., and S. W. Nielsen. 1980. Environmental survey of methylmercury levels in wild mink *(Mustela vison)* and otter *(Lutra canadensis)* from the northeastern United States and experimental pathology of methylmercurialism in the otter. p. 1728-1745. *In* J. A. Chapman and D. Pursley (Eds.). Proc. Worldwide Furbearers Conf., Frostburg, Md., 1980.
Odsjo, T., and J. Sondell. 1977. Population development and breeding success in the marsh harrier *Circus aeruginosus* in relation to levels of DDT, PCB, and mercury. (In Swedish with English summary.) Var Fagelvarld 36:152-160.
Oehme, G. 1981. Zur Quecksilberrückstandsbelastung tot aufgefundener Seeadler, *Haliaeetus albicilla*, in den Jahren 1967-1978. (In German with English summary.) Hercynia 18:353-364.
Parkhurst, C. R., and P. Thaxton. 1973. Toxicity of mercury to young chickens. 1. Effect on growth and mortality. Poult. Sci. 52:273-276.
Peakall, D. B., and J. L. Lincer. 1972. Methylmercury: its effect on eggshell thickness. Bull. Environ. Contam. Toxicol. 8:89-90.
Roelke, M. E., D. P. Schultz, C. F. Facemire, and S. F. Sundlof. 1991. Mercury contamination in the free-ranging endangered Florida panther *(Felis concolor coryi)*. Proc. Am. Assoc. Zoo Vet. 20:277-283.
Scheuhammer, A. M. 1987. The chronic toxicity of aluminium, cadmium, mercury, and lead in birds: a review. Environ. Pollut. 46:263-295.
Scott, M. L., J. R. Zimmerman, S. Marinsky, P. A. Mullenhoff, G. L. Rumsey, and R. W. Rice. 1975. Effects of PCBs, DDT, and mercury compounds upon egg production, hatchability, and shell quality in chickens and Japanese quail. Poult. Sci. 54:350-368.
Spann, J. W., R. G. Heath, J. F. Kreitzer, and L. N. Locke. 1972. Ethyl mercury *p*-toluene sulfonanilide: lethal and reproductive effects on pheasants. Science (Washington, D.C.) 175:328-331.
Stoewsand, G. S., J. L. Anderson, W. H. Gutenmann, C. A. Bache, and D. J. Lisk. 1971. Eggshell thinning in Japanese quail fed mercuric chloride. Science (Washington, D.C.) 173:1030-1031.
Stoewsand, G. S., C. A. Bache, and D. J. Lisk. 1978. Dietary selenium protection of methylmercury intoxication of Japanese quail. Bull. Environ. Contam. Toxicol. 11:152-156.
Takeuchi, T., F. M. D'Itri, P. V. Fischer, C. S. Annet, and M. Okabe. 1977. The outbreak of Minamata disease (methylmercury poisoning) in cats on northwestern Ontario reserves. Environ. Res. 13:215-228.
Thaxton, P., and C. R. Parkhurst. 1973. Abnormal mating behavior and reproductive dysfunction caused by mercury in Japanese quail. Proc. Soc. Exp. Biol. Med. 144:252-255.
Thompson, D. R., and R. W. Furness. 1989. The chemical form of mercury stored in south Atlantic seabirds. Environ. Pollut. 60:305-317.
Thompson, D. R., K. C. Hamer, and R. W. Furness. 1991. Mercury accumulation in great skuas *Catharacta skua* of known age and sex, and its effect upon breeding and survival. J. Appl. Ecol. 28:672-684.
Vermeer, K., F. A. J. Armstrong, and D. R. M. Hatch. 1973. Mercury in aquatic birds at Clay Lake, western Ontario. J. Wildl. Manage. 37:58-61.
Wiemeyer, S. N., T. G. Lamont, C. M. Bunck, C. R. Sindelar, F. J. Gramlich, J. D. Fraser, and M. A. Byrd. 1984. Organochlorine pesticide, polychlorobiphenyl and mercury residues in bald eagle eggs — 1969-79 — and their relationships to shell thinning and reproduction. Arch. Environ. Contam. Toxicol. 13:529-549.

Wobeser, G., N. O. Nielsen, and B. Schiefer. 1976a. Mercury and mink. I. Mercury contaminated fish as food for ranch mink. Can. J. Comp. Med. 40:30-33.

Wobeser, G., N. O. Nielsen, and B. Schiefer. 1976b. Mercury and mink. II. Experimental methylmercury intoxication. Can. J. Comp. Med. 40:34-45.

Wobeser, G., and M. Swift. 1976. Mercury poisoning in a wild mink. J. Wildl. Dis. 12:335-340.

Wren, C. D. 1985. Probable case of mercury poisoning in a wild otter, *Lutra canadensis*, in northwestern Ontario. Can. Field Nat. 99:112-114.

Wren, C. D. 1986. A review of metal accumulation and toxicity in wild mammals. I. Mercury. Environ. Res. 40:210-244.

CHAPTER 15

Metals in Marine Mammals*

Robin J. Law

Interpreting the significance of the concentrations of metals determined in the tissues of marine mammals is more problematic than for many other animals, largely because of the relative paucity of toxicological information. The purpose of this chapter is to review the literature (published to the beginning of 1994) regarding the occurrence of metals in marine mammal tissues and their effects. In areas where the literature for marine mammals is inadequate, I have drawn parallels with other mammalian studies to aid the interpretation.

INTRODUCTION

Metals enter the marine environment both naturally (for instance, as a result of the weathering of rocks or volcanic activity) and as a consequence of anthropogenic discharges. Many of these "trace elements" are essential for the health and growth of animals, including marine mammals. Unlike their exposure to modern synthetic organic chemicals, such as polychlorinated biphenyls (PCBs), the exposure of marine mammals to metals has occurred throughout history, during which they may have developed mechanisms either to control the internal concentrations of certain elements or to mitigate their toxic effects. A review of data published to mid-1983 on heavy metal concentrations in cetaceans and pinnipeds was prepared by Wagemann and Muir (1984), with the aim of establishing "concentration norms" for toxicant burdens. Data were presented for selenium, copper, zinc, mercury, and lead; however, in only two studies were elements other than selenium and mercury determined. Only for mercury in pinnipeds were the data sufficiently numerous to approach the authors' aim, but variations in sampling and reporting procedures hampered statistical treatment of even those data. Apart from investigations concerning the body distribution of metals, most determinations made subsequently have been in liver or

---

* This chapter is copyrighted by the British Crown, 1995.

kidney tissue. The concentrations of heavy metals are higher in these than most other tissues, and turnover rates are much higher in these tissues than in muscle. Concentrations of metals in the bones of striped dolphins have also been shown to vary with age and with biological factors such as pregnancy, parturition, lactation and suckling, and growth rate (Honda et al., 1986).

Data are sparse for many metals, and this chapter will be based primarily on information for mercury; copper, zinc, and cadmium will also be discussed in some detail, and some information will be provided for chromium, nickel, arsenic, and lead. Toxicological studies on the effects of ingested metals on marine mammals have essentially been limited to methylmercury in seals. Extrapolations will therefore be made from other mammalian data where they seem appropriate, but even so the picture is rather sketchy. All concentrations given in this chapter are related to wet weight of tissue; where concentrations in the original literature were related to dry weight of tissue, these have been converted to wet weight values, wherever possible, using information on the moisture content presented in each report for individual samples. Where only a mean value was given, this has been used to convert the reported concentrations and, in two cases, an average value for a species from our own data has been used.

As marine mammals do not breathe by means of gills, their uptake of metals directly from water is assumed to be negligible, although there remains the possibility of some slight intake from ingested seawater, by absorption through the skin, or from the atmosphere via the lungs (Augier et al., 1993). No estimates of the possible magnitude of uptake via these routes has been made, but they are probably of only minor importance. There are, thus, three major routes of uptake to be considered in each case:

1. Across the placenta before birth,
2. In milk during the suckling period, and
3. From food.

It is apparent, therefore, that once a marine mammal has been weaned, its uptake is predominantly from food, and its tissue burden will reflect the balance between ingestion and elimination. Areas of the world in which the concentrations of metals in food species are elevated, because of inputs from either anthropogenic or natural sources, may therefore show elevated concentrations in marine mammal tissues.

Of especial importance in understanding the dynamics of the uptake and release of metals is the role of metallothioneins. In fish, aquatic invertebrates, and mammals, metallothioneins play an important role in the transport and storage of metals. Metallothionein is a low-molecular-weight protein, the apoprotein (thionein) of which is induced (in fish) by both acute and chronic exposure to cadmium, copper, mercury, and zinc and which provides protection against the toxic effects of those metals by sequestering and thus reducing the amount of free metal ions (Hamilton and Mehrle, 1986). Saturation of metallothioneins by metals allows excess ions to spill over into other cellular compartments, resulting in pathological lesions. While control is maintained, the function of metallothioneins is to regulate the availability of metals for metal-dependent functions, thus controlling the accessibility of metals to specific intracellular sites. The role of metallothioneins in essential metal regulation (homeostasis) and detoxification consists of two parts of a common function; i.e., essential metals are sequestered or donated to biochemical reactions as metabolic needs dictate, while nonessential metals are sequestered into a nonavailable cellular compartment (Roesijadi, 1992). Metallothionein levels in common (harbor) seals *(Phoca vitulina)* have been correlated with age, and a statistically significant relation between these levels and the concentrations of cadmium and zinc in the liver, and of inorganic mercury in the kidney, suggests that the protein is indeed responsible for the sequestration of these metals in a marine mammal species (Tohyama et al., 1986).

Table 1 gives the ranges of concentrations observed in the liver for eight metals (chromium, nickel, copper, zinc, arsenic, cadmium, mercury, and lead) in a number of marine mammal species sampled in various parts of the world. This is not meant as an exhaustive review of the literature, but as more of an indication of the similarities and contrasts between species and locations. As most data are available for mercury, this element will be considered first.

MERCURY

No vital function for mercury in living organisms has yet been found. Manifestations of subacute mercury poisoning are primarily neurological; when mercury is ingested as methylmercury, the symptoms include loss of coordination, loss of vision and hearing, and mental deterioration (Clarkson, 1987). Methylmercury also impairs both the primary and secondary immune response and gives rise to fetal malformation and behavioral deficits in offspring (Von Burg and Greenwood, 1991). In humans sensitive to methylmercury, clinical manifestations of poisoning may occur at an intake of 4 µg of Hg per kilogram of body weight (Clarkson, 1987).

Whether released naturally (primarily from the degassing of the earth's crust and oceans) or from anthropogenic sources, mercury is largely in the inorganic form, and it may subsequently be methylated by microorganisms in freshwater and marine sediments, first to monomethyl mercury and then to the dimethyl form. Methylmercury is both more toxic than inorganic mercury and more readily bioaccumulated. Mercury in the muscle tissue of fish appears to be virtually all (~99%) in the form of methylmercury, and many other aquatic animals carry almost their entire muscle tissue mercury burden in the methylated form (Bloom, 1992). In fish, bioconcentration factors are also greater for methylmercury than for inorganic mercury, of the order of 10^6 to 10^7, rather than $<10^4$ for nonmethylmercury.

High concentrations of mercury in the livers of adult marine mammals have been reported from both areas contaminated by industrial discharges (up to 430 mg/kg in a gray seal *[Halichoerus grypus]* from Liverpool Bay, an area of northwest England that has received discharges from the chloralkali industry [Law et al., 1992]) and those receiving large inputs from natural sources (up to 1544 mg/kg in an adult female striped dolphin *[Stenella coeruleoalba]* from the Mediterranean Sea (André et al., 1991b) and up to 4400 mg/kg of dry weight in a striped dolphin and

Table 1 Ranges of Concentrations Observed for Eight Metals (mg/kg of wet weight) in the Livers of Marine Mammals

Element	Species[a]	Range (mg/kg)	Location	Sampling date	Ref.
Chromium	G	<0.5 to 2.0	Wales and N.W. England	1988-1991	Law et al. (1992)
	P	<0.5 to 1.8	S. and W. Wales	1988-1991	Law et al. (1992)
	CD	<0.5 to 1.6	S. and W. Wales	1990-1991	Law et al. (1992)
Nickel	C	<0.006 to 0.033	Skagerrak and Kattegat	1988	Frank et al. (1992)
	G	<0.5 to 2.1	Wales and N.W. England	1988-1991	Law et al. (1992)
	SD	0.05 to 0.49	Japan	1977-1980	Honda et al. (1983b)
Copper	C	2.3 to 20	Norway	1988	Skaare et al. (1990)
	G	2.2 to 28	N.W. England	1989-1991	Law et al. (1992)
	P	2.7 to 13	E. Scotland	1974	Falconer et al. (1983)
	SD	3.6 to 15	Japan	1977-1980	Honda et al. (1983b)
	PW	3.0 to 8.9	Faeroe Islands	1978	Julshamn et al. (1987)
Zinc	C	19 to 99	Norway	1988	Skaare et al. (1990)
	C	25 to 46	Skagerrak and Kattegat	1988	Frank et al. (1992)
	G	22 to 89	N.W. England	1989-1991	Law et al. (1992)
	P	18 to 68	E. Scotland	1974	Falconer et al. (1983)
	SD	27 to 109	Japan	1977-1980	Honda et al. (1983b)
	PW	41 to 117	Faeroe Islands	1978	Julshamn et al. (1987)
Arsenic	C	ND[b] to 0.9[c]	Alaska	1976-1978	Miles et al. (1992)
	C	0.03 to 0.42	Norway	1988	Skaare et al. (1990)
	CD	0.51 to 2.6	S.W. England	1992	Law (unpublished data)
	PW	0.28 to 1.8	Faeroe Islands	1978	Julshamn et al. (1987)
	PW	0.01 to 2.9[c]	Newfoundland	1980	Muir et al. (1988)
Cadmium	C	<0.1 to 1.0	Norway	1988	Skaare et al. (1990)
	C	<0.02 to 0.1	Skagerrak and Kattegat	1988	Frank et al. (1992)
	C	<0.06 to 0.5	E. England	1988	Law et al. (1991)
	C	<0.06 to 0.94	N. Ireland	1988-1989	Law et al. (1991)
	C	<0.002 to 0.006	Puget Sound, U.S.	1990	Calambokidis et al. (1991)
	G	<0.06 to 1.2	N.W. England	1989-1991	Law et al. (1992)
	G	0.06 to 2.9	N. Ireland	1988-1989	Law et al. (1991)
	R	0.31 to 21[c]	Arctic Canada	1983	Wagemann (1989)
	P	<0.05 to 0.94	E. Scotland	1974	Falconer et al. (1983)
	SD	<0.07 to 11	U.K. waters	1990-1991	Law et al. (1992), Law (unpublished data)
	SD	0.04 to 11	Japan	1977-1980	Honda et al. (1983b)
	WBD	0.06 to 2.3[c]	Newfoundland	1982	Muir et al. (1988)

Table 1 (continued) Ranges of Concentrations Observed for Eight Metals (mg/kg of wet weight) in the Livers of Marine Mammals

Element	Species[a]	Range (mg/kg)	Location	Sampling date	Ref.
	PW	0.74 to 125	Faeroe Islands	1978	Julshamn et al. (1987)
	PW	0.01 to 43[c]	Newfoundland	1980	Muir et al. (1988)
	PW	0.34 to 65	N.E. and S.W. England	1990-1991	Law (unpublished data)
	MW	2.2 to 33[c]	Antarctic	1980-1985	Honda et al. (1987)
Mercury	C	0.1 to 89	Norway	1988	Skaare et al. (1990)
	C	0.72 to 7.7	Skagerrak and Kattegat	1988	Frank et al. (1992)
	G	1.5 to 430	N.W. England	1989-1991	Law et al. (1992)
	R	0.71 to 22[c]	Arctic Canada	1983	Wagemann (1989)
	P	0.28 to 16	E. Scotland	1974	Falconer et al. (1983)
	P	0.6 to 190	W. Wales	1988-1991	Law et al. (1992)
	P	0.2 to 132	Belgium and Denmark	1987-1990	Joiris et al. (1991)
	SD	1.7 to 485	Japan	1977-1980	Honda et al. (1983b)
	SD	1.2 to 1544	Mediterranean France	1972-1980	André et al. (1991b)
	SpD	0.18 to 218	E. Tropical Pacific Ocean	1977-1983	André et al. (1990)
	WBD	0.13 to 1.6[c]	Newfoundland	1982	Muir et al. (1988)
	PW	0.91 to 150	Faeroe Islands	1978	Julshamn et al. (1987)
	PW	0.08 to 83[c]	Newfoundland	1980	Muir et al. (1988)
	MW	0.02 to 0.13[c]	Antarctic	1980-1985	Honda et al. (1987)
Lead	C	0.2 to 2.1	Alaska	1976-1978	Miles et al. (1992)
	C	0.03 to 0.91	Skagerrak and Kattegat	1988	Frank et al. (1992)
	G	0.38 to 1.9	N.W. England	1989-1991	Law et al. (1992)
	R	0.006 to 0.33[c]	Arctic Canada	1983	Wagemann (1989)
	P	0.08 to 0.37	W. Wales	1990-1991	Law et al. (1992)
	SD	0.03 to 0.64	Japan	1977-1980	Honda et al. (1983b)
	WBD	0.008 to 0.34[c]	Newfoundland	1982	Muir et al. (1988)
	PW	0.019 to 0.4[c]	Newfoundland	1980	Muir et al. (1988)
	MW	0.03 to 0.6[c]	Antarctic	1980-1985	Honda et al. (1987)

[a] Species: G, gray seal *(Halichoerus grypus)*; P, common (harbor) porpoise *(Phocoena phocoena)*; CD, common dolphin *(Delphinus delphis)*; C, common (harbor) seal *(Phoca vitulina)*; SD, striped dolphin *(Stenella coeruleoalba)*; PW, long-finned pilot whale *(Globicephala melas)*; R, ringed seal *(Phoca hispida)*; WBD, white-beaked dolphin *(Lagenorhyncus albirostris)*; MW, southern minke whale *(Balaenoptera acutorostrata)*; SpD, pantropical spotted dolphin *(Stenella attenuata)*.
[b] ND, not detected (limit of detection not reported).
[c] Converted from dry weight values using data on moisture content given in the paper.

13,156 mg/kg of dry weight in a bottle-nosed dolphin *[Tursiops truncatus]* from the northern Tyrrhenian Sea [Leonzio et al., 1992]). Leonzio et al. did not present data for the moisture content of their samples, and so no conversion to wet weight by this means was possible. For comparative purposes, therefore, I have undertaken a conversion using information derived from our own program. Average values for moisture content were 71.3% in striped dolphins ($n = 7$) and 70.9% in bottle-nosed dolphins ($n = 6$), leading to wet weight concentrations of 1263 and 3828 mg/kg, respectively, for the two animals. Much lower concentrations have been reported for essentially offshore animals from the Arctic region, such as the white-beaked dolphin *(Lagenorhyncus albirostris)* from Newfoundland (Table 1; 0.13 to 1.6 mg/kg; Muir et al., 1988) and the narwhal *(Monodon monoceros)* from Baffin Island (mean value, 6.1 mg/kg; standard deviation, 3.1 mg/kg; Wagemann et al., 1983).

Concentrations of mercury in fetal and neonatal marine mammals are normally low (<1 mg/kg in livers) (e.g., van de Ven et al., 1979), since only methylmercury is transferred across the placenta and the majority of mercury in marine mammals is stored in the inorganic form, as a result of the ability of marine mammals to demethylate mercury within their bodies. Harp seal *(Phoca groenlandica*, formerly *Pagophilus groenlandica)* pups acquired most (75 to 80%) of their mercury during gestation by transplacental transfer of methylmercury that they were unable to demethylate efficiently (Wagemann et al., 1988). The methylmercury concentrations in the livers of neonatal walruses *(Odobenus rosmarus rosmarus)* from northern Greenland and their mothers were comparable, and there were indications of some transfer occurring in milk as well as via the placenta (Born et al., 1981). Gray seal pups from the Farne Islands (United Kingdom), sampled in 1977, were found to have liver concentrations of methylmercury only around one tenth of those found in their mothers (van de Ven et al., 1979). Subsequently, accumulation of mercury through the local marine trophic network is the major contamination route for spotted *(Stenella attenuata)* and striped dolphins (André et al., 1991a) and, indeed, for marine mammals in general.

The trophic supply of mercury increases progressively with the mammal's growth, as the amount of food eaten and the preferred prey size increase (André et al., 1990). Thus, the body burden of marine mammals increases with age, as has been reported (for instance) for porpoises *(Phocoena phocoena)* (Gaskin et al., 1979; Falconer et al., 1983), striped dolphins (Itano et al., 1984a), common (harbor) seals (Reijnders, 1980; Miles et al., 1992), and gray seals (Sergeant and Armstrong, 1973). The liver is the most important accumulator of mercury in both pinnipeds and cetaceans (Wagemann and Muir, 1984; Marcovecchio et al., 1990) and, in spotted dolphins, three tissues (liver, skeletal muscle, and blubber) contain 95% of the mercury burden of the 18 tissues and organs analyzed (André et al., 1990). The proportion of organic mercury in the livers of pilot whales aged between 3 and 41 years ranged from 3 to 62%. This was negatively correlated with age ($r = -0.81$, $P < 0.01$), indicating that lower proportions were found in older individuals that, moreover, contained higher total mercury levels (Schintu et al., 1992). In fin whales *(Balaenoptera physalus*, a mysticete), the proportion of organic mercury was highest in the muscle tissue (~80%), whereas in the liver the mean percentage was about 40%. A high proportion of methylmercury is found in the muscle tissue of marine mammals generally (Dietz et al., 1990). A slight increase in concentration with age was observed for total mercury in both the muscle and liver and for organic mercury in the liver, which suggests a low rate of excretion for this metal even at low concentrations. No differences between sexes were found regarding mercury concentrations (Sanpera et al., 1993).

A number of experiments have been conducted in which seals have been fed diets including methylmercury (the form in which mercury occurs in their food), but none have been reported either with cetaceans as subjects or for other metals. Four harp seals were exposed to daily oral doses of methyl mercuric chloride (Ronald et al., 1977). Two fed doses of 0.25 mg/kg of body weight per day for 60 and 90 days did not show abnormal blood concentrations; two others fed 25 mg/kg of body weight per day died on days 20 and 26 of the experiment. The low-dosage seals exhibited a reduction in appetite and consequent weight loss, whereas both high-dosage seals became more lethargic and lost weight from day 3 onward. The measurement of blood parameters in these animals indicated toxic hepatitis, uremia, and renal failure. The observed failure in renal and hepatic functions was related to the high accumulation of mercury in these tissues. The concentrations of both total and methylmercury in the blood of all these animals rose linearly with time during the experiment. Neither control nor low-dosage seals showed overt clinical signs of mercury poisoning. In the high-dosage seals, the cause of death was ascribed to chronic renal failure, with their behavior prior to death resulting from uremia (Tessaro and Ronald, 1976). The concentrations of methylmercury in the livers of the high-dosage seals at death were 127 and 125 mg/kg, corresponding to total mercury concentrations of 134 and 142 mg/kg, respectively; i.e., essentially all the ingested mercury was still in the methyl form. The low-dosage seals exhibited somewhat lower concentrations; in the seal sacrificed after 60 days, the liver concentrations of methyl and total mercury were 18 and 64 mg/kg, respectively, and, after 90 days in the second seal, 76 and 83 mg/kg. The corresponding values in the control seals were 0.16 and 26 mg/kg and 0.2 and 1.7 mg/kg. Damage to the sensory hair cells of the cochlea was also found in the seals fed methylmercury in this experiment, with the degree of damage observed being related to the dose administered (Ramprashad and Ronald, 1977).

An adult female harp seal was fed methyl mercuric chloride at 0.25 mg of mercury per kilogram of body weight per day for 61 days; high concentrations of mercury accumulated in most tissues (Freeman et al., 1975). High concentrations of mercury (>75% of it as methylmercury) were found in the ovaries and adrenal glands. These organs were examined under a light microscope and appeared normal, but in vitro incubations of tissue from these organs indicated a marked alteration of steroid biosynthesis in tissue from the treated seal. At death, the concentration of mercury in the blood of the treated seal had risen to 9.9 mg/kg, compared with 0.14 mg/kg at the beginning of the experiment, whereas the concentration in the blood of the control seal was unchanged at the same level. The liver concentrations of total mercury in the treated and untreated seals were 64 and 25.8 mg/kg, respectively, while the equivalent methylmercury concentrations were 18.5 and 0.16 mg/kg.

In order to set these dietary intakes in context with respect to seals feeding in the wild, it is necessary to make some estimates of likely intake of mercury from fish alone. In the last experiment, the treated seal weighed 184 kg and was fed herring at 3 to 5% of body weight once or twice daily, i.e., 5.5 to 18 kg of fish per day. The mercury content of the herring provided (essentially all as methylmercury) was 0.1 mg/kg, so the seal ingested 0.55 to 1.8 mg of mercury daily from the fish. This corresponds to a dietary intake of 0.003 to 0.01 mg of mercury per kilogram of body weight per day. Only in an area such as the Mediterranean Sea where mercury concentrations in fish are particularly high (up to 7 mg/kg in edible tissues in some species; Bernhard, 1988) could the dietary intake of wild marine mammals approach even the lower doses used in these experiments.

Selenium has been shown to function as an antagonist to counteract the toxicity of a number of heavy metals (Levander, 1986; Fishbein, 1991) in both rats (Potter and Matrone, 1974) and Japanese quail (El Begearmi et al., 1977), for example. In gray seals fed low levels of methylmercury (~0.2 mg of methylmercury per kg of body weight per day), the concentrations of total mercury and selenium in both the liver and kidney increased in parallel, while in the other tissues examined, e.g., brain and blood, only the mercury concentration increased (van de Ven et al., 1979). Selenium was not administered along with methylmercury but was taken up from the diet, presumably in the form of organic selenium. The atomic ratio of mercury and selenium was found to be close to 1 in the livers of wild gray seals from the U.K. (van de Ven et al., 1979), ringed seals *(Phoca hispida)* and bearded seals *(Erignathus barbatus)* from Arctic Canada (Smith and Armstrong, 1978), and other pinnipeds and cetaceans from around the world (Koeman et al., 1975). An equimolar mercury:selenium ratio is not apparent in a large number of marine fish that make up a significant part of the diet of marine mammals; thus, the establishment of the 1:1 ratio must occur in the mammals themselves (Koeman et al., 1975; Luten et al., 1980; Cuvin-Aralar and Furness, 1991). Study by neutron activation analysis of subcellular fractions of a fresh liver homogenate from a seal showed that over 50% of the mercury and selenium occurred in the 700-g fraction (chiefly nuclei and cell wall material) with an equimolar ratio (Koeman et al., 1973). Histological studies of livers from specimens of Cuvier's beaked whale *(Ziphius cavirostris)* and bottle-nosed dolphin showed the presence of irregular black particles (1 to 5 µm) located in the connective tissue of the portal vessels (Martoja and Viale, 1977). Diffraction studies showed that these consisted of pure mercuric selenide (HgSe or tiemannite) deposited as the final stage of a slow mineralization and detoxification process (Martoja and Berry, 1980; Joiris et al., 1991). These particles are not attacked by proteolytic enzymes and are therefore inert (André et al., 1991b). Thus, marine mammals are apparently able to tolerate extremely high body burdens without ill effects. No indication of mercury intoxication was seen in 454 ringed and bearded seals with liver mercury concentrations up to 420 mg/kg (Smith and Armstrong, 1978). Also, no symptoms of mercury poisoning were evident in a mature striped dolphin in which the total mercury burden was 2.5 times as high as the fatal concentration reported for humans (Itano et al., 1984b) and in which the methylmercury burden also reached the fatal level for humans (Itano and Kawai, 1980). Only one case indicating possible toxic effects has been reported; that was a ringed seal from an area of heavy industrial mercury disposal (Helminen et al., 1968). Because marine mammals can tolerate high concentrations of mercury immobilized as the selenide, higher indeed than the concentrations estimated to be lethal to other mammals, methylmercury poisoning (as in the acute feeding experiments described above) must result from a high intake rate of methylmercury, i.e., with a threshold somewhere in the range of 0.25 to 25 mg/kg of body weight per day in seals, that accumulates in the body faster than the detoxification and mineralization process can cope with. This occurs despite reports that in vitro demethylation of methylmercury in liver and kidney tissues of seals showed about twice the activity seen in rats or mice (Himeno et al., 1989). A higher accumulation of methylmercury has also been observed in sick animals, suggesting that when weakened by disease, animals may be unable to detoxify organic mercury as efficiently as when they are well (Dietz et al., 1990).

A recent report (Rawson et al., 1993) has linked liver anomalies in bottle-nosed dolphins with chronic accumulation of mercury. Pigmented deposits were found in the livers and kidneys of nine animals with high hepatic mercury concentrations (61 to 443 mg/kg of wet weight); these deposits were not present in nine other animals with lower concentrations (<0.01 to 50 mg/kg of wet weight). The pigmented deposits were found to contain mercury and lipofuscin, which is believed to derive from damaged subcellular membranes and forms an indigestible residue within lysosomes (Robbins et al., 1981). Four of the nine animals with pigmented deposits also showed active liver damage, but only one from the other group; this was also shown not to be a simple relation with age (Rawson et al., 1993).

COPPER

Copper is an essential trace element, and copper deficiency anemia occurs in all species of animals (Davis and Mertz, 1987). Very young animals and neonates are normally richer in copper than are adults, and this has proved to be the case in four species of cetaceans, with the concentrations in the livers of a neonatal common porpoise and fetuses of common dolphin *(Delphinus delphis)*, Dall's porpoise *(Phocoenoides dalli)*, and a long-finned pilot whale *(Globicephala melas*, formerly *Globicephala melaena)* being 5 to 10 times higher than those found in their mothers, thus reflecting a considerable transplacental transfer of copper (Julshamn et al., 1987; Fujise et al., 1988; Law et al., 1992). Levels of copper in the tissues of young harp seals are also generally similar to or higher (up to a factor of 4) than those seen in their mothers (Ronald et al., 1984; Wagemann et al., 1988), so it is probably true for marine mammals in general. In animals generally, the newborn concentrations are largely maintained throughout the suckling period, followed by a steady fall during growth to adult values (Davis and Mertz, 1987). Copper concentrations in the mature striped dolphin (Honda et al., 1983b) and harbor porpoise (Falconer et al., 1983) have been shown not to increase with the length (age) of the animal. Falconer et al. (1983) found that liver concentrations of copper in 23 porpoises collected from the east coast of Scotland in 1974 ranged from 2.7 to 13 mg/kg, with very similar concentrations in both males and females and with no relation to age. In the majority of cases, copper concentrations in the livers of adult marine mammals are within the range of 3 to 30 mg/kg, and this may represent the normal range of homeostatic control in marine mammals (Law et al., 1991). Animal species differ greatly in their tolerance of high levels of copper in the diet, and metabolic interactions with other metals, such as cadmium, iron, and zinc, further complicate the interpretation (Davis and Mertz, 1987).

ZINC

Zinc is relatively nontoxic to mammals, and more than 200 zinc-based enzymes or other proteins have been identified from various sources encompassing all phyla. Zinc deficiency reduces the activity of certain enzymes, affects the production of hormones (especially testosterone and adrenal corticosteroids) and reproduction, and is essential to the integrity of the immune system. In growing animals, the first effects of zinc deficiency are reduced food intake and growth rate. The liver of the adult human contains 30 to 100 mg/kg of zinc and, in most other mammals, reported concentrations are in the lower part of this range (Hambidge et al., 1986). Based on their own results and a study of reported literature data, Law et al. (1991, 1992) have postulated that zinc concentrations in the liver of marine mammals are maintained within the range of approximately 20 to 100 mg/kg. There is little fetal storage of zinc in mammals generally (Underwood, 1977) and, in a neonatal porpoise and a fetal common dolphin liver, zinc concentrations were only one half and one third of those found in their mothers (Law et al., 1992). The transplacental transfer of zinc is therefore less than that for copper. Subsequently, zinc is taken in via milk and from food, but concentrations are not correlated with age in porpoises (Falconer et al., 1983), striped dolphins (Honda et al., 1983b), or northern fur seals *(Callorhinus ursinus)* (Goldblatt and Anthony, 1983). In mammals generally, the absorption of zinc appears readily influenced by a wide range of factors, with the supply and status of zinc major factors in homeostatic control. The liver is the major organ involved in zinc metabolism and, in rats, fluctuations in the levels of ^{65}Zn were accompanied by concomitant changes in the rate of synthesis of metallothionein and of zinc binding. In humans, the elimination of a ^{65}Zn tracer best fit a two-component model, the initial rapid phase having a half-life of 12.5 days and the subsequent phase, one of approximately 300 days, although in rats and mice the initial phase was much more rapid (Hambidge et al., 1986).

It has been suggested above that the ranges for homeostatic control of copper and zinc in marine mammal livers may lie close to 3 to 30 mg/kg and 20 to 100 mg/kg, respectively. Figure 1 shows the liver concentrations of copper and zinc in 243 marine mammals (11 species consisting of two seal and nine cetacean) plotted against one another (Law et al., 1991, 1992; unpublished data). It can be seen that, while some individuals exhibit copper concentrations of >30 mg/kg or zinc concentrations of >100 mg/kg, only one individual shows both together. This was a porpoise stranded on the Isle of Man, Irish Sea (reference number SMRU91-16; Law, 1994). Thirty-nine animals showed concentrations of either copper or zinc above these threshold values. If overloading of the control mechanism occurs, and if metallothionein is not

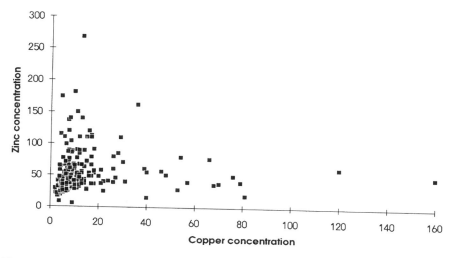

Figure 1 Concentrations of copper and zinc in the livers of 243 marine mammals (mg/kg of wet weight).

being synthesized sufficiently quickly to maintain homeostasis, then one could expect concentrations of both these elements to move out of control. In a study of polar bears *(Ursus maritimus)* from the Canadian Arctic, Norstrom et al. (1986) concluded that elevated levels of copper in their diet could disturb the homeostatic equilibrium of zinc in their livers and give rise to high concentrations of both elements concurrently in some animals. It may be that the individuals with the highest concentrations of either copper or zinc in our studies were fulfilling some specific metabolic need, and that this was the reason for the high values observed. Further study is needed in this area before such fluctuations can be fully explained.

CADMIUM

In humans, cadmium is virtually absent from the body at birth and accumulates with age up to ~50 years of age. Only 6% of dietary cadmium is absorbed, and the rest passes into the feces. In long-term, low-level exposure, approximately 33% of the body burden is localized in the kidneys and 17% in the liver. This retention results from sequestration by metallothionein and accounts for the very long half-life in the body (of the order of 26 years in humans but much shorter in other animals, from several weeks in mice to 2 years in monkeys). The biological half-life may, in fact, be dependent on the dose received. Aquatic food species including fish, crabs, oysters, and shrimp bioaccumulate cadmium. In marine invertebrates, this is largely dependent on the concentration of cadmium leached from sediment into water, but bioavailability is low as a result of the formation of stable complexes. Cadmium is toxic to virtually every system in the human body. The degree of the toxic effect may be determined by the capacity of the tissues, e.g., the kidneys, to synthesize metallothionein. High dietary concentrations of cadmium can lead to a wide range of effects in animals, including depressed growth, enteropathy, anemia, poor bone mineralization, severe kidney damage, cardiac enlargement, hypertension, and fetal malformation (Kostial, 1986). Moreover, a potential for cadmium to cause cancer has been shown recently in animal experiments (Stoeppler, 1991).

The level of cadmium in the livers of neonatal porpoises and of fetuses of a common dolphin and a Dall's porpoise, for instance, was very low, suggesting that transplacental transfer is negligible (Fujise et al., 1988; Law et al., 1992), although some transfer to fetal kidney has been reported to occur in pilot whales (Meador et al., 1993). The mother of the Dall's porpoise had a very high concentration of cadmium, 20.6 mg/kg in the liver and 34 mg/kg in the kidneys; such high concentrations are generally thought to relate to dietary preference. High concentrations of cadmium are observed in marine mammals that inhabit areas remote from pollution sources but that eat a large proportion of food species high in cadmium. High concentrations of cadmium are accumulated in the liver and gonads of cephalopods (Hamanaka et al., 1982; Pena et al., 1988). Concentrations are also high in pilot whales from the Faeroe Islands (Julshamn et al., 1987), Ross seals *(Ommatophoca rossi)* from the pack ice of Queen Maud Land (off Antarctica) (McClurg, 1984), and crabeater seals *(Lobodon carcinophagus)* from the Antarctic (Szefer et al., 1994), all of which feed on squid. High concentrations of cadmium in Antarctic krill also lead to high concentrations in southern minke whales *(Balaenoptera acutorostrata)* (Honda et al., 1987) that feed on them. High concentrations of cadmium were also seen in the liver and kidney of beluga *(Delphinapteras leucas)* and narwhal taken from West Greenland in 1984 to 1986 (Hansen et al., 1990). An increasing trend in cadmium concentrations in livers of polar bears from the west to the east of the Canadian Arctic also suggests that natural sources are the dominant ones for the marine food chain in that area (Norstrom et al., 1986; Muir et al., 1992).

Cadmium is transferred via milk from mother to calf in the striped dolphin, as indicated by a rapid increase in concentration from birth up to 1 year of age (which is also apparent for nickel and lead) (Honda et al., 1983b, 1986). Concentrations continue to increase in many cases with the length (age) of the animal, for example, in the common porpoise (Falconer et al., 1983) and the Steller sea lion *(Eumetopias jubatus)* (Hamanaka et al., 1982).

The magnitude of uptake differs among species, probably reflecting differences in habitat and diet. Pilot whales exhibited liver cadmium concentrations 20 times higher than those seen in white-beaked dolphins (both from Newfoundland), while porpoises from the Bay of Fundy and the Baltic Sea yielded concentrations 100 times lower than those in the dolphins (Muir et al., 1988). High concentrations were reported for livers of Pacific walruses *(Odobenus rosmarus divergens)* from the Bering Sea (up to 50 mg/kg) with no obvious signs of health problems apparent from gross examination of the harvested animals, although histological studies were not conducted (Taylor et al., 1989). Concentrations of cadmium and selenium were correlated in the livers of minke whales, belugas and narwhals from West Greenland, although it is not clear whether this represents a simple accumulation of both elements with age or the presence of some detoxifying mechanism, as is the case for mercury (Hansen et al., 1990). In humans, cadmium is stored in the kidneys to a concentration of 200 to 400 mg/kg, above which level renal damage prevents further accumulation (Piotrowski and Coleman, 1980). In the absence of specific information relating to marine mammals, it is reasonable to consider this as a tentative threshold level. As concentrations of cadmium are consistently higher in renal tissue than in hepatic tissue of both cetaceans and seals (Wagemann and Muir, 1984), this then corresponds to a liver concentration in the range of 40 to 200 mg/kg. In addition, Fujise et al. (1988) reported a study by Honda in which the renal cadmium concentrations in marine mammals increased with an increase of the hepatic cadmium concentration up to about 20 mg/kg, and above this value the renal cadmium concentration decreased. This was taken to indicate that renal dysfunction attributable to cadmium can occur in marine mammals with liver concentrations higher than 20 mg/kg. It can be seen from Table 1 that a number of the maximum concentrations fall within this tentative threshold area of 20 to 200 mg/kg, so further study of the dynamics and possible effects of cadmium in marine mammals would be of merit.

LEAD

On the basis of laboratory experiments, scientists suggest that lead is an essential element in rats, but the evidence so far is incomplete, as is the understanding of lead metabolism and the means by which it produces its toxic effects. Many enzymes, membranes, and biochemical processes have been shown to be affected by lead, but none has been shown to be both sensitive and important in explaining the manifestations of toxicity, especially at low levels of the metal. Lead has a multiplicity of biochemical effects, many of which involve the inhibition of enzyme systems, such as the cytochrome P-450-linked mixed-function oxidase system. The most common consequence of lead poisoning is anemia. The toxic effects of lead include renal damage, hypertension and cardiac disease, reduced antibody synthesis, and neurological effects (Quarterman, 1986). The absorption of lead varies widely, with factors such as species, age (young animals absorb a much greater amount of lead than do adults [Ewers and Schlipköter, 1991]), physiological state, and the nature of the diet. In humans, the vast majority of body lead resides in the skeleton and has a very long mean half-life. In Dall's porpoise and other marine mammals also, high lead concentrations were observed in skin and bone (Fujise et al., 1988). Lead is readily transferred across the placenta in humans, rats, and goats (Quarterman, 1986), and this also seems to be the case for marine mammals, since the concentrations of lead in the livers of a neonatal porpoise and a fetal common dolphin were approximately 35% of those seen in their mothers (Law et al., 1992); the hepatic lead concentrations in matured female southern minke whales decreased with the progress of gestation (Honda et al., 1987). Lead is also transferred from mother to calf via milk in the striped dolphin (Honda et al., 1983a). No correlation between tissue lead concentrations and age was observed for northern fur seals from Alaska (Goldblatt and Anthony, 1983). Liver concentrations are generally low (<1 mg/kg; Table 1) except in areas such as Liverpool Bay (N.W. England), where elevated concentrations (up to 4.3 mg/kg) in gray seals and cetaceans are thought to result from industrial activities, which have included the manufacture of alkyllead antiknock additives for gasoline (Law et al., 1992).

ARSENIC

Arsenic has been determined in only a few instances in marine mammal tissues. Almost all the data currently available are included in Table 1, with the exception of narwhal from Baffin Island sampled in 1978 to 1979, for which a range of concentration was not reported. The average concentration of arsenic in those animals was 0.33 mg/kg (S.D., 0.18 mg/kg) (Wagemann et al., 1983), i.e., within the range of values reported for the other animals (common seals, common dolphins, and pilot whales).

In pilot whales from waters of the Faeroe Islands, arsenic concentrations in liver and kidney tissues of two fetuses examined were lower than those seen in adults, suggesting that the degree of transplacental transfer is low. Marine fish and invertebrates (especially crustaceans), however, can contain quite high arsenic concentrations, often as much as 100 mg/kg. Almost all of this arsenic is present in the form of organoarsenical compounds, especially arsenobetaine (Phillips, 1990), which are relatively nontoxic and can be readily eliminated via the kidneys in mammals generally. Arsenic is an essential element, although the nutritional requirement for marine mammals is unknown. In humans, a normal arsenic supply corresponds to 12 to 25 µg per day (Anke, 1986).

CHROMIUM

Inorganic chromium compounds are poorly absorbed in animals, regardless of dose and dietary status, although there is some evidence that the natural complexes in the diet may be more available. Hexavalent chromium is better absorbed than trivalent chromium and is also much more toxic, an oral dose of 0.5 to 1 g of potassium dichromate being fatal to humans (Gauglhofer and Bianchi, 1991). Glucose intolerance is usually one of the first signs of chromium deficiency, followed by further abnormalities in glucose and lipid metabolism and nerve disorders (Anderson, 1987). Some chromium compounds are recognized as human carcinogens (Gauglhofer and Bianchi, 1991), although the cancers followed inhalation in industrial situations and may not therefore be relevant to marine mammals. In humans, chromium is generally recognized as an essential trace element, and a tentative dietary allowance for adults of 50 to 200 µg per day has been recommended (National Academy of Sciences, 1980). An assessment of the significance of the chromium concentrations reported in marine mammal tissues would require further study of chromium speciation, in both prey species and the mammals themselves.

NICKEL

In mammals generally, dietary nickel is poorly absorbed and relatively nontoxic. As yet, no biological function of nickel in animals has been firmly established. However, some findings indicate a role in the production of specific metalloenzymes or in the intestinal absorption of the ferric ion (Nielsen, 1987). Very few studies of marine mammals have included the analysis of tissues for nickel, and little is known of the form in which nickel is ingested from fish and marine invertebrates, although shellfish and crustacea generally contain higher concentrations of nickel in their tissues than do fish (Sunderman and Oskarsson, 1991). Higher concentrations may therefore be expected in tissues of marine mammals that eat euphausiids, such as krill, in large quantity (Martin, 1990), blue whales *(Balaenoptera musculus),* or fin whales *(Balaenoptera physalus),* for instance, than in fish-eating dolphins. No meaningful assessment of either the toxicity or deficiency of nickel in marine mammals can be made as yet.

SUMMARY

Metals enter the marine environment as a result of both natural processes and anthropogenic inputs. Once they are weaned, the primary route of uptake by marine mammals is from their food. In the case of essential trace elements such as copper and zinc, the concentrations in marine mammal tissues seem to be regulated. Toxic metals, such as cadmium, can be sequestered by metallothioneins into cellular compartments in which they are no longer available. The total concentrations of cadmium in liver and kidney tissues of some individuals may be close to or higher than tentatively established threshold effect levels, however, and closer scrutiny of the effectiveness of this mechanism would be advisable. Methylmercury, the form in which mercury is essentially present in muscle tissue of fish, has been shown to be toxic to seals in short-term (<100 days) experiments. Marine mammals have the ability to demethylate mercury within their bodies and to immobilize it as the selenide. When mercury is stored in this way, marine mammals may accumulate mercury to concentrations higher than those estimated to be lethal to other mammals without apparent harm, although recent data suggest that chronic toxicity may result in increased active liver disease. For other metals (chromium, nickel, arsenic, and lead), some data are available, but our knowledge is insufficient at this time to allow even tentative conclusions to be drawn. With the increase in the use of inductively coupled plasma/mass spectrometry as an alternative analytical technique to atomic absorption spectrophotometry, it is likely that future studies will routinely cover a wider range of elements than is presently the case, allowing a better assessment of the significance of metal concentrations in marine mammal tissues than can be made at present.

REFERENCES

Anderson, R. A. 1987. Chromium. *In* W. Mertz (Ed.). Trace elements in human and animal nutrition. 5th ed. Vol. 1. Academic Press, Orlando, Fla.

André, J. M., A. Boudou, and F. Ribeyre. 1991a. Mercury accumulation in delphinidae. Water Air Soil Pollut. 56:187-201.

André, J. M., A. Boudou, F. Ribeyre, and M. Bernhard. 1991b. Comparative study of mercury accumulation in dolphins *(Stenella coeruleoalba)* from French Atlantic and Mediterranean coasts. Sci. Total Environ. 104:191-209.

André, J. M., F. Ribeyre, and A. Boudou. 1990. Mercury contamination levels and distribution in tissues and organs of delphinids *(Stenella attenuata)* from the eastern tropical Pacific, in relation to biological and ecological factors. Mar. Environ. Res. 30:43-72.

Anke, M. 1986. Arsenic. *In* W. Mertz (Ed.). Trace elements in human and animal nutrition. 5th ed. Vol. 2. Academic Press, Orlando, Fla.

Augier, H., W. K. Park, and C. Ronneau. 1993. Mercury contamination of the striped dolphin *Stenella coeruleoalba* Meyen from the French Mediterranean coast. Mar. Pollut. Bull. 26:306-311.

Bernhard, M. 1988. Mercury in the Mediterranean. UNEP Reg. Seas Rep. Studies, No. 98. 141 pp.

Bloom, N. S. 1992. On the chemical form of mercury in edible fish and marine invertebrate tissue. Can. J. Fish. Aquat. Sci. 49:1010-1017.

Born, E. W., I. Kraul, and T. Kristensen. 1981. Mercury, DDT, and PCB in the Atlantic walrus *(Odobenus rosmarus rosmarus)* from the Thule district, north Greenland. Arctic 34:255-260.

Calambokidis, J., G. H. Steiger, L. J. Lowenstine, and D. S. Becker. 1991. Chemical contamination of harbor seal pups in Puget Sound. Report No. 910/9-91-032. U.S. Environmental Protection Agency, Seattle, Wash. 43 pp.

Clarkson, T. W. 1987. Mercury. *In* W. Mertz (Ed.). Trace elements in human and animal nutrition. 5th ed. Vol. 1. Academic Press, Orlando, Fla.

Cuvin-Aralar, M. L. A., and R. W. Furness. 1991. Mercury and selenium interaction: a review. Ecotoxicol. Environ. Saf. 21:348-364.

Davis, G. K., and W. Mertz. 1987. Copper. *In* W. Mertz (Ed.). Trace elements in human and animal nutrition. 5th ed. Vol. 1. Academic Press, Orlando, Fla.

Dietz, R., C. O. Nielsen, M. M. Hansen, and C. T. Hansen. 1990. Organic mercury in Greenland birds and mammals. Sci. Total Environ. 95:41-51.

El-Begearmi, M. M., M. L. Sunde, and H. E. Ganther. 1977. A mutual protective effect of mercury and selenium in Japanese quail. Poult. Sci. 56:313-322.

Ewers, U., and H.-W. Schlipköter. 1991. Lead. *In* E. Merian (Ed.). Metals and their compounds in the environment. VCH Publishers, Weinheim, Germany.

Falconer, C. R., I. M. Davies, and G. Topping. 1983. Trace metals in the common porpoise, *Phocoena phocoena*. Mar. Environ. Res. 8:119-127.

Fishbein, L. 1991. Selenium. *In* E. Merian (Ed.). Metals and their compounds in the environment. VCH Publishers, Weinheim, Germany.

Frank, A., V. Galgan, A. Roos, M. Olsson, L. R. Petersson, and A. Bignert. 1992. Metal concentrations in seals from Swedish waters. Ambio 21:529-538.

Freeman, H. C., G. Sangalang, J. F. Uthe, and K. Ronald. 1975. Steroidogenesis in vitro in the harp seal *(Pagophilus groenlandicus)* without and with methylmercury treatment in vivo. Environ. Physiol. Biochem. 5:428-439.

Fujise, Y., K. Honda, R. Tatsukawa, and S. Mishima. 1988. Tissue distribution of heavy metals in Dall's porpoise in the northwestern Pacific. Mar. Pollut. Bull. 19:226-230.

Gaskin, D. E., K. I. Stonefield, P. Suda, and R. Frank. 1979. Changes in mercury levels in harbour porpoises from the Bay of Fundy, Canada, and adjacent waters during 1969-1977. Arch. Environ. Contam. Toxicol. 8:733-762.

Gauglhofer, J., and V. Bianchi. 1991. Chromium. In E. Merian (Ed.). Metals and their compounds in the environment. VCH Publishers, Weinheim, Germany.

Goldblatt, C. J., and R. G. Anthony. 1983. Heavy metals in northern fur seals *(Callorhinus ursinus)* from the Pribilof Islands, Alaska. J. Environ. Qual. 12:478-482.

Hamanaka, T., T. Itoo, and S. Mishima. 1982. Age-related change and distribution of cadmium and zinc concentrations in the Steller sea lion *(Eumetopias jubata)* from the coast of Hokkaido, Japan. Mar. Pollut. Bull. 13:57-61.

Hambidge, K. M., C. E. Casey, and N. F. Krebs. 1986. Zinc. In W. Mertz (Ed.). Trace elements in human and animal nutrition. 5th ed. Vol. 2. Academic Press, Orlando, Fla.

Hamilton, S. J., and P. M. Mehrle. 1986. Metallothionein in fish: review of its importance in assessing stress from metal contaminants. Trans. Am. Fish. Soc. 115:596-609.

Hansen, C. T., C. O. Nielsen, R. Dietz, and M. M. Hansen. 1990. Zinc, cadmium, mercury, and selenium in minke whales, belugas, and narwhals from West Greenland. Polar Biol. 10:529-539.

Helminen, M., E. Karppanen, and J. I. Koivisto. 1968. Mercury content of the ringed seal of Lake Saama. Suom. Laaklehti 74:87-89.

Himeno, S., C. Watanabe, T. Hongo, T. Suzuki, A. Naganuma, and N. Imura. 1989. Body size and organ accumulation of mercury and selenium in young harbour seals *(Phoca vitulina)*. Bull. Environ. Contam. Toxicol. 42:503-509.

Honda, K., Y. Fujise, R. Tatsukawa, K. Itano, and N. Miyazaki. 1986. Age-related accumulation of heavy metals in bone of the striped dolphin, *Stenella coeruleoalba*. Mar. Environ. Res. 20:143-160.

Honda, K., R. Tatsukawa, and T. Fujiyama. 1983a. Distribution characteristics of heavy metals in the organs and tissues of striped dolphin, *Stenella coeruleoalba*. Agric. Biol. Chem. 46:3011-3021.

Honda, K., R. Tatsukawa, K. Itano, N. Miyazaki, and T. Fujiyama. 1983b. Heavy metal concentrations in muscle, liver, and kidney tissue of striped dolphin, *Stenella coeruleoalba*, and their variations with body length, weight, sex, and age. Agric. Biol. Chem. 47:1219-1228.

Honda, K., Y. Yamamoto, H. Kato, and R. Tatsukawa. 1987. Heavy metal accumulations and their recent changes in southern minke whales, *Balaenoptera acutorostrata*. Arch. Environ. Contam. Toxicol. 16:209-216.

Itano, K., and S. Kawai. 1980. Changes of mercury and selenium contents and biological half-life of mercury in the striped dolphins. p. 49-72. University of the Ryukyus, Okinawa, Japan.

Itano, K., S. Kawai, N. Miyazaki, R. Tatsukawa, and T. Fujiyama. 1984a. Mercury and selenium levels in striped dolphins caught off the Pacific coast of Japan. Agric. Biol. Chem. 48:1109-1116.

Itano, K., S. Kawai, N. Miyazaki, R. Tatsukawa, and T. Fujiyama. 1984b. Body burdens and distribution of mercury and selenium in striped dolphins. Agric. Biol. Chem. 48:1117-1121.

Joiris, C. R., L. Holsbeek, J. M. Bouquegneau, and M. Bossicart. 1991. Mercury contamination of the harbour porpoise *(Phocoena phocoena)* and other cetaceans from the North Sea and the Kattegat. Water Air Soil Pollut. 56:283-293.

Julshamn, K., A. Andersen, O. Ringdal, and J. Morkore. 1987. Trace elements intake in the Faeroe Islands. I. Element levels in edible parts of pilot whales *(Globicephalus meleanus)*. Sci. Total Environ. 65:53-62.

Koeman, J. H., W. H. M. Peeters, C. H. M. Koudstaal-Hol, P. S. Tjioe, and J. J. M. de Goeij. 1973. Mercury-selenium correlations in marine mammals. Nature (Lond.) 245:385-386.

Koeman, J. H., W. S. M. van de Ven, J. J. M. de Goeij, P. S. Tjioe, and J. L. van Haaften. 1975. Mercury and selenium in marine mammals and birds. Sci. Total Environ. 3:279-287.

Kostial, K. 1986. Cadmium. *In* W. Mertz (Ed.). Trace elements in human and animal nutrition. 5th ed. Vol. 2. Academic Press, Orlando, Fla.

Law, R. J. (compiler). 1994. Collaborative UK marine mammal project: summary of data produced 1988-1992. Fish. Res. Tech. Rep., MAFF Direct. Fish. Res., Lowestoft, 97:42.

Law, R. J., C. F. Fileman, A. D. Hopkins, J. R. Baker, J. Harwood, D. B. Jackson, S. Kennedy, A. R. Martin, and R. J. Morris. 1991. Concentrations of trace metals in the livers of marine mammals (seals, porpoises, and dolphins) from waters around the British Isles. Mar. Pollut. Bull. 22:183-191.

Law, R. J., B. R. Jones, J. R. Baker, S. Kennedy, R. Milne, and R. J. Morris. 1992. Trace metals in the livers of marine mammals from the Welsh coast and the Irish Sea. Mar. Pollut. Bull. 24:296-304.

Leonzio, C., S. Focardi, and C. Fossi. 1992. Heavy metals and selenium in stranded dolphins of the northern Tyrrhenian (NW Mediterranean). Sci. Total Environ. 119:77-84.

Levander, O. A. 1986. Selenium. *In* W. Mertz (Ed.). Trace elements in human and animal nutrition. 5th ed. Vol. 2. Academic Press, Orlando, Fla.

Luten, J. B., A. Ruiter, T. M. Ritskes, A. B. Rauchbaar, and G. Riekwel-Booy. 1980. Mercury and selenium in marine and freshwater fish. J. Food Sci. 45:416-419.

Marcovecchio, J. E., V. J. Moreno, R. O. Bastida, M. S. Gerpe, and D. H. Rodriguez. 1990. Tissue distribution of heavy metals in small cetaceans from the southwestern Atlantic Ocean. Mar. Pollut. Bull. 21:299-304.

Martin, A. R. 1990. Whales and dolphins. Salamander Books, London.

Martoja, R., and J.-P. Berry. 1980. Identification of tiemannite as a probable product of demethylation of mercury by selenium in cetaceans. A complement to the scheme of the biological cycle of mercury. Vie Milieu 30:7-10.

Martoja, R., and D. Viale. 1977. Accumulation de granules de seleniure mercurique dans le foie d'Odontocetes (Mammiferes, Cetaces): un mecanisme possible de detoxication du methylmercure par le selenium. (In French.) C. R. Hebd. Seances Acad. Sci. Ser. D Sci. Nat. 285:109-112.

McClurg, T. P. 1984. Trace metals and chlorinated hydrocarbons in Ross seals from Antarctica. Mar. Pollut. Bull. 15:384-389.

Meador, J. P., U. Varanasi, P. A. Robisch, and S.-L. Chan. 1993. Toxic metals in pilot whales *(Globicephala melaena)* from strandings in 1986 and 1990 on Cape Cod, Massachusetts. Can. J. Fish. Aquat. Sci. 50:2698-2706.

Miles, A. K., D. G. Calkins, and N. C. Coon. 1992. Toxic elements and organochlorines in harbor seals *(Phoca vitulina richardsi)*, Kodiak, Alaska, USA. Bull. Environ. Contam. Toxicol. 48:727-732.

Muir, D. C. G., R. Wagemann, N. P. Grift, R. J. Norstrom, M. Simon, and J. Lien. 1988. Organochlorine chemical and heavy metal contaminants in white-beaked dolphins *(Lagenorhyncus albirostris)* and pilot whales *(Globicephala melaena)* from the coast of Newfoundland, Canada. Arch. Environ. Contam. Toxicol. 17:613-629.

Muir, D. C. G., R. Wagemann, B. T. Hargrave, D. J. Thomas, D. B. Peakall, and R. J. Norstrom. 1992. Arctic marine ecosystem contamination. Sci. Total Environ. 122:75-134.

National Academy of Sciences. 1980. Recommended dietary allowances. 9th rev. ed. National Academy of Sciences, Washington, D.C.

Nielsen, F. H. 1987. Nickel. *In* W. Mertz (Ed.). Trace elements in human and animal nutrition. 5th ed. Vol. 1. Academic Press, Orlando, Fla.

Norstrom, R. J., R. E. Schweinsberg, and B. T. Collins. 1986. Heavy metals and essential elements in livers of the polar bear *(Ursus maritimus)* in the Canadian Arctic. Sci. Total Environ. 48:195-212.

Pena, N. I., V. J. Moreno, J. E. Marcovecchio, and A. Perez. 1988. Total mercury, cadmium, and lead distribution in tissues of the southern sea lion *(Otario flavescens)* in the ecosystem of Mar del Plata, Argentina. p. 140-146. *In* U. Seelinger, L. D. de Lacerda, and S. R. Patchineelam (Ed.). Metals in coastal environments of Latin America. Springer-Verlag, Heidelberg.

Phillips, D. J. H. 1990. Arsenic in aquatic organisms: a review, emphasizing chemical speciation. Aquat. Toxicol. (Amst.) 16:151-186.

Piotrowski, J. K., and D. O. Coleman. 1980. Environmental hazards of heavy metals: summary evaluation of lead, cadmium, and mercury. Report No. 20. Monitoring and Assessment Research Centre, University of London, London.

Potter, S., and G. Matrone. 1974. Effect of selenite on the toxicity of dietary methylmercury and mercuric chloride in the rat. J. Nutr. 104:638-647.

Quarterman, J. 1986. Lead. *In* W. Mertz (Ed.). Trace elements in human and animal nutrition. 5th ed. Vol. 2. Academic Press, Orlando, Fla.

Ramprashad, F., and K. Ronald. 1977. A surface preparation study on the effect of methylmercury on the sensory hair cell population in the cochlea of the harp seal *(Pagophilus groenlandicus* Erxleben, 1777). Can. J. Zool. 55:223-230.

Rawson, A. J., G. W. Patton, S. Hofmann, G. G. Pietra, and L. Johns. 1993. Liver abnormalities associated with chronic mercury accumulation in stranded Atlantic bottlenose dolphins. Exotoxicol. Environ. Saf. 25:41-47.

Reijnders, P. J. H. 1980. Organochlorine and heavy metal residues in harbour seals from the Wadden Sea and their possible effects on reproduction. Neth. J. Sea Res. 14:30-65.

Robbins, S. L., M. Angell, and V. Kumar. 1981. Basic pathology. 3rd ed. Saunders, Philadelphia, Pa.

Roesijadi, G. 1992. Metallothioneins in metal regulation and toxicity in aquatic animals. Aquat. Toxicol. (Amst.) 22:81-114.

Ronald, K., R. J. Frank, and J. Dougan. 1984. Pollutants in harp seals *(Pagophilus groenlandicus)*. II. Heavy metals and selenium. Sci. Total Environ. 38:153-166.

Ronald, K., S. V. Tessaro, J. F. Uthe, H. C. Freeman, and R. Frank. 1977. Methylmercury poisoning in the harp seal *(Pagophilus groenlandicus)*. Sci. Total Environ. 8:1-11.

Sanpera, C., R. Capelli, V. Minganti, and L. Jover. 1993. Total and organic mercury in North Atlantic fin whales: distribution pattern and biological related changes. Mar. Pollut. Bull. 26:135-139.

Schintu, M., F. Jean-Caurant, and J.-C. Amiard. 1992. Organomercury determination in biological reference materials: application to a study on mercury speciation in marine mammals off the Faröe Islands. Ecotoxicol. Environ. Saf. 24:95-101.

Sergeant, D. E., and F. A. J. Armstrong. 1973. Mercury in seals from Eastern Canada. J. Fish. Res. Board Can. 30:843-846.

Skaare, J. U., N. H. Markussen, G. Norheim, S. Haugen, and G. Holt. 1990. Levels of polychlorinated biphenyls, organochlorine pesticides, mercury, cadmium, copper, selenium, arsenic, and zinc in the harbour seal, *Phoca vitulina*, in Norwegian waters. Environ. Pollut. 66:309-324.

Smith, T. G., and F. A. J. Armstrong. 1978. Mercury and selenium in ringed and bearded seal tissues from Arctic Canada. Arctic 31:75-84.

Stoeppler, M. 1991. Cadmium. *In* E. Merian (Ed.). Metals and their compounds in the environment. VCH Publishers, Weinheim, Germany.

Sunderman, F. W., Jr., and A. Oskarsson. 1991. Nickel. *In* E. Merian (Ed.). Metals and their compounds in the environment. VCH Publishers, Weinheim, Germany.

Szefer, P., K. Szefer, J. Pempkowiak, B. Skwarzec, R. Bojanowski, and E. Holm. 1994. Distribution and coassociations of selected metals in seals of the Antarctic. Environ. Pollut. 83:341-349.

Taylor, D. L., S. Schliebe, and H. Metsker. 1989. Contaminants in blubber, liver, and kidney tissue of Pacific walruses. Mar. Pollut. Bull. 20:465-468.

Tessaro, S. V., and K. Ronald. 1976. The lesions of chronic methylmercury poisoning in the harp seal *(Pagophilus groenlandicus)*. ICES CM1976/N:7.

Tohyama, C., S.-I. Himeno, C. Watanabe, T. Suzuki, and M. Morita. 1986. The relationship of the increased level of metallothionein with heavy metal levels in the tissue of the harbor seal *(Phoca vitulina)*. Ecotoxicol. Environ. Saf. 12:85-94.

Underwood, E. J. 1977. Trace elements in human and animal nutrition. 4th ed. Academic Press, London.

van de Ven, W. S. M., J. H. Koeman, and A. Svenson. 1979. Mercury and selenium in wild and experimental seals. Chemosphere 8:539-555.

Von Burg, R., and M. R. Greenwood. 1991. Mercury. *In* E. Merian (Ed.). Metals and their compounds in the environment. VCH Publishers, Weinheim, Germany.

Wagemann, R. 1989. Comparison of heavy metals in two groups of ringed seals *(Phoca hispida)* from the Canadian Arctic. Can. J. Fish. Aquat. Sci. 46:1558-1563.

Wagemann, R., and D. C. G. Muir. 1984. Concentrations of heavy metals and organochlorine concentrations in marine mammals of northern waters: overview and evaluation. Can. Tech. Rep. Fish. Aquat. Sci. 1279:v + 97 pp.

Wagemann, R., N. B. Snow, A. Lutz, and D. P. Scott. 1983. Heavy metals in tissues and organs of the narwhal *(Monodon monoceros)*. Can. J. Fish. Aquat. Sci. 40 (Suppl. 2):206-214.

Wagemann, R., R. E. A. Stewart, W. L. Lockhart, B. E. Stewart, and M. Povoledo. 1988. Trace metals and methylmercury: associations and transfer in harp seal *(Phoca groenlandica)* mothers and their pups. Mar. Mammal Sci. 4:339-355.

CHAPTER 16

Cadmium in Small Mammals

John A. Cooke and Michael S. Johnson

INTRODUCTION

There has been much research into the toxicity of cadmium over the last three decades. Controlled studies involving inbred populations of laboratory rats, in particular, have provided data concerning many health effects of cadmium. Many of the studies have involved the parenteral administration of soluble cadmium salts, but these lack relevance to the long-term oral exposure that characterizes more realistic environmental conditions. A number of field-based investigations of wild populations of small mammals have been conducted over the last 15 years that have focused on the cadmium concentrations in tissue. Few of these studies, however, have provided evidence relating these tissue concentrations of cadmium to toxicological effects or possible dose-response characteristics. None has yielded results that show the ecotoxicological effects at the population level. Most of these field studies have been at sites that contain elevated levels of other metals, especially copper, lead, and zinc. Because interactions between these metals and cadmium are likely, the ecological and toxicological significance of such interactions is not well understood.

This review attempts to identify the critical concentrations in tissues of wild small mammals and their potential toxicological significance, mainly from the two aforementioned sources of information: studies on laboratory rats and field studies on wild small mammals. It is unfortunate that studies attempting to bridge the gap between laboratory and field approaches, e.g., laboratory experiments using wild species of small mammals, are rare.

METABOLISM AND TOXICITY

Cadmium has no known biological function. Its toxicity may originate through exposure via respiration or ingestion. The incidence of acute toxicity is rare, although the inhalation of cadmium fume is a well-known industrial hazard (Fielder and Dale, 1983). Fifty percent lethal dose values for acute toxicity of about 100 to 300 mg of cadmium per kilogram have been obtained for a range of soluble cadmium compounds in the rat when administered p.o. (Fielder and Dale, 1983).

Environmental cadmium exposure is predominantly a chronic problem reflected in the gradual accumulation of the metal in target organs with eventual tissue dysfunction. The metabolism of cadmium is broadly similar for most mammalian species. Absorption from the diet is low, <5% of the ingested cadmium, but diets low in calcium and iron can increase the percentage retained. Low protein, zinc, copper, and vitamin D may have similar effects (Bremner, 1979). Excretion of cadmium in feces will include intestinal epithelium, with bound cadmium, sloughed into the intestine (McLellan et al., 1978). However, this bound cadmium may also be reabsorbed from the gut lumen and lead to delayed absorption long after the cadmium was originally ingested (McLellan et al., 1978). Biliary excretion and urinary excretion of assimilated cadmium can occur, but the amounts are normally very small (<1% of absorbed cadmium), leading to long biological half-lives (100 to 300 days in rats) (Friberg et al., 1974; Bremner, 1979).

Cadmium is transported bound to protein, particularly metallothionein, and the main sites of accumulation are the intestinal mucosa, kidney, and liver (Bremner, 1979). Cadmium retention times are usually quoted as 10 to 30 years in humans, resulting in age accumulation wherein the concentrations in the kidney renal cortex have been shown to increase up to the age of 50 years and then to decline (Ryan et al., 1982). This decline may be associated with renal tubule dysfunction, which leads to increased urinary excretion of cadmium. The most usual sign of nephrotoxicity in proximal tubules is the presence of low molecular weight serum proteins in urine, especially β_2-microglobulin (Copius Peereboom and Copius Peereboom-Stegeman, 1981). Hyperexcretion of calcium and phosphorus may then occur together, with the disturbance of metabolism of these bone minerals leading to osteomalacia and osteoporosis. "Itai-itai" disease is the Japanese name for this syndrome among people (mostly multiparous women over 50 years) living on a rice diet high in cadmium. However, it is unlikely that cadmium exposure alone can be sufficient cause, and "itai-itai" disease is also usually associated with multiple dietary deficiencies (Friberg et al., 1986).

Exposure to cadmium (usually as cadmium chloride) through the diet has been extensively investigated in experimental studies with rats. The major effects including the following: reduction of food and water intake; growth depression; hypertension (increased blood pressure); anemia and bleaching of incisors; renal dysfunction (proteinuria); histological lesions in the kidney; and reduced calcification of bone and osteoporosis (Samarawickrama, 1979; Fielder and Dale, 1983). Hepatotoxicity has been noted in rats but usually occurs only at dose levels higher than those producing nephrotoxicity (Fielder and Dale, 1983). It is possible that mild testicular effects can occur, but the oral doses required are likely to be high, about 50 mg of cadmium per kilogram per day in rats (Fielder and Dale, 1983). Standardization of dose-response data and tissue concentrations is difficult, because of the influence of animal age and sex, other dietary components, the length of the experiment, and whether given in drinking water or food. However, it would seem that, with increasing chronic oral exposure, there would be an increase in the severity of effects ranging from mild symptoms of renal dysfunction and reduced calcification of bone at dietary levels of 10 ppm (dry weight) to more severe tubular lesions in the kidney at dietary concentrations of 50 ppm (Fielder and Dale, 1983). The equivalent range of dose is probably between 1 and 7 mg of cadmium per kilogram per day.

CADMIUM IN TERRESTRIAL HABITATS

Cadmium concentrations tend to be very low in most environmental media. Typical levels in uncontaminated soils are <1 mg/kg, although naturally higher values can occur when the soils are derived from the weathering of parent materials high in cadmium, such as black shales (Jackson and Alloway, 1992). Environmental contamination through atmospheric deposition can lead to much higher soil concentrations. Comparison of the major natural atmospheric source of cadmium, volcanic action (Nriago, 1979), with the major anthropogenic sources, nonferrous metal production, iron and steel making, fuel combustion (coal, oil, gas), and refuse incineration, for the European Community gave relative annual emission figures of 20 metric tons (t)/year, 33 t/year, 34 t/year, 8.5 t/year, and 31 t/year, respectively (Hutton, 1983). These figures indicate the importance of human activity in the global cycling of cadmium. Such emissions over the last 100 years have increased cadmium levels in soils of urban, rural, and industrial sites (Korte, 1982).

The interception and retention of cadmium-bearing atmospheric particulates by leaves, other trapping surfaces (bark, stems, surface litter), and soil with its subsequent uptake through plant roots (Martin et al., 1982) will lead to circulation of the element within the terrestrial ecosystem (Hughes et al., 1980). In this way, many plant-derived substrates with high cadmium concentrations will be provided for animal feeding, leading to elevated concentrations in most invertebrates and small mammals within the contaminated ecosystems. Martin and Coughtrey (1982), although primarily concerned with the biological monitoring of heavy metal pollution, review many examples of the concentrations of cadmium in both invertebrates (especially earthworms, wood lice, snails, and slugs) and small mammals from many different polluted habitats. Hunter et al. (1987a, 1987b, 1987c, 1989) described, in considerable detail, the distribution of cadmium in grassland ecosystems near a copper-cadmium refinery in northwest England. Some of their results are summarized in Table 1 and are both illustrative and representative of the ecological consequences of environmental cadmium contamination. Aerial deposition of cadmium induced proportional increases in most biotic components of the grassland ecosystem studied. Moreover, it is generally regarded as one of the most mobile toxic metals, because its transfer occurs along most terrestrial food chains, and because its transference potential is much greater than those of lead and zinc, for example (Roberts and Johnson, 1978). Cadmium can often be measured in equal or higher concentrations in animal consumers relative to the concentrations in their diet. For example, from Table 1, earthworms had higher concentrations of cadmium than those from the surface soil, and isopods had concentrations greater than those in litter (senescent leaf tissue).

Table 1   Cadmium (mg/kg of Dry Weight) in Grasslands from a Reference (Clean) Site and from Sites Near a Copper-Cadmium Refinery in Northwest England[a]

	No. of samples analyzed from each site	Reference site	1-km site	Refinery sites
Surface soil	40	0.8 ± 0.08[b]	6.9 ± 0.6[b]	15.4 ± 2.3[b]
Creeping bent grass (leaf) (*Agrostis stolonifera*)	100			
Live		0.63 ± 0.06	1.32 ± 0.07	3.3 ± 0.36
Senescent		0.68 ± 0.06	2.06 ± 0.13	10.4 ± 1.7
Earthworms (*Oligochaeta*)	60	4.1 ± 0.3	34.0 ± 2.8	107 ± 24.6
Wood lice (*Isopoda*)	60	14.7 ± 1	130 ± 45.9	231 ± 131
Spiders (*Lycosidae*)	400	2.6 ± 0.3	34.5 ± 5	102 ± 7.5
Beetles (*Carabidae*)	400	0.7 ± 0.1	5.6 ± 0.9	15.1 ± 1.1
Field vole (kidney) (*Microtus agrestis*)	19-23	1.7 ± 0.2	23.9 ± 5.6	88.8 ± 23.3
Common shrew (kidney) (*Sorex araneus*)	20-25	20.5 ± 1.6	156 ± 25	253 ± 75

[a] From Hunter et al., 1987a, 1987b, 1989.
[b] Mean ± standard error.

Significant accumulation can occur in wild small mammal kidney tissue. This is evident from a comparison between the values for the herbivore, field vole (*Microtus agrestis*), and those for live creeping bent grass (*Agrostis stolonifera*) leaf material (Table 1). Insectivorous small mammals, such as the common shrew (*Sorex araneus*), usually contain higher concentrations in the kidney than do herbivorous or omnivorous species in the same habitat (Table 1).

It is also clear that the application of phosphatic fertilizers and sewage sludges and other organic wastes to agricultural land can increase the soil cadmium levels considerably. A recent survey of sludge-treated soils in the United Kingdom showed cadmium concentrations ranging from 0.27 to 158.7 mg/kg (Jackson and Alloway, 1992). Thus, extremely high local concentrations can occur because of the high variability in the cadmium content of sludges and sludge application rates. Dried sludge on vegetation and soil surfaces can also be directly ingested by grazing animals (Chaney et al., 1987). It is possible for cadmium in sludge-treated grasslands to accumulate in the kidneys and livers of small mammals as field experiments with the meadow vole (*Microtus pennsylvanicus*) in Ohio have shown (Anderson et al., 1982; Maly, 1984).

Nonferrous metal mines, especially those exploiting lead-zinc ores, and the activities involved in processing the ores can be a significant source of environmental contamination through the disposal of cadmium-rich wastes (Roberts and Johnson, 1978). Wind dispersal of waste from spoil tips at a derelict mine complex was shown to lead to considerable increases in the cadmium concentrations in herbivorous and carnivorous invertebrates and indigenous small mammals (Roberts and Johnson, 1978). Considerable mobility of cadmium within the soil-plant-animal system can also occur after the reclamation of industrial waste disposal sites to grassland (Andrews and Cooke, 1984).

TISSUE CONCENTRATIONS IN WILD SPECIES

In a recent review that combined data from 13 studies between 1974 and 1987, the average concentrations (on a dry weight basis) from reference ("clean") sites for the whole body, liver, and kidney were 0.1 to 1.4 mg/kg, 0.2 to 1.5 mg/kg, and <0.1 to 5.6 mg/kg, respectively (Talmage and Walton, 1991). This data set included mice, rats, and voles but excluded moles and shrews that accumulate higher cadmium levels from their insectivorous diet. For example, shrews from uncontaminated sites had concentrations of 1.2 to 4 mg/kg (whole body), 2.9 to 25.4 mg/kg (liver), and 4.1 to 25.7 mg/kg (kidney) (Talmage and Walton, 1991). When small mammals are found in contaminated sites, the increased dietary concentrations lead to high concentrations in a range of tissues and organs, particularly the kidney and liver, regardless of whether the contamination is primarily derived from polluted soil or atmospheric deposition. Lesser increases (but often statistically significantly higher than at reference sites) are found in the heart, femur, muscle, pancreas, and hair (Andrews et al., 1984; Hunter et al., 1989). There seems little evidence of increased cadmium in the brain, lung, and testis, even at highly contaminated sites (Hunter et al., 1989).

Nearly all measurements of cadmium in the tissues of small mammals caught in the wild have shown that the kidney has the highest concentrations, with the liver being the next most important (Table 2). Table 2 also shows that, for most species, the kidney/liver concentration ratios are between 2.3 and 8.4 The obvious exceptions in Table 2 are the common shrew and European mole (both insectivores), where the ratio is below 1.0, with both the kidney and liver concentrations considerably higher than those for the other five species listed in Table 2.

It is probable that there is a strong relation between the intake of cadmium in the diet and the levels of this element in the kidney and liver, both responding by accumulating and storing the cadmium. However, there is a paucity of reliable data available that relate the dietary concentrations and dose to the tissue concentrations in wild small mammals. Table 3 summarizes some recent data and the response of three species found in contaminated grasslands near a copper-cadmium refinery, compared with a clean reference site (Hunter et al., 1987c, 1989). In two species, *Apodemus sylvaticus* and *M. agrestis,* exposed to a similar increase in dietary concentrations (0.7 to 3.3 mg/kg of dry weight) or daily dose (0.2 to 2 mg/kg of body weight), a considerable increase in the kidney and liver concentrations occurred. *S. araneus* received a much higher level of dietary cadmium than did the

Table 2 Kidney and Liver Cadmium Concentrations in Various Species of Small Mammals from Contaminated Habitats

Species	Contaminated site	Concentration (mg/kg of dry weight) Kidney	Liver	Ratio kidney/liver	Ref.
Bank vole (Clethrionomys glareolus)	Dulowa Forest, Poland	29.6	12.8	2.3	Sawicka-Kapusta et al. (1990)
	Y Fan, Pb/Zn mine, Wales	16.8	5.1	3.3	Johnson et al. (1978)
Wood mouse (Apodemus sylvaticus)	Smelter waste, Wales	18.0	5.5	3.3	Johnson et al. (1978)
	Minera, Pb/Zn mine, Wales	39.7	9.8	4.1	Johnson et al. (1978)
	Y Fan, Pb/Zn mine, Wales	10.3	2.49	4.1	Johnson et al. (1978)
	Cu/Cd refinery, England	41.7	18.2	2.3	Hunter et al. (1989)
	Fluorspar waste, England	1.78	0.71	2.5	Cooke et al. (1990)
Short-tailed field vole (Microtus agrestis)	Y Fan, Pb/Zn mine, Wales	8.91	1.06	8.4	Johnson et al. (1978)
	Cu/Cd refinery, England	88.8	22.7	3.9	Hunter et al. (1989)
	Fluorspar waste, England	5.3	1.8	2.9	Andrews et al. (1984)
	Budel, the Netherlands	2.7	0.57	4.7	Ma et al. (1991)
Meadow vole (Microtus pennsylvanicus)	Sludge-treated fields, USA	23[a]	7.9[a]	2.9	Anderson et al. (1982)
White-footed mouse (Peromyscus leucopus)	Wastewater-irrigated site, USA	2.3[a]	0.5[a]	5.0	Anthony and Kozlowski (1982)
Common shrew (Sorex araneus)	Cu/Cd refinery, England	253	578	0.43	Hunter et al. (1987)
	Fluorspar waste, England	158	236	0.67	Andrews et al. (1984)
	Budel, the Netherlands (Oct.-Nov.)	126	180	0.7	Ma et al. (1991)
	Budel, the Netherlands (Feb.-Mar.)	200	268	0.75	Ma et al. (1991)
European mole (Talpa europea)	Budel, the Netherlands	224	227	0.99	Ma (1987)

[a] Values changed from wet weight (dry weight = wet weight × 3.5).

other two species occupying the same grasslands in terms of both concentration and estimated dose (Table 3). The difference in the dietary concentration between *Sorex* and the other two species increased from 2 times at the clean reference site to 18 times at the site closest to the refinery. This is because of the contrasting nature of feeding habits between the species and in the relatively high concentrations reached in ground-dwelling invertebrates, the prey of the shrew (Hunter et al., 1987b) as shown in Table 1. The response by the shrew to the higher cadmium intake rates is to accumulate higher concentrations in the liver compared with those in the kidney, so that the kidney/liver ratios are above one at the reference site but well below one at the two contaminated sites (Table 3). A very similar pattern in the distribution of cadmium between the kidney and the liver was found for these three species occupying a cadmium-contaminated grassland established on fluorspar tailings (Cooke et al., 1990).

Table 3 Estimated Dietary Concentrations and Daily Cadmium Intake, as well as Concentrations in Kidney and Liver, for Three Species of Small Mammals[a]

Species	Site	Dietary concentration (mg/kg of dry weight)	Cadmium dose (mg/kg body weight per day)	Kidney concentration (mg/kg of dry weight)	Liver concentration (mg/kg of dry weight)	Kidney/liver ratio
Wood mouse (Apodemus sylvaticus)	1	0.89	0.22	2.0	0.4	5
	2	1.4		8.5	1.8	4.7
	3	3.01	0.94	41.7	18.2	2.3
Short-tailed field vole (Microtus agrestis)	1	0.67	0.35	1.7	0.7	2.4
	2	1.3		23.9	8.7	2.7
	3	3.3	2.11	88.8	22.7	3.9
Common shrew (Sorex araneus)	1	1.78	0.58	20.5	13.6	1.5
	2	16.0		156	245	0.63
	3	55.0	25.0	253	578	0.43

From Hunter et al., 1987c, 1989.
Site 1, reference site; site 2, 1 km from Cu/Cd refinery; site 3, close to Cu/Cd refinery.

The long biological half-lives of cadmium in the kidney and liver (Bremner, 1979) mean that cadmium tissue concentrations do depend on the age of the animal. Age accumulation has been suggested by the significant positive relations between total body burdens and body weight for a number of small mammal species from clean and contaminated sites (Schlesinger and Potter, 1974; Cooke et al., 1990). Shrews from contaminated sites show highly significant differences in the kidney and liver cadmium concentrations between juveniles and adults and significant positive relations between the concentrations in these organs and body weight (Hunter et al., 1989). The relation between age and cadmium concentrations must be recognized when interpreting tissue concentrations such as those given in Table 2, where sample populations are not separated into age groups.

In young animals and individuals with low dietary cadmium, the kidney has higher concentrations than does the liver and can be regarded as the target organ. If cadmium oral exposure is continual, then this probably remains the case for most small mammal species. However, this may change with high exposures, with kidney concentrations tending to reach a plateau level with the liver concentrations still increasing (Bremner, 1979). It has been shown in field-caught shrews (*S. araneus*) that, with increasing total body burdens of cadmium, the kidney/liver concentration ratios decrease, because the proportion of the body burden in the kidney decreases while that of the liver increases as the total body burden rises (Andrews and Cooke, 1984; Hunter et al., 1989; Ma et al., 1991). This has recently been confirmed in feeding trials of a laboratory population of *S. araneus* fed cadmium chloride-contaminated diets, where the proportion of the total body burden (over the range of 60 to 400 μg of cadmium) declined in the kidney from 17 to 8% and increased in the liver from about 60 to 80% (Dodds-Smith et al., 1992). As well as being a response to the magnitude of dietary cadmium, it may be that species differences highlighted in Tables 2 and 3 reflect differences in the cadmium bioavailability between a plant-based diet as compared with one containing invertebrate prey. Differences in dietary cadmium speciation (metallothionein or phytochelatin bound or as inorganic cadmium) could lead to differences in the tissue levels of cadmium and the relative tissue distribution in relation to the increasing dietary dose of cadmium (Jackson and Alloway, 1992). However, similarities in the tissue levels of cadmium and in the relative distribution between the kidney and the liver in field-caught shrews feeding on invertebrates compared with animals from laboratory experiments using cadmium chloride diets might suggest that basic physiological differences between species are more important than the dietary form of cadmium (Dodds-Smith et al., 1992).

CRITICAL TISSUE CONCENTRATIONS

It is broadly accepted that the kidney is the critical organ in mammalian cadmium toxicity in that it is the first organ in which damage is observed or adverse functional changes start to occur. In humans, one third to one half of the total body burden of cadmium may be in the kidney, with the concentration in the cortex being 1.25 times that of the whole organ (Friberg et al., 1986). The critical concentration of cadmium in the human kidney cortex at which tubular dysfunction and/or morphological kidney changes occur has been given as 200 mg/kg of wet weight (Friberg et al., 1974). This threshold has been reevaluated using in vivo neutron activation analysis and various metabolic models and is thought to remain the best estimate for the critical concentration at which renal dysfunction is likely to occur in 10% of the exposed human population (Friberg et al., 1986). In summaries of critical concentrations of cadmium in the renal cortex in experimental animals, Nomiyama (1986) cites similar values of 200 to 300 mg/kg of wet weight for proteinuria in mice, rats, and rabbits, and Fielder and Dale (1983) give 150 mg of cadmium per kilogram of wet weight as the critical concentration for tubular dysfunction following p.o. exposure. This number of 150 mg/kg can be approximated to 120 mg of cadmium per kilogram in the whole kidney on a wet weight basis or 420 mg of cadmium per kilogram on a dry weight basis. However, in relation to the other cadmium-induced effects as reviewed below, it can be proposed that these concentrations are perhaps a little too high and should be rounded downward to 100 mg/kg of wet weight and 350 mg/kg of dry weight.

A number of experimental studies on laboratory rodents exposed via the p.o. or s.c. route would suggest that evidence of toxicity can be found at lower kidney concentrations. Values as low as 30 to 60 mg/kg of wet weight in the rat kidney have been associated with proteinuria (Prigge, 1978), changes in urinary excretion of trace elements (Chmielnicka et al., 1989), and cell necrosis and degenerative changes in the proximal tubules (Itokawa et al., 1978; Aughey et al., 1984). At similar kidney concentrations in mice, loss of ionic regulation in proximal tubules has been indicated (Kendall et al., 1984), and severe tissue damage has been shown (Nicholson et al., 1983). In the latter study, although the animals were outwardly healthy, there were marked changes in the proximal tubules including cell necrosis, nuclear pyknosis, and mitochondrial swelling.

In one of the few electron microscopy studies of wild animals, *S. araneus* from a polluted smelter site was shown to have widespread damage of kidney and liver tissue (Hunter et al., 1984). Over the range of 150 to 560 mg/kg of dry weight in the kidney, the degree of ultrastructural damage in proximal tubule cells, including enlargement of apical cytoplasmic vesicles and scattered cell necrosis, was found to correlate with the tissue concentration of cadmium. Limited glomerular damage also occurred throughout the full age range of the population; i.e., this damage did not correlate with the kidney concentration. Corresponding liver concentrations of cadmium (300 to 1000 mg/kg) were associated with damage to hepatocytes that showed disrupted rough endoplasmic reticulum, swollen mitochondria, dilation of the smooth endoplasmic reticulum, and invagination of nuclei. Hepatocytes from adults contained numerous electron-dense cytoplasmic inclusion bodies with much of the cadmium bound to metallothionein (Hunter et al., 1984). Notwithstanding the tissue damage to both kidneys and liver, no evidence for clinical renal dysfunction could be found from analysis of urine, and the animals were seemingly in good condition when caught in the field (Hunter, 1984).

One of the major difficulties in understanding the effects of cadmium in wildlife is whether kidney damage (and that to liver and other tissues) is related to ecological fitness. The kidney does have spare functional capacity, and its regenerative capacity is great (Nicholson and Osborn, 1983). Similarly, cadmium-induced proteinuria, although it may persist throughout life, may be a condition that can be tolerated and is not necessarily indicative of progression to a more serious condition, such as renal failure. Because of these uncertainties, it has recently been concluded, in terms of environmental exposure, that cadmium is an element "looking for a disease to cause" (Davies, 1992). However, it can be argued with equal validity that tissue damage in wild small mammals subject to predation, food shortages, low temperatures, etc., can be of much greater significance than equivalent damage in relatively inactive, disease-free, well-fed laboratory rodents. It is well known that dietary deficiencies can considerably exacerbate cadmium toxicity (Bremner, 1979). If the proposal of 100 mg of cadmium per kilogram of wet weight or 350 mg of cadmium per kilogram of dry weight in the whole kidney is taken as an appropriate critical cadmium concentration in wild mammals, it can be seen from Table 2 that few wild small mammal species will attain these levels, even in highly contaminated habitats. It is the insectivorous species, such as the shrews (Soricidae) and moles (Talpidae), that would seem to be the important indicators and at the greatest toxicological risk.

SUMMARY

Cadmium is a widespread element in the environment, but at low concentrations. Increased cadmium in the terrestrial habitats of small mammals is derived from a variety of anthropogenic sources including atmospheric deposition, the application of phosphatic fertilizers and sewage sludges to land, and disused mine waste. Field studies have shown that cadmium, whether derived from the atmosphere or soil, is commonly found in most biotic components within a terrestrial ecosystem.

The concentrations measured in wild small mammals caught in contaminated sites show elevated cadmium in many tissues and organs, but most of the body burden is in the kidney and liver. Kidney concentrations ranged from 2 to 90 mg of cadmium per kilogram of dry weight in mice and voles and were 2 to 8 times the corresponding liver value. In shrews, liver values were usually higher, at 200 to 600 mg of cadmium per kilogram of dry weight as against kidney concentrations of 100 to 250 mg of cadmium per kilogram.

Using response criteria similar to those for human cadmium exposure and using primarily laboratory experiments with rats and mice, we suggest that 100 mg of cadmium per kilogram of wet weight or 350 mg of cadmium per kilogram of dry weight could be considered as the critical kidney concentration on a whole-organ basis. However, from field-based studies to date, it would seem that these concentrations will not be reached in species other than insectivores, such as shrews and moles. However, there are too few laboratory studies on wild species to enable any firm conclusions regarding dose-response characteristics or real hypotheses concerning the broader ecological consequences of environmental cadmium.

REFERENCES

Anderson, T. J., G. W. Barrett, C. S. Clark, V. J. Elia, and V. A. Majeti. 1982. Metal concentrations in tissues of meadow voles from sewage sludge-treated fields. J. Environ. Qual. 11:272-277.

Andrews, S. M., and J. A. Cooke. 1984. Cadmium within a contaminated grassland ecosystem established on metalliferous mine waste. p. 11-15. In D. Osborn (Ed.). Metals in animals. Inst. Terrestrial Ecol. Publ. No. 12. Natural Environment Research Council, Cambridge, U.K.

Andrews, S. M., M. S. Johnson, and J. A. Cooke. 1984. Cadmium in small mammals from grassland established on metalliferous mine waste. Environ. Pollut. Ser. A Ecol. Biol. 33:153-162.

Anthony, R. G., and R. Kozlowski. 1982. Heavy metals in tissues of small mammals inhabiting waste-water-irrigated habitats. J. Environ. Qual. 11:20-22.

Aughey, E., G. S. Fell, R. Scott, and M. Black. 1984. Histopathology of early effects of oral cadmium in the rat kidney. Environ. Health Perspect. 54:153-161.

Bremner, I. 1979. Mammalian absorption, transport, and excretion of cadmium. p. 175-193. In M. Webb (Ed.). The chemistry, biochemistry, and biology of cadmium. Elsevier/North-Holland Biomedical Press, Amsterdam.

Chaney, R. L., R. J. Bruins, D. E. Baker, R. F. Korcak, J. E. Smith, and D. Cole. 1987. Transfer of sludge-applied trace elements to the food chain. In A. L. Page, T. J. Logan, and J. A. Ryan (Eds.). Land application of sludge: food chain implications. Lewis Publishers, Chelsea, Mich.

Chmielnicka, J., T. Halatek, and U. Jedlinska. 1989. Correlation of cadmium-induced nephropathy and the metabolism of endogenous copper and zinc in rats. Ecotoxicol. Environ. Saf. 18:268-276.

Cooke, J. A., S. M. Andrews, and M. S. Johnson. 1990. Lead, zinc, cadmium, and fluoride in small mammals from contaminated grassland established on fluorspar tailings. Water Air Soil Pollut. 51:43-54.

Copius Peereboom, J. W., and J. H. J. Copius Peereboom-Stegeman. 1981. Exposure and health effects of cadmium. Part 2. Toxic effects of cadmium to animals and man. Toxicol. Environ. Chem. Rev. 4:67-178.

Davies, B. E. 1992. Trace metals in the environment: retrospect and prospect. p. 1-17. In D. C. Adriano (Ed.). Biogeochemistry of trace metals. Lewis Publishers, Boca Raton, Fla.

Dodds-Smith, M. E., M. S. Johnson, and D. J. Thompson. 1992. Trace metal accumulation by the shrew, *Sorex araneus*. II. Tissue distribution in kidney and liver. Ecotoxicol. Environ. Saf. 24:118-130.

Fielder, R. J., and E. A. Dale. 1983. Cadmium and its compounds. Toxicity Review No. 7. Health and safety executive. Her Majesty's Stationery Office, London.

Friberg, L., T. Kjellstrom, and G. F. Nordberg. 1986. Cadmium. In L. Friberg, G. F. Nordberg, and V. Vouk (Eds.). Handbook on the toxicology of metals. 2nd ed. Elsevier Science Publishers, Amsterdam.

Friberg, L., M. Piscator, G. F. Nordberg, and T. Kjellstrom. 1974. Cadmium in the environment. 2nd ed. CRC Press, Cleveland, Ohio.

Hughes, M. K., N. W. Lepp, and D. A. Phipps. 1980. Aerial heavy metal pollution in terrestrial ecosystems. Adv. Ecol. Res. 11:218-327.

Hunter, B. A. 1984. The Ecology and Toxicology of Trace Metals in Contaminated Grasslands. Ph.D. thesis. University of Liverpool, Liverpool, England.

Hunter, B. A., M. S. Johnson, and D. J. Thompson. 1984. Cadmium induced lesions in tissues of *Sorex araneus* from metal refinery grasslands. p. 39-44 In D. Osborn (Ed.). Metals in animals. Inst. Terrestrial Ecol. Publ. No. 12. Natural Environment Research Council, Cambridge, U.K.

Hunter, B. A., M. S. Johnson, and D. J. Thompson. 1987a. Ecotoxicology of copper and cadmium in a contaminated grassland ecosystem. I. Soil and vegetation contamination. J. Appl. Ecol. 24:573-586.

Hunter, B. A., M. S. Johnson, and D. J. Thompson. 1987b. Ecotoxicology of copper and cadmium in a contaminated grassland ecosystem. II. Invertebrates. J. Appl. Ecol. 24:587-599.

Hunter, B. A., M. S. Johnson, and D. J. Thompson. 1987c. Ecotoxicology of copper and cadmium in a contaminated grassland ecosystem. III. Small mammals. J. Appl. Ecol. 24:601-614.

Hunter, B. A., M. S. Johnson, and D. J. Thompson. 1989. Ecotoxicology of copper and cadmium in a contaminated grassland ecosystem. IV. Tissue distribution and age accumulation in small mammals. J. Appl. Ecol. 26:89-99.

Hutton, M. 1983. Sources of cadmium in the environment. Ecotoxicol. Environ. Saf. 7:9-24.

Itokawa, Y., K. Nishino, M. Takashima, T. Nakata, H. Kaito, E. Okamoto, K. Daijo, and J. Kawamura. 1978. Renal and skeletal lesions in experimental cadmium poisoning of rats. Histology and renal function. Environ. Res. 15:206-217.

Jackson, A. P., and B. J. Alloway. 1992. The transfer of cadmium from agricultural soils to the human food chain. p. 109-158. In D. C. Adriano (Ed.). Biogeochemistry of trace metals. Lewis Publishers, Boca Raton, Fla.

Johnson, M. S., R. D. Roberts, M. Hutton, and M. J. Inskip. 1978. Distribution of lead, zinc, and cadmium in small mammals from polluted environments. Oikos 30:153-159.

Kendall, M. D., J. K. Nicholson, and A. Warley. 1984. The effects of cadmium on the concentrations of some naturally occurring intracellular elements in the kidney. p. 50-54. In D. Osborn (Ed.). Metals in animals. Inst. Terrestrial Ecol. Publ. No. 12. Natural Environment Research Council, Cambridge, U.K.

Korte, F. 1982. Ecotoxicology of cadmium. Regul. Toxicol. Pharmacol. 2:184-208.

Ma, W. C. 1987. Heavy metal accumulation in the mole *Talpa europea* and earthworms as an indicator of metal bioavailability in terrestrial environments. Bull. Environ. Contam. Toxicol. 39:933-938.

Ma, W. C., W. Denneman, and J. Faber. 1991. Hazardous exposure of ground-living small mammals to cadmium and lead in contaminated terrestrial ecosystems. Arch. Environ. Contam. Toxicol. 20:266-270.

Maly, M. S. 1984. Survivorship of meadow voles *Microtus pennsylvanicus* from sewage sludge-treated fields. Bull. Environ. Contam. Toxicol. 32:724-731.

Martin, M. H. and P. J. Coughtrey. 1982. Biological monitoring of heavy metal pollution. Applied Science Publishers, London.

Martin, M. H., E. M. Duncan, and P. J. Coughtrey. 1982. The distribution of heavy metals in a contaminated woodland ecosystem. Environ. Pollut. Ser. B Chem. Phys. 3:147-157.

McLellan, J. S., P. R. Flanagan, M. J. Chamberlain, and L. S. Valberg. 1978. Measurement of dietary cadmium absorption in humans. J. Toxicol. Environ. Health 4:131-138.

Nicholson, J. K., M. D. Kendall, and D. Osborn. 1983. Cadmium and nephrotoxicity. Nature (Lond.) 304:633-635.

Nicholson, J. K., and D. Osborn. 1983. Kidney lesions in pelagic seabirds with high tissue levels of cadmium and mercury. J. Zool. (Lond.) 200:99-118.

Nomiyama, K. 1986. The chronic toxicity of cadmium: influence of environmental and other variables. p. 101-133. In E. C. Foulkes (Ed.). Handbook of experimental pharmacology. Vol. 80. Springer-Verlag, Berlin.

Nriago, J. O. 1979. Global inventory of natural and anthropogenic sources of metals in the atmosphere. Nature (Lond.) 279:409-411.

Prigge, E. 1978. Early signs of oral and inhalative cadmium uptake in rats. Arch Toxicol. 40:231-247.

Roberts, R. D., and M. S. Johnson. 1978. Dispersal of heavy metals from abandoned mine workings and their transference through terrestrial food chains. Environ. Pollut. 16:293-310.

Ryan, J. A., H. R. Pahren, and J. B. Lucas. 1982. Controlling cadmium in the human food chain: a review and rationale based on health effects. Environ. Res. 28:251-302.

Samarawickrama, G. P. 1979. Biological effects of cadmium in mammals. p. 341-421. In M. Webb (Ed.). The chemistry, biochemistry, and biology of cadmium. Elsevier/North-Holland Biomedical Press, Amsterdam.

Sawicka-Kapusta, K., R. Swiergosz, and M. Zakrzewska. 1990. Bank voles as monitors of environmental contamination by heavy metals. A remote wilderness area in Poland imperilled. Environ. Pollut. 67:315-324.

Schlesinger, W. H., and G. L. Potter. 1974. Lead, copper, and cadmium concentrations in small mammals in the Hubbard Brook Experimental Forest. Oikos 25:148-152.

Talmage, S. S., and B. T. Walton. 1991. Small mammals as monitors of environmental contaminants. Rev. Environ. Contam. Toxicol. 119:47-145.

CHAPTER 17

Cadmium in Birds

Robert W. Furness

INTRODUCTION

Cadmium, in association with zinc, is widely distributed in the earth's crust. Cadmium and zinc are commercially produced by smelting their mixed ores and, since the 1950s, about 18,000 tons of cadmium per year have been used in industrial processes such as plastic production, electroplating, and the manufacture of alloys and batteries. Almost all of this cadmium is eventually released into the environment. When added together with the dust and wastewater from smelting and refining, the cadmium from phosphate rocks, fertilizers, coal, and oils that is introduced into the environment by human activities totals 30,000 tons per year (Nriagu and Pacyna, 1988).

Cadmium deposits may be found on the land and in the air and waters. Some cadmium is deposited on terrestrial environments, especially those close to smelters. It also is deposited on land from the wear of vehicle tires, which contain cadmium at concentrations of about 50 mg/kg. Deposits of cadmium in the air are less important than those of mercury and lead, because cadmium is less volatile than these metals. Hence, atmospheric transport is less important for cadmium, but riverine transport is relatively more important (Furness and Rainbow, 1990). Sewage sludge may contain cadmium at concentrations up to 10 to 30 mg/kg. Much cadmium is deposited in bottom sediments on the continental shelf, but levels of dissolved cadmium in seas and oceans may be increasing (Nriagu and Pacyna, 1988). In addition, acidification of freshwaters will increase the concentrations of dissolved cadmium, resulting in greater amounts of cadmium being available for incorporation into the food chain (Scheuhammer, 1991).

Cadmium is not a nutritionally essential element for animals. Exposure to acutely high, or chronically low, levels may induce intracellular production of metallothionein, a low-molecular-weight protein rich in sulfur amino acids to which cadmium can be bound and, hence, rendered less toxic; metallothionein also has an important function as a store in zinc metabolism (Chakraborty et al., 1987; Fernando et al., 1989). A high accumulation of cadmium can, but does not always, lead to food chain amplification (increases in cadmium concentrations in animals at each step in the food chain), because metallothionein-bound cadmium has a long biological half-life in animals and because concentrations tend to increase with age. Molluscs have a particular tendency to accumulate large amounts of cadmium (Furness and Rainbow, 1990; Vermeer and Castilla, 1991), so it may be anticipated that, in unpolluted environments, long-lived birds feeding on molluscs (including cephalopods) may accumulate high concentrations of cadmium. Cadmium pollution can thus be anticipated to most likely affect mollusc-eating birds in enclosed coastal areas with high inputs of cadmium in sewage sludge or from smelter or refinery discharges (Vermeer and Castilla, 1991).

However, one may also anticipate that birds naturally exposed to high levels of cadmium through a diet of molluscs may have evolved greater tolerance to cadmium than that found in other bird species. Thus, it would be difficult to suggest critical levels of cadmium in diets or in tissues that are applicable to all birds. For this reason, I consider in this chapter the range of cadmium levels found in a wide variety of free-living birds, so that the influences of pollution and toxic effects can be set in the context of an enormous natural variation in cadmium accumulation. It is likely that a similar pattern and extent of variation in sensitivity to the toxic effects of cadmium also exist.

TISSUE DISTRIBUTION IN BIRDS

All cadmium concentrations reported in this chapter are given as parts per million (mg/kg), wet weight, of tissue, unless explicitly stated otherwise alongside the value cited.

Cadmium concentrations in birds are almost always highest in the kidney, lower in the liver, and very low in muscle (Nicholson, 1981; Thompson, 1990). Concentrations in eggs tend to be extremely low (Burger and Gochfeld, 1991), often below most atomic absorption spectrophotometry analytical detection limits. It is often not clear whether cadmium in feathers is derived from deposition into growing feathers from circulating cadmium in the blood, or whether it is all from atmospheric or aqueous deposition onto feather surfaces (Hahn, 1991), so feathers provide little or no information on cadmium accumulation from food. Mayack et al. (1981) found a highly significant correlation between the concentration of cadmium in the diet and the concentration of cadmium in feathers grown by captive young wood ducks, but these authors did not find any increase in blood cadmium in the dosed groups and did not consider the possibility (which seems likely) that feather cadmium concentrations were increased as a result of deposition of cadmium dust from the food onto the feather surface.

For healthy adult birds in wild populations, the concentration of cadmium in the liver is usually between one half and one tenth of the concentration in the kidney of the same bird (Lee et al., 1987; Thompson, 1990; Lock et al., 1992). Although cadmium concentrations are usually highest in kidney tissue, Scheuhammer (1987) advocated using the liver in monitoring biological exposure to cadmium, since the liver accumulates about one half of the total body burden of cadmium and since the cadmium content of the liver is extremely stable, probably because the liver is generally resistant to the toxic effects of cadmium. By contrast, concentrations of cadmium in the kidney can fall considerably after cadmium-induced tubular dysfunction (White et al., 1978; Goyer et al., 1984), so that low concentrations of cadmium in the kidney may indicate low exposure but may also result from the toxic effects of high exposure. Scheuhammer (1987) suggested that comparison of liver and kidney concentrations of cadmium may indicate cases of acute exposure to cadmium at high levels, because birds in such a situation would be expected to have liver concentrations of cadmium as high as or higher than those found in the kidney.

Concentrations of cadmium in bird muscle are so low that they are rarely reported. A few cases in the literature show muscle cadmium concentrations in adult birds to be around 0.2 to 5% of kidney concentrations (e.g., Osborn et al., 1979; Cheney et al., 1981; Leonzio et al., 1986; Nielsen and Dietz, 1989). Laboratory experiments with dosing show that a high intake of cadmium does result in elevated concentrations of cadmium in muscle, though to a much smaller extent than in the kidney and liver (Leach et al., 1979).

UPTAKE OF CADMIUM

The few studies of cadmium uptake by birds that have been conducted have used birds in controlled laboratory conditions and are, thus, of uncertain value as a model for wild birds, because the range of bird species used in laboratory studies is limited and particularly because the assimilation and toxicity of cadmium vary considerably according to the nutritional status of the bird. Stoewsand et al. (1986) showed that Japanese quail fed a diet containing powdered cadmium-contaminated earthworms accumulated cadmium in their liver and kidney but that concentrations of cadmium did not increase in the eggs. Intestinal uptake of cadmium (as Cd^{2+}) by Japanese quail in laboratory conditions was dose-dependent and represented about 0.4 to 2% of the dose (Scheuhammer, 1987, 1988). However, Koo et al. (1978) found that the intestinal uptake of cadmium was affected by calcium, cholecalciferol, and phosphorus. Rambeck and Kollmer (1990) found that the accumulated concentration of cadmium could vary 16-fold in the liver and 11-fold in the kidney of chickens according to the amount of iron, copper, vitamin C, or selenium in the diet.

Much of the ingested cadmium becomes bound in the intestinal epithelium, especially at low exposure. In laboratory experiments with Japanese quail, Scheuhammer (1988) found that, 4 days after p.o. administration of four daily doses totaling less than 20 mg/kg, about 0.7% of the dose could be found in the liver + kidneys + duodenum, while doses totaling 200 mg/kg led to 2% of the dose being in the liver + kidneys + duodenum after 4 days. Only at the highest dose did the liver cadmium concentration exceed the kidney cadmium concentration, and only at this dose did metallothionein concentrations in the liver and kidney increase above control concentrations. The increased absorption of cadmium at the highest dose may reflect direct toxic effects on the intestinal epithelium (Richardson and Fox, 1975), allowing cadmium to pass into the blood, to become associated with plasma albumen, and to be transported primarily to the liver rather than to the kidney. Scheuhammer (1988) found that the normal laboratory diet of the quail had a high zinc content and that this probably induced metallothionein production in the duodenum, which may in turn have influenced cadmium uptake. In wild birds, cadmium uptake could also be influenced by injury to the intestinal epithelium or by deficiencies in the dietary levels of calcium, zinc, or iron. These could all cause increased cadmium uptake (Scheuhammer, 1987). Thus, with only very limited information on laboratory uptake, it would be unwise to suggest a dietary threshold at which cadmium damage to the intestinal epithelium of wild birds would be expected. However, the laboratory studies suggest that a dietary intake of less than 1 mg of cadmium per kg of food would be most unlikely to lead to damage to intestinal epithelium or to any other toxic effects.

CONCENTRATIONS IN WILD BIRDS

Cadmium concentrations in healthy wild birds vary widely among species and among populations within species, with mean levels of <0.1 to 32 mg/kg in the liver and <0.3 to 137 mg/kg in the kidney (Walsh, 1990) (Table 1). In supposedly unpolluted areas, cadmium concentrations are consistently several orders of magnitude higher in pelagic seabirds than in terrestrial birds and, among seabirds, concentrations in squid-eating or insect-eating species are generally highest (Bull et al., 1977; Cheng et al., 1984; Muirhead and Furness, 1988; Honda et al., 1990; Elliott et al. 1992). The highest individual concentrations of cadmium measured in the kidneys of apparently healthy adult birds were 275 mg/kg in a Greenland kittiwake, 183 mg/kg in a Greenland glaucous gull (Nielsen and Dietz, 1989), 166 mg/kg in a macaroni penguin (Norheim, 1987), and 148 mg/kg in a wandering albatross (Muirhead and Furness, 1988). Levels in sea ducks and mollusc-eating waders are higher than in terrestrial birds, but much lower than in many seabirds. Pigeons from close to Heathrow, presumably exposed to cadmium from tire vaporization from landing aircraft, had cadmium concentrations in their kidneys considerably higher than in control pigeons from a less polluted area, but still far less than found in most seabirds from unpolluted sites. Similarly, passerines from areas of Korea close to heavy industry had higher cadmium levels than did birds from rural sites, but levels in the polluted birds were still orders of magnitude less than found in many seabirds.

Despite enormous variation in cadmium levels among species, intraspecific variation and skewness tend to be lower than for other nonessential metals, such as mercury and lead, for populations of seabirds though not for waders, wildfowl, or passerines (Karlog et al., 1983; Di Giulio and Scanlon, 1984b; Lee et al., 1989; Walsh, 1990). Walsh (1990) suggests that this may be an evolutionary consequence of long-term exposure to high natural levels of cadmium in oceanic food chains leading to seabirds evolving greater ability to regulate tissue concentrations of cadmium.

Di Giulio and Scanlon (1984b) found that carnivorous ducks had higher concentrations of cadmium than did herbivorous or omnivorous species, while Lee et al. (1989) found cadmium concentrations to increase in the sequence: terrestrial carnivores

Table 1 Levels of Cadmium Found in Adults in Some Populations of Wild Birds Considered to be Healthy[a]

Species	Considered polluted or not	Cadmium level in Kidney (mg/kg)	Liver (mg/kg)	Ref.
Wandering albatross	No	137	32	Muirhead and Furness (1988)
Sooty albatross	No	76	26	Muirhead and Furness (1988)
Kittiwake	No	76	11	Nielsen and Dietz (1989)
Great shearwater	No	74	15	Muirhead and Furness (1988)
Fulmar	No	55	17	Norheim (1987)
Macaroni penguin	No	49	9	Norheim (1987)
Sooty shearwater	No	41	7	Lee et al. (1987)
Long-tailed duck	?	28	4	Di Giulio and Scanlon (1984b)
Eider	?	12	5	Karlog et al. (1983)
Feral pigeon (Heathrow)	Yes	11	3	Hutton and Goodman (1980)
Oystercatcher	?	6	2	Stock et al. (1989)
"Korean passerines"	Yes	2.4	0.8	Lee et al. (1989)
"Korean waterfowl"	?	0.8	0.2	Lee et al. (1989)
Feral pigeon (Mortlake)	No	0.3	0.1	Hutton and Goodman (1980)

[a] All data are arithmetic means for samples, given as mg/kg of wet weight of tissue. Where the authors stated explicitly that they considered the birds they analyzed to be in an environment with cadmium pollution, the entry in the table notes this. Where cadmium pollution was considered to be negligible, this is recorded. Where cadmium pollution may be present but is not certain, a ? is recorded.

waterfowl, passerines, waders, seabirds. Although seabirds tend to live longer than most other birds, the high cadmium concentrations in seabirds appear to be a consequence of diet as well as age, since cadmium concentrations are generally low in long-lived terrestrial birds and higher among squid-eating seabirds than among fish-eating seabirds of similar longevity (Muirhead and Furness, 1988; Thompson, 1990).

Cadmium concentrations in the liver and kidney of breeding adult birds can vary by a factor of 4 between different stages in the breeding season as a result of changes in tissue physiology (Walsh, 1990). Although cadmium concentrations are usually the same in males and females, a few studies have suggested differences in levels between the sexes. Hutton (1981) and Stock et al. (1989) both found differences in cadmium concentrations between male and female oystercatchers but in opposite directions, so it is not clear if these differences were biologically meaningful.

Not surprisingly, since cadmium concentrations are negligible in egg contents and the biological half-life of cadmium is several years (Blomqvist et al., 1987; Scheuhammer, 1987), cadmium concentrations increase with age in birds. Many studies have shown that cadmium concentrations are higher in adults than in juveniles or immatures, often 10 times and sometimes as much as 100 times higher in adults than in chicks (Hulse et al., 1980; Stoneburner et al., 1980; Cheney et al., 1981; Maedgen et al., 1982; Stock et al., 1989; Lock et al., 1992). This large age effect would suggest that toxic effects would be more likely to arise among adults than among chicks. Among adults, a few studies have measured cadmium concentrations in birds of known age (due to banding of chicks) and found evidence of an age related accumulation in some species though not in others (Furness and Hutton 1979; Walsh, 1990).

It has been surprisingly difficult to show an elevation in cadmium concentrations in birds inhabiting environments supposedly polluted by cadmium. No differences in cadmium concentrations were found between gulls from garbage dumps and those from coastal sites (Leonzio et al., 1986) or between gulls or terns from industrialized sites and those from rural sites in New Zealand (Turner et al., 1978), while Howarth et al. (1981), contrary to their expectations, found significantly higher cadmium concentrations in terns breeding in a nonindustrialized than in an industrialized region in Australia. Pinowska et al. (1981) found higher cadmium concentrations in house sparrows (analyzed by homogenizing whole dried birds) from an industrialized region than from an agricultural region, but this could have been due to atmospheric deposition of cadmium onto the plumage (Hahn, 1991). Reid and Hacker (1982) found higher cadmium concentrations in laughing gulls from an industrialized bay in Texas than in conspecifics from a more rural bay. Brothers and Brown (1987) found a correlation between the cadmium concentrations in prions and the distance of their breeding colony from a site of ocean dumping of over 2 million tons of jarosite (which contains 200 ppm of cadmium). However, the mean concentration was only twice as high in the colony closer to the dump site, and the ranges of concentrations overlapped almost completely. Feral pigeons showed a very clear difference in cadmium concentrations between sites with differing levels of pollution (Table 1), the highest concentrations of cadmium reaching 70 mg/kg in the kidney (Hutton and Goodman, 1980).

TOXIC EFFECTS IN BIRDS

Pritzl et al. (1974) estimated the 50% lethal dose (LD_{50}) for 2-week-old Leghorn chicks to be about 565 mg/kg of dietary cadmium (as powdered cadmium carbonate in a commercial corn-soybean diet).

ALTERED BEHAVIOR

Heinz et al. (1983) showed that elevated dietary cadmium levels could affect the avoidance behavioral response of ducklings.

METALLOTHIONEIN INDUCTION

Cadmium concentrations above a certain tissue threshold level induce production of metallothionein, a low-molecular-weight, sulfhydryl-rich protein (Yamamura and Suzuki, 1984). Accumulation of cadmium bound to metallothionein tends also to lead to accumulation of zinc, since this metal also binds to metallothionein; concentrations of the two metals in the liver and kidney tend to correlate in birds.

cadmium- and zinc-binding protein was isolated by Osborn (1978) from pelagic seabirds with high concentrations of cadmium. With a molecular weight of about 10,000, this was believed to be a metallothionein. Elliott et al. (1992) reported a highly significant positive correlation between the renal cadmium and metallothionein concentration in seabirds from eastern Canada.

Avian metallothionein (molecular weight of approximately 10,000 to 12,000) differs from that found in mammals (molecular weight of 6000 to 7000); chicken liver metallothionein contains a single isoform with one histidine residue, whereas mammalian metallothionein contains two isoforms with no histidine. Metallothioneins in Japanese quail contain two isoforms, the major one appearing similar to that of chickens (Yamamura and Suzuki, 1984), and this is also the case in the pigeon (Lin et al., 1990). Scheuhammer and Templeton (1990) showed that, by giving p.o. doses of cadmium to Barbary doves, metallothionein production is regulated in a similar manner in liver and kidney and that the effects of cadmium, zinc, and copper are additive and noncompetitive in terms of the maintenance of metallothionein levels during chronic exposure to cadmium. Their control birds were fed a diet containing (on a dry weight basis) 0.2 mg/kg of cadmium, 7.4 mg/kg of zinc, and 4 mg/kg of copper. These birds produced only small amounts of metallothionein. Increasing cadmium in the diet to 2 mg/kg approximately doubled the metallothionein concentrations, and increasing cadmium in the diet to 20 mg/kg resulted in a 6-fold increase in metallothionein concentrations compared with those of controls.

Although it is easy to measure cadmium concentrations in the kidneys of dead birds, it is not so easy to assess the toxic hazard that a particular concentration of cadmium represents for a particular species of bird. Interspecific variation in tolerance is probably considerable though largely unknown. Metallothionein concentrations might provide a measure of the bird's response to cadmium in its tissues, though this may be complicated by the presence of other metals (zinc, copper, and to a lesser extent mercury) that may also induce metallothionein synthesis. Metallothionein concentrations might be used as an assay of the toxic potential of cadmium levels in different species of wild birds (Peakall, 1992). Low concentrations of metallothionein should indicate that cadmium concentrations in the bird are not sufficient to cause toxic effects, whereas high concentrations of metallothionein associated with high concentrations of cadmium would imply that metallothionein has been induced to detoxify cadmium. Whether the measurement of metallothionein concentrations can indicate the threshold level of cadmium that is of toxic significance remains to be tested. Elliott et al. (1992) presented data suggesting that seabirds of four species followed a common relation between renal metallothionein and cadmium concentrations.

DISTURBANCES OF IRON, ZINC, AND CALCIUM METABOLISM

It has long been known that intestinal absorption of iron is reduced in birds with an increased dietary intake of cadmium, and that elevated concentrations of cadmium affect iron, zinc, and copper concentrations in tissues (Freeland and Cousins, 1973).

Metallothionein has a role in the binding of zinc, which may be stored for essential functions in reproduction and molt. Induction of metallothionein may cause interactions between cadmium and zinc. Osborn (1978) suggested that the binding of both cadmium and zinc to metallothionein indicates that, rather than metallothionein functioning primarily to detoxify cadmium, cadmium may interfere with normal zinc metabolism through its binding to metallothionein. However, the zinc concentration is homeostatically controlled in liver and kidney tissues of birds, so binding of zinc to metallothionein will simply lead to an increased uptake of zinc or reduced excretion to compensate. Toxic effects due to a deficiency of available zinc seem unlikely. Since zinc is found in many enzymes, cadmium could interfere with zinc by binding to them, and this may cause toxic effects that mimic zinc deficiency. Cosson (1989) concluded that zinc was the most important metal influencing the production of metallothionein-like proteins in wild flamingos and egrets, although copper, zinc, mercury, and cadmium concentrations all showed correlations with metallothionein concentrations in the kidney of flamingos. He supported the earlier suggestion that metallothionein represents a zinc-storage protein rather than a cadmium detoxification protein. However, the relative importance of zinc and cadmium in determining metallothionein concentrations may be different in wild birds exposed to high cadmium levels, as in many seabirds. Elliott et al. (1992) suggested that the cadmium concentration was the main determinant of the metallothionein concentration in seabird kidneys.

Disturbances in iron, zinc, and calcium metabolism can be caused by a chronic dietary exposure to cadmium, and many of the histopathological effects induced in cadmium-dosed Japanese quail were closely similar to the effects induced by dietary deficiencies of iron (anemia, bone marrow hyperplasia, cardiac hypertrophy) or zinc (testicular hypoplasia) (Scheuhammer, 1987).

EFFECTS ON BONE DEVELOPMENT

Cadmium and copper have been shown in tissue culture experiments to induce bone damage through causing osteoporosis, inhibition of calcification, and inhibition of collagen synthesis (Miyahara et al., 1982; Kaji et al., 1986, 1988). The relevance of this for wild birds is not clear, though cadmium concentrations tend to be very low in chicks because virtually none is present in eggs and accumulation from food occurs over a much longer period than chick development.

SUPPRESSION OF EGG PRODUCTION

Cadmium in the diet suppressed egg production by mallards at 200 mg/kg (White and Finley, 1978) and by chickens at 60 mg/kg (Sell, 1975) and 12 or 48 mg/kg (Leach et al., 1979). In the last study, egg production was halved among birds given the higher dose, and these birds had accumulated about 100 mg/kg of cadmium in the kidney and about 40 mg/kg of cadmium in the liver. Such concentrations are similar to the highest found in breeding adult pelagic seabirds showing no signs of impaired egg production (Table 1), suggesting that seabirds chronically exposed to high natural levels of cadmium may be less sensitive to cadmium than are chickens dosed over short periods of time.

EGG SHELL THINNING

Leach et al. (1979) found that a diet containing 48 mg/kg of cadmium resulted in shell thinning of eggs laid by chickens (which accumulated kidney cadmium concentrations of about 110 mg/kg) in one experiment, but not in another where the same dose was given but where kidney concentrations were less elevated (to about 80 mg/kg). In the group given 48 mg/kg of cadmium and reaching the higher kidney concentrations, shell thickness was reduced and the cadmium content of the egg was elevated compared with eggs produced by control birds; however, there were no differences in these parameters in the replicate experiment compared with those of controls. There is no evidence from field studies of pelagic seabird species, with many individuals having kidney cadmium concentrations of 100 mg/kg or more, that such effects arise in wild birds, though no studies have been directed at assessing this. Thus, small effects might not have been noticed. Studies using in vitro preparations of the calcium-secreting part of the eggshell gland mucosa of chickens have shown that several metals, including mercury and cadmium, reduce the rate of ATP-dependent calcium binding and thereby can influence eggshell quality (Lundholm and Mathson, 1986). However, I am not aware of any studies of shell structure or quality in populations of birds with high burdens of cadmium.

KIDNEY DAMAGE

Although cadmium can cause damage to the liver at very high concentrations (Richardson et al., 1974), the critical organ in chronic toxicity is generally considered to be the kidney. Cadmium nephropathy is indicated by proximal tubule cell necrosis, proteinuria, glycosuria, increased urinary cadmium, decreased cadmium content in the kidney, and the appearance of metallothionein in the plasma (White et al., 1978; Mayack et al., 1981; Nicholson and Osborn, 1983; Nicholson et al., 1983; Scheuhammer, 1987). Similar damage can be caused by mercury (Nicholson and Osborn, 1984), and it is difficult to attribute damage in wild birds to any specific metal (Nicholson et al., 1983).

Experimental dosing has been used to cause kidney damage in a number of birds. Adult mallards given dietary cadmium at 20 mg/kg for up to 90 days (kidney accumulation, 50 to 60 mg/kg) showed no kidney damage, but 200 mg/kg in the diet (kidney accumulation, 130 to 140 mg/kg) caused renal tubular necrosis (White et al., 1978). Some mallard ducklings given 20 mg/kg of dietary cadmium did develop kidney damage, suggesting that young birds may be more sensitive than adults (Cain et al., 1983). Starlings given s.c. injections of cadmium chloride accumulated renal cadmium concentrations of 20 to 55 mg/kg and showed extensive kidney damage (podocyte vacuolation, mesangial matrix proliferation of renal corpuscles, cell necrosis, nuclear pyknosis, mitochondrial swelling, tubulorrhexis and some regenerative activity in the proximal tubules, hydropic changes in the distal tubules, chronic cortical inflammation, and cellular debris in the distal nephron lumen) (Nicholson and Osborn, 1983; Nicholson et al., 1983). None of these pathological changes were evident in kidneys of starlings that had not been given metal supplements. These authors also examined kidneys from apparently healthy adult seabirds collected from breeding colonies at a site relatively distant from sources of pollution. Fulmars, Manx shearwaters, and puffins contained cadmium concentrations in the kidneys in the range of 10 to 120 mg/kg and showed very similar kidney damage to that of the starlings experimentally dosed with cadmium, though the damage to seabird kidneys was generally less than that in the starling kidneys with a similar kidney cadmium concentration. Examination of the kidneys of a range of pelagic seabirds from a variety of sites in the North and South Atlantic and New Zealand (Furness, unpublished) found similar kidney damage in pelagic seabirds, in association with high concentrations of cadmium and mercury, as reported by Nicholson and Osborn, but Elliott et al. (1992) reported that Leach's storm petrels and puffins from eastern Canada with similar renal cadmium concentrations but containing rather low concentrations of mercury showed no signs of kidney damage. The apparent contradiction between these studies may be a consequence of the way in which histopathological changes are caused by combinations of metals. Rao et al. (1989) found that the combined administration of methylmercury and cadmium increased the severity of degenerative lesions in kidney proximal tubules.

The birds studied by Nicholson and Osborn (1983) contained renal cadmium concentrations below those generally shown to affect kidney structure in experimental studies using cadmium alone, but they contained rather high concentrations of mercury that may have contributed substantially to the kidney damage.

The birds sampled by Nicholson and Osborn and by Furness were outwardly healthy, in good body condition in terms of lipid and protein reserves, and apparently breeding successfully, so it is not clear whether the kidney damage and consequent tissue regeneration imposes a significant overall fitness cost to the individual. Nicholson and Osborn (1983) suggested that the metal concentrations in the seabirds they studied were natural, and that the damage they observed to the kidneys had no influence on fitness. Such a conclusion seems difficult to accept given the evident structural damage to kidney tissues, but it is equally difficult to believe that these birds would suffer a fitness reduction caused by naturally occurring metal concentrations to which the birds have been exposed through their evolutionary development. This problem remains to be resolved.

TESTICULAR DAMAGE

Testicular atrophy appears to be caused by exposure to cadmium concentrations similar to those causing damage to the kidneys. However, while kidney damage appears to have little or no influence on the survival of the individual, damage to testis function may have a more drastic impact through reduced fertility. White et al. (1978) found that a small proportion of mallards fed dietary cadmium at 20 mg/kg (accumulating cadmium to about 50 mg/kg in the kidney) developed slight gonadal alterations, while testes of those fed 200 mg/kg (accumulating cadmium to about

100 mg/kg in the kidney) atrophied and sperm production ceased. Richardson et al. (1974) found that testicular development in 4-week-old quail was retarded by high levels of cadmium or a dietary deficiency of zinc. However, development by 6 weeks of age was normal in cadmium-dosed quail, and sperm production was as high as in controls, even though cadmium-fed birds accumulated cadmium in the liver to concentrations of about 30 mg/kg, which represents about the same level of contamination as the 100 mg/kg in the sterile duck kidneys. Some seabirds, collected from nests with fertile eggs or chicks, have had cadmium concentrations in excess of 100 mg/kg in the kidney and 30 mg/kg in the liver, so it would appear that these concentrations do not cause sterility in seabirds, though a reduction in fertility cannot be ruled out.

ANEMIA, CARDIAC HYPERTROPHY, AND BONE MARROW HYPERPLASIA

Significant decreases in packed cell volume and hemoglobin concentration and significant increases in serum glutamic pyruvic transaminase were found by Cain et al. (1983) in mallard ducklings fed 20 mg/kg of cadmium for 12 weeks, which resulted in cadmium accumulations of about 60 mg/kg in the liver. Severe anemia, hypertrophy of the ventricles, and bone marrow hyperplasia were induced in Japanese quail fed diets with elevated cadmium (75 mg/kg) and which accumulated liver cadmium concentrations of about 30 mg/kg. Feeding a diet deficient in iron caused the same symptoms (Richardson et al., 1974), suggesting that cadmium may cause these changes by creating a deficiency of iron. Ascorbic acid added to the high-cadmium diet significantly alleviated or prevented almost all aspects of cadmium toxicity in this study. Schafer and Strugala (1986) concluded that cadmium-induced anemia is due to competition between iron and cadmium at the intestinal binding sites of mucosal transferrin, resulting in reduced iron absorption.

DUODENAL EPITHELIUM DAMAGE

Extensive damage to the absorptive epithelium of the duodenum may be a characteristic toxic effect of high levels of cadmium that does not arise as a result of dietary deficiencies in iron or zinc (Scheuhammer, 1987). Such damage was found in quail with liver cadmium concentrations of about 30 mg/kg (Richardson et al., 1974), but damage was not found in mallard ducklings with liver cadmium concentrations of about 60 mg/kg (Cain et al., 1983). Scheuhammer (1988) suggested that a daily dose of cadmium of more than 5 but less than 50 mg/kg causes damage to the intestinal epithelium of Japanese quail.

ALTERED ENERGY METABOLISM

Di Giulio and Scanlon (1984a) fed young mallards for 42 days on diets containing cadmium at concentrations of 0, 50, 150, or 450 mg/kg. Birds given the highest dose accumulated kidney and liver cadmium concentrations of about 120 mg/kg. The approximate 1:1 ratio of liver:kidney cadmium suggests that the cadmium represented an acute dose. The birds given the highest dose showed a significant loss in body mass and a lower daily food consumption (though not statistically different from other groups). Ducks on the highest cadmium diet showed reduced liver mass but increased kidney and adrenal mass. The authors concluded that cadmium interference with carbohydrate metabolism was much less than has been found in similar studies with mammals, but cadmium toxicity may be exacerbated by food shortage and vice versa (Di Giulio and Scanlon, 1985).

TERATOGENIC EFFECTS

Cadmium salts deposited on the blastoderm of chicken eggs caused teratogenic effects (Schowing, 1984). However, since birds apparently do not transfer cadmium into eggs, it is unlikely that this observation will be of any relevance to environmental studies of birds.

SUMMARY

Cadmium is a heavy metal released into the environment from smelting, from burning coal and oils, and from using phosphate rock fertilizers. It is accumulated by most organisms, especially molluscs. Levels in bird eggs are negligible, so embryotoxic effects are unlikely. Cadmium concentrations increase with age, being some 10 to 100 times higher in adults than in chicks. About half the body burden is stored in the liver, but the highest concentrations almost always occur in the kidney; in both tissues, metallothionein binds cadmium.

Only 0.4 to 2% of ingested cadmium is assimilated. Cadmium concentrations are <1 mg/kg in the liver of most passerines and birds of prey, 1 to 5 mg/kg in many waders and waterfowl that eat molluscs, and 5 to 35 mg/kg in many seabirds. Within this broad range, increases attributable to pollution tend to be small. The liver is the preferred organ for investigating cadmium concentrations in birds suspected of cadmium poisoning.

Toxic effects have been studied principally by laboratory dosing of ducks, chickens, and quail. Dietary cadmium concentrations in excess of 2 mg/kg induced increased synthesis of metallothionein, accumulation of cadmium and zinc, and possible disturbance of the metabolism of iron, zinc, and calcium. Toxic effects of cadmium (altered behavioral responses, suppression of egg production, egg shell thinning, kidney damage, testicular damage, duodenal epithelium damage, altered energy metabolism, anemia, bone marrow hyperplasia, and cardiac and adrenal hypertrophy) have been reported in laboratory studies with ducks, chickens, quail, and starlings.

After a review of the literature, it is evident that the liver and kidney are the most useful organs for analyzing cadmium, and I suggest that about 40 mg/kg in the liver or 100 mg/kg in the kidney should be considered tentative threshold tissue concentrations, above which cadmium poisoning in birds might be expected.

However, among wild birds, such high tissue concentrations of cadmium occur apparently naturally in a small proportion of individuals in some seabird populations and are believed by those reporting them to be harmless. Threshold toxic concentrations of cadmium may be rather higher among pelagic seabirds. No toxic effects of cadmium have been reported in wild bird populations apart from kidney damage similar to that in dosed birds, clearly evident at the ultrastructural level. Whether this damage has consequences for the fitness of birds with high burdens of cadmium and mercury remains unresolved.

REFERENCES

Blomqvist, S., A. Frank, and L. R. Petersson. 1987. Metals in liver and kidney tissues of autumn-migrating dunlin *Calidris alpina* and curlew sandpiper *Calidris ferruginea* staging at the Baltic Sea. Mar. Ecol. Prog. Ser. 35:1-13.

Brothers, N. P., and M. J. Brown. 1987. The potential use of fairy prions *(Pachyptila turtur)* as monitors of heavy metal levels in Tasmanian waters. Mar. Pollut. Bull. 18:132-134.

Bull, K. R., R. K. Murton, D. Osborn, P. Ward, and L. Cheng. 1977. High levels of cadmium in Atlantic seabirds and sea-skaters. Nature (Lond.) 269:507-509.

Burger, J., and M. Gochfeld. 1991. Cadmium and lead in common terns (Aves *Sterna hirundo*): relationships between levels in parents and eggs. Environ. Monit. Assess. 16:253-258.

Cain, B. W., L. Sileo, J. C. Franson, and J. Moore. 1983. Effects of dietary cadmium on mallard ducklings. Environ. Res. 32:286-297.

Chakraborty, T., I. B. Maiti, and B. B. Biswas. 1987. A single form of metallothionein is present in both heavy metal induced and neonatal chicken liver. J. Biosci. (Bangalore) 11:379-390.

Cheney, M. A., C. S. Hacker, and G. D. Schroder. 1981. Bioaccumulation of lead and cadmium in the Louisiana heron *(Hydranassa tricolor)* and the cattle egret *(Bubulcus ibis)*. Ecotoxicol. Environ. Saf. 5:211-224.

Cheng, L., M. Schulz-Baldes, and C. S. Harrison. 1984. Cadmium in ocean-skaters, *Halobates sericeus* (Insecta), and in their seabird predators. Mar. Biol. (N.Y.) 79:321-324.

Cosson, R. P. 1989. Relationships between heavy metal and metallothionein-like protein levels in the liver and kidney of two birds: the greater flamingo and the little egret. Comp. Biochem. Physiol. C Comp. Pharmacol. 94:243-248.

Di Giulio, R. T., and P. F. Scanlon. 1984a. Sublethal effects of cadmium ingestion on mallard ducks. Arch. Environ. Contam. Toxicol. 13:765-771.

Di Giulio, R. T., and P. F. Scanlon. 1984b. Heavy metals in tissues of waterfowl from the Chesapeake Bay, USA. Environ. Pollut. Ser. A Ecol. Biol. 35:29-48.

Di Giulio, R. T., and P. F. Scanlon. 1985. Effects of cadmium ingestion and food restriction on energy metabolism and tissue metal concentrations in mallard ducks. Environ. Res. 37:433-444.

Elliott, J. E., A. M. Scheuhammer, F. A. Leighton, and P. A. Pearce. 1992. Heavy metal and metallothionein concentrations in Atlantic Canadian seabirds. Arch. Environ. Contam. Toxicol. 22:63-73.

Fernando, L. P., D. Wei, and G. K. Andrews. 1989. Structure and expression of chicken metallothionein. J. Nutr. 119:309-318.

Freeland, J. H., and R. J. Cousins. 1973. Effect of dietary cadmium on anaemia, iron absorption, and cadmium binding protein in the chick. Nutr. Rep. Int. 8:337-347.

Furness, R. W., and M. Hutton. 1979. Pollutant levels in the great skua *Catharacta skua*. Environ. Pollut. 19:261-268.
Furness, R. W., and P. S. Rainbow. 1990. Heavy metals in the marine environment. CRC Press, Boca Raton, Fla.
Goyer, R. A., M. G. Cherian, and L. Delaquerriere-Richardson. 1984. Correlation of parameters of cadmium exposure with onset of cadmium-induced nephropathy in rats. J. Environ. Pathol. Toxicol. Oncol. 5:89-100.
Hahn, E. 1991. Schwermetallgehalte in Vogelfedern — ihre Ursache und der Einsatz von Federn standorttreuer Vogelarten im Rahmen von Bioindikationsverfahren. (In German.) Ber. Forschungszentrums Julich.
Heinz, G. H., S. D. Haseltine, and L. Sileo. 1983. Altered avoidance behavior of young black ducks fed cadmium. Environ. Toxicol. Chem. 2:419-421.
Honda, K., J. E. Marcovecchio, S. Kan, R. Tatsukawa, and H. Ogi. 1990. Metal concentrations in pelagic seabirds from the North Pacific Ocean. Arch. Environ. Contam. Toxicol. 19:704-711.
Howarth, D. M., A. J. Hulbert, and D. Horning. 1981. A comparative study of heavy metal accumulation in tissues of the crested tern, *Sterna bergii*, breeding near industrialized and non-industrialized areas. Aust. Wildl. Res. 8:665-672.
Hulse, M., J. S. Mahoney, G. D. Schroder, C. S. Hacker, and S. M. Pier. 1980. Environmentally acquired lead, cadmium, and manganese in the cattle egret, *Bubulcus ibis*, and the laughing gull, *Larus atricilla*. Arch. Environ. Contam. Toxicol. 9:65-78.
Hutton, M. 1981. Accumulation of heavy metals and selenium in three seabird species from the United Kingdom. Environ. Pollut. Ser. A Ecol. Biol. 26:129-145.
Hutton, M., and G. T. Goodman. 1980. Metal contamination of feral pigeons *Columba livia* from the London area. Part 1. Tissue accumulation of lead, cadmium, and zinc. Environ. Pollut. Ser. A Ecol. Biol. 22:207-217.
Kaji, T., R. Kawatani, M. Takata, T. Hoshino, T. Miyahara, H. Kozuka, and F. Koizumi. 1988. The effects of cadmium, copper, or zinc on formation of embryonic chick bone in tissue culture. Toxicology 50:303-316.
Kaji, T., H. Yamada, T. Hoshino, T. Miyahara, H. Kozuka, and Y. Naruse. 1986. A possible mechanism of cadmium-copper interaction in embryonic chick bone in tissue culture. Toxicol. Appl. Pharmacol. 86:243-252.
Karlog, O., K. Elvestad, and B. Clausen. 1983. Heavy metals (cadmium, copper, lead, and mercury) in common eiders *(Somateria mollissima)* from Denmark. Nord. Veterinaermed. 35:448-451.
Koo, S. I., C. S. Fullmer, and R. H. Wasserman. 1978. Intestinal absorption and retention of ^{109}Cd: effects of cholecalciferol, calcium status, and other variables. J. Nutr. 108:1812-1822.
Leach, R. M., K. W.-L. Wang, and D. E. Baker. 1979. Cadmium and the food chain: the effect of dietary cadmium on tissue composition in chicks and laying hens. J. Nutr. 109:437-443.
Lee, D. P., K. Honda, and R. Tatsukawa. 1987. Comparison of tissue distributions of heavy metals in birds in Japan and Korea. J. Yamashina Inst. Ornithol. 19:103-116.
Lee, D. P., K. Honda, R. Tatsukawa, and P.-O. Won. 1989. Distribution and residue level of mercury, cadmium, and lead in Korean birds. Bull. Environ. Contam. Toxicol. 43:550-555.
Leonzio, C., C. Fossi, and S. Focardi. 1986. Lead, mercury, cadmium, and selenium in two species of gull feeding on inland dumps, and in marine areas. Sci. Total Environ. 57:121-127.

Lin, L.-Y., W. C. Lin, and P. C. Huang. 1990. Pigeon metallothionein consists of two species. Biochim. Biophys. Acta 1037:248-255.

Lock, J. W., D. R. Thompson, R. W. Furness, and J. A. Bartle. 1992. Metal concentrations in seabirds of the New Zealand region. Environ. Pollut. 75:289-300.

Lundholm, C. E., and K. Mathson. 1986. Effect of some metal compounds on the Ca^{2+} binding and Ca^{2+}-Mg^{2+}-ATPase activity of eggshell gland mucosa homogenate from the domestic fowl. Acta Pharmacol. Toxicol. 59:410-415.

Maedgen, J. L., C. S. Hacker, G. D. Schroder, and F. W. Weir. 1982. Bioaccumulation of lead and cadmium in the royal tern and Sandwich tern. Arch. Environ. Contam. Toxicol. 11:99-102.

Mayack, L. A., P. B. Bush, O. J. Fletcher, R. K. Page, and T. T. Fendley. 1981. Tissue residues of dietary cadmium in wood ducks. Arch. Environ. Contam. Toxicol. 10:637-645.

Miyahara, T., Y. Oh-e, T. Komurasaki, and H. Kozuka. 1982. The effect of cadmium on the collagen metabolism of embryonic chicken bone in tissue culture and zinc-cadmium interaction. Eisei Kagaku 28:52. (Abstr.).

Muirhead, S. J., and R. W. Furness. 1988. Heavy metal concentrations in the tissues of seabirds from Gough Island, South Atlantic Ocean. Mar. Pollut. Bull. 19:278-283.

Nicholson, J. K. 1981. The comparative distribution of zinc, cadmium, and mercury in selected tissues of the herring gull *(Larus argentatus)*. Comp. Biochem. Physiol. C Comp. Pharmacol. 68:91-94.

Nicholson, J. K., M. D. Kendall, and D. Osborn. 1983. Cadmium and mercury nephrotoxicity. Nature (Lond.) 304:633-635.

Nicholson, J. K., and D. Osborn. 1983. Kidney lesions in pelagic seabirds with high tissue levels of cadmium and mercury. J. Zool. (Lond.) 200:99-118.

Nicholson, J. K., and D. Osborn. 1984. Kidney lesions in juvenile starlings *Sturnus vulgaris* fed on a mercury-contaminated synthetic diet. Environ. Pollut. Ser. A Ecol. Biol. 33:195-206.

Nielsen, C. O., and R. Dietz. 1989. Heavy metals in Greenland seabirds. Medd. Gronl.-Biosci. Rep. No. 29.

Norheim, N. 1987. Levels and interactions of heavy metals in sea birds from Svalbard and the Antarctic. Environ. Pollut. 47:83-94.

Nriagu, J. O., and J. M. Pacyna. 1988. Quantitative assessment of worldwide contamination of air, water, and soils by trace metals. Nature (Lond.) 333:134-139.

Osborn, D. 1978. A naturally occurring cadmium and zinc binding protein from the liver and kidney of *Fulmarus glacialis*, a pelagic North Atlantic seabird. Biochem. Pharmacol. 27:822-824.

Osborn, D., M. P. Harris, and J. K. Nicholson. 1979. Comparative tissue distribution of mercury, cadmium, and zinc in three species of pelagic seabirds. Comp. Biochem. Physiol. C Comp. Pharmacol. 64:61-67.

Peakall, D. 1992. Animal biomarkers as pollution indicators. Chapman and Hall, London.

Pinowska, B., K. Krasnicki, and J. Pinowski. 1981. Estimation of the degree of contamination of granivorous birds with heavy metals in agricultural and industrial landscapes. Ekol. Pol. 29:137-149.

Pritzl, M. C., Y. H. Lie, E. W. Kienholz, and C. E. Whiteman. 1974. The effect of dietary cadmium on development of young chickens. Poult. Sci. 53:2026-2029.

Rambeck, W. A., and W. E. Kollmer. 1990. Modifying cadmium retention in chickens by dietary supplements. J. Anim. Physiol. Anim. Nutr. 63:66-74.

Rao, P. V. V. P., S. A. Jordan, and M. K. Bhatnagar. 1989. Combined nephrotoxicity of methylmercury, lead, and cadmium in pekin ducks: metallothionein, metal interactions, and histopathology. J. Toxicol. Environ. Health 26:327-348.

Reid, M., and C. S. Hacker. 1982. Spatial and temporal variation in lead and cadmium in the laughing gull, *Larus atricilla*. Mar. Pollut. Bull. 13:387-389.

Richardson, M. E., and M. R. S. Fox. 1975. Dietary cadmium and enteropathy in the Japanese quail — histochemical and ultrastructural studies. Lab. Invest. 31:722-731.

Richardson, M. E., M. R. S. Fox, and B. E. Fry. 1974. Pathological changes produced in Japanese quail by ingestion of cadmium. J. Nutr. 104:323-338.

Schafer, S. G., and G. J. Strugala. 1986. Dietary cadmium and the development of anaemia. Trends Pharmacol. Sci. 7:430-432.

Scharenberg, W. 1989. Schwermetalle in Organen und Federn von Graureihern *(Ardea cinerea)* und Kormoranen *(Phalacrocorax carbo)*. (In German.) J. Ornithol. 130:25-33.

Scheuhammer, A. M. 1987. The chronic toxicity of aluminium, cadmium, mercury, and lead in birds: a review. Environ. Pollut. 46:263-295.

Scheuhammer, A. M. 1988. The dose-dependent deposition of cadmium into organs of Japanese quail following oral administration. Toxicol. Appl. Pharmacol. 95:153-161.

Scheuhammer, A. M. 1991. Effects of acidification on the availability of toxic metals and calcium to wild birds and mammals. Environ. Pollut. 71:329-375.

Scheuhammer, A. M., and D. M. Templeton. 1990. Metallothionein production: similar responsiveness of avian liver and kidney to chronic cadmium administration. Toxicology 60:151-160.

Schowing, J. 1984. Teratogenic effects of cadmium acetate and sulphate upon development of the chick embryo. Acta Morphol. Hung. 32:37-46.

Sell, J. L. 1975. Cadmium and the laying hen: apparent absorption, tissue distribution, and virtual absence of transfer into eggs. Poult. Sci. 54:1674-1678.

Stock, M., R. F. M. Herber, and H. M. A. Geron. 1989. Cadmium levels in oystercatcher *Haematopus ostralegus* from the German Wadden Sea. Mar. Ecol. Prog. Ser. 53:227-234.

Stoewsand, G. S., C. A. Bache, W. H. Gutenmann, and D. J. Lisk. 1986. Concentration of cadmium in *Coturnix* quail fed earthworms. J. Toxicol. Environ. Health 18:369-376.

Stoneburner, D. L., P. C. Patty, and W. B. Robertson. 1980. Evidence of heavy metal accumulations in sooty terns. Sci. Total Environ. 14:147-152.

Thompson, D. R. 1990. Metal levels in marine vertebrates. p. 143-182. *In* R. W. Furness and P. S. Rainbow (Eds.). Heavy metals in the marine environment. CRC Press, Boca Raton, Fla.

Turner, J. C., S. R. B. Solly, J. C. M. Mol-Krijnen, and V. Shanks. 1978. Organochlorine, fluorine, and heavy metal levels in some birds from New Zealand estuaries. N.Z. J. Sci. 21:99-102.

Vermeer, K., and J. C. Castilla. 1991. High cadmium residues observed during a pilot study in shorebirds and their prey downstream from the El Salvador copper mine, Chile. Bull. Environ. Contam. Toxicol. 46:242-248.

Walsh, P. M. 1990. The use of seabirds as monitors of heavy metals in the marine environment. p. 183-204. *In* R. W. Furness and P. S. Rainbow (Eds.). Heavy metals in the marine environment. CRC Press, Boca Raton, Fla.

White, D. H., and M. T. Finley. 1978. Uptake and retention of dietary cadmium in mallard ducks. Environ. Res. 17:53-59.

White, D. H., M. T. Finley, and J. F. Ferrell. 1978. Histopathologic effects of dietary cadmium on kidneys and testes of mallard ducks. J. Toxicol. Environ. Health 4:551-558.

Yamamura, M., and K. T. Suzuki. 1984. Induction and characterization of metallothionein in the liver and kidney of Japanese quail. Comp. Biochem. Physiol. B Comp. Biochem. 77:101-106.

CHAPTER 18

Heavy Metals in Aquatic Invertebrates

Philip S. Rainbow

INTRODUCTION

Although the term "heavy metal" is widely used in the biological literature, it is rarely defined by objective chemical criteria such as those proposed by Nieboer and Richardson (1980). This chapter uses the term to describe metals with ions of class B or intermediate character as described by Nieboer and Richardson (1980). Heavy metals will therefore include metals such as copper, iron, manganese, and zinc, known to be essential to organisms, as well as nonessential metals such as cadmium, lead, and mercury. It is impossible in the space available here to give equal attention and justice to all or many heavy metals, and therefore this text concentrates particularly on zinc, and to a lesser extent on copper, as examples of heavy metals.

In this chapter, the term "metal content" will refer to the total amount (e.g., micrograms) of metal in a body or part thereof, while "concentration" (typically, µg/g of dry weight) will refer to content per unit of weight, dry weight unless specified otherwise.

ACCUMULATION OF HEAVY METALS

As a result of their chemical reactivity all heavy metals, whether essential or not, have the potential to be toxic when taken up in excess by aquatic invertebrates. The major routes of metal uptake (Bryan, 1979; Luoma, 1983) into aquatic invertebrates are from solution (Simkiss and Taylor, 1989a) and from food (e.g., Timmermans et al., 1992). More esoteric routes are also present in some organisms (Rainbow, 1990a), such as pinocytosis of metal-rich particles in the gills of molluscs (Hobden, 1967; George et al., 1976) and the pharynx of ascidians (Kalk, 1963) or the entry of seawater into the pedal sinuses of particular gastropods (Depledge and Phillips, 1986).

The rates of uptake of heavy metals from solution by marine invertebrates depend greatly on external physicochemical factors (Sunda et al., 1978; Engel and Fowler, 1979; Luoma, 1983) and are beyond the short-term physiological control of the organism (Nugegoda and Rainbow, 1988a, 1989a, 1989b; Simkiss and Taylor, 1989a; Phillips and Rainbow, 1993), although in rare circumstances some estuarine organisms may make a physiological change to their overall permeability (apparent water permeability; Mantel and Farmer, 1983; Campbell and Jones, 1990) in response to changed external osmotic pressure, possibly with a subsequent effect on the metal uptake rate (Chan et al., 1992; Rainbow et al., 1993). There are nevertheless long-term interspecific differences among metal uptake rates of aquatic organisms (Nugegoda and Rainbow, 1988b; Rainbow and White, 1989, 1990). Generally, heavy metals enter aquatic organisms from solution in proportion to their ambient bioavailability correlated, for example, with the dissolved concentration, suggesting that any dissolved heavy metal will enter an aquatic organism at some rate and, indeed, that essential metals typically enter at rates in excess of metabolic requirements. On the other hand, it may be the case that, for some metals, the barriers to uptake are formidable at natural ambient bioavailabilities (Depledge and Rainbow, 1990). This is unlikely to be the case for any aquatic invertebrate passing water over a permeable respiratory surface, but may be the situation, for example, in an aquatic arthropod like an air-breathing insect or spider physiologically isolated from the aquatic medium.

The typical aquatic invertebrate can therefore be considered to be taking up heavy metals from solution (and food), with the potential to accumulate them to high internal concentrations, given the large number of potential metal-binding sites internally. Uptake of heavy metals from solution can usually be explained by a passive facilitated diffusion process (Bryan, 1971, 1979; Simkiss and Taylor, 1989a), as metals bind passively to a membrane transport protein and are passed thence down a gradient of metal-binding ligands of increasing metal affinity (Williams, 1981a, 1981b). (Other routes of uptake of dissolved heavy metals, for example, cadmium entry through active transport pumps for calcium [e.g., Wright, 1977, 1980], also occur to differing degrees of importance in different aquatic invertebrates in different situations; see Phillips and Rainbow [1993] and Rainbow and Dallinger [1993] for more detailed discussion.) In the facilitated diffusion model, significant back diffusion is prevented by binding of the heavy metals to large intracellular proteins before metals are transferred elsewhere in the body. The subsequent accumulation of a heavy metal by an invertebrate then depends on its particular accumulation strategy for that metal (Rainbow et al., 1990; Phillips and Rainbow, 1993).

In contrast to the situation for dissolved organic compounds (Phillips and Rainbow, 1993), the net accumulation of heavy metals by aquatic invertebrates is not predictable by simple chemical partition coefficients describing the relative solubilities of an element between, for example, water and lipid. Uptake of dissolved heavy metals by facilitated diffusion follows simple diffusion laws responding to changes in external dissolved concentrations (Depledge and Rainbow, 1990), but there is no corresponding automatic loss of metal from the body when the external concentration falls. The key to understanding this process is the high affinity of heavy metals for organic molecules, often via binding to nitrogen and sulfur contained therein (Nieboer and Richardson, 1980). Once a heavy metal has entered the body of an aquatic invertebrate, it is typically bound to high-molecular-weight ligands (usually proteins) and unavailable for significant back diffusion out of the body. Its subsequent fate depends on how it is handled physiologically by the organism, in effect the accumulation strategy of that invertebrate for that metal (Phillips and Rainbow, 1989, 1993; Rainbow et al., 1990). As a general rule, heavy metals reach body concentrations in aquatic invertebrates (Eisler, 1981) that are orders of magnitude greater on a wet weight basis than external dissolved concentrations, while diffusible metal continues to enter passively. If this accumulated metal is maintained in a metabolically available form, it has the potential to play a metabolic role (in the case of an essential heavy metal) but also to be toxic, since binding to intracellular molecules interferes with their metabolic functioning. There are physiological processes in aquatic invertebrates that detoxify (render metabolically unavailable) body metal levels in excess of metabolic requirements, if any, and in some circumstances excrete them from the body.

Accumulation strategies vary interspecifically for the same metal and intraspecifically for different metals (Rainbow and White, 1989; Rainbow et al., 1990). At one extreme, the aquatic invertebrate accumulates all metal taken up, with no significant excretion (strong net accumulation if the uptake rate is high). Under these circumstances, the metal must be stored in detoxified form. In the case of an essential metal, the majority of accumulated metal is in this stored detoxified form, leaving only some available for essential requirements, but in the case of a nonessential metal, presumably all the accumulated metal is detoxified. At the other extreme, an aquatic invertebrate may excrete all the metal that is entering in excess of metabolic requirements, thereby maintaining a relatively constant body content (regulation of body metal content), presumably equivalent to physiological need. In this case, the excretion rate is increased to match the uptake rate as it increases with the raised external metal concentration, but ultimately the uptake rate may be so high that the excretion rate fails to match it. This is the point of regulation breakdown where net accumulation occurs (White and Rainbow, 1982, 1984; Nugegoda and Rainbow, 1987, 1989c).

There is a gradient of accumulation strategies between the two extremes of strong net accumulation and body content regulation (Bryan, 1979; Phillips and Rainbow, 1989; Rainbow et al., 1990). Weak net accumulation may occur as a result of a very low metal uptake rate with no significant excretion, or it may be brought about if the uptake rate is high but is almost matched by significant excretion. Weak net accumulation approaches body content regulation with little increase in body metal content with an increase in metal uptake rate as external concentrations are increased, and weak net accumulation may be referred to as "partial regulation" (Phillips and Rainbow, 1989; Rainbow et al., 1990).

The crustaceans provide specific examples of different metal accumulation strategies. Barnacles are strong net accumulators of both essential metals like copper and zinc and nonessential metals like cadmium (Rainbow and White, 1989). Zinc body concentrations in barnacles typically reach 10,000 µg/g and may even exceed

100,000 μg/g (Walker et al., 1975a, 1975b; Rainbow, 1987). Zinc is taken up by the barnacle *Elminius modestus* at a high rate with no apparent significant excretion (Rainbow and White, 1989); therefore, all the zinc that is taken up is subsequently accumulated, and the barnacle is a strong net accumulator of zinc. The vast majority of this accumulated zinc is necessarily stored in a detoxified form, being bound in metabolically inert granules based predominantly on the anion pyrophosphate with other metals, particularly iron, also present (Walker et al., 1975a, 1975b; Pullen and Rainbow, 1991).

The binding of accumulated heavy metals in relatively inert granules is a common detoxification mechanism in invertebrates (Simkiss, 1977, 1981; Brown, 1982; George, 1982, 1990; Simkiss et al., 1982; Taylor and Simkiss, 1984; Viarengo, 1989), the implication being that such granules are not redissolved, although they may be excreted whole in some cases (George et al., 1980; Moore and Rainbow, 1984). Other granule types, including one based on the carbonate anion, may be present, for example, in aquatic molluscs (Mason and Nott, 1981; Simkiss, 1981). These carbonate-based granules act as temporary reusable metabolic reserves, for example, of calcium, and do not contain a detoxified store of heavy metals that would be potentially toxic if released (Simkiss, 1980, 1981; Mason and Nott, 1981). Another physiological process of detoxification of accumulated heavy metals involves metallothioneins, low-molecular-weight sulfur-rich proteins with a high affinity for certain heavy metals including cadmium, copper, mercury, and zinc (Roesijadi, 1980-1981; Engel and Brouwer, 1989; George, 1990). The products of lysosomal breakdown of metallothioneins may also be found in intracellular deposits (George, 1990), and some metals initially bound to metallothioneins may eventually be found in calcium-based or pyrophosphate-type granules (Nott and Langston, 1989; Simkiss and Taylor, 1989b; Pullen and Rainbow, 1991).

At the other end of the spectrum of accumulation strategies, the littoral decapod *Palaemon elegans* regulates its body content of zinc (White and Rainbow, 1982, 1984). The uptake rate of zinc into the decapod rises with the increased bioavailability of zinc in solution, caused (for example) by an increased dissolved zinc concentration (White and Rainbow, 1984), decreased salinity (Nugegoda and Rainbow, 1989a, 1989b), or the absence of a metal chelating agent (Nugegoda and Rainbow, 1988a). However, the uptake rate is matched by an increased zinc excretion rate, and the body zinc content stays approximately constant. The regulated body zinc content is presumed to be equivalent to that content required metabolically (White and Rainbow, 1985), although this may change intraspecifically with temperature (Nugegoda and Rainbow, 1987) and will vary interspecifically (Nugegoda and Rainbow, 1988b). This body zinc would therefore be in a metabolically available form. Eventually at a threshold of zinc bioavailability, the uptake rate of zinc is so great that it can no longer be matched by the excretion rate, the body content regulation breaks down, and net accumulation occurs. Death of *P. elegans* occurs when the zinc body content approximately doubles to about 200 μg of zinc per gram (White and Rainbow, 1982), indicating that much of the extra accumulated zinc is still in metabolically available form. This lethal accumulated body concentration contrasts markedly with the accumulated zinc concentrations in barnacles that typically exceed 10,000 µg of zinc per gram but are known to be in a detoxified form and therefore not metabolically available (Rainbow, 1987).

P. elegans also appears to regulate its body concentration of copper to about 100 µg/g (White and Rainbow, 1982), but any regulation mechanism needs confirmation by independent identification of uptake and accumulation rates. After breakdown of the apparent body concentration regulation, *P. elegans* survives net accumulation of copper to a body concentration of about 700 µg/g (White and Rainbow, 1982), suggesting that a portion of the newly accumulated copper is in a detoxified form. Indeed, copper-rich deposits are found in the hepatopancreatic cells of the decapod under these circumstances (Rainbow, 1988). In the case of the nonessential metal cadmium, *P. elegans* is a net accumulator at all external bioavailabilities (White and Rainbow, 1982, 1986; Rainbow and White, 1989). Indeed, there are no confirmed reports of the body content regulation of nonessential metals by aquatic invertebrates, such regulation being an accumulation strategy apparently restricted to essential metals, particularly zinc (Bryan, 1979; Rainbow, 1988, 1990a, 1993; Rainbow and Dallinger, 1993).

Among crustaceans, examples of weak net accumulators of dissolved zinc are the littoral amphipods *Echinogammarus pirloti* (Rainbow and White, 1989) and *Orchestia gammarellus* (Weeks and Rainbow, 1991). Both amphipods show no significant excretion of zinc taken up from solution, but their absolute zinc uptake rates are so low that net accumulation is small. The bivalve mussels *Mytilus edulis* and *Perna viridis* are also weak net accumulators (partial regulators) of zinc (George and Pirie, 1980; Phillips and Yim, 1981; Phillips, 1985; Chan, 1988). *M. edulis* excretes zinc in granular form from the kidney (George and Pirie, 1980), so weak net accumulation is apparently the net difference between significant uptake and significant excretion. Whereas barnacles store detoxified zinc-rich granules permanently, the storage of zinc-rich granules in the mytilid kidney is an example of the temporary storage of a detoxified metal.

Freshwater insect larvae appear to be net accumulators of heavy metals (Rainbow and Dallinger, 1993), with larvae of different insect families showing a range of accumulated metal concentrations under the same environmental conditions (Cain et al., 1992; Hare and Campbell, 1992; Timmermans, 1993). Some insect larvae can excrete metals (Hare et al., 1991), often via moulted exuvia (Timmermans and Walker, 1989), but accumulated body concentrations do change with metal exposure before reaching a new concentration (in some cases in a new apparent steady state) (Hare et al., 1991; Rainbow and Dallinger, 1993).

REQUIREMENTS OF ESSENTIAL METALS

It has been emphasized above that a component of the body content of an essential metal must be in a metabolically available form in order to play required roles in metabolism. Such a role would be as a vital part of a metal-associated enzyme, zinc being associated, for example, with carbonic anhydrase, carboxypeptidases, alkaline phosphatase, and alcohol dehydrogenase, among others (Pequegnat et al, 1969; Williams, 1984). It is possible to make theoretical estimates of the enzymatic requirements of tissues for essential metals (Pequegnat et al., 1969; White and Rainbow, 1985, 1987) and, in spite of the many assumptions made, such estimates do provide a baseline concentration for comparative purposes to perhaps within an order of magnitude. White and Rainbow (1985, 1987) drew upon the work of Pequegnat et al. (1969) with a more extensive enzyme database to conclude that the enzymatic requirements of crustacean soft tissues for copper and zinc approximated 26 µg of Cu per gram and 35 µg of Zn per gram (White and Rainbow, 1985) and that those for iron and manganese approximated 27 µg of Fe per gram and 4 µg of Mn per gram (White and Rainbow, 1987). Decapod crustaceans and certain molluscs have an additional nonenzymatic copper requirement to synthesize the blood pigment hemocyanin, that of the crab *Carcinus maenas* being an additional 57 µg of Cu per gram in the whole body. Depledge (1989a) justifiably criticized some of the assumptions and conclusions of White and Rainbow (1985) but, nevertheless, came up with the estimated total metabolic requirements of decapod crustaceans for copper and zinc of the same order (namely, 83 µg of Cu per gram and 68 µg of Zn per gram for *C. maenas*). Rainbow (1993) has pursued further such theoretical estimates of metabolic requirements for essential heavy metals, concluding that they are useful guidelines to minimum concentrations of essential metals to be expected in metabolically active invertebrate soft tissues. Moreover, it appears that body concentrations of essential metals observed in body content regulators usually exceed slightly the theoretical estimates, perhaps reflecting the presence of some small storage capacity above immediate requirements.

This discussion of the required concentrations of essential heavy metals has considered metabolically available metal. In some invertebrates, heavy metals may play other roles. The metals may be used in a form not immediately available to metabolism, as in the case of structural roles for metals in polychaete jaws. For example, the zinc concentrations of the jaws of the nereid polychaete worms *Nereis virens* and *Nereis (Hediste) diversicolor* fall in the range of 10,000 to 20,000 µg/g, the jaw zinc content therefore approximating 1 to 2% of the jaw dry weight and 20 to 50% of the zinc body content; this zinc is believed to contribute to the mechanical properties of the jaws (Bryan and Gibbs, 1979, 1980). Similarly, copper appears to be a structural feature of the jaws of glycerid polychaete worms; copper in the jaws of *Glycera gigantea* reaches concentrations of about 15,000 µg of Cu per gram, accounting for 1.5% of jaw dry weight and up to 67% of the total body burden of copper (Gibbs and Bryan, 1980). Accumulated vanadium in the tunic of some sea squirts (Carlisle, 1968; Michibata et al., 1986) may play antifouling and antipredatory roles (Stoecker, 1978, 1980), and high concentrations of copper in the branchiae of the polychaete worm *Melinna palmata* may serve to reduce their palatability to fish predators (Gibbs et al., 1981).

DEFICIENCY AND EXCESS OF ESSENTIAL METALS

It is possible, therefore, that, under circumstances of very low essential metal bioavailability, the body content of metal in an aquatic invertebrate may be too low to meet essential requirements, metabolic or otherwise. There is presumably an absolute minimum body content of an essential metal below which life is impossible, but there will also be a range of metal contents where the invertebrate is alive but suffering from sublethal nutrient deficiency.

Concrete examples of such cases are difficult to find and, indeed, to verify. Juveniles of the mesopelagic deep sea decapod *Systellaspis debilis* have body concentrations of copper that are low in comparison with those of adults of the same species, those of related carideans from coastal waters and, indeed, theoretical metabolic requirements (White and Rainbow, 1987; Rainbow and Abdennour, 1989). These low body contents of copper are correlated with low levels of the copper-bearing respiratory pigment hemocyanin (Rainbow and Abdennour, 1989). It remains to be shown, however, whether these juvenile decapods are suffering from copper deficiency or whether the relative lack of hemocyanin in juveniles is of no functional disadvantage, diffusion satisfying all requirements for oxygen transport. Hopkin (1993) considers the possibility of copper deficiency in invertebrates in a review of the biology of copper in (albeit terrestrial) isopod crustaceans. Some individual woodlice may in rare circumstances exhibit copper deficiency, particularly in the presence of a high bioavailability of cadmium acting antagonistically.

At the other end of the essential metal bioavailability spectrum, a metal may enter an aquatic invertebrate at a rate faster than it can be excreted or detoxified physiologically. The pool of metabolically available metal will then increase until it reaches a lethal content, although the total body metal content may be far below levels reached over a longer period in other individuals able to match the detoxification rate to the uptake rate. Good field data illustrating the phenomenon again derive from a terrestrial invertebrate, the woodlouse *Porcellio scaber* (Hopkin, 1990a). Populations of this woodlouse within a 3-km radius of a smelting works at Avonmouth, England, contained individuals that were moribund, apparently as a result of the zinc-detoxified storage capacity of the hepatopancreas being exceeded (Hopkin, 1990a, 1990b). This manifestation of the accumulation of a lethal content of the metabolically available component of total body metal content occurred when the mean zinc concentration of the hepatopancreas exceeded 8000 μg of Zn per gram (equivalent to a whole body concentration of about 800 μg of Zn per gram) (Hopkin, 1990b).

COMPONENTS OF THE METAL CONTENT OF AN AQUATIC INVERTEBRATE

Figure 1 summarizes many of the points discussed above, crucial to the understanding of the significance of a heavy metal concentration in an aquatic invertebrate. The ordinate represents the invertebrate's total body metal content that will increase with increasing ambient metal bioavailability. The sizes of the different components shown are of no quantitative significance, for these will vary hugely interspecifically for the same metal and intraspecifically for different metals. There will almost certainly be intraspecific variation for one metal with the physiological state of the animal.

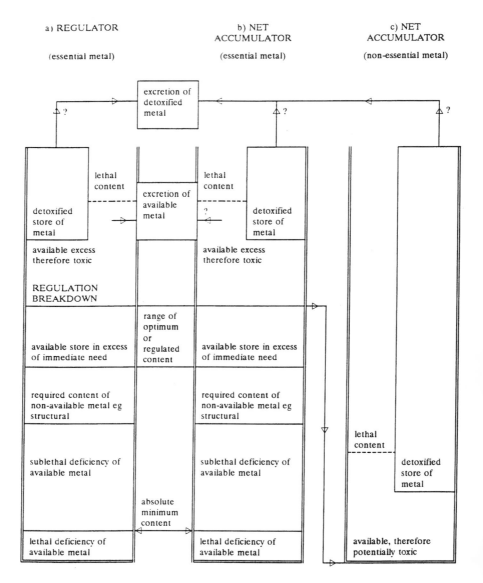

Figure 1 Schematic representation of the different components comprising the heavy metal body content (ordinate) of an aquatic invertebrate of (a) a body content regulator of an essential metal; (b) a net accumulator of an essential metal; or (c) a net accumulator of a nonessential metal. Total body metal contents will increase on the ordinate with increased ambient metal bioavailability (see text for details).

The first example (Figure 1) is that of an invertebrate regulating its total body content of an essential metal over a wide range of metal bioavailabilities until, ultimately, at a threshold bioavailability, body content regulation breaks down and net accumulation begins. This net accumulation may be in the form of a detoxified store or, if the uptake rate is high enough (or in the unlikely event of the invertebrate lacking physiological processes of detoxification), in a metabolically available and potentially toxic form. Thus, as cited above, the decapod crustacean *P. elegans* accumulates copper (at least partially) in a detoxified form after regulation breakdown, but metabolically available concentrations of zinc reach a lethal level with only an approximate doubling of the total metal content (White and Rainbow, 1982; Rainbow, 1988).

P. elegans excretes zinc (presumably from the metabolically available component) both before and after body content regulation breakdown (White and Rainbow, 1982; Nugegoda and Rainbow, 1989c). Excreted metal may also be in a detoxified form, for example, as granules from a crustacean hepatopancreas, an insect malpighian tubule, or a bivalve mollusc kidney as in the case of zinc in the mussel *M. edulis* (George et al., 1980), a partial regulator of zinc.

Net accumulation of essential metals (Figure 1) will necessarily contain detoxified stores of accumulated metals even if these stores are temporary, part or all of which might be excreted (in pulses?) to cause variation in the total body metal content above an optimum (required?) content. The amphipod *Stegocephaloides christianiensis* excretes, via the gut, crystals of iron-rich ferritin elaborated in cells of the ventral caeca (Moore and Rainbow, 1984), but the zinc-rich pyrophosphate-based granules of barnacles are stored permanently in tissues below the midgut (Walker et al., 1975a, 1975b; Rainbow, 1987; Pullen and Rainbow, 1991). In the latter case, the body zinc content of the barnacle increases over the barnacle's lifetime, although some zinc is lost periodically in the emission of gametes.

In the final example of a net accumulator of a nonessential metal such as cadmium or lead, there is by definition no requirement for the metal, and any accumulated metal is potentially toxic unless detoxified. As for essential metals, a lethal content of metabolically available metal will be reached if the metal uptake rate exceeds the rate at which the entering metal can be detoxified. Under such conditions of high metal bioavailability, an invertebrate may die from metal toxicity even if the total metal content is well below body contents reached in other individuals over longer time periods, the slower rate of uptake in the latter being matched by the rate of detoxification. It is possible that some net accumulators may excrete nonessential metals whether in detoxified form or from metabolically available pools (not shown in Figure 1), but such excretion rates will not match uptake rates.

Figure 1 does not show the possible release of essential metals from detoxified stores back into the metabolically available pool. Metallothioneins, for example, probably act as temporary stores of the essential metal copper in a detoxified form in decapod crustaceans, the copper being required for the synthesis of hemocyanin (Engel and Brouwer, 1987, 1989). Metallothioneins may also transfer zinc and copper to apoenzymes (Brady, 1982).

Although discussion of Figure 1 has been concerned with whole body contents, the same concepts can be applied to the compartmentalization of the metal contents of individual tissues or organs. Indeed, total body metal contents are a summation of the metal contents of the separate tissues; many tissues in net accumulators will be regulating metal contents, while one or more organs act as sites of metal accumulation for the whole body (Bryan, 1968, 1979; Mason and Nott, 1981; George, 1982; Viarengo, 1989; Depledge and Rainbow, 1990). Similarly, it is easier to understand how feedback loops allowing regulation of metal contents can be present at the level of tissues, as opposed to the level of the whole body (Depledge and Rainbow, 1990).

PHYSIOLOGICAL VARIATION OF METAL CONCENTRATIONS

Discussion so far has referred extensively to body metal contents, but it will apply also to metal concentrations (contents per unit of weight) in the absence of weight changes. Physiological variation will, however, alter dry weights, not only of a whole invertebrate but also differentially of the separate organs of the body, for example, with storage or emission of gametes or brooded juveniles and with the buildup or loss of energy reserves in times of high or low food availabilities. Thus, when considering metal concentrations as opposed to metal contents, there is another layer of intraspecific variability to be superimposed onto the division of the body metal into components as depicted in Figure 1.

Such variability will correlate with parameters that include size (Boyden, 1974, 1977), season (Gault et al., 1983; Amiard et al., 1986), and sex (Orren et al., 1980; Latouche and Mix, 1982) and may be inconsistent between populations. Thus, Goldberg et al. (1978) and Boalch et al. (1981) found no seasonal variation in soft tissue concentrations of heavy metals in *M. edulis* in the United States and southern England, respectively, while Gault et al. (1983) and Amiard et al. (1986) did detect significant seasonal variability in the same species from Northern Ireland and France.

Growth rates will vary between individuals with consequent effects on the relative dilution of metal content by body tissue. The tropical penaeid decapod *Metapenaeopsis palmensis* has a low body concentration of cadmium (Rainbow, 1990b) in comparison with temperate caridean decapods (White and Rainbow, 1982), probably as a result of a rapid growth rate.

Other evidence of the importance of the physiological state of an invertebrate on heavy metal content is now emerging (Depledge and Bjerregaard, 1989a; Hopkin, 1990b). Changes in temperature and the osmotic environment of the decapods *Palaemon serratus*, *Penaeus japonicus*, and *Carcinus maenas* are associated with changes in whole body loads and organ distributions of copper and iron (Spaargaren, 1983). The tissue distributions of copper and iron in *C. maenas* are also strongly influenced by the nutritional state (Depledge, 1989b). *P. elegans* regulates its body concentration of zinc to different levels at different temperatures (Nugegoda and Rainbow, 1987). The moult cycle of decapod crustaceans causes considerable changes in the tissue and body concentrations of copper (Engel, 1987; Depledge and Bjerregaard, 1989b), the moult cycle in turn being affected by seasonal temperature changes (Engel and Brouwer, 1987). Thus, a complicated relation exists between the total amounts of metals in tissues and the physiological state of the invertebrate, with variation between individual "physiotypes" discussed in the case of crustaceans by Depledge and Bjerregaard (1989a).

INTERPRETATION OF METAL CONCENTRATIONS

Where does this leave the chances of establishing background or baseline concentrations of trace metals in aquatic invertebrates? Clearly, it is impossible to define such a concentration of even a single metal interspecifically. Closely related species may differ dramatically in tissue and body contents of a heavy metal in the absence of differences in external bioavailabilities. Thus, the ampharetid polychaete worm *M. palmata* has a copper concentration of 760 μg of Cu per gram, while another ampharetid polychaete *Amphareta acutifrons* contains only 33 μg of Cu per gram (Gibbs et al., 1981). Similarly, the cirratulid polychaete *Tharyx marioni* contains an arsenic concentration of between 1000 and 3000 μg of As per gram, and another cirratulid *Cirrifomia tentaculata* has only 84 μg of As per gram (Gibbs et al., 1983).

Even intraspecifically, the concentrations of a metal may vary greatly as a result of inherent variability, not accountable by environmental or physiological factors. Thus, Lobel (1987a) found high variability in whole soft tissue concentrations of zinc in the mussel *M. edulis*, resulting from high variability in kidney zinc concentrations caused by biochemical differences between individual mussels of a single population.

Is it possible, therefore, to interpret the significance of a heavy metal concentration in an aquatic invertebrate body or tissue? Variability in metal accumulation strategies among organisms and in the relative sizes of the different components of the metal contents of organisms sharing the same metal accumulation strategy makes it difficult to make interspecific comparisons of metal concentrations even between closely related species with any confidence (e.g., Moore and Rainbow, 1987; Phillips and Rainbow, 1988). Intraspecifically, in the absence of differences in ambient metal bioavailability, individual variability with physiological state and further (often unexplained) inherent individual variability remain to confound interpretations of metal concentrations. It is difficult, then, to define absolutely a body concentration range reflecting "normal" conditions because of such variability, but some intraspecific comparisons are possible, particularly when the comparisons include populations exposed to unusually high metal bioavailabilities. Such interpretations are the basis of any heavy metal biomonitoring program that would necessarily involve the use of net accumulators (Phillips and Rainbow, 1993).

Typical biomonitoring data sets might be expected to fall conveniently into a group of "background" samples of approximately equal metal concentration, with remaining samples occupying a gradient of increased concentrations indicative of sites exposed to a range of increased metal bioavailabilities. In fact, physiological and other inherent individual variability often causes samples of the first group to fall along a gradient of concentrations themselves (Rainbow, 1993). In effect, it is impossible to define a point along the complete gradient of samples where increased metal bioavailability supersedes physiological or inherent variability as the primary determinant of a particular concentration. The presence of this gray area that is difficult to interpret does not, however, necessarily prevent clear conclusions to be drawn concerning samples at the top of any series of metal concentrations. It is often possible, therefore, to conclude that the metal concentration of a particular sample indicates the presence of atypically high metal bioavailability, significantly raised above those at the other sites monitored.

The data sets in Table 1 illustrate the point. Zinc concentrations in both the barnacle *Balanus amphitrite* and the mussel *Perna viridis* from Hong Kong fall along intraspecific gradients with no grouping of samples into background concentration sets. It is impossible to define the exact points where raised environmental bioavailability of zinc lifts the metal concentrations above the range controlled by physiological variability, although this may have occurred at Tung Chung in each case. Nevertheless, it is clear that the availability of zinc to the barnacle is significantly raised at Hang Hau and Rennies Mill (Phillips and Rainbow, 1988). The identification of sites of increased zinc bioavailability to the mussel is more difficult because *P. viridis* is a partial regulator of zinc, with relatively little effect of raised zinc bioavailability on soft tissue concentrations.

Table 1 Concentrations of Zinc (μg/g) of Dry Weight in the Bodies of the Barnacle *(Balanus amphitrite)* of Standardized Dry Weight (0.004 g) as Estimated from Best-Fit Double-Log Regressions Between Accumulated Body Metal Concentration *(y)* and Body Dry Weight *(x)*, with 95% Confidence Interval (CI). Also Shown are the Mean Concentrations of Zinc (μg/g of Dry Weight) in Soft Tissues of the Green-Lipped Mussel *Perna viridis* with 95% CI.[a,b]

Hong Kong site	Balanus amphitrite			Perna viridis		
	[Zn] (μg/g)	95% CI	ANCOVA	[Zn] (μg/g)	95% CI	ANOVA
Causeway Bay				150	14-286	A
Kennedy Town				146	4-288	A
Queens Pier				141	74-208	A
Hung Hom				118	68-168	A
Reef Island				114	67-161	A
Hang Hau	11,990	10,220-14,070	A	111	75-147	A
Rennies Mill	11,820	9,547-14,640	B	109	65-153	A
Chai Wan Kok	9,353	7,411-11,800	C	153	9-247	A
Kowloon Pier				108	39-177	A
Kwun Tong	7,276	5,269-10,050	C	115	79-151	A
North Point	7,870	6,070-10,210	C	96	38-154	B
Tung Chung	6,491	102-41,600	D	86	70-103	B
Wu Kai Sha	4,671	4,195-5,201	D	65	46-85	C
Tai Po Kau	4,381	3,415-5,620	E	61	42-80	C
Sha Tin	3,214	2,887-3,578	F	66	38-94	C
Lai Chi Chong	2,726	967-7,688	F	53	39-67	D

[a] Sites are ranked in approximate order of decreasing zinc contamination. Different letters in the ANCOVA column indicate significant differences ($P < 0.05$) in metal concentrations between samples by analysis of covariance of best-fit regressions. Different letters in the ANOVA column denote significant differences in metal concentrations between samples by analysis of variance. All samples were collected in April 1986 in Hong Kong.
[b] From Phillips, D. J. H. and P. S. Rainbow. 1988. *Mar. Ecol. Prog. Ser.* 49:83-93. With permission.

ZINC IN THE MUSSEL *MYTILUS EDULIS*: A MODEL INTERPRETATION

The interpretation of the significance of a metal concentration in an aquatic invertebrate does depend heavily on a detailed study of the biology of the metal in that invertebrate. George and Pirie (1980) have produced such a study for zinc in the mussel *M. edulis*, and data are also available on field concentrations of zinc in the soft parts of the mussel (see Table 2), including data on the variability of zinc concentrations in mussels from a single site (Lobel, 1987b). It is possible, then, to attempt to distinguish some of the components of body metal content identified in Figure 1.

Table 2 lists zinc concentrations in the soft tissues of mussels collected in Europe and North America (some from Mussel Watch Programs), including data from sites known or deduced to be zinc polluted (e.g., Hawke's Bay, Newfoundland; Lobel, 1987b). The presence of such zinc-contaminated sites is indicated by raised mean zinc concentrations in the mussels, *M. edulis*, that only partially regulate zinc (vide supra). Most samples in Table 2 are from unpolluted sites and, in order to avoid distortion of any averaging by the inclusion of data from polluted sites, a median zinc concentration is quoted. These median concentrations show remarkable consistency, particularly in light of the physiological variability discussed above. Typical mean zinc concentrations in the soft tissues of *M. edulis* fall in the range of 80 to 110 μg/g of dry weight.

Indeed, George and Pirie (1980) investigated zinc in Scottish mussels containing about 90 μg of Zn per gram in their soft tissues. They concluded that about 30% of body zinc was present as zinc-rich granules in the kidney. These zinc-rich granules are available for excretion (George and Pirie, 1980) and represent the detoxified store of metal illustrated in Figure 1, contributing 27 μg/g to the total soft tissue zinc concentration. The soft tissues of *M. edulis* would have approximately 35 μg/g of zinc associated with enzymes (White and Rainbow, 1985). Thus, about 62 μg of Zn per gram of a total of about 90 μg of Zn per gram in the soft tissues might be accounted for by these two components alone in a mussel from an unpolluted site.

When a mussel is exposed to a raised zinc bioavailability, the total soft tissue concentration of zinc rises; soft tissue concentrations above 200 μg of Zn per gram certainly indicate the presence of extra ambient zinc. Concentrations between 110 and 200 μg of Zn per gram represent a gray area under the influence of physiological variability, with evidence for high environmental zinc increasing as mussel soft tissue concentrations approach 200 μg of Zn per gram. Most of the extra zinc in the whole soft tissues under such circumstances is attributable to increased zinc in the detoxified store in the kidney, awaiting excretion (Lobel, 1987b). This component now represents a higher proportion of a higher total zinc content (see Figure 1).

Similarly detailed studies have been made of the biology of zinc in oysters, notoriously strong accumulators of the metal. For example, Thomson et al. (1985) showed that more than 90% of body zinc in the Pacific oyster *Crassostrea gigas* from the metal-rich Derwent Estuary, Tasmania, is found in detoxified form in granules in amebocyte cells in the blood. In the case of the nonessential metal cadmium and *M. edulis*, 85% of body cadmium in mussels chronically exposed to the metal was detoxified in granules, the rest being present in the cytoplasmic fraction of tissues, usually bound to proteins including metallothioneins (George and Pirie, 1979; George, 1982).

Table 2 Soft Tissue Concentrations of Zinc (µg/g) in the Mussel *(Mytilus edulis)*

Location	Date	Mean(s)	Median of means of samples	Median of samples	No. of means or samples	Sample range	Standard deviation	No. of samples	Ref.
Europe									
Solent, England		91							Segar et al. (1971)
South Devon, England	1977-1979	77.9-285	97.6		14	16.1-634			Boalch et al. (1981)
Northern Ireland	1980-1981	66.1-213	91.8		11	50.6-312			Gault et al. (1983)
Bourgneuf, France	1982-1984					39.8-112			Amiard et al. (1986)
North and Baltic Seas	1973	66-169	83		6	38-575			Karbe et al. (1977)
Scandinavia	1976	14-460	97		54				Phillips (1977)
North America									
East Coast, U.S.	1976			106	19	67-189		19	Goldberg et al. (1978)
Bodega Bay, U.S.	1976-1978	135					32	29	Lauenstein et al. (1990)
	1986-1988	110					42	9	Lauenstein et al. (1990)
Narragansett Bay, U.S.	1976-1978	120					38	20	Lauenstein et al. (1990)
	1986-1988	100					11	15	Lauenstein et al. (1990)
Bellevue, Newfoundland	1988	80					26.8	94	Lobel et al. (1989)
Hawke's Bay, Newfoundland (zinc polluted)	1985	211		184	39	105-593		39	Lobel (1987b)

SUMMARY

Heavy metals that are toxic, whether essential or not, are taken up and accumulated by aquatic invertebrates, usually to concentrations greatly in excess of those in an equivalent amount of the surrounding medium. Accumulation strategies of invertebrates for metals vary greatly between species and intraspecifically between metals, an absolute body concentration that is high for one species being low for another.

Toxic effects are not related to absolute body concentrations but are manifest only when the rate of uptake of a toxic metal exceeds the rates of physiological/biochemical detoxification and/or excretion. An invertebrate with a low total metal concentration may be suffering sublethal toxic effects, resulting from a recent increase in metal uptake rate, while other conspecifics may be free from toxicity, although containing much higher metal concentrations accumulated in detoxified form over an extended time period.

The identification and quantification of different components of the total metal content of an invertebrate (e.g., metabolically available levels, temporary or permanent detoxified metal stores) offer scope for the interpretation of the significance of the metal concentration accumulated in that invertebrate. Furthermore, the comparison of intraspecific metal concentrations of aquatic invertebrates in a biomonitoring program does allow the identification of sites of raised toxic metal bioavailability.

Therefore, the measurement of metal concentrations in aquatic invertebrates cannot tell us directly whether that metal is poisoning the organism. Nevertheless, in situations of metal contamination, the measurement of metal concentrations in a suite of well-researched biomonitors does allow us to recognize whether accumulated concentrations are atypically high, with a real possibility that toxic effects may be present, a vital step in any recognition of potential ecotoxicological effects in the environment.

REFERENCES

Amiard, J. C., C. Amiard-Triquet, B. Berthet, and C. Metayer. 1986. Contribution to the ecotoxicological study of cadmium, lead, copper, and zinc in the mussel *Mytilus edulis*. I. Field study. Mar. Biol. (Berl.) 90:425-431.

Boalch, R., S. Chan, and D. Taylor. 1981. Seasonal variation in the trace metal content of *Mytilus edulis*. Mar. Pollut. Bull. 12:276-280.

Boyden, C. R. 1974. Trace element content and body size in molluscs. Nature (Lond.) 251:311-314.

Boyden, C. R. 1977. Effect of size upon metal content of shellfish. J. Mar. Biol. Assoc. U.K. 57:675-714.

Brady, F. O. 1982. The physiological function of metallothionein. Trends Biochem. Sci. 7:143-145.

Brown, B. E. 1982. The form and function of metal-containing "granules" in invertebrate tissues. Biol. Rev. Camb. Philos. Soc. 57:621-667.

Bryan, G. W. 1968. Concentrations of zinc and copper in the tissues of decapod crustaceans. J. Mar. Biol. Assoc. U.K. 48:303-321.

Bryan, G. W. 1971. The effects of heavy metals (other than mercury) on marine and estuarine organisms. Proc. R. Soc. Lond. Ser. B Biol. Sci. 177:389-410.

Bryan, G. W. 1979. Bioaccumulation of marine pollutants. Philos. Trans. R. Soc. Lond. B Biol. Sci. 286:483-505.

Bryan, G. W., and P. E. Gibbs. 1979. Zinc — a major inorganic component of nereid polychaete jaws. J. Mar. Biol. Assoc. U.K. 59:969-973.

Bryan, G. W., and P. E. Gibbs. 1980. Metals in nereid polychaetes: the contributions of metal in the jaws to the total body burden. J. Mar. Biol. Assoc. U.K. 60:641-654.

Cain, D. J., S. N. Luoma, J. L. Carter, and S. V. Fend. 1992. Aquatic insects as bioindicators of trace element contamination in cobble-bottom rivers and streams. Can. J. Fish. Aquat. Sci. 49:2141-2154.

Campbell, R. J., and M. B. Jones. 1990. Water permeability of *Palaemon longirostris* and other euryhaline caridean prawns. J. Exp. Biol. 150:145-158.

Carlisle, D. B. 1968. Vanadium and other metals in ascidians. Proc. R. Soc. Lond. Ser. B Biol. Sci. 171:31-41.

Chan, H. M. 1988. Accumulation and tolerance to cadmium, copper, lead, and zinc by the green mussel *Perna viridis*. Mar. Ecol. Prog. Ser. 48:295-303.

Chan, H. M., P. Bjerregaard, P. S. Rainbow, and M. H. Depledge. 1992. The uptake of zinc and cadmium by two populations of shore crabs *(Carcinus maenas)* at different salinities. Mar. Ecol. Prog. Ser. 86:91-97.

Depledge, M. H. 1989a. Re-evaluation of metabolic requirements for copper and zinc in decapod crustaceans. Mar. Environ. Res. 27:115-126.

Depledge, M. H. 1989b. Studies on copper and iron concentrations, distributions, and uptake in the brachyuran, *Carcinus maenas* (L.), following starvation. Ophelia 30:187-197.

Depledge, M. H., and P. Bjerregaard. 1989a. Explaining individual variation in trace metal concentrations in selected marine invertebrates: the importance of interactions between physiological state and environmental factors. p. 121-126. *In* J. C. Aldrich (Ed.). Phenotypic responses and individuality in aquatic ecototherms. JAPAGA, Ashford, Ireland.

Depledge, M. H., and P. Bjerregaard. 1989b. Haemolymph protein composition and copper levels in decapod crustaceans. Helgol. Meeresunters. 43:207-223.

Depledge, M. H., and D. J. Phillips. 1986. Circulation, respiration, and fluid dynamics in the gastropod mollusc, *Hemifusus tuba* (Gmelin). J. Exp. Mar. Biol. Ecol. 95:1-13.

Depledge, M. H., and P. S. Rainbow. 1990. Models of regulation and accumulation of trace metals in marine invertebrates. Comp. Biochem. Physiol. C Comp. Pharmacol. 97:1-7.

Eisler, R. 1981. Trace metal concentrations in marine organisms. Pergamon Press, New York. 696 pp.

Engel, D. W. 1987. Metal regulation and molting in the blue crab *Callinectes sapidus*: zinc and metallothionein. Biol. Bull. (Woods Hole) 172:69-82.

Engel, D. W., and M. Brouwer. 1987. Metal regulation and molting in the blue crab, *Callinectes sapidus*: metallothionein function in metal metabolism. Biol. Bull. (Woods Hole) 173:239-251.

Engel, D. W., and M. Brouwer. 1989. Metallothionein and metallothionein-like proteins: physiological importance. Adv. Comp. Environ. Physiol. 5:53-75.

Engel, D. W., and B. A. Fowler. 1979. Factors influencing cadmium accumulation and its toxicity to marine organisms. Environ. Health Perspect. 28:81-88.

Gault, N. F. S., E. L. C. Tolland, and J. G. Parker. 1983. Spatial and temporal trends in heavy metal concentrations in mussels from Northern Island coastal waters. Mar. Biol. (Berl.) 77:307-316.

George, S. G. 1982. Subcellular accumulation and detoxication of metals in aquatic animals. p. 3-52. *In* W. B. Vernberg, A. Calabrese, F. P. Thurberg, and J. F. Vernberg (Eds.). Physiological mechanisms of marine pollutant toxicity. Academic Press, New York.

George, S. G. 1990. Biochemical and cytological assessments of metal toxicity in marine animals. p. 123-142. *In* R. W. Furness and P. S. Rainbow (Eds.). Heavy metals in the marine environment. CRC Press, Boca Raton, Fla.

George, S. G., and B. J. S. Pirie. 1979. The occurrence of cadmium in sub-cellular particles in the kidney of the marine mussel, *Mytilus edulis*, exposed to cadmium. The use of electron microprobe analysis. Biochim. Biophys. Acta 580:234-244.

George, S. G., and B. J. S. Pirie. 1980. Metabolism of zinc in the mussel *Mytilus edulis* (L.): a combined ultrastructural and biochemical study. J. Mar. Biol. Assoc. U.K. 60:575-590.

George, S. G., B. J. S. Pirie, and T. L. Coombs. 1976. The kinetics of accumulation and excretion of ferric hydroxide in *Mytilus edulis* (L.) and its distribution in the tissues. J. Exp. Mar. Biol. Ecol. 23:71-84.

George, S. G., B. J. S. Pirie, and T. L. Coombs. 1980. Isolation and elemental analysis of metal-rich granules from the kidney of the scallop, *Pecten maximus* (L.). J. Exp. Mar. Biol. Ecol. 42:143-156.

Gibbs, P. E., and G. W. Bryan. 1980. Copper — the major metal component of glycerid polychaete jaws. J. Mar. Biol. Assoc. U.K. 60:205-214.

Gibbs, P. E., G. W. Bryan, and K. P. Ryan. 1981. Copper accumulation by the polychaete *Melinna palmata*: an antipredation mechanism? J. Mar. Biol. Assoc. U.K. 61:707-722.

Gibbs, P. E., W. J. Langston, G. R. Burt, and P. L. Pascoe. 1983. *Tharyx marioni* (Polychaeta): a remarkable accumulator of arsenic. J. Mar. Biol. Assoc. U.K. 63:313-325.

Goldberg, E. D., V. T. Bowen, J. W. Farrington, G. Harvey, J. H. Martin, P. L. Parker, R. W. Risebrough, W. Robertson, E. Schneider, and E. Gamble. 1978. The mussel watch. Environ. Conserv. 5:101-125.

Hare, L., and P. G. C. Campbell. 1992. Temporal variations of trace metals in aquatic insects. Freshwater Biol. 27:13-27.

Hare, L., E. Saouter, P. G. C. Campbell, A. Tessier, F. Ribeyre, and A. Boudou. 1991. Dynamics of cadmium, lead, and zinc exchange between nymphs of the burrowing mayfly *Hexagenia rigida* (Ephemeroptera) and the environment. Can. J. Fish. Aquat. Sci. 48:39-47.

Hobden, D. J. 1967. Iron metabolism in *Mytilus edulis*. I. Variation in total content and distribution. J. Mar. Biol. Assoc. U.K. 47:597-606.

Hopkin, S. P. 1990a. Species-specific differences in the net assimilation of zinc, cadmium, lead, copper, and iron by the terrestrial isopods *Oniscus asellus* and *Porcellio scaber*. J. Appl. Ecol. 27:460-474.

Hopkin, S. P. 1990b. Critical concentrations, pathways of detoxification, and cellular ecotoxicology of metals in terrestrial arthropods. Funct. Ecol. 4:321-327.

Hopkin, S. P. 1993. Deficiency and excess of copper in terrestrial isopods. p. 359-382. *In* R. Dallinger and P. S. Rainbow (Eds.). Ecotoxicology of metals in invertebrates. Lewis Publishers, Boca Raton, Fla.

Kalk, M. 1963. Absorption of vanadium by tunicates. Nature (Lond.) 198:1010-1011.

Karbe, L., C. Schnier, and H. O. Siewers. 1977. Trace elements in mussels *(Mytilus edulis)* from coastal areas of the North Sea and the Baltic. Multielement analyses using instrumental neutron activation analysis. J. Radioanal. Chem. 37:927-943.

Latouche, Y. D., and M. C. Mix. 1982. The effects of depuration, size, and sex on trace metal levels in bay mussels. Mar. Pollut. Bull. 13:27-29.

Lauenstein, G. G., A. Robertson, and T. P. O'Connor. 1990. Comparison of trace metal data in mussels and oysters from a mussel watch programme of the 1970s with those from a 1980s programme. Mar. Pollut. Bull. 21:440-447.

Lobel, P. B. 1987a. Short-term and long-term uptake of zinc by the mussel, *Mytilus edulis*: a study in individual variability. Arch. Environ. Contam. Toxicol. 16:723-732.

Lobel, P. B. 1987b. Intersite, intrasite, and inherent variability of the whole soft tissue zinc concentrations of individual mussels *Mytilus edulis:* importance of the kidney. Mar. Environ. Res. 21:59-71.

Lobel, P. B., S. P. Belkhode, S. E. Jackson, and H. P. Longerich. 1989. A universal method for quantifying and comparing the residual variability of element concentrations in biological tissues using 25 elements in the mussel *Mytilus edulis* as a model. Mar. Biol. (Berl.) 102:513-518.

Luoma, S. N. 1983. Bioavailability of trace metals to aquatic organisms — a review. Sci. Total Environ. 28:1-22.

Mantel, L. H., and L. L. Farmer. 1983. Osmotic and ionic regulation. p. 53-161. *In* L. H. Mantel (Ed.). The biology of crustacea. Vol. 5. Academic Press, New York.

Mason, A. Z., and J. A. Nott. 1981. The role of intracellular biomineralised granules in the regulation and detoxification of metals in gastropods with special reference to the marine prosobranch *Littorina littorea*. Aquat. Toxicol. (Amst.) 1:239-256.

Michibata, H., T. Terada, N. Anada, K. Yamakara, and T. Namakurai. 1986. The accumulation and distribution of vanadium, iron, and manganese in solitary ascidians. Biol. Bull. (Woods Hole) 171:672-681.

Moore, P. G., and P. S. Rainbow. 1984. Ferritin crystals in the gut caeca of *Stegocephaloides christianiensis* Boeck and other Stegocephalidae (Amphipoda: Gammaridea): a functional interpretation. Philos. Trans. R. Soc. Lond. B Biol. Sci. 306:219-245.

Moore, P. G., and P. S. Rainbow. 1987. Copper and zinc in an ecological series of talitroidean Amphipoda (Crustacea). Oecologia (Heidelb.) 73:120-126.

Nieboer, E., and D. H. S. Richardson. 1980. The replacement of the nondescript term "heavy metals" by a biologically and chemically significant classification of metal ions. Environ. Pollut. Ser. B Chem. Phys. 1:3-26.

Nott, J. A., and W. J. Langston. 1989. Cadmium and the phosphate granules in *Littorina littorea*. J. Mar. Biol. Assoc. U.K. 69:219-227.

Nugegoda, D., and P. S. Rainbow. 1987. The effect of temperature on zinc regulation by the decapod crustacean *Palaemon elegans* Rathke. Ophelia 27:17-30.

Nugegoda, D., and P. S. Rainbow. 1988a. Effect of a chelating agent (EDTA) on zinc uptake and regulation by *Palaemon elegans* (Crustacea: Decapoda). J. Mar. Biol. Assoc. U.K. 68:25-40.

Nugegoda, D., and P. S. Rainbow. 1988b. Zinc uptake and regulation by the sublittoral prawn *Pandalus montagui* (Crustacea: Decapoda). Estuarine Coastal Shelf Sci. 26:619-632.

Nugegoda, D., and P. S. Rainbow. 1989a. Effects of salinity changes on zinc uptake and regulation by the decapod crustaceans *Palaemon elegans* and *Palaemonetes varians*. Mar. Ecol. Prog. Ser. 51:57-75.

Nugegoda, D., and P. S. Rainbow. 1989b. Salinity, osmolality, and zinc uptake in *Palaemon elegans* (Crustacea: Decapoda). Mar. Ecol. Prog. Ser. 55:149-157.

Nugegoda, D., and P. S. Rainbow. 1989c. Zinc uptake rate and regulation breakdown in the decapod crustacean *Palaemon elegans* Rathke. Ophelia 30:199-212.

Orren, M. J., G. A. Eagle, H. F-K. O. Hennig, and A. Green. 1980. Variations in trace metal content of the mussel *Chloromytilus meridionalis* (Kr.) with season and sex. Mar. Pollut. Bull. 11:253-257.

Pequegnat, J. E., S. W. Fowler, and L. F. Small. 1969. Estimates of the zinc requirements of marine organisms. J. Fish. Res. Board Can. 26:145-150.

Phillips, D. J. H. 1977. The common mussel *Mytilus edulis* as an indicator of trace metals in Scandinavian waters. I. Zinc and cadmium. Mar. Biol. (Berl.) 43:283-291.

Phillips, D. J. H. 1985. Organochlorines and trace metals in green-lipped mussels *Perna viridis* from Hong Kong waters: a test of indicator ability. Mar. Ecol. Prog. Ser. 21:251-258.

Phillips, D. J. H., and P. S. Rainbow. 1988. Barnacles and mussels as biomonitors of trace elements: a comparative study. Mar. Ecol. Prog. Ser. 49:83-93.

Phillips, D. J. H., and P. S. Rainbow. 1989. Strategies of trace metal sequestration in aquatic organisms. Mar. Environ. Res. 28:207-210.

Phillips, D. J. H., and P. S. Rainbow. 1993. Biomonitoring of trace aquatic contaminants. Elsevier Applied Science, London. 371 pp.

Phillips, D. J. H., and W. W.-S. Yim. 1981. A comparative evaluation of oysters, mussels, and sediments as indicators of trace metals in Hong Kong waters. Mar. Ecol. Prog. Ser. 6:285-293.

Pullen, J. S. H., and P. S. Rainbow. 1991. The composition of pyrophosphate heavy metal detoxification granules in barnacles. J. Exp. Mar. Biol. Ecol. 150:249-266.

Rainbow, P. S. 1987. Heavy metals in barnacles. p. 405-417. *In* A. J. Southward (Ed.). Barnacle biology. A. A. Balkema, Rotterdam.

Rainbow, P. S. 1988. The significance of trace metal concentrations in decapods. Symp. Zool. Soc. Lond. 59:291-313.

Rainbow, P. S. 1990a. Heavy metal levels in marine invertebrates. p. 67-79. *In* R. W. Furness and P. S. Rainbow (Eds.). Heavy metals in the marine environment. CRC Press, Boca Raton, Fla.

Rainbow, P. S. 1990b. Trace metal concentrations in a Hong Kong penaeid prawn, *Metapenaeopsis palmensis* (Haswell). p. 1221-1228. *In* B. Morton (Ed.). Proc. 2nd Int. Mar. Biol. Workshop: the marine flora and fauna of Hong Kong and southern China, Hong Kong, 1986. Hong Kong University Press, Hong Kong.

Rainbow, P. S. 1993. The significance of trace metal concentrations in marine invertebrates. p. 3-23. *In* R. Dallinger and P. S. Rainbow (Eds.). Ecotoxicology of metals in invertebrates. Lewis Publishers, Boca Raton, Fla.

Rainbow, P. S., and C. Abdennour. 1989. Copper and haemocyanin in the mesopelagic decapod crustacean *Systellaspis debilis*. Oceanol. Acta 12:91-94.

Rainbow, P. S., and R. Dallinger. 1993. Metal uptake, regulation, and excretion in freshwater invertebrates. p. 119-131. *In* R. Dallinger and P. S. Rainbow (Eds.). Ecotoxicology of metals in invertebrates. Lewis Publishers, Boca Raton, Fla.

Rainbow, P. S., I. Malik, and P. O'Brien. 1993. Physicochemical and physiological effects on the uptake of dissolved zinc and cadmium by the amphipod crustacean *Orchestia gammarellus*. Aquat. Toxicol. 25:15-30.

Rainbow, P. S., D. J. H. Phillips, and M. H. Depledge. 1990. The significance of trace metal concentrations in marine invertebrates. Mar. Pollut. Bull. 21:321-324.

Rainbow, P. S., and S. L. White. 1989. Comparative strategies of heavy metal accumulation by crustaceans: zinc, copper, and cadmium in a decapod, an amphipod, and a barnacle. Hydrobiologia 174:245-262.

Rainbow, P. S., and S. L. White. 1990. Comparative accumulation of cobalt by three crustaceans: a decapod, an amphipod, and a barnacle. Aquat. Toxicol. (Amst.) 16:113-126.

Roesijadi, G. 1980-1981. The significance of low molecular weight, metallothionein-like proteins in marine invertebrates: current status. Mar. Environ. Res. 4:167-179.

Segar, D. A., J. D. Collins, and J. P. Riley. 1971. The distribution of the major and some minor elements in marine animals. II. Molluscs. J. Mar. Biol. Assoc. U.K. 51:131-136.

Simkiss, K. 1977. Biomineralisation and detoxification. Calcif. Tissue Res. 24:199-200.

Simkiss, K. 1980. Detoxification, calcification, and intracellular storage of ions. p. 13-18. *In* M. Omori and N. Watabe (Eds.). The mechanisms of biomineralisation in animals and plants. Tokai University Press, Tokyo.

Simkiss, K. 1981. Calcium, pyrophosphate, and cellular pollution. Trends Biochem. Sci. 6:III-V.
Simkiss, K., and M. G. Taylor. 1989a. Metal fluxes across the membranes of aquatic organisms. CRC Rev. Aquat. Sci. 1:173-189.
Simkiss, K., and M. G. Taylor. 1989b. Convergence of cellular systems of metal detoxification. Mar. Environ. Res. 28:211-214.
Simkiss, K., M. Taylor, and A. Z. Mason. 1982. Metal detoxification and bioaccumulation in molluscs. Mar. Biol. Lett. 3:187-201.
Spaargaren, D. H. 1983. Osmotically induced changes in copper and iron concentrations in three euryhaline crustacean species. Neth. J. Sea Res. 17:96-105.
Stoecker, D. 1978. Resistance of a tunicate to fouling. Biol. Bull. (Woods Hole) 155:615-626.
Stoecker, D. 1980. Relationships between chemical defence and ecology in benthic ascidians. Mar. Ecol. Prog. Ser. 3:257-265.
Sunda, W. G., D. W. Engel, and R. M. Thuotte. 1978. Effects of chemical speciation on the toxicity of cadmium to grass shrimp *Palaemonetes pugio*: importance of free cadmium ion. Environ. Sci. Technol. 12:409-413.
Taylor, M. G., and K. Simkiss. 1984. Inorganic deposits in invertebrate tissues. Environ. Chem. 3:102-138.
Thomson, J. D., B. J. S. Pirie, and S. G. George. 1985. Cellular metal distribution in the Pacific oyster, *Crassostrea gigas* (Thun.) determined by quantitative X-ray microprobe analysis. J. Exp. Mar. Biol. Ecol. 85:37-45.
Timmermans, K. R. 1993. Accumulation and effects of trace metals in freshwater invertebrates. p. 133-148. *In* R. Dallinger and P. S. Rainbow (Eds.). Ecotoxicology of metals. Lewis Publishers, Boca Raton, Fla.
Timmermans, K. R., E. Spijkerman, M. Tonkes, and H. Govers. 1992. Cadmium and zinc uptake by two species of aquatic invertebrate predators from dietary and aqueous sources. Can. J. Fish. Aquat. Sci. 49:655-662.
Timmermans, K. R., and P. A. Walker. 1989. The fate of trace metals during the metamorphosis of chironomids (Diptera, Chironomidae). Environ. Pollut. 62:73-85.
Viarengo, A. 1989. Heavy metals in marine invertebrates: mechanisms of regulation toxicity at the cellular level. CRC Crit. Rev. Aquat. Sci. 1:295-317.
Walker, G., P. S. Rainbow, P. Foster, and D. J. Crisp. 1975a. Barnacles: possible indicators of zinc pollution? Mar. Biol. (Berl.) 30:57-65.
Walker, G., P. S. Rainbow, P. Foster, and D. L. Holland. 1975b. Zinc phosphate granules in tissue surrounding the midgut of the barnacle *Balanus balanoides*. Mar. Biol. (Berl.) 33:161-166.
Weeks, J. M., and P. S. Rainbow. 1991. The uptake and accumulation of zinc and copper from solution by two species of talitrid amphipods (Crustacea). J. Mar. Biol. Assoc. U.K. 71:811-826.
White, S. L., and P. S. Rainbow. 1982. Regulation and accumulation of copper, zinc, and cadmium by the shrimp *Palaemon elegans*. Mar. Ecol. Prog. Ser. 8:95-101.
White, S. L., and P. S. Rainbow. 1984. Regulation of zinc concentrations by *Palaemon elegans* (Crustacea: Decapoda): zinc flux and effects of temperature, zinc concentration, and moulting. Mar. Ecol. Prog. Ser. 16:135-147.
White, S. L., and P. S. Rainbow. 1985. On the metabolic requirements for copper and zinc in molluscs and crustaceans. Mar. Environ. Res. 16:215-229.
White, S. L., and P. S. Rainbow. 1986. Accumulation of cadmium by *Palaemon elegans* (Crustacea: Decapoda). Mar. Ecol. Prog. Ser. 32:17-25.
White, S. L., and P. S. Rainbow. 1987. Heavy metal concentrations and size effects in the mesopelagic decapod crustacean *Systellaspis debilis*. Mar. Ecol. Prog. Ser. 37:147-151.

Williams, R. J. P. 1981a. Physico-chemical aspects of inorganic element transfer through membranes. Philos. Trans. R. Soc. Lond. B Biol. Sci. 294:57-74.
Williams, R. J. P. 1981b. Natural selection of the chemical elements. Proc. R. Soc. Lond. Ser. B Biol. Sci. 213:361-397.
Williams, R. J. P. 1984. Zinc: what is its role in biology? Endeavour (Oxf.) 8:65-70.
Wright, D. A. 1977. The effect of salinity on cadmium uptake by the tissues of the shore crab *Carcinus maenas*. J. Exp. Biol. 67:137-146.
Wright, D. A. 1980. Cadmium and calcium interactions in the freshwater amphipod *Gammarus pulex*.. Freshwater Biol. 10:123-133.

CHAPTER **19**

Selenium in Aquatic Organisms

A. Dennis Lemly

INTRODUCTION

The role and importance of selenium as an environmental contaminant have gained widespread attention among research scientists, natural resource managers, and federal and state regulatory agencies during the past decade. Although the basic toxicological symptoms and paradox of selenium (nutritionally required in small amounts but highly toxic in slightly greater amounts) have been known for many years (Draize and Beath, 1935; Ellis et al., 1937; Rosenfeld and Beath, 1946; Hartley and Grant, 1961), it was not until the late 1970s and early 1980s that the potential for widespread contamination of aquatic ecosystems due to human activities was recognized (Andren et al., 1975; Cherry and Guthrie, 1977; Evans et al., 1980; National Research Council, 1980; Braunstein et al., 1981). In fact, as recently as 1970, selenium was being called the "unknown pollutant" with respect to what was known about its cycling and toxicity in the aquatic environment (Copeland, 1970).

It is now evident that two factors stand apart as the major human-related causes of selenium mobilization and introduction into aquatic systems in the United States. The first major factor is the procurement, processing, and combustion of fossil fuels. Selenium is an important trace element present in coal, crude oil, oil shale, coal conversion materials (liquefaction oils and synthetic gases), and their waste by-products (Pillay et al., 1969; American Petroleum Institute, 1978; Fruchter and Petersen, 1979; Clark et al., 1980; Cowser and Richmond, 1980; Schlinger and Richter, 1980; U.S. Environmental Protection Agency, 1980; Nystrom and Post, 1982). It can be leached directly from coal and oil-shale mining, preparation, and storage sites (Davis and Boegly, 1981; Heaton et al., 1982; Jones, 1990) and, more important, is highly concentrated in the mineral fraction (fly ash and bottom ash) remaining after coal is burned (Kaakinen et al., 1975; Klein et al., 1975). Over 70 million tons of coal fly ash are produced annually in the U.S., and most of it is disposed by dumping into wet-slurry or dry-ash basins (Murtha et al., 1983). Selenium-laden leachate and overflow from these basins often make their way into rivers, streams, and impoundments. Selenium concentrations can rapidly increase in fish and aquatic organisms in the receiving water, ultimately resulting in tissue damage, reproductive failure, and elimination of entire fish communities (Cumbie and Van Horn, 1978; Garrett and Inman, 1984; Lemly, 1985a; Sorensen, 1986).

The second major factor is the irrigation of seleniferous soils for crop production in arid and semiarid regions of the country. Deposits of cretaceous marine shales have weathered to produce high-selenium soils in many areas of the western U.S., notably the San Joaquin Valley of California and certain parts of Wyoming, Colorado, Nevada, North and South Dakota, Montana, New Mexico, Arizona, Utah, Nebraska, and Kansas (Kubota, 1980; Tanji et al., 1986; Presser and Ohlendorf, 1987; Allen and Wilson, 1990; Presser et al., 1990; Severson et al., 1991). These areas usually require substantial irrigation for agricultural crop production, and the drainage that results may be highly contaminated with dissolved selenium salts that have leached from the soil (Presser and Barnes, 1985; Saiki, 1986a; Summers and Anderson, 1986; Fuji, 1988; Deverel et al., 1989). Selenium in agricultural irrigation drainwater was responsible for massive poisoning of fish and wildlife at Kesterson National Wildlife Refuge, California, in the early to mid-1980s (Marshall, 1985; Hoffman et al., 1986; Saiki, 1986a, 1986b; Saiki and Lowe, 1987; Ohlendorf et al., 1986, 1988). Subsequent studies have shown irrigation-related selenium contamination to be a threat to aquatic systems and wildlife refuges in many western states (Summers and Anderson, 1986; U.S. Fish and Wildlife Service, 1986; Ohlendorf et al., 1987; Sylvester et al., 1991).

Federal and state regulatory agencies have become actively involved with the environmental selenium issue. In 1987, the U.S. Environmental Protection Agency lowered the permissible concentration of waterborne selenium from 35 µg/l (parts per billion) to 5 µg/l to provide increased protection for fish and aquatic life, citing environmental damage caused by coal fly ash (U.S. Environmental Protection Agency, 1987). Some states have enacted or proposed their own local standards for selenium that are even more restrictive than the U.S. Environmental Protection Agency national freshwater criterion (North Carolina Division of Environmental Management, 1986; California State Water Resources Control Board, 1987; California Environmental Protection Agency, 1992). The case histories, research studies, and regulatory responses all indicate selenium to be a highly hazardous contaminant with the ability to quickly affect fish and aquatic life. Natural resource managers are now very aware of this threat, and most have developed and implemented aquatic monitoring programs to detect and evaluate selenium contamination. Central to this evaluation process is the interpretation of selenium concentrations in tissues of fish and aquatic organisms. It is critical to know the biological importance of this trace element to forage fish and food-chain organisms themselves and to predatory species of fish and the wildlife that feed on them.

INTERPRETING SELENIUM CONCENTRATIONS

All values for tissue concentrations in this chapter are given as dry weight. Data from references that reported only wet weights were converted to dry weights by assuming 75% moisture, i.e., 4 times wet weight concentration.

FISH TISSUES

Salmonids are very sensitive to selenium contamination, and they exhibit toxic symptoms even when tissue concentrations are quite low. In laboratory studies, Hunn et al. (1987) exposed rainbow trout fry *(Oncorhynchus mykiss)* to waterborne sodium selenite and found that significant mortality occurred when whole-body concentrations exceeded 1 µg/g (parts per million). Hodson et al. (1980) and Hilton et al. (1980) exposed juvenile rainbow trout to waterborne and dietary sodium selenite and found that significant changes in blood chemistry occurred when whole-body tissue concentrations reached about 2 µg/g (liver tissues contained 51 µg/g). Survival was reduced when whole-body concentrations exceeded 5 µg/g. Hamilton et al. (1986, 1989, 1990) exposed juvenile chinook salmon *(Oncorhynchus tshawytscha)* to combinations of waterborne and dietary selenium (waterborne sodium selenate/sodium selenite ratio, 6:1; field-source selenium diet and seleno-DL-methionine-spiked commercial diet) and observed that smoltification and seawater migration were impaired when whole-body tissue concentrations reached about 20 µg/g (Table 1). Mortality occurred when concentrations exceeded 5 µg/g. However, growth was impaired at whole-body tissue concentrations of only 2 to 3 µg/g; these concentrations were only 2 to 3 times those of the controls (0.8 to 1.0 µg/g).

In studies of juvenile and adult fathead minnows *(Pimephales promelas,* a cyprinid), Bennett et al. (1986) and Ogle and Knight (1989) reported that growth was inhibited at whole-body tissue concentrations of 5 to 8 µg/g or greater (selenium administered as waterborne sodium selenate and a dietary mixture of 25% sodium selenate, 50% sodium selenite, and 25% seleno-L-methionine). Schultz and Hermanutz (1990) dosed outdoor experimental streams with 10 µg/l of sodium selenite and observed the effects on reproduction in fathead minnows. Reproductive success (survival of fry to swim-up) was impaired when ovarian tissue of spawning females contained about 24 µg/g and resultant fry contained about 16 µg/g on a whole-body basis (Table 1).

Coughlan and Velte (1989) fed selenium-laden red shiners *(Notropis lutrensis)* collected from a contaminated power plant reservoir to juvenile striped bass *(Morone saxatilis,* a percichthyid) and found that the fish accumulated 14 to 16 µg/g in skeletal muscle tissue, did not gain weight, and died within 78 days. There was also extensive tissue damage in the liver and trunk kidney of these fish. Selenium in agricultural irrigation drainwater has been implicated in the decline of striped bass in California (Greenberg and Kopec, 1986). Whole-body concentrations in juvenile striped bass collected from areas impacted by irrigation drainage ranged from about 5 to 8 µg/g (Greenberg and Kopec, 1986; Saiki and Palawski, 1990). Selenium concentrations in striped bass that succumbed to irrigation drainage in laboratory studies were about 2 µg/g (Saiki et al., 1992).

Several field and laboratory studies have been conducted that describe the effects of selenium on bluegill *(Lepomis macrochirus)* and other centrarchids (Table 1). Cumbie and Van Horn (1978) and Lemly (1985a, 1985b) found that selenium concentrations of 12 to 16 µg/g in skeletal muscle and 40 to 60 µg/g in ovaries were associated with reproductive failure and mortality in nine species of centrarchids present in a power plant cooling reservoir in North Carolina. Similar effects were

Table 1 Selenium Concentrations Associated with Toxic Effects in Fish and Aquatic Organisms

Species	Tissue	Selenium concentration[a] (μg/g)	Effect	Ref.
Rainbow trout (Oncorhynchus mykiss)	Whole body	2	Blood changes	Hodson et al. (1980)
	Liver	51	Blood changes	Hodson et al. (1980)
	Whole body	5	Mortality	Hilton et al. (1980)
	Whole body	1	Mortality	Hunn et al. (1987)
Chinook salmon (Oncorhynchus tshawytscha)	Whole body	20	Reduced smolting	Hamilton et al. (1986)
	Whole body	2	Reduced growth	Hamilton et al. (1990)
	Whole body	5	Mortality	Hamilton et al. (1990)
Fathead minnow (Pimephales promelas)	Whole body	5	Reduced growth	Ogle and Knight (1989)
	Ovaries	24	Reproductive failure	Schultz and Hermanutz (1990)
	Whole body	16	Reproductive failure	Schultz and Hermanutz (1990)
Striped bass (Morone saxatilis)	Skeletal muscle	14	Mortality	Coughlan and Velte (1989)
	Whole body	2	Mortality	Saiki et al. (1992)
Bluegill sunfish (Lepomis macrochirus)	Skeletal muscle	20	Mortality	Finley (1985)
	Liver	34	Mortality	Finley (1985)
	Carcass	24	Reproductive failure	Gillespie and Baumann (1986)
	Ovaries	23	Reproductive failure	Gillespie and Baumann (1986)
	Whole body	5	Mortality	U.S. Fish and Wildlife Service (1990)
	Whole body	19	Reproductive failure	Coyle et al. (1993)
	Ovaries	34	Reproductive failure	Coyle et al. (1993)
	Eggs	42	Reproductive failure	Coyle et al. (1993)
	Ovaries	18	Reproductive failure	Hermanutz et al. (1992)
	Skeletal muscle	16	Reproductive failure	Hermanutz et al. (1992)
	Liver	29	Reproductive failure	Hermanutz et al. (1992)
	Whole body	18	Reproductive failure	Hermanutz et al. (1992)
	Whole body	15	Teratogenic defects	Lemly (1993b)
Green alga (Selenastrum capricornutum)	Whole organism	20	Reduced cell replication	Foe and Knight (1986)
Cyanobacterium (Anabaena flos-aquae)	Whole organism	394	Reduced chlorophyll a	Kiffney and Knight (1990)
Cladoceran (Daphnia magna)	Whole organism	15	Reduced weight	Ingersoll et al. (1990)
	Whole organism	32	Reproductive failure	Ingersoll et al. (1990)

[a] Selenium concentrations in parts per million (μg/g) on a dry weight basis.

reported for centrarchids in a selenium-contaminated reservoir in Texas, where skeletal muscle concentrations were in the 8- to 36-μg/g range (Garrett and Inman, 1984). Several physiologically important changes in blood parameters, tissue structure in major organs (ovary, kidney, liver, heart, gills), and organ weight-body weight relations have also been described for centrarchids in these contaminated reservoirs (Sorenson et al., 1984; Sorensen, 1986, 1988). Selenium concentrations of 20 to 80 μg/g in the various tissues were associated with pathological conditions.

Finley (1985) fed selenium-laden invertebrates from a contaminated reservoir to juvenile bluegill and found that mortality occurred when skeletal muscle tissues contained 20 to 32 μg/g and liver tissue contained 32 to 86 μg/g. Whole-body concentrations of only 4 to 6 μg/g were associated with mortality when juvenile bluegill were fed selenomethionine-spiked commercial diets in the laboratory (U.S. Fish and Wildlife Service, 1990).

Gillespie and Baumann (1986) brought selenium-laden adult bluegill from a contaminated power plant reservoir into the laboratory and spawned them artificially to produce crosses between clean and contaminated parents. The results showed that contaminated females (selenium concentrations of 8 to 36 μg/g in carcasses and 12 to 55 μg/g in ovaries) did not produce viable offspring. The fertility and hatchability of the eggs were not affected, but the high concentration of selenium (12 to 55 μg/g) transferred from the eggs to developing embryos during yolk-sac absorption resulted in edema, morphological deformities, and death prior to the swim-up stage. Similar findings were reported by Woock et al. (1987).

In a laboratory study, Lemly (1993a) examined the effect of reductions in water temperature and photoperiod, mimicking winter conditions, on the toxicity of combined dietary (5.1 μg/g of dry weight) and waterborne (4.8 μg/l) selenium to juvenile bluegill. Elevated selenium caused hematological changes and gill damage that reduced respiratory capacity while increasing respiratory demand and oxygen consumption. Elevated selenium in combination with low water temperature (4°C) caused reduced activity and feeding, depletion of 50 to 80% of body lipid, and significant mortality within 60 days. Fish in warm-water selenium exposures continued to actively feed, and lipid depletion did not occur despite increased oxygen consumption. The combination of selenium-induced elevation in energy demand and reductions in feeding due to cold temperature and short photoperiod was designated "winter stress syndrome." This syndrome caused young bluegill to undergo a drain in energy that resulted in the death of about one third of the fish. Selenium concentrations (whole-body) of 5-8 μg/g were associated with these effects.

In another laboratory study, Coyle et al. (1993) evaluated the effects of waterborne and dietary selenium (waterborne sodium selenate:sodium selenite ratio, 6:1; seleno-L-methionine-spiked commercial diet) on the reproductive success of bluegill. The results showed that offspring from females containing whole-body selenium concentrations of 16 to 18 μg/g (30 to 38 μg/g in ovaries; 40 to 45 μg/g in eggs) failed to survive beyond the swim-up stage (5 to 7 days post hatch). No effect was observed on the health or survival of adult fish, spawning frequency, eggs per spawn, or hatchability of eggs. The authors recommended that gravid ovarian tissue be used to monitor the selenium concentrations and the potential effects on bluegill populations, because this tissue delivers the toxic "dose" to the developing fry. They concluded that ovarian selenium concentrations in excess of 13 µg/g may result in reproductive impairment. This number agrees very well with the findings of Gillespie and Baumann (1986), who observed that feral bluegill with ovarian selenium concentrations of 12 µg/g or greater failed to produce viable offspring. It also agrees with the findings of Hermanutz et al. (1992), who dosed outdoor experimental streams with sodium selenite, allowed natural cycling and bioaccumulation to occur in the food chain and adult bluegill, and then measured spawning success. They found that ovarian selenium concentrations of 10 to 24 µg/g in the skeletal muscle, 22 to 85 µg/g in the liver, and 12 to 35 µg/g in the whole body were associated with a 15% decrease in survival of adult fish and almost complete failure of reproduction, with larvae exhibiting edema, lordosis (spinal deformities), and hemorrhaging.

Lemly (1993b) studied the teratogenic effects of selenium in natural populations of centrarchids and other warm-water fish species. He determined the prevalence of abnormalities and associated tissue selenium concentrations in a contaminated lake and two reference lakes over a period of 17 years. Whole-body selenium concentrations of 15 µg/g were associated with a 10-fold higher incidence of teratogenic defects in centrarchid populations in the contaminated lake (tissue selenium was 1 to 3 µg/g in the reference lakes). The relation between tissue selenium residues and the prevalence of malformations approximated an exponential function over the range of 1 to 80 µg/g of selenium and 0 to 70% deformities ($R^2 = 0.881$, $P < 0.01$). Lemly concluded that this relation could be used to predict the role of teratogenic defects in warm-water fish populations suspected of having selenium-related reproductive failure.

Depending on the specific tissue (i.e., skeletal muscle, ovary, liver, whole body, etc.), selenium in fish from control test groups or habitats with low ambient selenium concentrations usually ranges from about 1 to 5 µg/g (Baumann and May, 1984; Lemly, 1985a; Gillespie and Baumann, 1986; Hermanutz et al., 1992; Coyle et al., 1993). However, reduced growth, tissue damage in major organs, reproductive impairment, and mortality begin to occur when concentrations reach 4 to 16 µg/g (Table 1). The margin between "normal" and toxic concentrations in tissues is thus extremely narrow. A small margin of safety, along with the propensity of selenium to bioaccumulate in aquatic food chains and become toxic in the diet (Table 2), underscores the biological importance of even slight increases in environmental selenium.

I recommend that the following tissue concentrations be taken as concentrations of concern, i.e., toxic effects thresholds, for the overall health and reproductive vigor of freshwater and anadromous fish: whole body, 4 µg/g; skeletal muscle (skinless fillets), 8 µg/g; liver, 12 µg/g; ovary and eggs, 10 µg/g (Table 3). Laboratory and field studies show that the most sensitive indicator of selenium impacts on centrarchid populations is reproductive success (Cumbie and Van Horn, 1978; Lemly, 1985a; Gillespie and Baumann, 1986; Woock et al., 1987; Hermanutz et al., 1992; Coyle et al., 1993). Reproductive success was also the most sensitive indicator for fathead minnows (Pyron and Beitinger, 1989; Schultz and Hermanutz, 1990) and, although reproductive studies are lacking, probably holds for salmonids and

Table 2 Concentrations of Selenium Known to be Toxic in the Diets of Fish and Wildlife

Species	Dietary selenium concentration[a] (μg/g)	Effect	Ref.
Rainbow trout	9	Mortality	Goettl and Davies (1978)
(Oncorhynchus mykiss)	13	Mortality	Hilton et al. (1980)
	11	Kidney damage	Hilton and Hodson (1983)
Chinook salmon	6.5	Mortality	Hamilton et al. (1989)
(Oncorhynchus tshawytscha)	5	Reduced growth	Hamilton et al. (1990)
Fathead minnow (Pimephales promelas)	20	Reduced growth	Ogle and Knight (1989)
Striped bass (Morone saxatilis)	39	Mortality	Coughlan and Velte (1989)
Bluegill (Lepomis	54	Mortality	Finley (1985)
macrochirus)	6.5	Mortality	U.S. Fish and Wildlife Service (1990)
	5	Mortality	Lemly (1993a)
	13	Reproductive failure	Woock et al. (1987)
	33	Reproductive failure	Coyle et al. (1993)
Mallard duck (Anas	11	Reproductive failure	Heinz et al. (1987)
platyrhynchos)	9	Reproductive failure	Heinz et al. (1989)

[a] Selenium concentrations in parts per million (μg/g) on a dry weight basis.

Table 3 Toxic Effects Thresholds for Selenium Concentrations in Water, Food Chain Organisms, and Fish Tissues

Selenium source	Selenium concentration[a]	Effect
Water		
Inorganic selenium	2 μg/l	Food-chain bioaccumulation and reproductive failure in fish and wildlife
Organic selenium	<1 μg/l	Food-chain bioaccumulation and reproductive failure in fish and wildlife
Food-chain organisms	3 μg/g	Reproductive failure in fish and wildlife
Fish tissues		
Whole body	4 μg/g	Mortality of juveniles and reproductive failure
Skeletal muscle (skinless fillets)	8 μg/g	Reproductive failure
Liver	12 μg/g	Reproductive failure
Ovary and eggs	10 μg/g	Reproductive failure

[a] Selenium concentrations in parts per billion for water; parts per million (μg/g) on a dry weight basis for food-chain organisms and fish tissues.

percichthyids because the selenium concentrations at which tissue damage and mortality of fry and juveniles occur are very similar to those for centrarchids (Lemly, 1985a; Hamilton et al., 1986, 1989, 1990; Coughlan and Velte, 1989). The most precise way to assess the selenium status and potential reproductive viability of adult fish is to measure the selenium concentrations in gravid ovaries. Measuring concentrations in ovaries integrates waterborne and dietary exposure and allows evaluation of the most sensitive biological endpoint. Resource managers and fisheries biologists should consider evaluation of gravid ovaries when designing aquatic monitoring studies for selenium.

AQUATIC FOOD ORGANISMS OF FISH AND WILDLIFE

Although data on the toxicity or bioaccumulation of selenium have been reported for invertebrates and other aquatic food organisms of fish and wildlife (see reviews by Eisler, 1985; Maier et al., 1988; Ogle et al., 1988; Ohlendorf, 1989), few studies have reported tissue concentrations associated with toxic effects. Foe and Knight (1986) cultured the green alga (*Selenastrum capricornutum*) in solutions of sodium selenite and found that chlorophyll a concentrations, dry weight, and cell replication were reduced when tissue concentrations reached about 20 µg/g (Table 1). Cell division was completely stopped when concentrations reached 100 to 500 µg/g. Kiffney and Knight (1990) exposed the cyanobacterium (*Anabaena flos-aquae*) to solutions of sodium selenate, sodium selenite, and seleno-L-methionine and found that chlorophyll a concentrations were unaffected until tissue concentrations reached 394 µg/g, regardless of the chemical form of selenium used. Ingersoll et al. (1990) exposed the cladoceran *Daphnia magna* to a 6:1 ratio of waterborne sodium selenate and sodium selenite and observed that concentrations of 15 µg/g or greater were associated with reduced weight of adults. The production of young cladocerans was significantly reduced when tissue concentrations in adults reached about 32 µg/g (Table 1).

Field studies show that benthic invertebrates and certain forage fishes (mosquitofish *[Gambusia affinis]*, red shiners *[N. lutrensis]*, fathead minnows) can accumulate 20 to 370 µg/g of selenium and still maintain stable, reproducing populations (Woock and Summers, 1984; Lemly, 1985a, 1985b; Saiki, 1986a, 1986b; Saiki and Lowe, 1987; Barnum and Gilmer, 1988; Roth and Horne, 1988; Schuler, 1989). Plankton and aquatic plants seem to be largely unaffected with concentrations of 30 µg/g or more (Woock and Summers, 1984; Lemly, 1985a; Saiki, 1986a, 1986b; Roth and Horne, 1988; Schuler et al., 1990).

The most important aspect of selenium accumulation in aquatic food chains is not direct toxicity to the organisms themselves but, rather, the dietary source of selenium that they provide to fish and wildlife. The consensus among researchers is that most of the selenium in fish tissues results from uptake through diet rather than through water (Cumbie and Van Horn, 1978; Finley, 1985; Lemly, 1982, 1985a; Hamilton et al., 1986, 1990; Woock et al., 1987; Besser et al., 1993; Coyle et al., 1993). The environmental selenium cycle includes strong bioaccumulation steps in the aquatic food-chain that greatly increase the dietary concentrations of selenium available to the fish and birds that consume aquatic organisms (Lemly, 1985a, 1989; Saiki, 1986a, 1986b; Lemly and Smith, 1987, 1991; Saiki and Lowe, 1987; Hothem and Ohlendorf, 1989). Thus, a small increase in waterborne selenium will result in a disproportionately large elevation of selenium concentrations in fish and wildlife tissues. Moreover, selenium is efficiently transferred from parents to offspring through the eggs (Gillespie and Baumann, 1986; Heinz et al., 1987, 1989; Schultz and Hermanutz, 1990; Coyle et al., 1993). A contaminated aquatic food-chain can leave a legacy of selenium poisoning in fish and wildlife populations for many generations (Duke Power Company, 1980; Lemly, 1985a; Sorensen, 1988; Coughlan and Velte, 1989; Ohlendorf, 1989).

The toxic effect thresholds for selenium impacts on food-chain organisms (20 to 394 µg/g) are much higher than the dietary effect thresholds for fish and wildlife. Dietary concentrations of 6.5 µg/g or greater (as selenomethionine) caused mortality and reproductive failure in centrarchids (Woock et al., 1987; U.S. Fish and Wildlife Service, 1990; Coyle et al., 1993), whereas dietary concentrations of 6.5 to 10 µg/g reduced survival of juvenile salmonids (Goettl and Davies, 1978; Hilton et al., 1980; Hilton and Hodson, 1983; Hamilton et al., 1986, 1989, 1990) (Table 2). The uptake kinetics and effects concentrations for natural field-source selenium diets and seleno-L-methionine-spiked commercial diets are nearly identical (Woock et al., 1985, 1987; Hamilton et al., 1986, 1989), indicating that the results of dietary toxicity studies with seleno-L-methionine in the laboratory can be used to accurately evaluate the hazard of food-chain concentrations in the field. Hilton et al. (1980) suggested that diets containing in excess of 3 µg/g may ultimately be toxic to rainbow trout. Similar dietary effect concentrations have been found for wildlife. Studies by Heinz et al. (1987, 1989) showed that the reproduction of mallard ducks *(Anas platyrhynchos)* was impaired at a dietary concentration of 9 µg/g, with the effects threshold falling between 4 and 9 µg/g. These toxicity values contrast sharply with those for aquatic invertebrates, which can tolerate up to 300 µg/g in the diet without effects on reproduction, growth, or survival (Foe and Knight, 1986). Food-chain organisms can thus build up tissue concentrations of selenium that are toxic to predators while remaining unaffected themselves. I recommend 3 µg/g as the toxic threshold for selenium in aquatic food-chain organisms consumed by fish and wildlife (Table 3).

WATER

Selenium is strongly bioaccumulated in aquatic habitats. This characteristic results in a marked elevation of concentrations in food-chain organisms as compared with waterborne concentrations (Lemly, 1985b; Maier et al., 1988; Ogle et al., 1988; Ohlendorf, 1989). It is critical to know how much bioaccumulation can be expected for a given aqueous concentration of selenium in order to evaluate the potential for dietary toxicity and reproductive effects in predatory species of fish and wildlife. Laboratory studies show that seleno-L-methionine (an organic selenium compound) is bioconcentrated over 200,000 times by zooplankton when water concentrations average 0.5 to 0.8 µg/l (Besser et al., 1989, 1993). Resultant selenium concentrations in zooplankton were over 100 µg/g, a concentration that far exceeds the dietary toxicity threshold for fish and wildlife (3 µg/g). Organoselenium compounds can comprise a substantial portion of the total waterborne selenium concentration in aquatic environments (Chau et al., 1976; Cutter, 1982, 1986, 1991; Cooke and

Bruland, 1987), although the complete range of chemical species is poorly described. The potential for bioaccumulation and toxicity due to organic selenium is very high.

Inorganic selenium (selenate, selenite) bioconcentrates more readily in phytoplankton than zooplankton, and tissue concentrations as high as 18 μg/g can result when waterborne concentrations are in the 7- to 10-μg/l range (Besser et al., 1993), resulting in bioconcentration factors of about 3000. It is at the primary producer and primary consumer levels of the food-chain (phytoplankton and zooplankton) that most of the bioconcentration (waterborne uptake) and bioaccumulation (combined waterborne and dietary uptake) occur. The biomagnification of selenium (progressively higher concentrations in successive trophic levels of the food chain) is not clearly indicated in laboratory studies (Bennett et al., 1986; Besser et al., 1989, 1993). However, some field studies have found that the body burdens continue to rise from 2 to 6 times through the food chain in a pattern suggestive of biomagnification (Woock and Summers, 1984; Lemly, 1985a, 1986; Saiki, 1986a; Lemly and Smith, 1987; Saiki and Lowe, 1987; Barnum and Gilmer, 1988; Hothem and Ohlendorf, 1989).

Field studies have documented selenium bioaccumulation factors of 500 to 35,000 in contaminated aquatic habitats where concentrations of waterborne selenium were in the 2- to 16-μg/l range (Sager and Cofield, 1984; Woock, 1984; Woock and Summers, 1984; Lemly, 1985a, 1985b; Baumann and Gillespie, 1986; Barnum and Gilmer, 1988). These waterborne concentrations resulted in food-chain concentrations of 10 to 60 μg/g that, again, far exceed the 3-μg/g dietary toxicity threshold for fish and wildlife. Until about 1986, most published literature on the selenium concentrations in aquatic food chains and on the associated impacts upon predatory species pertained to fisheries in power plant cooling reservoirs in the southeastern U.S. (Duke Power Company, 1980; Garrett and Inman, 1984; Woock and Summers, 1984; Lemly, 1985a, 1985b; Gillespie and Baumann, 1986). In these aquatic systems, the threshold for significant bioaccumulation was in the range of 2 to 5 μg/l (Lemly, 1985a, 1985b, 1986; Lemly and Smith, 1987).

Several recent studies of selenium in agricultural irrigation drainwater in the western U.S. show that selenium may accumulate to toxic concentrations (10 to 20 μg/g) in the food chain when waterborne concentrations are in the 0.5- to 3-μg/l range (Barnum and Gilmer, 1988; Schroeder et al., 1988; Stevens et al., 1988; Hoffman et al., 1990; Saiki, 1990; Skorupa and Ohlendorf, 1991; Hallock et al., 1993). This degree of bioaccumulation seems to be related to environmental conditions that favor the formation and uptake of biologically active organic selenium compounds. One scenario is that these organic forms are released by algal cells during bloom conditions and are present for very short periods of time (hours or days) and in ultratrace amounts (<1 μg/l). These organic forms may be taken up by food-chain organisms in a manner similar to free seleno-amino acids, such as selenomethionine, and result in disproportionately high tissue concentrations as compared with inorganic selenate and selenite (Besser et al., 1989, 1993; Sanders et al., 1992). Another possibility is that organic selenium present in decaying plant and animal matter is recycled into the food-chain and is responsible for unexpectedly high concentrations in fish and wildlife. Field observations indicate that detrital material may contain higher concentrations of selenium than underlying sediment or associated biota (Saiki and Lowe, 1987; Parker and Knight, 1989; Hallock et al., 1993). Microbial communities living in or on detritus can bioconcentrate selenium from the benthic microlayer as dissolved organoselenium is released from decaying organic matter (Sanders et al., 1992). The detrital food-chain may thus be exposed to a highly concentrated source of selenium in some cases. In these situations, the concentrations in fish and wildlife tissues may be much higher than would be predicted or anticipated solely on the basis of total waterborne selenium.

The environmental speciation of selenium is complex, and several chemical forms are likely to be present in solution at a given location and time (Cutter, 1982, 1986, 1991; McKeown and Marinas, 1986; Cooke and Bruland, 1987). However, the patterns and magnitude of bioaccumulation are similar enough among aquatic systems impacted by power production wastes, agricultural irrigation drainwater, and other selenium sources that a common number can be given as the threshold for detrimental effects. I recommend that waterborne selenium concentrations of 2 μg/l or greater (total recoverable basis in 0.45-μm filtered samples) be considered highly hazardous to the health and long-term survival of fish and wildlife (Table 3). Some species will be relatively unaffected at these concentrations, but sensitive species, many of which are the most important in terms of ecological integrity and public recreational value, are likely to be seriously impacted. It should also be recognized that, under certain environmental conditions, 1 μg/l of selenium or less has the potential to bioaccumulate to concentrations in the food-chain that are toxic to predatory species.

The concentrations of waterborne selenium should be evaluated along with the concentrations present in food chain organisms and fish and wildlife tissues to form an integrated and conclusive assessment of the selenium status and health of the aquatic ecosystem under study. An approach that measures both water and tissue concentrations will combine the strengths of chemical and biological monitoring to elucidate cause-and-effect relations (Hodson, 1990).

SUMMARY

The most widespread human-caused sources of selenium mobilization and introduction into aquatic ecosystems in the U.S. today are the extraction and utilization of coal for generating electric power and the irrigation of high-selenium soils for agricultural production. Once in aquatic systems, selenium is readily taken up from solution by food-chain organisms and can quickly reach concentrations that are toxic to the fish and wildlife that consume them. Selenium is efficiently transferred in eggs from parents to offspring, where it can cause edema, hemorrhaging, spinal deformities, and death. Reproductive success is more sensitive to selenium toxicity than are the growth and survival of juvenile and adult fish. Waterborne concentrations in the low-μg/l range can bioaccumulate in the food-chain and result in an elevated dietary selenium intake and the reproductive failure of adult fish with little or no additional symptoms of selenium poisoning in the entire aquatic system. Thus, the impact of selenium can be subtle and yet very dramatic to fish populations as reproductive failure occurs.

It is now possible to formulate diagnostic selenium concentrations for three distinct ecosystem-level components: water, the food-chain, and predatory fish (consuming fish or invertebrate prey). Waterborne selenium concentrations of 2 µg/l or greater (on a total recoverable basis in 0.45-µm filtered samples) should be considered hazardous to the health and long-term survival of fish and wildlife populations because of the high potential for food-chain bioaccumulation, dietary toxicity, and reproductive effects. In some cases, ultratrace amounts of soluble organic selenium might lead to bioaccumulation and toxicity even when the total waterborne concentrations are less than 1 µg/l.

Food-chain organisms such as zooplankton, benthic invertebrates, and certain forage fishes can accumulate up to 30 µg/g of selenium, dry weight (some taxa up to 370 µg/g) with no apparent effect on survival or population levels. However, the dietary toxicity threshold for fish and wildlife is only 3 µg/g; these food organisms would supply a toxic dose of selenium while being unaffected themselves. Consequently, food-chain organisms containing 3 µg/g of dry weight or more should be viewed as potentially lethal to the fish and aquatic birds that consume them.

Thresholds for tissue concentrations that affect the health and reproductive success of freshwater and anadromous fish are as follows: whole-body, 4 µg/g; skeletal muscle (skinless fillets), 8 µg/g; liver, 12 µg/g; and ovaries and eggs, 10 µg/g. The most precise way to evaluate the potential reproductive viability of adult fish populations is to measure the selenium concentrations in gravid ovaries. This single measure integrates waterborne and dietary exposure and allows an evaluation based on the most sensitive biological endpoint. Resource managers and aquatic biologists should obtain measurements of the selenium concentrations in water, food-chain organisms, and fish and wildlife tissues in order to formulate a comprehensive and conclusive assessment of the overall selenium status and health of aquatic ecosystems.

REFERENCES

Allen, G. T., and R. M. Wilson. 1990. Selenium in the aquatic environment of Quivira National Wildlife Refuge. Prairie Nat. 22:129-135.

American Petroleum Institute. 1978. Analysis of refinery wastewaters for the EPA priority pollutants. Publ. 4296. American Petroleum Institute, Washington, D.C., 36 pp.

Andren, A. W., D. H. Klein, and Y. Talmi. 1975. Selenium in coal-fired steam plant emissions. Environ. Sci. Technol. 9:856-858.

Barnum, D. A., and D. S. Gilmer. 1988. Selenium levels in biota from irrigation drainwater impoundments in the San Joaquin Valley, California. Lake Reservoir Manage. 4:181-186.

Baumann, P. C., and R. B. Gillespie. 1986. Selenium bioaccumulation in gonads of largemouth bass and bluegill from three power plant cooling reservoirs. Environ. Toxicol. Chem. 5:695-701.

Baumann, P. C., and T. W. May. 1984. Selenium residues in fish from inland waters of the United States. p. 7-1 to 7-16. *In* Workshop proceedings: the effects of trace elements on aquatic ecosystems. Tech. Rep. EA-3329. Electric Power Research Institute, Palo Alto, Calif.

Bennett, W. N., A. S. Brooks, and M. E. Boraas. 1986. Selenium uptake and transfer in an aquatic food chain and its effects on fathead minnow larvae. Arch. Environ. Contam. Toxicol. 15:513-517.

Besser, J. M., T. J. Canfield, and T. W. La Point. 1993. Bioaccumulation of organic and inorganic selenium in a laboratory food chain. Environ. Toxicol. Chem. 12:57-72.

Besser, J. M., J. N. Huckins, E. E. Little, and T. W. La Point. 1989. Distribution and bioaccumulation of selenium in aquatic microcosms. Environ. Pollut. 62:1-12.

Braunstein, H. M., E. D. Copenhaver, and H. A. Pfuderer (Eds.). 1981. Environmental, health, and control aspects of coal conversion — an information overview. Vol. 2. Ann Arbor Science Publishers, Ann Arbor, Mich., 430 pp.

California Environmental Protection Agency (EPA). 1992. Derivation of site-specific water quality for selenium in San Francisco Bay. Tech. Rep. California EPA, Oakland, Calif., 37 pp.

California State Water Resources Control Board. 1987. Regulation of agricultural drainage to the San Joaquin River. Tech. Rep. W.Q. 85-1. State Water Resources Control Board, Sacramento, Calif., 76 pp.

Chau, Y. K., P. T. S. Wong, B. A. Silverberg, P. L. Luxon, and G. A. Bengert. 1976. Methylation of selenium in the aquatic environment. Science (Washington, D.C.) 192:1130-1131.

Cherry, D. S., and R. K. Guthrie. 1977. Toxic metals in surface waters from coal ash. Water Resour. Bull. 13:1227-1236.

Clark, P. J., R. A. Zingaro, K. J. Irgolic, and A. N. McGinley. 1980. Arsenic and selenium in Texas lignite. Int. J. Environ. Anal. Chem. 7:295-314.

Cooke, T. D., and K. W. Bruland. 1987. Aquatic chemistry of selenium: evidence of biomethylation. Environ. Sci. Technol. 21:1214-1219.

Copeland, R. 1970. Selenium: the unknown pollutant. Limnos 3:7-9.

Coughlan, D. J., and J. S. Velte. 1989. Dietary toxicity of selenium-contaminated red shiners to striped bass. Trans. Am. Fish. Soc. 118:400-408.

Cowser, K. E., and C. R. Richmond (Eds.). 1980. Synthetic fossil fuel technology: potential health and environmental effects. Ann Arbor Science Publishers, Ann Arbor, Mich., 511 pp.

Coyle, J. J., D. R. Buckler, C. G. Ingersoll, J. F. Fairchild, and T. W. May. 1993. Effect of dietary selenium on the reproductive success of bluegill sunfish *(Lepomis macrochirus)*. Environ. Toxicol. Chem. 12:551-565.

Cumbie, P. M., and S. L. Van Horn. 1978. Selenium accumulation associated with fish mortality and reproductive failure. Proc. Annu. Conf. Southeast. Assoc. Fish Wildl. Agencies 32:612-624.

Cutter, G. A. 1982. Selenium in reducing waters. Science (Washington, D.C.) 217:829-831.

Cutter, G. A. 1986. Speciation of selenium and arsenic in natural waters and sediments. *In* Vol. 1. Selenium speciation. Publ. EA-4641. Electric Power Research Institute, Palo Alto, Calif., 75 pp.

Cutter, G. A. 1991. Selenium biogeochemistry in reservoirs. *In* Vol. 1. Time series and mass balance results. Publ. EN-7281. Electric Power Research Institute, Palo Alto, Calif., 67 pp.

Davis, E. C., and W. J. Boegly, Jr. 1981. Coal pile leachate quality. J. Environ. Eng. Div. Am. Soc. Civil Eng. 107:399-417.

Deverel, S. J., J. L. Fio, and R. J. Gilliom. 1989. Selenium in tile drain water. p. 81-91. *In* Preliminary assessment of sources, distribution, and mobility of selenium in the San Joaquin Valley, California. Water Resour. Invest. Rep. 88-4186. U.S. Geological Survey, Sacramento, Calif.

Draize, J. H., and O. A. Beath. 1935. Observations on the pathology of blind staggers and alkali disease. Am. Vet. Med. Assoc. J. 86:753-763.

Duke Power Company. 1980. Toxic effects of selenium on stocked bluegill (*Lepomis macrochirus*) in Belews Lake, North Carolina, April-September 1979. Tech. Rep. Duke Power Company, Charlotte, N.C., 18 pp.

Eisler, R. 1985. Selenium hazards to fish, wildlife, and invertebrates: a synoptic review. Biol. Rep. 85(1.5), Contam. Hazard Rev. No. 5. U.S. Fish and Wildlife Service, Washington, D.C., 57 pp.

Ellis, M. M., H. L. Motley, M. D. Ellis, and R. O. Jones. 1937. Selenium poisoning in fishes. Proc. Soc. Exp. Biol. Med. 36:519-522.

Evans, D. W., J. G. Wiener, and J. H. Horton. 1980. Trace element inputs from a coal burning power plant to adjacent terrestrial and aquatic environments. Air Pollut. Control Assoc. J. 30:567-573.

Finley, K. A. 1985. Observations of bluegills fed selenium-contaminated *Hexagenia* nymphs collected from Belews Lake, North Carolina. Bull. Environ. Contam. Toxicol. 35:816-825.

Foe, C., and A. W. Knight. 1986. Selenium bioaccumulation, regulation, and toxicity in the green alga, *Selenastrum capricornutum*, and dietary toxicity of the contaminated alga to *Daphnia magna*. p. 77-88. *In* Proc. First Environ. Symp.: selenium in the environment. Calif. Agric. Technol. Inst. Publ. No. CAT1/860201. California State University, Fresno, Calif.

Fruchter, J. S., and M. R. Petersen. 1979. Environmental characterization of products and effluents from coal conversion processes. p. 247-275. *In* Analytical methods for coal and coal products. Vol. 3. Academic Press, New York.

Fuji, R. 1988. Water-quality and sediment-chemistry data of drain water and evaporation ponds from Tulare Lake Drainage District, Kings County, California, March 1985 to March 1986. Open-File Rep. 87-700. U.S. Geological Survey, Sacramento, Calif., 19 pp.

Garrett, G. P., and C. R. Inman. 1984. Selenium-induced changes in fish populations of a heated reservoir. Proc. Annu. Conf. Southeast. Assoc. Fish Wildl. Agencies 38:291-301.

Gillespie, R. B., and P. C. Baumann. 1986. Effects of high tissue concentrations of selenium on reproduction by bluegills. Trans. Am. Fish. Soc. 115:208-213.

Goettl, J. P., Jr., and P. H. Davies. 1978. Water pollution studies. Job Prog. Rep., Federal Aid Project F-33-R-13. Colorado Department of Natural Resources, Division of Wildlife, Denver, Colo., 45 pp.

Greenberg, A. J., and D. Kopec. 1986. Decline of Bay-Delta fisheries and increased selenium loading: possible correlation? p. 69-81. *In* Selenium and agricultural drainage: implications for San Francisco Bay and the California environment. Proc. Sec. Selenium Symp. The Bay Institute of San Francisco, Tiburon, Calif.

Hallock, R. J., H. L. Burge, S. P. Thompson, and R. J. Hoffman. 1993. Irrigation drainage in and near Stillwater, Humboldt, and Fernley wildlife management areas and Carson Lake, west-central Nevada, 1988-1990. II. Effects on wildife. Water Resour. Invest. Rep. 91-401. U.S. Geological Survey, Carson City, Nev., 115 pp.

Hamilton, S. J., K. J. Buhl, and N. L. Faerber. 1989. Toxicity of selenium in the diet to chinook salmon. p. 22-34. *In* A. Q. Howard (Ed.). Selenium and agricultural drainage: implications for San Francisco Bay and the California environment. Proc. 4th Selenium Symp. The Bay Institute of San Francisco, Tiburon, Calif.

Hamilton, S. J., K. J. Buhl, N. L. Faerber, R. H. Wiedmeyer, and F. A. Bullard. 1990. Toxicity of organic selenium in the diet to chinook salmon. Environ. Toxicol. Chem. 9:347-358.

Hamilton, S. J., A. N. Palmisano, G. W. Wedemeyer, and W. T. Yasutake. 1986. Impacts of selenium on early life stages and smoltification of fall chinook salmon. Trans. N. Am. Wildl. Nat. Resour. Conf. 51:343-356.

Hartley, W. J., and A. B. Grant. 1961. A review of selenium responsive diseases of New Zealand livestock. Fed. Proc. 20:679-688.

Heaton, R. C., J. M. Williams, J. P. Bertino, L. E. Wangen, A. M. Nyitray, M. M. Jones, P. L. Wanek, and P. Wagner. 1982. Leaching behaviors of high-sulfur coal wastes from two Appalachian coal preparation plants. Publ. LA-9356-MS. U.S. Dept. Energy, Los Alamos National Laboratory, Los Alamos, N.M., 78 pp.

Heinz, G. H., D. J. Hoffman, and L. G. Gold. 1989. Impaired reproduction of mallards fed an organic form of selenium. J. Wildl. Manage. 53:418-428.

Heinz, G. H., D. J. Hoffman, A. J. Krynitsky, and D. M. G. Weller. 1987. Reproduction in mallards fed selenium. Environ. Toxicol. Chem. 6:423-433.

Hermanutz, R. O., K. N. Allen, T. H. Roush, and S. F. Hedtke. 1992. Effects of elevated selenium concentrations on bluegills *(Lepomis macrochirus)* in outdoor experimental streams. Environ. Toxicol. Chem. 11:217-224.

Hilton, J. W., and P. V. Hodson. 1983. Effect of increased dietary carbohydrate on selenium metabolism and toxicity in rainbow trout *(Salmo gairdneri)*. J. Nutr. 113:1241-1248.

Hilton, J. W., P. V. Hodson, and S. J. Slinger. 1980. The requirement and toxicity of selenium in rainbow trout *(Salmo gairdneri)*. J. Nutr. 110:2527-2535.

Hodson, P. V. 1990. Indicators of ecosystem health at the species level and the example of selenium effects on fish. Environ. Monit. Assess. 15:241-254.

Hodson, P. V., D. J. Spry, and B. R. Blunt. 1980. Effects on rainbow trout *(Salmo gairdneri)* of a chronic exposure to waterborne selenium. Can. J. Fish. Aquat. Sci. 37:233-240.

Hoffman, R. J., R. J. Hallock, T. G. Rowe, M. S. Lico, H. L. Burge, and S. P. Thompson. 1990. Reconnaissance investigation of water quality, bottom sediment, and biota associated with irrigation drainage in and near Stillwater Wildlife Management Area, Churchill County, Nevada, 1986-87. Water Resour. Invest. Rep. 89-4105. U.S. Geological Survey, Carson City, Nev., 150 pp.

Hoffman, S. E., N. J. Williams, and A. I. Herson. 1986. The Kesterson story. p. 1-9. *In* Proc. First Biennial Conf.: Is current technology the answer? National Water Supply Improvement Association, Springfield, Va.

Hothem, R. L., and H. M. Ohlendorf. 1989. Contaminants in foods of aquatic birds at Kesterson Reservoir, California, 1985. Arch. Environ. Contam. Toxicol. 18:773-786.

Hunn, J. B., S. J. Hamilton, and D. R. Buckler. 1987. Toxicity of sodium selenite to rainbow trout fry. Water Res. 21:233-238.

Ingersoll, C. G., F. J. Dwyer, and T. W. May. 1990. Toxicity of inorganic and organic selenium to *Daphnia magna* (Cladocera) and *Chironomus riparius* (Diptera). Environ. Toxicol. Chem. 9:1171-1181.

Jones, D. R. 1990. Batch leaching studies of Rundle oil shale. J. Environ. Qual. 19:408-413.

Kaakinen, J. W., R. M. Jorden, M. H. Lawasani, and R. E. West. 1975. Trace element behavior in coal-fired power plant. Environ. Sci. Technol. 9:862-869.

Kiffney, P., and A. Knight. 1990. The toxicity and bioaccumulation of selenate, selenite, and seleno-L-methionine in the cyanobacterium *Anabaena flos-aquae*. Arch. Environ. Contam. Toxicol. 19:488-494.

Klein, D. H., A. W. Andren, and N. E. Bolton. 1975. Trace element discharges from coal combustion for power production. Water Air Soil Pollut. 5:71-77.

Kubota, J. 1980. Regional distribution of trace element problems in North America. p. 441-466. *In* B. E. Davies (Ed.). Applied soil trace elements. John Wiley & Sons, New York.

Lemly, A. D. 1982. Response of juvenile centrarchids to sublethal concentrations of waterborne selenium. I. Uptake, tissue distribution, and retention. Aquat. Toxicol. (N.Y.) 2:235-252.

Lemly, A. D. 1985a. Toxicology of selenium in a freshwater reservoir: implications for environmental hazard evaluation and safety. Ecotoxicol. Environ. Saf. 10:314-338.

Lemly, A. D. 1985b. Ecological basis for regulating aquatic emissions from the power industry: the case with selenium. Regul. Toxicol. Pharmacol. 5:465-486.

Lemly, A. D. 1986. Effects of selenium on fish and other aquatic life. p. 153-162. In J. B. Summers and S. S. Anderson (Eds.). Toxic substances in agricultural water supply and drainage: defining the problems. U.S. Committee on Irrigation and Drainage, Denver, Colo.

Lemly, A. D. 1989. Cycling of selenium in the environment. p. 113-123. In A. Q. Howard (Ed.). Selenium and agricultural drainage: implications for San Francisco Bay and the California environment. Proc. 4th Selenium Symp. The Bay Institute of San Francisco, Tiburon, Calif.

Lemly, A. D. 1993a. Metabolic stress during winter increases the toxicity of selenium to fish. Aquat. Toxicol. (Amst.) 27:133-158.

Lemly, A. D. 1993b. Teratogenic effects of selenium in natural populations of freshwater fish. Ecotoxicol. Environ. Saf. 26:181-204.

Lemly, A. D., and G. J. Smith. 1987. Aquatic cycling of selenium: implications for fish and wildlife. Fish Wildl. Leaflet 12. U.S. Fish and Wildlife Service, Washington, D.C., 10 pp.

Lemly, A. D., and G. J. Smith. 1991. Aquatic cycling of selenium: implications for fish and wildlife. p. 43-53. In R. C. Severson, S. E. Fisher, Jr., and L. P. Gough (Eds.). Proc. 1990 Billings Land Reclamation Symp. Selenium in arid and semiarid environments, western United States. Circ. 1064. U.S. Geological Survey, Denver, Colo.

Maier, K. J., C. Foe, R. S. Ogle, M. J. Williams, A. W. Knight, P. Kiffney, and L. A. Melton. 1988. The dynamics of selenium in aquatic ecosystems. p. 361-408. In D. D. Hemphill (Ed.). Trace substances in environmental health. Vol. 21. University of Missouri, Columbia, Mo.

Marshall, E. 1985. Selenium poisons refuge, California politics. Science (Washington, D.C.) 229:144-146.

McKeown, B., and B. Marinas. 1986. The chemistry of selenium in an aqueous environment. p. 7-15. In Proc. First Environ. Symp.: selenium in the environment. Calif. Agric. Technol. Inst. Publ. No. CAT1/860201. California State University, Fresno, Calif.

Murtha, M. J., G. Burnet, and N. Harnby. 1983. Power plant fly ash — disposal and utilization. Environ. Prog. 2:193-198.

National Research Council. 1980. Energy and the fate of ecosystems. National Academy Press, Washington, D.C., 399 pp.

North Carolina Division of Environmental Management. 1986. North Carolina water quality standards documentation: the freshwater chemistry and toxicity of selenium with an emphasis on its effects in North Carolina. Rep. No. 86-02. North Carolina Dept. Natural Resources and Community Development, Raleigh, N.C., 52 pp.

Nystrom, R. R., and G. Post. 1982. Chronic effects of ammonia-stripped oil shale retort water on fishes, birds, and mammals. Bull. Environ. Contam. Toxicol. 28:271-276.

Ogle, R. S., and A. W. Knight. 1989. Effects of elevated foodborne selenium on growth and reproduction of the fathead minnow *(Pimephales promelas)*. Arch. Environ. Contam. Toxicol. 18:795-803.

Ogle, R. S., K. J. Maier, P. Kiffney, M. J. Williams, A. Brasher, L. A. Melton, and A. W. Knight. 1988. Bioaccumulation of selenium in aquatic ecosystems. Lake Reservoir Manage. 4:165-173.

Ohlendorf, H. M. 1989. Bioaccumulation and effects of selenium in wildlife. p. 133-177. In Selenium in agriculture and the environment. Special Publ. No. 23. Soil Science Society of America, Madison, Wisc.

Ohlendorf, H. M., D. J. Hoffman, M. K. Saiki, and T. W. Aldrich. 1986. Embryonic mortality and abnormalities of aquatic birds: apparent impacts of selenium from irrigation drainwater. Sci. Total Environ. 52:49-63.

Ohlendorf, H. M., R. L. Hothem, T. W. Aldrich, and A. J. Krynitski. 1987. Selenium contamination of the grasslands, a major California waterfowl area. Sci. Total Environ. 66:169-183.

Ohlendorf, H. M., A. W. Kilness, J. L. Simmons, R. K. Stroud, D. J. Hoffman, and J. F. Moore. 1988. Selenium toxicosis in wild aquatic birds. J. Toxicol. Environ. Health 24:67-92.

Parker, M. S., and A. W. Knight. 1989. Biological characterization of agricultural drainage evaporation ponds. Water Sci. Eng. Paper No. 4521. University of California, Davis, Calif., 52 pp.

Pillay, K. K. S., C. C. Thomas, Jr., and J. W. Kaminski. 1969. Neutron activation analysis of the selenium content of fossil fuels. Nucl. Appl. Technol. 7:478-483.

Presser, T. S., and I. Barnes. 1985. Dissolved constituents including selenium in waters in the vicinity of Kesterson National Wildlife Refuge and the west grassland, Fresno and Merced counties, California. Water Resour. Invest. Rep. 85-4220. U.S. Geological Survey, Menlo Park, Calif., 73 pp.

Presser, T. S., and H. M. Ohlendorf. 1987. Biogeochemical cycling of selenium in the San Joaquin Valley, California, USA. Environ. Manage. 11:805-821.

Presser, T. S., W. C. Swain, R. R. Tidball, and R. C. Severson. 1990. Geologic sources, mobilization, and transport of selenium from the California coast ranges to the western San Joaquin Valley: a reconnaissance study. Water Resour. Invest. Rep. 90-4070. U.S. Geological Survey, Menlo Park, Calif., 66 pp.

Pyron, M., and T. L. Beitinger. 1989. Effect of selenium on reproductive behavior and fry of fathead minnows. Bull. Environ. Contam. Toxicol. 42:609-613.

Rosenfeld, I., and O. A. Beath. 1946. Pathology of selenium poisoning. Wyo. Agric. Exp. Stn. Bull. 275:1-27.

Roth, J. C., and A. J. Horne. 1988. Quantitative studies of the biota of Kesterson Reservoir: numbers and biomass of food-chain organisms in several habitats in the deep ponds during the first two years after the cessation of drainwater inputs. Tech. Rep. Sanitary Engineering and Environmental Health Research Laboratory, University of California, Berkeley, Calif., 57 pp.

Sager, D. R., and C. R. Cofield. 1984. Differential accumulation of selenium among axial muscle, reproductive, and liver tissues of four warm water fish species. Water Res. Bull. 20:359-363.

Saiki, M. K. 1986a. Concentrations of selenium in aquatic food-chain organisms and fish exposed to agricultural tile drainage water. p. 25-33. In Selenium and agricultural drainage: implications for San Francisco Bay and the California environment. Proc. 2nd. Selenium Symp. The Bay Institute of San Francisco, Tiburon, Calif.

Saiki, M. K. 1986b. A field example of selenium contamination in an aquatic food chain. p. 67-76. In Proc. First Environ. Symp.: selenium in the environment. Calif. Agric. Technol. Inst. Publ. No. CAT1/860201. California State University, Fresno, Calif.

Saiki, M. K. 1990. Elemental concentrations in fishes from the Salton Sea, southeastern California. Water Air Soil Pollut. 52:41-56.

Saiki, M. K., and T. P. Lowe. 1987. Selenium in aquatic organisms from subsurface agricultural drainage water, San Joaquin Valley, California. Arch. Environ. Contam. Toxicol. 16:657-670.

Saiki, M. K., and D. U. Palawski. 1990. Selenium and other elements in juvenile striped bass from the San Joaquin Valley and San Francisco Estuary, California. Arch. Environ. Contam. Toxicol. 19:717-730.

Saiki, M. K., M. R. Jennings, and R. H. Wiedmeyer. 1992. Toxicity of agricultural subsurface drainwater from the San Joaquin Valley, California, to juvenile chinook salmon and striped bass. Trans. Am. Fish. Soc. 121:78-93.

Sanders, J. G., G. F. Riedel, C. C. Gilmour, R. W. Osman, C. E. Goulden, and R. W. Sanders. 1992. Selenium cycling and impact in aquatic systems. p. 1-228. *In* Modeling toxic effects on aquatic ecosystems. Prog. Rep. RP2020-11. Electric Power Research Institute, Palo Alto, Calif.

Schlinger, W. G., and G. N. Richter. 1980. An environmental evaluation of the Texaco coal gasification process. p. 150-168. *In* Proc. First Int. Gas Res. Conf. University of Chicago Press, Chicago, Ill.

Schroeder, R. A., D. U. Palawski, and J. P. Skorupa. 1988. Reconnaissance investigation of water quality, bottom sediment, and biota associated with irrigation drainage in the Tulare Lake bed area, southern San Joaquin Valley, California, 1986-87. Water Resour. Invest. Rep. 88-4001. U.S. Geological Survey, Sacramento, Calif., 86 pp.

Schuler, C. 1989. Selenium and boron accumulation in wetlands and waterfowl food at Kesterson Reservoir. p. 91-101. *In* A. Q. Howard (Ed.). Selenium and agricultural drainage: implications for San Francisco Bay and the California environment. Proc. 4th Selenium Symp. The Bay Institute of San Francisco, Tiburon, Calif.

Schuler, C. A., R. G. Anthony, and H. M. Ohlendorf. 1990. Selenium in wetlands and waterfowl foods at Kesterson Reservoir, California, 1984. Arch. Environ. Contam. Toxicol. 19:845-853.

Schultz, R., and R. Hermanutz. 1990. Transfer of toxic concentrations of selenium from parent to progeny in the fathead minnow *(Pimephales promelas)*. Bull. Environ. Contam. Toxicol. 45:568-573.

Severson, R. C., S. E. Fisher, Jr., and L. P. Gough (Eds.). 1991. Proc. 1990 Billings Land Reclamation Symp. Selenium in arid and semiarid environments, western United States. Circ. 1064. U.S. Geological Survey, Denver, Colo., 146 pp.

Skorupa, J. P., and H. M. Ohlendorf. 1991. Contaminants in drainage water and avian risk thresholds. p. 345-368. *In* A. Dinar and D. Zilberman (Eds.). The economics and management of water and drainage in agriculture. Kluwer Academic Publishers, New York.

Sorensen, E. M. B. 1986. The effects of selenium on freshwater teleosts. Rev. Environ. Toxicol. 2:59-116.

Sorensen, E. M. B. 1988. Selenium accumulation, reproductive status, and histopathological changes in environmentally exposed redear sunfish. Arch. Toxicol. 61:324-329.

Sorensen, E. M. B., P. M. Cumbie, T. L. Bauer, J. S. Bell, and C. W. Harlan. 1984. Histopathological, hematological, condition-factor, and organ weight changes associated with selenium accumulation in fish from Belews Lake, North Carolina. Arch. Environ. Contam. Toxicol. 13:153-162.

Stevens, D. W., B. Waddell, and J. B. Miller. 1988. Reconnaissance investigation of water quality, bottom sediment, and biota associated with irrigation drainage in the Middle Green River Basin, Utah, 1986-87. Water Resour. Invest. Rep. 88-4011. U.S. Geological Survey, Salt Lake City, Utah, 70 pp.

Summers, J. B., and S. S. Anderson (Eds.). 1986. Toxic substances in agricultural water supply and drainage: defining the problems. U.S. Comm. on Irrigation and Drainage, Denver, Colo., 358 pp.

Sylvester, M. A., J. P. Deason, H. R. Feltz, and R. A. Engberg. 1991. Preliminary results of the Department of the Interior's irrigation drainage studies. p. 115-122. *In* R. C. Severson, S. E. Fisher, Jr., and L. P. Gough (Eds.). Proc. 1990 Billings Land Reclamation Symp. Selenium in arid and semiarid environments, western United States. Circ. 1064. U.S. Geological Survey, Denver, Colo.

Tanji, K., A. Lauchli, and J. Meyer. 1986. Selenium in the San Joaquin Valley. Environment (Washington, D.C.) 28:6-11, 34-39.

U.S. Environmental Protection Agency (EPA). 1980. Oil Shale Symp.: sampling, analysis, and quality assurance. Tech. Rep. EPA-600-9/80/022. U.S. EPA, Environmental Research Laboratory, Cincinnati, Ohio, 586 pp.

U.S. Environmental Protection Agency (EPA). 1987. Ambient water quality criteria for selenium — 1987. Tech. Rep. EPA-440/5-87-006. Office of Water Regulations and Standards, U.S. EPA, Washington, D.C., 121 pp.

U.S. Fish and Wildlife Service. 1986. Preliminary survey of contaminant issues of concern on national wildlife refuges. Division of Refuge Management, U.S. Fish and Wildlife Service, Washington, D.C., 176 pp.

U.S. Fish and Wildlife Service. 1990. Agricultural irrigation drainwater studies in support of the San Joaquin Valley drainage program — final report. U.S. Fish and Wildlife Service, National Fisheries Contaminant Research Center, Columbia, Mo., 309 pp.

Woock, S. E. 1984. Accumulation of selenium by golden shiners *Notemigonus crysoleucas* — Hyco Reservoir, N.C. Cage Study 1981-1982. Tech. Rep. Carolina Power and Light Company, New Hill, N.C., 19 pp.

Woock, S. E., W. T. Bryson, K. A. MacPherson, M. A. Mallin, and W. E. Partin. 1985. Roxboro steam electric plant Hyco Reservoir 1984 bioassay report. Tech. Rep. Carolina Power and Light Company, New Hill, N.C., 48 pp.

Woock, S. E., W. R. Garrett, W. E. Partin, and W. T. Bryson. 1987. Decreased survival and teratogenesis during laboratory selenium exposures to bluegill, *Lepomis macrochirus*. Bull. Environ. Contam. Toxicol. 39:998-1005.

Woock, S. E., and P. B. Summers. 1984. Selenium monitoring in Hyco Reservoir (N.C.) waters (1977-1981) and biota (1977-1980). p. 6-1 to 6-27. *In* Workshop proceedings: the effects of trace elements on aquatic ecosystems. Tech. Rep. EA-3329. Electric Power Research Institute, Palo Alto, Calif.

CHAPTER 20

Selenium in Birds

Gary H. Heinz

INTRODUCTION

Selenium is a semimetallic trace element that birds and other wildlife need in small amounts for good health. Selenium deficiencies in livestock occur in some parts of the world and must be corrected by additions of selenium to the diet. However, the range of dietary concentrations that provides adequate but nontoxic amounts of selenium is narrow compared with the ranges for some other elements.

As early as the 13th century, Marco Polo described what was much later determined to be selenium poisoning in horses in Northern China (Spallholz and Raftery, 1984). In the 1930s, grains grown on seleniferous soils in South Dakota caused reproductive failure when fed to chickens (Poley and Moxon, 1938). The most drastic incident of selenium poisoning in wild birds occurred at the Kesterson National Wildlife Refuge in California during the early 1980s. Water used to irrigate crops in the San Joaquin Valley of California dissolved naturally occurring selenium salts from the soil, and when the selenium-laden subsurface water was drained from the agricultural fields and into Kesterson Reservoir on the refuge, dangerous levels of selenium accumulated in plants and animals used as foods by birds (Ohlendorf et al., 1986a; Ohlendorf et al., 1988; Ohlendorf, 1989). Reproductive failure and adult mortality occurred. Similar problems have been discovered elsewhere in the western United States, most notably in the Tulare Basin in California (Skorupa and Ohlendorf, 1991).

High concentrations of selenium in foods of wildlife are not limited to areas where soils are naturally high in selenium. Areas contaminated with sewage sludge, fly ash, and emissions from metal smelting plants may build up high levels of selenium (Robberecht et al., 1983; Wadge and Hutton, 1986; Cappon, 1991).

An assessment of the toxicity of selenium is complicated by its occurrence in many different chemical forms, some differing greatly in their toxicity to birds. The four common oxidation states are selenide (–2), elemental selenium (0), selenite (+4), and selenate (+6). Elemental selenium is virtually insoluble in water and presents little risk to birds. Both selenite and selenate are toxic to birds, but organic selenides pose the greatest hazard. Among the organic selenides, selenomethionine was shown to be highly toxic to birds and seems to be the most likely form to harm wild birds.

Selenium's ability to interact with other environmental contaminants, especially other elements, also sometimes complicates an interpretation of toxic thresholds in tissues of birds. Although I will not attempt to interpret critical levels of selenium in the presence of elevated levels of other pollutants, the reader needs to be aware that such interactions exist. The most studied of these interactions is between selenium and mercury, when each counteracts the toxicity of the other (Cuvin-Aralar and Furness, 1991), but selenium toxicity has also been reported to be diminished by elevated levels of lead (Donaldson and McGowan, 1989), copper and cadmium (Hill, 1974), silver (Jensen, 1975), and arsenic (Thapar et al., 1969).

The purpose of this chapter is to identify the concentrations of selenium that are toxic in avian tissues and eggs and to discuss how different chemical forms of selenium and its interactions with other environmental contaminants can alter toxicity.

INTERPRETING SELENIUM CONCENTRATIONS IN EGGS AND TISSUES

EGGS

The embryo is the avian life stage most sensitive to selenium poisoning (Poley et al., 1937; Poley and Moxon, 1938; Heinz et al., 1987; Hoffman and Heinz, 1988; Heinz et al., 1989). Because it is the selenium in the egg, rather than in the parent bird, that causes developmental abnormalities and death of avian embryos, selenium in the egg gives the most sensitive measure for evaluating hazards to birds. Given the rapid accumulation and loss patterns of selenium in birds (Heinz et al., 1990), selenium concentrations in eggs also probably best represent contamination of the local environment. Additional advantages of measuring selenium in eggs are that eggs are frequently easier to collect than adult birds, the loss of one egg from a nest probably has little effect on a population, and the egg represents an integration of the exposure of the adult female over time.

The concentration detected in eggs and the toxicity of that concentration seem to depend on the chemical form of the ingested selenium. Organoselenium compounds are believed to be major forms in plants and animals. One organoselenium compound, selenomethionine, when fed to breeding mallards was more toxic to embryos than was selenocystine or sodium selenite (Heinz et al., 1989). In wheat seeds and soybean protein, selenomethionine is a major form of selenium (Olson et al., 1970; Yasumoto et al., 1988). Hamilton et al. (1990) found selenomethionine to be an excellent model for selenium poisoning in chinook salmon *(Oncorhynchus tshawytscha)* when compared with the toxicity of selenium that was biologically incorporated into mosquitofish *(Gambusia affinis)* collected at Kesterson Reservoir in California. Nevertheless, the chemical forms of selenium in aquatic foods of birds have received virtually no study. It is likely that other chemical forms of selenium are present to some degree in plants and animals eaten by birds, yet the toxic concentrations of few selenium compounds have been determined in birds.

Information from the studies cited below regarding dietary selenomethionine leads me to conclude that about 3 ppm of selenium on a wet-weight basis in bird eggs should be considered the threshold of reproductive impairment.

In a laboratory study, mallards *(Anas platyrhynchos)* were fed 0, 1, 2, 4, 8, or 16 ppm of selenium as selenomethionine on what were close to dry-weight concentrations (Heinz et al., 1989). The reproductive success of the groups fed 1, 2, or 4 ppm of selenium did not significantly differ from that of controls; mean selenium concentrations in a sample of 15 eggs from each of these groups were 0.83, 1.6, and 3.4 ppm on a wet-weight basis. The group fed 8 ppm produced 57% as many healthy ducklings as the controls; the reduction in numbers was caused mainly by hatching failure and the early death of those that did hatch. A sample of 15 eggs from this group contained 11 ppm of selenium. The group fed 16 ppm failed to produce any healthy young, and a sample of 10 of their eggs contained an average of 18 ppm of selenium. Therefore, based on this study, the highest mean selenium concentration in eggs not associated with reproductive impairment was 3.4 ppm, and the lowest mean toxic concentration was 11 ppm.

In another laboratory study, mallards were fed 10 ppm of selenium as selenomethionine (Heinz et al., 1987). Reproductive success was significantly lower in the treated ducks than in controls, and a small sample of five eggs from the treated birds contained a mean of 4.6 ppm of selenium on a wet-weight basis. Because mallards were fed only one dietary concentration of selenium in the form of selenomethionine (10 ppm), no safe level was established in this experiment. All that can be said is that the safe level in eggs was below 4.6 ppm of selenium.

In a laboratory study designed to measure the lingering effects of an overwinter exposure to selenomethionine on reproduction, mallards were fed 15 ppm of selenium for 21 weeks prior to the onset of laying (Heinz and Fitzgerald, 1993b). Females began laying after various lengths of time off treatment. This experimental design was not ideal for determining the lowest concentration of selenium in eggs associated with reproductive impairment, but the authors were able to make some general conclusions. Some fertile eggs hatched when selenium in eggs was as high as 6 to 9 ppm on a wet-weight basis, but other eggs failed to hatch when concentrations of selenium were estimated to be between 3 and 5 ppm. The authors concluded that the most logical reason why some embryos die while others survive when exposed to a given concentration of selenium is that mallard embryos vary in their sensitivity to selenium.

In a laboratory study with black-crowned night herons *(Nycticorax nycticorax)*, 10 ppm of selenium as selenomethionine in the diet (on close to a dry-weight basis) did not reduce the hatching success of fertile eggs (Smith et al., 1988). The eggs of treated herons contained a mean concentration of 3.3 ppm of selenium on a wet-weight basis. The results from this study must be taken with some caution, however, because sample sizes were small and hatching of control eggs was poor.

Martin (1988) fed Japanese quail *(Coturnix coturnix japonica)* diets containing 5 or 8 ppm and chickens *(Gallus domesticus)* 10 ppm of selenium as selenomethionine, respectively. At 5 ppm of selenium, the hatching success of fertile quail eggs (56.4%) was lower than that of controls (76.4%); eggs from treated females contained 7.1 ppm of selenium on a wet-weight basis. At 8 ppm of selenium, the hatching of quail eggs was further decreased to 10.4% (compared with 75.1% for controls in that trial), and selenium in eggs averaged 12 ppm. The hatching success of the chickens fed 10 ppm of selenium also was depressed (23.2% compared with 84.5% for controls), and selenium in eggs averaged 9.6 ppm. No-effect concentrations in the diet or eggs were not determined.

In another study with chickens, diets were supplemented with seleniferous grains in amounts to produce dietary concentrations of 2.5, 5, and 10 ppm of selenium (Poley and Moxon, 1938; Moxon and Poley, 1938). Selenomethionine is reported to be a dominant form of selenium in grains (Olson et al., 1970). Modern statistical techniques were not applied to these data, and chemical analyses were different from those used today, but at 2.5 ppm of selenium in the diet, the hatching success of fertile eggs was no different from that of controls, and a sample of eggs contained 1.75 ppm of selenium in egg white and 1.67 ppm in egg yolk, both on a wet-weight basis. At 5 ppm of selenium in the diet, the hatching of eggs was "slightly reduced," and selenium in egg whites and yolks averaged 2.95 and 2.73 ppm, respectively. At 10 ppm of selenium, hatching decreased to zero, and egg whites and yolks contained 6.40 and 3.92 ppm of selenium, respectively. A selenium threshold of about 3 ppm in whole eggs, therefore, was associated with reproductive impairment, revealing similar results between this early study and more rigorous recent studies.

Harmful concentrations of selenium in eggs may be of a different magnitude when another chemical form of selenium, sodium selenite, is fed to birds. A diet containing 7 ppm of selenium as sodium selenite caused reproductive impairment in chickens and resulted in only 0.87 and 2.02 ppm (wet weight) of selenium in egg white and yolk (Ort and Latshaw, 1978).

In another study with chickens, a diet containing 8 ppm of selenium as sodium selenite impaired reproduction, and whole eggs contained from 1.46 to 1.86 ppm of selenium on a wet-weight basis (Arnold et al., 1973). The chemical form of selenium in chicken eggs seems to be different when sodium selenite rather than selenomethionine is fed (Latshaw, 1975; Latshaw and Osman, 1975).

In mallards, a dietary concentration of 25 ppm of selenium as sodium selenite impaired reproduction, but resulted in a mean of only 1.3 ppm of selenium (wet weight) in eggs (Heinz et al., 1987). Therefore, although higher dietary concentrations of sodium selenite than selenomethionine must be fed to mallards to harm reproduction, lower concentrations of selenium in eggs are associated with harm.

In nature, birds may eat foods containing several different chemical forms of selenium, each form having its own propensity to accumulate in eggs and its own toxicity to embryos once in the egg. Unfortunately, little is known about the kinds and percentages of chemical forms in different foods, their ability to be transferred by the female to her eggs, or their toxicity to embryos. It is known that selenomethionine is a major chemical form in some plants and a highly toxic form to embryos, causing, in the laboratory, the kinds of deformities seen in nature. Furthermore, the toxic thresholds in eggs, estimated from feeding selenomethionine in laboratory studies, are very close to the thresholds derived from field studies. Therefore, until more is learned about the prevalence and toxicity of the many different chemical forms of selenium in avian foods, it seems best to have based the threshold level that affects avian reproduction at 3 ppm of selenium (wet weight) in eggs on a combination of field data and feeding studies with selenomethionine.

As mentioned, the concentrations of selenium estimated from laboratory studies to be toxic in eggs are similar to those estimated from field studies. Studying birds at the Kesterson Reservoir in California, Ohlendorf et al. (1986b) used logistic regression to estimate a 50% chance of embryo death or deformity in American coots *(Fulica americana)* when selenium concentrations in eggs reached about 4.9 ppm on a wet-weight basis. The estimated selenium concentration causing the same effect in black-necked stilts *(Himantopus mexicanus)* was 7.1 ppm (wet weight). The value for eggs of eared grebes *(Podiceps nigricollis)* could not be calculated because even the lowest selenium concentration detected in eggs, about 11 ppm (wet weight), was embryotoxic. The logistic approach is best suited to estimate the 50% effect concentration, not the concentrations of selenium in eggs at which embryo deaths and deformities begin for each species. These concentrations would obviously be somewhat lower than the 50% effect levels.

Skorupa and Ohlendorf (1991) examined the relation between selenium concentrations in eggs and reproductive impairment at the population level. Embryo deformities were detected in only 3 of 55 populations of birds that had a mean selenium concentration of less than about 0.9 ppm (wet weight) in eggs; this is a concentration of selenium judged to represent a background level. However, deformities were detected in 9 of 10 populations of birds in which the mean selenium concentration in eggs exceeded about 14 ppm (wet weight). Their data suggested that a teratogenic threshold at the population level existed between about 4 and 7 ppm of selenium on a wet-weight basis. These authors also discussed some preliminary data on threshold concentrations of selenium affecting the hatching success (which is more sensitive to selenium than are teratogenic effects) of black-necked stilt and American avocet *(Recurvirostra americana)* eggs, again at the population level. They concluded that significantly reduced hatching success was associated with mean selenium concentrations in eggs exceeding about 2.4 ppm (wet weight) in a population of birds. Because the threshold of 2.4 ppm of selenium in eggs was based on means from populations, some of the eggs from those affected populations contained more than 2.4 ppm of selenium.

Based on laboratory studies in which the organic form of selenium, selenomethionine, was fed and on field studies in which the dietary forms of selenium were unknown, the designation of 3 ppm of selenium in eggs as a threshold for reproductive impairment seems warranted. A much lower threshold would be in the range of background levels in nature and at a lower concentration than was shown safe when mallards were fed selenomethionine in the laboratory. However, readers should understand that setting the threshold at 3 ppm leaves only a narrow margin of safety, especially because so few species have been tested under controlled laboratory conditions. The threshold concentration of 3 ppm of selenium in eggs should not be confused with a *safe* level that might be set for regulatory purposes, a level incorporating a safety factor to account for unknowns.

LIVER

Based on the laboratory and field data discussed below, I recommend that concentrations of greater than 10 ppm of selenium in the liver on a wet-weight basis be considered possibly harmful to the health of young and adult birds and that concentrations above about 3 ppm in the liver of laying females may be associated with reproductive impairment.

Selenium concentrations in the liver have been used to estimate both exposure and effects on birds. Although accumulation in the liver is dose dependent (Hoffman et al., 1991), the hepatic concentration is only an imprecise estimator of the pathological condition of a bird. The cutoff is not clear between selenium concentrations in the livers of birds killed by selenium poisoning and others exposed to high concentrations but collected alive. The livers of birds found dead at the Kesterson Reservoir contained 8 to 26 ppm (wet weight) of selenium, whereas the livers of birds shot there contained 11 to 26 ppm of selenium (Ohlendorf et al., 1988).

In a laboratory study, surviving mallard ducklings fed 40 ppm of selenium as selenomethionine had a mean of 68 ppm of selenium in the liver (wet weight), whereas ducklings that died had a mean of 60 ppm (Heinz et al., 1988). In another laboratory study, this time with adult male mallards fed 100 ppm of selenium as selenomethionine, the livers of survivors contained a mean of 43 ppm of selenium (wet weight), and the livers of birds that died contained a mean of 38 ppm of selenium (Heinz, 1993).

When adult male mallards were fed 32 ppm of selenium as selenomethionine, they accumulated an average of 29 ppm (wet weight) in their livers (Hoffman et al., 1991). One of 10 birds fed 32 ppm of selenium died, and others had hyperplasia of the bile duct and hemosiderin pigmentation of the liver and spleen. Various other sublethal effects, such as elevated plasma alkaline phosphatase activity and a change in the ratio of hepatic oxidized glutathione to reduced glutathione, were observed in ducks with lower hepatic concentrations. At a dietary concentration of 8 ppm of selenium, which caused several of the physiological effects mentioned above, the mean concentration of selenium in the liver was about 12.5 ppm of selenium on a wet-weight basis.

Based on these laboratory studies, in which selenium was present as selenomethionine and the only element fed at toxic concentrations, mortality of young and adult mallards could occur when hepatic concentrations of selenium reach roughly 20 or more ppm on a wet-weight basis, and important sublethal effects are likely when the concentrations exceed about 10 ppm.

Using selenium concentrations in adult female livers to predict when reproductive impairment occurs in birds is not nearly as good as using selenium concentrations in eggs, because it is the selenium in the egg that actually harms the embryo. Extrapolating from liver to egg will introduce additional uncertainty above that already existing for the egg. However, in a controlled laboratory study, the correlation between selenium concentrations in eggs and in the livers of laying females was demonstrated by feeding mallards selenomethionine (Heinz et al., 1989). Therefore, when selenium concentrations in eggs are not available, the concentrations in the livers of females during the breeding season can be used to estimate whether reproduction might be impaired. When selenium concentrations are known for both the eggs and livers of breeding females, judgments on the hazards of selenium to reproduction should be based on selenium in the egg.

Because selenium concentrations quickly build up or decline in the liver when birds are introduced to or removed from a selenium-contaminated diet (Heinz et al., 1990), measurements of selenium in the livers of females collected outside the breeding season are probably not very useful in predicting reproductive effects. In laboratory studies of reproduction, the livers of male mallards contained more selenium than did the livers of females fed the same diets (Heinz et al., 1987; Heinz et al., 1989). Because females use the egg as a route of selenium excretion unavailable to males, one would expect that, in the field, the lowest reproductive effect threshold of selenium would be in the livers of laying females and that the livers of males would be less useful in predicting effects on reproduction, even if the males were collected during the breeding season and from the area where reproduction is of concern. The advantage of sampling laying females, however, may be more academic than practical. In nature, it is easier and more likely that a female would be collected before or after egg laying, at which time the concentration of selenium in her liver should be the same as in the liver of a male. If one collects breeding males in the wild or has reason to believe that the collected females were not collected during egg laying, the 3-ppm threshold concentration of selenium in the liver would be on the low side; a value of 4 to 6 ppm might be more appropriate.

The reasoning behind the 3-ppm threshold of selenium in the livers of egg-laying females is based on information from the laboratory and field studies discussed below. In a laboratory study with mallards, a dietary concentration of 8 ppm of selenium as selenomethionine significantly reduced reproductive success, and livers of the treated females contained a mean of 3.5 ppm of selenium on a wet-weight basis (Heinz et al., 1989). In the same study, reproductive success was not significantly different between females fed 4 ppm of selenium and controls, and livers contained a mean of 2.4 ppm (wet weight) of selenium. Based on a regression equation of selenium concentrations in female livers vs. their eggs (Heinz et al., 1989), the threshold concentration of 3 ppm of selenium in eggs corresponds to a value of 1.6 ppm of selenium (wet weight) in the liver. However, I am uncertain whether the data for this regression were linear in the lower end of the selenium range. If the data are curvilinear, a value of 3 ppm of selenium in eggs may correspond to a value of roughly 3 ppm for the liver.

In a related study with mallards, females fed 10 ppm of selenium as selenomethionine had reduced reproductive success and a mean of 4.7 ppm (wet weight) of selenium in the liver (Heinz et al., 1987). Because no dietary concentrations below 10 ppm were used, a no-effect level of selenium in the liver was not determined.

In these laboratory studies with mallards, between 16 and 31 eggs were laid before each female was sacrificed. Depletion of selenium through egg laying, therefore, may have been greater in the laboratory than in nature where birds lay fewer eggs. If depletion of selenium is greater by females in a laboratory study, the selenium concentrations in the liver associated with reproductive impairment could be on the low side.

Based on field data, a very high risk of embryonic deformity exists when the mean selenium concentration (wet weight) in the liver of a population of birds (both sexes included and females not necessarily laying) exceeded about 9 ppm (U.S. Fish and Wildlife Service, 1990). Populations with means below about 3 ppm of selenium on a wet-weight basis generally did not have many deformed embryos.

The suggested threshold of reproductive impairment of 3 ppm of selenium on a wet-weight basis in the livers of laying females is admittedly a rough approximation. In laboratory and field studies, some birds or populations of birds had normal reproductive success when selenium in the liver exceeded 3 ppm. The 3-ppm value is influenced by the laboratory-derived regression equation between eggs and livers that I discussed earlier. The uncertainty surrounding this threshold for the liver reinforces the need for determining selenium levels in eggs to assess risks to reproductive success.

OTHER TISSUES

Tissues other than eggs and liver have been analyzed in laboratory studies of selenium toxicity to birds and in field studies in selenium-contaminated areas. Although the toxic thresholds have not been studied as well for these other tissues as for eggs and liver, pertinent results from field studies and experimental studies with selenomethionine are important if these other tissues are the only ones for which selenium concentrations are reported.

Kidney

Ohlendorf et al. (1990) reported a significant correlation ($r = 0.98$) between selenium in the liver and kidney of American coots from clean and contaminated areas in California; both tissues had similar amounts of selenium. In a separate report, Ohlendorf et al. (1988) concluded that, for freshwater aquatic birds in selenium-contaminated areas, the liver/kidney ratio for selenium is about 1. In an experimental study with chickens fed a range of from 0.1 to 6 ppm of selenium as selenomethionine, Moksnes (1983) also found nearly the same concentrations of selenium in the kidney as in the liver.

Breast Muscle

Ohlendorf et al. (1990) also reported a significant correlation ($r = 0.69$) between selenium in the liver and in the breast muscle of juvenile ducks from the Kesterson Reservoir in California. The predictive equation was

$$\text{Log Se in muscle} = 0.22 + 0.65 \log \text{Se in liver}.$$

When mallards in a laboratory study were fed 10 ppm of selenium as selenomethionine, females had nearly identical wet-weight concentrations of selenium in the liver (4.7 ppm) and breast muscle (4.9 ppm), whereas males had more in the liver (8.6 ppm) than in breast muscle (3.1 ppm) (Heinz et al., 1987). Because the females were laying eggs, they may have been using stores of selenium from the liver to incorporate into eggs.

Fairbrother and Fowles (1990) reported more selenium in breast muscle (about 6.7 ppm on a wet-weight basis) than in the liver (about 4.5 ppm) of male mallards given drinking water containing 2.2 mg of selenium per liter for 12 weeks. When chickens were fed from 0.1 to 6 ppm of selenium as selenomethionine for 18 weeks, selenium concentrations in breast muscle (0.29 ppm) ranged from about half of those in the liver (0.60 ppm) at 0.1 ppm in the diet to nearly equal concentrations (5.4 and 6.6 ppm) at 6 ppm in the diet (Moksnes, 1983).

Heinz et al. (1990) fed female mallards 10 ppm of selenium as selenomethionine for 6 weeks, followed by 6 weeks off treatment, and measured selenium in the liver and breast muscle. By 6 weeks, selenium in the liver averaged about 7.4 ppm (wet weight) and in breast muscle, about 6.3 ppm. However, selenium in the liver had nearly peaked after about 1 week, whereas muscle was projected to reach a peak of about 8 ppm after 81 days. Likewise, selenium was lost faster from the liver than from breast muscle, indicating that the two tissues contain similar concentrations of selenium, but only after both reach an equilibrium.

Blood

In the same study by Heinz et al. (1990), female mallards were fed increasingly high diets of selenium as selenomethionine (from 10 ppm to 160 ppm over a period of 31 days). At the end of the 31-day exposure, birds began to die. Survivors contained means of about 22.6 ppm of selenium in the liver and 12 ppm in the blood. However, as with muscle, the rate of loss of selenium was faster from the liver than from the blood, suggesting that consistent relations between selenium in the blood and in the liver may occur only at equilibrium levels.

In another study, adult male mallards that were fed 20, 40, or 80 ppm of selenium as selenomethionine began dying when samples of blood from surviving ducks in the same pens contained means of about 5 to 14 ppm of selenium (wet weight) (Heinz and Fitzgerald, 1993a). However, samples of blood were not taken from any of the birds that died. Therefore, comparisons of selenium concentrations between the dead and the survivors were not possible.

SUMMARY

Selenium is an environmental contaminant from soils naturally high in selenium, smelting operations, and other industrial activities. Although selenium occurs in many different chemical forms, most of the laboratory data used to derive toxic threshold concentrations were based on one highly toxic form, selenomethionine. Data were also used from the field, where selenomethionine is believed to be a major, but probably not the sole, chemical form in the foods of birds.

Reproductive success is more sensitive to selenium poisoning than are the health and survival of young and adult birds. The threshold of reproductive problems, primarily deformities of embryos and hatching failure, is believed to occur when selenium concentrations in eggs exceed about 3 ppm on a wet-weight basis. Liver residues also can be used, although not as successfully, to predict whether selenium will impair reproduction in birds. When the livers of egg-laying females contain more than about 3 ppm of selenium on a wet-weight basis, reproductive impairment is possible.

When the livers of young or adult birds contain more than 10 ppm of selenium on a wet-weight basis, important sublethal effects may occur, and when wet-weight concentrations exceed 20 ppm, the survival of young and adults may be jeopardized.

No other tissues are presently as suitable as eggs and livers for predicting harm to birds, but approximate thresholds of toxicity exist for the kidney, breast muscle, and blood under certain conditions.

REFERENCES

Arnold, R. L., O. E. Olson, and C. W. Carlson. 1973. Dietary selenium and arsenic additions and their effects on tissue and egg selenium. Poult. Sci. 52:847-854.

Cappon, C. J. 1991. Sewage sludge as a source of environmental selenium. Sci. Total Environ. 100:177-205.

Cuvin-Aralar, M. L. A., and R. W. Furness. 1991. Mercury and selenium interaction: a review. Ecotoxicol. Environ. Saf. 21:348-364.

Donaldson, W. E., and C. McGowan. 1989. Lead toxicity in chickens: interaction with toxic dietary levels of selenium. Biol. Trace Elem. Res. 20:127-133.

Fairbrother, A., and J. Fowles. 1990. Subchronic effects of sodium selenite and selenomethionine on several immune functions in mallards. Arch. Environ. Contam. Toxicol. 19:836-844.

Hamilton, S. J., K. J. Buhl, N. L. Faerber, R. H. Wiedmeyer, and F. A. Bullard. 1990. Toxicity of organic selenium in the diet to chinook salmon. Environ. Toxicol. Chem. 9:347-358.

Heinz, G. H. 1993. Re-exposure of mallards to selenium after chronic exposure. Environ. Toxicol. Chem. 12:1691-1694.

Heinz, G. H., and M. A. Fitzgerald. 1993a. Overwinter survival of mallards fed selenium. Arch. Environ. Contam. Toxicol. 25:90-94.

Heinz, G. H., and M. A. Fitzgerald. 1993b. Reproduction of mallards following overwinter exposure to selenium. Environ. Pollut. 81:117-122.

Heinz, G. H., D. J. Hoffman, and L. G. Gold. 1988. Toxicity of organic and inorganic selenium to mallard ducklings. Arch. Environ. Contam. Toxicol. 17:561-568.

Heinz, G. H., D. J. Hoffman, and L. G. Gold. 1989. Impaired reproduction of mallards fed an organic form of selenium. J. Wildl. Manage. 53:418-428.

Heinz, G. H., D. J. Hoffman, A. J. Krynitsky, and D. M. G. Weller. 1987. Reproduction in mallards fed selenium. Environ. Toxicol. Chem. 6:423-433.

Heinz, G. H., G. W. Pendleton, A. J. Krynitsky, and L. G. Gold. 1990. Selenium accumulation and elimination in mallards. Arch. Environ. Contam. Toxicol. 19:374-379.

Hill, C. H. 1974. Reversal of selenium toxicity in chicks by mercury, copper, and cadmium. J. Nutr. 104:593-598.
Hoffman, D. J., and G. H. Heinz. 1988. Embryotoxic and teratogenic effects of selenium in the diet of mallards. J. Toxicol. Environ. Health 24:477-490.
Hoffman, D. J., G. H. Heinz, L. J. LeCaptain, C. M. Bunck, and D. E. Green. 1991. Subchronic hepatotoxicity of selenomethionine ingestion in mallard ducks. J. Toxicol. Environ. Health 32:449-464.
Jensen, L. S. 1975. Modification of a selenium toxicity in chicks by dietary silver and copper. J. Nutr. 105:769-775.
Latshaw, J. D. 1975. Natural and selenite selenium in the hen and egg. J. Nutr. 105:32-37.
Latshaw, J. D., and M. Osman. 1975. Distribution of selenium in egg white and yolk after feeding natural and synthetic selenium compounds. Poult. Sci. 54:1244-1252.
Martin, P. F. 1988. The Toxic and Teratogenic Effects of Selenium and Boron on Avian Reproduction. M.S. thesis. University of California, Davis, Calif.
Moksnes, K. 1983. Selenium deposition in tissues and eggs of laying hens given surplus of selenium as selenomethionine. Acta Vet. Scand. 24:34-44.
Moxon, A. L., and W. E. Poley. 1938. The relation of selenium content of grains in the ration to the selenium content of poultry carcass and eggs. Poult. Sci. 17:77-80.
Ohlendorf, H. M. 1989. Bioaccumulation and effects of selenium in wildlife, p. 133-177. *In* L. W. Jacobs (Ed.). Selenium in agriculture and the environment. Special Publication 23. American Society of Agronomy and Soil Science Society of America, Madison, Wisc.
Ohlendorf, H. M., D. J. Hoffman, M. K. Saiki, and T. W. Aldrich. 1986a. Embryonic mortality and abnormalities of aquatic birds: apparent impacts of selenium from irrigation drainwater. Sci. Total Environ. 52:49-63.
Ohlendorf, H. M., R. L. Hothem, C. M. Bunck, T. W. Aldrich, and J. F. Moore. 1986b. Relationships between selenium concentrations and avian reproduction. Trans. N. Am. Wildl. Nat. Resour. Conf. 51:330-342.
Ohlendorf, H. M., R. L. Hothem, C. M. Bunck, and K. C. Marois. 1990. Bioaccumulation of selenium in birds at Kesterson Reservoir, California. Arch. Environ. Contam. Toxicol. 19:495-507.
Ohlendorf, H. M., A. W. Kilness, J. L. Simmons, R. K. Stroud, D. J. Hoffman, and J. F. Moore. 1988. Selenium toxicosis in wild aquatic birds. J. Toxicol. Environ. Health 24:67-92.
Olson, O. E., E. J. Novacek, E. I. Whitehead, and I. S. Palmer. 1970. Investigations on selenium in wheat. Phytochemistry (Oxf.) 9:1181-1188.
Ort, J. F., and J. D. Latshaw. 1978. The toxic level of sodium selenite in the diet of laying chickens. J. Nutr. 108:1114-1120.
Poley, W. E., and A. L. Moxon. 1938. Tolerance levels of seleniferous grains in laying rations. Poult. Sci. 17:72-76.
Poley, W. E., A. L. Moxon, and K. W. Franke. 1937. Further studies of the effects of selenium poisoning on hatchability. Poult. Sci. 16:219-225.
Robberecht, H., H. Deelstra, D. Vanden Berghe, and R. Van Grieken. 1983. Metal pollution and selenium distributions in soils and grass near a non-ferrous plant. Sci. Total Environ. 29:229-241.
Skorupa, J. P., and H. M. Ohlendorf. 1991. Contaminants in drainage water and avian risk thresholds, p. 345-368. *In* A. Dinar and D. Zilberman (Eds.). The economics and management of water and drainage in agriculture. Kluwer Academic Publishers, Norwell, Mass.
Smith, G. J., G. H. Heinz, D. J. Hoffman, J. W. Spann, and A. J. Krynitsky. 1988. Reproduction in black-crowned night-herons fed selenium. Lake Reservoir Manage. 4:175-180.

Spallholz, J. E., and A. Raftery. 1984. Nutritional, chemical, and toxicological evaluation of a high-selenium yeast, p. 516-529. *In* G. F. Combs, Jr., J. E. Spallholz, O. A. Levander, and J. E. Oldfield (Eds.). Selenium in biology and medicine, part A. Van Nostrand Reinhold, New York, NY.

Thapar, N. T., E. Guenthner, C. W. Carlson, and O. E. Olson. 1969. Dietary selenium and arsenic additions to diets for chickens over a life cycle. Poult. Sci. 48:1988-1993.

U.S. Fish and Wildlife Service. 1990. Summary report: effects of irrigation drainwater contaminants on wildlife, p. 1-38. U.S. Fish and Wildlife Service, Patuxent Wildlife Research Center, Laurel, Md.

Wadge, A., and M. Hutton. 1986. The uptake of cadmium, lead, and selenium by barley and cabbage grown on soils amended with refuse incinerator fly ash. Plant Soil 96:407-412.

Yasumoto, K., T. Suzuki, and M. Yoshido. 1988. Identification of selenomethionine in soybean protein. J. Agric. Food Chem. 36:463-467.

CHAPTER 21

Fluoride in Birds

W. James Fleming

INTRODUCTION

The availability of fluoride to animals is a result of both natural and anthropogenic processes. Some bedrock formations and overlying soils, as well as geothermal waters, enrich local and regional levels of fluoride. Industrial releases of fluorides are most often as gases or airborne particulates (Smith and Hodge, 1979), although waterborne industrial releases also occur (Andreasen and Stroud, 1987). Industrial fluoride by-products are associated with the mining or production of aluminum, phosphates, steel, copper, petroleum, bricks, and ceramics, as well as coal-fired power generation (Smith and Hodge, 1979). Emissions of fluorides during the late 1960s were estimated to be 1.36×10^7 kg per year and are expected to more than double by the year 2000 (Smith and Hodge, 1979).

Some wild bird populations are exposed to high amounts of fluoride and, as a result, fluoride concentrations in tissues may become elevated. Bone and eggshell are the most commonly assayed tissues for assessing fluoride exposure. It is to be expected that fluoride will almost always be detected in these tissues, even for samples collected from unpolluted sites. However, there is no indication in the literature that fluoride concentrations in biological tissues can be clearly related to biological effects.

Fluoride has rarely been implicated in the demise of either individuals or populations of wild birds. There is only a single, published report of possible fluoride-induced mortality of wildfowl (Andreasen and Stroud, 1987). Descriptive reports suggest that community and population effects may occur, but experimental evidence is lacking. Van Toledo (1978) speculated that the number of avian species was depressed near fluoride-emitting aluminum factories in Europe. Newman (1977) believed that the nesting density of house martins *(Delichon urbica)* was reduced near an aluminum plant with high fluoride emissions in Czechoslovakia. Further studies suggest that the low density of birds was a generalized response to air pollutants, of which fluoride was one (J. R. Newman, personal communication).

Consumption of foods or water high in fluoride content is probably the most important route of avian exposure to fluoride. Upon ingestion, the availability of other elements (e.g., calcium, magnesium, and aluminum) in the ingesta may influence the amount of fluoride absorption and storage of fluoride in tissues (Gardiner et al., 1961; Rogler and Parker, 1972; Cukir et al., 1978; Hahn and Guenter, 1986).

Once ingested and absorbed, fluoride may be stored or excreted. Excretion of fluoride in the blood is primarily through the cloaca, with the salt gland, an accessory ion excretion site in many species of birds, playing only a minor role (Culik, 1987). Fluoride that is not excreted becomes incorporated in bone or eggshells.

Fluoride is not listed as an essential element for birds (Scott et al., 1976). However, as in mammals, it may provide benefits to some species under specific nutritional circumstances. The addition of fluoride to poultry diets has been repeatedly investigated as a means of strengthening bone and eggshell. Such dietary supplementation has proven useful, especially when dietary calcium is limited.

The sensitivity of mammals to fluoride varies widely. Ungulates are particularly sensitive, with both laboratory and field investigations demonstrating debilitating effects on dentition and bone. The range of sensitivity in birds is not known. Most experimental feeding studies to date have been conducted with galliforms. Field investigations have not revealed any avian species that are peculiarly sensitive to fluoride.

FLUORIDE INGESTION

It is reasonable to expect that the ingestion of fluoride is strongly related to the food habits of individual species. Airborne particulates containing fluoride settle on leaf surfaces, and fluoride gases are taken into photosynthetic surfaces through leaf stomates. Fluorides are not readily transported through plant tissues; therefore, herbivorous species that consume leafy materials are expected to have a higher exposure than grainivores. Exposure of raptors also seems low because of the generally low fluoride content reported in soft tissues. For raptors that consume bone while ingesting prey, those that do not digest the bone might be expected to have lower fluoride exposure than species that digest bone. However, there is no clear-cut relation between fluoride concentrations in raptors and their bone consumption and digestion characteristics (Seel and Thomson, 1984). Many fish-eating birds digest the bones of small prey and therefore may absorb fluoride from ingested bone. Because pollinating and leaf-feeding insects accumulate fluoride (Dewey, 1973), insectivorous birds feeding on these groups of insects in fluoride-contaminated environments should accumulate relatively high fluoride levels.

In general, field data support these expectations for differences in fluoride exposure that are based on the food habits of individual species. Surveys of fluoride concentrations in birds across a variety of habitats suggest that short-lived, seed-eating birds have lower bone fluoride concentrations than long-lived, omnivorous species (Stewart et al., 1974; Kay et al., 1975), although it is unclear whether diet or age is the more important factor. Seel and Thomson (1984) reported that avian predators that feed mainly on birds have higher fluoride concentrations in femurs than species that feed mainly on small mammals. In five species of owls, the fluoride concentrations were highest in the two species that consumed both small mammals and earthworms. These authors speculated that earthworms contained ingested soil-borne fluoride, which in turn increased fluoride exposure in the owls that consumed them (Seel and Thomson, 1984).

RESIDUES IN BONE

Fluoride readily accumulates in bone, where it is incorporated into the hydroxylapatite lattice. Accumulation is strongly related to exposure although age, growth rates, and interactions with other elements may affect intestinal adsorption, excretion, and deposition in bone. Accumulation is especially rapid in actively growing young animals, in which bone fluoride quickly reaches equilibrium with dietary fluoride content (Suttie et al., 1964, 1984). Once incorporated in bone, fluoride becomes relatively unavailable and inactive in biological processes. The biological availability of bone fluoride in domestic chickens was well illustrated by Merkley (1981), who supplied chickens (beginning at 1 day of age) with calcium-adequate diets and fluoride-supplemented water for 20 weeks, followed by 25 weeks on untreated water. The chickens were actively laying eggs during the posttreatment period, which probably maximized the mobilization of both calcium and fluoride. However, the fluoride concentrations in bone did not decrease during the posttreatment period.

The relation of sex and age to the fluoride content of bone is somewhat variable. Fluoride concentrations in male and female sparrowhawks *(Accipiter nisus)* and in female Eurasian kestrels *(Falco tinnunculus)* increased significantly with age (Seel et al., 1987). They did not increase with age in male Eurasian kestrels or in combined sex groups of tawny owls *(Strix aluco)* or barn owls *(Tyto alba)* (Seel et al., 1987). Male sparrowhawks and Eurasian kestrels had 1.3 and 1.5 times more fluoride in femurs than did females. For two owl species, both having considerably less fluoride than the two accipiters, there were no differences in fluoride concentrations between sexes (Seel et al., 1987). Fluoride in tibiotarsi from black-crowned night herons *(Nycticorax nycticorax)* showed a strong, positive correlation with age, but there were no differences between sexes (Henny and Burke, 1990). The fluoride content of magpie *(Pica pica)* nestlings' bone (24 days old) did not differ significantly from that of adults (Seel, 1983).

Seel et al. (1987) examined the hypothesis that fluoride concentrations in bone increase with body size. Using size differences among species ($n = 4$) and between sexes, they found no link with fluoride concentrations.

EXPRESSION OF FLUORIDE CONTENT OF BONE

The fluoride content of bone is normally expressed as ppm ($\mu g/g$), dry weight, but can be stated as fat-free dry weight, ash weight, or fat-free ash weight. I found no published conversion factors to adjust for the fat content of bone. However, for long bones of birds, the fat content of bone after drying, and especially after ashing, should not dramatically affect the values of residues. For black-crowned night herons, a regression equation ($r = 0.95$) can be used to predict dry weight and ash weight (Henny and Burke, 1990):

$$\mu g \text{ of fluoride}/g \text{ (dry weight)} = -340.228 + 0.515 \, [\mu g \text{ of fluoride}/g \text{ (ash weight)}]$$

Herein, I have presented fluoride residue data as reported in the cited publications.

REPRODUCTION

High dietary fluoride has been linked with poor reproductive performance in both domestic and wild birds. However, the fluoride content of bone does not appear to be closely related to productivity. This is probably because dietary fluoride is more biologically available, and thus biologically active, than is fluoride sequestered in bone.

Screech owls *(Otus asio)* fed a meat diet supplemented with 200 ppm of fluoride produced fewer young than did owls fed the same diet without added fluoride (Pattee et al., 1988). Although the mean fluoride levels in femurs were numerically greater in the 200-ppm (dietary) group (geometric mean = 1134 ppm [dry weight] for males and 1720 ppm for females) compared with those levels in femurs from controls, the differences were not significant. High variability in residues within experimental groups contributed to the failure to detect statistical differences in bone fluoride. The number of fertile eggs was not significantly correlated with the fluoride content of femurs for either sex, even though significantly fewer fertile eggs were produced in the 200-ppm group compared with those from the control group.

Fluoride concentrations in the femurs of adult starlings *(Sturnus vulgaris)* were significantly increased with dietary fluoride supplementation (W. J. Fleming, unpublished data). Starlings fed the highest dietary concentration (1080 ppm of fluoride) produced significantly fewer hatchlings than did groups fed diets supplemented with 0, 120, or 360 ppm of fluoride. The major factors contributing to the lower number of young were not significantly correlated with the fluoride content of the femurs of either males or females (extreme fluoride concentrations of 910 to 4500 ppm [dry weight] for males and 980 to 4700 ppm for females). Adult laying hens *(Gallus domesticus)* fed 1300 ppm of fluoride for 16 weeks exhibited a reduced food intake and lower weight gain and required more feed per egg produced than did hens fed 0 or 100 ppm of fluoride or combinations of fluoride plus aluminum. The fluoride concentration in tibiae of the 1300-ppm fluoride group averaged 2600 ppm (fat-free dry weight) and was not different from that of other groups with normal weight gains, food intake, and efficiency in converting calories to egg production (Hahn and Guenter, 1986).

GROWTH RATES

Nestling altricial species, fledging at less than normal body weights, may have lower survivorship probabilities (Perrins, 1963; Loman, 1977). Fledging weights were reduced in nestling starlings administered daily oral doses of 23 mg of fluoride per kilogram of body weight (Fleming et al., 1987). The fluoride content of femurs in this group averaged 39 times that of controls (geometric mean = 3600 vs. 93 ppm, dry weight). To test the influence of the method of fluoride administration, young coturnix *(Coturnix japonica)* were fed diets supplemented with fluoride, and a second group was given daily oral doses of fluoride by gavage (Fleming and Schuler, 1988). Coturnix consuming up to 1000 mg of fluoride per day experienced no mortality and averaged 43 times more fluoride in femurs than did controls (geometric mean = 13,400 vs. 312 ppm, dry weight). Coturnix dosed daily with 54 mg of fluoride per kilogram per day experienced 73% mortality through 16 days of dosing. The fluoride content of femurs in survivors was not available, but 36 mg of fluoride per kilogram per day (the next lower dose group) resulted in no mortality and an average of 5045 ppm (dry weight) of fluoride in femurs. There were no differences in the growth of coturnix fed diets resulting in average tibial fluoride concentrations of 13 to 2223 ppm (fat-free ash weight) (Chan et al., 1973).

Diets supplemented with 400 ppm of fluoride were fed to domestic turkey poults *(Meleagris gallopavo)*. At 18 weeks, treated turkeys weighed significantly less than did controls. Fluoride concentrations in femurs averaged 7770 ppm (dry weight) in these poults vs. 30 ppm in controls (Nahorniak et al., 1983). Reduced growth was also reported for domestic chicks fed diets supplemented with 850 ppm of fluoride (Rogler and Parker, 1972). The fluoride content of bone averaged 15,610 ppm (dry weight) at 4 days of age and 24,704 ppm at 14 days of age. Even greater reductions in growth occurred in chicks fed diets containing fluoride plus high magnesium, although the fluoride content in bone was about one third less than for high fluoride alone. Magnesium alone did not affect growth rates. The addition of calcium to diets decreased the amount of fluoride in tissues and improved growth rates of fluoride-alone and fluoride-plus-magnesium treatment groups, but growth rates still remained below those in controls. Thus, the interaction of fluoride with other elements can influence the concentration of fluoride in tissues as well as complicate the interpretation of fluoride residues in tissues.

BONE STRENGTH

Henny and Burke (1990) examined bone strength in relation to bone fluoride content in black-crowned night herons. Their work was prompted by the observation of a heron that snapped its tibia in what appeared to be a minor collision. They found that the fluoride content of bone increased with age (up to 11,000 ppm, ash weight). However, the fluoride could not be clearly implicated in reducing the bone strength in these herons.

Bone strength has always been of great interest in the domestic poultry industry, where maximizing production and minimizing bone breakage during processing are goals. The bone strength was not reduced in turkey poults fed diets supplemented with 400 ppm of fluoride (Nahorniak et al., 1983). Average fluoride residues in tibiae of these poults at 18 weeks were 7770 ppm (ash weight) compared with 30 ppm for controls. Suttie et al. (1984) reported no difference in tibial breaking strength for young chickens and turkeys fed up to 800 ppm of fluoride. After 6 weeks on experimental diets, the high-dose groups averaged fluoride residues in femora from 10,863 to 15,560 ppm (ash weight). There was no difference in bone morphology among groups.

The breaking strength of humeri and tibiae was about twice as great in experimental groups of laying hens with tibial fluoride content averaging 64,500 to 95,500 ppm (fat-free ash weight) compared with 13,000 to 15,100 ppm in controls (Merkley, 1981). The breaking strength of bones from laying hens with an average 32,700 to 53,300 ppm of fluoride in tibiae did not differ from that of bones with only 13,000 to 15,100 ppm of fluoride. The load strength of humeri of young poultry provided fluoridated water was increased at ≥100 ppm in the diet, which resulted in 38,200 to 42,000 ppm (fat-free ash weight) of humeral fluoride (Merkley, 1976). Tibial torsion strength was reduced in coturnix fed fluoride, in which tibial fluoride concentrations averaged 1963 to 2223 ppm (ash weight) (Chan et al., 1973).

RESIDUES IN SOFT TISSUES

Fluoride can be found in some soft tissues, but not at the high concentrations found in bone. Significant concentrations in soft tissues occur primarily after the renal capacity for fluoride excretion is exceeded. The interpretive link between soft tissue fluoride accumulation and fluoride toxicosis, unfortunately, has not been defined. At present, it can even be questioned whether fluoride accumulates in soft tissues or simply represents fluoride in plasma, urine, or other body fluids contained in tissues and organs. In the following discussion, the fluoride content of soft tissues is presented as ppm of wet weight.

Liver, spleen, pancreas, proventriculus, duodenum, and plasma fluoride concentrations increased 8 to 21 times over those of controls in poultry fed 600 ppm of fluoride for 4 to 6 weeks. For example, the average liver fluoride concentrations increased from 0.15 ppm to a maximum of 2.25 ppm (Suttie et al., 1984). Hahn and Guenter (1986) suggested that fluoride is readily accumulated in bone and kidney, but not in liver and muscle. In general, this is in agreement with the findings of Haman et al. (1936). Fluoride also accumulates in the pituitary gland (Demole and Held, 1963; Zhavoronkov and Edemskii, 1974, as cited by Walash and Ridha, 1980) and probably in some other soft tissues.

The plasma fluoride concentration in animals is normally less than 1 ppm. In rats fed fluoride, the food consumption decreased when plasma fluoride increased to 3 ppm (Simon and Suttie, 1968). Chickens are notably resistant to fluoride compared with mammals (Suttie et al., 1964). The mean plasma fluoride concentrations ranged from 3.84 (after 4 days) to 8.20 ppm (after 14 days) in groups of chicks fed 800 ppm of fluoride. These chicks grew more slowly than did birds fed diets without added fluoride (Rogler and Parker, 1972). Diets containing ≥400 ppm of fluoride reduce the growth of domestic chicks (Gardiner et al., 1961, 1968; Van Toledo and Combs, 1984). The earlier in life that chicks are exposed to fluoride, the more pronounced is this reduction (Gerry et al., 1947; Griffith et al., 1963). A portion of the reduced growth is due to diminished feed consumption (Phillips et al., 1935; Gardiner et al., 1968). However, decreased metabolic efficiency also contributes to the reduction in weight gains in growing chicks (Gardiner et al., 1968).

Adult laying hens fed 1300 ppm of fluoride for 16 weeks had a reduced food intake and smaller weight gain, and they required more feed per egg produced than did hens fed 0 or 100 ppm of fluoride or combinations of fluoride plus aluminum. The concentrations of fluoride in the liver (19.2 ppm), kidney (31.8 ppm), and plasma (10.1 ppm) from the high-dose group were different from those of most other groups in which food consumption and weight gains were normal (Hahn and Guenter, 1986). However, the fluoride content of tibiae (2600 ppm of dry weight) of the 1300-ppm group was not different from that of other groups with normal weight gains, food intake, and efficiency in converting calories to egg production. High fluoride concentrations circulating in the blood may have an effect on yolk synthesis (Guenter and Hahn, 1986).

Fluoride concentrations in soft tissues have been used to describe the single avian die-off attributed to fluoride poisoning. A subsample of 97 snow geese *(Chen caerulescens)* that died from apparent fluoride poisoning had fluoride levels of 32 to 129 ppm in the liver and of 6.8 to 25 ppm in the brain. Fluoride concentrations in the liver and brain of three snow geese that served as reference specimens were less than 1 ppm (Andreasen and Stroud, 1987). Experimental data are not available to substantiate a clinical diagnosis of fluoride poisoning based on fluoride residues in birds.

RESIDUES IN EGGS

EGGSHELLS

Eggshells appear to be the standard substrate for assessing fluoride residues in eggs. The fluoride content of eggshells is expressed as ppm of dry weight. In the following presentation of research summaries, it will become obvious that fluoride is not a potent teratogen, and its presence in the eggshell does not predict reproductive success. Rather, any fluoride-induced reproductive effects seem to be more closely related to the physiology and behaviors of breeding adults. Eggshell integrity may be an exception because of the propensity of fluoride to accumulate in eggshells.

Does the fluoride content of eggshells reflect the degree of exposure to fluoride? The answer is a qualified "yes." In all species of birds studied, whether domestic or wild, the fluoride content of the eggshell increases with experimental or environmental exposure to elevated fluoride levels. However, the fluoride content of individual eggshells, even in controlled experiments, varies widely. For example, within a single clutch of screech owl eggs ($n = 4$), fluoride concentrations ranged from 74 to 220 ppm (Pattee et al., 1988). The within-clutch fluoride content of eggshells from starlings fed fluoride-supplemented diets throughout the breeding season commonly differed by factors of 5 to 10 (W. J. Fleming, unpublished data).

Seel (1983) speculated that the variability of fluoride residues in eggshells of magpies might be due to the short-term availability of fluoride during the brief period of shell formation. Merkley (1981) clearly demonstrated that for laying hens, which are indeterminate layers, the fluoride content of eggshells reflects the dietary intake of fluoride during the days immediately preceding the laying of individual eggs. The mobilization of fluoride from bone had minimal influence on the fluoride content of eggshells. Whether this is the case for determinate layers is not known. However, medullary bone is reportedly a major source of calcium for the eggshells of birds and, as such, may contribute fluoride to the eggshell. Screech owls and American kestrels *(Falco sparverius)*, which are determinate layers, produced eggshells that displayed a trend toward progressively greater concentrations of fluoride in relation to egg order. This was apparent even though the fluoride content of diets remained the same on all days of the study (Bird and Massari, 1983; Pattee et al., 1988). As calcium and, thus fluoride, are depleted from medullary bone, one could speculate that dietary fluoride may contribute more to the fluoride content of eggshells. This shift in importance of dietary fluoride could yield a consistent pattern of within-clutch variability in the fluoride content of eggshells. However, two points argue against this. First, in screech owls, the fluoride content of femora did not correlate well ($r = 0.51$) with that of eggshells (Pattee et al., 1988). Second, starlings, which are determinate layers, did not consistently display an increasing trend of fluoride content of eggshells in relation to egg order (W. J. Fleming, unpublished data).

Unhatched, indented eggshells from two tawny owl nests found near an aluminum factory contained an average of 240 ppm of fluoride (Van Toledo, 1978). To explore the potential effect of fluoride, Pattee et al. (1988) fed screech owls a commercial diet supplemented with fluoride. Sodium fluoride was added to the diets at 0, 40, and 200 mg/kg on a wet-weight basis, which is equivalent to about 0, 100, and 500 mg/kg on a dry-weight basis. The control diet contained 27 ppm of fluoride (wet weight) and adequate nutrient content, including 1.74% calcium and 1.05% phosphorus (dry weight). Control birds produced 4.2 young per clutch and averaged 6.4 ppm of fluoride in the eggshell of the third egg per clutch. The 200-ppm dietary group produced significantly fewer young per clutch (2.6 young) than did the controls and had significant higher (87.2 ppm) fluoride in eggshells. The 40-ppm group produced 3.2 young per clutch, a number not significantly different from that of controls. Fluoride concentrations in the shells of the third eggs of the 40-ppm group averaged 53.3 ppm, significantly different from controls but not from the 200-ppm dietary group. The fluoride content of eggshells of the third eggs was not significantly correlated with the number of fertile eggs per clutch. The eggshell fluoride content of clutches in the 200-ppm group, in which one or more eggs failed to hatch, ranged from 30 to 220 ppm. That range excluded fluoride values for unhatched controls but included residues in eggs from an intermediate dietary group in which hatching did not differ from that of controls.

There was no difference in the hatching success of American kestrels fed diets containing fluoride (Bird and Massari, 1983). The fluoride content of eggshells in the high-dose group (50 ppm) averaged 53 ppm in fourth eggs, which were the eggs within each clutch that tended to have the highest fluoride content.

I fed starlings diets supplemented with 0, 120, 360, or 1080 ppm of fluoride as NaF during the breeding season (unpublished data). The mean fluoride concentrations in eggshells increased in relation to the amount of fluoride added to the feed. Starlings fed the 1080-ppm diet initiated nest building, but few nested. Of those that did nest, egg laying began later than in other groups. The number of eggs laid per clutch was lower, and a significantly greater proportion of the eggs were removed from the nest by the parents. The fluoride content of the eggshells of this dietary group had extreme values of 367 and 770 ppm. Reproduction in the starlings fed the 120- and 320-ppm diets was similar to that of controls. The fluoride content of eggshells in the 120-ppm dietary group had extreme values of 14 and 420 ppm for first eggs and 22 and 130 ppm for fourth eggs. For the 320-ppm dietary group, comparable fluoride concentrations were 64 and 380 ppm for first eggs and 120 and 320 ppm for fourth eggs, respectively. Of the eggs incubated to term, there were no differences in hatching success among experimental groups.

Laying hens provided drinking water containing 100 ppm of fluoride suffered no reduction in egg fertility or hatching success of fertile eggs (Merkley and Sexton, 1982). The fluoride content of eggshells averaged 87 ppm (Merkley, 1981). Though fluoride residues in eggshells were not reported, diets containing 1000 and 1300 ppm of fluoride had no effect on the fertility or hatching success of chicken eggs (Guenter, 1979).

There were no differences in chicken egg weight, eggshell thickness, and breaking strength for eggshells averaging 19.9 to 87 ppm of fluoride (Merkley, 1981). Laying hens fed diets supplemented with 1300 ppm of fluoride produced eggs of similar weight compared with those of controls. The fluoride content of eggshells averaged 307 ppm (Hahn and Guenter, 1986). However, laying hens fed 1000 and 1300 ppm of fluoride produced smaller eggs with thinner shells and more deformities (cracked and thin-shelled eggs) compared with eggs of controls (Guenter and Hahn, 1986). Kuhl and Sullivan (1976) fed laying hens 500 ppm of fluoride in combination with monosodium phosphate for 16 weeks. Hens laid heavier eggs with a reduced shell-breaking strength. The fluoride content of these eggshells averaged 150 ppm. A level of 50 ppm of fluoride in the drinking water of coturnix yielded no difference in eggshell thickness.

Genetic differences among strains of laying hens may contribute to differences in sensitivities to fluoride, according to Van Toledo and Combs (1984). They reported that a strain with low eggshell strength was less sensitive to fluoride-induced changes in eggshells than was a strain selected for high eggshell strength.

Diets supplemented with up to 200 ppm of fluoride produced eggshells of normal thickness in screech owls (Pattee et al., 1988), but egg weights and hatchling size were reduced (Hoffman et al., 1985). Fluoride residues averaged >10 times the control concentrations in these screech owl eggs (geometric mean = 87.2 vs. 6.4 ppm) (Pattee et al., 1988). Starling diets supplemented with up to 320 ppm of fluoride yielded no differences in an eggshell thickness index or in egg volume (W. J. Fleming, unpublished data). Eggshell thickness for eggs produced by a group of starlings fed a 1080-ppm fluoride-supplemented diet appeared to be reduced slightly, but the sample size was too small for statistical testing. Bird and Massari (1983) reported a reduction of eggshell thickness in American kestrels fed fluoride, but interpretation of their data was complicated by other factors that I believe invalidated their report of shell thinning.

YOLK AND ALBUMEN

There is disagreement in the literature about whether fluoride will accumulate in yolk and albumen. Said et al. (1979) reported that laying hens fed 648 ppm of fluoride produced eggs with up to 0.48 and 0.66 ppm of fluoride in the albumen and yolk. Comparable values for controls were 0.23 and 0.27 ppm. At these levels of fluoride, there were no significant differences in egg weight, interior quality of the shell, shell quality, fertility, or hatching success. Laying hens provided fluoride-treated drinking water produced eggshells averaging 54 to 87 ppm; in contrast the yolk and albumen averaged less than 1 ppm of fluoride (Merkley, 1981). Seel (1983) reported that fluoride concentrations in eggshells of magpies in a fluoride-contaminated environment were twice those of controls, but that there were no differences in fluoride levels in egg contents (yolk plus albumen). Kuhl and Sullivan (1976) reported no increase in the fluoride content of egg yolk and albumen in chickens fed high-phosphorus diets supplemented with fluoride (the mean values for yolk and albumen were 0.55 and 0.35 ppm). However, fluoride supplementation increased the fluoride concentrations in eggshells up to an average of 150 ppm. The fluoride content of eggshells and eggshell weights and breaking strength depended on both the phosphate and fluoride in the diet.

RECOMMENDATIONS FOR ASSESSING FLUORIDE CONTAMINATION IN BIRDS

Fluoride readily accumulates in the eggshell and calcium-containing tissues of birds. The amount of fluoride in bone and eggshell may be influenced by the food habits, age, sex, and dietary concentrations of fluoride and other elements. Young, rapidly growing birds quickly accumulate fluoride in bone. Several studies indicated that the fluoride content of bone continues to increase with age in birds, especially when exposure levels are not extremely high. Other studies suggest that an equilibrium is reached that is dependent on dietary exposures. However, once fluoride is sequestered in either bone or eggshell, it appears not to be readily available to affect homeostatic processes.

Because of the biological inactivity of fluoride in bone and the tendency of bone to reflect cumulative exposure, it seems unlikely that the fluoride content of bone can be linked to biologically significant endpoints, such as growth and reproduction. Existing literature tends to support this view. However, bone density, strength, and other variables related to bone integrity may be affected by fluoride exposure. I conclude that the fluoride content of avian bone may be useful in assessing fluoride exposure, but that it has little value for predicting biological effects in birds.

Dietary fluoride may affect the egg size, number of eggs laid, number of eggs removed from the nest prior to hatching, and perhaps hatching success. However, the fluoride content of eggshells has not been significantly linked to these variables in a manner that can be used to interpret field data. Basic to this problem is the high variability in fluoride concentrations in both experimental and field situations. Because of the high variability in the fluoride content of eggshells, even within clutches, it is difficult to conclude that the fluoride content of an eggshell of a single egg can be reliable in predicting or assessing nesting outcome.

As with bone, the physical characteristics of the eggshell (thickness, for example) are the variables most likely to be linked to the fluoride content of eggshells. However, a definitive link between the physical characteristics and fluoride content of eggshells is neither proven nor available for interpreting field data.

Experimental studies and field reports from fluoride-contaminated areas indicate that the fluoride content of bone and eggshells increases with fluoride exposure. Perhaps, however, we have emphasized fluoride-rich tissues too much. Soft tissues or body fluids (plasma), though having far less potential to accumulate fluoride, may provide better links to biological effects. The two obvious reasons for this statement are the following: (1) the fluoride content of soft tissues probably reflects current, rather than cumulative, exposure; and (2) fluoride in soft tissues has not been sequestered into biologically inactive matrices. Future work should examine the potential for the fluoride content of soft tissues to be linked with biologically significant endpoints in birds.

SUMMARY

Fluoride residues in bone and eggshell are good indicators of the cumulative exposure of birds to fluoride. As such, they are useful for detecting exposure and for monitoring geographical and temporal trends. However, they cannot be definitely linked to growth, reproductive outcome, or mortality in a manner suitable for interpreting field residue data. Fluoride concentrations in soft tissues are generally low and are probably transient. Concentrations in soft tissues probably reflect the levels of fluorides in plasma or other body fluids and, as such, represent recent exposure. Soft tissues are more likely to reflect biologically active concentrations of fluorides. However, few studies have reported the fluoride levels in soft tissues of birds, and no studies have attempted to link biological outcomes with residues in soft tissues of birds.

ACKNOWLEDGMENTS

I thank Jim Newman and John Cooke for their reviews of the manuscript. Medhat Mohamed collected and organized the literature I used, and Dorothy Wright typed and formatted several drafts of the manuscript.

REFERENCES

Andreasen, J. K., and R. K. Stroud. 1987. Industrial halide wastes cause acute mortality of snow geese in Oklahoma. Environ. Toxicol. Chem. 6:291-293.

Bird, D. M., and C. Massari. 1983. Effects of dietary sodium fluoride on bone fluoride levels and reproductive performance of captive American kestrels. Environ. Pollut. Ser. A Ecol. Biol. 31:67-76.

Chan, M. M., R. B. Rucker, F. Zeman, and R. S. Riggins. 1973. Effect of fluoride on bone formation and strength in Japanese quail. J. Nutr. 103:1431-1440.

Cukir, A., T. W. Sullivan, and F. B. Mather. 1978. Alleviation of fluoride toxicity in starting turkeys and chicks with aluminum. Poult. Sci. 57:498-505.

Culik, B. 1987. Fluoride excretion in Adelie penguins *(Pygoscelis adeliae)* and mallard ducks *(Anas platyrhynchos)*. Comp. Biochem. Physiol. A Comp. Physiol. 88:229-233.

Demole, V., and A. J. Held. 1963. State of health of the population of the region of Mohlin-Rheinfelden, a suspected region of fluorosis. Bull. Schweiz. Akad. Med. Wiss. 19:375-390.

Dewey, J. F. 1973. Accumulation of fluorides by insects near an emission source in Western Montana. Environ. Entomol. 2:179-182.

Fleming, W. J., C. E. Grue, C. A. Schuler, and C. M. Bunck. 1987. Effects of oral doses of fluoride on nestling European starlings. Arch. Environ. Contam. Toxicol. 16:483-489.

Fleming, W. J., and C. A. Schuler. 1988. Influence of the method of fluoride administration on toxicity and fluoride concentrations in Japanese quail. Environ. Toxicol. Chem. 7:841-845.

Gardiner, E. E., J. C. Rogler, and H. E. Parker. 1961. Interrelationships between magnesium and fluoride in chicks. J. Nutr. 75:270-274.

Gardiner, E. E., K. S. Winchell, and R. Hironaka. 1968. The influence of dietary sodium fluoride on the utilization and metabolizable energy value of a poultry diet. Poult. Sci. 47:1241-1244.

Gerry, R. W., C. W. Carrick, R. E. Roberts, and S. M. Hauge. 1947. Phosphate supplements of different fluorine content as sources of phosphorus for chickens. Poult. Sci. 26:323-334.

Griffith, F. D., H. E. Parker, and J. C. Rogler. 1963. Observations on a magnesium-fluoride interrelationship in chicks. J. Nutr. 79:251-256.

Guenter, W. 1979. Fluorine toxicity and laying hen performance. Poult. Sci. 58:1063. (Abstr.).

Guenter, W., and P. H. B. Hahn. 1986. Fluorine toxicity and laying hen performance. Poult. Sci. 65:769-778.

Hahn, P. H. B., and W. Guenter. 1986. Effect of dietary fluoride and aluminum on laying hen performance and fluoride concentration in blood, soft tissue, bone, and egg. Poult. Sci. 65:1343-1349.

Haman, K., P. H. Phillips, and J. G. Halpin. 1936. The distribution and storage of fluoride in the tissue of the laying hen. Poult. Sci. 15:154-157.

Henny, C. J., and P. M. Burke. 1990. Fluoride accumulation and bone strength in wild black-crowned night-herons. Arch. Environ. Contam. Toxicol. 19:132-137.

Hoffman, D. J., O. H. Pattee, and S. N. Wiemeyer. 1985. Effects of fluoride on screech owl reproduction: teratological evaluation, growth, and blood chemistry in hatchlings. Toxicol. Lett. (Amst.) 26:19-24.

Kay, C. E., P. C. Toruangeau, and C. C. Gordon. 1975. Fluoride levels in indigenous animals and plants collected from uncontaminated ecosystems. Fluoride 8:125-133.

Kuhl, H. J., and T. W. Sullivan. 1976. Effect of sodium fluoride and high fluorine fertilizer phosphates on performance of laying chickens and egg shell quality. Poult. Sci. 55:2055. (Abstr.).

Loman, J. 1977. Factors affecting clutch size and brood size in the crow, *Corvus cornix*. Oikos 29:294-301.
Merkley, J. W. 1976. Increased bone strength in coop-reared broilers provided fluoridated water. Poult. Sci. 55:1313-1319.
Merkley, J. W. 1981. The effect of sodium fluoride on egg production, egg quality, and bone strength of caged layers. Poult. Sci. 60:771-776.
Merkley, J. W., and T. J. Sexton. 1982. Reproductive performance of white leghorns provided fluoride. Poult. Sci. 61:52-56.
Nahorniak, N. A., P. E. Waibel, W. G. Olson, M. M. Walser, and H. E. Dziuk. 1983. Effect of dietary sodium fluoride on growth and bone development in growing turkeys. Poult. Sci. 62:2048-2055.
Newman, J. R. 1977. Sensitivity of the house martin, *Delichon urbica*, to fluoride emissions. Fluoride 10:73-76.
Pattee, O. H., S. N. Wiemeyer, and D. M. Swineford. 1988. Effects of dietary fluoride on reproduction in eastern screech-owls. Arch. Environ. Contam. Toxicol. 17:213-218.
Perrins, C. 1963. Survival of the great tit *Parus major*. Proc. Int. Ornithol. Congr. 13:717-728.
Phillips, P. H., H. English, and E. B. Hart. 1935. The augmentation of the toxicity of fluorosis in the chick by feeding desiccated thyroid. J. Nutr. 10:399-407.
Rogler, J. C., and H. E. Parker. 1972. Effects of excess calcium on a fluoride-magnesium interrelationship in chicks. J. Nutr. 102:1699-1708.
Said, N. W., M. L. Sunde, H. R. Bird, and J. W. Suttie. 1979. Raw rock phosphate as a phosphorus supplement for growing pullets and layers. Poult. Sci. 58:1557-1563.
Scott, M. L., M. C. Mesheim, and R. J. Young. 1976. Nutrition of the chicken. M. L. Scott and Associates, Ithaca, N.Y. 555 pp.
Seel, D. C. 1983. Fluoride in the magpie. Annu. Rep. Inst. Terrestrial Ecol. 1982:45-48.
Seel, D. C., and A. G. Thomson. 1984. Bone fluoride in predatory birds in the British Isles. Environ. Pollut. Ser. A Ecol. Biol. 36:367-374.
Seel, D. C., A. G. Thomson, and R. E. Bryant. 1987. Bone fluoride in four species of predatory bird in the British Isles. p. 211-221. *In* P. J. Coughtrey, M. H. Martin, and M. H. Unsworth (Eds.). Pollutant transport and fate in ecosystems. Blackwell Scientific Publications, Oxford, England.
Simon, G., and J. W. Suttie. 1968. Effect of dietary fluoride on food intake and plasma fluoride concentrations in the rat. J. Nutr. 96:152-156.
Smith, F. A., and H. C. Hodge. 1979. Airborne fluorides and man. Part I. Crit. Rev. Environ. Control 8:293-371.
Stewart, D. J., T. R. Manley, D. A. White, K. L. Harrison, and E. A. Stringer. 1974. National fluoride levels in Bluff area, New Zealand. I. Concentrations in wildlife and domestic animals. N.Z. J. Sci. 17:105-113.
Suttie, J. W., D. L. Kolstad, and M. L. Sunde. 1984. Fluoride tolerance of the young chick and turkey poult. Poult. Sci. 63:738-743.
Suttie, J. W., P. H. Phillips, and E. C. Faltin. 1964. Serum fluoride in the chick. Proc. Soc. Exp. Biol. Med. 115:575-577.
Van Toledo, B. 1978. Fluoride content in eggs of wild birds (*Parus major* L. and *Strix aluco* L.) and the common house-hen *(Gallus domesticus)*. Fluoride 11:198-207.
Van Toledo, B., and G. F. Combs, Jr. 1984. Fluorosis in the laying hen. Poult. Sci. 63:1543-1552.
Walash, M. N., and M. T. Ridha. 1980. Preliminary investigation of the influence of sodium fluoride on the developing chick embryo. Zool. Soc. Egypt Bull. 30:29-36.
Zhavoronkov, A. A., and A. I. Edemskii. 1974. Histopathology of the adenohypophysis in experimental fluorosis. Byull. Eksp. Biol. Med. 78:108-111.

CHAPTER 22

Fluoride in Small Mammals

John A. Cooke, Iain C. Boulton, and Michael S. Johnson

INTRODUCTION

Fluoride is a halogen, estimated as the 13th most abundant element in the earth's crust (0.065%). It is widely distributed in nature (rocks, soils, water, vegetation, and animals) and, because of its high reactivity, fluorine occurs predominantly as inorganic fluoride compounds (EPA, 1980). Although controversy exists as to whether it is an essential element in animals (Hodge and Smith, 1981), studies in humans and laboratory animals have shown that fluoride can have beneficial effects. The regular consumption of small amounts of fluoride has been advocated for the prevention of dental caries in humans (WHO, 1984) and has been prescribed for the treatment of the bone disease osteoporosis (Hodge and Smith, 1968).

The hazards of excessive fluoride intake in animals have been known for a long time. Ingestion of large quantities can lead to acute poisoning (Hodge and Smith, 1965). Chronic toxicity results from prolonged exposure to relatively lower doses (WHO, 1984). The chronic disease has been called fluorosis, although it is not a clearly defined condition as its symptoms can range from mild to extremely severe (Suttie, 1977). Of the domestic animals, including sheep, horses, and pigs, dairy cattle are believed to be the most sensitive. Because of their economic importance, fluorosis in animals is best known from field observation and long-term experiments on cattle (NAS, 1974). Poultry are regarded as the most fluoride-tolerant domestic animal, with safe levels in the diet being 2 to 6 times higher than those of domestic mammals (NAS, 1974).

Damaging fluorosis in cattle is associated with the accumulation of fluoride in the hard tissues and includes the following: dental fluorosis (mottling, erosion of enamel, and excessive wear) that can interfere with feeding; bone damage that results in lameness; and systemic fluorosis in which the fluoride intake causes loss of appetite, stunted growth, and reduced milk yield (Allcroft et al., 1965). The most practical environmental standard to prevent chronic fluorosis in cattle is based upon the fluoride content of forage, whereby the fluoride concentration of grass should not exceed an average annual concentration of 40 mg of fluoride per kilogram (dry weight), with no 2 consecutive months exceeding 60 mg of fluoride per kilogram or 80 mg/kg for more than 1 month (Suttie, 1969).

The earliest accounts of fluoride poisoning among livestock appear to be associated with volcanic eruptions (Fridriksson, 1983). In 1845, the eruption of Hekla in Iceland was followed by emaciation, decreased milk yields, and lameness in sheep, which condition was associated with dental lesions such as abrasion of molar teeth, pitting of incisors, and bone changes including the thickening of joints (Roholm, 1937). Many areas of the world, particularly tropical countries such as India, have naturally high concentrations of fluoride in rocks and water supplies that lead to endemic fluorosis in humans as well as domestic animals (WHO, 1984).

During the latter half of the 20th century, the most significant inputs of fluoride into the environment are from industrial activities. Fluorides are emitted into the air from industries producing steel, aluminum, bricks, glass, fertilizers, and chemical fluoride compounds, together with power stations and oil refineries (NAS, 1971). Gases (particularly hydrogen fluoride) and particulates are deposited onto the vegetation and soils surrounding these sources through dry deposition, including uptake by plant stomata and wet deposition during rainfall (Davison, 1987). Soils and groundwater can also become contaminated by fluorides through the agricultural application of sewage sludges (Davis, 1980) and phosphate fertilizers. Such environmental contamination may also occur as a result of the extraction and processing of fluoride-containing minerals such as fluorspar (CaF_2). A particular concern is the often large areas of high-fluoride waste materials derived from mining industries that are frequently used for agricultural and amenity purposes following reclamation and the establishment of grassland (Johnson, 1980).

Whether by direct deposition from the atmosphere and uptake through plant leaves or via uptake through roots from soils, fluoride accumulation may be substantial by vegetation in contaminated habitats (Cooke et al., 1976; Davison, 1987). Herbivorous animals consuming this plant material may absorb and accumulate fluoride in quantities sufficient to exert a toxic effect. In extensive studies within specific contaminated habitats or defined geographical areas near a fluoride source, researchers have shown that most animal species including invertebrates, birds, and small mammals accumulate fluoride often greatly in excess of background levels (Walton, 1986; Andrews et al., 1989). However, compared with domestic animals, comparatively little is known about the effects and dietary tolerances of fluoride in wild animals (WHO, 1984). Walton (1988) reviewed the literature on the levels of fluoride in bone of wild mammals, but the experimental toxicology or dose-response aspects were not covered by this author.

Most experimental studies on the toxic effects of fluoride in small mammals have been performed in laboratory species (especially the rat, white mouse, rabbit, and guinea pig) using highly soluble sodium fluoride (Whitford, 1989). This chapter includes interpretations of these data together with results from recent laboratory studies on wild species of small mammals common to the seminatural grasslands in Europe. In wild species, the potential toxicity of ingested fluoride compounds will be strongly influenced by factors affecting their absorption, such as the pH of the gastrointestinal fluids (Whitford, 1989) and the presence of components shown to reduce absorption, such as calcium and aluminum ions (Harrison et al., 1984; Spencer et al., 1985).

Organic fluoride compounds are not common in nature, although some tropical plants are able to synthesize certain compounds such as monofluoroacetate (Hall, 1972). There is relatively little information available concerning the environmental toxicity of the numerous organic fluorine compounds used in industry and the household. However, some compounds, particularly sodium fluoroacetate and fluoracetamide, have been used as poisons in rodent control. There is no conclusive evidence that plants or animals synthesize the more acutely poisonous organofluorides from inorganic fluorides that result from environmental pollution. Thus, it is assumed in this chapter that all the examples described relate to the environmental effects of inorganic fluoride.

ACUTE AND SUBACUTE LETHAL POISONING

Following a single exposure to a relatively large dose of fluoride, small mammals exhibit a series of symptoms that are classified as an acute toxicosis (Davis, 1961; Hodge and Smith, 1965). Fluorides ingested p.o. have been demonstrated to elicit some or all of the following responses in laboratory animals: a severe irritation of the linings of the gastrointestinal tract, whether a localized inflammation (Pashley et al., 1984) or an acute hemorrhagic enteritis (Davis, 1961); salivation and vomiting; diarrhea; and respiratory, cardiac, and nervous depression. Following a fatal dose, all or most of the above symptoms are enhanced, followed by collapse, convulsions, coma, and death. The anatomical lesions of acute and lethal fluoride intoxication are usually confined to an acute hyperemia of the viscera, although numerous studies have shown that the most prominent changes occur in the kidney and liver. These changes may include a cloudy swelling of the organs, disorganization of tissue layers, and necrotic lesions in the deeper regions (Davis, 1961; Hodge and Smith, 1965).

The overall effects of acute poisoning are related to the peak concentrations of fluoride in the plasma and soft tissues (Singer et al., 1978). Plasma fluoride is generally regarded as a reliable indicator of recent exposure (Whitford, 1989). However, with a single acute dose, the peak concentration declines rapidly to normal levels over the following 24 hours. A major factor in acute toxicity is the removal rate of fluoride from the blood by the skeletal system. This means that young, growing animals remove more fluoride from the blood plasma than do older, mature animals (Wallace-Durban, 1954; Whitford et al., 1990). Laboratory results for rats are given by De Lopez et al. (1976). The p.o. 24-hour 50% lethal dose (LD_{50}) values were 31 mg/kg for 250-g rats and 52 mg/kg for 80-g animals. Plasma fluoride concentrations of 8 to 10 mg/l were often associated with death regardless of dose or body weight (age), but these concentrations were more rapidly attained in the older, heavier animals. The p.o. 24-hour LD_{50} for mice is of a similar order, 45 mg/kg (Lim et al., 1978).

Fatal intoxication can occur within days or a few weeks of repeated subacute doses of inorganic fluorides with the progression of intoxication varying with the magnitude of the dose (Davis, 1961). In an experiment in which rats were given drinking water containing sodium fluoride, 5 of 30 animals had died after 5 days with 200 ppm; with 300 ppm, 12 of 16 died in 5 days, and the 30-day 50% lethal concentration (LC_{50}) was calculated to be 205 ppm (Taylor et al., 1961). This represented 40 mg/kg/day of fluoride when the water consumption was normal. Data for other species are sparse, but Davis (1961) notes that fatal intoxication occurred in the guinea pig after doses of 20 mg of fluoride per kilogram for 21 days and 3 doses of 100 mg/kg killing rabbits in 6 days. In a laboratory study (Boulton, 1992) with 80 ppm of fluoride as sodium fluoride in drinking water, 50% of the 24 experimental field voles *(Microtus agrestis)* were dead in 25 days, whereas no deaths occurred in laboratory mice treated in the same manner. This represents a dose of 83 mg/kg/day for field voles, because of their high water consumption, compared with a dose of 17 mg/kg/day in mice. These data highlight the importance of differences between species, not just in terms of susceptibility to fluorides but in basic rates of feeding or water consumption and, therefore, the fluoride dose acquired.

CHRONIC TOXICITY

Lethal poisoning in experimental animals can occur through prolonged exposure to dietary fluoride for weeks or months, and it is somewhat arbitrary to separate subacute from chronic toxicity. In acute lethal poisoning, it is difficult to relate specific tissue levels to the cause of death, whereas in the prolonged exposure to relatively lower dietary levels of fluoride, characterization is possible because of the accumulation of fluoride by mineralized tissues. The ease with which fluoride ions can substitute for hydroxyl ions in the lattice structure of the apatitic component of hard tissue mineral is the most probable explanation of the ready uptake of fluoride by these tissues (Eanes, 1983; WHO, 1984). Increases in the concentration in the skeleton and teeth relate to the duration and level of exposure to fluoride. Concentrations of fluoride in soft tissues generally remain low unless associated with ectopic calcification, which can occur in the aorta, tendons, cartilage, and placenta, or with high rates of urinary excretion, which can leave higher residue concentrations in the kidneys (WHO, 1984).

TEETH

The most important signs of chronic fluorosis in small mammals are visible and microscopical damage to the teeth (Fejerskov et al., 1983; Abe et al., 1986). Lindemann (1967) reviewed laboratory rat experiments conducted between the years 1925 and 1962 for the effects of fluoride on incisor enamel. In rats, low doses lead to a striation or banding of the enamel of the cutting incisors. Further progressive changes occur with the accumulation of fluoride and are typified by total pigment loss (from orange/yellow to white), chalkiness of the enamel, hypoplasia (pitting and loss of enamel), and increased fragility of the teeth.

Recently a scoring system based upon the observable morphological changes in the incisors and molars of wild and experimental rodents has been devised (Table 1), adapted from one used to clinically classify the various degrees of dental fluorosis in cattle and other domestic and wild large mammals (Shupe and Olson, 1983). In an experimental study conducted over 60 years ago, a similar grading of alterations in rat incisor enamel was related to dietary concentrations (Smith and Leverton, 1934).

Table 1 Summary of the Scoring System for the Classification of Fluoride-induced Lesions in the Teeth of Small Mammals[a]

Score	Appearance	Incisor	Molar
0	Normal	Enamel smooth, glossy orange-yellow color, normal shape	White-enamel, solid creamy dentine on grinding face, solid enamel "ridge"
1	Questionable	Slight deviation from normal	Slight deviation from normal
2	Slight	Faint horizontal banding of enamel, mottling, chalky spots, slight erosion	Slight increase in erosion of grinding face, dentine darkened
3	Moderate	Whitened enamel, mottling, incisor banding, erosion of tips, staining	Increased erosion of grinding face, dentinal cavities, enamel chipped
4	Marked	Enamel hypoplasia pitting and staining, heavy erosion of cutting tips, loss of enamel color	Grinding surface nearly worn flat, dentinal cavities, severe staining, abnormal wear
5	Severe	Lesions much worse than for score 4, cutting tips splayed and eroded to blunt "stubs", total color loss, abnormal curvature	Grinding surface worn flat, heavily cavitated dentine, severe staining, shrunken profile, deformation of tooth roots

[a] From Boulton et al. (1994a); in part devised by Andrews (1985).

Tooth lesions may have a profound effect on the normal operation of teeth in preparing food prior to digestion, leading to reduced food intake and perhaps starvation. Also, rodent incisors require even wear to maintain their cutting edges and counteract the constant growth from open roots. It is proposed here that animals with at least moderate dental fluorosis (score 3, Table 1), where some tooth erosion has occurred, could be regarded as being likely to suffer chronic nutritional problems. This level of dental fluorosis could be associated with about 2000 mg of fluoride per kilogram (dry weight) in incisor and molar teeth and 2500 mg of fluoride per kilogram in the femur and whole skeleton.

In a study of field voles caught from various fluoride-contaminated sites (Boulton, 1992), whole-tooth fluoride concentrations corresponding to particular lesion scores were measured (see Table 2). The severity of incisor dental lesions can also be related quite well to femur (or whole skeleton) fluoride levels. In this field vole study, normal (score 0), questionable (score 1), slight (score 2), moderate (score 3), marked (score 4), and severe (score 5) dental fluorosis corresponded to mean femur fluoride concentrations of 189 ± 45 (standard error, SE), 314 ± 67, 1866 ± 298, 4619 ± 403, 6175 ± 364, and 5722 ± 364 mg of fluoride per kilogram, respectively.

Table 2 Changes in the Dental Lesion Score with Fluoride Concentration (mg/kg of dry weight) in the Incisor and Molar of the Short-tailed Field Vole *(M. agrestis)*[a]

Lesion score	No. in sample	Incisor fluoride[b]	No.	Molar fluoride[b]
0	2	291 ± 88	15	295 ± 42
1	14	323 ± 61	30	975 ± 195
2	29	1,172 ± 154	20	2,394 ± 306
3	19	2,484 ± 191	17	4,368 ± 372
4	19	3,474 ± 236	8	5,058 ± 434
5	8	4,192 ± 414	3	6,299 ± 150

[a] Mean ± standard error of combined results from four field sites 1988 and 1990.
[b] From Boulton, I. C., J. A. Cooke, and M. S. Johnson. 1994a. Environ. Pollut. 85:161-167. With permission.

In an experimental study with field voles, diets were prepared as pellets made from dried grass from different fluoride-contaminated habitats (Boulton et al., 1994b). With a diet of 100 mg of fluoride per kilogram (on a dry weight basis giving an overall dose of 25 mg/kg/day), composed of grass from near a fluorochemical works, all the animals survived 84 days but showed moderate to marked dental lesions (scores 3 and 4), with mean fluoride concentrations in incisors of 3022 ± 107 (SE), in molars of 3883 ± 155, and in femurs of 3201 ± 130 mg of fluoride per kilogram.

A second group of 10 animals with 100 mg of fluoride per kilogram in a diet using grass from near an aluminum smelter all died by day 77, also with moderate to marked dental fluorosis, but mean fluoride concentrations that were somewhat lower: in incisors, 2157 ± 99 (SE), in molars, 2673 ± 160, and in femurs, 2120 ± 407 mg of fluoride per kilogram.

A third experimental group of 10 animals was given 100 mg of fluoride per kilogram as pellets made from grass from a fluorspar tailings dam. In marked contrast to the other groups with the same fluoride concentration, but where the fluoride had been derived from atmospheric pollution, there was only slight (score 2) damage to the teeth. There were correspondingly much lower mean fluoride concentrations in the hard tissues: in incisors, 478 ± 40 (SE), in molars, 496 ± 42, and in femurs, 909 ± 40 mg of fluoride per kilogram. This was because of reduced absorption by the vole gut from a diet in which the fluoride from the plant had a much lower solubility (Boulton et al., 1994b).

In a similar study with field voles, but with a higher dietary concentration of 300 mg of fluoride per kilogram (dry weight) as pellets made from dried grass collected from near the fluorochemical works, all the animals showed a rapid weight loss indicative of malnutrition and death within 28 days (Boulton et al., 1994b). These animals had marked (score 4) dental fluorosis of both the incisors and molars similar to that of the animals fed 100 mg of fluoride/kg and scored after 84 days of exposure. However, they had much higher mean fluoride concentrations (mg of fluoride per kilogram) in incisors (5543 ± 413 [SE]) and molars (5932 ± 405) but not in femurs (2930 ± 152).

Similar marked-to-severe dental lesions were observed in incisors from experimental deer mice *(Peromyscus maniculatus)* with femur fluoride concentrations of

3106 to 3975 mg of fluoride per kilogram, but, surprisingly, differences in structure or wear were not observed in molar teeth (Newman and Markey, 1976). In a recent comparative experimental study of laboratory mice, wood mice *(Apodemus sylvaticus)*, field voles, and bank voles *(Clethyrionomys glareolus)* with fluoride in the drinking water, marked dental effects (score 4, Table I) on the upper incisors and molars were noted with mean femur fluoride concentrations of 3105 ± 116 (SE), 2412 ± 90, 2960 ± 155, and 4278 ± 578 mg/kg, respectively (Boulton, 1992).

Walton (1987) described similar dental lesions occurring in wild field voles and wood mice caught near an aluminum smelter in Wales when whole-skeleton concentrations were greater than 2500 mg/kg. Moles *(Talpa europaea)* caught near the same smelter also showed some tooth wear, with one individual with 11,100 mg of fluoride per kilogram in the whole skeleton showing severe attrition of some teeth and accretions of hard material on several others (Walton, 1986). Common shrews *(Sorex araneus)* from the same location were not reported as showing any abnormal tooth wear (Walton, 1986). Similarly, shrews caught on a reclaimed tailings dam (Andrews et al., 1989) and near another British aluminum smelter (Andrews, 1985) also did not show any changes in teeth from normal, even though the femur fluoride concentrations exceeded 6000 mg/kg in some animals. Although insectivorous shrews accumulate fluoride in hard tissues to a considerably high level, they appear to be relatively resistant to dental fluorosis compared with rodents.

Tooth and femur fluoride levels are only a guide to the severity of dental fluorosis, because tooth lesions result from the effect of fluoride on enamel development (especially the enamel-forming cells or ameloblasts) and other mineralizing processes (Fejerskov et al., 1983). Thus, the pathogenic mechanisms depend upon fluoride levels in extracellular fluid at the time of tooth formation and may reflect, therefore, past high dietary concentrations that were transient. In wild small mammals, this is significant when there can be considerable spatial and temporal variation in vegetation fluoride in polluted sites near aerial sources of fluorides (Davison, 1987).

SKELETON

A series of biochemical, functional, and morphological changes have been observed in the skeletons of large domestic animals, particularly cattle (Shupe and Olson, 1983) and wild ungulates (Shupe et al., 1984). Following prolonged intake of fluoride, typical changes include the development of exostotic lesions (bony outgrowths) and periosteal hyperostosis (general thickening of some bones), which gives a rough and porous appearance to the structure compared with normal, hard, smooth bones (NAS, 1971). In laboratory experiments with rats and mice and in accounts of field-caught wild small mammals, these gross morphological changes have not usually been observed. Changes that do occur are usually confined to histological and radiological disturbances in long limb bones, ribs, and flat bones of the skull and pelvis (Chavassieux, 1990).

Walton (1987) reports taking radiographs of skeletons of field voles and wood mice caught near an aluminum smelter from 1977 to 1985 with skeleton concentrations up to 15,000 mg/kg, but no changes could be detected when compared with animals from unpolluted sites. Outgrowths (exostoses) and extraskeletal mineralization were absent. Similarly, no gross morphological changes in the bones of field voles, wood mice, bank voles, or common shrews have been recorded in recent studies (Andrews, 1985; Boulton, 1992).

Present evidence suggests that fluoride has little effect on the skeletons of wild small mammals, compared with the severe damage it causes to their teeth. Femur or skeletal fluoride concentrations are only useful in indicating the chronic toxicity associated with dental fluorosis. Evidence for the importance of tissues other than teeth, bones, or plasma as indicators of critical levels associated with fluoride toxicity is also sparse.

SUMMARY

Fluoride is an environmental contaminant associated mainly with atmospheric pollution from specific industries and some agricultural practices. It can occur in many different chemical forms, but most concern is for the toxicity of relatively soluble inorganic forms. It is a general toxin affecting many biochemical and physiological processes and can cause acute and subacute lethal poisoning of small mammals. Such general systemic toxic effects are difficult to assess in terms of specific tissue levels, although 10 mg of fluoride per liter in the blood plasma may be a reasonable guide.

Environmental fluoride is of particular concern because of its chronic toxicity to vertebrates. It accumulates in the bones and teeth, and high concentrations during their mineralization can lead to abnormal development. In small mammals, dental rather than skeletal fluorosis is of particular importance, with extreme attrition of incisor and molar teeth of rodents commonly observed. It is proposed here that 2000 mg of fluoride per kilogram (on a dry weight basis) in whole teeth or 2500 mg of fluoride per kilogram in femur or whole-skeleton samples would be indicative of sublethal effects and a shortened life span. No other tissues seem to be suitable in predicting chronic fluoride toxicity in wild small mammals.

The critical concentration of fluoride in grass, 40 mg of fluoride per kilogram of dry weight, which is regarded as providing safeguards for continual cattle grazing, should also prevent chronic toxicity in herbivorous small mammals, such as the *Microtus* species. A concentration of 100 mg/kg causes marked dental fluorosis and has been shown to lead to death after 2 to 3 months of dietary exposure in experimental field voles *(M. agrestis)*.

REFERENCES

Abe, T., M. Masuoka, M. Nomura, and H. Miyajima. 1986. Effects of fluoride on developing enamel and dentine of rat incisors. p. 299-305. *In* H. Tsunoda and M. H. Yu (Eds.). Fluoride research 1985. Elsevier Science Publishers, Amsterdam.

Allcroft, R., K. N. Burns, and C. N. Herbert. 1965. Fluorosis in cattle. II. Development and alleviation: experimental studies. Anim. Dis. Surv. Rep. No. 2. Part II. Her Majesty's Stationery Office, London.

Andrews, S. M. 1985. Aspects of the Ecology of Fluoride and Heavy Metals in Contaminated Grassland. Ph.D. thesis. Sunderland Polytechnic, Sunderland, U.K. 221 pp.

Andrews, S. M., J. A. Cooke, and M. S. Johnson. 1989. Distribution of trace element pollutants in a contaminated ecosystem established on metalliferous fluorspar tailings. III. Fluoride. Environ. Pollut. 60:165-179.

Boulton, I. C. 1992. Environmental and Experimental Toxicology of Fluoride in Wild Small Mammals. Ph.D. thesis. University of Sunderland, Sunderland, U.K. 146 pp.

Boulton, I. C., J. A. Cooke, and M. S. Johnson. 1994a. Fluoride accumulation and toxicity in wild small mammals. Environ. Pollut. 85:161-167.

Boulton, I. C., J. A. Cooke, and M. S. Johnson. 1994b. Experimental fluoride accumulation and toxicity in the short-tailed field vole *(Microtus agrestis)*. J. Zool. (Lond.) 234:409-421.

Chavassieux, P. 1990. Bone effects of fluoride in animal models *in vivo*. A review and a recent study. J. Bone Miner. Res. 5 (Suppl.):S95-S99.

Cooke, J. A., M. S. Johnson, A. W. Davison, and A. D. Bradshaw. 1976. Fluoride in plants colonising fluorspar wastes in the Peak District and Weardale. Environ. Pollut. 11:9-23.

Davis, R. D. 1980. Uptake of fluoride by ryegrass in soil treated with sewage sludge. Environ. Pollut. Ser. B Chem. Phys. 1:277-284.

Davis, R. K. 1961. Fluorides: a critical review. V. Fluoride intoxication in laboratory animals. J. Occup. Med. 3:593-601.

Davison, A. W. 1987. Pathways of fluoride transfer in terrestrial ecosystems. p. 193-210. *In* P. J. Coughtrey, M. H. Martin, and M. H. Unsworth (Eds.). Pollutant transport and fate in ecosystems. Blackwell Scientific Publications, Oxford.

De Lopez, O. H., F. A. Smith, and H. C. Hodge. 1976. Plasma fluoride concentrations in rats acutely poisoned with sodium fluoride. Toxicol. Appl. Pharmacol. 37:75-83.

Eanes, E. D. 1983. Effect of fluoride on mineralization of teeth and bones. p. 195-198. *In* J. L. Shupe, H. B. Peterson, and N.C. Leone (Eds.). Fluorides: effects on vegetation, animals, and humans. Paragon Press, Salt Lake City, Utah.

Environmental Protection Agency (EPA). 1980. Reviews of the environmental effects of pollutants. IX. Fluoride. EPA, Cincinnati, Ohio. 441 pp.

Fejerskov, O., A. Richards, and K. Josephsen. 1983. Pathogenesis and biochemical findings of dental fluorosis in various species. p. 305-317. *In* J. L. Shupe, H. B. Peterson, and N. C. Leone (Eds.). Fluorides: effects on vegetation, animals, and humans. Paragon Press, Salt Lake City, Utah.

Fridriksson, S. 1983. Fluoride problems following volcanic eruptions. p. 339-344. *In* J. L. Shupe, H. B. Peterson, and N. C. Leone (Eds.). Fluorides: effects on vegetation, animals, and humans. Paragon Press, Salt Lake City, Utah.

Hall, R. J. 1972. The distribution of organic fluoride in some toxic tropical plants. New Phytol. 71:855-871.

Harrison, J. E., A. J. W. Hitchman, S. A. Hasany, A. Hitchman, and C. S. Tam. 1984. The effect of diet calcium on fluoride toxicity in growing rats. Can. J. Physiol. Pharmacol. 62:259-265.

Hodge, H. C., and F. A. Smith. 1965. Biological properties of inorganic fluorides. p. 27-33. *In* J. H. Simons (Ed.). Fluorine chemistry. Vol. 4. Academic Press, New York.

Hodge, H. C., and F. A. Smith. 1968. Fluorides and man. Annu. Rev. Pharmacol. 8:395-408.

Hodge, H. C., and F. A. Smith. 1981. Fluoride disorders of mineral metabolism. p. 439-483. Vol. 1. Academic Press, New York.

Johnson, M. S. 1980. Revegetation and the development of wildlife interest on disused fluorspar tailings dams in Great Britain. Reclam. Rev. 3:209-216.

Lim, J. K., G. J. Renaldo, and P. Chapman. 1978. LD_{50} of SnF_2, NaF, and Na_2PO_3F in the mouse compared to the rat. Caries Res. 12:177-179.

Lindemann, G. 1967. Pigment alterations and other disturbances in rat incisor enamel in chronic fluorosis and recovery. Acta Odontol. Scand. 25:525-539.

National Academy of Sciences (NAS). 1971. Biological effects of atmospheric pollutants: fluorides. NAS, Washington, D.C. 295 pp.

National Academy of Sciences (NAS). 1974. Effects of fluorides in animals. NAS, Washington, D.C. 70 pp.

Newman, J. R., and D. Markey. 1976. Effects of elevated levels of fluoride on deer mice *(Peromyscus maniculatus)*. Fluoride 9:47-53.

Pashley, D. H., N. B. Allison, R. P. Easman, R. V. N. McKinney, J. A. Horner, and G. M. Whitford. 1984. The effects of fluoride on the gastric mucosa of the rat. J. Oral Pathol. 13:535-545.

Roholm, K. 1937. Fluoride intoxication: a clinical-hygienic study. H. K. Lewis and Co., London. 364 pp.

Shupe, J. L., and A. E. Olson. 1983. Clinical and pathological aspects of fluoride toxicosis in animals. p. 319-338. *In* J. L. Shupe, H. B. Peterson, and N. C. Leone (Eds.). Fluorides: effects on vegetation, animals, and humans. Paragon Press, Salt Lake City, Utah.

Shupe, J. L., A. E. Olson, H. B. Peterson, and J. B. Low. 1984. Fluoride toxicosis in wild ungulates. J. Am. Vet. Med. Assoc. 185:1295-1300.

Singer, L., W. D. Armstrong, and R. H. Ophaug. 1978. Effects of acute fluoride intoxication on rats. Proc. Soc. Exp. Biol. Med. 157:363-368.

Smith, M. C., and R. M. Leverton. 1934. Comparative toxicity of fluorine compounds. Ind. Eng. Chem. 26:791-797.

Spencer, H., L. Kramer, D. Osis, and E. Wiatrowski. 1985. Effects of aluminum hydroxide on fluoride and calcium metabolism. J. Environ. Pathol. Toxicol. Oncol. 6:33-41.

Suttie, J. W. 1969. Air quality standards for the protection of farm animals from fluorides. J. Air Pollut. Assoc. 19:239-242.

Suttie, J. W. 1977. Effects of fluoride on livestock. J. Occup. Med. 19:40-48.

Taylor, J. M., D. E. Gardner, J. K. Scott, E. A. Maynard, W. L. Downs, F. A. Smith, and H. C. Hodge. 1961. Toxic effects of fluoride on the rat kidney. XI. Chronic effects. Toxicol. Appl. Pharmacol. 3:290-314.

Wallace-Durban, P. 1954. The metabolism of fluorine in the rat using F^{18} as a tracer. J. Dent. Res. 33:789-800.

Walton, K. C. 1986. Fluoride in moles, shrews, and earthworms near an aluminum reduction plant. Environ. Pollut. Ser. A Ecol. Biol. 42:361-371.

Walton, K. C. 1987. Tooth damage in field voles, wood mice, and moles in areas polluted by fluoride from an aluminum reduction plant. Sci. Total Environ. 65:257-260.

Walton, K. C. 1988. Environmental fluoride and fluorosis in mammals. Mammal Rev. 18:77-90.

Whitford, G. M. 1989. The metabolism and toxicity of fluoride. Monogr. Oral Sci. No. 13. Karger Press, Basel, Switzerland. 154 pp.

Whitford, G. M., N. L. Birdsong-Whitford, and C. Finidori. 1990. Acute oral toxicity of sodium fluoride and monofluorophosphate separately or in combination in rats. Caries Res. 24:121-126.

World Health Organization (WHO). 1984. Fluorine and fluorides. Environ. Health Criter. No. 36. WHO, Geneva. 136 pp.

Index

Acetylcholinesterase inhibition, 22
Acid lakes, 307-311, 329-330
Adriatic Sea, 121
African clawed frog, 77
AHH induction, 184-185
Air, 12
ALAD (delta-aminovulinic acid dehydratase), 258-259, 268, 290-291, see also Lead
ALA synthase activity, 174
Albatross, Laysan, 276
Alberta, Canada, 64
Aldrin, 10, 73, 74, 79, see also Cyclodienes; Dieldrin
Algae, 120, 121, 128, 129, 131, 436, 439, see also Aquatic organisms
Alkylmercury fungicides, 299, see also Mercury
Alligator, 353
American avocet, 457
American black duck, 111
American coot, 460
American kestrel, 53, 85, 88, 100-101, 105, 106, 109, 174, 175, 177, 178, 187, 271, 352, 472-474
delta-Aminovulinic acid dehydratase (ALAD), 258-259, 268, 290-291, see also Lead
Analytical issues if residue studies, 1-24, 233-236
Andean condor, 270
Antarctica, 119, 160, 162
Antarctic Ocean, 119
Aquatic birds, 15, 18, see Birds and individual species
Aquatic organisms
  heavy metals in, 411-425, see also Metals
  PCBs and, 118-126
    feral organisms, 136-139
    health effects, 122-126
    laboratory studies, 127-136
Aquatic PAHs, 231-245, see also PAHs
Arabian Gulf, 239
Arctic Ocean, 119, 120
Arkansas, 195-196
Aroclors, 167-174
Arsenic, 4, 364, 374
Arsenobetaine, 374
Arthropods, 120, 127, 132
Atlantic Ocean, 119, 120, 161, 239
Atlantic puffin, 173, 176
Avian species, see Birds and individual species
Avocet, American, 457

Baja California, 58, 65
Bald eagle, 14, 52, 59, 64-65, 82-83, 106, 169, 187-188, 193, 270, 347-348
Ball Lake (Ontario), 347
Baltic Sea, 161-162, 188, 348, 424
Bank vole, 386, 485
Barnacle, 414, 415, 422
Barn owl, 56, 61, 81, 84, 87, 467
Bass, see also Fish; Mercury
  largemouth, 307-308, 325
  rock, 323
  striped, 435, 436, 439
Bat, 13, 81, 84, 158-159
Bear, polar, 371
Bearded seal, 368
Behavioral effects, see also Neurotoxicity
  cadmium in birds, 398
  cyclodienes, 91-92
  mercury in fish, 324-325
  PCBs, 128, 130
  selenium in fish, 437
Beluga, 372
Bengalese finch, 170
Benthic species, see Aquatic organisms
Bering Sea, 119, 120
Bill deformities, 186, 188, see also Teratogenicity
Bioaccumulation, 9-13, 66
  fluoride in birds, 467
  heavy metals in aquatic invertebrates, 419-422
  metal in aquatic invertebrates, 414-415
  PAHs, 243-244
  selenium and methylmercury, 316
Bioavailability
  of cadmium, 388
  of PAHs, 243
Biocentration factor for dioxins in fish, 218-223
Biogenic amines, 91
Biomagnification, 169, 199
Biomagnification factors (BMF) for PCBs, 196
Biomarkers, 21-22
  for cyclodienes, 89-91
  lead in mammals, 290-293
Bird eggs, see Eggs; Eggshell thinning
Birds, 5, 6, 11-12, see also individual species
  aquatic, 15, 18
  cadmium in, 393-404, see also Cadmium
  dioxins in
    field studies of TCDD, 191-195
    forage residue studies, 196

laboratory studies, 189-191
endangered, 14-15, 184-185, 199, see also individual species
fluoride in, 466-474, see also Fluoride
lead in, see also Lead
  birds other than waterfowl, 267-277
  waterfowl, 253-263
mercury in, 345-352, see also Mercury
organochlorines in, 49-67, 99-112
  chlordane, 101-103
  chlorodecone, 105-106
  DDT/DDD/DDE, 49-67
  dicofol, 108-109
  endosulfan, 103
  heptachlor, 100-101
  hexachlorobenzine, 106-108
  hexachlorocyclohexane, 106
  methoxychlor, 110
  mirex, 103-105
  toxaphene, 110-111
PCBs in
  egg injection studies, 173-179
  field studies, 182-189
  forage residue studies, 196
  pen and laboratory studies, 169-173
  summary of findings, 196-199
  toxic equivalency factors and, 179-182
representative studies, 16-17, 20-21
selenium in, 453-462, see also Selenium
Blackbird, red-winged, 77, 85, 88, 100-102, 171-172, 274, 350
Black-crowned night heron, 52-53, 59, 64-65, 67, 188-189, 193, 455-456, 467, 469-470
Black duck, 56, 61, 111, 351
Black-headed gull, 179, 274
Black skimmer, 348
Blood residues/effects
  DDT/DDE/DDD in birds, 55
  lead, 256, 260, 268, 285-286
  PCBs in mammals, 157, 161, 163
  selenium in birds, 460
Bluebird, eastern, 191, 199
Bluegill sunfish, 75, 436, 437, 439
Blue whale, 375
Bobwhite quail, 51, 75, 76, 91, 103, 104, 169, 177, 273
Bodega Bay, 424
Bone development and cadmium in birds, 400-401
Bone residues
  fluoride
    in birds, 467-468, 470-471
    in small mammals, 485-486
  lead, 287-289

Bottle-nosed dolphin, 363, 366
Brain function and lead toxicity, 286
Brain residues, 8-10
  DDT/DDE/DDD in birds, 50-53
  endrin lethal levels, 78
  lead, 286-287, 289
  mercury in fish, 317, 324
  PCBs, 157, 159, 163, 196-197
British Columbia, 194-195
Brook trout, 317-318, 322, 327
Brown-headed cowbird, 51, 52, 54, 55, 100-102, 170-172
Brown pelican, 7, 14, 58, 59, 64-67, 88, 89
Bullfrog, 77
Buzzard, 81, 347
  common, 270
  honey, 271

Cadmium
  bioavailability, 388
  in birds
    concentrations in wild birds, 396-398
    mechanism of toxicity, 393-394
    tissue distribution, 394-395
    toxic effects, 398-404
    uptake, 395-396
  in invertebrates, 383, 385
  in marine mammals, 364, 371-373
  and mercury toxicity, 347
  in small mammals
    critical tissue concentrations, 388-389
    metabolism and toxicity, 376-383
    in terrestrial habitats, 383-385
    tissue concentrations in wild species, 385-388
Calcium metabolism and cadmium in birds, 400
California, 58, 65, 455
California condor, 59, 270
California quail, 75, 82
Canada, 64, 120, 124, 160, 185-186, 194-195, 311-312, 322-323, 353, 424
Canada geese, 100, 106, 108
Carcinogenicity of PCBs, 122, 136-137
Caspian tern, 185, 192
Cat, 353, 354
Catfish, channel, 78
Cattle, 74, 285-288, 479-480, 485
Cattle egret, 275
Cephalopods, 372, see also individual species
Channel catfish, 78
Chicken, 52, 56-57, 65, 66, 74, 85, 87, 105-106, 108, 110, 170-172, 175-179, 190-191, 199, 273, 351-352, 354, 456, 460, 468-470, 473
Chinook salmon, 435, 436, 439

Chloralkali manufacture, 307, 329
Chlordane, 6, 101-103, see also Organochlorines
Chlorinated camphenes, 110, see also Toxaphene
Chlorodecone, 103, 105-106
Chromium in marine mammals, 364, 374-375
Chukar, 82
Churchill River reservoir, 311-312
Civier's beaked whale, 368
Clams, see Molluscs
Clawed frogs, 75
Clay Lake (Ontario), 322-323, 353
Clophan, 171, see also PCBs
Cockerel, 52, 54, see also Chicken
Coelenterata, 129
Colorado, 82
Columbia River, 188
Common buzzard, 270
Common dolphin, 372
Common gallinule, 85
Common loon, 53, 347, 348
Common murres, 170, 174
Common shrew, 386, 387
Common tern, 177, 185, 192, 347
Condor
  Andean, 270
  California, 59, 270
Congeners
  dioxins/furans in fish, 202-215
  PCBs, 179
    in birds, 174-175, 198-199
    in mammals, 161-162
Contaminant dispersal, 11-13
Coot, American, 460
Copper, 416, 417, 419
  in aquatic invertebrates, 415
  in marine mammals, 364, 369-370
Cormorant
  double-breasted, 58
  double-crested, 185-196, 192-193
  great, 170, 171, 186
Cotton rat, 79, 81
Coturnix, 469, 470
Cowbird, 77, 78, 170, 350
  brown-headed, 51, 52, 54, 55, 100-102, 170-172, 274
Crabeater seal, 372
Crane
  sandhill, 275
  whooping, 275
Crude oil contamintion, 242, 433, see also PAHs
Crustacea, 120, 121, 129, 130-132, see also Aquatic organisms; Invertebrates
Cuckoo, 274

Cyanobacterium, 436, 439
Cyclodienes
  behavioral changes and, 91-92
  biogenic amines and, 91
  biomarkers for, 89-91
  dietary toxicity, 73-75
  endosulfan, 103
  lethality studies, 75-78
  liver/hepatic enzyme effects, 90
  mortality
    bald eagle, 82-83
    barn owl, 84
    bat, 84
    environmental, 79-80
    experimental, 75-79
    otter, 84-85
    peregrine falcon, 83-84
    wildlife incidents, 80-82
  reproductive effects, 85-89
  thyroid effects, 90-91
Cytochrome P-450, 180, 213, 373

Dacthal, 107
Dall's porpoise, 360-370, 372
Daphnia sp., 127-128, see also Aquatic organisms
DDD
  in birds, 49-67, see also DDT; Organochlorines
  characteristics, 10
DDE, 7, 13, 15, 23-24, 182-183
  in birds, 49-67, see also DDT; Organochlorines
  reproductive effects, 83, 88
DDT, 4, 7-9, 11, 83
  in birds
    brain, 50-53
    eggs, 56-61
    food levels, 66
    lethal residues, 50-55
    liver, 53-55
    other tissues, 55
    reproductive activity and, 61-65
    sublethal residues, 56-66
  dieldrin storage and, 78-79
  reproductive impairment, 88
Decapods, 420-421
Dechlorodicofol, 108
Deer, 75, 76, 86, 90-91
Deer mouse, 485
Deer mule, 75
Derwent Estuary (Tasmania), 425
Diagnostic lethality, see under Lethality
Diagnostic residue levels, 7-9
Dichlorobenzohydrol, 108
Dichlorobenzophenone, 108

Dichlorodiphenyltrichloroethane, see DDT; Organochlorines
Dicofol, 108-109
Dieldrin, 79, 83, 85, 88, 92, see also Organochlorines
 mortality field studies, 81-82
Dietary toxicity
 cadmium, 382-383
 cyclodienes, 73-80
 fluoride
  in birds, 469-470
  in small mammals, 483-485
 mercury, 353-355, see also Mercury
  in fish, 301-304
  in marine mammals, 368-369
 PCBs, 124
 PCBs/dioxin in forage, 196
 selenium in birds, 456-458
Dioxins
 in birds
  field studies of TCDD, 191-195
  forage residue studies, 196
  laboratory studies, 189-191
 in fish
  effect concentrations, 214-223
  environmental/effect concentrations compared, 223-225
  TECs, 212-214
Dispersal of contamination, 11-13
Dog, 76, 289
Dolphin, 363, 366
Double-crested cormorant, 58, 185-186, 192-193
Dove, 271-273
 mourning, 172-173, 272
 ring, 90, 91, 352
 ringed turtle-, 108-109, 172-173, 272
 ring-necked turtle-, 176
 rock, 106
 stock, 80
Duck, 455
 American black, 111
 black, 56, 61, 351
 goldeneye, 179
 mallard, 75, 78, 85, 86-91, 109, 169, 173, 176, 178, 179, 256-263, 351, 439, 440, 457-461
 wood, 193, 195-196
Dutch Wadden Sea, 119, 121, 160-161

Eagle
 bald, 14, 52, 59, 64-65, 82-83, 106, 169, 187-188, 193, 270, 347-348
 golden, 88, 271, 347-348
 white-tailed, 346-347, 348
 white-tailed sea, 188, 193
Eastern bluebird, 191, 199
Eastern screech owl, 109
Egg injection studies of PCBs, 176-179
Eggs, 7, 13, see also Eggshell thinning; Reproductive effects
 bird
  DDT/DDE/DDD, 56-65
  fluoride in, 471-474
  mercury in, 347-350
  selenium in, 454-458
 fish, 326-327
Eggshell thickness/thinning, 14, 15, 83, 85, 108-109, 471-473
 cadmium and, 401
 DDT/DDE/DDD and, 50, 56-61
 mercury and, 352
Eggshell weight, 14
Egret, 64-65, 275
Egypt, 119
Elbe River (Germany), 121
Endosulfan, 103
Endrin, 2, 4, 6, 7, 78, 82, 88, 89, 92, see also Cyclodienes; Dieldrin; Organochlorines
 12-keto-, 74
English Channel, 119
English River (Ontario), 353
Environmental Pollution by Pesticides (Edwards), 13
Eurasian kestrel, 54, 467
Eurasian sparrowhawk, 54, 56, 59
European kestrel, 80, 81
European mole, 385, 386
European sparrowhawk, 270
European starling, 100-102

Falcon
 peregrine, 14, 56, 58-60, 66, 67, 81, 83-84, 89, 194, 270-271, 348-349
 prairie, 64, 101, 270-271, 348
Fathead minnow, 322, 326-327, 435, 436, 439
Feral pigeon, 272
Ferret, 354
Ferrochelatase inhibition, 291
Field vole, 384, 387, 483-486
Finch, 170, 171
Finland, 120
Finnish Archipelago, 239
Fin whale, 366
Fire ant control, 5-6, 9-10
Fish, 11-12, 75, 120, 121, 125-127, 129-132, see also Aquatic organisms and individual species
 dioxins/furans in

effect concentrations, 214-223
environmental/effect concentrations compared, 223-225
TECs, 212-214
PCB effects, 125-127
selenium in, 435-440
Fishing sinkers, 275
Fish predators, 12, 13, 89, 105, 182, 466-467, see also individual species
Flamingo, 275
Flooding and mercury poisoning, 312
Florida, 58
Florida panther, 353
Fluoride
  in birds
    assessment recommendations, 474-475
    bone residues and expression, 467-468
    bone strength and, 470-471
    egg residues, 471-474
    growth rates and, 470
    ingestion, 466-467
    reproductive effects, 468-469
    soft-tissue residues, 470-471
  in livestock, 479-480
  in small mammals
    acute and subacute lethality, 481-482
    chronic toxicity, 482
    dental, 482-485
    skeletal, 485-486
Food chains, 7, 9-12, 23-24
  aquatic, 439-440
Food residues, see Dietary toxicity
Forster's tern, 182, 184-185, 192, 348
Fowler's toad, 75
Fox, 80, 353
France, 119, 120, 424
Free erythrocyte porphyrin activity, 291
Frog, 87, 89
  African clawed, 77
  bull-, 77
  clawed, 75
  leopard, 77
  Southern leopard, 78
Fulmar, 347
Fungicides, 107, 299, see also Mercury
Furans in fish, 211-225, see also Dioxins

Gallinule, 85
Genelly, R.E., 2
German Wadden Sea, 160-161
Germany, 121
Gizzard shad, 78
GLEMEDS, 180-181
Glomma estuary (Norway), 121

Golden eagle, 88, 271, 347-348
Goldeneye duck, 179
Goose
  Canada, 100, 106, 108
  domestic, 179
  pink-footed, 81
  snow, 10, 81, 471
Goshawk, 350
Grackle, 78, 100-102, 170-172, 274, 350
Gray bat, 81
Gray partridge, 101, 273
Gray seal, 363
Great blue heron, 52-53, 65, 193-195
Great Britain, 84-85, 417, 424
Great cormorant, 171, 186
Great Lakes, 119-121, 123, 124, 183, 185-188, 191-192, 199
Great Lakes embryo mortality, edema, and deformities syndrome (GLEMEDS), 180-181
Great skuas, 349
Green Bay, 185, 186
Grouse, sharp-tailed, 76
Growth effects
  dioxins in fish, 222-223
  fluoride in birds, 469
  PCBs, 129-132, 173-175
  selenium in fish, 438
Gulf of Bothria, 160
Gulf of Finland, 160
Gulf of Mexico, 119
Gull
  black-headed, 179, 274
  herring, 179, 182-183, 191-192, 196-197, 274, 347, 348
  laughing, 274
  lesser brown-backed, 174
Gull Island, 347
Gunshot, 253-263, 270-276, see also Lead

Halogens, see Fluoride
Hamilton Harbor, 347
Harbor porpoise, 366
Harbor seal, 366
Harp seal, 366
Harrier, marsh, 348
Hawk
  red-tailed, 270, 350
  sparrow, see Sparrowhawk
Hawke's Bay, 424
Heavy metals, see Metals and individual types
Hepatoma cell line studies, 180, 199, 213
Heptachlor, 4, 5, 100-101, see also Organochlorines

Herbicides, see individual chemicals
Heron
  black-crowned night, 52-53, 59, 64-65, 67, 188-189, 193, 455-456, 467, 469-470
  great, 170
  great blue, 52-53, 65, 193-195
Herring gull, 179, 182-183, 191-192, 196-197, 274, 347, 348
Hexachlorobenzine, 103, 106-108
Hexachlorocyclohexane, 106
History of studies, 1-24, see also Residue studies
Honey buzzard, 271
Hong Kong, 423
Horse, 285-286
Hot spots, 12
House martin, 465-466
House sparrow, 50
Hudson River, 119, 121
Humic waters, 307-311, 329-330
Hydrocarbons, see PAHs

Ibis, white-faced, 61, 64-65, 349
Iceland, 162
Immune function
  PCBs and, 128, 133, 137-139, 179
  zinc and, 370-371
Indian Ocean, 120
Insecticides, see individual chemicals
Insect larvae, 415
Invertebrates, heavy metals in aquatic, 411-425, see also Metals
Ireland, 424
Irrigation water, 11, 434, 435
Isodrin, see Cyclodienes; Dieldrin
Itai-itai disease, 382

James River (Virginia), 105, 106
Japan, 121, 322, 329, 345
Japanese quail, 75-79, 85, 90, 101, 107-108, 169, 170, 172-174, 198-199, 273-274, 352, 456

Kapone, see Chlorodecone
Kelthane, 108, see also Dicofol
Kesterson National Wildlife Refuge (California), 434, 455, 458
Kestrel, 89, 107, 347
  American, 53, 85, 88, 100-101, 105, 106, 109, 174, 175, 177, 187, 271, 352, 472-474
  Eurasian, 54, 467
  European, 80, 81
  sparrowhead, 6
12-Ketoendrin, 74, see also Cyclodienes; Endrin
Kidney residues/effects
  cadmium, 385-389, 401-402
  DDT/DDE/DDD, 55
  lead, 256, 260, 268, 274, 287-289, 292
  mercury, 347
  PCBs, 157
  selenium, 459
King vulture, 270
Krill, 372, 375

La Grande hydroelectric complex (Canada), 312
Lake Butte des Morts (Wisconsin), 119
Lake Geneva, 120, 121
Lake Huron, 120
Lake Michigan, 119, 121, 183, 186
Lake Ontario, 119-121, 183, 185-186, 212
Lake pollution, 311, 329-330, see also Mercury and individual lakes
Lake Superior, 119
Lake trout, 307
Lamb, 286
Largemouth bass, 307-308, 325
Lark, meadow, 81
Laughing gull, 274
Laysan albatross, 276
Lead
  in birds other than waterfowl
    charadiformes, 274
    ciconiformes, 275
    columbiformes, 271-273
    falconiformes, 270-271
    galliformes, 273-274
    gaviformes, 275
    gruiformes, 275
    mechanisms of toxicity, 267-270
    passeriformes, 274
    procelariiformes, 276
    strigiformes, 275-276
  clinical symptoms of poisoning, 285
  in mammals
    biochemical biomarkers, 290-292
    blood concentration, 284-287
    interpretation of tissue concentrations, 284-290
    mechanism of absorption and toxicity, 283-284
    morphological biomarkers, 292-293
    organ and bone concentrations, 287-290
  in marine mammals, 365, 373-374
  subclinical effects, 268, 286-287
  in waterfowl
    background and terminology, 253-254
    factors influencing concentrations, 255-256
    interpretive recommendations, 261-263
    sublethal effects, 256-259

INDEX

489

threshold concentrations, 259-261
tissue distribution, 254-255
Lead acetate, 270-273, 276
Lead powder, 271
Lead shot, see Gunshot; Lead
Leopard frog, 77, 78
Lesser brown-backed gull, 174
Lesser scaup, 81, 255, 260
Lethal body burden of dioxins in fish, 222, 224-225
Lethality, see also Mortality
  aldrin/endrin/dieldrin, 75
  DDT/DDD/DDE in birds, 50-55
  diagnostic
    chlordane, 101-102
    chlorodecone, 105
    heptachlor, 100
    hexachlorobenzene, 107
    hexachlorohexane, 106
    mirex, 103-104
  dieldrin, 76-77
  dioxins in fish, 223
  endrin in brain, 78
  fluoride acute and subacute, 481-482
  mercury in birds, 351
  PCBs, 129, 131, 140, 157, 159, 163, 169-172
Lindane, 106, see also Hexachlorocyclohexane
Lipophilicity, 66
Liver residues/effects
  cadmium, 385-388
  cyclodienes, 90
  DDT/DDE/DDD, 53-55
  lead, 256, 260, 268, 274, 287-289
  mercury, 346-347, 350
  PCBs
    in birds, 198-199
    in mammals, 157, 159, 163
  selenium, 458-459
  TCDD, 191
Livestock, fluoride in, 479-480
Long-eared owl, 347
Long Island Sound, 121
Loon, common, 53, 347, 348
Louisiana, 89
Louisiana pelican, 7

Malformations, see Teratogenicity
Mallard duck, 75, 78, 85-91, 109, 169, 173, 176, 178, 179, 256-263, 351, 439, 440, 457-461
Mammals, 15, 82, see also Marine mammals
  cadmium in small, 375-389
  fluoride in small, 479-486
  lead in, 283-293, see also Lead

mercury detoxification in, 312-316
mercury in, 352-355
PCBs in, 122-123, 155-164
  bat, 158-159
  congener-specific studies, 161-163
  mink, 156-157
  seal, 160-161
  sea lion, 159-160
  total PCB studies, 155-161
Manx shearwater, 347
Marine mammals
  metals in, 361-376, see also Metals and individual metals and species
  PCBs in, 159-163
Marsh harrier, 348
Marten, 353
Martin, house, 465-466
Maryland, 158
Massachusetts, 119
Meadowlark, 81
Meadow vole, 384, 386
Mealworms, 158-159
Mediterranean Sea, 119
Mercury, 80
  biotransformation, 349
  in birds
    in eggs, 347-350
    free-living populations and individuals, 346-350
    laboratory-based studies, 350-352
  cadmium and, 347
  in freshwater fish
    atmospheric deposition to low pH and humic lakes, 306-311
    critical tissue concentrations, 321-325
    detoxification mechanisms, 313-317
    diet, trophic structure, and longevity, 305-306
    elevated concentrations, 304-305
    field studies, 322-323
    human exposure to, 300-301
    laboratory studies, 317-322
    mode of action, 313
    newly flooded reservoirs, 311-312
    point-source pollution, 306
    reproductive/embryonic effects, 325-328
    residues in mercury-intoxicated adult fish, 312-313
    tissue distribution and retention, 304
    toxicity and associated residues, 317-325
    uptake from food and water, 301-304
  in mammals
    free-living populations and individuals, 353
    laboratory-based studies, 353-355
  in marine mammals, 363-369

selenium and, 346, 368, 453
Merlin, 64, 101, 348
Metallothioneins, 316, 362, 371, 399, 414, 420,
    see also Mercury; Metals
Metals, see also individual metals
  heavy in aquatic invertebrates
    accumulation, 411-415
    body requirements, 415-416
    components of content, 416-420
    deficiency and excess, 416
    interpretation, 421-422
    physiological variations, 420-421
    uptake, 412-413
    zinc in *Mytilus edulis*, 422-425
  lead
    in birds other than waterfowl, 267-277
    in mammals, 283-293
    in marine mammals, 373-374
    in waterfowl, 253-263
  in marine mammals
    arsenic, 364, 374
    cadmium, 364, 371-373
    chromium, 364, 374-375
    copper, 364, 369-370
    historical background, 361-363
    lead, 365, 373-374
    mercury, 363-369
    nickel, 364, 375
    zinc, 364, 370-371
  mercury
    in birds and terrestrial mammals, 345-355
    in freshwater fish, 299-330
    in marine mammals, 363-369
  selenium, see also Selenium
    in aquatic organisms, 433-444
Methoxychlor, 110
Methylmercury, see Mercury
MFO activity, 138, 174, 178-179, 191, 240
Michigan, 119, 185
Migrant exposure, 15
Minimata Bay (Japan), 322, 329, 345
Mink, 85, 156-157, 163, 353, 354
Minke whale, 372
Minnow, fathead, 322, 326-327, 435, 436, 439
Mirex, 103-105
Missouri, 84
Mole, 289, 385, 386
Molluscs, 120, 121, 131, see also Aquatic
    organisms; Invertebrates
  heavy metals in, 411-425, see also Metals
  PAHs in, 240-244
Monitoring programs, 18-21
Monkey, 284
Mortality

cyclodienes, 76-78
  bald eagle, 82-83
  barn owl, 84
  bat, 84
  environmental, 79-80
  experimental, 75-79
  otter, 84-85
  peregrine falcon, 83-84
  wildlife incidents, 80-85
DDT/DDD/DDE in birds, 8-10, 50-55
dieldrin field studies, 81-82
direct, 6-7, 14, 83
Mosquitofish, 325, 439
Mourning dove, 172-173
Mouse, 78, 287-288
  deer, 485
  white-footed, 386
  wood, 284, 386, 387, 485, 486
Murres, common, 170, 174
Muscle residues/effects
  mercury
    in birds, 350
    in fish, 324
  selenium in birds, 460
Mussel, 415, 422-425, see also Aquatic
    organisms; Invertebrates; Molluscs
Mute swan, 255-256, 260

Nagaragawa River (Japan), 121
Narwhal, 366, 372
National Contaminant Biomonitoring Program,
    19
Netherlands, 186
Neurotoxicity
  of lead, 286
  of mercury, 312, 324-325, 329
New Bedford Harbor (Massachusetts), 119, 121
Newfoundland, 424
New York state, 121
Nickel in marine mammals, 364, 375
Nile River, 119
NOAELs, 89, 184, 188
NOELs, 89, 324
Nonachlor, 102, see also Chlordane
North America, 84
North Atlantic Ocean, 239
North Carolina, 82
North Sea, 119, 121, 161, 424
Norway, 121, 162

Oceanic contamination, 12
Onondaga Lake (New York), 304
Orchard treatments, 5-6, 82
Organochlorine index, 53

INDEX

491

Organochlorines, 49, 99, 111-112, see also DDD; DDE; DDT
  in birds, 49-67, 99-112
  chlordane, 101-103
  chlorodecone, 105-106
  DDT, DDD, and DDE, 49-67
  dicofol, 108-109
  and endosulfan, 103
  endosulfan, 103
  heptachlor, 100-101
  hexachlorobenzine, 106-108
  hexachlorocyclohexane, 106
  methoxychlor, 110
  mirex, 103-105
  toxaphene, 110-111
  vs. dieldrin, 88
Osprey, 14, 65, 106, 271
Otter, 84-85, 353, 355
Owl, 81, 467
  barn, 56, 61, 81, 84, 87, 467
  long-eared, 347
  screech, 88, 109, 173, 176, 275-276, 468, 472-474
  snowy, 275-276
  tawny, 467, 472-473
Oyster, 425, see also Aquatic organisms; Invertebrates; Molluscs

Pacific Ocean, 119
Pacific Sea, 119, 120
PAHs
  analytic considerations, 233-236
  in aquatic environment, 231-233
  detection limits, 236
  fate in vertebrates, 237-238
  in feral fish, 237-240
  in finfish and marine mammals, 237-240
  free, 239-240
  individual analysis, 235-236
  low molecular weight, 240, 242
  in molluscs, 240-243
  sediment quality criteria, 243-244
  threshold values, 244-245
Palaemon elegans, 414-415
Panther, Florida, 353
Partridge, gray, 101, 273
Passerines, see individual species
PCBs, 7, 84, see also Dioxins
  in aquatic environments, 118-122
  in aquatic organisms, 136-139
  in birds
    black-crowned night heron, 188-189
    cormorant, 185-187
    eagle, 187-188
    egg injection studies, 173-179
    field studies, 182-189
    forage residue studies, 196
    herring gull, 182-183
    pen and laboratory studies, 169-173
    tern, 183-185
    toxic equivalency factors and, 179-182
  and health effects
    in aquatic organisms, 123-126
    in mammals, 122
  laboratory studies, 127-136
  in mammals, 155-164
    bat, 158-159
    congener-specific studies, 161-163
    mink, 156-157
    seal, 160-161
    sea lion, 159-160
    total PCB studies, 155-161
  vector effects and, 123
  vs. dieldrin, 85
  waterborne concentrations, 118
PCDDs, 169, see also PCBs
  environmental/effect concentrations compared, 223-225
  in fish
    effect concentrations, 214-223
    TECs, 212-214
PCDFs
  environmental/effect concentrations compared, 223-225
  in fish
    effect concentrations, 214-223
    TECs, 212-214
Pelican
  brown, 7, 14, 58, 59, 64-67, 88, 89
  Louisiana, 7
Pentachloronitrobenzene, 107
Peregrine falcon, 14, 56, 58-60, 66, 81, 83-84, 89, 194, 270-271, 348-349
Peripheral neuropathy, 286
pH and mercury toxicity, 307-311, 329-330
Pheasant, 56, 65, 76, 78, 80, 85, 106, 169-170, 172-173, 177, 179, 190, 273, 350, 351
Phenoclor, 171, see also PCBs
Phytoplankton, 128, 129, 131, see also Aquatic organisms
Pigeon, 77, 90, 271-273
  feral, 272
  wood, 80, 272
Pike, 317, 322-323, 327-328
Pilot whale, 366
Pink-footed goose, 81
Plankton, 120, 121, 439
Plumbism, 289

Polar bear, 371
Polychlorinated biphenyls, see PCBs
Polychlorinated dibenzodioxins, see PCBs; PCDDs
Polychlorinated dibenzofurans, see Dioxins; PCDFs
Polychlorinated dibenzo-*p*-dioxins, see Dioxins; PCDDs
Polycyclic aromatic hydrocarbons, 231-248, see also PAHs
Population decline, 14-15, 56, see also Reproductive effects
Porphyria, 174
Porpoise, 162
  Dall's, 360-370, 372
  harbor, 366
Poultry, see individual species
Prairie falcon, 64, 101, 270-271, 348
Principles developed, 4, 5, 7, 10, 14, 21, 23-24, see also Residue studies
Priority pollutants, 232
Protozoa, 129, 131
Ptarmigan, willow, 273
Puffin, 173, 176, 347
Puget Sound, 119, 121, 124
Purple gallinule, 85

Quail
  bobwhite, 51, 75, 76, 91, 103, 104, 169, 177, 273
  California, 75, 82
  Japanese, 75-79, 85, 101, 107-108, 169, 170, 172-174, 198-199, 273, 352, 456
  Japanese, 90

Rabbit, 74, 79
Raccoon, 353
Rail, sora, 275
Rainbow trout, 75, 307, 320-322, 324, 326, 435, 436
Raptors, 15, 17, 64, 81-84, 106, 107, 109, 193-194, 270-271, 346-347, 348, 350, 466, see also individual species
Rat, 75, 76, 286, 482-483, 486
  cotton, 79, 81
  laboratory, 284, 288-289
Redshank, 274
Red shiner, 435
Red-tailed hawk, 270, 350
Red-winged blackbird, 77, 85, 88, 100, 101-102, 171-172, 274
Reproductive effects, 13-15, 14-15, 23-24, 53, 83
  cadmium in birds, 401, 403

cyclodienes, 85-89
DDT/DDD/DDE, 50, 61-62, 65
DDT/DDE/DDE, 88
dioxins in fish, 219-222
fluoride in birds, 468-469
lead, 285, 290
mercury
  in birds, 346-347
  in fish, 325-328, 330
organochlorines
  chlordane, 103
  chlorodecone, 105-106
  dicofol, 108-109
  heptachlor, 100-101
  hexachlorobenzene, 107-108
  hexachlorohexane, 106
  methoxychlor, 109
  mirex, 104
  toxaphene, 109-110
PAHs, 239
PCBs
  in aquatic organisms, 125, 127-128, 132-133, 135
  in birds, 172-173, 176-179, 198
  in mammals, 155-157, 159-160, 163
selenium
  in birds, 455-460
  in fish, 434, 437-439
Reptiles, 13
Reservoir pollution, 311-312
Reservoirs, 329
Residue studies
  accomplishments of, 22-24
  biomarkers, 21-22
  criticisms of, 6, 7
  diagnostic residue levels, 7-9
  dispersal of contamination, 11-13
  early history and principles, 1-3
  food chains, 9-11
  genesis of wildlife studies, 3-4
  monitoring programs, 18-21
  nontreated areas, 6-7
  principles developed, 23-24
  threatened avian species, 13-18
  treated areas, 4-6
Richardson's merlin, 348
Ring dove, 90, 91, 352
Ringed seal, 368
Ringed turtledove, 108-109, 172-173, 176, 272
Ring-necked pheasant, 56, 65, 169-170, 172-173, 273, 351
Robin, 50, 51, 81, 83, 274
Rock bass, 323
Rock dove, 106

INDEX

Rodents, 74, 79, 82, 84, 286-288, see also individual species
  fluoride in small mammals, 479-487
Rook, 80
Royal tern, 274
Russ seal, 372

Saginaw Bay, 185
Salmon, 324, 435, 436, 439
Sample egg technique, 61-66, see also Eggshell thinning; Eggs
Sandhill crane, 275
Sandwich tern, 80, 81, 274
San Francisco Bay, 121, 188-189
Saskatchewan River, 353
Scandinavia, 424
Scaup, lesser, 81, 255, 260
Scotland, 88, 353, 424
Screech owl, 88, 109, 173, 176, 275-276, 468, 472, 474
Sculpin, slimy, 307
Seal, 6, 160-161, 366-369
  bearded, 368
  crabeater, 372
  gray, 363
  harbor, 366
  harp, 366
  ringed, 368
  Russ, 372
Sea lion, 159-160
  Stellar, 372
Sediment quality criteria, 242-244
Sediment residues, 212, 242-244
Selenium
  in aquatic organisms
    background, 433-435
    in dietary sources, 440-441
    in fish species, 435-440
    interpretation, 435
    in water, 441-443
  in birds
    blood, 460
    breast muscle, 460
    eggs, 454-458
    kidneys, 459
    liver, 458-459
  and mercury toxicity, 346, 354, 368, 453
  and methylmercury bioaccumulation, 316
Sewage sludge, 453
Shad, gizzard, 78
Sharp-tailed finch, 171
Sharp-tailed grouse, 76
Shearwater, Manx, 347
Shellfish, 411-426, see also Aquatic organisms

Shiawassee River, 119
Shiner, red, 435
Short-tailed field vole, 386, 387
Short-tailed shrew, 76
Shrew, 287-288, 292
  common, 385-387
  short-tailed, 76, 78
Skimmer, black, 348
Skin, DDT/DDE/DDD in birds, 55
Skuas, great, 349
Slimy sculpin, 307
Smelt, 307
Snow goose, 10, 81, 471
Snowy egret, 64-65
Snowy owl, 275-276
Sodium selenite, see Selenium
Somatic organ index, 290
Sora rail, 275
South Caroline, 58
Spain, 239
Sparrow, house, 50, 274
Sparrowhawk, 80, 81, 89, 467
  Eurasian, 54, 56, 59
  European, 270
Sparrowhead kestrel, 6
Squid, 372
Starling, 78, 170-172, 274, 350, 352, 468, 473
  European, 100-102
Stellar sea lion, 372
Stock dove, 80
Striped bass, 435, 436, 439
Striped dolphin, 363
Sunfish, bluegill, 75, 436, 437, 439
Swan, mute, 255-256, 260
Sweden, 85, 119-121, 345, 348, 353, 424
Swedish Baltic Sea, 161-162
Switzerland, 120

Talpid mole, 289
Tasmania, 425
Tawny owl, 467, 472-473
TCDD, 189-191, see also Dioxins; PCBs
TCDD equivalents, 179-182
TEFs, 179-182
Telodrin, see Cyclodienes; Dieldrin
Teratogenicity, 186, 188, see also Reproductive effects
  mercury, 325-328
  PCBs, 176-179
  selenium, 438, 460
Tern, 183-185
  Caspian, 185, 192
  common, 177, 185, 192, 347
  Forster's, 182, 184-185, 192, 199, 348

royal, 274
sandwich, 80, 81, 274
Testicular effects, see Reproductive effects
Tetrachlorobenzodioxin, see Dioxins; PCBs; TCDD
2,3,7,8-Tetrachlorodibenzo-*p*-dioxin, see Dioxins; TCDD
Texas, 58
Thickness index, 58, see also Eggshell thickness
Threshold concentrations of lead, 258-259
Thyroid effects
 cyclodienes, 90-91
 PCBs, 135, 174, 184
 TCDD, 194
Toad, Fowler's, 75
Tolerances, 2-4
Total equivalency concentration, PCDDs and PCDFs in fish, 212-214
Toxaphene, 4, 110-111
Toxic effects thresholds of selenium
 in aquatic food chain, 440
 in fish, 438, 439
Toxic equivalency factors (TEFs), 162, 179-182, 212-214
Trophic levels, 10
Trophic structure and mercury toxicity, 305-306
Trout
 brook, 317-318, 322, 327
 lake, 307
 rainbow, 75, 307, 320-322, 324, 326, 435, 436
Turkey, 239, 469, 470
 wild, 273
Turkey (country), 179
Turkey vulture, 271, 347
Turtle, yellow mud, 73
Turtledove, ringed, 108-109, 172-173, 176, 272

United Kingdom, 79, 82-84, 353
U.S. Environmental Protection Agency (EPA), 3
U.S. Fish and Wildlife Service, 3-4, 75-76, 82

Vanadium, 416
Visceral mass PAH elevation, 242
Vitamin A imbalance, 191
Vitellogenin transfer, 211-212
Vole, 74, 82, 284, 287-288
 bank, 386, 485
 field, 384, 483-486
 meadow, 384, 386

 short-tailed field, 386, 387
Vulture, 270, 271

Wabigoon Lake, 347
Walleye, 317, 327-328
Walrus, 366, 372-373
Washington state, 82
Waterfowl, lead in, 253-263, see also Lead and individual species
Water pollution, 11, 12, 212, see also Aquatic invertebrates; Aquatic organisms; Fish
 mercury, see also Mercury
 mercury in fish, 299-325
 PAHs, 231-245, see also PAHs
 PCBs, 118-123
 selenium, 433-434, 441-443
Whale, 162
 blue, 375
 Civier's beaked, 368
 fin, 366
 minke, 372
 pilot, 366
White-beaked dolphin, 366
White-faced ibis, 61, 64-65, 349
White-footed mouse, 386
White-tailed deer, 76, 86
White-tailed eagle, 188, 193, 346-348
Whooping crane, 275
Wildlife, 5
Wild turkey, 273
Willow ptarmigan, 273
Winter stress syndrome in fish, 437
Wisconsin, 119, 185
Wood duck, 193, 195-196
Woodlouse, 417
Wood mouse, 284, 386, 387, 485, 486
Wood pigeon, 80, 272
Worms, 416, 421

Yellow mud turtle, 73

Zinc, 415-417, 419, 422
 and cadmium in birds, 400
 in marine mammals, 364, 370-371
 mussel *Mytilus edulis* model, 422-425
Zinc porphyrin activity, 291
Zooplankton, 127-129, 131, 132, see also Aquatic organisms